BUSINESS AND ECONOMICS

(continued)

(*continued on back endpaper*)

Finite Mathematics

for the Managerial, Life,
and Social Sciences

Eighth Edition

Finite Mathematics

for the Managerial, Life, and Social Sciences

Eighth Edition

S. T. TAN

STONEHILL COLLEGE

THOMSON

BROOKS/COLE

Australia • Canada • Mexico • Singapore • Spain
United Kingdom • United States

THOMSON

BROOKS/COLE

Executive Publisher: Curt Hinrichs
Development Editor: Danielle Derbenti
Senior Assistant Editor: Ann Day
Editorial Assistant: Fiona Chong
Technology Project Manager: Earl Perry
Marketing Manager: Tom Ziolkowski
Marketing Assistant: Jessica Bothwell
Advertising Project Manager: Nathaniel Bergson-Michelson
Project Manager, Editorial Production: Sandra Craig
Art Director: Lee Friedman
Print/Media Buyer: Doreen Suruki

Permissions Editor: Sarah Harkrader
Production: Cecile Joyner, The Cooper Company
Text Designer: Diane Beasley
Photo Researcher: Stephen Forsling
Copy Editor: Betty Duncan
Illustrator: Better Graphics, Inc.; Accurate Art
Cover Designer: Irene Morris
Cover Image: Portrait of Chris Shannon by Peter Kiar
Cover Printer: Phoenix Color Corp
Compositor: Better Graphics, Inc.
Printer: R. R. Donnelley/Willard

For more information about our products, contact us at:
Thomson Learning Academic Resource Center
1-800-423-0563

For permission to use material from this text or product, submit a request online at **http://www.thomsonrights.com**. Any additional questions about permissions can be submitted by email to **thomsonrights@thomson.com.**

Library of Congress Control Number: 2004114812

Student Edition: ISBN 0-534-49214-2

Instructor's Edition: ISBN 0-495-01028-6

International Student Edition: ISBN: 0-495-01510-5
(Not for sale in the United States)

Thomson Higher Education
10 Davis Drive
Belmont, CA 94002-3098
USA

Asia (including India)
Thomson Learning
5 Shenton Way
#01-01 UIC Building
Singapore 068808

Australia/New Zealand
Thomson Learning Australia
102 Dodds Street
Southbank, Victoria 3006
Australia

Canada
Thomson Nelson
1120 Birchmount Road
Toronto, Ontario M1K 5G4
Canada

UK/Europe/Middle East/Africa
Thomson Learning
High Holborn House
50/51 Bedford Row
London WC1R 4LR
United Kingdom

Latin America
Thomson Learning
Seneca, 53
Colonia Polanco
11560 Mexico
D.F. Mexico

Spain (including Portugal)
Thomson Paraninfo
Calle Magallanes, 25
28015 Madrid, Spain

TO PAT, BILL, AND MICHAEL

Contents

*Sections marked with an asterisk are not prerequisites for later material.

CHAPTER 3

Linear Programming: A Geometric Approach 167

CHAPTER 4

Linear Programming: An Algebraic Approach 215

CHAPTER 5

Mathematics of Finance 277

Preface

Math is an integral part of our increasingly complex daily life. *Finite Mathematics for the Managerial, Life, and Social Sciences, Eighth Edition*, attempts to illustrate this point with its applied approach to mathematics. Our objective for this Eighth Edition is threefold: (1) to write an applied text that motivates students while providing the background in the quantitative techniques necessary to better understand and appreciate the courses normally taken in undergraduate training, (2) to lay the foundation for more advanced courses, such as statistics and operations research, and (3) to make the text a useful tool for instructors. The only prerequisite for understanding this text is 1 to 2 years, or the equivalent, of high school algebra.

Features of the Eighth Edition

Coverage of Topics This text offers more than enough material for a one-semester or two-quarter course. Optional sections have been marked with an asterisk in the table of contents, thereby allowing the instructor to be flexible in choosing the topics most suitable for his or her course. The following chart on chapter dependency is provided to help the instructor design a course that is most suitable for the intended audience.

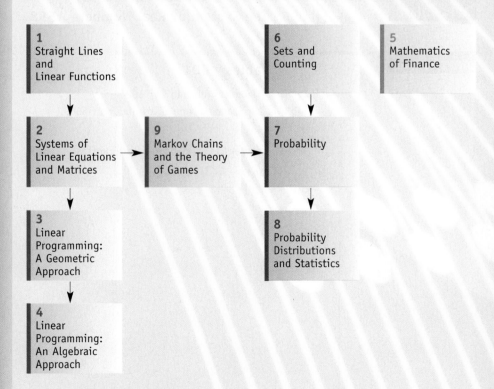

- **Approach** A problem-solving approach is stressed throughout the book. Numerous examples and solved problems are used to amplify each new concept or result in order to facilitate students' comprehension of the material. Graphs and pictures are used extensively to help students visualize the concepts and ideas being presented.
- **Level of Presentation** Our approach is intuitive, and we state the results informally. However, we have taken special care to ensure that this approach does not compromise the mathematical content and accuracy.

Applications The applications provide another opportunity to show the student the connection between mathematics and the real world.

- **Current and Relevant Examples and Exercises** are drawn from the fields of business, economics, social and behavioral sciences, life sciences, physical sciences, and other fields of general interest. In the examples, these are highlighted with new icons that illustrate the various applications.

APPLIED EXAMPLE 4   Financing a Car   After making a down payment of $2000 for an automobile, Murphy paid $200 per month for 36 months with interest charged at 12% per year compounded monthly on the unpaid balance. What was the original cost of the car? What portion of Murphy's total car payments went toward interest charges?

Solution   The loan taken up by Murphy is given by the present value of the annuity

$$P = \frac{200[1 - (1.01)^{-36}]}{0.01} = 200a_{\overline{36}|0.01}$$

- **New Applications** Many new real-life applications have been introduced. Among these applications are sales of GPS Equipment, Broadband Internet Households, Switching Internet Service Providers, Digital vs. Film Cameras, Online Sales of Used Autos, Financing College Expenses, Balloon Payment Mortgages; Nurses Salaries, Revenue Growth of a Home Theater Business, Same-Sex Marriage, Rollover Deaths, Switching Jobs, Downloading Music, Americans without Health Insurance, Access to Capital, and Volkswagen's Revenue.

---

75. **SALES OF GPS EQUIPMENT**   The annual sales (in billions of dollars) of global positioning systems (GPS) equipment from 2000 through 2006 follow. (Sales in 2004 through 2006 are projections.) Here, $x = 0$ corresponds to 2000.

Year x	0	1	2	3	4	5	6
Annual Sales, y	7.9	9.6	11.5	13.3	15.2	17	18.8

a. Plot the annual sales (y) versus the year (x).
b. Draw a straight line L through the points corresponding to 2000 and 2006.
c. Derive an equation of the line L.

▪ **New Portfolios** are designed to convey to the student the real-world experiences of professionals who have a background in mathematics and use it in their daily business interactions.

PORTFOLIO   Morgan Wilson

TITLE   Land Use Planner
INSTITUTION   City of Burien

As a Land Use Planner for the city of Burien, Washington, I assist property owners every day in the development of their land. By definition, land use planners develop plans and recommend policies for managing land use. To do this, I must take into account many existing and potential factors, such as public transportation, zoning laws, and other municipal laws. By using the basic ideas of linear programming, I work with the property owners to figure out maximum and minimum use requirements for each individual situation. Then, I am able to review and evaluate proposals for land use plans and prepare recommendations. All this is necessary to process an application for a land development permit.

Here's how it works. A property owner will come to me who wants to start a business on a vacant commercially zoned piece of property. First, we would have a discussion to find out what type of commercial zoning the property is in and whether or not the use is permitted or would require additional land use review. If the use is permitted and no further land use review is required, I would let the applicant know what criteria would have to be met and shown on building plans. At this point the applicant will begin working with their building contractor, architect, or engineer and landscape architect to meet the zoning code criteria. Once the

applicant has worked with one or more of these professionals, building plans can be submitted for review. Then, they are routed to several different departments (building, engineer, public works, and the fire department). Because I am the land use planner for the project, one set of plans is routed to my desk for review.

During this review, I determine whether or not the zoning requirements have been met in order to make a final determination of the application. These zoning requirements are assessed by asking the applicant to give us a site plan showing lot area measurements, building and impervious surface coverage calculations, and building setbacks, just to name a few. Additionally, I would have to determine the parking requirements. How many off-street parking spaces are required? What are the isle widths? Is there enough room for backing space? Then, I would look at the landscaping requirements. Plans would need to be drawn up by a landscape architect and list specifics about the location, size, and types of plants that will be used.

By weighing all of these factors and measurements, I am able to determine the viability of a land development project. The basic ideas of linear programming are, fundamentally, at the heart of this determination and are key to the day-to-day choices I must make in my profession.

▪ **Explore & Discuss** boxes, appearing throughout the main body of the text, offer optional questions that can be discussed in class or assigned as homework. These questions generally require more thought and effort than the usual exercises. They may also be used to add a writing component to the class, giving students opportunities to articulate what they have learned. Complete solutions to these exercises are given in the *Instructor's Solutions Manual*.

EXPLORE & DISCUSS

1. Consider the amortization Formula (13):

$$R = \frac{Pi}{1 - (1 + i)^{-n}}$$

Suppose you know the values of R, P, and n and you wish to determine i. Explain why you can accomplish this task by finding the point of intersection of the graphs of the functions

$$y_1 = R \quad \text{and} \quad y_2 = \frac{Pi}{1 - (1 + i)^{-n}}$$

Real-Life Data   Many of the applications are based on mathematical models (functions) that the author has constructed using data drawn from various sources including current newspapers and magazines, and data obtained through the Internet. Sources are given in the text for these applied problems. In Functions and Linear Models (Section 1.3), the modeling process is discussed and students are asked to use a model (function) constructed from real-life data to answer questions about the Market for Cholesterol-Reducing Drugs. Then in Section 1.5, students learn how to construct the function used in that model by using the least-squares method. Hands-on experience constructing models from other real-life data is provided by the exercises that follow.

Exercise Sets   The exercise sets are designed to help students understand and apply the concepts developed in each section. Three types of exercises are included in these sets:

- **Self-Check Exercises** offer students immediate feedback on key concepts with worked-out solutions following the section exercises.
- **New Concept Questions** are designed to test students' understanding of the basic concepts discussed in the section and at the same time encourage students to explain these concepts in their own words.
- **Exercises** provide an ample set of problems of a routine computational nature followed by an extensive set of application-oriented problems.

5.3   Self-Check Exercises

1. The Mendozas wish to borrow $100,000 from a bank to help finance the purchase of a house. Their banker has offered the following plans for their consideration. In plan I, the Mendozas have 30 yr to repay the loan in monthly installments with interest on the unpaid balance charged at 10.5%/year compounded monthly. In plan II, the loan is to be repaid in monthly installments over 15 yr with interest on the unpaid balance charged at 9.75%/year compounded monthly.
   a. Find the monthly repayment for each plan.
   b. What is the difference in total payments made under each plan?

2. Harris, a self-employed individual who is 46 yr old, is setting up a defined-benefit retirement plan. If he wishes to have $250,000 in this retirement account by age 65, what is the size of each yearly installment he will be required to make into a savings account earning interest at $8\frac{1}{4}$%/year? (Assume that Harris is eligible to make each of the 20 required contributions.)

Solutions to Self-Check Exercises 5.3 can be found on page 318.

5.3   Concept Questions

1. Write the amortization formula.
   a. If P and i are fixed and n is allowed to increase, what will happen to R?
   b. Interpret the result of part (a).

2. Using the formula for computing a sinking fund payment, show that if the number of payments into a sinking fund increases, then the size of the periodic payment into the sinking fund decreases.

5.3   Exercises

In Exercises 1–8, find the periodic payment R required to amortize a loan of P dollars over t years with interest earned at the rate of r%/year compounded m times a year

12. $S = 120,000$, $r = 4.5$, $t = 30$, $m = 6$

13. $S = 250,000$, $r = 10.5$, $t = 25$, $m = 12$

5.3   Solutions to Self-Check Exercises

1. a. We use Equation (13) in each instance. Under plan I,

$$P = 100,000 \qquad i = \frac{r}{m} = \frac{0.105}{12} = 0.00875$$

$$n = (30)(12) = 360$$

Therefore, the size of each monthly repayment under plan I is

$$R = \frac{100,000(0.00875)}{1 - (1.00875)^{-360}}$$

$$\approx 914.74$$

or $914.74.
Under plan II,

$$P = 100,000 \qquad i = \frac{r}{m} = \frac{0.0975}{12} = 0.008125$$

$$n = (15)(12) = 180$$

Therefore, the size of each monthly repayment under plan II is

b. Under plan I, the total amount of repayments will be

$$(360)(914.74) = 329,306.40 \qquad \text{Number of payments} \\ \times \text{ the size of each installment}$$

or $329,306.40. Under plan II, the total amount of repayments will be

$$(180)(1059.36) = 190,684.80$$

or $190,684.80. Therefore, the difference in payments is

$$329,306.40 - 190,684.80 = 138,621.60$$

or $138,621.60.

2. We use Equation (14) with

$$S = 250,000$$

$$i = r = 0.0825 \qquad \text{Since } m = 1$$

$$n = 20$$

giving the required size of each installment as

Review Sections   These sections are designed to help students review the material in each section and assess their understanding of basic concepts as well as problem-solving skills.

- **Summary of Principal Formulas and Terms** highlights important equations and terms with page numbers given for quick review.
- **New Concept Review Questions** give students a chance to check their knowledge of the basic definitions and concepts given in each chapter.
- **Review Exercises** offer routine computational exercises followed by applied problems.
- **New Before Moving On . . . Exercises** give students a chance to see if they have mastered the basic computational skills developed in each chapter. If they solve a problem incorrectly, they can go to the companion Web site and try again. In fact, they can keep on trying until they get it right. If students need step-by-step help, they can utilize the *iLrn* Tutorials that are keyed to the text and work out similar problems at their own pace.

CHAPTER 4   Summary of Principal Terms

  TERMS

standard maximization problem (216)	pivot column (220)	standard minimization problem (244)
slack variable (217)	pivot row (220)	primal problem (244)
basic variable (218)	pivot element (220)	dual problem (244)
nonbasic variable (218)	simplex tableau (220)	nonstandard problem (260)

CHAPTER 4 **Before Moving On . . .**

1. Consider the following linear programming problem:

$$\text{Maximize} \quad P = x + 2y - 3z$$
$$\text{subject to} \quad 2x + y - z \leq 3$$
$$x - 2y + 3z \leq 1$$
$$3x + 2y + 4z \leq 17$$
$$x \geq 0, y \geq 0, z \geq 0$$

Write the initial simplex tableau for the problem and identify the pivot element to be used in the first iteration of the simplex method.

2. The following simplex tableau is in final form. Find the solution to the linear programming problem associated with this tableau.

x	y	z	u	v	w	P	Constant
0	$\frac{1}{2}$	0	1	$-\frac{1}{2}$	0	0	2
0	$\frac{1}{4}$	1	0	$\frac{5}{4}$	$-\frac{1}{2}$	0	11
1	$\frac{1}{4}$	0	0	$-\frac{3}{4}$	$\frac{1}{2}$	0	2
0	$\frac{13}{4}$	0	0	$\frac{1}{4}$	$\frac{1}{2}$	1	28

3. Using the simplex method, solve the following linear programming problem:

$$\text{Maximize} \quad P = 5x + 2y$$
$$\text{subject to} \quad 4x + 3y \leq 30$$
$$2x - 3y \leq 6$$
$$x \geq 0, y \geq 0$$

Technology   Throughout the text, opportunities to explore mathematics through technology are given.

▪ **Exploring with Technology Questions** appear throughout the main body of the text and serve to enhance the student's understanding of the concepts and theory presented. Complete solutions to these exercises are given in the *Instructor's Solutions Manual*.

 EXPLORING WITH TECHNOLOGY

Investments allowed to grow over time can increase in value surprisingly fast. Consider the potential growth of $10,000 if earnings are reinvested. More specifically, suppose $A_1(t)$, $A_2(t)$, $A_3(t)$, $A_4(t)$, and $A_5(t)$ denote the accumulated values of an investment of $10,000 over a term of t years, and earning interest at the rate of 4%, 6%, 8%, 10%, and 12% per year compounded annually.

1. Find expressions for $A_1(t)$, $A_2(t)$, . . . , $A_5(t)$.

2. Use a graphing utility to plot the graphs of A_1, A_2, . . . , A_5 on the same set of axes, using the viewing window $[0, 20] \times [0, 100{,}000]$.

3. Use TRACE to find $A_1(20)$, $A_2(20)$, . . . , $A_5(20)$ and interpret your results.

▪ **Using Technology Subsections** that offer optional material explaining the use of graphing calculators as a tool to solve problems in finite mathematics and to construct and analyze mathematical models are placed at the end of appropriate sections. Once again many relevant applications with sourced data are introduced here. These subsections are written in the traditional example–exercise format, with answers given at the back of the book. They may be used in the classroom if desired or as material for self-study by the student. Illustrations showing graphing calculator screens and Microsoft Excel 2003 are extensively used. In many instances there are alternative ways of entering data onto a spreadsheet and/or dialog box, but only one method is presented here. Step-by-step instructions (including keystrokes) for many popular calculators are now given on the disc that accompanies the text. Written instructions are also given at the Web site.

USING TECHNOLOGY

■ Amortizing a Loan

Graphing Utility

Here we use the TI-83 TVM SOLVER function to help us solve problems involving amortization and sinking funds.

EXAMPLE 1 Finding the Payment to Amortize a Loan The Wongs are considering obtaining a preapproved 30-year loan of $120,000 to help finance the purchase of a house. The mortgage company charges interest at the rate of 8% per year on the unpaid balance, with interest computations made at the end of each month. What will be the monthly installments if the loan is amortized at the end of the term?

Solution We use the TI-83 TVM SOLVER with the following inputs:

$$N = 360 \qquad (30)(12)$$

■ TECHNOLOGY EXERCISES

1. Find the periodic payment required to amortize a loan of $55,000 over 120 periods with interest earned at the rate of $6\frac{5}{8}\%$/period.

2. Find the periodic payment required to amortize a loan of $178,000 over 180 periods with interest earned at the rate of

8. Find the periodic payment required to accumulate $144,000 over 120 periods with interest earned at the rate of $\frac{5}{8}\%$/period.

9. A loan of $120,000 is to be repaid over a 10-yr period through equal installments made at the end of each year. If

■ **New Interactive Video Skillbuilder CD**, in the back of every new text, contains hours of video instruction from award-winning teacher Deborah Upton of Stonehill College. Watch as she walks you through key examples from the text, step by step—giving you a foundation in the skills that you need to know. Each example found on the CD is identified by the video icon located in the margin.

APPLIED EXAMPLE 3 Saving for a College Education As a savings program toward Alberto's college education, his parents decide to deposit $100 at the end of every month into a bank account paying interest at the rate of 6% per year compounded monthly. If the savings program began when Alberto was 6 years old, how much money would have accumulated by the time he turns 18?

■ **New Graphing Calculator Tutorial**, by Larry Schroeder of Carl Sandburg College, can also be found on the *Interactive Video Skillbuilder CD* and includes step-by-step instructions, as well as video lessons.

■ **Student Resources on the Web** Students and instructors will now have access to the following additional materials at the Companion Web site: http://series.brookscole.com/tans

  ▪ Review material and practice chapter quizzes and tests
  ▪ Group projects and extended problems for each chapter
  ▪ Instructions, including keystrokes, for the procedures referenced in the text for specific calculators (TI-82, TI-83, TI-85, TI-86, and other popular models)

Other Changes in the Eighth Edition

■ **Expanded Coverage of Mathematical Modeling** In Linear Functions and Mathematical Modeling, a discussion of the mathematical modeling process has been added followed by a new applied example. Here students are asked to draw conclusions from a model constructed from real-life data.

■ **Using Technology subsections have been updated for Office 2003** and new dialog boxes are now shown.

■ **Other Changes** Continuous compound interest is now covered in Section 5.1. A discussion of the median and the mode has been added to Section 8.3.

■ **A Revised Student Solutions Manual** Problem-solving strategies and additional algebra steps and review for selected problems (identified in the *Instructor's Solutions Manual*) have been added to this supplement.

Teaching Aids

■ **Instructor's Solutions Manual** includes solutions to all exercises. ISBN 0-534-49215-0

■ **Instructor's Suite CD** contains complete solutions to all exercises, along with PowerPoint slide presentations and test items for every chapter, in formats compatible with Microsoft Office. ISBN 0-534-49291-6

■ **Printed Test Bank**, by Tracy Wang, is available to adopters of the book. ISBN 0-534-49216-9

■ **iLrn Testing**, available online or on CD-ROM. *iLrn Testing* is browser-based, fully integrated testing and course management software. With no need for plugins or downloads, *iLrn* offers algorithmically generated problem values and machine-graded free response mathematics. ISBN 0-534-49217-7

Learning Aids

■ **Student Solutions Manual**, available to both students and instructors, includes the solutions to odd-numbered exercises. ISBN 0-534-49218-5

■ **WebTutor Advantage for WebCT & Blackboard**, by Larry Schroeder, Carl Sandburg College, contains expanded online study tools including: step-by-step lecture notes; student study guide with step-by-step TI-89/92/83/86 and Microsoft Excel explanations; a quick check interactive student problem for each online example, with accompanying step-by-step solution and step-by-step TI-89/92/83/86 solution; practice quizzes by chapter sections that can be used as electronically graded online exercises, and much more. ISBN for *WebCT* 0-534-49219-3 and ISBN for *Blackboard* 0-534-49211-8

Acknowledgments

I wish to express my personal appreciation to each of the following reviewers of this Eighth Edition, whose many suggestions have helped make a much improved book.

Ronald Barnes
University of Houston

Larry Blaine
Plymouth State College

Candy Giovanni
Michigan State University

Joseph Macaluso
DeSales University

Marna Mozeff
Drexel University

Deborah Primm
Jacksonville State University

Michael Sterner
University of Montevallo

I also thank those previous edition reviewers whose comments and suggestions have helped to get the book this far.

Daniel D. Anderson
University of Iowa

Randy Anderson
California State University—Fresno

Ronald D. Baker
University of Delaware

Ronald Barnes
University of Houston—Downtown

Frank E. Bennett
Mount Saint Vincent University

Teresa L. Bittner
Canada College

Michael Button
San Diego City College

Frederick J. Carter
St. Mary's University

Charles E. Cleaver
The Citadel

Leslie S. Cobar
University of New Orleans

Matthew P. Coleman
Fairfield University

William Coppage
Wright State University

Jerry Davis
Johnson State College

Michael W. Ecker
Pennsylvania State University, Wilkes-Barre Campus

Bruce Edwards
University of Florida—Gainesville

Robert B. Eicken
Illinois Central College

Charles S. Frady
Georgia State University

Howard Frisinger
Colorado State University

William Geeslin
University of New Hampshire

Larry Gerstein
University of California—Santa Barbara

David Gross
University of Connecticut

Murli Gupta
George Washington University

John Haverhals
Bradley University

Yvette Hester
Texas A & M University

Sharon S. Hewlett
University of New Orleans

Patricia Hickey
Baylor University

Xiaoming Huang
Heidelberg College

Harry C. Hutchins
Southern Illinois University

Frank Jenkins
John Carroll University

Bruce Johnson
University of Victoria

David E. Joyce
Clark University

Martin Kotler
Pace University

John Kutzke
University of Portland

Paul E. Long
University of Arkansas

Larry Luck
Anoka-Ramsey Community College

Sandra Wray McAfee
University of Michigan

Gary MacGillivray
University of Victoria

Gary A. Martin
University of Massachusetts—Dartmouth

Norman R. Martin
Northern Arizona University

Ruth Mikkelson
University of Wisconsin—Stout

Maurice Monahan
South Dakota State University

John A. Muzzey
Lyndon State College

James D. Nelson
Western Michigan University

Ralph J. Neuhaus
University of Idaho

Richard J. O'Malley
University of Wisconsin—Milwaukee

Lloyd Olson
North Dakota State University

Wesley Orser
Clark College

Lavon B. Page
North Carolina State University

James Perkins
Piedmont Virginia Community College

Richard D. Porter
Northeastern University

Sandra Pryor Clarkson
Hunter College—SUNY

Richard Quindley
Bridgewater State College

C. Rao
University of Wisconsin

Chris Rodger
Auburn University

Robert H. Rodine
Northern Illinois University

Thomas N. Roe
South Dakota State University

Arnold Schroeder
Long Beach City College

Donald R. Sherbert
University of Illinois

Ron Smit
University of Portland

John St. Clair
Matlow State Community College

Lowell Stultz
Texas Township Campus

Francis J. Vlasko
Kutztown University

Lawrence V. Welch
Western Illinois University

I also wish to thank my colleague, Deborah Upton, who did a great job preparing the videos that now accompany the text and who helped with the accuracy check of the text. Special thanks also go to Tracy Wang for preparing the PowerPoint slides and the test bank, and to Tau Guo for his many helpful suggestions for improving the text.

My thanks also go to the editorial, production, and marketing staffs of Brooks/Cole: Curt Hinrichs, Danielle Derbenti, Ann Day, Sandra Craig, Tom Ziolkowski, Doreen Suruki, Fiona Chong, Earl Perry, Jessica Bothwell, and Sarah Harkrader for all of their help and support during the development and production of this edition. Finally, I wish to thank Cecile Joyner of The Cooper Company and Betty Duncan for doing an excellent job ensuring the accuracy and readability of this Eighth Edition, Diane Beasley for the design of the interior of the book, and Irene Morris for the cover design. Simply stated, the team I have been working with is outstanding, and I truly appreciate all of their hard work and effort.

S. T. Tan

About the Author

SOO T. TAN received his S.B. degree from Massachusetts Institute of Technology, his M.S. degree from the University of Wisconsin-Madison, and his Ph.D. from the University of California at Los Angeles. He has published numerous papers in Optimal Control Theory, Numerical Analysis, and Mathematics of Finance. He is currently a Professor of Mathematics at Stonehill College.

"By the time I started writing the first of what turned out to be a series of textbooks in mathematics for students in the managerial, life, and social sciences, I had quite a few years of experience teaching mathematics to non-mathematics majors. One of the most important lessons I learned from my early experience teaching these courses is that many of the students come into these courses with some degree of apprehension. This awareness led to the intuitive approach I have adopted in all of my texts. As you will see, I try to introduce each abstract mathematical concept through an example drawn from a common, real-life experience. Once the idea has been conveyed, I then proceed to make it precise, thereby assuring that no mathematical rigor is lost in this intuitive treatment of the subject. Another lesson I learned from my students is that they have a much greater appreciation of the material if the applications are drawn from their fields of interest and from situations that occur in the real world. This is one reason you will see so many exercises in my texts that are modeled on data gathered from newspapers, magazines, journals, and other media. Whether it be the market for cholesterol-reducing drugs, financing a home, bidding for cable rights, broadband Internet households, or Starbuck's annual sales, I weave topics of current interest into my examples and exercises, to keep the book relevant to all of my readers."

1 Straight Lines and Linear Functions

Which process should the company use? Robertson Controls Company must decide between two manufacturing processes for its Model C electronic thermostats. In Example 4, page 44, you will see how to determine which process will be more profitable.

© Jim Arbogast/PhotoDisc

THIS CHAPTER INTRODUCES the Cartesian coordinate system, a system that allows us to represent points in the plane in terms of ordered pairs of real numbers. This in turn enables us to compute the distance between two points algebraically. We also study straight lines. *Linear functions*, whose graphs are straight lines, can be used to describe many relationships between two quantities. These relationships can be found in fields of study as diverse as business, economics, the social sciences, physics, and medicine. In addition, we see how some practical problems can be solved by finding the point(s) of intersection of two straight lines. Finally, we learn how to find an algebraic representation of the straight line that "best" fits a set of data points that are scattered about a straight line.

1.1 The Cartesian Coordinate System

■ The Cartesian Coordinate System

The real number system is made up of the set of real numbers together with the usual operations of addition, subtraction, multiplication, and division. We assume that you are familiar with the rules governing these algebraic operations (see Appendix B).

Real numbers may be represented geometrically by points on a line. This line is called the **real number,** or **coordinate, line.** We can construct the real number line as follows: Arbitrarily select a point on a straight line to represent the number 0. This point is called the **origin.** If the line is horizontal, then choose a point at a convenient distance to the right of the origin to represent the number 1. This determines the scale for the number line. Each positive real number x lies x units to the right of 0, and each negative real number x lies $-x$ units to the left of 0.

In this manner, a one-to-one correspondence is set up between the set of real numbers and the set of points on the number line, with all the positive numbers lying to the right of the origin and all the negative numbers lying to the left of the origin (Figure 1).

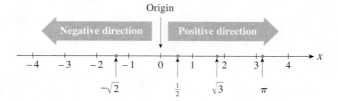

FIGURE 1
The real number line

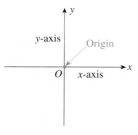

FIGURE 2
The Cartesian coordinate system

In a similar manner, we can represent points in a plane (a two-dimensional space) by using the **Cartesian coordinate system,** which we construct as follows: Take two perpendicular lines, one of which is normally chosen to be horizontal. These lines intersect at a point O, called the **origin** (Figure 2). The horizontal line is called the ***x*-axis,** and the vertical line is called the **y-axis.** A number scale is set up along the x-axis, with the positive numbers lying to the right of the origin and the negative numbers lying to the left of it. Similarly, a number scale is set up along the y-axis, with the positive numbers lying above the origin and the negative numbers lying below it.

Note   The number scales on the two axes need not be the same. Indeed, in many applications different quantities are represented by x and y. For example, x may represent the number of cell phones sold and y the total revenue resulting from the sales. In such cases it is often desirable to choose different number scales to represent the different quantities. Note, however, that the zeros of both number scales coincide at the origin of the two-dimensional coordinate system.   ■

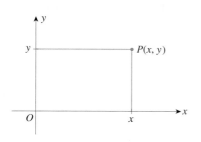

FIGURE 3
An ordered pair in the coordinate plane

We can represent a point in the plane uniquely in this coordinate system by an **ordered pair** of numbers—that is, a pair (x, y), where x is the first number and y the second. To see this, let P be any point in the plane (Figure 3). Draw perpendiculars from P to the x-axis and y-axis, respectively. Then the number x is precisely the number that corresponds to the point on the x-axis at which the perpendicular through P hits the x-axis. Similarly, y is the number that corresponds to the point on the y-axis at which the perpendicular through P crosses the y-axis.

Conversely, given an ordered pair (x, y), with x as the first number and y the second, a point P in the plane is uniquely determined as follows: Locate the point on the x-axis represented by the number x and draw a line through that point parallel to the y-axis. Next, locate the point on the y-axis represented by the number y and draw a line through that point parallel to the x-axis. The point of intersection of these two lines is the point P (Figure 3).

In the ordered pair (x, y), x is called the **abscissa,** or **x-coordinate,** y is called the **ordinate,** or **y-coordinate,** and x and y together are referred to as the **coordinates** of the point P. The point P with x-coordinate equal to a and y-coordinate equal to b is often written $P(a, b)$.

The points $A(2, 3)$, $B(-2, 3)$, $C(-2, -3)$, $D(2, -3)$, $E(3, 2)$, $F(4, 0)$, and $G(0, -5)$ are plotted in Figure 4.

Note   In general, $(x, y) \neq (y, x)$. This is illustrated by the points A and E in Figure 4.

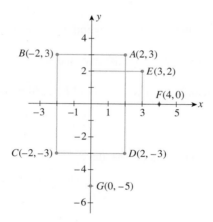

FIGURE 4
Several points in the coordinate plane

The axes divide the plane into four quadrants. Quadrant I consists of the points P with coordinates x and y, denoted by $P(x, y)$, satisfying $x > 0$ and $y > 0$; Quadrant II, the points $P(x, y)$, where $x < 0$ and $y > 0$; Quadrant III, the points $P(x, y)$, where $x < 0$ and $y < 0$; and Quadrant IV, the points $P(x, y)$, where $x > 0$ and $y < 0$ (Figure 5).

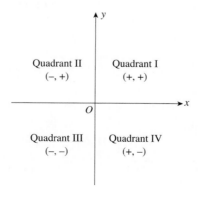

FIGURE 5
The four quadrants in the coordinate plane

FIGURE 6
The distance between two points in the coordinate plane

The Distance Formula

One immediate benefit that arises from using the Cartesian coordinate system is that the distance between any two points in the plane may be expressed solely in terms of the coordinates of the points. Suppose, for example, (x_1, y_1) and (x_2, y_2) are any two points in the plane (Figure 6). Then the distance d between these two points is, by the Pythagorean theorem,

$$d = \sqrt{(x_2 - x_1)^2 + (y_2 - y_1)^2}$$

For a proof of this result, see Exercise 45, page 9.

Distance Formula

The distance d between two points $P_1(x_1, y_1)$ and $P_2(x_2, y_2)$ in the plane is given by

$$d = \sqrt{(x_2 - x_1)^2 + (y_2 - y_1)^2} \qquad \textbf{(1)}$$

In what follows, we give several applications of the distance formula.

EXAMPLE 1   Find the distance between the points $(-4, 3)$ and $(2, 6)$.

Solution   Let $P_1(-4, 3)$ and $P_2(2, 6)$ be points in the plane. Then, we have

$$x_1 = -4 \quad \text{and} \quad y_1 = 3$$
$$x_2 = 2 \qquad\qquad y_2 = 6$$

Using Formula (1), we have

$$d = \sqrt{[2 - (-4)]^2 + (6 - 3)^2}$$
$$= \sqrt{6^2 + 3^2}$$
$$= \sqrt{45}$$
$$= 3\sqrt{5}$$

EXPLORE & DISCUSS

Refer to Example 1. Suppose we label the point $(2, 6)$ as P_1 and the point $(-4, 3)$ as P_2.
(1) Show that the distance d between the two points is the same as that obtained earlier.
(2) Prove that, in general, the distance d in Formula (1) is independent of the way we label the two points.

VIDEO

$ **APPLIED EXAMPLE 2   The Cost of Laying Cable**   In Figure 7, S represents the position of a power relay station located on a straight coastal highway, and M shows the location of a marine biology experimental station on a nearby island. A cable is to be laid connecting the relay station with the experimental station. If the cost of running the cable on land is \$1.50 per running foot and the cost of running the cable underwater is \$2.50 per running foot, find the total cost for laying the cable.

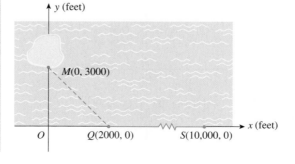

FIGURE 7
The cable will connect the relay station S to the experimental station M.

Solution The length of cable required on land is given by the distance from S to Q. This distance is (10,000 − 2000), or 8000 feet. Next, we see that the length of cable required underwater is given by the distance from Q to M. This distance is

$$\begin{aligned}\sqrt{(0-2000)^2+(3000-0)^2} &= \sqrt{2000^2+3000^2}\\ &= \sqrt{13{,}000{,}000}\\ &\approx 3605.55\end{aligned}$$

or approximately 3605.55 feet. Therefore, the total cost for laying the cable is

$$1.5(8000) + 2.5(3605.55) = 21{,}013.875$$

or approximately \$21,014.

VIDEO

EXAMPLE 3 Let $P(x, y)$ denote a point lying on the circle with radius r and center $C(h, k)$ (Figure 8). Find a relationship between x and y.

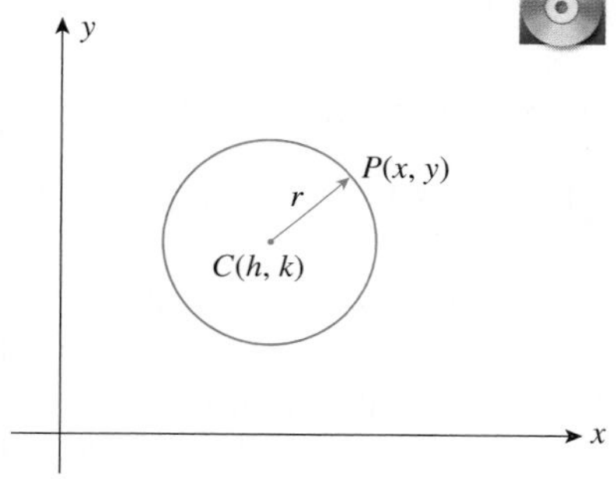

FIGURE 8
A circle with radius r and center $C(h, k)$

Solution By the definition of a circle, the distance between $C(h, k)$ and $P(x, y)$ is r. Using Formula (1), we have

$$\sqrt{(x-h)^2+(y-k)^2} = r$$

which, upon squaring both sides, gives the equation

$$(x-h)^2+(y-k)^2 = r^2$$

that must be satisfied by the variables x and y.

A summary of the result obtained in Example 3 follows.

Equation of a Circle
An equation of the circle with center $C(h, k)$ and radius r is given by

$$(x-h)^2+(y-k)^2 = r^2 \tag{2}$$

EXAMPLE 4 Find an equation of the circle with (a) radius 2 and center (−1, 3) and (b) radius 3 and center located at the origin.

Solution

a. We use Formula (2) with $r = 2$, $h = -1$, and $k = 3$, obtaining

$$\begin{aligned}[x-(-1)]^2+(y-3)^2 &= 2^2\\ (x+1)^2+(y-3)^2 &= 4\end{aligned}$$

(Figure 9a).

b. Using Formula (2) with $r = 3$, $h = k = 0$, we obtain

$$\begin{aligned}x^2+y^2 &= 3^2\\ x^2+y^2 &= 9\end{aligned}$$

(Figure 9b).

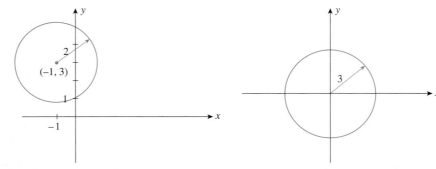

FIGURE 9

(a) The circle with radius 2 and center (−1, 3)

(b) The circle with radius 3 and center (0, 0)

EXPLORE & DISCUSS

1. Use the distance formula to help you describe the set of points in the xy-plane satisfying each of the following inequalities.

   a. $(x - h)^2 + (y - k)^2 \leq r^2$ **c.** $(x - h)^2 + (y - k)^2 \geq r^2$

   b. $(x - h)^2 + (y - k)^2 < r^2$ **d.** $(x - h)^2 + (y - k)^2 > r^2$

2. Consider the equation $x^2 + y^2 = 4$.

   a. Show that $y = \pm\sqrt{4 - x^2}$.

   b. Describe the set of points (x, y) in the xy-plane satisfying the equation

   $$\text{(i)} \ \ y = \sqrt{4 - x^2} \qquad \text{(ii)} \ \ y = -\sqrt{4 - x^2}$$

1.1 Self-Check Exercises

1. **a.** Plot the points $A(4, -2)$, $B(2, 3)$, and $C(-3, 1)$.

   b. Find the distance between the points A and B, between B and C, and between A and C.

   c. Use the Pythagorean theorem to show that the triangle with vertices A, B, and C is a right triangle.

2. The accompanying figure shows the location of cities A, B, and C. Suppose a pilot wishes to fly from city A to city C but must make a mandatory stopover in city B. If the single-engine light plane has a range of 650 mi, can the pilot make the trip without refueling in city B?

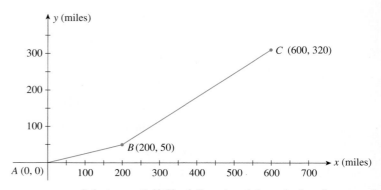

Solutions to Self-Check Exercises 1.1 can be found on page 9.

1.1 Concept Questions

1. What can you say about the signs of a and b if the point $P(a, b)$ lies in (a) the second quadrant? (b) The third quadrant? (c) The fourth quadrant?

2. **a.** What is the distance between $P_1(x_1, y_1)$ and $P_2(x_2, y_2)$?
   b. When you use the distance formula, does it matter which point is labeled P_1 and which point is labeled P_2? Explain.

1.1 Exercises

In Exercises 1–6, refer to the accompanying figure and determine the coordinates of the point and the quadrant in which it is located.

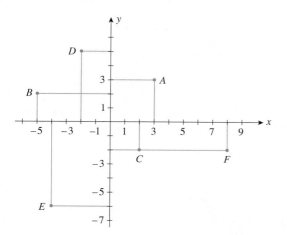

1. A **2.** B **3.** C

4. D **5.** E **6.** F

In Exercises 7–12, refer to the accompanying figure.

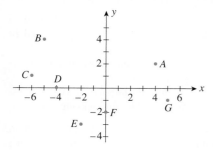

7. Which point has coordinates $(4, 2)$?

8. What are the coordinates of point B?

9. Which points have negative y-coordinates?

10. Which point has a negative x-coordinate and a negative y-coordinate?

11. Which point has an x-coordinate that is equal to zero?

12. Which point has a y-coordinate that is equal to zero?

In Exercises 13–20, sketch a set of coordinate axes and plot the point.

13. $(-2, 5)$ **14.** $(1, 3)$

15. $(3, -1)$ **16.** $(3, -4)$

17. $(8, -7/2)$ **18.** $(-5/2, 3/2)$

19. $(4.5, -4.5)$ **20.** $(1.2, -3.4)$

In Exercises 21–24, find the distance between the points.

21. $(1, 3)$ and $(4, 7)$ **22.** $(1, 0)$ and $(4, 4)$

23. $(-1, 3)$ and $(4, 9)$

24. $(-2, 1)$ and $(10, 6)$

25. Find the coordinates of the points that are 10 units away from the origin and have a y-coordinate equal to -6.

26. Find the coordinates of the points that are 5 units away from the origin and have an x-coordinate equal to 3.

27. Show that the points $(3, 4)$, $(-3, 7)$, $(-6, 1)$, and $(0, -2)$ form the vertices of a square.

28. Show that the triangle with vertices $(-5, 2)$, $(-2, 5)$, and $(5, -2)$ is a right triangle.

In Exercises 29–34, find an equation of the circle that satisfies the conditions.

29. Radius 5 and center $(2, -3)$

30. Radius 3 and center $(-2, -4)$

31. Radius 5 and center at the origin

32. Center at the origin and passes through $(2, 3)$

33. Center $(2, -3)$ and passes through $(5, 2)$

34. Center $(-a, a)$ and radius $2a$

35. DISTANCE TRAVELED A grand tour of four cities begins at city *A* and makes successive stops at cities *B*, *C*, and *D* before returning to city *A*. If the cities are located as shown in the accompanying figure, find the total distance covered on the tour.

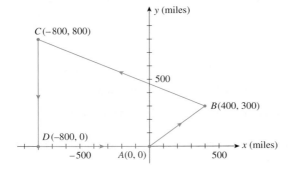

36. DELIVERY CHARGES A furniture store offers free setup and delivery services to all points within a 25-mi radius of its warehouse distribution center. If you live 20 mi east and 14 mi south of the warehouse, will you incur a delivery charge? Justify your answer.

37. OPTIMIZING TRAVEL TIME Towns *A*, *B*, *C*, and *D* are located as shown in the accompanying figure. Two highways link town *A* to town *D*. Route 1 runs from town *A* to town *D* via town *B*, and Route 2 runs from town *A* to town *D* via town *C*. If a salesman wishes to drive from town *A* to town *D* and traffic conditions are such that he could expect to average the same speed on either route, which highway should he take in order to arrive in the shortest time?

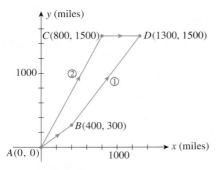

38. MINIMIZING SHIPPING COSTS Refer to the figure for Exercise 37. Suppose a fleet of 100 automobiles are to be shipped from an assembly plant in town *A* to town *D*. They may be shipped either by freight train along Route 1 at a cost of 22¢/mile/automobile or by truck along Route 2 at a cost of 21¢/mile/automobile. Which means of transportation minimizes the shipping cost? What is the net savings?

39. CONSUMER DECISIONS Will Barclay wishes to determine which antenna he should purchase for his home. The TV store has supplied him with the following information:

| Range in miles | | | |
VHF	UHF	Model	Price
30	20	A	$40
45	35	B	50
60	40	C	60
75	55	D	70

Will wishes to receive Channel 17 (VHF), which is located 25 mi east and 35 mi north of his home, and Channel 38 (UHF), which is located 20 mi south and 32 mi west of his home. Which model will allow him to receive both channels at the least cost? (Assume that the terrain between Will's home and both broadcasting stations is flat.)

40. COST OF LAYING CABLE In the accompanying diagram, *S* represents the position of a power relay station located on a straight coastal highway, and *M* shows the location of a marine biology experimental station on a nearby island. A cable is to be laid connecting the relay station with the experimental station. If the cost of running the cable on land is $1.50/running foot and the cost of running cable underwater is $2.50/running foot, find an expression in terms of *x* that gives the total cost of laying the cable. Use this expression to find the total cost when $x = 1500$ and when $x = 2500$.

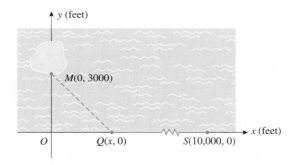

41. Two ships leave port at the same time. Ship *A* sails north at a speed of 20 mph while ship *B* sails east at a speed of 30 mph.
   a. Find an expression in terms of the time *t* (in hours) giving the distance between the two ships.
   b. Using the expression obtained in part (a), find the distance between the two ships 2 hr after leaving port.

42. Sailing north at a speed of 25 mph, ship *A* leaves a port. A half hour later, ship *B* leaves the same port, sailing east at a speed of 20 mph. Let *t* (in hours) denote the time ship *B* has been at sea.
   a. Find an expression in terms of *t*, giving the distance between the two ships.
   b. Use the expression obtained in part (a) to find the distance between the two ships 2 hr after ship *A* has left the port.

In Exercises 43 and 44, determine whether the statement is true or false. If it is true, explain why it is true. If it is false, give an example to show why it is false.

43. If the distance between the points $P_1(a, b)$ and $P_2(c, d)$ is D, then the distance between the points $P_1(a, b)$ and $P_3(kc, kd)$ ($k \neq 0$) is given by $|k|D$.

44. The circle with equation $kx^2 + ky^2 = a^2$ lies inside the circle with equation $x^2 + y^2 = a^2$, provided $k > 1$.

45. Let (x_1, y_1) and (x_2, y_2) be two points lying in the xy-plane. Show that the distance between the two points is given by

$$d = \sqrt{(x_2 - x_1)^2 + (y_2 - y_1)^2}$$

Hint: Refer to the accompanying figure and use the Pythagorean theorem.

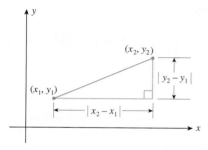

46. In the Cartesian coordinate system, the two axes are perpendicular to each other. Consider a coordinate system in which the x- and y-axis are noncollinear (that is, the axes do not lie along a straight line) and are not perpendicular to each other (see the accompanying figure).

a. Describe how a point is represented in this coordinate system by an ordered pair (x, y) of real numbers. Conversely, show how an ordered pair (x, y) of real numbers uniquely determines a point in the plane.

b. Suppose you want to find a formula for the distance between two points, $P_1(x_1, y_1)$ and $P_2(x_2, y_2)$, in the plane. What advantage does the Cartesian coordinate system have over the coordinate system under consideration? Comment on your answer.

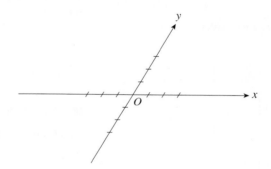

1.1 Solutions to Self-Check Exercises

1. a. The points are plotted in the accompanying figure.

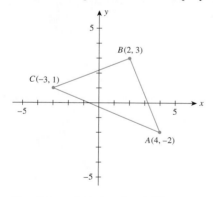

b. The distance between A and B is

$$d(A, B) = \sqrt{(2 - 4)^2 + [3 - (-2)]^2}$$
$$= \sqrt{(-2)^2 + 5^2} = \sqrt{4 + 25} = \sqrt{29}$$

The distance between B and C is

$$d(B, C) = \sqrt{(-3 - 2)^2 + (1 - 3)^2}$$
$$= \sqrt{(-5)^2 + (-2)^2} = \sqrt{25 + 4} = \sqrt{29}$$

The distance between A and C is

$$d(A, C) = \sqrt{(-3 - 4)^2 + [1 - (-2)]^2}$$
$$= \sqrt{(-7)^2 + 3^2} = \sqrt{49 + 9} = \sqrt{58}$$

c. We will show that

$$[d(A, C)]^2 = [d(A, B)]^2 + [d(B, C)]^2$$

From part (b), we see that $[d(A, B)]^2 = 29$, $[d(B, C)]^2 = 29$, and $[d(A, C)]^2 = 58$, and the desired result follows.

2. The distance between city A and city B is

$$d(A, B) = \sqrt{200^2 + 50^2} \approx 206$$

or 206 mi. The distance between city B and city C is

$$d(B, C) = \sqrt{(600 - 200)^2 + (320 - 50)^2}$$
$$= \sqrt{400^2 + 270^2} \approx 483$$

or 483 mi. Therefore, the total distance the pilot would have to cover is 689 mi, so she must refuel in city B.

1.2   Straight Lines

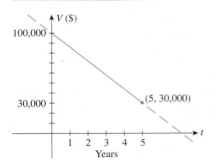

FIGURE 10
Linear depreciation of an asset

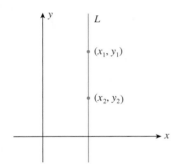

FIGURE 11
The slope is undefined.

In computing income tax, business firms are allowed by law to depreciate certain assets such as buildings, machines, furniture, automobiles, and so on, over a period of time. *Linear depreciation*, or the *straight-line method*, is often used for this purpose. The graph of the straight line shown in Figure 10 describes the book value V of a computer that has an initial value of \$100,000 and that is being depreciated linearly over 5 years with a scrap value of \$30,000. Note that only the solid portion of the straight line is of interest here.

The book value of the computer at the end of year t, where t lies between 0 and 5, can be read directly from the graph. But there is one shortcoming in this approach: The result depends on how accurately you draw and read the graph. A better and more accurate method is based on finding an *algebraic* representation of the depreciation line. (We will continue our discussion of the linear depreciation problem in Section 1.3.)

To see how a straight line in the xy-plane may be described algebraically, we need to first recall certain properties of straight lines.

■ Slope of a Line

Let L denote the unique straight line that passes through the two distinct points (x_1, y_1) and (x_2, y_2). If $x_1 = x_2$, then L is a vertical line, and the slope is undefined (Figure 11). If $x_1 \neq x_2$, we define the slope of L as follows.

Slope of a Nonvertical Line

If (x_1, y_1) and (x_2, y_2) are any two distinct points on a nonvertical line L, then the slope m of L is given by

$$m = \frac{\Delta y}{\Delta x} = \frac{y_2 - y_1}{x_2 - x_1} \tag{3}$$

(Figure 12).

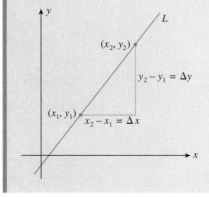

FIGURE 12

Thus, the slope of a straight line is a constant whenever it is defined.

The number $\Delta y = y_2 - y_1$ (Δy is read "delta y") is a measure of the vertical change in y, and $\Delta x = x_2 - x_1$ is a measure of the horizontal change in x as shown in Figure 12. From this figure we can see that the slope m of a straight line L is a

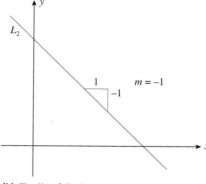

FIGURE 13

(a) The line rises ($m > 0$).   **(b)** The line falls ($m < 0$).

measure of the *rate of change of y with respect to x*. Furthermore, the slope of a non-vertical straight line is constant, and this tells us that this rate of change is constant.

Figure 13a shows a straight line L_1 with slope 2. Observe that L_1 has the property that a 1-unit increase in x results in a 2-unit increase in y. To see this, let $\Delta x = 1$ in Equation (3) so that $m = \Delta y$. Since $m = 2$, we conclude that $\Delta y = 2$. Similarly, Figure 13b shows a line L_2 with slope -1. Observe that a straight line with positive slope slants upward from left to right (y increases as x increases), whereas a line with negative slope slants downward from left to right (y decreases as x increases). Finally, Figure 14 shows a family of straight lines passing through the origin with indicated slopes.

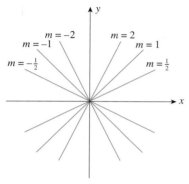

FIGURE 14
A family of straight lines

EXPLORE & DISCUSS

Show that the slope of a nonvertical line is independent of the two distinct points used to compute it.

Hint: Suppose we pick two other distinct points, $P_3(x_3, y_3)$ and $P_4(x_4, y_4)$ lying on L. Draw a picture and use similar triangles to demonstrate that using P_3 and P_4 gives the same value as that obtained using P_1 and P_2.

EXAMPLE 1   Sketch the straight line that passes through the point $(-2, 5)$ and has slope $-\frac{4}{3}$.

Solution   First, plot the point $(-2, 5)$ (Figure 15). Next, recall that a slope of $-\frac{4}{3}$ indicates that an increase of 1 unit in the x-direction produces a *decrease* of $\frac{4}{3}$ units in the y-direction, or equivalently, a 3-unit increase in the x-direction produces a $3(\frac{4}{3})$, or 4-unit, decrease in the y-direction. Using this information, we plot the point $(1, 1)$ and draw the line through the two points.

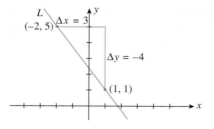

FIGURE 15
L has slope $-\frac{4}{3}$ and passes through $(-2, 5)$.

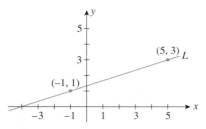

FIGURE 16
L passes through (5, 3) and (−1, 1).

EXAMPLE 2   Find the slope *m* of the line that passes through the points (−1, 1) and (5, 3).

Solution   Choose (x_1, y_1) to be the point (−1, 1) and (x_2, y_2) to be the point (5, 3). Then, with $x_1 = -1$, $y_1 = 1$, $x_2 = 5$, and $y_2 = 3$, we find, using Equation (3),

$$m = \frac{y_2 - y_1}{x_2 - x_1} = \frac{3 - 1}{5 - (-1)} = \frac{2}{6} = \frac{1}{3}$$

Figure 16). Try to verify that the result obtained would have been the same had we chosen the point (−1, 1) to be (x_2, y_2) and the point (5, 3) to be (x_1, y_1). ■

EXAMPLE 3   Find the slope of the line that passes through the points (−2, 5) and (3, 5).

Solution   The slope of the required line is given by

$$m = \frac{5 - 5}{3 - (-2)} = \frac{0}{5} = 0$$

(Figure 17).

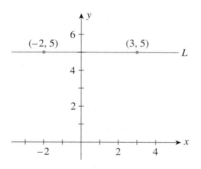

FIGURE 17
The slope of the horizontal line *L* is zero.

Note   In general, the slope of a horizontal line is zero. ■

We can use the slope of a straight line to determine whether a line is parallel to another line.

Parallel Lines
Two distinct lines are **parallel** if and only if their slopes are equal or their slopes are undefined.

EXAMPLE 4   Let L_1 be a line that passes through the points (−2, 9) and (1, 3) and let L_2 be the line that passes through the points (−4, 10) and (3, −4). Determine whether L_1 and L_2 are parallel.

Solution   The slope m_1 of L_1 is given by

$$m_1 = \frac{3 - 9}{1 - (-2)} = -2$$

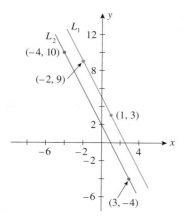

FIGURE 18
L_1 and L_2 have the same slope and hence are parallel.

The slope m_2 of L_2 is given by

$$m_2 = \frac{-4 - 10}{3 - (-4)} = -2$$

Since $m_1 = m_2$, the lines L_1 and L_2 are in fact parallel (Figure 18). ∎

◼ Equations of Lines

We now show that every straight line lying in the xy-plane may be represented by an equation involving the variables x and y. One immediate benefit of this is that problems involving straight lines may be solved algebraically.

Let L be a straight line parallel to the y-axis (perpendicular to the x-axis) (Figure 19). Then L crosses the x-axis at some point $(a, 0)$ with the x-coordinate given by $x = a$, where a is some real number. Any other point on L has the form $(a, \bar{y})$, where $\bar{y}$ is an appropriate number. Therefore, the vertical line L is described by the sole condition

$$x = a$$

and this is accordingly an equation of L. For example, the equation $x = -2$ represents a vertical line 2 units to the left of the y-axis, and the equation $x = 3$ represents a vertical line 3 units to the right of the y-axis (Figure 20).

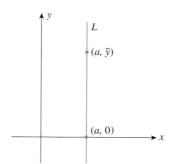

FIGURE 19
The vertical line $x = a$

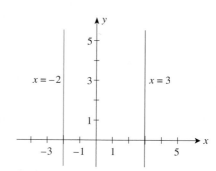

FIGURE 20
The vertical lines $x = -2$ and $x = 3$

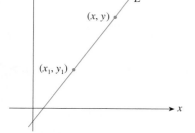

FIGURE 21
L passes through (x_1, y_1) and has slope m.

Next, suppose L is a nonvertical line so that it has a well-defined slope m. Suppose (x_1, y_1) is a fixed point lying on L and (x, y) is a variable point on L distinct from (x_1, y_1) (Figure 21). Using Equation (3) with the point $(x_2, y_2) = (x, y)$, we find that the slope of L is given by

$$m = \frac{y - y_1}{x - x_1}$$

Upon multiplying both sides of the equation by $x - x_1$, we obtain Equation (4).

Point-Slope Form
An equation of the line that has slope m and passes through the point (x_1, y_1) is given by

$$y - y_1 = m(x - x_1) \tag{4}$$

Equation (4) is called the *point-slope form* of the equation of a line since it utilizes a given point (x_1, y_1) on a line and the slope m of the line.

EXAMPLE 5   Find an equation of the line that passes through the point $(1, 3)$ and has slope 2.

Solution   Using the point-slope form of the equation of a line with the point $(1, 3)$ and $m = 2$, we obtain

$$y - 3 = 2(x - 1) \qquad {\scriptstyle y - y_1 = m(x - x_1)}$$

which, when simplified, becomes

$$2x - y + 1 = 0$$

(Figure 22). ∎

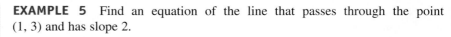

EXAMPLE 6   Find an equation of the line that passes through the points $(-3, 2)$ and $(4, -1)$.

Solution   The slope of the line is given by

$$m = \frac{-1 - 2}{4 - (-3)} = -\frac{3}{7}$$

Using the point-slope form of the equation of a line with the point $(4, -1)$ and the slope $m = -\frac{3}{7}$, we have

$$y + 1 = -\frac{3}{7}(x - 4) \qquad {\scriptstyle y - y_1 = m(x - x_1)}$$

$$7y + 7 = -3x + 12$$

$$3x + 7y - 5 = 0$$

(Figure 23).

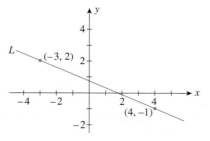

FIGURE 22
L passes through (1, 3) and has slope 2.

FIGURE 23
L passes through (−3, 2) and (4, −1).

∎

We can use the slope of a straight line to determine whether a line is perpendicular to another line.

Perpendicular Lines

If L_1 and L_2 are two distinct nonvertical lines that have slopes m_1 and m_2, respectively, then L_1 is **perpendicular** to L_2 (written $L_1 \perp L_2$) if and only if

$$m_1 = -\frac{1}{m_2}$$

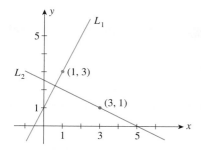

FIGURE 24
L_2 is perpendicular to L_1 and passes through (3, 1).

If the line L_1 is vertical (so that its slope is undefined), then L_1 is perpendicular to another line, L_2, if and only if L_2 is horizontal (so that its slope is zero). For a proof of these results, see Exercise 90, page 23.

EXAMPLE 7 Find an equation of the line that passes through the point (3, 1) and is perpendicular to the line of Example 5.

Solution Since the slope of the line in Example 5 is 2, the slope of the required line is given by $m = -\frac{1}{2}$, the negative reciprocal of 2. Using the point-slope form of the equation of a line, we obtain

$$y - 1 = -\frac{1}{2}(x - 3) \quad \text{\small $y - y_1 = m(x - x_1)$}$$
$$2y - 2 = -x + 3$$
$$x + 2y - 5 = 0$$

(Figure 24). ∎

 EXPLORING WITH TECHNOLOGY

1. Use a graphing utility to plot the straight lines L_1 and L_2 with equations $2x + y - 5 = 0$ and $41x + 20y - 11 = 0$ on the same set of axes, using the standard viewing window.
   a. Can you tell if the lines L_1 and L_2 are parallel to each other?
   b. Verify your observations by computing the slopes of L_1 and L_2 algebraically.

2. Use a graphing utility to plot the straight lines L_1 and L_2 with equations $x + 2y - 5 = 0$ and $5x - y + 5 = 0$ on the same set of axes, using the standard viewing window.
   a. Can you tell if the lines L_1 and L_2 are perpendicular to each other?
   b. Verify your observation by computing the slopes of L_1 and L_2 algebraically.

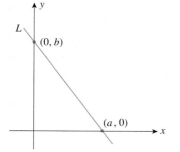

FIGURE 25
The line L has x-intercept a and y-intercept b.

A straight line L that is neither horizontal nor vertical cuts the x-axis and the y-axis at, say, points $(a, 0)$ and $(0, b)$, respectively (Figure 25). The numbers a and b are called the **x-intercept** and **y-intercept,** respectively, of L.

Now, let L be a line with slope m and y-intercept b. Using Equation (4), the point-slope form of the equation of a line, with the point given by $(0, b)$ and slope m, we have

$$y - b = m(x - 0)$$
$$y = mx + b$$

This is called the *slope-intercept form* of an equation of a line.

Slope-Intercept Form
The equation of the line that has slope m and intersects the y-axis at the point $(0, b)$ is given by

$$y = mx + b \tag{5}$$

EXAMPLE 8　Find an equation of the line that has slope 3 and y-intercept -4.

Solution　Using Equation (5) with $m = 3$ and $b = -4$, we obtain the required equation:

$$y = 3x - 4$$

EXAMPLE 9　Determine the slope and y-intercept of the line whose equation is $3x - 4y = 8$.

Solution　Rewrite the given equation in the slope-intercept form and obtain

$$y = \frac{3}{4}x - 2$$

Comparing this result with Equation (5), we find $m = \frac{3}{4}$ and $b = -2$, and we conclude that the slope and y-intercept of the given line are $\frac{3}{4}$ and -2, respectively.

EXPLORING WITH TECHNOLOGY

1. Use a graphing utility to plot the straight lines with equations $y = -2x + 3$, $y = -x + 3$, $y = x + 3$, and $y = 2.5x + 3$ on the same set of axes, using the standard viewing window. What effect does changing the coefficient m of x in the equation $y = mx + b$ have on its graph?

2. Use a graphing utility to plot the straight lines with equations $y = 2x - 2$, $y = 2x - 1$, $y = 2x$, $y = 2x + 1$, and $y = 2x + 4$ on the same set of axes, using the standard viewing window. What effect does changing the constant b in the equation $y = mx + b$ have on its graph?

3. Describe in words the effect of changing both m and b in the equation $y = mx + b$.

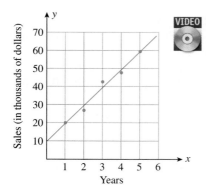

FIGURE 26
Sales of a sporting goods store

APPLIED EXAMPLE 10　**Predicting Sales Figures**　The sales manager of a local sporting goods store plotted sales versus time for the last 5 years and found the points to lie approximately along a straight line (Figure 26). By using the points corresponding to the first and fifth years, find an equation of the *trend line*. What sales figure can be predicted for the sixth year?

Solution　Using Equation (3) with the points $(1, 20)$ and $(5, 60)$, we find that the slope of the required line is given by

$$m = \frac{60 - 20}{5 - 1} = 10$$

Next, using the point-slope form of the equation of a line with the point $(1, 20)$ and $m = 10$, we obtain

$$y - 20 = 10(x - 1)$$
$$y = 10x + 10$$

as the required equation.

The sales figure for the sixth year is obtained by letting $x = 6$ in the last equation, giving

$$y = 10(6) + 10 = 70$$

or $70,000. ∎

APPLIED EXAMPLE 11 Predicting the Value of Art Suppose an art object purchased for $50,000 is expected to appreciate in value at a constant rate of $5000 per year for the next 5 years. Use Equation (5) to write an equation predicting the value of the art object in the next several years. What will be its value 3 years from the purchase date?

EXPLORE & DISCUSS

Refer to Example 11. Can the equation predicting the value of the art object be used to predict long-term growth?

Solution Let x denote the time (in years) that has elapsed since the purchase date and let y denote the object's value (in dollars). Then, $y = 50,000$ when $x = 0$. Furthermore, the slope of the required equation is given by $m = 5000$ since each unit increase in x (1 year) implies an increase of 5000 units (dollars) in y. Using Equation (5) with $m = 5000$ and $b = 50,000$, we obtain

$$y = 5000x + 50,000$$

Three years from the purchase date, the value of the object will be given by

$$y = 5000(3) + 50,000$$

or $65,000. ∎

▬ General Form of an Equation of a Line

We have considered several forms of the equation of a straight line in the plane. These different forms of the equation are equivalent to each other. In fact, each is a special case of the following equation.

General Form of a Linear Equation
The equation

$$Ax + By + C = 0 \tag{6}$$

where A, B, and C are constants and A and B are not both zero, is called the general form of a linear equation in the variables x and y.

We now state (without proof) an important result concerning the algebraic representation of straight lines in the plane.

An equation of a straight line is a linear equation; conversely, every linear equation represents a straight line.

This result justifies the use of the adjective *linear* in describing Equation (6).

EXAMPLE 12 Sketch the straight line represented by the equation

$$3x - 4y - 12 = 0$$

Solution   Since every straight line is uniquely determined by two distinct points, we need find only two such points through which the line passes in order to sketch it. For convenience, let's compute the points at which the line crosses the x- and y-axes. Setting $y = 0$, we find $x = 4$, so the line crosses the x-axis at the point $(4, 0)$. Setting $x = 0$ gives $y = -3$, so the line crosses the y-axis at the point $(0, -3)$. A sketch of the line appears in Figure 27.

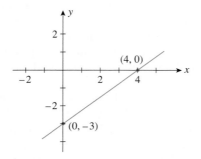

FIGURE 27
The straight line $3x - 4y = 12$

Following is a summary of the common forms of the equations of straight lines discussed in this section.

Equations of Straight Lines

Vertical line: $x = a$

Horizontal line: $y = b$

Point-slope form: $y - y_1 = m(x - x_1)$

Slope-intercept form: $y = mx + b$

General form: $Ax + By + C = 0$

1.2   Self-Check Exercises

1. Determine the number a so that the line passing through the points $(a, 2)$ and $(3, 6)$ is parallel to a line with slope 4.

2. Find an equation of the line that passes through the point $(3, -1)$ and is perpendicular to a line with slope $-\frac{1}{2}$.

3. Does the point $(3, -3)$ lie on the line with equation $2x - 3y - 12 = 0$? Sketch the graph of the line.

4. The percent of people over age 65 who have high school diplomas is summarized in the following table:

Year, x	1960	1965	1970	1975	1980	1985	1990
Percent with Diplomas, y	20	25	30	36	42	47	52

Source: U.S. Department of Commerce

a. Plot the percent of people over age 65 who have high school diplomas (y) versus the year (x).

b. Draw the straight line L through the points $(1960, 20)$ and $(1990, 52)$.

c. Find an equation of the line L.

d. Assume the trend continued and estimate the percent of people over age 65 who had high school diplomas by the year 1995.

Solutions to Self-Check Exercises 1.2 can be found on page 23.

1.2 Concept Questions

1. What is the slope of a nonvertical line? What can you say about the slope of a vertical line?

2. Give (a) the point-slope form, (b) the slope-intercept form, and (c) the general form of an equation of a line.

3. Let L_1 have slope m_1 and let L_2 have slope m_2. State the conditions on m_1 and m_2 if (a) L_1 is parallel to L_2 and (b) L_1 is perpendicular to L_2.

1.2 Exercises

In Exercises 1–4, find the slope of the line shown in each figure.

1.

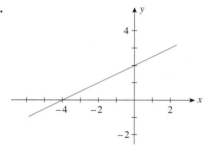

2.

3.

4.

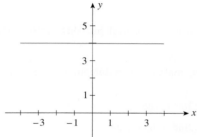

In Exercises 5–10, find the slope of the line that passes through the pair of points.

5. $(4, 3)$ and $(5, 8)$

6. $(4, 5)$ and $(3, 8)$

7. $(-2, 3)$ and $(4, 8)$

8. $(-2, -2)$ and $(4, -4)$

9. (a, b) and (c, d)

10. $(-a + 1, b - 1)$ and $(a + 1, -b)$

11. Given the equation $y = 4x - 3$, answer the following questions.
    a. If x increases by 1 unit, what is the corresponding change in y?
    b. If x decreases by 2 units, what is the corresponding change in y?

12. Given the equation $2x + 3y = 4$, answer the following questions.
    a. Is the slope of the line described by this equation positive or negative?
    b. As x increases in value, does y increase or decrease?
    c. If x decreases by 2 units, what is the corresponding change in y?

In Exercises 13 and 14, determine whether the lines through the pairs of points are parallel.

13. $A(1, -2)$, $B(-3, -10)$ and $C(1, 5)$, $D(-1, 1)$

14. $A(2, 3)$, $B(2, -2)$ and $C(-2, 4)$, $D(-2, 5)$

In Exercises 15 and 16, determine whether the lines through the pairs of points are perpendicular.

15. $A(-2, 5)$, $B(4, 2)$ and $C(-1, -2)$, $D(3, 6)$

16. $A(2, 0)$, $B(1, -2)$ and $C(4, 2)$, $D(-8, 4)$

17. If the line passing through the points $(1, a)$ and $(4, -2)$ is parallel to the line passing through the points $(2, 8)$ and $(-7, a + 4)$, what is the value of a?

18. If the line passing through the points $(a, 1)$ and $(5, 8)$ is parallel to the line passing through the points $(4, 9)$ and $(a + 2, 1)$, what is the value of a?

19. Find an equation of the horizontal line that passes through $(-4, -3)$.

20. Find an equation of the vertical line that passes through $(0, 5)$.

In Exercises 21–26, match the statement with one of the graphs a–f.

21. The slope of the line is zero.

22. The slope of the line is undefined.

23. The slope of the line is positive, and its y-intercept is positive.

24. The slope of the line is positive, and its y-intercept is negative.

25. The slope of the line is negative, and its x-intercept is negative.

26. The slope of the line is negative, and its x-intercept is positive.

a.

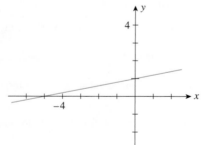

b.

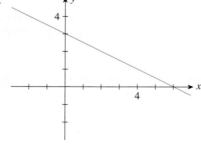

c.

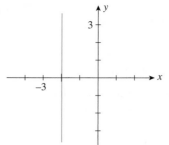

d.

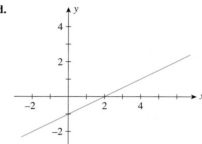

e.

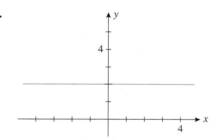

f.

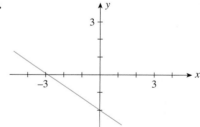

In Exercises 27–30, find an equation of the line that passes through the point and has the indicated slope m.

27. $(3, -4)$; $m = 2$ **28.** $(2, 4)$; $m = -1$

29. $(-3, 2)$; $m = 0$ **30.** $(1, 2)$; $m = -\dfrac{1}{2}$

In Exercises 31–34, find an equation of the line that passes through the points.

31. $(2, 4)$ and $(3, 7)$ **32.** $(2, 1)$ and $(2, 5)$

33. $(1, 2)$ and $(-3, -2)$ **34.** $(-1, -2)$ and $(3, -4)$

In Exercises 35–38, find an equation of the line that has slope m and y-intercept b.

35. $m = 3$; $b = 4$

36. $m = -2$; $b = -1$

37. $m = 0$; $b = 5$

38. $m = -\dfrac{1}{2}$; $b = \dfrac{3}{4}$

In Exercises 39–44, write the equation in the slope-intercept form and then find the slope and y-intercept of the corresponding line.

39. $x - 2y = 0$

40. $y - 2 = 0$

41. $2x - 3y - 9 = 0$

42. $3x - 4y + 8 = 0$

43. $2x + 4y = 14$

44. $5x + 8y - 24 = 0$

45. Find an equation of the line that passes through the point $(-2, 2)$ and is parallel to the line $2x - 4y - 8 = 0$.

46. Find an equation of the line that passes through the point $(2, 4)$ and is perpendicular to the line $3x + 4y - 22 = 0$.

In Exercises 47–52, find an equation of the line that satisfies the condition.

47. The line parallel to the x-axis and 6 units below it

48. The line passing through the origin and parallel to the line passing through the points $(2, 4)$ and $(4, 7)$

49. The line passing through the point (a, b) with slope equal to zero

50. The line passing through $(-3, 4)$ and parallel to the x-axis

51. The line passing through $(-5, -4)$ and parallel to the line passing through $(-3, 2)$ and $(6, 8)$

52. The line passing through (a, b) with undefined slope

53. Given that the point $P(-3, 5)$ lies on the line $kx + 3y + 9 = 0$, find k.

54. Given that the point $P(2, -3)$ lies on the line $-2x + ky + 10 = 0$, find k.

In Exercises 55–60, sketch the straight line defined by the linear equation by finding the x- and y-intercepts.

Hint: See Example 12.

55. $3x - 2y + 6 = 0$

56. $2x - 5y + 10 = 0$

57. $x + 2y - 4 = 0$

58. $2x + 3y - 15 = 0$

59. $y + 5 = 0$

60. $-2x - 8y + 24 = 0$

61. Show that an equation of a line through the points $(a, 0)$ and $(0, b)$ with $a \neq 0$ and $b \neq 0$ can be written in the form

$$\frac{x}{a} + \frac{y}{b} = 1$$

(Recall that the numbers a and b are the x- and y-intercepts, respectively, of the line. This form of an equation of a line is called the **intercept form**.)

In Exercises 62–65, use the results of Exercise 61 to find an equation of a line with the x- and y-intercepts.

62. x-intercept 3; y-intercept 4

63. x-intercept -2; y-intercept -4

64. x-intercept $-\dfrac{1}{2}$; y-intercept $\dfrac{3}{4}$

65. x-intercept 4; y-intercept $-\dfrac{1}{2}$

In Exercises 66 and 67, determine whether the points lie on a straight line.

66. $A(-1, 7)$, $B(2, -2)$, and $C(5, -9)$

67. $A(-2, 1)$, $B(1, 7)$, and $C(4, 13)$

68. TEMPERATURE CONVERSION The relationship between the temperature in degrees Fahrenheit ($°F$) and the temperature in degrees Celsius ($°C$) is

$$F = \frac{9}{5}C + 32$$

a. Sketch the line with the given equation.

b. What is the slope of the line? What does it represent?

c. What is the F-intercept of the line? What does it represent?

69. NUCLEAR PLANT UTILIZATION The United States is not building many nuclear plants, but the ones it has are running at nearly full capacity. The output (as a percent of total capacity) of nuclear plants is described by the equation

$$y = 1.9467t + 70.082$$

where t is measured in years, with $t = 0$ corresponding to the beginning of 1990.

a. Sketch the line with the given equation.

b. What is the slope and the y-intercept of the line found in part (a)?

c. Give an interpretation of the slope and the y-intercept of the line found in part (a).

d. If the utilization of nuclear power continues to grow at the same rate and the total capacity of nuclear plants in the United States remains constant, by what year can the plants be expected to be generating at maximum capacity?

Source: Nuclear Energy Institute

70. SOCIAL SECURITY CONTRIBUTIONS For wages less than the maximum taxable wage base, Social Security contributions by employees are 7.65% of the employee's wages.

a. Find an equation that expresses the relationship between the wages earned (x) and the Social Security taxes paid (y) by an employee who earns less than the maximum taxable wage base.

b. For each additional dollar that an employee earns, by how much is his or her Social Security contribution increased? (Assume that the employee's wages are less than the maximum taxable wage base.)

c. What Social Security contributions will an employee who earns $35,000 (which is less than the maximum taxable wage base) be required to make?

Source: Social Security Administration

71. **COLLEGE ADMISSIONS**   Using data compiled by the Admissions Office at Faber University, college admissions officers estimate that 55% of the students who are offered admission to the freshman class at the university will actually enroll.

a. Find an equation that expresses the relationship between the number of students who actually enroll (y) and the number of students who are offered admission to the university (x).

b. If the desired freshman class size for the upcoming academic year is 1100 students, how many students should be admitted?

72. **WEIGHT OF WHALES**   The equation $W = 3.51L - 192$, expressing the relationship between the length L (in feet) and the expected weight W (in British tons) of adult blue whales, was adopted in the late 1960s by the International Whaling Commission.

a. What is the expected weight of an 80-ft blue whale?

b. Sketch the straight line that represents the equation.

73. **THE NARROWING GENDER GAP**   Since the founding of the Equal Employment Opportunity Commission and the passage of equal-pay laws, the gulf between men's and women's earnings has continued to close gradually. At the beginning of 1990 ($t = 0$), women's wages were 68% of men's wages, and by the beginning of 2000 ($t = 10$), women's wages were 80% of men's wages. If this gap between women's and men's wages continued to narrow *linearly*, what percent of men's wages were women's wages at the beginning of 2004?

Source: Journal of Economic Perspectives

74. **SALES OF NAVIGATION SYSTEMS**   The projected number of navigation systems (in millions) installed in vehicles in North America, Europe, and Japan from 2002 through 2006 follow. Here, $x = 0$ corresponds to 2002.

Year x	0	1	2	3	4
Systems Installed, y	3.9	4.7	5.8	6.8	7.8

a. Plot the annual sales (y) versus the year (x).

b. Draw a straight line L through the points corresponding to 2002 and 2006.

c. Derive an equation of the line L.

d. Use the equation found in part (c) to estimate the number of navigation systems installed in 2005. Compare this figure with the projected sales for that year.

Source: ABI Research

75. **SALES OF GPS EQUIPMENT**   The annual sales (in billions of dollars) of global positioning systems (GPS) equipment from 2000 through 2006 follow. (Sales in 2004 through 2006 are projections.) Here, $x = 0$ corresponds to 2000.

Year x	0	1	2	3	4	5	6
Annual Sales, y	7.9	9.6	11.5	13.3	15.2	17	18.8

a. Plot the annual sales (y) versus the year (x).

b. Draw a straight line L through the points corresponding to 2000 and 2006.

c. Derive an equation of the line L.

d. Use the equation found in part (c) to estimate the annual sales of GPS equipment in 2005. Compare this figure with the projected sales for that year.

Source: ABI Research

76. **IDEAL HEIGHTS AND WEIGHTS FOR WOMEN**   The Venus Health Club for Women provides its members with the following table, which gives the average desirable weight (in pounds) for women of a certain height (in inches):

Height, x	60	63	66	69	72
Weight, y	108	118	129	140	152

a. Plot the weight (y) versus the height (x).

b. Draw a straight line L through the points corresponding to heights of 5 ft and 6 ft.

c. Derive an equation of the line L.

d. Using the equation of part (c), estimate the average desirable weight for a woman who is 5 ft, 5 in. tall.

77. **COST OF A COMMODITY**   A manufacturer obtained the following data relating the cost y (in dollars) to the number of units (x) of a commodity produced:

Units Produced, x	0	20	40	60	80	100
Cost in Dollars, y	200	208	222	230	242	250

a. Plot the cost (y) versus the quantity produced (x).

b. Draw a straight line through the points (0, 200) and (100, 250).

c. Derive an equation of the straight line of part (b).

d. Taking this equation to be an approximation of the relationship between the cost and the level of production, estimate the cost of producing 54 units of the commodity.

78. **DIGITAL TV SERVICES**   The percent of homes with digital TV services stood at 5% at the beginning of 1999 ($t = 0$) and was projected to grow linearly so that at the beginning of 2003 ($t = 4$) the percent of such homes would be 25%.

a. Derive an equation of the line passing through the points $A(0, 5)$ and $B(4, 25)$.

b. Plot the line with the equation found in part (a).

c. Using the equation found in part (a), find the percent of homes with digital TV services at the beginning of 2001.

Source: Paul Kagan Associates

79. SALES GROWTH Metro Department Store's annual sales (in millions of dollars) during the past 5 yr were

Annual Sales, y	5.8	6.2	7.2	8.4	9.0
Year, x	1	2	3	4	5

a. Plot the annual sales (y) versus the year (x).
b. Draw a straight line L through the points corresponding to the first and fifth years.
c. Derive an equation of the line L.
d. Using the equation found in part (c), estimate Metro's annual sales 4 yr from now ($x = 9$).

80. Is there a difference between the statements "The slope of a straight line is zero" and "The slope of a straight line does not exist (is not defined)"? Explain your answer.

81. Consider the slope-intercept form of a straight line $y = mx + b$. Describe the family of straight lines obtained by keeping
a. The value of m fixed and allowing the value of b to vary.
b. The value of b fixed and allowing the value of m to vary.

In Exercises 82–88, determine whether the statement is true or false. If it is true, explain why it is true. If it is false, give an example to show why it is false.

82. Suppose the slope of a line L is $-\frac{1}{2}$ and P is a given point on L. If Q is the point on L lying 4 units to the left of P, then Q is situated 2 units above P.

83. The point $(-1, 1)$ lies on the line with equation $3x + 7y = 5$.

84. The point $(1, k)$ lies on the line with equation $3x + 4y = 12$ if and only if $k = \frac{9}{4}$.

85. The line with equation $Ax + By + C = 0$ $(B \neq 0)$ and the line with equation $ax + by + c = 0$ $(b \neq 0)$ are parallel if $Ab - aB = 0$.

86. If the slope of the line L_1 is positive, then the slope of a line L_2 perpendicular to L_1 may be positive or negative.

87. The lines with equation $ax + by + c_1 = 0$ and $bx - ay + c_2 = 0$, where $a \neq 0$ and $b \neq 0$, are perpendicular to each other.

88. If L is the line with equation $Ax + By + C = 0$, where $A \neq 0$, then L crosses the x-axis at the point $(-C/A, 0)$.

89. Show that two distinct lines with equations $a_1x + b_1y + c_1 = 0$ and $a_2x + b_2y + c_2 = 0$, respectively, are parallel if and only if $a_1b_2 - b_1a_2 = 0$.
Hint: Write each equation in the slope-intercept form and compare.

90. Prove that if a line L_1 with slope m_1 is perpendicular to a line L_2 with slope m_2, then $m_1m_2 = -1$.
Hint: Refer to the accompanying figure. Show that $m_1 = b$ and $m_2 = c$. Next, apply the Pythagorean theorem and the distance formula to the triangles OAC, OCB, and OBA to show that $1 = -bc$.

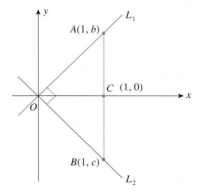

1.2 Solutions to Self-Check Exercises

1. The slope of the line that passes through the points $(a, 2)$ and $(3, 6)$ is

$$m = \frac{6 - 2}{3 - a} = \frac{4}{3 - a}$$

Since this line is parallel to a line with slope 4, m must be equal to 4; that is,

$$\frac{4}{3 - a} = 4$$

or, upon multiplying both sides of the equation by $3 - a$,

$$4 = 4(3 - a)$$
$$4 = 12 - 4a$$
$$4a = 8$$
$$a = 2$$

2. Since the required line L is perpendicular to a line with slope $-\frac{1}{2}$, the slope of L is

$$m = -\frac{1}{-\frac{1}{2}} = 2$$

Next, using the point-slope form of the equation of a line, we have

$$y - (-1) = 2(x - 3)$$
$$y + 1 = 2x - 6$$
$$y = 2x - 7$$

3. Substituting $x = 3$ and $y = -3$ into the left-hand side of the given equation, we find

$$2(3) - 3(-3) - 12 = 3$$

which is not equal to zero (the right-hand side). Therefore, $(3, -3)$ does not lie on the line with equation $2x - 3y - 12 = 0$.

Setting $x = 0$, we find $y = -4$, the y-intercept. Next, setting $y = 0$ gives $x = 6$, the x-intercept. We now draw the line passing through the points $(0, -4)$ and $(6, 0)$, as shown in the accompanying figure.

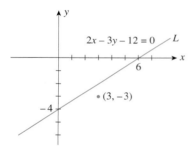

4. a and **b.** See the accompanying figure.

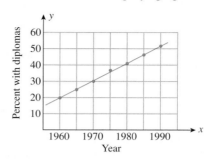

c. The slope of L is

$$m = \frac{52 - 20}{1990 - 1960} = \frac{32}{30} = \frac{16}{15}$$

Using the point-slope form of the equation of a line with the point $(1960, 20)$, we find

$$y - 20 = \frac{16}{15}(x - 1960) = \frac{16}{15}x - \frac{(16)(1960)}{15}$$
$$y = \frac{16}{15}x - \frac{6272}{3} + 20$$
$$= \frac{16}{15}x - \frac{6212}{3}$$

d. To estimate the percent of people over age 65 who had high school diplomas by the year 1995, let $x = 1995$ in the equation obtained in part (c). Thus, the required estimate is

$$y = \frac{16}{15}(1995) - \frac{6212}{3} \approx 57.33$$

or approximately 57%.

USING TECHNOLOGY

■ Graphing a Straight Line

Graphing Utility

The first step in plotting a straight line with a graphing utility is to select a suitable viewing window. We usually do this by experimenting. For example, you might first plot the straight line using the **standard viewing window** $[-10, 10] \times [-10, 10]$. If necessary, you then might adjust the viewing window by enlarging it or reducing it to obtain a sufficiently complete view of the line or at least the portion of the line that is of interest.

EXAMPLE 1   Plot the straight line $2x + 3y - 6 = 0$ in the standard viewing window.

Solution   The straight line in the standard viewing window is shown in Figure T1.

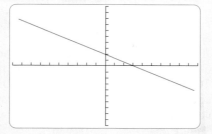

FIGURE T1
The straight line $2x + 3y - 6 = 0$ in the standard viewing window.

EXAMPLE 2   Plot the straight line $2x + 3y - 30 = 0$ in (a) the standard viewing window and (b) the viewing window $[-5, 20] \times [-5, 20]$.

Solution

a. The straight line in the standard viewing window is shown in Figure T2a.
b. The straight line in the viewing window $[-5, 20] \times [-5, 20]$ is shown in Figure T2b. This figure certainly gives a more complete view of the straight line.

FIGURE T2

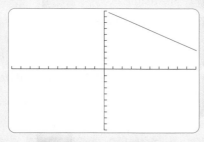

(a) The graph of $2x + 3y - 30 = 0$ in the standard viewing window

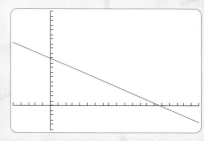

(b) The graph of $2x + 3y - 30 = 0$ in the viewing window $[-5, 20] \times [-5, 20]$

Excel

In the examples and exercises that follow, we assume that you are familiar with the basic features of Microsoft Excel. Please consult your Excel manual or use Excel's Help features to answer questions regarding the standard commands and operating instructions for Excel.

EXAMPLE 3   Plot the graph of the straight line $2x + 3y - 6 = 0$ over the interval $[-10, 10]$.

Solution

1. *Write the equation in the slope-intercept form:*

$$y = -\frac{2}{3}x + 2$$

2. *Create a table of values.* First, enter the input values: Enter the values of the endpoints of the interval over which you are graphing the straight line. (Recall that

(*continued*)

we only need two distinct data points to draw the graph of a straight line. In general, we select the endpoints of the interval over which the straight line is to be drawn as our data points.) In this case, we enter -10 in cell A2 and 10 in cell A3.

Second, enter the formula for computing the y-values: Here we enter

$$= -(2/3)*A2+2$$

in cell B2 and then press Enter .

Third, evaluate the function at the other input value: To extend the formula to cell B3, move the pointer to the small black box at the lower right corner of cell B2 (the cell containing the formula). Observe that the pointer now appears as a black $+$ (plus sign). Drag this pointer through cell B3 and then release it. The y-value, -4.66667, corresponding to the x-value in cell A3(10) will appear in cell B3 (Figure T3).

	A	B
1	x	y
2	−10	8.666667
3	10	−4.66667

FIGURE T3
Table of values for x and y

3. *Graph the straight line determined by these points.* First, highlight the numerical values in the table. Here we highlight cells A2:A3 and B2:B3. Next, click the Chart Wizard button on the toolbar.

Step 1   In the Chart Type dialog box that appears, select **XY(Scatter)** . Next, select the second chart in the first column under Chart sub-type. Then click **Next** at the bottom of the dialog box.

Step 2   Click **Columns** next to Series in: Then click **Next** at the bottom of the dialog box.

Step 3   Click the **Titles** tab. In the Chart title: box, enter y = −(2/3)x + 2. In the Value (X) axis: box, type x. In the Value (Y) axis: box, type y. Click the **Legend** tab. Next, click the **Show Legend** box to remove the check mark. Click **Finish** at the bottom of the dialog box.

The graph shown in Figure T4 will appear.

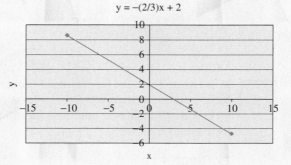

FIGURE T4
The graph of $y = -\frac{2}{3}x + 2$ over the interval $[-10, 10]$

If the interval over which the straight line is to be plotted is not specified, then you may have to experiment to find an appropriate interval for the x-values in your graph. For example, you might first plot the straight line over the interval $[-10, 10]$.

Note: Boldfaced words/characters enclosed in a box (for example, **Enter**) indicate that an action (click, select, or press) is required. Words/characters printed blue (for example, Chart Type) indicate words/characters that appear on the screen. Words/characters printed in a typewriter font (for example, = (−2/3)*A2+2) indicate words/characters that need to be typed and entered.

If necessary you then might adjust the interval by enlarging it or reducing it to obtain a sufficiently complete view of the line or at least the portion of the line that is of interest.

EXAMPLE 4 Plot the straight line $2x + 3y - 30 = 0$ over the intervals (a) $[-10, 10]$, and (b) $[-5, 20]$.

Solution a and b. We first cast the equation in the slope-intercept form, obtaining $y = -\frac{2}{3}x + 10$. Following the procedure given in Example 3, we obtain the graphs shown in Figure T5.

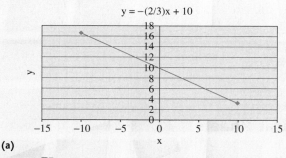

(a) (b)

FIGURE T5
The graph of $y = -\frac{2}{3}x + 10$ over the intervals (a) $[-10, 10]$ and (b) $[-5, 20]$

Observe that the graph in Figure T5b includes the x- and y-intercepts. This figure certainly gives a more complete view of the straight line.

TECHNOLOGY EXERCISES

Graphing Utility

In Exercises 1–6, plot the straight line with the equation in the standard viewing window.

1. $3.2x + 2.1y - 6.72 = 0$

2. $2.3x - 4.1y - 9.43 = 0$

3. $1.6x + 5.1y = 8.16$

4. $-3.2x + 2.1y = 6.72$

5. $2.8x = -1.6y + 4.48$

6. $3.3y = 4.2x - 13.86$

In Exercises 7–10, plot the straight line with the equation in (a) the standard viewing window and (b) the indicated viewing window.

7. $12.1x + 4.1y - 49.61 = 0$; $[-10, 10] \times [-10, 20]$

8. $4.1x - 15.2y - 62.32 = 0$; $[-10, 20] \times [-10, 10]$

9. $20x + 16y = 300$; $[-10, 20] \times [-10, 30]$

10. $32.2x + 21y = 676.2$; $[-10, 30] \times [-10, 40]$

In Exercises 11–18, plot the straight line with the equation in an appropriate viewing window. (*Note:* The answer is *not* unique.)

11. $20x + 30y = 600$

12. $30x - 20y = 600$

13. $22.4x + 16.1y - 352 = 0$

14. $18.2x - 15.1y = 274.8$

15. $1.2x + 20y = 24$

16. $30x - 2.1y = 63$

17. $-4x + 12y = 50$

18. $20x - 12.2y = 240$

(continued)

Excel

In Exercises 1–6, plot the straight line with the equation over the interval [−10, 10].

1. $3.2x + 2.1y − 6.72 = 0$ **2.** $2.3x − 4.1y − 9.43 = 0$

3. $1.6x + 5.1y = 8.16$ **4.** $−3.2x + 2.1y = 6.72$

5. $2.8x = −1.6y + 4.48$ **6.** $3.3y = 4.2x − 13.86$

In Exercises 7–10, plot the straight line with the equation over the indicated interval.

7. $12.1x + 4.1y − 49.61 = 0$; $[−10, 10]$

8. $4.1x − 15.2y − 62.32 = 0$; $[−10, 20]$

9. $20x + 16y = 300$; $[−10, 20]$

10. $32.2x + 21y = 676.2$; $[−10, 30]$

In Exercises 11–18, plot the straight line with the equation. (*Note*: The answer is *not* unique.)

11. $20x + 30y = 600$ **12.** $30x − 20y = 600$

13. $22.4x + 16.1y − 352 = 0$

14. $18.2x − 15.1y = 274.8$ **15.** $1.2x + 20y = 24$

16. $30x − 2.1y = 63$ **17.** $−4x + 12y = 50$

18. $20x − 12.2y = 240$

1.3   Linear Functions and Mathematical Models

▬ Mathematical Models

Regardless of the field from which a real-world problem is drawn, the problem is solved by analyzing it through a process called **mathematical modeling.** The four steps in this process, as illustrated in Figure 28, follow.

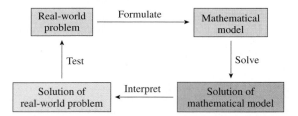

FIGURE 28

1. **Formulate** Given a real-world problem, our first task is to formulate the problem, using the language of mathematics. The many techniques used in constructing mathematical models range from theoretical consideration of the problem on the one extreme to an interpretation of data associated with the problem on the other. For example, the mathematical model giving the accumulated amount at any time when a certain sum of money is deposited in the bank can be derived theoretically (see Chapter 5). On the other hand, many of the mathematical models in this book are constructed by studying the data associated with the problem. In Section 1.5, we will see how linear equations (models) can be constructed from a given set of data points. Also, in the ensuing chapters, we will see how other mathematical models, including statistical and probability models, are used to describe and analyze real-world situations.

2. **Solve** Once a mathematical model has been constructed, we can use the appropriate mathematical techniques, which we will develop throughout the book, to solve the problem.

3. **Interpret** Bearing in mind that the solution obtained in step 2 is just the solution of the mathematical model, we need to interpret these results in the context of the original real-world problem.

4. **Test** Some mathematical models of real-world applications describe the situations with complete accuracy. For example, the model describing a deposit in a bank account gives the exact accumulated amount in the account at any time. But other mathematical models give, at best, an approximate description of the real-world problem. In this case we need to test the accuracy of the model by observing how well it describes the original real-world problem and how well it predicts past and/or future behavior. If the results are unsatisfactory, then we may have to reconsider the assumptions made in the construction of the model or, in the worst case, return to step 1.

We now look at an important way of describing the relationship between two quantities using the notion of a function. As you will see subsequently, many mathematical models are represented by functions.

▬ Functions

A manufacturer would like to know how his company's profit is related to its production level; a biologist would like to know how the population of a certain culture of bacteria will change with time; a psychologist would like to know the relationship between the learning time of an individual and the length of a vocabulary list; and a chemist would like to know how the initial speed of a chemical reaction is related to the amount of substrate used. In each instance, we are concerned with the same question: How does one quantity depend on another? The relationship between two quantities is conveniently described in mathematics by using the concept of a function.

> **Function**
> A **function** f is a rule that assigns to each value of x one and only one value of y.

The number y is normally denoted by $f(x)$, read "f of x," emphasizing the dependency of y on x.

An example of a function may be drawn from the familiar relationship between the area of a circle and its radius. Letting x and y denote the radius and area of a circle, respectively, we have, from elementary geometry,

$$y = \pi x^2$$

This equation defines y as a function of x, since for each admissible value of x (a nonnegative number representing the radius of a certain circle) there corresponds precisely one number $y = \pi x^2$ giving the area of the circle. This *area function* may be written as

$$f(x) = \pi x^2 \tag{7}$$

For example, to compute the area of a circle with a radius of 5 inches, we simply replace x in Equation (7) by the number 5. Thus, the area of the circle is

$$f(5) = \pi 5^2 = 25\pi$$

or 25π square inches.

Suppose we are given the function $y = f(x)$.* The variable x is referred to as the **independent variable,** and the variable y is called the **dependent variable.** The set of all values that may be assumed by x is called the **domain** of the function f, and the set comprising all the values assumed by $y = f(x)$ as x takes on all possible values in its domain is called the **range** of the function f. For the area function (7), the domain of f is the set of all nonnegative numbers x, and the range of f is the set of all nonnegative numbers y.

We now focus our attention on an important class of functions known as linear functions. Recall that a linear equation in x and y has the form $Ax + By + C = 0$, where A, B, and C are constants and A and B are not both zero. If $B \neq 0$, the equation can always be solved for y in terms of x; in fact, as we saw in Section 1.2, the equation may be cast in the slope-intercept form:

$$y = mx + b \qquad (m, b \text{ constants}) \tag{8}$$

Equation (8) defines y as a function of x. The domain and range of this function is the set of all real numbers. Furthermore, the graph of this function, as we saw in Section 1.2, is a straight line in the plane. For this reason, the function $f(x) = mx + b$ is called a linear function.

> **Linear Function**
> The function f defined by
>
> $$f(x) = mx + b$$
>
> where m and b are constants, is called a **linear function.**

Linear functions play an important role in the quantitative analysis of business and economic problems. First, many problems arising in these and other fields are *linear* in nature or are *linear* in the intervals of interest and thus can be formulated in terms of linear functions. Second, because linear functions are relatively easy to work with, assumptions involving linearity are often made in the formulation of problems. In many cases these assumptions are justified, and acceptable mathematical models are obtained that approximate real-life situations.

APPLIED EXAMPLE 1   Market for Cholesterol-Reducing Drugs   In a study conducted in early 2000, experts projected a rise in the market for cholesterol-reducing drugs. The U.S. market (in billions of dollars) for such drugs from 1999 through 2004 is given in the following table:

Year	1999	2000	2001	2002	2003	2004
Market	12.07	14.07	16.21	18.28	20	21.72

A mathematical model giving the approximate U.S. market over the period in question is given by

$$M(t) = 1.95t + 12.19$$

where t is measured in years, with $t = 0$ corresponding to 1999.

---

*It is customary to refer to a function f as $f(x)$.

a. Sketch the graph of the function M and the given data on the same set of axes.
b. Assuming that the projection held and the trend continued, what was the market for cholesterol-reducing drugs in 2005 ($t = 6$)?
c. What was the rate of increase of the market for cholesterol-reducing drugs over the period in question?

Source: S. G. Cowen

Solution
a. The graph of M is shown in Figure 29.

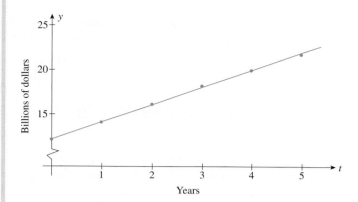

FIGURE 29
Projected U.S. market for cholesterol-reducing drugs

b. The projected market in 2005 for cholesterol-reducing drugs was approximately

$$M(6) = 1.95(6) + 12.19$$
$$= 23.89$$

or $23.89 billion.
c. The function M is linear, and so we see that the rate of increase of the market for cholesterol-reducing drugs is given by the slope of the straight line represented by M, which is approximately $1.95 billion per year.   ■

In Section 1.5, we will show how the function in Example 1 is actually constructed using the method of least squares.

In the rest of this section, we look at several applications that can be modeled using linear functions.

■ Simple Depreciation

We first discussed linear depreciation in Section 1.2 as a real-world application of straight lines. The following example illustrates how to derive an equation describing the book value of an asset being depreciated linearly.

APPLIED EXAMPLE 2   Linear Depreciation   A printing machine has an original value of $100,000 and is to be depreciated linearly over 5 years with a $30,000 scrap value. Find an expression giving the book value at the end of year t. What will be the book value of the machine at the end of the second year? What is the rate of depreciation of the printing machine?

Solution   Let V denote the printing machine's book value at the end of the tth year. Since the depreciation is linear, V is a linear function of t. Equivalently, the graph of the function is a straight line. Now, to find an equation of the straight line, observe that $V = 100{,}000$ when $t = 0$; this tells us that the line passes through the point $(0, 100{,}000)$. Similarly, the condition that $V = 30{,}000$ when $t = 5$ says that the line also passes through the point $(5, 30{,}000)$. The slope of the line is given by

$$m = \frac{100{,}000 - 30{,}000}{0 - 5} = -\frac{70{,}000}{5} = -14{,}000$$

Using the point-slope form of the equation of a line with the point $(0, 100{,}000)$ and the slope $m = -14{,}000$, we have

$$V - 100{,}000 = -14{,}000(t - 0)$$
$$V = -14{,}000t + 100{,}000$$

the required expression. The book value at the end of the second year is given by

$$V = -14{,}000(2) + 100{,}000 = 72{,}000$$

or \$72,000. The rate of depreciation of the machine is given by the negative of the slope of the depreciation line. Since the slope of the line is $m = -14{,}000$, the rate of depreciation is \$14,000 per year. The graph of $V = -14{,}000t + 100{,}000$ is sketched in Figure 30.   ∎

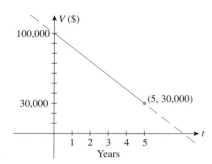

FIGURE 30
Linear depreciation of an asset

Linear Cost, Revenue, and Profit Functions

Whether a business is a sole proprietorship or a large corporation, the owner or chief executive must constantly keep track of operating costs, revenue resulting from the sale of products or services, and perhaps most important, the profits realized. Three functions provide management with a measure of these quantities: the total cost function, the revenue function, and the profit function.

> **Cost, Revenue, and Profit Functions**
> Let x denote the number of units of a product manufactured or sold. Then, the **total cost function** is
>
> $$C(x) = \text{Total cost of manufacturing } x \text{ units of the product}$$
>
> The **revenue function** is
>
> $$R(x) = \text{Total revenue realized from the sale of } x \text{ units of the product}$$
>
> The **profit function** is
>
> $$P(x) = \text{Total profit realized from manufacturing and selling } x \text{ units of the product}$$

Generally speaking, the total cost, revenue, and profit functions associated with a company are more likely than not to be nonlinear (these functions are best studied using the tools of calculus). But *linear* cost, revenue, and profit functions do arise in practice, and we will consider such functions in this section. Before deriving explicit forms of these functions, we need to recall some common terminology.

The costs incurred in operating a business are usually classified into two categories. Costs that remain more or less constant regardless of the firm's activity level are called **fixed costs.** Examples of fixed costs are rental fees and executive salaries. Costs that vary with production or sales are called **variable costs.** Examples of variable costs are wages and costs for raw materials.

Suppose a firm has a fixed cost of F dollars, a production cost of c dollars per unit, and a selling price of s dollars per unit. Then the *cost function* $C(x)$, the *revenue function* $R(x)$, and the *profit function* $P(x)$ for the firm are given by

$$C(x) = cx + F$$
$$R(x) = sx$$
$$P(x) = R(x) - C(x) \qquad \text{Revenue} - \text{cost}$$
$$= (s - c)x - F$$

where x denotes the number of units of the commodity produced and sold. The functions C, R, and P are linear functions of x.

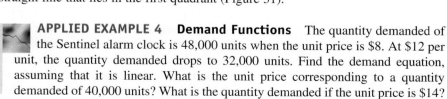 **APPLIED EXAMPLE 3 Profit Functions** Puritron, a manufacturer of water filters, has a monthly fixed cost of $20,000, a production cost of $20 per unit, and a selling price of $30 per unit. Find the cost function, the revenue function, and the profit function for Puritron.

Solution Let x denote the number of units produced and sold. Then,

$$C(x) = 20x + 20{,}000$$
$$R(x) = 30x$$
$$P(x) = R(x) - C(x)$$
$$= 30x - (20x + 20{,}000)$$
$$= 10x - 20{,}000$$

Linear Demand and Supply Curves

In a free-market economy, consumer demand for a particular commodity depends on the commodity's unit price. A **demand equation** expresses this relationship between the unit price and the quantity demanded. The corresponding graph of the demand equation is called a **demand curve.** In general, the quantity demanded of a commodity decreases as its unit price increases, and vice versa. Accordingly, a *demand function,* defined by $p = f(x)$, where p measures the unit price and x measures the number of units of the commodity, is generally characterized as a decreasing function of x; that is, $p = f(x)$ decreases as x increases.

The simplest demand function is defined by a linear equation in x and p, where both x and p assume only nonnegative values. Its graph is a straight line having a negative slope. Thus, the demand curve in this case is that part of the graph of a straight line that lies in the first quadrant (Figure 31).

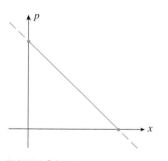

FIGURE 31
A graph of a linear demand function

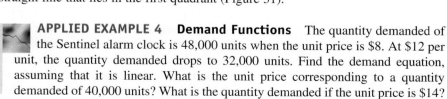 **APPLIED EXAMPLE 4 Demand Functions** The quantity demanded of the Sentinel alarm clock is 48,000 units when the unit price is $8. At $12 per unit, the quantity demanded drops to 32,000 units. Find the demand equation, assuming that it is linear. What is the unit price corresponding to a quantity demanded of 40,000 units? What is the quantity demanded if the unit price is $14?

Solution   Let p denote the unit price of an alarm clock (in dollars) and let x (in units of 1000) denote the quantity demanded when the unit price of the clocks is $\$p$. When $p = 8$, $x = 48$ and the point $(48, 8)$ lies on the demand curve. Similarly, when $p = 12$, $x = 32$ and the point $(32, 12)$ also lies on the demand curve. Since the demand equation is linear, its graph is a straight line. The slope of the required line is given by

$$m = \frac{12 - 8}{32 - 48} = \frac{4}{-16} = -\frac{1}{4}$$

So, using the point-slope form of an equation of a line with the point $(48, 8)$, we find that

$$p - 8 = -\frac{1}{4}(x - 48)$$

$$p = -\frac{1}{4}x + 20$$

is the required equation. The demand curve is shown in Figure 32. If the quantity demanded is 40,000 units ($x = 40$), the demand equation yields

$$y = -\frac{1}{4}(40) + 20 = 10$$

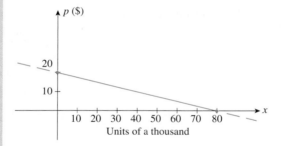

FIGURE 32
The graph of the demand equation
$p = -\frac{1}{4}x + 20$

and we see that the corresponding unit price is $10. Next, if the unit price is $14 ($p = 14$), the demand equation yields

$$14 = -\frac{1}{4}x + 20$$

$$\frac{1}{4}x = 6$$

$$x = 24$$

so the quantity demanded will be 24,000 units.   ■

In a competitive market, a relationship also exists between the unit price of a commodity and its availability in the market. In general, an increase in a commodity's unit price will induce the manufacturer to increase the supply of that commodity. Conversely, a decrease in the unit price generally leads to a drop in the supply. An equation that expresses the relationship between the unit price and the quantity supplied is called a **supply equation,** and the corresponding graph is called a **supply curve.** A supply function, defined by $p = f(x)$, is generally characterized by an increasing function of x; that is, $p = f(x)$ increases as x increases.

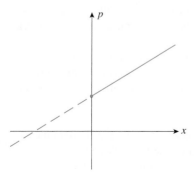

FIGURE 33
A graph of a linear supply function

As in the case of a demand equation, the simplest supply equation is a linear equation in p and x, where p and x have the same meaning as before but the line has a positive slope. The supply curve corresponding to a linear supply function is that part of the straight line that lies in the first quadrant (Figure 33).

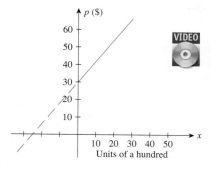

FIGURE 34
The graph of the supply equation
$4p - 5x = 120$

APPLIED EXAMPLE 5 Supply Functions The supply equation for a commodity is given by $4p - 5x = 120$, where p is measured in dollars and x is measured in units of 100.

a. Sketch the corresponding curve.
b. How many units will be marketed when the unit price is $55?

Solution
a. Setting $x = 0$, we find the p-intercept to be 30. Next, setting $p = 0$, we find the x-intercept to be -24. The supply curve is sketched in Figure 34.
b. Substituting $p = 55$ in the supply equation, we have $4(55) - 5x = 120$, or $x = 20$, so the amount marketed will be 2000 units. ■

1.3 Self-Check Exercises

1. A manufacturer has a monthly fixed cost of $60,000 and a production cost of $10 for each unit produced. The product sells for $15/unit.
   a. What is the cost function?
   b. What is the revenue function?
   c. What is the profit function?
   d. Compute the profit (loss) corresponding to production levels of 10,000 and 14,000 units.

2. The quantity demanded for a certain make of 30-in. × 52-in. area rug is 500 when the unit price is $100. For each $20 decrease in the unit price, the quantity demanded increases by 500 units. Find the demand equation and sketch its graph.

Solutions to Self-Check Exercises 1.3 can be found on page 38.

1.3 Concept Questions

1. **a.** What is a *function*? Give an example.
   b. What is a *linear function*? Give an example.
   c. What is the domain of a linear function? The range?
   d. What is the graph of a linear function?

2. What is the general form of a linear cost function? A linear revenue function? A linear profit function?

3. Is the slope of a linear demand curve positive or negative? The slope of a linear supply curve?

1.3 Exercises

In Exercises 1–10, determine whether the equation defines y as a linear function of x. If so, write it in the form $y = mx + b$.

1. $2x + 3y = 6$

2. $-2x + 4y = 7$

3. $x = 2y - 4$

4. $2x = 3y + 8$

5. $2x - 4y + 9 = 0$

6. $3x - 6y + 7 = 0$

7. $2x^2 - 8y + 4 = 0$

8. $3\sqrt{x} + 4y = 0$

9. $2x - 3y^2 + 8 = 0$ **10.** $2x + \sqrt{y} - 4 = 0$

11. A manufacturer has a monthly fixed cost of $40,000 and a production cost of $8 for each unit produced. The product sells for $12/unit.
   a. What is the cost function?
   b. What is the revenue function?
   c. What is the profit function?
   d. Compute the profit (loss) corresponding to production levels of 8000 and 12,000 units.

12. A manufacturer has a monthly fixed cost of $100,000 and a production cost of $14 for each unit produced. The product sells for $20/unit.
   a. What is the cost function?
   b. What is the revenue function?
   c. What is the profit function?
   d. Compute the profit (loss) corresponding to production levels of 12,000 and 20,000 units.

13. Find the constants m and b in the linear function $f(x) = mx + b$ so that $f(0) = 2$ and $f(3) = -1$.

14. Find the constants m and b in the linear function $f(x) = mx + b$ so that $f(2) = 4$ and the straight line represented by f has slope -1.

15. LINEAR DEPRECIATION   An office building worth $1 million when completed in 2003 is being depreciated linearly over 50 yr. What will be the book value of the building in 2008? In 2013? (Assume the scrap value is $0.)

16. LINEAR DEPRECIATION   An automobile purchased for use by the manager of a firm at a price of $24,000 is to be depreciated using the straight-line method over 5 yr. What will be the book value of the automobile at the end of 3 yr? (Assume the scrap value is $0.)

17. CONSUMPTION FUNCTIONS   A certain economy's consumption function is given by the equation

$$C(x) = 0.75x + 6$$

where $C(x)$ is the personal consumption expenditure in billions of dollars and x is the disposable personal income in billions of dollars. Find $C(0)$, $C(50)$, and $C(100)$.

18. SALES TAX   In a certain state, the sales tax T on the amount of taxable goods is 6% of the value of the goods purchased (x), where both T and x are measured in dollars.
   a. Express T as a function of x.
   b. Find $T(200)$ and $T(5.60)$.

19. SOCIAL SECURITY BENEFITS   Social Security recipients receive an automatic cost-of-living adjustment (COLA) once each year. Their monthly benefit is increased by the same percent that consumer prices have increased during the preceding year. Suppose consumer prices have increased by 5.3% during the preceding year.
   a. Express the adjusted monthly benefit of a Social Security recipient as a function of his or her current monthly benefit.

   b. If Carlos Garcia's monthly Social Security benefit is now $620, what will be his adjusted monthly benefit?

20. PROFIT FUNCTIONS   AutoTime, a manufacturer of 24-hr variable timers, has a monthly fixed cost of $48,000 and a production cost of $8 for each timer manufactured. The timers sell for $14 each.
   a. What is the cost function?
   b. What is the revenue function?
   c. What is the profit function?
   d. Compute the profit (loss) corresponding to production levels of 4000, 6000, and 10,000 timers, respectively.

21. PROFIT FUNCTIONS   The management of TMI finds that the monthly fixed costs attributable to the company's blank-tape division amount to $12,100.00. If the cost for producing each reel of tape is $.60 and each reel of tape sells for $1.15, find the company's cost function, revenue function, and profit function.

22. LINEAR DEPRECIATION   In 2003, National Textile installed a new machine in one of its factories at a cost of $250,000. The machine is depreciated linearly over 10 yr with a scrap value of $10,000.
   a. Find an expression for the machine's book value in the tth year of use $(0 \le t \le 10)$.
   b. Sketch the graph of the function of part (a).
   c. Find the machine's book value in 2007.
   d. Find the rate at which the machine is being depreciated.

23. LINEAR DEPRECIATION   A server purchased at a cost of $60,000 in 2002 has a scrap value of $12,000 at the end of 4 yr. If the straight-line method of depreciation is used,
   a. Find the rate of depreciation.
   b. Find the linear equation expressing the server's book value at the end of t yr.
   c. Sketch the graph of the function of part (b).
   d. Find the server's book value at the end of the third year.

24. LINEAR DEPRECIATION   Suppose an asset has an original value of $C and is depreciated linearly over N yr with a scrap value of $S. Show that the asset's book value at the end of the tth year is described by the function

$$V(t) = C - \left(\frac{C - S}{N}\right)t$$

Hint: Find an equation of the straight line passing through the points $(0, C)$ and (N, S). (Why?)

25. Rework Exercise 15 using the formula derived in Exercise 24.

26. Rework Exercise 16 using the formula derived in Exercise 24.

27. DRUG DOSAGES   A method sometimes used by pediatricians to calculate the dosage of medicine for children is based on the child's surface area. If a denotes the adult dosage (in milligrams) and if S is the child's surface area (in square meters), then the child's dosage is given by

$$D(S) = \frac{Sa}{1.7}$$

a. Show that D is a linear function of S.
Hint: Think of D as having the form $D(S) = mS + b$. What is the slope m and the y-intercept b?
b. If the adult dose of a drug is 500 mg, how much should a child whose surface area is 0.4 m² receive?

28. **DRUG DOSAGES** Cowling's rule is a method for calculating pediatric drug dosages. If a denotes the adult dosage (in milligrams) and if t is the child's age (in years), then the child's dosage is given by

$$D(t) = \left(\frac{t+1}{24}\right)a$$

a. Show that D is a linear function of t.
Hint: Think of $D(t)$ as having the form $D(t) = mt + b$. What is the slope m and the y-intercept b?
b. If the adult dose of a drug is 500 mg, how much should a 4-yr-old child receive?

29. **BROADBAND INTERNET HOUSEHOLDS** The number of U.S. broadband Internet households stood at 20 million at the beginning of 2002 and is projected to grow at the rate of 7.5 million households per year for the next 6 years.
a. Find a linear function $f(t)$ giving the projected U.S. broadband Internet households (in millions) in year t, where $t = 0$ corresponds to the beginning of 2002.
b. What is the projected size of U.S. broadband Internet households at the beginning of 2008?

Source: Strategy Analytics Inc.

30. **DIAL-UP INTERNET HOUSEHOLDS** The number of U.S. dial-up Internet households stood at 42.5 million at the beginning of 2004 and is projected to decline at the rate of 3.9 million households per year for the next 4 yr.
a. Find a linear function f giving the projected U.S. dial-up Internet households (in millions) in year t, where $t = 0$ corresponds to the beginning of 2004.
b. What is the projected size of U.S. dial-up Internet households at the beginning of 2008?

Source: Strategy Analytics Inc.

31. **CELSIUS AND FAHRENHEIT TEMPERATURES** The relationship between temperature measured in the Celsius scale and the Fahrenheit scale is linear. The freezing point is 0°C and 32°F, and the boiling point is 100°C and 212°F, respectively.
a. Find an equation giving the relationship between the temperature F measured in the Fahrenheit scale and the temperature C measured in the Celsius scale.
b. Find F as a function of C and use this formula to determine the temperature in Fahrenheit corresponding to a temperature of 20°C.
c. Find C as a function of F and use this formula to determine the temperature in Celsius corresponding to a temperature of 70°F.

32. **CRICKET CHIRPING AND TEMPERATURE** Entomologists have discovered that a linear relationship exists between the number of chirps of crickets of a certain species and the air temperature. When the temperature is 70°F, the crickets chirp at the rate of 120 chirps/min, and when the temperature is 80°F, they chirp at the rate of 160 chirps/min.
a. Find an equation giving the relationship between the air temperature T and the number of chirps/minute N of the crickets.
b. Find N as a function of T and use this formula to determine the rate at which the crickets chirp when the temperature is 102°F.

33. **DEMAND FOR CLOCK RADIOS** In the accompanying figure, L_1 is the demand curve for the model A clock radio manufactured by Ace Radio, and L_2 is the demand curve for the model B clock radio. Which line has the greater slope? Interpret your results.

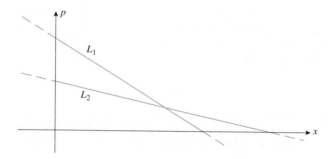

34. **SUPPLY OF CLOCK RADIOS** In the accompanying figure, L_1 is the supply curve for the model A clock radio manufactured by Ace Radio, and L_2 is the supply curve for the model B clock radio. Which line has the greater slope? Interpret your results.

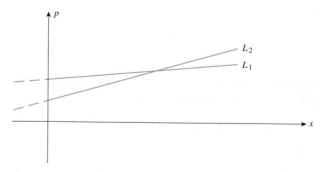

For each demand equation in Exercises 35–38, where x represents the quantity demanded in units of 1000 and p is the unit price in dollars, (a) sketch the demand curve and (b) determine the quantity demanded corresponding to the given unit price p.

35. $2x + 3p - 18 = 0$; $p = 4$

36. $5p + 4x - 80 = 0; p = 10$

37. $p = -3x + 60; p = 30$ **38.** $p = -0.4x + 120; p = 80$

39. DEMAND FUNCTIONS   At a unit price of $55, the quantity demanded of a certain commodity is 1000 units. At a unit price of $85, the demand drops to 600 units. Given that it is linear, find the demand equation. Above what price will there be no demand? What quantity would be demanded if the commodity were free?

40. DEMAND FUNCTIONS   The quantity demanded for a certain brand of portable CD players is 200 units when the unit price is set at $90. The quantity demanded is 1200 units when the unit price is $40. Find the demand equation and sketch its graph.

41. DEMAND FUNCTIONS   Assume that a certain commodity's demand equation has the form $p = ax + b$, where x is the quantity demanded and p is the unit price in dollars. Suppose the quantity demanded is 1000 units when the unit price is $9.00 and 6000 when the unit price is $4.00. What is the quantity demanded when the unit price is $7.50?

42. DEMAND FUNCTIONS   The demand equation for the Sicard wristwatch is

$$p = -0.025x + 50$$

where x is the quantity demanded per week and p is the unit price in dollars. Sketch the graph of the demand equation. What is the highest price (theoretically) anyone would pay for the watch?

For each supply equation in Exercises 43–46, where x is the quantity supplied in units of 1000 and p is the unit price in dollars, (a) sketch the supply curve and (b) determine the number of units of the commodity the supplier will make available in the market at the given unit price.

43. $3x - 4p + 24 = 0; p = 8$

44. $\frac{1}{2}x - \frac{2}{3}p + 12 = 0; p = 24$

45. $p = 2x + 10; p = 14$ **46.** $p = \frac{1}{2}x + 20; p = 28$

47. SUPPLY FUNCTIONS   Suppliers of a certain brand of digital voice recorders will make 10,000 available in the market if the unit price is $45. At a unit price of $50, 20,000 units will be made available. Assuming that the relationship between the unit price and the quantity supplied is linear, derive the supply equation. Sketch the supply curve and determine the quantity suppliers will make available when the unit price is $70.

48. SUPPLY FUNCTIONS   Producers will make 2000 refrigerators available when the unit price is $330. At a unit price of $390, 6000 refrigerators will be marketed. Find the equation relating the unit price of a refrigerator to the quantity supplied if the equation is known to be linear. How many refrigerators will be marketed when the unit price is $450? What is the lowest price at which a refrigerator will be marketed?

In Exercises 49 and 50, determine whether the statement is true or false. If it is true, explain why it is true. If it is false, give an example to show why it is false.

49. Suppose $C(x) = cx + F$ and $R(x) = sx$ are the cost and revenue functions of a certain firm. Then, the firm is making a profit if its level of production is less than $F/(s - c)$.

50. If $p = mx + b$ is a linear demand curve, then it is generally true that $m < 0$.

1.3   Solutions to Self-Check Exercises

1. Let x denote the number of units produced and sold. Then
   a. $C(x) = 10x + 60,000$
   b. $R(x) = 15x$
   c. $P(x) = R(x) - C(x) = 15x - (10x + 60,000)$
   $$= 5x - 60,000$$
   d. $P(10,000) = 5(10,000) - 60,000$
   $$= -10,000$$
   or a loss of $10,000 per month.
   $P(14,000) = 5(14,000) - 60,000$
   $$= 10,000$$
   or a profit of $10,000 per month.

2. Let p denote the price of a rug (in dollars) and let x denote the quantity of rugs demanded when the unit price is p. The condition that the quantity demanded is 500 when the unit price is $100 tells us that the demand curve passes through the point $(500, 100)$. Next, the condition that for each $20 decrease in the unit price the quantity demanded increases by 500 tells us that the demand curve is linear and that its slope is given by $-\frac{20}{500}$, or $-\frac{1}{25}$. Therefore, letting $m = -\frac{1}{25}$ in the demand equation

$$p = mx + b$$

we find

$$p = -\frac{1}{25}x + b$$

To determine b, use the fact that the straight line passes through $(500, 100)$ to obtain

$$100 = -\frac{1}{25}(500) + b$$

or $b = 120$. Therefore, the required equation is

$$p = -\frac{1}{25}x + 120$$

The graph of the demand curve $p = -\frac{1}{25}x + 120$ is sketched in the accompanying figure.

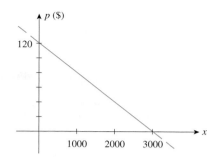

USING TECHNOLOGY

▬ Evaluating a Function

Graphing Utility

A graphing utility can be used to find the value of a function f at a given point with minimal effort. However, to find the value of y for a given value of x in a linear equation such as $Ax + By + C = 0$, the equation must first be cast in the slope-intercept form $y = mx + b$, thus revealing the desired rule $f(x) = mx + b$ for y as a function of x.

EXAMPLE 1 Consider the equation $2x + 5y = 7$.

a. Plot the straight line with the given equation in the standard viewing window.
b. Find the value of y when $x = 2$ and verify your result by direct computation.
c. Find the value of y when $x = 1.732$.

Solution

a. The straight line with equation $2x + 5y = 7$ or, equivalently, $y = -\frac{2}{5}x + \frac{7}{5}$ in the standard viewing window, is shown in Figure T1.
b. Using the evaluation function of the graphing utility and the value of 2 for x, we find $y = 0.6$. This result is verified by computing

$$y = -\frac{2}{5}(2) + \frac{7}{5} = -\frac{4}{5} + \frac{7}{5} = \frac{3}{5} = 0.6$$

when $x = 2$.
c. Once again using the evaluation function of the graphing utility, this time with the value 1.732 for x, we find $y = 0.7072$. The efficacy of the graphing utility is clearly demonstrated here!

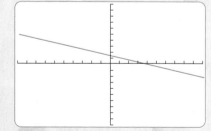

FIGURE T1
The straight line $2x + 5y = 7$ in the standard viewing window

⚠ When evaluating $f(x)$ at $x = a$, remember that the number a must lie between xMin and xMax.

(continued)

EXAMPLE 2   According to Pacific Gas and Electric, the nation's largest utility company, the demand for electricity (in megawatts) in year t is approximately

$$D(t) = 295t + 328 \qquad (0 \le t \le 10)$$

with $t = 0$ corresponding to 1990.

a. Plot the graph of D in the viewing window $[0, 10] \times [0, 3500]$.
b. What was the demand for electricity in 1999?
Source: Pacific Gas and Electric

Solution

a. The graph of D is shown in Figure T2.

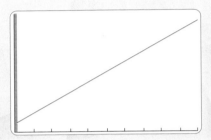

FIGURE T2
The graph of D in the viewing window
$[0, 10] \times [0, 3500]$

b. Evaluating the function at $t = 9$, we find $y = 2983$. Therefore, the demand for electricity in 1999 was approximately 2983 megawatts.

Excel

Excel can be used to find the value of a function at a given value with minimal effort. However, to find the value of y for a given value of x in a linear equation such as $Ax + By + C = 0$, the equation must first be cast in the slope-intercept form $y = mx + b$, thus revealing the desired rule $f(x) = mx + b$ for y as a function of x.

EXAMPLE 3   Consider the equation $2x + 5y = 7$.

a. Find the value of y for $x = 0$, 5, and 10.
b. Plot the straight line with the given equation over the interval $[0, 10]$.

Solution

a. Since this is a linear equation, we first cast the equation in slope-intercept form:

$$y = -\frac{2}{5}x + \frac{7}{5}$$

Next, we create a table of values (Figure T3), following the same procedure outlined in Example 3, pages 25–26. In this case we use the formula `=(-2/5)*A2+7/5` for the y-values.

	A	B
1	x	y
2	0	1.4
3	5	-0.6
4	10	-2.6

FIGURE T3
Table of values for x and y

Note: Words/characters printed in a typewriter font (for example, `=(-2/3)*A2+2`) indicate words/characters that need to be typed and entered.

b. Following the procedure outlined in Example 3, page 25, we obtain the graph shown in Figure T4.

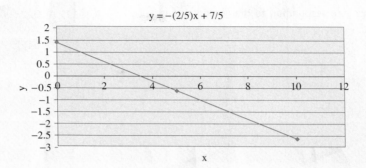

$$y = -(2/5)x + 7/5$$

FIGURE T4
The graph of $y = -\frac{2}{5}x + \frac{7}{5}$ over the interval [0, 10]

EXAMPLE 4 According to Pacific Gas and Electric, the nation's largest utility company, the demand for electricity (in megawatts) in year t is approximately

$$D(t) = 295t + 328 \qquad (0 \leq t \leq 10)$$

with $t = 0$ corresponding to 1990.

a. Plot the graph of D over the interval [0, 10].
b. What was the demand for electricity in 1999?
Source: Pacific Gas and Electric

Solution

a. Following the instructions given in Example 3, page 25, we obtain the following spreadsheet and graph shown in Figure T5. [*Note:* We have added the value 9 to the table because we are asked to compute the value of $D(t)$ when $t = 9$ in part (b). Also, we have made the appropriate entries for the title and x- and y-axis labels.]

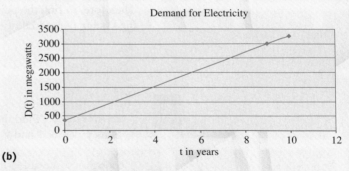

	A	B
1	t	D(t)
2	0	328
3	9	2983
4	10	3278

(a)

(b)

FIGURE T5
(a) The table of values for t and $D(t)$ and (b) the graph showing demand for electricity

b. From the table of values, we see that

$$D(9) = 2983$$

or 2983 megawatts.

(*continued*)

TECHNOLOGY EXERCISES

Find the value of y corresponding to the given value of x.

1. $3.1x + 2.4y - 12 = 0$; $x = 2.1$

2. $1.2x - 3.2y + 8.2 = 0$; $x = 1.2$

3. $2.8x + 4.2y = 16.3$; $x = 1.5$

4. $-1.8x + 3.2y - 6.3 = 0$; $x = -2.1$

5. $22.1x + 18.2y - 400 = 0$; $x = 12.1$

6. $17.1x - 24.31y - 512 = 0$; $x = -8.2$

7. $2.8x = 1.41y - 2.64$; $x = 0.3$

8. $0.8x = 3.2y - 4.3$; $x = -0.4$

1.4   Intersection of Straight Lines

▬ Finding the Point of Intersection

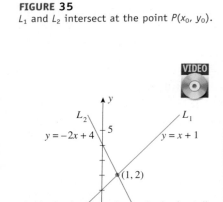

FIGURE 35
L_1 and L_2 intersect at the point $P(x_0, y_0)$.

The solution of certain practical problems involves finding the point of intersection of two straight lines. To see how such a problem may be solved algebraically, suppose we are given two straight lines L_1 and L_2 with equations

$$y = m_1x + b_1 \quad \text{and} \quad y = m_2x + b_2$$

(where m_1, b_1, m_2, and b_2 are constants) that intersect at the point $P(x_0, y_0)$ (Figure 35).

The point $P(x_0, y_0)$ lies on the line L_1 and so satisfies the equation $y = m_1x + b_1$. It also lies on the line L_2 and so satisfies the equation $y = m_2x + b_2$. Therefore, to find the point of intersection $P(x_0, y_0)$ of the lines L_1 and L_2, we solve the system composed of the two equations

$$y = m_1x + b_1 \quad \text{and} \quad y = m_2x + b_2$$

for x and y.

VIDEO

EXAMPLE 1   Find the point of intersection of the straight lines that have equations $y = x + 1$ and $y = -2x + 4$.

Solution   We solve the given simultaneous equations. Substituting the value y as given in the first equation into the second, we obtain

$$x + 1 = -2x + 4$$
$$3x = 3$$
$$x = 1$$

FIGURE 36
The point of intersection of L_1 and L_2 is $(1, 2)$.

Substituting this value of x into either one of the given equations yields $y = 2$. Therefore, the required point of intersection is $(1, 2)$ (Figure 36).   ▪

EXPLORING WITH TECHNOLOGY

1. Use a graphing utility to plot the straight lines L_1 and L_2 with equations $y = 3x - 2$ and $y = -2x + 3$, respectively, on the same set of axes, using the standard viewing window. Then use TRACE and ZOOM to find the point of intersection of L_1 and L_2. Repeat using the "intersection" function of your graphing utility.

2. Find the point of intersection of L_1 and L_2 algebraically.

3. Comment on the effectiveness of each method.

We now turn to some applications involving the intersections of pairs of straight lines.

Break-Even Analysis

Consider a firm with (linear) cost function $C(x)$, revenue function $R(x)$, and profit function $P(x)$ given by

$$C(x) = cx + F$$
$$R(x) = sx$$
$$P(x) = R(x) - C(x) = (s - c)x - F$$

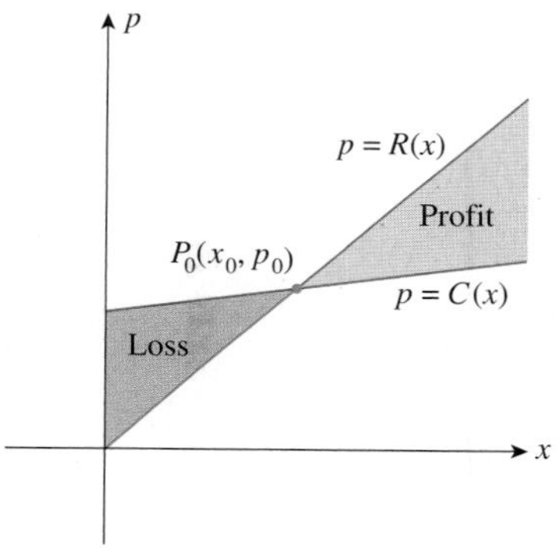

where c denotes the unit cost of production, s denotes the selling price per unit, F denotes the fixed cost incurred by the firm, and x denotes the level of production and sales. The level of production at which the firm neither makes a profit nor sustains a loss is called the **break-even level of operation** and may be determined by solving the equations $p = C(x)$ and $p = R(x)$ simultaneously. For at this level of production, x_0, the profit is zero, so

$$P(x_0) = R(x_0) - C(x_0) = 0$$
$$R(x_0) = C(x_0)$$

The point $P_0(x_0, p_0)$, the solution of the simultaneous equations $p = R(x)$ and $p = C(x)$, is referred to as the break-even point; the number x_0 and the number p_0 are called the **break-even quantity** and the **break-even revenue,** respectively.

Geometrically, the break-even point $P_0(x_0, p_0)$ is just the point of intersection of the straight lines representing the cost and revenue functions, respectively. This follows because $P_0(x_0, p_0)$, being the solution of the simultaneous equations $p = R(x)$ and $p = C(x)$, must lie on both these lines simultaneously (Figure 37).

Note that if $x < x_0$, then $R(x) < C(x)$ so that $P(x) = R(x) - C(x) < 0$, and thus the firm sustains a loss at this level of production. On the other hand, if $x > x_0$, then $P(x) > 0$, and the firm operates at a profitable level.

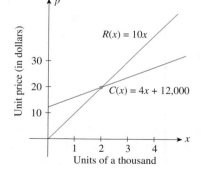

FIGURE 37
P_0 is the break-even point.

FIGURE 38
The point at which $R(x) = C(x)$ is the break-even point.

APPLIED EXAMPLE 2   Break-Even Level   Prescott manufactures its products at a cost of $4 per unit and sells them for $10 per unit. If the firm's fixed cost is $12,000 per month, determine the firm's break-even point.

Solution   The cost function C and the revenue function R are given by $C(x) = 4x + 12,000$ and $R(x) = 10x$, respectively (Figure 38).

Setting $R(x) = C(x)$, we obtain

$$10x = 4x + 12{,}000$$
$$6x = 12{,}000$$
$$x = 2000$$

Substituting this value of x into $R(x) = 10x$ gives

$$R(2000) = (10)(2000) = 20{,}000$$

So, for a break-even operation, the firm should manufacture 2000 units of its product, resulting in a break-even revenue of $20,000 per month.   ▪

APPLIED EXAMPLE 3   Break-Even Analysis   Using the data given in Example 2, answer the following questions:

a. What is the loss sustained by the firm if only 1500 units are produced and sold each month?
b. What is the profit if 3000 units are produced and sold each month?
c. How many units should the firm produce in order to realize a minimum monthly profit of $9000?

Solution   The profit function P is given by the rule

$$P(x) = R(x) - C(x)$$
$$= 10x - (4x + 12{,}000)$$
$$= 6x - 12{,}000$$

a. If 1500 units are produced and sold each month, we have

$$P(1500) = 6(1500) - 12{,}000 = -3000$$

so the firm will sustain a loss of $3000 per month.
b. If 3000 units are produced and sold each month, we have

$$P(3000) = 6(3000) - 12{,}000 = 6000$$

or a monthly profit of $6000.
c. Substituting 9000 for $P(x)$ in the equation $P(x) = 6x - 12{,}000$, we obtain

$$9000 = 6x - 12{,}000$$
$$6x = 21{,}000$$
$$x = 3500$$

Thus, the firm should produce at least 3500 units in order to realize a $9000 minimum monthly profit.   ▪

APPLIED EXAMPLE 4 Decision Analysis The management of Robertson Controls must decide between two manufacturing processes for its model C electronic thermostat. The monthly cost of the first process is given by $C_1(x) = 20x + 10{,}000$ dollars, where x is the number of thermostats produced; the monthly cost of the second process is given by $C_2(x) = 10x + 30{,}000$ dollars. If the projected monthly sales are 800 thermostats at a unit price of $40, which process should management choose in order to maximize the company's profit?

Solution The break-even level of operation using the first process is obtained by solving the equation

$$40x = 20x + 10,000$$
$$20x = 10,000$$
$$x = 500$$

giving an output of 500 units. Next, we solve the equation

$$40x = 10x + 30,000$$
$$30x = 30,000$$
$$x = 1000$$

giving an output of 1000 units for a break-even operation using the second process. Since the projected sales are 800 units, we conclude that management should choose the first process, which will give the firm a profit. ∎

 APPLIED EXAMPLE 5 Decision Analysis Referring to Example 4, decide which process Robertson's management should choose if the projected monthly sales are (a) 1500 units and (b) 3000 units.

Solution In both cases, the production is past the break-even level. Since the revenue is the same regardless of which process is employed, the decision will be based on how much each process costs.

a. If $x = 1500$, then

$$C_1(x) = (20)(1500) + 10,000 = 40,000$$
$$C_2(x) = (10)(1500) + 30,000 = 45,000$$

Hence, management should choose the first process.

b. If $x = 3000$, then

$$C_1(x) = (20)(3000) + 10,000 = 70,000$$
$$C_2(x) = (10)(3000) + 30,000 = 60,000$$

In this case, management should choose the second process. ∎

 EXPLORING WITH TECHNOLOGY

1. Use a graphing utility to plot the straight lines L_1 and L_2 with equations $y = 2x - 1$ and $y = 2.1x + 3$, respectively, on the same set of axes, using the standard viewing window. Do the lines appear to intersect?

2. Plot the straight lines L_1 and L_2, using the viewing window $[-100, 100] \times [-100, 100]$. Do the lines appear to intersect? Can you find the point of intersection using TRACE and ZOOM? Using the "intersection" function of your graphing utility?

3. Find the point of intersection of L_1 and L_2 algebraically.

4. Comment on the effectiveness of the solution methods in parts 2 and 3.

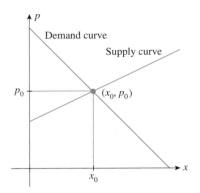

FIGURE 39
Market equilibrium is represented by
the point (x_0, p_0).

Market Equilibrium

Under pure competition, the price of a commodity eventually settles at a level dictated by the condition that the supply of the commodity be equal to the demand for it. If the price is too high, consumers will be more reluctant to buy, and if the price is too low, the supplier will be more reluctant to make the product available in the marketplace. Market equilibrium is said to prevail when the quantity produced is equal to the quantity demanded. The quantity produced at market equilibrium is called the equilibrium quantity, and the corresponding price is called the equilibrium price.

From a geometric point of view, market equilibrium corresponds to the point at which the demand curve and the supply curve intersect. In Figure 39, x_0 represents the equilibrium quantity and p_0 the equilibrium price. The point (x_0, p_0) lies on the supply curve and therefore satisfies the supply equation. At the same time, it also lies on the demand curve and therefore satisfies the demand equation. Thus, to find the point (x_0, p_0), and hence the equilibrium quantity and price, we solve the demand and supply equations simultaneously for x and p. For meaningful solutions, x and p must both be positive.

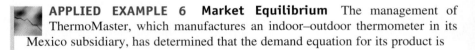 **APPLIED EXAMPLE 6 Market Equilibrium** The management of ThermoMaster, which manufactures an indoor–outdoor thermometer in its Mexico subsidiary, has determined that the demand equation for its product is

$$5x + 3p - 30 = 0$$

where p is the price of a thermometer in dollars and x is the quantity demanded in units of a thousand. The supply equation for these thermometers is

$$52x - 30p + 45 = 0$$

where x (measured in thousands) is the quantity ThermoMaster will make available in the market at p dollars each. Find the equilibrium quantity and price.

Solution   We need to solve the system of equations

$$\begin{aligned} 5x + 3p - 30 &= 0 \\ 52x - 30p + 45 &= 0 \end{aligned}$$

for x and p. Let us use the *method of substitution* to solve it. As the name suggests, this method calls for choosing one of the equations in the system, solving for one variable in terms of the other, and then substituting the resulting expression into the other equation. This gives an equation in one variable that can then be solved in the usual manner.

Let's solve the first equation for p in terms of x. Thus,

$$3p = -5x + 30$$

$$p = -\frac{5}{3}x + 10$$

Next, we substitute this value of p into the second equation, obtaining

$$52x - 30\left(-\frac{5}{3}x + 10\right) + 45 = 0$$

$$52x + 50x - 300 + 45 = 0$$

$$102x - 255 = 0$$

$$x = \frac{255}{102} = \frac{5}{2}$$

The corresponding value of p is found by substituting this value of x into the equation for p obtained earlier. Thus,

$$p = -\frac{5}{3}\left(\frac{5}{2}\right) + 10 = -\frac{25}{6} + 10$$

$$= \frac{35}{6} \approx 5.83$$

We conclude that the equilibrium quantity is 2500 units (remember that x is measured in units of a thousand) and the equilibrium price is $5.83 per thermometer.

■

VIDEO

APPLIED EXAMPLE 7 Market Equilibrium The quantity demanded of a certain model of DVD player is 8000 units when the unit price is $260. At a unit price of $200, the quantity demanded increases to 10,000 units. The manufacturer will not market any players if the price is $100 or lower. However, for each $50 increase in the unit price above $100, the manufacturer will market an additional 1000 units. Both the demand and the supply equations are known to be linear.

a. Find the demand equation.
b. Find the supply equation.
c. Find the equilibrium quantity and price.

Solution Let p denote the unit price in hundreds of dollars and let x denote the number of units of players in thousands.

a. Since the demand function is linear, the demand curve is a straight line passing through the points (8, 2.6) and (10, 2). Its slope is

$$m = \frac{2 - 2.6}{10 - 8} = -0.3$$

Using the point (10, 2) and the slope $m = -0.3$ in the point-slope form of the equation of a line, we see that the required demand equation is

$$p - 2 = -0.3(x - 10)$$

$$p = -0.3x + 5 \qquad \text{Figure 40}$$

b. The supply curve is the straight line passing through the points (0, 1) and (1, 1.5). Its slope is

$$m = \frac{1.5 - 1}{1 - 0} = 0.5$$

Using the point $(0, 1)$ and the slope $m = 0.5$ in the point-slope form of the equation of a line, we see that the required supply equation is

$$p - 1 = 0.5(x - 0)$$
$$p = 0.5x + 1 \qquad \text{Figure 40}$$

c. To find the market equilibrium, we solve simultaneously the system comprising the demand and supply equations obtained in parts (a) and (b)—that is, the system

$$p = -0.3x + 5$$
$$p = 0.5x + 1$$

Subtracting the first equation from the second gives

$$0.8x - 4 = 0$$

and $x = 5$. Substituting this value of x in the second equation gives $p = 3.5$. Thus, the equilibrium quantity is 5000 units and the equilibrium price is $350 (Figure 40).

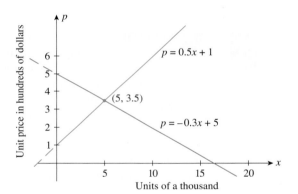

FIGURE 40
Market equilibrium occurs at the point (5, 3.5).

1.4   Self-Check Exercises

1. Find the point of intersection of the straight lines with equations $2x + 3y = 6$ and $x - 3y = 4$.

2. There is no demand for a certain make of one-time use camera when the unit price is $12. However, when the unit price is $8, the quantity demanded is 8000/wk. The suppliers will not market any cameras if the unit price is $2 or lower. At $4/camera, however, the manufacturer will make available

5000 cameras/wk. Both the demand and supply functions are known to be linear.
   a. Find the demand equation.
   b. Find the supply equation.
   c. Find the equilibrium quantity and price.

Solutions to Self-Check Exercises 1.4 can be found on page 51.

1.4   Concept Questions

1. Explain why the intersection of a linear demand curve and a linear supply curve must lie in the first quadrant.

2. Explain the meaning of each term:
   a. *Break-even point*
   b. *Break-even quantity*
   c. *Break-even revenue*

3. Explain the meaning of each term:
   a. *Market equilibrium*
   b. *Equilibrium quantity*
   c. *Equilibrium price*

1.4   Exercises

In Exercises 1–6, find the point of intersection of each pair of straight lines.

1. $y = 3x + 4$
   $y = -2x + 14$

2. $y = -4x - 7$
   $-y = 5x + 10$

3. $2x - 3y = 6$
   $3x + 6y = 16$

4. $2x + 4y = 11$
   $-5x + 3y = 5$

5. $y = \dfrac{1}{4}x - 5$
   $2x - \dfrac{3}{2}y = 1$

6. $y = \dfrac{2}{3}x - 4$
   $x + 3y + 3 = 0$

In Exercises 7–10, find the break-even point for the firm whose cost function C and revenue function R are given.

7. $C(x) = 5x + 10,000; R(x) = 15x$

8. $C(x) = 15x + 12,000; R(x) = 21x$

9. $C(x) = 0.2x + 120; R(x) = 0.4x$

10. $C(x) = 150x + 20,000; R(x) = 270x$

11. **BREAK-EVEN ANALYSIS** AutoTime, a manufacturer of 24-hr variable timers, has a monthly fixed cost of $48,000 and a production cost of $8 for each timer manufactured. The units sell for $14 each.
    a. Sketch the graphs of the cost function and the revenue function and hence find the break-even point graphically.
    b. Find the break-even point algebraically.
    c. Sketch the graph of the profit function.
    d. At what point does the graph of the profit function cross the x-axis? Interpret your result.

12. **BREAK-EVEN ANALYSIS** A division of Carter Enterprises produces "Personal Income Tax" diaries. Each diary sells for $8. The monthly fixed costs incurred by the division are $25,000, and the variable cost of producing each diary is $3.

    a. Find the break-even point for the division.
    b. What should be the level of sales in order for the division to realize a 15% profit over the cost of making the diaries?

13. **BREAK-EVEN ANALYSIS** A division of the Gibson Corporation manufactures bicycle pumps. Each pump sells for $9, and the variable cost of producing each unit is 40% of the selling price. The monthly fixed costs incurred by the division are $50,000. What is the break-even point for the division?

14. **LEASING** Ace Truck Leasing Company leases a certain size truck for $30/day and $.15/mi, whereas Acme Truck Leasing Company leases the same size truck for $25/day and $.20/mi.
    a. Find the functions describing the daily cost of leasing from each company.
    b. Sketch the graphs of the two functions on the same set of axes.
    c. If a customer plans to drive at most 70 mi, which company should he rent a truck from for 1 day?

15. **DECISION ANALYSIS** A product may be made using machine I or machine II. The manufacturer estimates that the monthly fixed costs of using machine I are $18,000, whereas the monthly fixed costs of using machine II are $15,000. The variable costs of manufacturing 1 unit of the product using machine I and machine II are $15 and $20, respectively. The product sells for $50 each.
    a. Find the cost functions associated with using each machine.
    b. Sketch the graphs of the cost functions of part (a) and the revenue functions on the same set of axes.
    c. Which machine should management choose to maximize their profit if the projected sales are 450 units? 550 units? 650 units?
    d. What is the profit for each case in part (c)?

16. The annual sales of Crimson Drug Store are expected to be given by $S = 2.3 + 0.4t$ million dollars t yr from now, whereas the annual sales of Cambridge Drug Store are expected to be given by $S = 1.2 + 0.6t$ million dollars t yr from now. When will Cambridge's annual sales first surpass Crimson's annual sales?

17. LCDs VERSUS CRTs The global shipments of traditional cathode-ray tube monitors (CRTs) is approximated by the equation

$$y = -12t + 88 \qquad (0 \le t \le 3)$$

where y is measured in millions and t in years, with $t = 0$ corresponding to 2001. The equation

$$y = 18t + 13.4 \qquad (0 \le t \le 3)$$

gives the approximate number (in millions) of liquid crystal displays (LCDs) over the same period. When did the global shipments of LCDs first overtake the global shipments of CRTs?

Source: IDC

18. DIGITAL VERSUS FILM CAMERAS The sales of digital cameras (in millions of units) in year t is given by the function

$$f(t) = 3.05t + 6.85 \qquad (0 \le t \le 3)$$

where $t = 0$ corresponds to 2001. Over that same period, the sales of film cameras (in millions of units) is given by

$$g(t) = -1.85t + 16.58 \qquad (0 \le t \le 3)$$

a. Show that more film cameras than digital cameras were sold in 2001.
b. When did the sales of digital cameras first exceed those of film cameras?

Source: Popular Science

19. U.S. FINANCIAL TRANSACTIONS The percent of U.S. transactions by check between 2001 ($t = 0$) and 2010 ($t = 9$) is projected to be

$$f(t) = -\frac{11}{9}t + 43 \qquad (0 \le t \le 9)$$

whereas the percent of transactions done electronically during the same period is projected to be

$$g(t) = \frac{11}{3}t + 23 \qquad (0 \le t \le 9)$$

a. Sketch the graphs of f and g on the same set of axes.
b. Find the time when transactions done electronically first exceeded those done by check.

Source: Foreign Policy

20. BROADBAND VERSUS DIAL-UP The number of U.S. broadband Internet households (in millions) between 2004 ($t = 0$) and 2008 ($t = 4$) is projected to be

$$f(t) = 7.5t + 35 \qquad (0 \le t \le 4)$$

Over the same period, the number of U.S. dial-up Internet households (in millions) is projected to be

$$g(t) = -3.9t + 42.5 \qquad (0 \le t \le 4)$$

a. Sketch the graphs of f and g on the same set of axes.
b. Solve the equation $f(t) = g(t)$ and interpret your result.

Source: Strategic Analytics Inc.

For each pair of supply-and-demand equations in Exercises 21–24, where x represents the quantity demanded in units of 1000 and p is the unit price in dollars, find the equilibrium quantity and the equilibrium price.

21. $4x + 3p - 59 = 0$ and $5x - 6p + 14 = 0$

22. $2x + 7p - 56 = 0$ and $3x - 11p + 45 = 0$

23. $p = -2x + 22$ and $p = 3x + 12$

24. $p = -0.3x + 6$ and $p = 0.15x + 1.5$

25. EQUILIBRIUM QUANTITY AND PRICE The quantity demanded of a certain brand of DVD player is 3000/wk when the unit price is \$485. For each decrease in unit price of \$20 below \$485, the quantity demanded increases by 250 units. The suppliers will not market any DVD players if the unit price is \$300 or lower. But at a unit price of \$525, they are willing to make available 2500 units in the market. The supply equation is also known to be linear.

a. Find the demand equation.
b. Find the supply equation.
c. Find the equilibrium quantity and price.

26. EQUILIBRIUM QUANTITY AND PRICE The demand equation for the Drake GPS Navigator is $x + 4p - 800 = 0$, where x is the quantity demanded per week and p is the wholesale unit price in dollars. The supply equation is $x - 20p + 1000 = 0$, where x is the quantity the supplier will make available in the market when the wholesale price is p dollars each. Find the equilibrium quantity and the equilibrium price for the GPS Navigators.

27. EQUILIBRIUM QUANTITY AND PRICE The demand equation for the Schmidt-3000 fax machine is $3x + p - 1500 = 0$, where x is the quantity demanded per week and p is the unit price in dollars. The supply equation is $2x - 3p + 1200 = 0$, where x is the quantity the supplier will make available in the market when the unit price is p dollars. Find the equilibrium quantity and the equilibrium price for the fax machines.

28. EQUILIBRIUM QUANTITY AND PRICE The quantity demanded each month of Russo Espresso Makers is 250 when the unit price is \$140. The quantity demanded each month is 1000 when the unit price is \$110. The suppliers will market 750 espresso makers if the unit price is \$60 or lower. At a unit price of \$80, they are willing to make available 2250 units in the market. Both the demand and supply equations are known to be linear.

a. Find the demand equation.
b. Find the supply equation.
c. Find the equilibrium quantity and the equilibrium price.

29. Suppose the demand-and-supply equations for a certain commodity are given by $p = ax + b$ and $p = cx + d$, respectively, where $a < 0$, $c > 0$, and $b > d > 0$ (see the accompanying figure).

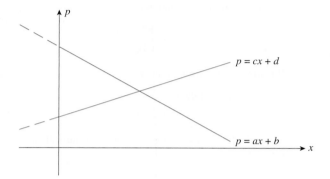

a. Find the equilibrium quantity and equilibrium price in terms of a, b, c, and d.
b. Use part (a) to determine what happens to the market equilibrium if c is increased while a, b, and d remain fixed. Interpret your answer in economic terms.
c. Use part (a) to determine what happens to the market equilibrium if b is decreased while a, c, and d remain fixed. Interpret your answer in economic terms.

30. Suppose the cost function associated with a product is $C(x) = cx + F$ dollars and the revenue function is $R(x) = sx$, where c denotes the unit cost of production, s denotes the unit selling price, F denotes the fixed cost incurred by the firm, and x denotes the level of production and sales. Find the break-even quantity and the break-even revenue in terms of the constants c, s, and F and interpret your results in economic terms.

In Exercises 31 and 32, determine whether the statement is true or false. If it is true, explain why it is true. If it is false, give an example to show why it is false.

31. Suppose $C(x) = cx + F$ and $R(x) = sx$ are the cost and revenue functions of a certain firm. Then, the firm is operating at a break-even level of production if its level of production is $F/(s - c)$.

32. If both the demand equation and the supply equation for a certain commodity are linear, then there must be at least one equilibrium point.

33. Let L_1 and L_2 be two nonvertical straight lines in the plane with equations $y = m_1x + b_1$ and $y = m_2x + b_2$, respectively. Find conditions on m_1, m_2, b_1, and b_2 so that (a) L_1 and L_2 do not intersect, (b) L_1 and L_2 intersect at one and only one point, and (c) L_1 and L_2 intersect at infinitely many points.

34. Find conditions on a_1, a_2, b_1, b_2, c_1, and c_2 so that the system of linear equations

$$a_1x + b_1y = c_1$$
$$a_2x + b_2y = c_2$$

has (a) no solution, (b) a unique solution, and (c) infinitely many solutions.
Hint: Use the results of Exercise 33.

1.4 Solutions to Self-Check Exercises

1. The point of intersection of the two straight lines is found by solving the system of linear equations

$$2x + 3y = 6$$
$$x - 3y = 4$$

Solving the first equation for y in terms of x, we obtain

$$y = -\frac{2}{3}x + 2$$

Substituting this expression for y into the second equation, we obtain

$$x - 3\left(-\frac{2}{3}x + 2\right) = 4$$
$$x + 2x - 6 = 4$$
$$3x = 10$$

or $x = \frac{10}{3}$. Substituting this value of x into the expression for y obtained earlier, we find

$$y = -\frac{2}{3}\left(\frac{10}{3}\right) + 2 = -\frac{2}{9}$$

Therefore, the point of intersection is $\left(\frac{10}{3}, -\frac{2}{9}\right)$.

2. **a.** Let p denote the price per camera and x the quantity demanded. The given conditions imply that $x = 0$ when $p = 12$ and $x = 8000$ when $p = 8$. Since the demand equation is linear, it has the form

$$p = mx + b$$

Now, the first condition implies that

$$12 = m(0) + b \quad \text{or} \quad b = 12$$

Therefore,

$$p = mx + 12$$

Using the second condition, we find

$$8 = 8000m + 12$$

$$m = -\frac{4}{8000} = -0.0005$$

Therefore, the required demand equation is

$$p = -0.0005x + 12$$

b. Let p denote the price per camera and x the quantity made available at that price. Then, since the supply equation is linear, it also has the form

$$p = mx + b$$

The first condition implies that $x = 0$ when $p = 2$, so we have

$$2 = m(0) + b \quad \text{or} \quad b = 2$$

Therefore,

$$p = mx + 2$$

Next, using the second condition, $x = 5000$ when $p = 4$, we find

$$4 = 5000m + 2$$

giving $m = 0.0004$. So the required supply equation is

$$p = 0.0004x + 2$$

c. The equilibrium quantity and price are found by solving the system of linear equations

$$p = -0.0005x + 12$$
$$p = 0.0004x + 2$$

Equating the two equations yields

$$-0.0005x + 12 = 0.0004x + 2$$
$$0.0009x = 10$$

or $x \approx 11{,}111$. Substituting this value of x into either equation in the system yields

$$p \approx 6.44$$

Therefore, the equilibrium quantity is 11,111, and the equilibrium price is $6.44.

USING TECHNOLOGY

■ Finding the Point(s) of Intersection of Two Graphs

Graphing Utility

A graphing utility can be used to find the point(s) of intersection of the graphs of two functions. Once again, it is important to remember that if the graphs are straight lines, the linear equations defining these lines must first be recast in the slope-intercept form.

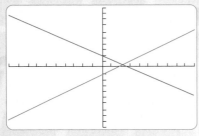

FIGURE T1
The straight lines $2x + 3y = 6$ and $3x - 4y - 5 = 0$

EXAMPLE 1   Find the points of intersection of the straight lines with equations $2x + 3y = 6$ and $3x - 4y - 5 = 0$.

Solution   Solving each equation for y in terms of x, we obtain

$$y = -\frac{2}{3}x + 2 \quad \text{and} \quad y = \frac{3}{4}x - \frac{5}{4}$$

as the respective equations in the slope-intercept form. The graphs of the two straight lines in the standard viewing window are shown in Figure T1.

Then, using TRACE and ZOOM or the function for finding the point of intersection of two graphs, we find that the point of intersection, accurate to four decimal places, is (2.2941, 0.4706).

TECHNOLOGY EXERCISES

In Exercises 1–6, find the point of intersection of the pair of straight lines with the given equations. Express your answers accurate to four decimal places.

1. $y = 2x + 5$ and $y = -3x + 8$

2. $y = 1.2x + 6.2$ and $y = -4.3x + 9.1$

3. $2x - 5y = 7$ and $3x + 4y = 12$

4. $1.4x - 6.2y = 8.4$ and $4.1x + 7.3y = 14.4$

5. $2.1x = 5.1y + 71$ and $3.2x = 8.4y + 16.8$

6. $8.3x = 6.2y + 9.3$ and $12.4x = 12.3y + 24.6$

7. BREAK-EVEN ANALYSIS PhotoMax makes single-use cameras that sell for $7.89 each and cost $3.24 each to produce. The weekly fixed cost for the company is $16,500.
   a. Plot the graphs of the cost function and the revenue function in the viewing window $[0, 6000] \times [0, 60,000]$.
   b. Find the break-even point, using the viewing window $[0, 6000] \times [-20,000, 20,000]$.
   c. Plot the graph of the profit function and verify the result of part (b) by finding the x-intercept.

8. BREAK-EVEN ANALYSIS The Monde Company makes a wine cooler with a capacity of 24 bottles. Each wine cooler sells for $245. The monthly fixed costs incurred by the company are $385,000, and the variable cost of producing each wine cooler is $90.50.
   a. Find the break-even point for the company.
   b. Find the level of sales needed to ensure that the company will realize a profit of 21% over the cost of producing the wine coolers.

9. LEASING Ace Truck Leasing Company leases a certain size truck for $34/day and $.18/mi, whereas Acme Truck Leasing Company leases the same size truck for $28/day and $.22/mi.
   a. Find the functions describing the daily cost of leasing from each company.
   b. Plot the graphs of the two functions, using the same viewing window.
   c. Find the point of intersection of the graphs of part (b).
   d. Use the result of part (c) to find a criterion that a customer can use to help her decide which company to rent the truck

from if she knows the maximum distance that she will drive on the day of rental.

10. BANK DEPOSITS The total deposits with a branch of Randolph Bank currently stand at $20.384 million and are projected to grow at the rate of $1.019 million/yr for the next 5 yr. The total deposits with a branch of Madison Bank, in the same shopping complex as the Randolph Bank, currently stands at $18.521 million and are expected to grow at the rate of $1.482 million/yr for the next 5 yr.
   a. Find the function describing the total deposits with each bank for the next 5 yr.
   b. Plot the graphs of the two functions found in part (a) using the same viewing window.
   c. Do the total deposits of Madison catch up to those of Randolph over the period in question? If so, at what time?

11. EQUILIBRIUM QUANTITY AND PRICE The quantity demanded of a certain brand of a two-way radio is 2000/wk when the unit price is $84. For each decrease in unit price of $5 below $84, the quantity demanded increases by 50 units. The supplier will not market any of the two-way radios if the unit price is $60 or less. But at a unit price of $90, the supplier is willing to make available 1800/wk in the market. The supply equation is also known to be linear.
   a. Find the demand and supply equations.
   b. Plot the graphs of the supply and demand equations and find their point of intersection.
   c. Find the equilibrium quantity and price.

12. EQUILIBRIUM QUANTITY AND PRICE The demand equation for the Miramar Heat Machine, a ceramic heater, is $1.1x + 3.8p - 901 = 0$, where x is the quantity demanded each week and p is the wholesale unit price in dollars. The corresponding supply equation is $0.9x - 20.4p + 1038 = 0$, where x is the quantity the supplier will make available in the market when the wholesale price is p dollars each.
   a. Plot the graphs of the demand and supply equations, using the same viewing window.
   b. Find the equilibrium quantity and the equilibrium price for the Miramar heaters.

1.5   The Method of Least Squares (Optional)

▬ The Method of Least Squares

In Example 10, Section 1.2, we saw how a linear equation may be used to approximate the sales trend for a local sporting goods store. The *trend line*, as we saw, may be used to predict the store's future sales. Recall that we obtained the trend line in Example 10 by requiring that the line pass through two data points, the rationale being that such a line seems to *fit* the data reasonably well.

In this section we describe a general method known as the **method of least squares** for determining a straight line that, in some sense, best fits a set of data points when the points are scattered about a straight line. To illustrate the principle behind the method of least squares, suppose, for simplicity, that we are given five data points,

$$P_1(x_1, y_1), \quad P_2(x_2, y_2), \quad P_3(x_3, y_3), \quad P_4(x_4, y_4), \quad P_5(x_5, y_5)$$

describing the relationship between the two variables x and y. By plotting these data points, we obtain a graph called a **scatter diagram** (Figure 41).

If we try to *fit* a straight line to these data points, the line will miss the first, second, third, fourth, and fifth data points by the amounts d_1, d_2, d_3, d_4, and d_5, respectively (Figure 42). We can think of the amounts $d_1, d_2, \ldots, d_5$ as the errors made when the values $y_1, y_2, \ldots, y_5$ are approximated by the corresponding values of y lying on the straight line L.

The **principle of least squares** states that the straight line L that fits the data points *best* is the one chosen by requiring that the sum of the squares of $d_1, d_2, \ldots, d_5$—that is,

$$d_1^2 + d_2^2 + d_3^2 + d_4^2 + d_5^2$$

be made as small as possible. In other words, the least-squares criterion calls for minimizing the sum of the squares of the errors. The line L obtained in this manner is called the **least-squares line,** or *regression line.*

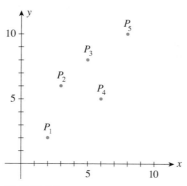

FIGURE 41
A scatter diagram

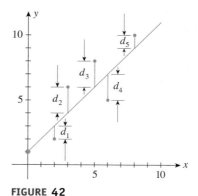

FIGURE 42
d_i is the vertical distance between the straight line and a given data point.

The method for computing the least-squares lines that best fits a set of data points is contained in the following result, which we state without proof.

The Method of Least Squares

Suppose we are given n data points

$$P_1(x_1, y_1), \quad P_2(x_2, y_2), \quad P_3(x_3, y_3), \ldots, P_n(x_n, y_n)$$

Then, the least-squares (regression) line for the data is given by the linear equation (function)

$$y = f(x) = mx + b$$

where the constants m and b satisfy the **normal equations**

$$nb + (x_1 + x_2 + \cdots + x_n)m = y_1 + y_2 + \cdots + y_n \tag{9}$$

$$(x_1 + x_2 + \cdots + x_n)b + (x_1^2 + x_2^2 + \cdots + x_n^2)m$$
$$= x_1y_1 + x_2y_2 + \cdots + x_ny_n \tag{10}$$

simultaneously.

EXAMPLE 1   Find the least-squares line for the data

$$P_1(1, 1), \quad P_2(2, 3), \quad P_3(3, 4), \quad P_4(4, 3), \quad P_5(5, 6)$$

Solution   Here, we have $n = 5$ and

$$x_1 = 1 \qquad x_2 = 2 \qquad x_3 = 3 \qquad x_4 = 4 \qquad x_5 = 5$$
$$y_1 = 1 \qquad y_2 = 3 \qquad y_3 = 4 \qquad y_4 = 3 \qquad y_5 = 6$$

Before using Equations (9) and (10), it is convenient to summarize this data in the form of a table:

	x	y	x^2	xy
	1	1	1	1
	2	3	4	6
	3	4	9	12
	4	3	16	12
	5	6	25	30
Sum	15	17	55	61

Using this table and (9) and (10), we obtain the normal equations

$$5b + 15m = 17 \tag{11}$$

$$15b + 55m = 61 \tag{12}$$

Solving Equation (11) for b gives

$$b = -3m + \frac{17}{5} \tag{13}$$

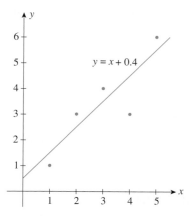

FIGURE 43
The least-squares line $y = x + 0.4$ and the given data points

which upon substitution into Equation (12) gives

$$15\left(-3m + \frac{17}{5}\right) + 55m = 61$$
$$-45m + 51 + 55m = 61$$
$$10m = 10$$
$$m = 1$$

Substituting this value of m into Equation (13) gives

$$b = -3 + \frac{17}{5} = \frac{2}{5} = 0.4$$

Therefore, the required least-squares line is

$$y = x + 0.4$$

The scatter diagram and the least-squares line are shown in Figure 43.   ■

APPLIED EXAMPLE 2 Advertising and Profit The proprietor of Leisure Travel Service compiled the following data relating the annual profit of the firm to its annual advertising expenditure (both measured in thousands of dollars):

Annual Advertising Expenditure, x	12	14	17	21	26	30
Annual Profit, y	60	70	90	100	100	120

a. Determine the equation of the least-squares line for these data.
b. Draw a scatter diagram and the least-squares line for these data.
c. Use the result obtained in part (a) to predict Leisure Travel's annual profit if the annual advertising budget is $20,000.

Solution

a. The calculations required for obtaining the normal equations are summarized in the accompanying table:

	x	y	x^2	xy
	12	60	144	720
	14	70	196	980
	17	90	289	1,530
	21	100	441	2,100
	26	100	676	2,600
	30	120	900	3,600
Sum	120	540	2,646	11,530

The normal equations are

$$6b + 120m = 540 \tag{14}$$
$$120b + 2646m = 11,530 \tag{15}$$

Solving Equation (14) for b gives

$$b = -20m + 90 \qquad \text{(16)}$$

which upon substitution into Equation (15) gives

$$120(-20m + 90) + 2646m = 11{,}530$$
$$-2400m + 10{,}800 + 2646m = 11{,}530$$
$$246m = 730$$
$$m \approx 2.97$$

Substituting this value of m into Equation (16) gives

$$b = -20(2.97) + 90$$
$$= 30.6$$

Therefore, the required least-squares line is given by

$$y = f(x) = 2.97x + 30.6$$

b. The scatter diagram and the least-squares line are shown in Figure 44.

c. Leisure Travel's predicted annual profit corresponding to an annual budget of $20,000 is given by

$$f(20) = 2.97(20) + 30.6 = 90$$

or $90,000.

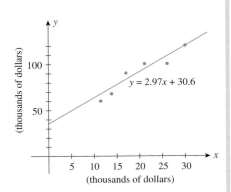

FIGURE 44
Profit versus advertising expenditure

APPLIED EXAMPLE 3 **Market for Cholesterol-Reducing Drugs**
Refer to Example 1 of Section 1.3. In a study conducted in early 2000, experts projected a rise in the market for cholesterol-reducing drugs. The U.S. market (in billions of dollars) for such drugs from 1999 through 2004 is given in the following table:

Year	1999	2000	2001	2002	2003	2004
Market	12.07	14.07	16.21	18.28	20	21.72

Find a function giving the U.S. market for cholesterol-reducing drugs between 1999 and 2004, using the least-squares technique. (Here, $t = 0$ corresponds to 1999.)

Solution The calculations required for obtaining the normal equations are summarized in the following table:

t	y	t^2	ty
0	12.07	0	0
1	14.07	1	14.07
2	16.21	4	32.42
3	18.28	9	54.84
4	20.0	16	80
5	21.72	25	108.6
Sum 15	102.35	55	289.93

The normal equations are

$$6b + 15m = 102.35 \tag{17}$$
$$15b + 55m = 289.93 \tag{18}$$

Solving Equation (17) for b gives

$$b = -2.5m + 17.058 \tag{19}$$

which upon substitution into Equation (18) gives

$$15(-2.5m + 17.058) + 55m = 289.93$$
$$-37.5m + 255.87 + 55m = 289.93$$
$$17.5m = 34.06$$
$$m \approx 1.946$$

Substituting this value of m into Equation (19) gives

$$b = -2.5(1.946) + 17.058 = 12.193$$

Therefore, the required function is

$$M(t) = 1.95t + 12.19$$

as obtained before.

1.5 Self-Check Exercises

1. Find an equation of the least-squares line for the data

x	1	3	4	5	7
y	4	10	11	12	16

2. In a market research study for Century Communications, the following data were provided based on the projected monthly sales x (in thousands) of a videocassette version of a box-office-hit adventure movie with a proposed wholesale unit price of p dollars.

x	2.2	5.4	7.0	11.5	14.6
p	38	36	34.5	30	28.5

Find the demand equation if the demand curve is the least-squares line for these data.

Solutions to Self-Check Exercises 1.5 can be found on page 62.

1.5 Concept Questions

1. Explain the terms (a) *scatter diagram* and (b) *least-squares line*.

2. State the principle of least squares in your own words.

1.5   Exercises

In Exercises 1–6, (a) find the equation of the least-squares line for the data and (b) draw a scatter diagram for the data and graph the least-squares line.

1.

x	1	2	3	4
y	4	6	8	11

2.

x	1	3	5	7	9
y	9	8	6	3	2

3.

x	1	2	3	4	4	6
y	4.5	5	3	2	3.5	1

4.

x	1	1	2	3	4	4	5
y	2	3	3	3.5	3.5	4	5

5. $P_1(1, 3), P_2(2, 5), P_3(3, 5), P_4(4, 7), P_5(5, 8)$

6. $P_1(1, 8), P_2(2, 6), P_3(5, 6), P_4(7, 4), P_5(10, 1)$

7. COLLEGE ADMISSIONS The accompanying data were compiled by the admissions office at Faber College during the past 5 yr. The data relate the number of college brochures and follow-up letters (x) sent to a preselected list of high school juniors who had taken the PSAT and the number of completed applications (y) received from these students (both measured in units of a thousand).

x	4	4.5	5	5.5	6
y	0.5	0.6	0.8	0.9	1.2

a. Determine the equation of the least-squares line for these data.

b. Draw a scatter diagram and the least-squares line for these data.

c. Use the result obtained in part (a) to predict the number of completed applications expected if 6400 brochures and follow-up letters are sent out during the next year.

8. STARBUCKS' STORE COUNT According to *Company Reports*, the number of Starbucks stores worldwide between 1999 and 2003 are as follows:

Year, x	0	1	2	3	4
Stores, y	2135	3501	4709	5886	7225

(Here, $x = 0$ corresponds to 1999.)

a. Find an equation of the least-squares line for these data.

b. Use the result of part (a) to estimate the rate at which new stores were opened annually in North America for the period in question.

Source: Company Reports

9. SAT VERBAL SCORES The accompanying data were compiled by the superintendent of schools in a large metropolitan area. The table shows the average SAT verbal scores of high school seniors during the 5 yr since the district implemented its "back-to-basics" program.

Year, x	1	2	3	4	5
Average Score, y	436	438	428	430	426

a. Determine the equation of the least-squares line for these data.

b. Draw a scatter diagram and the least-squares line for these data.

c. Use the result obtained in part (a) to predict the average SAT verbal score of high school seniors 2 yr from now ($x = 7$).

10. NET SALES The management of Kaldor, a manufacturer of electric motors, submitted the accompanying data in its annual stockholders report. The table shows the net sales (in millions of dollars) during the 5 yr that have elapsed since the new management team took over:

Year, x	1	2	3	4	5
Net Sales, y	426	437	460	473	477

(The first year the firm operated under the new management corresponds to the time period $x = 1$, and the four subsequent years correspond to $x = 2, 3, 4,$ and 5.)

a. Determine the equation of the least-squares line for these data.

b. Draw a scatter diagram and the least-squares line for these data.

c. Use the result obtained in part (a) to predict the net sales for the upcoming year.

11. MASS-TRANSIT SUBSIDIES The accompanying table gives the projected state subsidies (in millions of dollars) to the Massachusetts Bay Transit Authority (MBTA) over a 5-yr period.

Year, x	1	2	3	4	5
Subsidy, y	20	24	26	28	32

a. Find an equation of the least-squares line for these data.

b. Use the result of part (a) to estimate the state subsidy to the MBTA for the eighth year ($x = 8$).

Source: Massachusetts Bay Transit Authority

12. INFORMATION SECURITY SOFTWARE SALES As online attacks persist, spending on information security software continues to rise. The following table gives the forecast for the worldwide sales (in billions of dollars) of information security software through 2007:

Year, t	0	1	2	3	4	5
Spending, y	6.8	8.3	9.8	11.3	12.8	14.9

(Here, $t = 0$ corresponds to 2002.)
a. Find an equation of the least-squares line for these data.
b. Use the result of part (a) to forecast the spending on information security software in 2008, assuming that the trend continues.

Source: International Data Corp.

13. **U.S. DRUG SALES**   The following table gives the total sales of drugs (in billions of dollars) in the United States from 1999 ($t = 0$) through 2003:

Year, t	0	1	2	3	4
Sales, y	126	144	171	191	216

a. Find an equation of the least-squares line for these data.
b. Use the result of part (a) to predict the total sales of drugs in 2005, assuming that the trend continues.

Source: IMS Health

14. **NURSES' SALARIES**   The average hourly salary of hospital nurses in metropolitan Boston (in dollars) from 2000 through 2004 is given in the following table:

Year	2000	2001	2002	2003	2004
Hourly Salary, y	27	29	31	32	35

a. Find an equation of the least-squares line for these data. (Let $x = 0$ represent 2000.)
b. If the trend continues, what would you expect the average hourly salary of nurses to be in 2006?

Source: American Association of Colleges of Nursing

15. **NET-CONNECTED COMPUTERS IN EUROPE**   The projected number of computers (in millions) connected to the Internet in Europe from 1998 through 2002 is summarized in the accompanying table:

Year, x	0	1	2	3	4
Net-Connected Computers, y	21.7	32.1	45.0	58.3	69.6

(Here, $x = 0$ corresponds to the beginning of 1998.)
a. Find an equation of the least-squares line for these data.
b. Use the result of part (a) to estimate the projected number of computers connected to the Internet in Europe at the beginning of 2005, assuming the trend continued.

Source: Dataquest Inc.

16. **MALE LIFE EXPECTANCY AT 65**   The Census Bureau projections of male life expectancy at age 65 in the United States are summarized in the following table:

Year, x	0	10	20	30	40	50
Years Beyond 65, y	15.9	16.8	17.6	18.5	19.3	20.3

(Here, $x = 0$ corresponds to 2000.)
a. Find an equation of the least-squares line for these data.
b. Use the result of part (a) to estimate the life expectancy at 65 of a male in 2040. How does this result compare with the given data for that year?
c. Use the result of part (a) to estimate the life expectancy at 65 of a male in 2030.

Source: U.S. Census Bureau

17. **FEMALE LIFE EXPECTANCY AT 65**   The Census Bureau projections of female life expectancy at age 65 in the United States are summarized in the following table:

Year, x	0	10	20	30	40	50
Years Beyond 65, y	19.5	20.0	20.6	21.2	21.8	22.4

(Here, $x = 0$ corresponds to 2000.)
a. Find an equation of the least-squares line for these data.
b. Use the result of part (a) to estimate the life expectancy at 65 of a female in 2040. How does this result compare with the given data for that year?
c. Use the result of part (a) to estimate the life expectancy at 65 of a female in 2030.

Source: U.S. Census Bureau

18. **U.S. ONLINE BANKING HOUSEHOLDS**   The following table gives the projected U.S. online banking households as a percentage of all U.S. banking households from 2001 ($x = 1$) through 2007 ($x = 7$):

Year, x	1	2	3	4	5	6	7
Percent of Households, y	21.2	26.7	32.2	37.7	43.2	48.7	54.2

a. Find an equation of the least-squares line for these data.
b. Use the result of part (a) to estimate the projected percent of U.S. online banking households in 2008.

Source: Jupiter Research

19. **SALES OF GPS EQUIPMENT**   The annual sales (in billions of dollars) of global positioning system (GPS) equipment from the year 2000 through 2006 follow (sales in 2004 through 2006 are projections):

Year, x	0	1	2	3	4	5	6
Annual Sales, y	7.9	9.6	11.5	13.3	15.2	16	18.8

(Here, $x = 0$ corresponds to the year 2000.)
a. Find an equation of the least-squares line for these data.
b. Use the equation found in part (a) to estimate the annual sales of GPS equipment for 2005.

Source: ABI Research

20. ONLINE SALES OF USED AUTOS The amount (in millions of dollars) of used autos sold online in the United States is expected to grow in accordance with the figures given in the following table:

Year, x	0	1	2	3	4	5	6	7
Sales, y	1	1.4	2.2	2.8	3.6	4.2	5.0	5.8

(Here, x = 0 corresponds to 2000.)
a. Find an equation of the least-squares line for these data.
b. Use the result of part (a) to estimate the sales of used autos online in 2008, assuming that the predicted trend continues through that year.

Source: comScore Networks Inc.

21. WIRELESS SUBSCRIBERS The projected number of wireless subscribers y (in millions) from 2000 through 2006 is summarized in the accompanying table:

Year, x	0	1	2	3
Subscribers, y	90.4	100.0	110.4	120.4

Year, x	4	5	6
Subscribers, y	130.8	140.4	150.0

(Here, x = 0 corresponds to the beginning of 2000.)
a. Find an equation of the least-squares line for these data.
b. Use the result of part (a) to estimate the projected number of wireless subscribers at the beginning of 2006. How does this result compare with the given data for that year?

Source: BancAmerica Robertson Stephens

22. AUTHENTICATION TECHNOLOGY With computer security always a hot-button issue, demand is growing for technology that authenticates and authorizes computer users. The following table gives the authentication software sales (in billions of dollars) from 1999 through 2004:

Year, x	0	1	2	3	4	5
Sales, y	2.4	2.9	3.7	4.5	5.2	6.1

(Here, x = 0 represents 1999.)
a. Find an equation of the least-squares line for these data.
b. Use the result of part (a) to estimate the sales for 2007, assuming the trend continues.

Source: International Data Corporation

23. ONLINE SPENDING The convenience of shopping on the Web combined with high-speed broadband access services are spurring online spending. The projected online spending per buyer (in dollars) from 2002 (x = 0) through 2008 (x = 6) is given in the following table:

Year, x	0	1	2	3	4	5	6
Spending, y	501	540	585	631	680	728	779

a. Find an equation of the least-squares line for these data.
b. Use the result of part (a) to estimate the rate of change of spending per buyer between 2002 and 2008.

Source: Commerce Dept.

24. CALLING CARDS The market for prepaid calling cards is projected to grow steadily through 2008. The following table gives the projected sales of prepaid phone card sales (in billions of dollars) from 2002 through 2008:

Year, x	0	1	2	3	4	5	6
Sales, y	3.7	4.0	4.4	4.8	5.2	5.8	6.3

a. Find an equation of the least-squares line for these data.
b. Use the result of part (a) to estimate the rate at which the sales of prepaid phone cards will grow over the period in question.

Source: Atlantic-ACM

25. SOCIAL SECURITY WAGE BASE The Social Security (FICA) wage base (in thousands of dollars) from 1999 to 2004 is given in the accompanying table:

Year	1999	2000	2001
Wage Base, y	72.6	76.2	80.4

Year	2002	2003	2004
Wage Base, y	84.9	87	87.9

a. Find an equation of the least-squares line for these data. (Let x = 1 represent 1999.)
b. Use the result of part (a) to estimate the FICA wage base in 2006.

Source: The World Almanac

26. OUTPATIENT VISITS With an aging population, the demand for health care, as measured by outpatient visits, is steadily growing. The number of outpatient visits (in millions) from 1991 through 2001 is recorded in the following table:

Year, x	0	1	2	3	4	5
Number of Visits, y	320	340	362	380	416	440

Year, x	6	7	8	9	10
Number of Visits, y	444	470	495	520	530

(Here, x = 0 corresponds to 1991.)
a. Find an equation of the least-squares line for these data.
b. Use the result of part (a) to estimate the number of outpatient visits in 2004, assuming that the trend continued.

Source: PriceWaterhouse Coopers

In Exercises 27–30, determine whether the statement is true or false. If it is true, explain why it is true. If it is false, give an example to show why it is false.

27. The least-squares line must pass through at least one data point.

28. The error incurred in approximating n data points using the least-squares linear function is zero if and only if the n data points lie along a straight line.

29. If the data consist of two distinct points, then the least-squares line is just the line that passes through the two points.

30. A data point lies on the least-squares line if and only if the vertical distance between the point and the line is equal to zero.

1.5 Solutions to Self-Check Exercises

1. The calculations required for obtaining the normal equations may be summarized as follows:

	x	y	x^2	xy
	1	4	1	4
	3	10	9	30
	4	11	16	44
	5	12	25	60
	7	16	49	112
Sum	20	53	100	250

The normal equations are

$$5b + 20m = 53$$
$$20b + 100m = 250$$

Solving the first equation for b gives

$$b = -4m + \frac{53}{5}$$

which, upon substitution into the second equation, yields

$$20\left(-4m + \frac{53}{5}\right) + 100m = 250$$

$$-80m + 212 + 100m = 250$$

$$20m = 38$$

$$m = 1.9$$

Substituting this value of m into the expression for b found earlier, we find

$$b = -4(1.9) + \frac{53}{5} = 3$$

Therefore, an equation of the least-squares line is

$$y = 1.9x + 3$$

2. The calculations required for obtaining the normal equations may be summarized as follows:

	x	p	x^2	xp
	2.2	38.0	4.84	83.6
	5.4	36.0	29.16	194.4
	7.0	34.5	49.00	241.5
	11.5	30.0	132.25	345.0
	14.6	28.5	213.16	416.1
Sum	40.7	167.0	428.41	1280.6

The normal equations are

$$5b + 40.7m = 167$$
$$40.7b + 428.41m = 1280.6$$

Solving this system of linear equations simultaneously, we find that

$$m \approx -0.81 \quad \text{and} \quad b \approx 39.99$$

Therefore, an equation of the least-squares line is given by

$$p = f(x) = -0.81x + 39.99$$

which is the required demand equation provided

$$0 \leq x \leq 49.37$$

USING TECHNOLOGY

■ Finding an Equation of a Least-Squares Line

Graphing Utility

A graphing utility is especially useful in calculating an equation of the least-squares line for a set of data. We simply enter the given data in the form of lists into the calculator and then use the linear regression function to obtain the coefficients of the required equation.

EXAMPLE 1 Find an equation of the least-squares line for the data

x	1.1	2.3	3.2	4.6	5.8	6.7	8
y	−5.8	−5.1	−4.8	−4.4	−3.7	−3.2	−2.5

Plot the scatter diagram and the least-squares line for this data.

Solution First, we enter the data as

$$x_1 = 1.1 \qquad y_1 = -5.8 \qquad x_2 = 2.3 \qquad y_2 = -5.1 \qquad x_3 = 3.2$$
$$y_3 = -4.8 \qquad x_4 = 4.6 \qquad y_4 = -4.4 \qquad x_5 = 5.8 \qquad y_5 = -3.7$$
$$x_6 = 6.7 \qquad y_6 = -3.2 \qquad x_7 = 8 \qquad y_7 = -2.5$$

Then, using the linear regression function from the statistics menu, we obtain the output shown in Figure T1a. Therefore, an equation of the least-squares line ($y = a + bx$) is

$$y = -6.3 + 0.46x$$

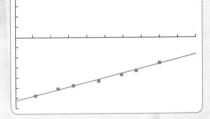

LinReg
y = ax+b
a = .4605609794
b = −6.299969007

(a) The TI-83 linear regression screen

(b) The scatter diagram and least-squares line for the data

FIGURE T1

The graph of the least-squares equation and the scatter diagram for the data are shown in Figure T1b.

EXAMPLE 2 According to Pacific Gas and Electric, the nation's largest utility company, the demand for electricity from 1990 through the year 2000 is summarized in the following table:

(continued)

t	0	2	4	6	8	10
y	333	917	1500	2117	2667	3292

Here $t = 0$ corresponds to 1990, and y gives the amount of electricity demanded in the year t, measured in megawatts. Find an equation of the least-squares line for these data.

Source: Pacific Gas and Electric

LinReg
 $y = ax+b$
 $a = 295.1714286$
 $b = 328.4761905$

∎

FIGURE T2
The TI-83 linear regression screen

Solution   First, we enter the data as

$$x_1 = 0 \qquad y_1 = 333 \qquad x_2 = 2 \qquad y_2 = 917 \qquad x_3 = 4 \qquad y_3 = 1500$$
$$x_4 = 6 \qquad y_4 = 2117 \qquad x_5 = 8 \qquad y_5 = 2667 \qquad x_6 = 10 \qquad y_6 = 3292$$

Then, using the linear regression function from the statistics menu, we obtain the output shown in Figure T2. Therefore, an equation of the least-squares line is

$$y = 328 + 295t$$

∎

Excel

Excel can be used to find an equation of the least-squares line for a set of data and to plot a scatter diagram and the least-squares line for the data.

EXAMPLE 3   Find an equation of the least-squares line for the data given in the following table:

x	1.1	2.3	3.2	4.6	5.8	6.7	8
y	−5.8	−5.1	−4.8	−4.4	−3.7	−3.2	−2.5

Plot the scatter diagram and the least-squares line for this data.

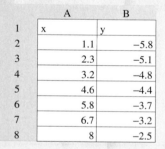

	A	B
1	x	y
2	1.1	−5.8
3	2.3	−5.1
4	3.2	−4.8
5	4.6	−4.4
6	5.8	−3.7
7	6.7	−3.2
8	8	−2.5

FIGURE T3
Table of values for *x* and *y*

Solution

1. *Set up a table of values on a spreadsheet* (Figure T3).
2. *Plot the scatter diagram.* Highlight the numerical values in the table of values. Click the | **Chart Wizard** | button on the toolbar. Follow the procedure given in Example 3, page 25, with these exceptions: In step 1, select the first chart in the first column under Chart sub-type:; in step 3, in the Chart title: box, type Scatter diagram and least-squares line. The scatter diagram will appear.
3. *Insert the least-squares line.* First, click | **Chart** | on the menu bar and then click | **Add Trendline** |. Next, select the | **Linear** | graph and then click the | **Options** | tab. Finally select | **Display equation on chart** | and then click | **OK** |. The equation

$$y = 0.4606x - 6.3$$

will appear on the chart (Figure T4).

Note: Boldfaced words/characters enclosed in a box (for example, | **Enter** |) indicate that an action (click, select, or press) is required. Words/characters printed blue (for example, Chart sub-type:) indicate words/characters that appear on the screen. Words/characters printed in a typewriter font (for example, =(−2/3)*A2+2) indicate words/characters that need to be typed and entered.

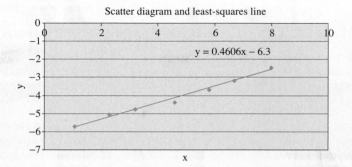

FIGURE T4
Scatter diagram and least-squares line
for the given data

EXAMPLE 4 According to Pacific Gas and Electric, the nation's largest utility company, the demand for electricity from 1990 through 2000 is summarized in the following table:

t	0	2	4	6	8	10
y	333	917	1500	2117	2667	3292

Here $t = 0$ corresponds to 1990, and y gives the amount of electricity in year t, measured in megawatts. Find an equation of the least-squares line for these data.

Solution

1. *Set up a table of values on a spreadsheet* (Figure T5).
2. *Find the equation of the least-squares line for this data.* Click ⃞Tools⃞ on the menu bar and then click ⃞Data Analysis⃞. From the Data Analysis dialog box that appears, select ⃞Regression⃞ and then click ⃞OK⃞. In the Regression dialog box that appears, select the ⃞Input Y Range:⃞ box and then enter the y-values by highlighting cells B2:B7. Next select the ⃞Input X Range:⃞ box and enter the x-values by highlighting cells A2:A7. Click ⃞OK⃞ and a SUMMARY OUTPUT box will appear. In the third table in this box, you will see the entries shown in Figure T6. These entries give the value of the y-intercept and the coefficient of x in the equation $y = mx + b$. In our example, we are using the variable t instead of x, so the required equation is

$$y = 328 + 295t$$

	A	B
	x	y
1		
2	0	333
3	2	917
4	4	1500
5	6	2117
6	8	2667
7	10	3292

FIGURE T5
Table of values for x and y

	Coefficients
Intercept	328.4762
X Variable 1	295.1714

FIGURE T6
Entries in the SUMMARY OUTPUT box

TECHNOLOGY EXERCISES

In Exercises 1–4, find an equation of the least-squares line for the data.

1.

x	2.1	3.4	4.7	5.6	6.8	7.2
y	8.8	12.1	14.8	16.9	19.8	21.1

2.

x	1.1	2.4	3.2	4.7	5.6	7.2
y	−0.5	1.2	2.4	4.4	5.7	8.1

3.

x	−2.1	−1.1	0.1	1.4	2.5	4.2	5.1
y	6.2	4.7	3.5	1.9	0.4	−1.4	−2.5

(continued)

4.

x	−1.12	0.1	1.24	2.76	4.21	6.82
y	7.61	4.9	2.74	−0.47	−3.51	−8.94

5. STARBUCKS' ANNUAL SALES According to company reports, Starbucks' annual sales (in billions of dollars) for 1998 through 2003 are as follows:

Year, x	0	1	2	3	4	5
Sales, y	1.28	1.73	2.18	2.65	3.29	4.08

(Here, $x = 0$ corresponds to 1998.)
a. Find an equation of the least-squares line for these data.
b. Use the result to estimate Starbucks' sales for 2006, assuming that the trend continues.
Source: Company reports

6. SALES OF DRUGS Sales of drugs called analeptics, which are used to treat attention-deficit disorders, were rising even before some companies began advertising them to parents. The following table gives the sales of analeptics (in millions of dollars) from 1995 through 2000:

Year, x	0	1	2	3	4	5
Sales, y	382	455	536	618	664	758

(Here, $x = 0$ represents 1995.)
a. Find an equation of the least-squares line for these data.
b. Use the result of part (a) to estimate the sales for 2002, assuming that the trend continued.
Source: IMS Health

7. WASTE GENERATION According to data from the Council on Environmental Quality, the amount of waste (in millions of tons per year) generated in the United States from 1960 to 1990 was:

Year	1960	1965	1970	1975
Amount, y	81	100	120	124

Year	1980	1985	1990
Amount, y	140	152	164

(Let x be in units of 5 and let $x = 1$ represent 1960.)
a. Find an equation of the least-squares line for these data.
b. Use the result of part (a) to estimate the amount of waste generated in the year 2000, assuming that the trend continued.
Source: Council on Environmental Quality

8. ONLINE TRAVEL More and more travelers are purchasing their tickets online. According to industry projections, the U.S. online travel revenue (in billions of dollars) from 2001 through 2005 is given in the following table:

Year, t	0	1	2	3	4
Revenue	16.3	21.0	25.0	28.8	32.7

(Here, $t = 0$ corresponds to 2001.)
a. Find an equation of the least-squares line for these data.
b. Use the result of part (a) to estimate the U.S. online travel revenue for 2006.
Source: Forrester Research Inc.

9. WORLD ENERGY CONSUMPTION According to the U.S. Department of Energy, the consumption of energy by countries of the industrialized world is projected to rise over the next 25 years. The consumption (in trillion cubic feet) from 2000 through 2025 is summarized in the following table.

Year	2000	2005	2010	2015	2020	2025
Consumption, y	214	225	240	255	270	285

(Let x be measured in 5-year intervals and $x = 0$ represent 2000.)
a. Find an equation of the least-squares line for these data.
b. Use the result of part (a) to estimate the energy consumption in the industrialized world in 2012.
Source: U.S. Department of Energy

10. ANNUAL COLLEGE COSTS The annual U.S. college costs (including tuition, room, and board) from 1991–1992 through 2001–2002 are given in the accompanying table:

Academic Year, x	0	1	2	3	4
Cost ($), y	7077	7452	7931	8306	8800

Academic Year, x	5	6	7	8	9	10
Cost ($), y	9206	9588	10,076	10,444	10,818	11,454

(Here, $x = 0$ corresponds to academic year 1991–1992.)
a. Find an equation of the least-squares line for these data.
b. Use the result of part (a) to plot the least-squares line.
c. Use the result of part (a) to determine the approximate average rate of increase of college costs per year for the period in question.
Source: National Center for Education Statistics

CHAPTER 1 Summary of Principal Formulas and Terms

FORMULAS

1. Distance between two points	$d = \sqrt{(x_2 - x_1)^2 + (y_2 - y_1)^2}$
2. Equation of a circle	$(x - h)^2 + (y - k)^2 = r^2$
3. Slope of a nonvertical line	$m = \dfrac{y_2 - y_1}{x_2 - x_1}$
4. Equation of a vertical line	$x = a$
5. Equation of a horizontal line	$y = b$
6. Point-slope form of the equation of a line	$y - y_1 = m(x - x_1)$
7. Slope-intercept form of the equation of a line	$y = mx + b$
8. General equation of a line	$Ax + By + C = 0$

TERMS

Cartesian coordinate system (2)
ordered pair (2)
coordinates (3)
parallel lines (12)
perpendicular lines (14)
function (29)
independent variable (30)

dependent variable (30)
domain (30)
range (30)
linear function (30)
total cost function (32)
revenue function (32)
profit function (32)

demand function (33)
supply function (34)
break-even point (43)
market equilibrium (46)
equilibrium quantity (46)
equilibrium price (46)

CHAPTER 1 Concept Review Questions

Fill in the blanks.

1. A point in the plane can be represented uniquely by a/an _____ pair of numbers. The first number of the pair is called the _____, and the second number of the pair is called the _____.

2. **a.** The point $P(a, 0)$ lies on the _____ axis, and the point $P(0, b)$ lies on the _____ axis.
   b. If the point $P(a, b)$ lies in the fourth quadrant, then the point $P(-a, b)$ lies in the _____ quadrant.

3. The distance between two points $P(a, b)$ and $P(c, d)$ is _____.

4. An equation of a circle with center $C(a, b)$ and radius r is given by _____.

5. **a.** If $P_1(x_1, y_1)$ and $P_2(x_2, y_2)$ are any two distinct points on a nonvertical line L, then the slope of L is $m =$ _____.

   b. The slope of a vertical line is _____.
   c. The slope of a horizontal line is _____.
   d. The slope of a line that slants upward is _____.

6. If L_1 and L_2 are nonvertical lines with slopes m_1 and m_2, respectively, then L_1 is parallel to L_2 if and only if _____ and L_1 is perpendicular to L_2 if and only if _____.

7. **a.** An equation of the line passing through the point $P(x_1, y_1)$ and having slope m is _____. It is called the _____ form of an equation of a line.
   b. An equation of the line that has slope m and y-intercept b is _____. It is called the _____ form of an equation of a line.

8. **a.** The general form of an equation of a line is _____.
   b. If a line has equation $ax + by + c = 0$ $(b \neq 0)$, then its slope is _____.

9. A linear function is a function of the form $f(x) = $ _____.

10. **a.** A demand function expresses the relationship between the unit _____ and the quantity _____ of a commodity. The graph of the demand function is called the _____ curve.

   b. A supply function expresses the relationship between the unit _____ and the quantity _____ of a commodity. The graph of the supply function is called the _____ curve.

11. If $R(x)$ and $C(x)$ denote the total revenue and the total cost incurred in manufacturing x units of a commodity, then the solution of the simultaneous equations $p = C(x)$ and $p = R(x)$ gives the _____ point.

12. The equilibrium quantity and the equilibrium price are found by solving the system composed of the _____ equation and the _____ equation.

CHAPTER 1   Review Exercises

In Exercises 1–4, find the distance between the two points.

1. (2, 1) and (6, 4)

2. (9, 6) and (6, 2)

3. $(-2, -3)$ and $(1, -7)$

4. $\left(\frac{1}{2}, \sqrt{3}\right)$ and $\left(-\frac{1}{2}, 2\sqrt{3}\right)$

In Exercises 5–10, find an equation of the line L that passes through the point $(-2, 4)$ and satisfies each condition.

5. L is a vertical line.

6. L is a horizontal line.

7. L passes through the point $\left(3, \frac{7}{2}\right)$.

8. The x-intercept of L is 3.

9. L is parallel to the line $5x - 2y = 6$.

10. L is perpendicular to the line $4x + 3y = 6$.

11. Find an equation of the line with slope $-\frac{1}{2}$ and y-intercept -3.

12. Find the slope and y-intercept of the line with equation $3x - 5y = 6$.

13. Find an equation of the line passing through the point (2, 3) and parallel to the line with equation $3x + 4y - 8 = 0$.

14. Find an equation of the line passing through the point $(-1, 3)$ and parallel to the line joining the points $(-3, 4)$ and (2, 1).

15. Find an equation of the line passing through the point $(-2, -4)$ that is perpendicular to the line with equation $2x - 3y - 24 = 0$.

In Exercises 16 and 17, sketch the graph of the equation.

16. $3x - 4y = 24$

17. $-2x + 5y = 15$

18. **SALES OF MP3 PLAYERS**   Sales of a certain brand of MP3 players are approximated by the relationship

$$S(x) = 6000x + 30{,}000 \qquad (0 \le x \le 5)$$

where $S(x)$ denotes the number of MP3 players sold in year x ($x = 0$ corresponds to the year 2000). Find the number of MP3 players expected to be sold in 2005.

19. **COMPANY SALES**   A company's total sales (in millions of dollars) are approximately linear as a function of time (in years). Sales in 1999 were $2.4 million, whereas sales in 2004 amounted to $7.4 million.

   a. Find an equation giving the company's sales as a function of time.

   b. What were the sales in 2002?

20. Show that the triangle with vertices $A(1, 1)$, $B(5, 3)$, and $C(4, 5)$ is a right triangle.

21. **CLARK'S RULE**   Clark's rule is a method for calculating pediatric drug dosages based on a child's weight. If a denotes the adult dosage (in milligrams) and if w is the child's weight (in pounds), then the child's dosage is given by

$$D(w) = \frac{aw}{150}$$

   a. Show that D is a linear function of w.

   b. If the adult dose of a substance is 500 mg, how much should a 35-lb child receive?

22. **LINEAR DEPRECIATION**   An office building worth $6 million when it was completed in 2000 is being depreciated linearly over 30 years.

   a. What is the rate of depreciation?

   b. What will be the book value of the building in 2010?

23. **LINEAR DEPRECIATION**   In 2000 a manufacturer installed a new machine in her factory at a cost of $300,000. The machine is depreciated linearly over 12 yr with a scrap value of $30,000.

   a. What is the rate of depreciation of the machine per year?

   b. Find an expression for the book value of the machine in year t ($0 \le t \le 12$).

24. **PROFIT FUNCTIONS**   A company has a fixed cost of $30,000 and a production cost of $6 for each disposable camera it manufactures. Each camera sells for $10.

a. What is the cost function?
b. What is the revenue function?
c. What is the profit function?
d. Compute the profit (loss) corresponding to production levels of 6000, 8000, and 12,000 units, respectively.

25. DEMAND EQUATIONS There is no demand for a certain commodity when the unit price is $200 or more, but for each $10 decrease in price below $200, the quantity demanded increases by 200 units. Find the demand equation and sketch its graph.

26. SUPPLY EQUATIONS Bicycle suppliers will make 200 bicycles available in the market per month when the unit price is $50 and 2000 bicycles available per month when the unit price is $100. Find the supply equation if it is known to be linear.

In Exercises 27 and 28, find the point of intersection of the lines with the given equations.

27. $3x + 4y = -6$ and $2x + 5y = -11$

28. $y = \frac{3}{4}x + 6$ and $3x - 2y + 3 = 0$

29. The cost function and the revenue function for a certain firm are given by $C(x) = 12x + 20,000$ and $R(x) = 20x$, respectively. Find the break-even point for the company.

30. Given the demand equation $3x + p - 40 = 0$ and the supply equation $2x - p + 10 = 0$, where p is the unit price in dollars and x represents the quantity demanded in units of a thousand, determine the equilibrium quantity and the equilibrium price.

31. COLLEGE ADMISSIONS The accompanying data were compiled by the Admissions Office of Carter College during the past 5 yr. The data relate the number of college brochures and follow-up letters (x) sent to a preselected list of high school juniors who took the PSAT and the number of completed applications (y) received from these students (both measured in thousands).

Brochures Sent, x	1.8	2	3.2
Applications Completed, y	0.4	0.5	0.7

Brochures Sent, x	4	4.8
Applications Completed, y	1	1.3

a. Derive an equation of the straight line L that passes through the points $(2, 0.5)$ and $(4, 1)$.
b. Use this equation to predict the number of completed applications that might be expected if 6400 brochures and follow-up letters are sent out during the next year.

32. EQUILIBRIUM QUANTITY AND PRICE The demand equation for the Edmund compact refrigerator is $2x + 7p - 1760 = 0$, where x is the quantity demanded each week and p is the unit price in dollars. The supply equation for these refrigerators is $3x - 56p + 2680 = 0$, where x is the quantity the supplier will make available in the market when the wholesale price is p dollars each. Find the equilibrium quantity and the equilibrium price for the Edmund compact refrigerators.

The problem-solving skills that you learn in each chapter are building blocks for the rest of the course. Therefore, it is a good idea to make sure that you have mastered these skills before moving on to the next chapter. The Before Moving On exercises that follow are designed for that purpose. After taking this test, you can see where your weaknesses, if any, are. Then you can go to http://series. brookscole.com/tan/ where you will find a link to our Companion Website. Here, you can click on the Before Moving On button, which will lead you to other versions of these tests. There you can re-test yourself on those exercises that you solved incorrectly. (You can also test yourself on these basic skills before taking your course quizzes and exams.)

If you feel that you need additional help with these exercises, you can use the *iLrn Tutorials*, as well as *vMentor*™ for live online help from a tutor.

CHAPTER 1 **Before Moving On . . .**

1. Plot the points $A(-2, 1)$ and $B(3, 4)$ on the same set of axes and find the distance between A and B.

2. Find an equation of the line passing through the point $(3, 1)$ and parallel to the line $3x - y - 4 = 0$.

3. Let L be the line passing through the points $(1, 2)$ and $(3, 5)$. Is L perpendicular to the line $2x + 3y = 10$?

4. The monthly total revenue function and total cost function for a company are $R(x) = 15x$ and $C(x) = 18x + 22,000$, respectively, where x is the number of units produced and both $R(x)$ and $C(x)$ are measured in dollars.

a. What is the unit cost for producing the product?
b. What is the monthly fixed cost for the company?
c. What is the selling price for each unit of the product?

5. Find the point of intersection of the lines $2x - 3y = -2$ and $9x + 12y = 25$.

6. The annual sales of Best Furniture Store are expected to be given by $S_1 = 4.2 + 0.4t$ million dollars t yr from now, whereas the annual sales of Lowe's Furniture Store are expected to be given by $S_2 = 2.2 + 0.8t$ million dollars t yr from now. When will Lowe's annual sales first surpass Best's annual sales?

Systems of Linear Equations and Matrices

How fast is the traffic moving? The flow of downtown traffic is controlled by traffic lights installed at the intersections. One of the roads is to be resurfaced. In Example 5, page 101, you will see how the flow patterns must be altered in order to ensure a smooth flow of traffic even during rush hour.

© Spike Mafford/PhotoDisc

THE LINEAR EQUATIONS in two variables studied in Chapter 1 are readily extended to the case involving more than two variables. For example, a linear equation in three variables represents a plane in three-dimensional space. In this chapter, we see how some real-world problems can be formulated in terms of systems of linear equations, and we also develop two methods for solving these equations.

In addition, we see how *matrices* (ordered rectangular arrays of numbers) can be used to write systems of linear equations in compact form. We then go on to consider some real-life applications of matrices. Finally, we show how matrices can be used to describe the Leontief input–output model, an important tool used by economists. For his work in formulating this model, Wassily Leontief was awarded the Nobel Prize in 1973.

2.1   Systems of Linear Equations: An Introduction

■ Systems of Equations

Recall that in Section 1.4 we had to solve two simultaneous linear equations in order to find the *break-even point* and the *equilibrium point*. These are two examples of real-world problems that call for the solution of a **system of linear equations** in two or more variables. In this chapter we take up a more systematic study of such systems.

We begin by considering a system of two linear equations in two variables. Recall that such a system may be written in the general form

$$ax + by = h$$
$$cx + dy = k \tag{1}$$

where a, b, c, d, h, and k are real constants and neither a and b nor c and d are both zero.

Now let's study the nature of the **solution of a system of linear equations** in more detail. Recall that the graph of each equation in System (1) is a straight line in the plane, so that geometrically the solution to the system is the point(s) of intersection of the two straight lines L_1 and L_2, represented by the first and second equations of the system.

Given two lines L_1 and L_2, then *one and only one* of the following may occur:

a. L_1 and L_2 intersect at exactly one point.
b. L_1 and L_2 are parallel and coincident.
c. L_1 and L_2 are parallel and distinct.

(See Figure 1.) In the first case, the system has a unique solution corresponding to the single point of intersection of the two lines. In the second case, the system has infinitely many solutions corresponding to the points lying on the same line. Finally, in the third case, the system has no solution, since the two lines do not intersect.

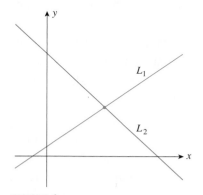

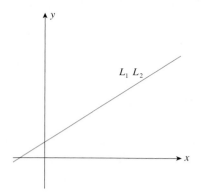

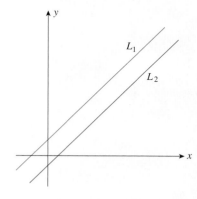

FIGURE 1
(a) Unique solution   **(b)** Infinitely many solutions   **(c)** No solution

---

EXPLORE & DISCUSS

Generalize the discussion on this page to the case where there are three straight lines in the plane defined by three linear equations. What if there are n lines defined by n equations?

Let's illustrate each of these possibilities by considering some specific examples.

1. A system of equations with exactly one solution Consider the system

$$2x - y = 1$$
$$3x + 2y = 12$$

Solving the first equation for y in terms of x, we obtain the equation

$$y = 2x - 1$$

Substituting this expression for y into the second equation yields

$$3x + 2(2x - 1) = 12$$
$$3x + 4x - 2 = 12$$
$$7x = 14$$
$$x = 2$$

Finally, substituting this value of x into the expression for y obtained earlier gives

$$y = 2(2) - 1 = 3$$

Therefore, the unique solution of the system is given by $x = 2$ and $y = 3$. Geometrically, the two lines represented by the two linear equations that make up the system intersect at the point $(2, 3)$ (Figure 2).

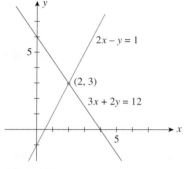

FIGURE 2
A system of equations with one solution

Note   We can check our result by substituting the values $x = 2$ and $y = 3$ into the equations. Thus,

$$2(2) - (3) = 1 \quad \checkmark$$
$$3(2) + 2(3) = 12 \quad \checkmark$$

From the geometric point of view, we have just verified that the point $(2, 3)$ lies on both lines.   ∎

2. A system of equations with infinitely many solutions Consider the system

$$2x - y = 1$$
$$6x - 3y = 3$$

Solving the first equation for y in terms of x, we obtain the equation

$$y = 2x - 1$$

Substituting this expression for y into the second equation gives

$$6x - 3(2x - 1) = 3$$
$$6x - 6x + 3 = 3$$
$$0 = 0$$

which is a true statement. This result follows from the fact that the second equation is equivalent to the first. (To see this, just multiply both sides of the first equation by 3.) Our computations have revealed that the system of two equations is equivalent to the single equation $2x - y = 1$. Thus, any ordered pair of numbers (x, y) satisfying the equation $2x - y = 1$ (or $y = 2x - 1$) constitutes a solution to the system.

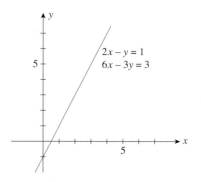

FIGURE 3
A system of equations with infinitely many solutions; each point on the line is a solution.

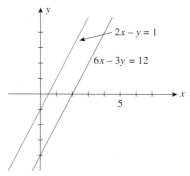

FIGURE 4
A system of equations with no solution

In particular, by assigning the value t to x, where t is any real number, we find that $y = 2t - 1$, so the ordered pair $(t, 2t - 1)$ is a solution of the system. The variable t is called a **parameter**. For example, setting $t = 0$ gives the point $(0, -1)$ as a solution, and setting $t = 1$ gives the point $(1, 1)$ as another solution of the system. Since t represents any real number, there are infinitely many solutions to the system. Geometrically, the two equations in the system represent the same line, and all solutions of the system are points lying on the line (Figure 3). Such a system is said to be **dependent**.

3. A system of equations that has no solution Consider the system

$$2x - \; y = 1$$
$$6x - 3y = 12$$

The first equation is equivalent to $y = 2x - 1$. Substituting this expression for y into the second equation gives

$$6x - 3(2x - 1) = 12$$
$$6x - 6x + 3 = 12$$
$$0 = 9$$

which is clearly impossible. Thus, there is no solution to the system of equations. To interpret this situation geometrically, cast both equations in the slope-intercept form, obtaining

$$y = 2x - 1$$
$$y = 2x - 4$$

We see at once that the lines represented by these equations are parallel (each has slope 2) and distinct since the first has y-intercept -1 and the second has y-intercept -4 (Figure 4). Systems with no solutions, such as this one, are said to be **inconsistent**.

EXPLORE & DISCUSS

1. Consider a system composed of two linear equations in two variables. Can the system have exactly two solutions? Exactly three solutions? Exactly a finite number of solutions?

2. Suppose at least one of the equations in a system composed of two equations in two variables is nonlinear. Can the system have no solution? Exactly one solution? Exactly two solutions? Exactly a finite number of solutions? Infinitely many solutions? Illustrate each answer with a sketch.

Note We have used the method of substitution in solving each of these systems. If you are familiar with the method of elimination, you might want to re-solve each of these systems using this method. We will study the method of elimination in detail in Section 2.2. ▪

In Section 1.4, we presented some real-world applications of systems involving two linear equations in two variables. Here is an example involving a system of three linear equations in three variables.

APPLIED EXAMPLE 1   Manufacturing: Production Scheduling   Ace Novelty wishes to produce three types of souvenirs: types A, B, and C. To manufacture a type-A souvenir requires 2 minutes on machine I, 1 minute on machine II, and 2 minutes on machine III. A type-B souvenir requires 1 minute on machine I, 3 minutes on machine II, and 1 minute on machine III. A type-C souvenir requires 1 minute on machine I and 2 minutes each on machines II and III. There are 3 hours available on machine I, 5 hours available on machine II, and 4 hours available on machine III for processing the order. How many souvenirs of each type should Ace Novelty make in order to use all of the available time? Formulate but do not solve the problem. (We will solve this problem in Example 7, Section 2.2.)

Solution   The given information may be tabulated as follows:

	Type A	Type B	Type C	Time Available (min)
Machine I	2	1	1	180
Machine II	1	3	2	300
Machine III	2	1	2	240

We have to determine the number of each of *three* types of souvenirs to be made. So, let x, y, and z denote the respective numbers of type-A, type-B, and type-C souvenirs to be made. The total amount of time that machine I is used is given by $2x + y + z$ minutes and must equal 180 minutes. This leads to the equation

$$2x + y + z = 180 \qquad \text{Time spent on machine I}$$

Similar considerations on the use of machines II and III lead to the following equations:

$$x + 3y + 2z = 300 \qquad \text{Time spent on machine II}$$
$$2x + y + 2z = 240 \qquad \text{Time spent on machine III}$$

Since the variables x, y, and z must satisfy simultaneously the three conditions represented by the three equations, the solution to the problem is found by solving the following system of linear equations:

$$
\begin{aligned}
2x + y + z &= 180 \\
x + 3y + 2z &= 300 \\
2x + y + 2z &= 240
\end{aligned}
$$

Solutions of Systems of Equations

We will complete the solution of the problem posed in Example 1 later on (page 89). For the moment, let's look at the geometric interpretation of a system of linear equations, such as the system in Example 1, in order to gain some insight into the nature of the solution.

A linear system composed of three linear equations in three variables x, y, and z has the general form

$$
\begin{aligned}
a_1 x + b_1 y + c_1 z &= d_1 \\
a_2 x + b_2 y + c_2 z &= d_2 \\
a_3 x + b_3 y + c_3 z &= d_3
\end{aligned} \tag{2}
$$

Just as a linear equation in two variables represents a straight line in the plane, it can be shown that a linear equation $ax + by + cz = d$ (a, b, and c not simultaneously equal to zero) in three variables represents a plane in three-dimensional space. Thus, each equation in System (2) represents a *plane* in three-dimensional space, and the *solution(s) of the system* is precisely the point(s) of intersection of the three planes defined by the three linear equations that make up the system. As before, the system has one and only one solution, infinitely many solutions, or no solution, depending on whether and how the planes intersect one another. Figure 5 illustrates each of these possibilities.

In Figure 5a, the three planes intersect at a point corresponding to the situation in which System (2) has a unique solution. Figure 5b depicts the situation in which there are infinitely many solutions to the system. Here, the three planes intersect along a line, and the solutions are represented by the infinitely many points lying on this line. In Figure 5c, the three planes are parallel and distinct, so there is no point in common to all three planes, and System (2) has no solution in this case.

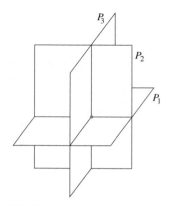

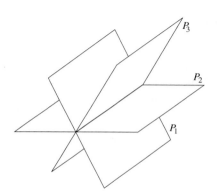

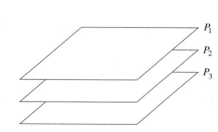

FIGURE 5
(a) A unique solution

(b) Infinitely many solutions

(c) No solution

Note The situations depicted in Figure 5 are by no means exhaustive. You may consider various other orientations of the three planes that would illustrate the three possible outcomes in solving a system of linear equations involving three variables.

■

Linear Equations in *n* Variables
A linear equation in n variables, $x_1, x_2, \ldots, x_n$ is one of the form

$$a_1x_1 + a_2x_2 + \cdots + a_nx_n = c$$

where $a_1, a_2, \ldots, a_n$ (not all zero) and c are constants.

For example, the equation

$$3x_1 + 2x_2 - 4x_3 + 6x_4 = 8$$

is a linear equation in the four variables, x_1, x_2, x_3, and x_4.

EXPLORE & DISCUSS

Refer to the Note on page 76.

Using the orientation of three planes, illustrate the outcomes in solving a system of three linear equations in three variables that result in no solution or infinitely many solutions.

When the number of variables involved in a linear equation exceeds three, we no longer have the geometric interpretation we had for the lower-dimensional spaces. Nevertheless, the algebraic concepts of the lower-dimensional spaces generalize to higher dimensions. For this reason, a linear equation in n variables, $a_1x_1 + a_2x_2 + \cdots + a_nx_n = c$, where $a_1, a_2, \ldots, a_n$ are not all zero, is referred to as an *n-dimensional hyperplane*. We may interpret the solution(s) to a system comprising a finite number of such linear equations to be the *point(s) of intersection* of the hyperplanes defined by the equations that make up the system. As in the case of systems involving two or three variables, it can be shown that only three possibilities exist regarding the nature of the solution of such a system: (1) a unique solution, (2) infinitely many solutions, and (3) no solution.

2.1 Self-Check Exercises

1. Determine whether the system of linear equations

$$2x - 3y = 12$$
$$x + 2y = 6$$

has (a) a unique solution, (b) infinitely many solutions, or (c) no solution. Find all solutions whenever they exist. Make a sketch of the set of lines described by the system.

2. A farmer has 200 acres of land suitable for cultivating crops A, B, and C. The cost per acre of cultivating crops A, B, and

C is $40, $60, and $80, respectively. The farmer has $12,600 available for cultivation. Each acre of crop A requires 20 labor-hours, each acre of crop B requires 25 labor-hours, and each acre of crop C requires 40 labor-hours. The farmer has a maximum of 5950 labor-hours available. If she wishes to use all of her cultivatable land, the entire budget, and all the labor available, how many acres of each crop should she plant? Formulate but do not solve the problem.

Solutions to Self-Check Exercises 2.1 can be found on page 79.

2.1 Concept Questions

1. Suppose you are given a system of two linear equations in two variables.
   a. What can you say about the solution(s) of the system of equations?
   b. Give a geometric interpretation of your answers to the question in part (a).

2. Suppose you are given a system of two linear equations in two variables.
   a. Explain what it means for the system to be dependent.
   b. Explain what it means for the system to be inconsistent.

2.1 Exercises

In Exercises 1–12, determine whether each system of linear equations has (a) one and only one solution, (b) infinitely many solutions, or (c) no solution. Find all solutions whenever they exist.

1. $x - 3y = -1$
   $4x + 3y = 11$

2. $2x - 4y = 5$
   $3x + 2y = 6$

3. $x + 4y = 7$
   $\frac{1}{2}x + 2y = 5$

4. $3x - 4y = 7$
   $9x - 12y = 14$

5. $x + 2y = 7$
   $2x - y = 4$

6. $\frac{3}{2}x - 2y = 4$
   $x + \frac{1}{3}y = 2$

7. $2x - 5y = 10$
$6x - 15y = 30$

8. $5x - 6y = 8$
$10x - 12y = 16$

9. $4x - 5y = 14$
$2x + 3y = -4$

10. $\dfrac{5}{4}x - \dfrac{2}{3}y = 3$
$\dfrac{1}{4}x + \dfrac{5}{3}y = 6$

11. $2x - 3y = 6$
$6x - 9y = 12$

12. $\dfrac{2}{3}x + y = 5$
$\dfrac{1}{2}x + \dfrac{3}{4}y = \dfrac{15}{4}$

13. Determine the value of k for which the system of linear equations

$$2x - y = 3$$
$$4x + ky = 4$$

has no solution.

14. Determine the value of k for which the system of linear equations

$$3x + 4y = 12$$
$$x + ky = 4$$

has infinitely many solutions. Then find all solutions corresponding to this value of k.

In Exercises 15–27, formulate but do not solve the problem. You will be asked to solve these problems in the next section.

15. AGRICULTURE   The Johnson Farm has 500 acres of land allotted for cultivating corn and wheat. The cost of cultivating corn and wheat (including seeds and labor) is $42 and $30/acre, respectively. Jacob Johnson has $18,600 available for cultivating these crops. If he wishes to use all the allotted land and his entire budget for cultivating these two crops, how many acres of each crop should he plant?

16. INVESTMENTS   Michael Perez has a total of $2000 on deposit with two savings institutions. One pays interest at the rate of 6%/year, whereas the other pays interest at the rate of 8%/year. If Michael earned a total of $144 in interest during a single year, how much does he have on deposit in each institution?

17. MIXTURES   The Coffee Shoppe sells a coffee blend made from two coffees, one costing $2.50/lb and the other costing $3.00/lb. If the blended coffee sells for $2.80/lb, find how much of each coffee is used to obtain the desired blend. (Assume the weight of the blended coffee is 100 lb.)

18. INVESTMENTS   Kelly Fisher has a total of $30,000 invested in two municipal bonds that have yields of 8% and 10% interest per year, respectively. If the interest Kelly receives from the bonds in a year is $2640, how much does she have invested in each bond?

19. RIDERSHIP   The total number of passengers riding a certain city bus during the morning shift is 1000. If the child's fare is $.25, the adult fare is $.75, and the total revenue from the fares in the morning shift is $650, how many children and how many adults rode the bus during the morning shift?

20. REAL ESTATE   Cantwell Associates, a real estate developer, is planning to build a new apartment complex consisting of one-bedroom units and two- and three-bedroom townhouses. A total of 192 units is planned, and the number of family units (two- and three-bedroom townhouses) will equal the number of one-bedroom units. If the number of one-bedroom units will be 3 times the number of three-bedroom units, find how many units of each type will be in the complex.

21. INVESTMENT   PLANNING   The   annual   interest   on   Sid Carrington's three investments amounted to $21,600: 6% on a savings account, 8% on mutual funds, and 12% on bonds. If the amount of Sid's investment in bonds was twice the amount of his investment in the savings account, and the interest earned from his investment in bonds was equal to the dividends he received from his investment in mutual funds, find how much money he placed in each type of investment.

22. INVESTMENT CLUB   A private investment club has $200,000 earmarked for investment in stocks. To arrive at an acceptable overall level of risk, the stocks that management is considering have been classified into three categories: high-risk, medium-risk, and low-risk. Management estimates that high-risk stocks will have a rate of return of 15%/year; medium-risk stocks, 10%/year; and low-risk stocks, 6%/year. The members have decided that the investment in low-risk stocks should be equal to the sum of the investments in the stocks of the other two categories. Determine how much the club should invest in each type of stock if the investment goal is to have a return of $20,000/year on the total investment. (Assume that all the money available for investment is invested.)

23. MIXTURE PROBLEM—FERTILIZER   Lawnco produces three grades of commercial fertilizers. A 100-lb bag of grade-A fertilizer contains 18 lb of nitrogen, 4 lb of phosphate, and 5 lb of potassium. A 100-lb bag of grade-B fertilizer contains 20 lb of nitrogen and 4 lb each of phosphate and potassium. A 100-lb bag of grade-C fertilizer contains 24 lb of nitrogen, 3 lb of phosphate, and 6 lb of potassium. How many 100-lb bags of each of the three grades of fertilizers should Lawnco produce if 26,400 lb of nitrogen, 4900 lb of phosphate, and 6200 lb of potassium are available and all the nutrients are used?

24. BOX-OFFICE RECEIPTS   A theater has a seating capacity of 900 and charges $2 for children, $3 for students, and $4 for adults. At a certain screening with full attendance, there were half as many adults as children and students combined. The receipts totaled $2800. How many children attended the show?

25. **MANAGEMENT DECISIONS** The management of Hartman Rent-A-Car has allocated $1.5 million to buy a fleet of new automobiles consisting of compact, intermediate, and full-size cars. Compacts cost $12,000 each, intermediate-size cars cost $18,000 each, and full-size cars cost $24,000 each. If Hartman purchases twice as many compacts as intermediate-size cars and the total number of cars to be purchased is 100, determine how many cars of each type will be purchased. (Assume that the entire budget will be used.)

26. **INVESTMENT CLUBS** The management of a private investment club has a fund of $200,000 earmarked for investment in stocks. To arrive at an acceptable overall level of risk, the stocks that management is considering have been classified into three categories: high-risk, medium-risk, and low-risk. Management estimates that high-risk stocks will have a rate of return of 15%/year; medium-risk stocks, 10%/year; and low-risk stocks, 6%/year. The investment in low-risk stocks is to be twice the sum of the investments in stocks of the other two categories. If the investment goal is to have an average rate of return of 9%/year on the total investment, determine how much the club should invest in each type of stock. (Assume that all the money available for investment is invested.)

27. **DIET PLANNING** A dietitian wishes to plan a meal around three foods. The percentage of the daily requirements of proteins, carbohydrates, and iron contained in each ounce of the three foods is summarized in the accompanying table:

	Food I	Food II	Food III
Proteins (%)	10	6	8
Carbohydrates (%)	10	12	6
Iron (%)	5	4	12

Determine how many ounces of each food the dietitian should include in the meal to meet exactly the daily requirement of proteins, carbohydrates, and iron (100% of each).

In Exercises 28–30, determine whether the statement is true or false. If it is true, explain why it is true. If it is false, give an example to show why it is false.

28. A system composed of two linear equations must have at least one solution if the straight lines represented by these equations are nonparallel.

29. Suppose the straight lines represented by a system of three linear equations in two variables are parallel to each other. Then, the system has no solution, or it has infinitely many solutions.

30. If at least two of the three lines represented by a system composed of three linear equations in two variables are parallel, then the system has no solution.

2.1 Solutions to Self-Check Exercises

1. Solving the first equation for y in terms of x, we obtain

$$y = \frac{2}{3}x - 4$$

Next, substituting this result into the second equation of the system, we find

$$x + 2\left(\frac{2}{3}x - 4\right) = 6$$

$$x + \frac{4}{3}x - 8 = 6$$

$$\frac{7}{3}x = 14$$

$$x = 6$$

Substituting this value of x into the expression for y obtained earlier, we have

$$y = \frac{2}{3}(6) - 4 = 0$$

Therefore, the system has the unique solution $x = 6$ and $y = 0$. Both lines are shown in the accompanying figure.

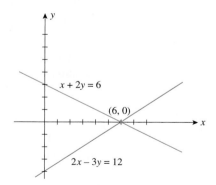

2. Let x, y, and z denote the number of acres of crop A, crop B, and crop C, respectively, to be cultivated. Then, the condition that all the cultivatable land be used translates into the equation

$$x + y + z = 200$$

Next, the total cost incurred in cultivating all three crops is $40x + 60y + 80z$ dollars, and since the entire budget is to be expended, we have

$$40x + 60y + 80z = 12,600$$

Finally, the amount of labor required to cultivate all three crops is $20x + 25y + 40z$ hr, and since all the available labor is to be used, we have

$$20x + 25y + 40z = 5950$$

Thus, the solution is found by solving the following system of linear equations:

$$
\begin{aligned}
x + \quad y + \quad z &= \quad 200 \\
40x + 60y + 80z &= 12{,}600 \\
20x + 25y + 40z &= \quad 5{,}950
\end{aligned}
$$

2.2 Systems of Linear Equations: Unique Solutions

▬ The Gauss–Jordan Method

The method of substitution used in Section 2.1 is well suited to solving a system of linear equations when the number of linear equations and variables is small. But for large systems, the steps involved in the procedure become difficult to manage.

The **Gauss–Jordan elimination method** is a suitable technique for solving systems of linear equations of any size. One advantage of this technique is its adaptability to the computer. This method involves a sequence of operations on a system of linear equations to obtain at each stage an **equivalent system**—that is, a system having the same solution as the original system. The reduction is complete when the original system has been transformed so that it is in a certain standard form from which the solution can be easily read.

The operations of the Gauss–Jordan elimination method are:

1. Interchange any two equations.
2. Replace an equation by a nonzero constant multiple of itself.
3. Replace an equation by the sum of that equation and a constant multiple of any other equation.

To illustrate the Gauss–Jordan elimination method for solving systems of linear equations, let's apply it to the solution of the following system:

$$
\begin{aligned}
2x + 4y &= 8 \\
3x - 2y &= 4
\end{aligned}
$$

We begin by working with the first, or x, column. First, we transform the system into an equivalent system in which the coefficient of x in the first equation is 1:

$$
\begin{aligned}
2x + 4y &= 8 \\
3x - 2y &= 4
\end{aligned}
\tag{3a}
$$

$$
\begin{aligned}
x + 2y &= 4 \\
3x - 2y &= 4
\end{aligned}
\qquad
\begin{array}{l}
\text{Multiply the first equation} \\
\text{in (3a) by } \tfrac{1}{2} \text{ (Operation 2).}
\end{array}
\tag{3b}
$$

Next, we eliminate x from the second equation:

$$x + 2y = 4$$
$$-8y = -8$$

Replace the second equation in (3b) by the sum of $-3 \times$ the first equation + the second equation (Operation 3): **(3c)**

$$\begin{array}{r} -3x - 6y = -12 \\ \underline{3x - 2y = 4} \\ -8y = -8 \end{array}$$

Then, we obtain the following equivalent system in which the coefficient of y in the second equation is 1:

$$x + 2y = 4$$
$$y = 1$$

Multiply the second equation in (3c) by $-\frac{1}{8}$ (Operation 2). **(3d)**

Next, we eliminate y in the first equation:

$$x = 2$$
$$y = 1$$

Replace the first equation in (3d) by the sum of $-2 \times$ the second equation + the first equation (Operation 3):

$$\begin{array}{r} x + 2y = 4 \\ \underline{- 2y = -2} \\ x = 2 \end{array}$$

This system is now in standard form, and we can read off the solution to (3a) as $x = 2$ and $y = 1$. We can also express this solution as (2, 1) and interpret it geometrically as the point of intersection of the two lines represented by the two linear equations that make up the given system of equations.

Let's consider another example involving a system of three linear equations and three variables.

EXAMPLE 1   Solve the following system of equations:

$$2x + 4y + 6z = 22$$
$$3x + 8y + 5z = 27$$
$$-x + y + 2z = 2$$

Solution   First, we transform this system into an equivalent system in which the coefficient of x in the first equation is 1:

$$2x + 4y + 6z = 22$$
$$3x + 8y + 5z = 27$$
$$-x + y + 2z = 2$$
(4a)

$$x + 2y + 3z = 11$$
$$3x + 8y + 5z = 27$$
$$-x + y + 2z = 2$$

Multiply the first equation in (4a) by $\frac{1}{2}$. **(4b)**

Next, we eliminate the variable x from all equations except the first:

$$x + 2y + 3z = 11$$
$$2y - 4z = -6$$
$$-x + y + 2z = 2$$

Replace the second equation in (4b) by the sum of $-3 \times$ the first equation + the second equation: **(4c)**

$$\begin{array}{r} -3x - 6y - 9z = -33 \\ \underline{3x + 8y + 5z = 27} \\ 2y - 4z = -6 \end{array}$$

$$x + 2y + 3z = 11$$
$$2y - 4z = -6$$
$$3y + 5z = 13$$

Replace the third equation in (4c)
by the sum of the first equation +
the third equation: **(4d)**

$$x + 2y + 3z = 11$$
$$\underline{-x + y + 2z = 2}$$
$$3y + 5z = 13$$

Then we transform System (4d) into yet another equivalent system, in which the coefficient of y in the second equation is 1:

$$x + 2y + 3z = 11$$
$$y - 2z = -3$$
$$3y + 5z = 13$$

Multiply the second equation
in (4d) by $\frac{1}{2}$.   **(4e)**

We now eliminate y from all equations except the second, using Operation 3 of the elimination method:

$$x + 7z = 17$$
$$y - 2z = -3$$
$$3y + 5z = 13$$

Replace the first equation in (4e)
by the sum of the first equation +
$(-2) \times$ the second equation: **(4f)**

$$x + 2y + 3z = 11$$
$$\underline{-2y + 4z = 6}$$
$$x + 7z = 17$$

$$x + 7z = 17$$
$$y - 2z = -3$$
$$11z = 22$$

Replace the third equation in (4f)
by the sum of $(-3) \times$ the second
equation + the third equation: **(4g)**

$$-3y + 6z = 9$$
$$\underline{3y + 5z = 13}$$
$$11z = 22$$

Finally, multiplying the third equation by $\frac{1}{11}$ in (4g) leads to the system

$$x + 7z = 17$$
$$y - 2z = -3$$
$$z = 2$$

Eliminating z from all equations except the third (try it!) then leads to the system

$$x = 3$$
$$y = 1$$
$$z = 2$$

(4h)

In its final form, the solution to the given system of equations can be easily read off! We have $x = 3$, $y = 1$, and $z = 2$. Geometrically, the point $(3, 1, 2)$ lies in the intersection of the three planes described by the three equations comprising the given system.   ∎

▬ Augmented Matrices

Observe from the preceding example that in each step of the reduction process the variables x, y, and z play no significant role except as a reminder of the position of each coefficient in the system. With the aid of **matrices,** which are rectangular arrays of numbers, we can eliminate writing the variables at each step of the reduction and thus save ourselves a great deal of work. For example, the system

$$2x + 4y + 6z = 22$$
$$3x + 8y + 5z = 27 \tag{5}$$
$$-x + y + 2z = 2$$

may be represented by the matrix

$$\begin{bmatrix} 2 & 4 & 6 & | & 22 \\ 3 & 8 & 5 & | & 27 \\ -1 & 1 & 2 & | & 2 \end{bmatrix} \tag{6}$$

The augmented matrix representing System (5)

The submatrix, consisting of the first three columns of Matrix (6), is called the **coefficient matrix** of System (5). The matrix itself, (6), is referred to as the **augmented matrix** of System (5) since it is obtained by joining the matrix of coefficients to the column (matrix) of constants. The vertical line separates the column of constants from the matrix of coefficients.

The next example shows how much work you can save by using matrices instead of the standard representation of the systems of linear equations.

EXAMPLE 2 Write the augmented matrix corresponding to each equivalent system given in (4a) through (4h).

Solution The required sequence of augmented matrices follows.

	Equivalent System	**Augmented Matrix**	
a.	$2x + 4y + 6z = 22$ $3x + 8y + 5z = 27$ $-x + y + 2z = 2$	$\begin{bmatrix} 2 & 4 & 6 & \| & 22 \\ 3 & 8 & 5 & \| & 27 \\ -1 & 1 & 2 & \| & 2 \end{bmatrix}$	**(7a)**
b.	$x + 2y + 3z = 11$ $3x + 8y + 5z = 27$ $-x + y + 2z = 2$	$\begin{bmatrix} 1 & 2 & 3 & \| & 11 \\ 3 & 8 & 5 & \| & 27 \\ -1 & 1 & 2 & \| & 2 \end{bmatrix}$	**(7b)**
c.	$x + 2y + 3z = 11$ $2y - 4z = -6$ $-x + y + 2z = 2$	$\begin{bmatrix} 1 & 2 & 3 & \| & 11 \\ 0 & 2 & -4 & \| & -6 \\ -1 & 1 & 2 & \| & 2 \end{bmatrix}$	**(7c)**
d.	$x + 2y + 3z = 11$ $2y - 4z = -6$ $3y + 5z = 13$	$\begin{bmatrix} 1 & 2 & 3 & \| & 11 \\ 0 & 2 & -4 & \| & -6 \\ 0 & 3 & 5 & \| & 13 \end{bmatrix}$	**(7d)**
e.	$x + 2y + 3z = 11$ $y - 2z = -3$ $3y + 5z = 13$	$\begin{bmatrix} 1 & 2 & 3 & \| & 11 \\ 0 & 1 & -2 & \| & -3 \\ 0 & 3 & 5 & \| & 13 \end{bmatrix}$	**(7e)**
f.	$x + 7z = 17$ $y - 2z = -3$ $3y + 5z = 13$	$\begin{bmatrix} 1 & 0 & 7 & \| & 17 \\ 0 & 1 & -2 & \| & -3 \\ 0 & 3 & 5 & \| & 13 \end{bmatrix}$	**(7f)**

g. $\begin{aligned} x \quad + 7z &= 17 \\ y - 2z &= -3 \\ 11z &= 22 \end{aligned}$ $\left[\begin{array}{ccc|c} 1 & 0 & 7 & 17 \\ 0 & 1 & -2 & -3 \\ 0 & 0 & 11 & 22 \end{array}\right]$ **(7g)**

h. $\begin{aligned} x \qquad &= 3 \\ y \quad &= 1 \\ z &= 2 \end{aligned}$ $\left[\begin{array}{ccc|c} 1 & 0 & 0 & 3 \\ 0 & 1 & 0 & 1 \\ 0 & 0 & 1 & 2 \end{array}\right]$ **(7h)** ■

The augmented matrix in (7h) is an example of a matrix in row-reduced form. In general, an augmented matrix with m rows and n columns (called an $m \times n$ matrix) is in **row-reduced form** if it satisfies the following conditions.

Row-Reduced Form of a Matrix

1. Each row consisting entirely of zeros lies below any other row having nonzero entries.
2. The first nonzero entry in each row is 1 (called a **leading 1**).
3. In any two successive (nonzero) rows, the leading 1 in the lower row lies to the right of the leading 1 in the upper row.
4. If a column contains a leading 1, then the other entries in that column are zeros.

EXAMPLE 3   Determine which of the following matrices are in row-reduced form. If a matrix is not in row-reduced form, state which condition is violated.

a. $\left[\begin{array}{ccc|c} 1 & 0 & 0 & 0 \\ 0 & 1 & 0 & 0 \\ 0 & 0 & 1 & 3 \end{array}\right]$ **b.** $\left[\begin{array}{ccc|c} 1 & 0 & 0 & 4 \\ 0 & 1 & 0 & 3 \\ 0 & 0 & 0 & 0 \end{array}\right]$ **c.** $\left[\begin{array}{ccc|c} 1 & 2 & 0 & 0 \\ 0 & 0 & 1 & 0 \\ 0 & 0 & 0 & 1 \end{array}\right]$

d. $\left[\begin{array}{ccc|c} 0 & 1 & 2 & -2 \\ 1 & 0 & 0 & 3 \\ 0 & 0 & 1 & 2 \end{array}\right]$ **e.** $\left[\begin{array}{ccc|c} 1 & 2 & 0 & 0 \\ 0 & 0 & 1 & 3 \\ 0 & 0 & 2 & 1 \end{array}\right]$ **f.** $\left[\begin{array}{cc|c} 1 & 0 & 4 \\ 0 & 3 & 0 \\ 0 & 0 & 0 \end{array}\right]$

g. $\left[\begin{array}{ccc|c} 0 & 0 & 0 & 0 \\ 1 & 0 & 0 & 3 \\ 0 & 1 & 0 & 2 \end{array}\right]$

Solution   The matrices in parts (a)–(c) are in row-reduced form.

d. This matrix is not in row-reduced form. Conditions 3 and 4 are violated: The leading 1 in row 2 lies to the left of the leading 1 in row 1. Also, column 3 contains a leading 1 in row 3 and a nonzero element above it.
e. This matrix is not in row-reduced form. Conditions 2 and 4 are violated: The first nonzero entry in row 3 is a 2, not a 1. Also, column 3 contains a leading 1 and has a nonzero entry below it.
f. This matrix is not in row-reduced form. Condition 2 is violated: The first nonzero entry in row 2 is not a leading 1.
g. This matrix is not in row-reduced form. Condition 1 is violated: Row 1 consists of all zeros and does not lie below the other nonzero rows.   ■

The foregoing discussion suggests the following adaptation of the Gauss–Jordan elimination method in solving systems of linear equations using matrices. First, the three operations on the equations of a system (see page 80) translate into the following row operations on the corresponding augmented matrices.

Row Operations

1. Interchange any two rows.
2. Replace any row by a nonzero constant multiple of itself.
3. Replace any row by the sum of that row and a constant multiple of any other row.

We obtained the augmented matrices in Example 2 by using the same operations that we used on the equivalent system of equations in Example 1.

To help us describe the Gauss–Jordan elimination method using matrices, let's introduce some terminology. We begin by defining what is meant by a unit column.

Unit Column

A column in a coefficient matrix is in unit form if one of the entries in the column is a 1 and the other entries are zeros.

For example, in the coefficient matrix of (7d), only the first column is in unit form; in the coefficient matrix of (7h), all three columns are in unit form. Now, the sequence of row operations that transforms the augmented matrix (7a) into the equivalent matrix (7d) in which the first column

$$2$$
$$3$$
$$-1$$

of (7a) is transformed into the unit column

$$1$$
$$0$$
$$0$$

is called pivoting the matrix about the element (number) 2. Similarly, we have pivoted about the element 2 in the second column of (7d), shown circled,

$$2$$
$$\text{②}$$
$$3$$

in order to obtain the augmented matrix (7g). Finally, pivoting about the element 11 in column 3 of (7g)

$$7$$
$$-2$$
$$\text{⑪}$$

leads to the augmented matrix (7h), in which all columns to the left of the vertical line are in unit form. The element about which a matrix is pivoted is called the *pivot element*.

Before looking at the next example, let's introduce the following notation for the three types of row operations.

Notation for Row Operations

Letting R_i denote the ith row of a matrix, we write:

Operation 1 $R_i \leftrightarrow R_j$ to mean: Interchange row i with row j.

Operation 2 cR_i to mean: Replace row i with c times row i.

Operation 3 $R_i + aR_j$ to mean: Replace row i with the sum of row i and a times row j.

EXAMPLE 4   Pivot the matrix about the circled element.

$$\begin{bmatrix} ③ & 5 & | & 9 \\ 2 & 3 & | & 5 \end{bmatrix}$$

Solution   Using the notation just introduced, we obtain

$$\begin{bmatrix} 3 & 5 & | & 9 \\ 2 & 3 & | & 5 \end{bmatrix} \xrightarrow{\frac{1}{3}R_1} \begin{bmatrix} 1 & \frac{5}{3} & | & 3 \\ 2 & 3 & | & 5 \end{bmatrix} \xrightarrow{R_2 - 2R_1} \begin{bmatrix} 1 & \frac{5}{3} & | & 3 \\ 0 & -\frac{1}{3} & | & -1 \end{bmatrix}$$

The first column, which originally contained the entry 3, is now in unit form, with a 1 where the pivot element used to be, and we are done.

Alternate Solution   In the first solution, we used Operation 2 to obtain a 1 where the pivot element was originally. Alternatively, we can use Operation 3 as follows:

$$\begin{bmatrix} 3 & 5 & | & 9 \\ 2 & 3 & | & 5 \end{bmatrix} \xrightarrow{R_1 - R_2} \begin{bmatrix} 1 & 2 & | & 4 \\ 2 & 3 & | & 5 \end{bmatrix} \xrightarrow{R_2 - 2R_1} \begin{bmatrix} 1 & 2 & | & 4 \\ 0 & -1 & | & -3 \end{bmatrix}$$ ∎

Note   In Example 4, the two matrices

$$\begin{bmatrix} 1 & \frac{5}{3} & | & 3 \\ 0 & -\frac{1}{3} & | & -1 \end{bmatrix} \quad \text{and} \quad \begin{bmatrix} 1 & 2 & | & 4 \\ 0 & -1 & | & -3 \end{bmatrix}$$

look quite different, but they are in fact equivalent. You can verify this by observing that they represent the systems of equations

$$x + \frac{5}{3}y = 3 \quad \text{and} \quad x + 2y = 4$$

$$-\frac{1}{3}y = -1 \qquad\qquad -y = -3$$

respectively, and both have the same solution: $x = -2$ and $y = 3$. Example 4 also shows that we can sometimes avoid working with fractions by using the appropriate row operation. ∎

A summary of the Gauss–Jordan method follows.

The Gauss–Jordan Elimination Method
1. Write the augmented matrix corresponding to the linear system.
2. Interchange rows (Operation 1), if necessary, to obtain an augmented matrix in which the first entry in the first row is nonzero. Then pivot the matrix about this entry.
3. Interchange the second row with any row below it, if necessary, to obtain an augmented matrix in which the second entry in the second row is nonzero. Pivot the matrix about this entry.
4. Continue until the final matrix is in row-reduced form.

Before writing the augmented matrix, be sure to write all equations with the variables on the left and constant terms on the right of the equal sign. Also, make sure that the variables are in the same order in all equations.

EXAMPLE 5   Solve the system of linear equations given by

$$3x - 2y + 8z = 9$$
$$-2x + 2y + z = 3 \tag{8}$$
$$x + 2y - 3z = 8$$

Solution   Using the Gauss–Jordan elimination method, we obtain the following sequence of equivalent augmented matrices:

$$\begin{bmatrix} ③ & -2 & 8 & | & 9 \\ -2 & 2 & 1 & | & 3 \\ 1 & 2 & -3 & | & 8 \end{bmatrix} \xrightarrow{R_1 + R_2} \begin{bmatrix} 1 & 0 & 9 & | & 12 \\ -2 & 2 & 1 & | & 3 \\ 1 & 2 & -3 & | & 8 \end{bmatrix}$$

$$\xrightarrow[R_3 - R_1]{R_2 + 2R_1} \begin{bmatrix} 1 & 0 & 9 & | & 12 \\ 0 & 2 & 19 & | & 27 \\ 0 & 2 & -12 & | & -4 \end{bmatrix}$$

$$\xrightarrow{R_2 \leftrightarrow R_3} \begin{bmatrix} 1 & 0 & 9 & | & 12 \\ 0 & ② & -12 & | & -4 \\ 0 & 2 & 19 & | & 27 \end{bmatrix}$$

$$\xrightarrow{\frac{1}{2}R_2} \begin{bmatrix} 1 & 0 & 9 & | & 12 \\ 0 & 1 & -6 & | & -2 \\ 0 & 2 & 19 & | & 27 \end{bmatrix}$$

$$\xrightarrow{R_3 - 2R_2} \begin{bmatrix} 1 & 0 & 9 & | & 12 \\ 0 & 1 & -6 & | & -2 \\ 0 & 0 & ㉛ & | & 31 \end{bmatrix}$$

$$\xrightarrow{\frac{1}{31}R_3}\begin{bmatrix} 1 & 0 & 9 & | & 12 \\ 0 & 1 & -6 & | & -2 \\ 0 & 0 & 1 & | & 1 \end{bmatrix}$$

$$\xrightarrow[R_2 + 6R_3]{R_1 - 9R_3}\begin{bmatrix} 1 & 0 & 0 & | & 3 \\ 0 & 1 & 0 & | & 4 \\ 0 & 0 & 1 & | & 1 \end{bmatrix}$$

The solution to System (8) is given by $x = 3$, $y = 4$, and $z = 1$ and may be verified by substitution into System (8) as follows:

$$3(3) - 2(4) + 8(1) = 9 \quad \checkmark$$
$$-2(3) + 2(4) + 1 = 3 \quad \checkmark$$
$$3 + 2(4) - 3(1) = 8 \quad \checkmark \qquad \blacksquare$$

 When searching for an element to serve as a pivot, it is important to keep in mind that you may work only with the row containing the potential pivot or any row *below* it. To see what can go wrong if this caution is not heeded, consider the following augmented matrix for some linear system:

$$\begin{bmatrix} 1 & 1 & 2 & | & 3 \\ 0 & 0 & 3 & | & 1 \\ 0 & 2 & 1 & | & -2 \end{bmatrix}$$

Observe that column 1 is in unit form. The next step in the Gauss–Jordan elimination procedure calls for obtaining a nonzero element in the second position of row 2. If you use row 1 (which is *above* the row under consideration) to help you obtain the pivot, you might proceed as follows:

$$\begin{bmatrix} 1 & 1 & 2 & | & 3 \\ 0 & 0 & 3 & | & 1 \\ 0 & 2 & 1 & | & -2 \end{bmatrix} \xrightarrow{R_2 \leftrightarrow R_1} \begin{bmatrix} 0 & 0 & 3 & | & 1 \\ 1 & 1 & 2 & | & 3 \\ 0 & 2 & 1 & | & -2 \end{bmatrix}$$

As you can see, not only have we obtained a nonzero element to serve as the next pivot, but it is already a 1, thus obviating the next step. This seems like a good move. But beware, we have undone some of our earlier work: Column 1 is no longer in the unit form where a 1 appears first. The correct move in this case is to interchange row 2 with row 3.

The next example illustrates how to handle a situation in which the entry in row 1 of the augmented matrix is zero.

EXPLORE & DISCUSS

1. Can the phrase "a nonzero constant multiple of itself" in a type-2 row operation be replaced by "any constant multiple of itself"? Explain.

2. Can a row of an augmented matrix be replaced by one obtained by adding a constant to every element in that row without changing the solution of the system of linear equations? Explain.

EXAMPLE 6 Solve the system of linear equations given by

$$
\begin{aligned}
2y + 3z &= 7 \\
3x + 6y - 12z &= -3 \\
5x - 2y + 2z &= -7
\end{aligned}
$$

Solution Using the Gauss–Jordan elimination method, we obtain the following sequence of equivalent augmented matrices:

$$
\left[\begin{array}{ccc|c}
0 & 2 & 3 & 7 \\
3 & 6 & -12 & -3 \\
5 & -2 & 2 & -7
\end{array}\right]
\xrightarrow{R_1 \leftrightarrow R_2}
\left[\begin{array}{ccc|c}
③ & 6 & -12 & -3 \\
0 & 2 & 3 & 7 \\
5 & -2 & 2 & -7
\end{array}\right]
\xrightarrow{\frac{1}{3}R_1}
\left[\begin{array}{ccc|c}
1 & 2 & -4 & -1 \\
0 & 2 & 3 & 7 \\
5 & -2 & 2 & -7
\end{array}\right]
\xrightarrow{R_3 - 5R_1}
$$

$$
\left[\begin{array}{ccc|c}
1 & 2 & -4 & -1 \\
0 & ② & 3 & 7 \\
0 & -12 & 22 & -2
\end{array}\right]
\xrightarrow{\frac{1}{2}R_2}
\left[\begin{array}{ccc|c}
1 & 2 & -4 & -1 \\
0 & 1 & \frac{3}{2} & \frac{7}{2} \\
0 & -12 & 22 & -2
\end{array}\right]
\xrightarrow[R_3 + 12R_2]{R_1 - 2R_2}
\left[\begin{array}{ccc|c}
1 & 0 & -7 & -8 \\
0 & 1 & \frac{3}{2} & \frac{7}{2} \\
0 & 0 & ㊵ & 40
\end{array}\right]
\xrightarrow{\frac{1}{40}R_3}
$$

$$
\left[\begin{array}{ccc|c}
1 & 0 & -7 & -8 \\
0 & 1 & \frac{3}{2} & \frac{7}{2} \\
0 & 0 & 1 & 1
\end{array}\right]
\xrightarrow[R_2 - \frac{3}{2}R_3]{R_1 + 7R_3}
\left[\begin{array}{ccc|c}
1 & 0 & 0 & -1 \\
0 & 1 & 0 & 2 \\
0 & 0 & 1 & 1
\end{array}\right]
$$

The solution to the system is given by $x = -1$, $y = 2$, and $z = 1$ and may be verified by substitution into the system. ■

 APPLIED EXAMPLE 7 Manufacturing: Production Scheduling Complete the solution to Example 1 in Section 2.1, page 75.

Solution To complete the solution of the problem posed in Example 1, recall that the mathematical formulation of the problem led to the following system of linear equations:

$$
\begin{aligned}
2x + y + z &= 180 \\
x + 3y + 2z &= 300 \\
2x + y + 2z &= 240
\end{aligned}
$$

where x, y, and z denote the respective numbers of type-A, type-B, and type-C souvenirs to be made.

Solving the foregoing system of linear equations by the Gauss–Jordan elimination method, we obtain the following sequence of equivalent augmented matrices:

$$
\left[\begin{array}{ccc|c}
② & 1 & 1 & 180 \\
1 & 3 & 2 & 300 \\
2 & 1 & 2 & 240
\end{array}\right]
\xrightarrow{R_1 \leftrightarrow R_2}
\left[\begin{array}{ccc|c}
1 & 3 & 2 & 300 \\
2 & 1 & 1 & 180 \\
2 & 1 & 2 & 240
\end{array}\right]
$$

$$
\xrightarrow[R_3 - 2R_1]{R_2 - 2R_1}
\left[\begin{array}{ccc|c}
1 & 3 & 2 & 300 \\
0 & ⑤ & -3 & -420 \\
0 & -5 & -2 & -360
\end{array}\right]
$$

$$\xrightarrow{-\frac{1}{5}R_2} \begin{bmatrix} 1 & 3 & 2 & | & 300 \\ 0 & 1 & \frac{3}{5} & | & 84 \\ 0 & -5 & -2 & | & -360 \end{bmatrix}$$

$$\xrightarrow[R_3 + 5R_2]{R_1 - 3R_2} \begin{bmatrix} 1 & 0 & \frac{1}{5} & | & 48 \\ 0 & 1 & \frac{3}{5} & | & 84 \\ 0 & 0 & ① & | & 60 \end{bmatrix}$$

$$\xrightarrow[R_2 - \frac{3}{5}R_3]{R_1 - \frac{1}{5}R_3} \begin{bmatrix} 1 & 0 & 0 & | & 36 \\ 0 & 1 & 0 & | & 48 \\ 0 & 0 & 1 & | & 60 \end{bmatrix}$$

Thus, $x = 36$, $y = 48$, and $z = 60$; that is, Ace Novelty should make 36 type-A souvenirs, 48 type-B souvenirs, and 60 type-C souvenirs in order to use all available machine time. ∎

2.2   Self-Check Exercises

1. Solve the system of linear equations

$$\begin{aligned} 2x + 3y + z &= 6 \\ x - 2y + 3z &= -3 \\ 3x + 2y - 4z &= 12 \end{aligned}$$

using the Gauss–Jordan elimination method.

2. A farmer has 200 acres of land suitable for cultivating crops A, B, and C. The cost per acre of cultivating crop A, crop B,

and crop C is $40, $60, and $80, respectively. The farmer has $12,600 available for land cultivation. Each acre of crop A requires 20 labor-hours, each acre of crop B requires 25 labor-hours, and each acre of crop C requires 40 labor-hours. The farmer has a maximum of 5950 labor-hours available. If he wishes to use all of his cultivatable land, the entire budget, and all the labor available, how many acres of each crop should he plant?

Solutions to Self-Check Exercises 2.2 can be found on page 94.

2.2   Concept Questions

1. **a.** Explain what it means for two systems of linear equations to be equivalent to each other.
   b. Give the meaning of the following notation used for row operations in the Gauss–Jordan elimination method:

   a. $R_i \leftrightarrow R_j$ **b.** cR_i **c.** $R_i + aR_j$

2. **a.** What is an *augmented matrix*? A *coefficient matrix*? A *unit column*?
   b. Explain what is meant by a *pivot operation*.

3. Suppose that a matrix is in row-reduced form.
   a. What is the position of a row consisting entirely of zeros relative to the nonzero rows?
   b. What is the first nonzero entry in each row?
   c. What is the position of the leading 1's in successive nonzero rows?
   d. If a column contains a leading 1, then what is the value of the other entries in that column?

2.2 Exercises

In Exercises 1–4, write the augmented matrix corresponding to each system of equations.

1. $2x - 3y = 7$
  $3x + y = 4$

2. $3x + 7y - 8z = 5$
  $x + 3z = -2$
  $4x - 3y = 7$

3. $-y + 2z = 6$
  $2x + 2y - 8z = 7$
  $3y + 4z = 0$

4. $3x_1 + 2x_2 = 0$
  $x_1 - x_2 + 2x_3 = 4$
  $2x_2 - 3x_3 = 5$

In Exercises 5–8, write the system of equations corresponding to each augmented matrix.

5. $\begin{bmatrix} 3 & 2 & | & -4 \\ 1 & -1 & | & 5 \end{bmatrix}$

6. $\begin{bmatrix} 0 & 3 & 2 & | & 4 \\ 1 & -1 & -2 & | & -3 \\ 4 & 0 & 3 & | & 2 \end{bmatrix}$

7. $\begin{bmatrix} 1 & 3 & 2 & | & 4 \\ 2 & 0 & 0 & | & 5 \\ 3 & -3 & 2 & | & 6 \end{bmatrix}$

8. $\begin{bmatrix} 2 & 3 & 1 & | & 6 \\ 4 & 3 & 2 & | & 5 \\ 0 & 0 & 0 & | & 0 \end{bmatrix}$

In Exercises 9–18, indicate whether the matrix is in row-reduced form.

9. $\begin{bmatrix} 1 & 0 & | & 3 \\ 0 & 1 & | & -2 \end{bmatrix}$

10. $\begin{bmatrix} 1 & 1 & | & 3 \\ 0 & 0 & | & 0 \end{bmatrix}$

11. $\begin{bmatrix} 0 & 1 & | & 3 \\ 1 & 0 & | & 5 \end{bmatrix}$

12. $\begin{bmatrix} 0 & 1 & | & 3 \\ 0 & 0 & | & 5 \end{bmatrix}$

13. $\begin{bmatrix} 1 & 0 & 0 & | & 3 \\ 0 & 1 & 0 & | & 4 \\ 0 & 0 & 1 & | & 5 \end{bmatrix}$

14. $\begin{bmatrix} 1 & 0 & 0 & | & -1 \\ 0 & 1 & 0 & | & -2 \\ 0 & 0 & 2 & | & -3 \end{bmatrix}$

15. $\begin{bmatrix} 1 & 0 & 1 & | & 3 \\ 0 & 1 & 0 & | & 4 \\ 0 & 0 & -1 & | & 6 \end{bmatrix}$

16. $\begin{bmatrix} 1 & 0 & | & -10 \\ 0 & 1 & | & 2 \\ 0 & 0 & | & 0 \end{bmatrix}$

17. $\begin{bmatrix} 0 & 0 & 0 & | & 0 \\ 0 & 1 & 2 & | & 4 \\ 0 & 0 & 0 & | & 0 \end{bmatrix}$

18. $\begin{bmatrix} 1 & 0 & 0 & | & 3 \\ 0 & 1 & 0 & | & 6 \\ 0 & 0 & 0 & | & 4 \\ 0 & 0 & 1 & | & 5 \end{bmatrix}$

In Exercises 19–26, pivot the system about the circled element.

19. $\begin{bmatrix} ② & 4 & | & 8 \\ 3 & 1 & | & 2 \end{bmatrix}$

20. $\begin{bmatrix} 3 & 2 & | & 6 \\ ④ & 2 & | & 5 \end{bmatrix}$

21. $\begin{bmatrix} ⟨-1⟩ & 2 & | & 3 \\ 6 & 4 & | & 2 \end{bmatrix}$

22. $\begin{bmatrix} ① & 3 & | & 4 \\ 2 & 4 & | & 6 \end{bmatrix}$

23. $\begin{bmatrix} ② & 4 & 6 & | & 12 \\ 2 & 3 & 1 & | & 5 \\ 3 & -1 & 2 & | & 4 \end{bmatrix}$

24. $\begin{bmatrix} 1 & 3 & 2 & | & 4 \\ ② & 4 & 8 & | & 6 \\ -1 & 2 & 3 & | & 4 \end{bmatrix}$

25. $\begin{bmatrix} 0 & 1 & 3 & | & 4 \\ 2 & 4 & ① & | & 3 \\ 5 & 6 & 2 & | & -4 \end{bmatrix}$

26. $\begin{bmatrix} 1 & 2 & 3 & | & 5 \\ 0 & ⟨-3⟩ & 3 & | & 2 \\ 0 & 4 & -1 & | & 3 \end{bmatrix}$

In Exercises 27–30, fill in the missing entries by performing the indicated row operations to obtain the row-reduced matrices.

27. $\begin{bmatrix} 3 & 9 & | & 6 \\ 2 & 1 & | & 4 \end{bmatrix} \xrightarrow{\frac{1}{3}R_1} \begin{bmatrix} \cdot & \cdot & | & \cdot \\ 2 & 1 & | & 4 \end{bmatrix} \xrightarrow{R_2 - 2R_1}$

$\begin{bmatrix} 1 & 3 & | & 2 \\ \cdot & \cdot & | & \cdot \end{bmatrix} \xrightarrow{-\frac{1}{5}R_2} \begin{bmatrix} 1 & 3 & | & 2 \\ \cdot & \cdot & | & \cdot \end{bmatrix} \xrightarrow{R_1 - 3R_2} \begin{bmatrix} 1 & 0 & | & 2 \\ 0 & 1 & | & 0 \end{bmatrix}$

28. $\begin{bmatrix} 1 & 2 & | & 1 \\ 2 & 3 & | & -1 \end{bmatrix} \xrightarrow{R_2 - 2R_1} \begin{bmatrix} 1 & 2 & | & 1 \\ \cdot & \cdot & | & \cdot \end{bmatrix} \xrightarrow{-R_2}$

$\begin{bmatrix} 1 & 2 & | & 1 \\ \cdot & \cdot & | & \cdot \end{bmatrix} \xrightarrow{R_1 - 2R_2} \begin{bmatrix} 1 & 0 & | & -5 \\ 0 & 1 & | & 3 \end{bmatrix}$

29. $\begin{bmatrix} 1 & 3 & 1 & | & 3 \\ 3 & 8 & 3 & | & 7 \\ 2 & -3 & 1 & | & -10 \end{bmatrix} \xrightarrow[R_3 - 2R_1]{R_2 - 3R_1} \begin{bmatrix} 1 & 3 & 1 & | & 3 \\ \cdot & \cdot & \cdot & | & \cdot \\ \cdot & \cdot & \cdot & | & \cdot \end{bmatrix} \xrightarrow{-R_2}$

$\begin{bmatrix} 1 & 3 & 1 & | & 3 \\ \cdot & \cdot & \cdot & | & \cdot \\ 0 & -9 & -1 & | & -16 \end{bmatrix} \xrightarrow[R_3 + 9R_2]{R_1 - 3R_2}$

$\begin{bmatrix} \cdot & \cdot & \cdot & | & \cdot \\ 0 & 1 & 0 & | & 2 \\ \cdot & \cdot & \cdot & | & \cdot \end{bmatrix} \xrightarrow[-R_3]{R_1 + R_3} \begin{bmatrix} 1 & 0 & 0 & | & -1 \\ 0 & 1 & 0 & | & 2 \\ 0 & 0 & 1 & | & -2 \end{bmatrix}$

30. $\begin{bmatrix} 0 & 1 & 3 & | & -4 \\ 1 & 2 & 1 & | & 7 \\ 1 & -2 & 0 & | & 1 \end{bmatrix} \xrightarrow{R_1 \leftrightarrow R_2} \begin{bmatrix} \cdot & \cdot & \cdot & | & \cdot \\ \cdot & \cdot & \cdot & | & \cdot \\ 1 & -2 & 0 & | & 1 \end{bmatrix}$

$\xrightarrow{R_3 - R_1} \begin{bmatrix} 1 & 2 & 1 & | & 7 \\ 0 & 1 & 3 & | & -4 \\ \cdot & \cdot & \cdot & | & \cdot \end{bmatrix} \xrightarrow[R_3 + 4R_2]{R_1 + \frac{1}{2}R_3} \begin{bmatrix} \cdot & \cdot & \cdot & | & \cdot \\ 0 & 1 & 3 & | & -4 \\ \cdot & \cdot & \cdot & | & \cdot \end{bmatrix}$

$\xrightarrow{\frac{1}{11}R_3} \begin{bmatrix} 1 & 0 & \frac{1}{2} & | & 4 \\ 0 & 1 & 3 & | & -4 \\ \cdot & \cdot & \cdot & | & \cdot \end{bmatrix} \xrightarrow[R_2 - 3R_3]{R_1 - \frac{1}{2}R_3} \begin{bmatrix} 1 & 0 & 0 & | & 5 \\ 0 & 1 & 0 & | & 2 \\ 0 & 0 & 1 & | & -2 \end{bmatrix}$

31. Write a system of linear equations for the augmented matrix of Exercise 27. Using the results of Exercise 27, determine the solution of the system.

32. Repeat Exercise 31 for the augmented matrix of Exercise 28.

33. Repeat Exercise 31 for the augmented matrix of Exercise 29.

34. Repeat Exercise 31 for the augmented matrix of Exercise 30.

In Exercises 35–50, solve the system of linear equations, using the Gauss–Jordan elimination method.

35. $\begin{aligned} x - 2y &= 8 \\ 3x + 4y &= 4 \end{aligned}$

36. $\begin{aligned} 3x + y &= 1 \\ -7x - 2y &= -1 \end{aligned}$

37. $\begin{aligned} 2x - 3y &= -8 \\ 4x + y &= -2 \end{aligned}$

38. $\begin{aligned} 5x + 3y &= 9 \\ -2x + y &= -8 \end{aligned}$

39. $\begin{aligned} x + y + z &= 0 \\ 2x - y + z &= 1 \\ x + y - 2z &= 2 \end{aligned}$

40. $\begin{aligned} 2x + y - 2z &= 4 \\ x + 3y - z &= -3 \\ 3x + 4y - z &= 7 \end{aligned}$

41. $\begin{aligned} 2x + 2y + z &= 9 \\ x \quad\;\; + z &= 4 \\ 4y - 3z &= 17 \end{aligned}$

42. $\begin{aligned} 2x + 3y - 2z &= 10 \\ 3x - 2y + 2z &= 0 \\ 4x - y + 3z &= -1 \end{aligned}$

43. $\begin{aligned} - x_2 + x_3 &= 2 \\ 4x_1 - 3x_2 + 2x_3 &= 16 \\ 3x_1 + 2x_2 + x_3 &= 11 \end{aligned}$

44. $\begin{aligned} 2x + 4y - 6z &= 38 \\ x + 2y + 3z &= 7 \\ 3x - 4y + 4z &= -19 \end{aligned}$

45. $\begin{aligned} x_1 - 2x_2 + x_3 &= 6 \\ 2x_1 + x_2 - 3x_3 &= -3 \\ x_1 - 3x_2 + 3x_3 &= 10 \end{aligned}$

46. $\begin{aligned} 2x + 3y - 6z &= -11 \\ x - 2y + 3z &= 9 \\ 3x + y &= 7 \end{aligned}$

47. $\begin{aligned} 2x \quad\;\; + 3z &= -1 \\ 3x - 2y + z &= 9 \\ x + y + 4z &= 4 \end{aligned}$

48. $\begin{aligned} 2x_1 - x_2 + 3x_3 &= -4 \\ x_1 - 2x_2 + x_3 &= -1 \\ x_1 - 5x_2 + 2x_3 &= -3 \end{aligned}$

49. $\begin{aligned} x_1 - x_2 + 3x_3 &= 14 \\ x_1 + x_2 + x_3 &= 6 \\ -2x_1 - x_2 + x_3 &= -4 \end{aligned}$

50. $\begin{aligned} 2x_1 - x_2 - x_3 &= 0 \\ 3x_1 + 2x_2 + x_3 &= 7 \\ x_1 + 2x_2 + 2x_3 &= 5 \end{aligned}$

The problems in Exercises 51–63 correspond to those in Exercises 15–27, Section 2.1. Use the results of your previous work to help you solve these problems.

51. AGRICULTURE   The Johnson Farm has 500 acres of land allotted for cultivating corn and wheat. The cost of cultivating corn and wheat (including seeds and labor) is $42 and $30/acre, respectively. Jacob Johnson has $18,600 available for cultivating these crops. If he wishes to use all the allotted land and his entire budget for cultivating these two crops, how many acres of each crop should he plant?

52. INVESTMENTS   Michael Perez has a total of $2000 on deposit with two savings institutions. One pays interest at the rate of 6%/year, whereas the other pays interest at the rate of 8%/year. If Michael earned a total of $144 in interest during a single year, how much does he have on deposit in each institution?

53. MIXTURES   The Coffee Shoppe sells a coffee blend made from two coffees, one costing $2.50/lb and the other costing $3.00/lb. If the blended coffee sells for $2.80/lb, find how much of each coffee is used to obtain the desired blend. (Assume the weight of the blended coffee is 100 lb.)

54. INVESTMENTS   Kelly Fisher has a total of $30,000 invested in two municipal bonds that have yields of 8% and 10% interest per year, respectively. If the interest Kelly receives from the bonds in a year is $2640, how much does she have invested in each bond?

55. RIDERSHIP   The total number of passengers riding a certain city bus during the morning shift is 1000. If the child's fare is $.25, the adult fare is $.75, and the total revenue from the fares in the morning shift is $650, how many children and how many adults rode the bus during the morning shift?

56. REAL ESTATE   Cantwell Associates, a real estate developer, is planning to build a new apartment complex consisting of one-bedroom units and two- and three-bedroom townhouses. A total of 192 units is planned, and the number of family units (two- and three-bedroom townhouses) will equal the number of one-bedroom units. If the number of one-bedroom units will be 3 times the number of three-bedroom units, find how many units of each type will be in the complex.

57. INVESTMENT PLANNING   The annual interest on Sid Carrington's three investments amounted to $21,600: 6% on a savings account, 8% on mutual funds, and 12% on bonds. If the amount of Sid's investment in bonds was twice the amount of his investment in the savings account, and the interest earned from his investment in bonds was equal to the dividends he received from his investment in mutual funds, find how much money he placed in each type of investment.

58. INVESTMENT CLUB   A private investment club has $200,000 earmarked for investment in stocks. To arrive at an acceptable overall level of risk, the stocks that management is considering have been classified into three categories: high-risk, medium-risk, and low-risk. Management estimates that high-risk stocks will have a rate of return of 15%/year; medium-risk stocks, 10%/year; and low-risk stocks, 6%/year. The members have decided that the investment in low-risk stocks should be equal to the sum of the investments in the stocks of the other two categories. Determine how much the club should invest in each type of stock if the investment goal is to have a return of $20,000/year on the total investment. (Assume that all the money available for investment is invested.)

59. MIXTURE PROBLEM—FERTILIZER   Lawnco produces three grades of commercial fertilizers. A 100-lb bag of grade-A fertilizer contains 18 lb of nitrogen, 4 lb of phosphate, and 5 lb of potassium. A 100-lb bag of grade-B fertilizer contains 20 lb of nitrogen and 4 lb each of phosphate and potassium. A 100-lb bag of grade-C fertilizer contains 24 lb of nitrogen, 3 lb of phosphate, and 6 lb of potassium. How many 100-lb bags of each of the three grades of fertilizers should Lawnco produce if 26,400 lb of nitrogen, 4900 lb of phosphate, and 6200 lb of potassium are available and all the nutrients are used?

60. **Box-Office Receipts**   A theater has a seating capacity of 900 and charges $2 for children, $3 for students, and $4 for adults. At a certain screening with full attendance, there were half as many adults as children and students combined. The receipts totaled $2800. How many children attended the show?

61. **Management Decisions**   The management of Hartman Rent-A-Car has allocated $1.5 million to buy a fleet of new automobiles consisting of compact, intermediate, and full-size cars. Compacts cost $12,000 each, intermediate-size cars cost $18,000 each, and full-size cars cost $24,000 each. If Hartman purchases twice as many compacts as intermediate-size cars and the total number of cars to be purchased is 100, determine how many cars of each type will be purchased. (Assume that the entire budget will be used.)

62. **Investment Clubs**   The management of a private investment club has a fund of $200,000 earmarked for investment in stocks. To arrive at an acceptable overall level of risk, the stocks that management is considering have been classified into three categories: high-risk, medium-risk, and low-risk. Management estimates that high-risk stocks will have a rate of return of 15%/year; medium-risk stocks, 10%/year; and low-risk stocks, 6%/year. The investment in low-risk stocks is to be twice the sum of the investments in stocks of the other two categories. If the investment goal is to have an average rate of return of 9%/year on the total investment, determine how much the club should invest in each type of stock. (Assume all of the money available for investment is invested.)

63. **Diet Planning**   A dietitian wishes to plan a meal around three foods. The percent of the daily requirements of proteins, carbohydrates, and iron contained in each ounce of the three foods is summarized in the accompanying table:

	Food I	Food II	Food III
Proteins (%)	10	6	8
Carbohydrates (%)	10	12	6
Iron (%)	5	4	12

Determine how many ounces of each food the dietitian should include in the meal to meet exactly the daily requirement of proteins, carbohydrates, and iron (100% of each).

64. **Investments**   Mr. and Mrs. Garcia have a total of $100,000 to be invested in stocks, bonds, and a money market account.

The stocks have a rate of return of 12%/year, while the bonds and the money market account pay 8% and 4%/year, respectively. They have stipulated that the amount invested in the money market account should be equal to the sum of 20% of the amount invested in stocks and 10% of the amount invested in bonds. How should the Garcias allocate their resources if they require an annual income of $10,000 from their investments?

65. **Box-Office Receipts**   For the opening night at the Opera House, a total of 1000 tickets were sold. Front orchestra seats cost $80 apiece, rear orchestra seats cost $60 apiece, and front balcony seats cost $50 apiece. The combined number of tickets sold for the front orchestra and rear orchestra exceeded twice the number of front balcony tickets sold by 400. The total receipts for the performance were $62,800. Determine how many tickets of each type were sold.

66. **Production Scheduling**   A manufacturer of women's blouses makes three types of blouses: sleeveless, short-sleeve, and long-sleeve. The time (in minutes) required by each department to produce a dozen blouses of each type is shown in the accompanying table:

	Sleeveless	Short-Sleeve	Long-Sleeve
Cutting	9	12	15
Sewing	22	24	28
Packaging	6	8	8

The cutting, sewing, and packaging departments have available a maximum of 80, 160, and 48 labor-hours, respectively, per day. How many dozens of each type of blouse can be produced each day if the plant is operated at full capacity?

67. **Business Travel Expenses**   An executive of Trident Communications recently traveled to London, Paris, and Rome. He paid $180, $230, and $160/night for lodging in London, Paris, and Rome, respectively, and his hotel bills totaled $2660. He spent $110, $120, and $90/day for his meals in London, Paris, and Rome, respectively, and his expenses for meals totaled $1520. If he spent as many days in London as he did in Paris and Rome combined, how many days did he stay in each city?

68. **Vacation Costs**   Joan and Dick spent 2 wk (14 nights) touring four cities on the East Coast—Boston, New York, Philadelphia, and Washington, D.C. They paid $120, $200, $80, and $100/night for lodging in each city, respectively, and their total hotel bill came to $2020. The number of days they spent in New York was the same as the total number of days they spent in Boston and Washington, D.C., and the couple spent 3 times as many days in New York as they did in Philadelphia. How many days did Joan and Dick stay in each city?

In Exercises 69 and 70, determine whether the statement is true or false. If it is true, explain why it is true. If it is false, give an example to show why it is false.

69. An equivalent system of linear equations can be obtained from a system of equations by replacing one of its equations by any constant multiple of itself.

70. If the augmented matrix corresponding to a system of three linear equations in three variables has a row of the form $[0 \ \ 0 \ \ 0 \ | \ a]$, where a is a nonzero number, then the system has no solution.

2.2 Solutions to Self-Check Exercises

1. We obtain the following sequence of equivalent augmented matrices:

$$\begin{bmatrix} 2 & 3 & 1 & | & 6 \\ 1 & -2 & 3 & | & -3 \\ 3 & 2 & -4 & | & 12 \end{bmatrix} \xrightarrow{R_1 \leftrightarrow R_2} \begin{bmatrix} ① & -2 & 3 & | & -3 \\ 2 & 3 & 1 & | & 6 \\ 3 & 2 & -4 & | & 12 \end{bmatrix}$$

$$\xrightarrow[R_3 - 3R_1]{R_2 - 2R_1} \begin{bmatrix} 1 & -2 & 3 & | & -3 \\ 0 & 7 & -5 & | & 12 \\ 0 & 8 & -13 & | & 21 \end{bmatrix} \xrightarrow{R_2 \leftrightarrow R_3}$$

$$\begin{bmatrix} 1 & -2 & 3 & | & -3 \\ 0 & ⑧ & -13 & | & 21 \\ 0 & 7 & -5 & | & 12 \end{bmatrix} \xrightarrow{R_2 - R_3} \begin{bmatrix} 1 & -2 & 3 & | & -3 \\ 0 & 1 & -8 & | & 9 \\ 0 & 7 & -5 & | & 12 \end{bmatrix}$$

$$\xrightarrow[R_3 - 7R_2]{R_1 + 2R_2} \begin{bmatrix} 1 & 0 & -13 & | & 15 \\ 0 & 1 & -8 & | & 9 \\ 0 & 0 & 51 & | & -51 \end{bmatrix} \xrightarrow{\frac{1}{51}R_3} \begin{bmatrix} 1 & 0 & -13 & | & 15 \\ 0 & 1 & -8 & | & 9 \\ 0 & 0 & ① & | & -1 \end{bmatrix}$$

$$\xrightarrow[R_2 + 8R_3]{R_1 + 13R_3} \begin{bmatrix} 1 & 0 & 0 & | & 2 \\ 0 & 1 & 0 & | & 1 \\ 0 & 0 & 1 & | & -1 \end{bmatrix}$$

The solution to the system is $x = 2$, $y = 1$, and $z = -1$.

2. Referring to the solution of Exercise 2, Self-Check Exercises 2.1, we see that the problem reduces to solving the following system of linear equations:

$$\begin{aligned} x + \quad y + \quad z &= \quad 200 \\ 40x + 60y + 80z &= 12{,}600 \\ 20x + 25y + 40z &= \quad 5{,}950 \end{aligned}$$

Using the Gauss–Jordan elimination method, we have

$$\begin{bmatrix} 1 & 1 & 1 & | & 200 \\ 40 & 60 & 80 & | & 12{,}600 \\ 20 & 25 & 40 & | & 5{,}950 \end{bmatrix} \xrightarrow[R_3 - 20R_1]{R_2 - 40R_1} \begin{bmatrix} 1 & 1 & 1 & | & 200 \\ 0 & 20 & 40 & | & 4600 \\ 0 & 5 & 20 & | & 1950 \end{bmatrix}$$

$$\xrightarrow{\frac{1}{20}R_2} \begin{bmatrix} 1 & 1 & 1 & | & 200 \\ 0 & 1 & 2 & | & 230 \\ 0 & 5 & 20 & | & 1950 \end{bmatrix} \xrightarrow[R_3 - 5R_2]{R_1 - R_2} \begin{bmatrix} 1 & 0 & -1 & | & -30 \\ 0 & 1 & 2 & | & 230 \\ 0 & 0 & 10 & | & 800 \end{bmatrix}$$

$$\xrightarrow{\frac{1}{10}R_3} \begin{bmatrix} 1 & 0 & -1 & | & -30 \\ 0 & 1 & 2 & | & 230 \\ 0 & 0 & 1 & | & 80 \end{bmatrix} \xrightarrow[R_2 - 2R_3]{R_1 + R_3} \begin{bmatrix} 1 & 0 & 0 & | & 50 \\ 0 & 1 & 0 & | & 70 \\ 0 & 0 & 1 & | & 80 \end{bmatrix}$$

From the last augmented matrix in reduced form, we see that $x = 50$, $y = 70$, and $z = 80$. Therefore, the farmer should plant 50 acres of crop A, 70 acres of crop B, and 80 acres of crop C.

USING TECHNOLOGY

■ Systems of Linear Equations: Unique Solutions

Solving a System of Linear Equations Using the Gauss–Jordan Method

The three matrix operations can be performed on a matrix, using a graphing utility. The commands are summarized below:

	Calculator Function		
Operation	**TI-83**	**TI-86**	
$R_i \leftrightarrow R_j$	**rowSwap([A], i, j)**	**rSwap(A, i, j)**	or equivalent
cR_i	***row**(c, [A], i)	**multR(c, A, i)**	or equivalent
$R_i + aR_j$	***row+**(a, [A], j, i)	**mRAdd(a, A, j, i)**	or equivalent

When a row operation is performed on a matrix, the result is stored as an answer in the calculator. If another operation is performed on this matrix, then the matrix is erased. Should a mistake be made in the operation, then the previous matrix is lost. For this reason, you should store the results of each operation. We do this by pressing **STO,** followed by the name of a matrix, and then **ENTER.** We use this process in the following example.

EXAMPLE 1 Use a graphing utility to solve the following system of linear equations by the Gauss–Jordan method (see Example 5 in Section 2.2):

$$3x - 2y + 8z = 9$$
$$-2x + 2y + z = 3$$
$$x + 2y - 3z = 8$$

Solution Using the Gauss–Jordan method, we obtain the following sequence of equivalent matrices.

$$\begin{bmatrix} 3 & -2 & 8 & | & 9 \\ -2 & 2 & 1 & | & 3 \\ 1 & 2 & -3 & | & 8 \end{bmatrix} \xrightarrow{\text{*row} + (1, [A], 2, 1) \blacktriangleright B}$$

$$\begin{bmatrix} 1 & 0 & 9 & | & 12 \\ -2 & 2 & 1 & | & 3 \\ 1 & 2 & -3 & | & 8 \end{bmatrix} \xrightarrow{\text{*row} + (2, [B], 1, 2) \blacktriangleright C}$$

$$\begin{bmatrix} 1 & 0 & 9 & | & 12 \\ 0 & 2 & 19 & | & 27 \\ 1 & 2 & -3 & | & 8 \end{bmatrix} \xrightarrow{\text{*row} + (-1, [C], 1, 3) \blacktriangleright B}$$

$$\begin{bmatrix} 1 & 0 & 9 & | & 12 \\ 0 & 2 & 19 & | & 27 \\ 0 & 2 & -12 & | & -4 \end{bmatrix} \xrightarrow{\text{*row}(\frac{1}{2}, [B], 2) \blacktriangleright C}$$

$$\begin{bmatrix} 1 & 0 & 9 & | & 12 \\ 0 & 1 & 9.5 & | & 13.5 \\ 0 & 2 & -12 & | & -4 \end{bmatrix} \xrightarrow{\text{*row} + (-2, [C], 2, 3) \blacktriangleright B}$$

$$\begin{bmatrix} 1 & 0 & 9 & | & 12 \\ 0 & 1 & 9.5 & | & 13.5 \\ 0 & 0 & -31 & | & -31 \end{bmatrix} \xrightarrow{\text{*row}(-\frac{1}{31}, [B], 3) \blacktriangleright C}$$

(continued)

$$\begin{bmatrix} 1 & 0 & 9 & | & 12 \\ 0 & 1 & 9.5 & | & 13.5 \\ 0 & 0 & 1 & | & 1 \end{bmatrix} \xrightarrow{\ *\text{row} + (-9, [C], 3, 1)\ \blacktriangleright B\ }$$

$$\begin{bmatrix} 1 & 0 & 0 & | & 3 \\ 0 & 1 & 9.5 & | & 13.5 \\ 0 & 0 & 1 & | & 1 \end{bmatrix} \xrightarrow{\ *\text{row} + (-9.5, [B], 3, 2)\ \blacktriangleright C\ } \begin{bmatrix} 1 & 0 & 0 & | & 3 \\ 0 & 1 & 0 & | & 4 \\ 0 & 0 & 1 & | & 1 \end{bmatrix}$$

The last matrix is in row-reduced form, and we see that the solution of the system is $x = 3$, $y = 4$, and $z = 1$.   ∎

Using rref (TI-83 and TI-86) to Solve a System of Linear Equations

The operation **rref** (or equivalent function in your utility, if there is one) will transform an augmented matrix into one that is in row-reduced form. For example, using **rref,** we find

$$\begin{bmatrix} 3 & -2 & 8 & | & 9 \\ -2 & 2 & 1 & | & 3 \\ 1 & 2 & -3 & | & 8 \end{bmatrix} \xrightarrow{\text{rref}} \begin{bmatrix} 1 & 0 & 0 & | & 3 \\ 0 & 1 & 0 & | & 4 \\ 0 & 0 & 1 & | & 1 \end{bmatrix}$$

as obtained earlier!

Using SIMULT (TI-86) to Solve a System of Equations

The operation **SIMULT** (or equivalent operation on your utility, if there is one) of a graphing utility can be used to solve a system of n linear equations in n variables, where n is an integer between 2 and 30.

EXAMPLE 2   Use the **SIMULT** operation to solve the system of Example 1.

Solution   Call for the **SIMULT** operation. Since the system under consideration has three equations in three variables, enter $n = 3$. Next, enter a1, 1 = 3, a1, 2 = -2, a1, 3 = 8, . . . , b1 = 9, a2, 1 = -2, . . . , b3 = 8. Select <SOLVE> and the display

$$\text{x1} = 3$$
$$\text{x2} = 4$$
$$\text{x3} = 1$$

appears on the screen, giving $x = 3$, $y = 4$, and $z = 1$ as the required solution.   ∎

TECHNOLOGY EXERCISES

Use a graphing utility to solve the system of equations (a) by the Gauss–Jordan method, (b) using the rref operation, and (c) using SIMULT.

1. $x_1 - 2x_2 + 2x_3 - 3x_4 = -7$
    $3x_1 + 2x_2 - x_3 + 5x_4 = 22$
    $2x_1 - 3x_2 + 4x_3 - x_4 = -3$
    $3x_1 - 2x_2 - x_3 + 2x_4 = 12$

2. $2x_1 - x_2 + 3x_3 - 2x_4 = -2$
$x_1 - 2x_2 + x_3 - 3x_4 = 2$
$x_1 - 5x_2 + 2x_3 + 3x_4 = -6$
$-3x_1 + 3x_2 - 4x_3 - 4x_4 = 9$

3. $2x_1 + x_2 + 3x_3 - x_4 = 9$
$-x_1 - 2x_2 \quad\quad - 3x_4 = -1$
$x_1 \quad\quad - 3x_3 + x_4 = 10$
$x_1 - x_2 - x_3 - x_4 = 8$

4. $x_1 - 2x_2 - 2x_3 + x_4 = 1$
$2x_1 - x_2 + 2x_3 + 3x_4 = -2$
$-x_1 - 5x_2 + 7x_3 - 2x_4 = 3$
$3x_1 - 4x_2 + 3x_3 + 4x_4 = -4$

5. $2x_1 - 2x_2 + 3x_3 - x_4 + 2x_5 = 16$
$3x_1 + x_2 - 2x_3 + x_4 - 3x_5 = -11$
$x_1 + 3x_2 - 4x_3 + 3x_4 - x_5 = -13$
$2x_1 - x_2 + 3x_3 - 2x_4 + 2x_5 = 15$
$3x_1 + 4x_2 - 3x_3 + 5x_4 - x_5 = -10$

6. $2.1x_1 - 3.2x_2 + 6.4x_3 + 7x_4 - 3.2x_5 = 54.3$
$4.1x_1 + 2.2x_2 - 3.1x_3 - 4.2x_4 + 3.3x_5 = -20.81$
$3.4x_1 - 6.2x_2 + 4.7x_3 + 2.1x_4 - 5.3x_5 = 24.7$
$4.1x_1 + 7.3x_2 + 5.2x_3 + 6.1x_4 - 8.2x_5 = 29.25$
$2.8x_1 + 5.2x_2 + 3.1x_3 + 5.4x_4 + 3.8x_5 = 43.72$

2.3 Systems of Linear Equations: Underdetermined and Overdetermined Systems

In this section, we continue our study of systems of linear equations. More specifically, we look at systems that have infinitely many solutions and those that have no solution. We also study systems of linear equations in which the number of variables is not equal to the number of equations in the system.

▬ Solution(s) of Linear Equations

Our first example illustrates the situation in which a system of linear equations has infinitely many solutions.

EXAMPLE 1 A System of Equations with an Infinite Number of Solutions Solve the system of linear equations given by

$$\begin{aligned} x + 2y - 3z &= -2 \\ 3x - y - 2z &= 1 \\ 2x + 3y - 5z &= -3 \end{aligned} \tag{9}$$

Solution Using the Gauss–Jordan elimination method, we obtain the following sequence of equivalent augmented matrices:

$$\left[\begin{array}{ccc|c} \textcircled{1} & 2 & -3 & -2 \\ 3 & -1 & -2 & 1 \\ 2 & 3 & -5 & -3 \end{array}\right] \xrightarrow[R_3 - 2R_1]{R_2 - 3R_1} \left[\begin{array}{ccc|c} 1 & 2 & -3 & -2 \\ 0 & \textcircled{-7} & 7 & 7 \\ 0 & -1 & 1 & 1 \end{array}\right] \xrightarrow{-\frac{1}{7}R_2}$$

$$\left[\begin{array}{ccc|c} 1 & 2 & -3 & -2 \\ 0 & 1 & -1 & -1 \\ 0 & -1 & 1 & 1 \end{array}\right] \xrightarrow[R_3 + R_2]{R_1 - 2R_2} \left[\begin{array}{ccc|c} 1 & 0 & -1 & 0 \\ 0 & 1 & -1 & -1 \\ 0 & 0 & 0 & 0 \end{array}\right]$$

The last augmented matrix is in row-reduced form. Interpreting it as a system of linear equations gives

$$x - z = 0$$
$$y - z = -1$$

a system of two equations in the three variables x, y, and z.

Let's now single out one variable—say, z—and solve for x and y in terms of it. We obtain

$$x = z$$
$$y = z - 1$$

If we assign a particular value to z—say, $z = 0$—we obtain $x = 0$ and $y = -1$, giving the solution $(0, -1, 0)$ to System (9). By setting $z = 1$, we obtain the solution $(1, 0, 1)$. In general, if we set $z = t$, where t represents some real number (called a parameter), we obtain a solution given by $(t, t - 1, t)$. Since the parameter t may be any real number, we see that System (9) has infinitely many solutions. Geometrically, the solutions of System (9) lie on the straight line in three-dimensional space given by the intersection of the three planes determined by the three equations in the system. ∎

Note   In Example 1 we chose the parameter to be z because it is more convenient to solve for x and y (both the x- and y-columns are in unit form) in terms of z. ∎

The next example shows what happens in the elimination procedure when the system does not have a solution.

EXAMPLE 2   A System of Equations That Has No Solution   Solve the system of linear equations given by

$$\begin{aligned} x + y + z &= 1 \\ 3x - y - z &= 4 \\ x + 5y + 5z &= -1 \end{aligned} \qquad (10)$$

Solution   Using the Gauss–Jordan elimination method, we obtain the following sequence of equivalent augmented matrices:

$$\begin{bmatrix} ① & 1 & 1 & | & 1 \\ 3 & -1 & -1 & | & 4 \\ 1 & 5 & 5 & | & -1 \end{bmatrix} \xrightarrow[R_3 - R_1]{R_2 - 3R_1} \begin{bmatrix} 1 & 1 & 1 & | & 1 \\ 0 & -4 & -4 & | & 1 \\ 0 & 4 & 4 & | & -2 \end{bmatrix}$$

$$\xrightarrow{R_3 + R_2} \begin{bmatrix} 1 & 1 & 1 & | & 1 \\ 0 & -4 & -4 & | & 1 \\ 0 & 0 & 0 & | & -1 \end{bmatrix}$$

Observe that row 3 in the last matrix reads $0x + 0y + 0z = -1$—that is, $0 = -1$! We conclude therefore that System (10) is inconsistent and has no solution. Geometrically, we have a situation in which two of the planes intersect in a straight line but the third plane is parallel to this line of intersection of the two planes and does not intersect it. Consequently, there is no point of intersection of the three planes.   ∎

Example 2 illustrates the following more general result of using the Gauss–Jordan elimination procedure.

Systems with No Solution
If there is a row in the augmented matrix containing all zeros to the left of the vertical line and a nonzero entry to the right of the line, then the system of equations has no solution.

It may have dawned on you that in all the previous examples we have dealt only with systems involving exactly the same number of linear equations as there are variables. However, systems in which the number of equations is different from the number of variables also occur in practice. Indeed, we will consider such systems in Examples 3 and 4.

The following theorem provides us with some preliminary information on a system of linear equations.

THEOREM 1
a. If the number of equations is greater than or equal to the number of variables in a linear system, then one of the following is true:
    i. The system has no solution.
    ii. The system has exactly one solution.
    iii. The system has infinitely many solutions.
b. If there are fewer equations than variables in a linear system, then the system either has no solution or it has infinitely many solutions.

Note   Theorem 1 may be used to tell us, before we even begin to solve a problem, what the nature of the solution may be.   ∎

Although we will not prove this theorem, you should recall that we have illustrated geometrically part (a) for the case in which there are exactly as many equations (three) as there are variables. To show the validity of part (b), let us once again consider the case in which a system has three variables. Now, if there is only one equation in the system, then it is clear that there are infinitely many solutions corresponding geometrically to all the points lying on the plane represented by the equation.

Next, if there are two equations in the system, then *only* the following possibilities exist:

1. The two planes are parallel and distinct.
2. The two planes intersect in a straight line.
3. The two planes are coincident (the two equations define the same plane) (Figure 6).

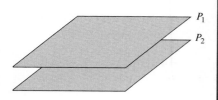

P_1
P_2

(a) No solution

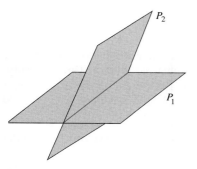

P_2

P_1

(b) Infinitely many solutions

P_1, P_2

(c) Infinitely many solutions
FIGURE 6

EXPLORE & DISCUSS

Give a geometric interpretation of Theorem 1 for a linear system composed of equations involving two variables. Specifically, illustrate what can happen if there are three linear equations in the system (the case involving two linear equations has already been discussed in Section 2.1). What if there are four linear equations? What if there is only one linear equation in the system?

Thus, either there is no solution or there are infinitely many solutions corresponding to the points lying on a line of intersection of the two planes or on a single plane determined by the two equations. In the case where two planes intersect in a straight line, the solutions will involve one parameter, and in the case where the two planes are coincident, the solutions will involve two parameters.

EXAMPLE 3   A System with More Equations Than Variables   Solve the following system of linear equations:

$$x + 2y = 4$$
$$x - 2y = 0$$
$$4x + 3y = 12$$

Solution   We obtain the following sequence of equivalent augmented matrices:

$$\begin{bmatrix} \textcircled{1} & 2 & 4 \\ 1 & -2 & 0 \\ 4 & 3 & 12 \end{bmatrix} \xrightarrow[R_3 - 4R_1]{R_2 - R_1} \begin{bmatrix} 1 & 2 & 4 \\ 0 & \textcircled{-4} & -4 \\ 0 & -5 & -4 \end{bmatrix} \xrightarrow{-\frac{1}{4}R_2}$$

$$\begin{bmatrix} 1 & 2 & 4 \\ 0 & 1 & 1 \\ 0 & -5 & -4 \end{bmatrix} \xrightarrow[R_3 + 5R_2]{R_1 - 2R_2} \begin{bmatrix} 1 & 0 & 2 \\ 0 & 1 & 1 \\ 0 & 0 & 1 \end{bmatrix}$$

The last row of the row-reduced augmented matrix implies that $0 = 1$, which is impossible, so we conclude that the given system has no solution. Geometrically, the three lines defined by the three equations in the system do not intersect at a point. (To see this for yourself, draw the graphs of these equations.) ∎

EXAMPLE 4   A System with More Variables Than Equations   Solve the following system of linear equations:

$$x + 2y - 3z + w = -2$$
$$3x - y - 2z - 4w = 1$$
$$2x + 3y - 5z + w = -3$$

Solution   First, observe that the given system consists of three equations in four variables, and so, by Theorem 1b, either the system has no solution or it has infinitely many solutions. To solve it we use the Gauss–Jordan method and obtain the following sequence of equivalent augmented matrices:

$$
\begin{bmatrix}
① & 2 & -3 & 1 & | & -2 \\
3 & -1 & -2 & -4 & | & 1 \\
2 & 3 & -5 & 1 & | & -3
\end{bmatrix}
\xrightarrow[R_3 - 2R_1]{R_2 - 3R_1}
\begin{bmatrix}
1 & 2 & -3 & 1 & | & -2 \\
0 & ⑦ & 7 & -7 & | & 7 \\
0 & -1 & 1 & -1 & | & 1
\end{bmatrix}
\xrightarrow{-\frac{1}{7}R_2}
$$

$$
\begin{bmatrix}
1 & 2 & -3 & 1 & | & -2 \\
0 & 1 & -1 & 1 & | & -1 \\
0 & -1 & 1 & -1 & | & 1
\end{bmatrix}
\xrightarrow[R_3 + R_2]{R_1 - 2R_2}
\begin{bmatrix}
1 & 0 & -1 & -1 & | & 0 \\
0 & 1 & -1 & 1 & | & -1 \\
0 & 0 & 0 & 0 & | & 0
\end{bmatrix}
$$

The last augmented matrix is in row-reduced form. Observe that the given system is equivalent to the system

$$
\begin{aligned}
x - z - w &= 0 \\
y - z + w &= -1
\end{aligned}
$$

of two equations in four variables. Thus, we may solve for two of the variables in terms of the other two. Letting $z = s$ and $w = t$ (s, t, parameters), we find that

$$
\begin{aligned}
x &= s + t \\
y &= s - t - 1 \\
z &= s \\
w &= t
\end{aligned}
$$

The solutions may be written in the form $(s + t, s - t - 1, s, t)$, where s and t are any real numbers. Geometrically, the three equations in the system represent three hyperplanes in four-dimensional space (since there are four variables) and their "points" of intersection lie in a two-dimensional subspace of four-space (since there are two parameters). ∎

Note In Example 4, we assigned parameters to z and w rather than to x and y because x and y are readily solved in terms of z and w. ∎

The following example illustrates a situation in which a system of linear equations has infinitely many solutions.

APPLIED EXAMPLE 5 Traffic Control Figure 7 shows the flow of downtown traffic in a certain city during the rush hours on a typical weekday. The arrows indicate the direction of traffic flow on each one-way road, and the average number of vehicles entering and leaving each intersection per hour appears beside each road. Fifth and Sixth Avenues can each handle up to 2000 vehicles per hour without causing congestion, whereas the maximum capacity of each of the two streets is 1000 vehicles per hour. The flow of traffic is controlled by traffic lights installed at each of the four intersections.

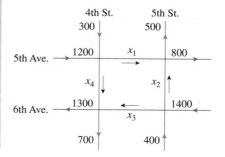

FIGURE 7

a. Write a general expression involving the rates of flow—x_1, x_2, x_3, x_4—and suggest two possible flow patterns that will ensure no traffic congestion.

b. Suppose the part of 4th Street between 5th Avenue and 6th Avenue is to be resurfaced and traffic flow between the two junctions has to be slowed to at most 300 vehicles per hour. Find two possible flow patterns that will result in a smooth flow of traffic.

Solution

a. To avoid congestion, all traffic entering an intersection must also leave that intersection. Applying this condition to each of the four intersections in a clockwise direction beginning with the 5th Avenue and 4th Street intersection, we obtain the following equations:

$$1500 = x_1 + x_4$$
$$1300 = x_1 + x_2$$
$$1800 = x_2 + x_3$$
$$2000 = x_3 + x_4$$

This system of four linear equations in the four variables x_1, x_2, x_3, x_4 may be rewritten in the more standard form

$$
\begin{aligned}
x_1 \quad\quad\quad\quad + x_4 &= 1500 \\
x_1 + x_2 \quad\quad\quad &= 1300 \\
x_2 + x_3 \quad &= 1800 \\
x_3 + x_4 &= 2000
\end{aligned}
$$

Using the Gauss–Jordan elimination method to solve the system, we obtain

$$
\begin{bmatrix}
1 & 0 & 0 & 1 & | & 1500 \\
1 & 1 & 0 & 0 & | & 1300 \\
0 & 1 & 1 & 0 & | & 1800 \\
0 & 0 & 1 & 1 & | & 2000
\end{bmatrix}
\xrightarrow{R_2 - R_1}
\begin{bmatrix}
1 & 0 & 0 & 1 & | & 1500 \\
0 & 1 & 0 & -1 & | & -200 \\
0 & 1 & 1 & 0 & | & 1800 \\
0 & 0 & 1 & 1 & | & 2000
\end{bmatrix}
$$

$$
\xrightarrow{R_3 - R_2}
\begin{bmatrix}
1 & 0 & 0 & 1 & | & 1500 \\
0 & 1 & 0 & -1 & | & -200 \\
0 & 0 & 1 & 1 & | & 2000 \\
0 & 0 & 1 & 1 & | & 2000
\end{bmatrix}
$$

$$
\xrightarrow{R_4 - R_3}
\begin{bmatrix}
1 & 0 & 0 & 1 & | & 1500 \\
0 & 1 & 0 & -1 & | & -200 \\
0 & 0 & 1 & 1 & | & 2000 \\
0 & 0 & 0 & 0 & | & 0
\end{bmatrix}
$$

The last augmented matrix is in row-reduced form and is equivalent to a system of three linear equations in the four variables x_1, x_2, x_3, x_4. Thus, we may express three of the variables—say, x_1, x_2, x_3—in terms of x_4. Setting $x_4 = t$ (t, a parameter), we may write the infinitely many solutions of the system as

$$
\begin{aligned}
x_1 &= 1500 - t \\
x_2 &= -200 + t \\
x_3 &= 2000 - t \\
x_4 &= t
\end{aligned}
$$

Observe that for a meaningful solution, $200 \le t \le 1000$, since x_1, x_2, x_3, and x_4 must all be nonnegative, and the maximum capacity of a street is 1000. For example, picking $t = 300$ gives the flow pattern

$$x_1 = 1200 \qquad x_2 = 100 \qquad x_3 = 1700 \qquad x_4 = 300$$

Selecting $t = 500$ gives the flow pattern

$$x_1 = 1000 \qquad x_2 = 300 \qquad x_3 = 1500 \qquad x_4 = 500$$

b. In this case, x_4 must not exceed 300. Again, using the results of part (a), we find, upon setting $x_4 = t = 300$, the flow pattern

$$x_1 = 1200 \qquad x_2 = 100 \qquad x_3 = 1700 \qquad x_4 = 300$$

obtained earlier. Picking $t = 250$ gives the flow pattern

$$x_1 = 1250 \qquad x_2 = 50 \qquad x_3 = 1750 \qquad x_4 = 250$$

2.3 Self-Check Exercises

1. The following augmented matrix in row-reduced form is equivalent to the augmented matrix of a certain system of linear equations. Use this result to solve the system of equations.

$$\begin{bmatrix} 1 & 0 & -1 & | & 3 \\ 0 & 1 & 5 & | & -2 \\ 0 & 0 & 0 & | & 0 \end{bmatrix}$$

2. Solve the system of linear equations

$$\begin{aligned} 2x - 3y + z &= 6 \\ x + 2y + 4z &= -4 \\ x - 5y - 3z &= 10 \end{aligned}$$

using the Gauss–Jordan elimination method.

3. Solve the system of linear equations

$$\begin{aligned} x - 2y + 3z &= 9 \\ 2x + 3y - z &= 4 \\ x + 5y - 4z &= 2 \end{aligned}$$

using the Gauss–Jordan elimination method.

Solutions to Self-Check Exercises 2.3 can be found on page 106.

2.3 Concept Questions

1. a. If a system of linear equations has the same number of equations or more equations than variables, what can you say about the nature of its solution(s)?

b. If a system of linear equations has fewer equations than variables, what can you say about the nature of its solution(s)?

2. A system consists of three linear equations in four variables. Can the system have a unique solution?

2.3   Exercises

In Exercises 1–12, given that the augmented matrix in row-reduced form is equivalent to the augmented matrix of a system of linear equations, (a) determine whether the system has a solution and (b) find the solution or solutions to the system, if they exist.

1. $\begin{bmatrix} 1 & 0 & 0 & | & 3 \\ 0 & 1 & 0 & | & -1 \\ 0 & 0 & 1 & | & 2 \end{bmatrix}$
2. $\begin{bmatrix} 1 & 0 & 0 & | & 3 \\ 0 & 1 & 0 & | & -2 \\ 0 & 0 & 1 & | & 1 \end{bmatrix}$

3. $\begin{bmatrix} 1 & 0 & | & 2 \\ 0 & 1 & | & 4 \\ 0 & 0 & | & 0 \end{bmatrix}$
4. $\begin{bmatrix} 1 & 0 & 0 & | & 3 \\ 0 & 1 & 0 & | & 1 \\ 0 & 0 & 0 & | & 0 \end{bmatrix}$

5. $\begin{bmatrix} 1 & 0 & 1 & | & 4 \\ 0 & 1 & 0 & | & -2 \end{bmatrix}$
6. $\begin{bmatrix} 1 & 0 & 0 & 0 & | & 3 \\ 0 & 1 & 1 & 0 & | & -1 \\ 0 & 0 & 0 & 1 & | & 2 \end{bmatrix}$

7. $\begin{bmatrix} 1 & 0 & 0 & 0 & | & 2 \\ 0 & 1 & 0 & 0 & | & 1 \\ 0 & 0 & 1 & 0 & | & 3 \\ 0 & 0 & 0 & 0 & | & 1 \end{bmatrix}$
8. $\begin{bmatrix} 1 & 0 & 0 & | & 4 \\ 0 & 1 & 0 & | & -1 \\ 0 & 0 & 1 & | & 3 \\ 0 & 0 & 0 & | & 1 \end{bmatrix}$

9. $\begin{bmatrix} 1 & 0 & 0 & 0 & | & 2 \\ 0 & 1 & 0 & 0 & | & -1 \\ 0 & 0 & 1 & 1 & | & 2 \\ 0 & 0 & 0 & 0 & | & 0 \end{bmatrix}$
10. $\begin{bmatrix} 0 & 1 & 0 & 1 & | & 3 \\ 0 & 0 & 1 & -2 & | & 4 \\ 0 & 0 & 0 & 0 & | & 0 \\ 0 & 0 & 0 & 0 & | & 0 \end{bmatrix}$

11. $\begin{bmatrix} 1 & 0 & 3 & 0 & | & 2 \\ 0 & 1 & -1 & 0 & | & 1 \\ 0 & 0 & 0 & 0 & | & 0 \\ 0 & 0 & 0 & 0 & | & 0 \end{bmatrix}$
12. $\begin{bmatrix} 1 & 0 & 3 & -1 & | & 4 \\ 0 & 1 & -2 & 3 & | & 2 \\ 0 & 0 & 0 & 0 & | & 0 \\ 0 & 0 & 0 & 0 & | & 0 \end{bmatrix}$

In Exercises 13–32, solve the system of linear equations, using the Gauss–Jordan elimination method.

13. $2x - y = 3$
    $x + 2y = 4$
    $2x + 3y = 7$

14. $x + 2y = 3$
    $2x - 3y = -8$
    $x - 4y = -9$

15. $3x - 2y = -3$
    $2x + y = 3$
    $x - 2y = -5$

16. $2x + 3y = 2$
    $x + 3y = -2$
    $x - y = 3$

17. $3x - 2y = 5$
    $-x + 3y = -4$
    $2x - 4y = 6$

18. $4x + 6y = 8$
    $3x - 2y = -7$
    $x + 3y = 5$

19. $x - 2y = 2$
    $7x - 14y = 14$
    $3x - 6y = 6$

20. $x + 2y + z = -2$
    $-2x - 3y - z = 1$
    $2x + 4y + 2z = -4$

21. $3x + 2y = 4$
    $-\frac{3}{2}x - y = -2$
    $6x + 4y = 8$

22. $3y + 2z = 4$
    $2x - y - 3z = 3$
    $2x + 2y - z = 7$

23. $2x_1 - x_2 + x_3 = -4$
    $3x_1 - \frac{3}{2}x_2 + \frac{3}{2}x_3 = -6$
    $-6x_1 + 3x_2 - 3x_3 = 12$

24. $x + y - 2z = -3$
    $2x - y + 3z = 7$
    $x - 2y + 5z = 0$

25. $x - 2y + 3z = 4$
    $2x + 3y - z = 2$
    $x + 2y - 3z = -6$

26. $x_1 - 2x_2 + x_3 = -3$
    $2x_1 + x_2 - 2x_3 = 2$
    $x_1 + 3x_2 - 3x_3 = 5$

27. $4x + y - z = 4$
    $8x + 2y - 2z = 8$

28. $x_1 + 2x_2 + 4x_3 = 2$
    $x_1 + x_2 + 2x_3 = 1$

29. $2x + y - 3z = 1$
    $x - y + 2z = 1$
    $5x - 2y + 3z = 6$

30. $3x - 9y + 6z = -12$
    $x - 3y + 2z = -4$
    $2x - 6y + 4z = 8$

31. $x + 2y - z = -4$
    $2x + y + z = 7$
    $x + 3y + 2z = 7$
    $x - 3y + z = 9$

32. $3x - 2y + z = 4$
    $x + 3y - 4z = -3$
    $2x - 3y + 5z = 7$
    $x - 8y + 9z = 10$

33. **MANAGEMENT DECISIONS**   The management of Hartman Rent-A-Car has allocated $1,008,000 to purchase 60 new automobiles to add to their existing fleet of rental cars. The company will choose from compact, mid-sized, and full-sized cars costing $12,000, $19,200, and $26,400 each, respectively. Find formulas giving the options available to the company. Give two specific options. (*Note:* Your answers will *not* be unique.)

34. **NUTRITION**   A dietitian wishes to plan a meal around three foods. The meal is to include 8800 units of vitamin A, 3380 units of vitamin C, and 1020 units of calcium. The number of units of the vitamins and calcium in each ounce of the foods is summarized in the accompanying table:

	Food I	Food II	Food III
Vitamin A	400	1200	800
Vitamin C	110	570	340
Calcium	90	30	60

Determine the amount of each food the dietitian should include in the meal in order to meet the vitamin and calcium requirements.

35. **NUTRITION**   Refer to Exercise 34. In planning for another meal, the dietitian changes the requirement of vitamin C to 2160 units instead of 3380 units. All other requirements remain the same. Show that such a meal cannot be planned around the same foods.

36. INVESTMENTS Mr. and Mrs. Garcia have a total of $100,000 to be invested in stocks, bonds, and a money market account. The stocks have a rate of return of 12%/year, while the bonds and the money market account pay 8% and 4%/year, respectively. They have stipulated that the amount invested in stocks should be equal to the sum of the amount invested in bonds and 3 times the amount invested in the money market account. How should the Garcias allocate their resources if they require an annual income of $10,000 from their investments?

37. TRAFFIC CONTROL The accompanying figure shows the flow of traffic near a city's Civic Center during the rush hours on a typical weekday. Each road can handle a maximum of 1000 cars/hour without causing congestion. The flow of traffic is controlled by traffic lights at each of the five intersections.

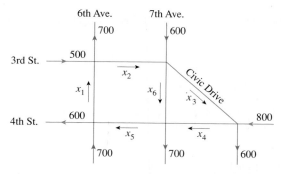

a. Set up a system of linear equations describing the traffic flow.

b. Solve the system devised in part (a) and suggest two possible traffic-flow patterns that will ensure no traffic congestion.

c. Suppose 7th Avenue between 3rd and 4th Streets is soon to be closed for road repairs. Find one possible flow pattern that will result in a smooth flow of traffic.

38. TRAFFIC CONTROL The accompanying figure shows the flow of downtown traffic during the rush hours on a typical weekday. Each avenue can handle up to 1500 vehicles/hour without causing congestion, whereas the maximum capacity of each street is 1000 vehicles/hour. The flow of traffic is controlled by traffic lights at each of the six intersections.

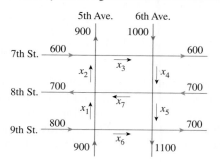

a. Set up a system of linear equations describing the traffic flow.

b. Solve the system devised in part (a) and suggest two possible traffic-flow patterns that will ensure no traffic congestion.

c. Suppose the traffic flow along 9th Street between 5th and 6th Avenues, x_6, is restricted due to sewer construction. What is the minimum permissible traffic flow along this road that will not result in traffic congestion?

39. Determine the value of k so that the following system of linear equations has a solution and then find the solution:

$$2x + 3y = 2$$
$$x + 4y = 6$$
$$5x + ky = 2$$

40. Determine the value of k so that the following system of linear equations has infinitely many solutions and then find the solutions:

$$3x - 2y + 4z = 12$$
$$-9x + 6y - 12z = k$$

In Exercises 41 and 42, determine whether the statement is true or false. If it is true, explain why it is true. If it is false, give an example to show why it is false.

41. A system of linear equations having fewer equations than variables has no solution, a unique solution, or infinitely many solutions.

42. A system of linear equations having more equations than variables has no solution, a unique solution, or infinitely many solutions.

2.3 Solutions to Self-Check Exercises

1. Let x, y, and z denote the variables. Then, the given row-reduced augmented matrix tells us that the system of linear equations is equivalent to the two equations

$$
\begin{aligned}
x \quad - z &= 3 \\
y + 5z &= -2
\end{aligned}
$$

Letting $z = t$, where t is a parameter, we find the infinitely many solutions given by

$$
\begin{aligned}
x &= t + 3 \\
y &= -5t - 2 \\
z &= t
\end{aligned}
$$

2. We obtain the following sequence of equivalent augmented matrices:

$$
\begin{bmatrix}
2 & -3 & 1 & | & 6 \\
1 & 2 & 4 & | & -4 \\
1 & -5 & -3 & | & 10
\end{bmatrix}
\xrightarrow{R_1 \leftrightarrow R_2}
$$

$$
\begin{bmatrix}
1 & 2 & 4 & | & -4 \\
2 & -3 & 1 & | & 6 \\
1 & -5 & -3 & | & 10
\end{bmatrix}
\xrightarrow[R_3 - R_1]{R_2 - 2R_1}
$$

$$
\begin{bmatrix}
1 & 2 & 4 & | & -4 \\
0 & -7 & -7 & | & 14 \\
0 & -7 & -7 & | & 14
\end{bmatrix}
\xrightarrow{-\frac{1}{7}R_2}
$$

$$
\begin{bmatrix}
1 & 2 & 4 & | & -4 \\
0 & 1 & 1 & | & -2 \\
0 & -7 & -7 & | & 14
\end{bmatrix}
\xrightarrow[R_3 + 7R_2]{R_1 - 2R_2}
\begin{bmatrix}
1 & 0 & 2 & | & 0 \\
0 & 1 & 1 & | & -2 \\
0 & 0 & 0 & | & 0
\end{bmatrix}
$$

The last augmented matrix, which is in row-reduced form, tells us that the given system of linear equations is equivalent to the following system of two equations:

$$
\begin{aligned}
x \quad + 2z &= 0 \\
y + z &= -2
\end{aligned}
$$

Letting $z = t$, where t is a parameter, we see that the infinitely many solutions are given by

$$
\begin{aligned}
x &= -2t \\
y &= -t - 2 \\
z &= t
\end{aligned}
$$

3. We obtain the following sequence of equivalent augmented matrices:

$$
\begin{bmatrix}
① & -2 & 3 & | & 9 \\
2 & 3 & -1 & | & 4 \\
1 & 5 & -4 & | & 2
\end{bmatrix}
\xrightarrow[R_3 - R_1]{R_2 - 2R_1}
$$

$$
\begin{bmatrix}
1 & -2 & 3 & | & 9 \\
0 & 7 & -7 & | & -14 \\
0 & 7 & -7 & | & -7
\end{bmatrix}
\xrightarrow{R_3 - R_2}
\begin{bmatrix}
1 & -2 & 3 & | & 9 \\
0 & 7 & -7 & | & -14 \\
0 & 0 & 0 & | & 7
\end{bmatrix}
$$

Since the last row of the final augmented matrix is equivalent to the equation $0 = 7$, a contradiction, we conclude that the given system has no solution.

USING TECHNOLOGY

■ **Systems of Linear Equations: Underdetermined and Overdetermined Systems**

We can use the row operations of a graphing utility to solve a system of m linear equations in n unknowns by the Gauss–Jordan method, as we did in the previous technology section. We can also use the **rref** or equivalent operation to obtain the row-reduced form without going through all the steps of the Gauss–Jordan method. The **SIMULT** function, however, cannot be used to solve a system where the number of equations and the number of variables are not the same.

EXAMPLE 1 Solve the system

$$\begin{aligned}
x_1 - 2x_2 + 4x_3 &= 2 \\
2x_1 + x_2 - 2x_3 &= -1 \\
3x_1 - x_2 + 2x_3 &= 1 \\
2x_1 + 6x_2 - 12x_3 &= -6
\end{aligned}$$

Solution First, we enter the augmented matrix A into the calculator as

$$A = \begin{bmatrix} 1 & -2 & 4 & 2 \\ 2 & 1 & -2 & -1 \\ 3 & -1 & 2 & 1 \\ 2 & 6 & -12 & -6 \end{bmatrix}$$

Then using the **rref** or equivalent operation, we obtain the equivalent matrix

$$\begin{bmatrix} 1 & 0 & 0 & 0 \\ 0 & 1 & -2 & -1 \\ 0 & 0 & 0 & 0 \\ 0 & 0 & 0 & 0 \end{bmatrix}$$

in reduced form. Thus, the given system is equivalent to

$$\begin{aligned}
x_1 &= 0 \\
x_2 - 2x_3 &= -1
\end{aligned}$$

Letting $x_3 = t$, where t is a parameter, we see that the solutions are $(0, 2t - 1, t)$. ∎

TECHNOLOGY EXERCISES

Use a graphing utility to solve the system of equations using the rref or equivalent operation.

1. $\begin{aligned}
2x_1 - x_2 - x_3 &= 0 \\
3x_1 - 2x_2 - x_3 &= -1 \\
-x_1 + 2x_2 - x_3 &= 3 \\
2x_2 - 2x_3 &= 4
\end{aligned}$

2. $\begin{aligned}
3x_1 + x_2 - 4x_3 &= 5 \\
2x_1 - 3x_2 + 2x_3 &= -4 \\
-x_1 - 2x_2 + 4x_3 &= 6 \\
4x_1 + 3x_2 - 5x_3 &= 9
\end{aligned}$

3. $\begin{aligned}
2x_1 + 3x_2 + 2x_3 + x_4 &= -1 \\
x_1 - x_2 + x_3 - 2x_4 &= -8 \\
5x_1 + 6x_2 - 2x_3 + 2x_4 &= 11 \\
x_1 + 3x_2 + 8x_3 + x_4 &= -14
\end{aligned}$

4. $\begin{aligned}
x_1 - x_2 + 3x_3 - 6x_4 &= 2 \\
x_1 + x_2 + x_3 - 2x_4 &= 2 \\
-2x_1 - x_2 + x_3 + 2x_4 &= 0
\end{aligned}$

5. $\begin{aligned}
x_1 + x_2 - x_3 - x_4 &= -1 \\
x_1 - x_2 + x_3 + 4x_4 &= -6 \\
3x_1 + x_2 - x_3 + 2x_4 &= -4 \\
5x_1 + x_2 - 3x_3 + x_4 &= -9
\end{aligned}$

6. $\begin{aligned}
1.2x_1 - 2.3x_2 + 4.2x_3 + 5.4x_4 - 1.6x_5 &= 4.2 \\
2.3x_1 + 1.4x_2 - 3.1x_3 + 3.3x_4 - 2.4x_5 &= 6.3 \\
1.7x_1 + 2.6x_2 - 4.3x_3 + 7.2x_4 - 1.8x_5 &= 7.8 \\
2.6x_1 - 4.2x_2 + 8.3x_3 - 1.6x_4 + 2.5x_5 &= 6.4
\end{aligned}$

2.4   Matrices

■ Using Matrices to Represent Data

Many practical problems are solved by using arithmetic operations on the data associated with the problems. By properly organizing the data into *blocks* of numbers, we can then carry out these arithmetic operations in an orderly and efficient manner. In particular, this systematic approach enables us to use the computer to full advantage.

Let's begin by considering how the monthly output data of a manufacturer may be organized. The Acrosonic Company manufactures four different loudspeaker systems in three separate locations. The company's May output is summarized in Table 1.

TABLE 1

	Model A	Model B	Model C	Model D
Location I	320	280	460	280
Location II	480	360	580	0
Location III	540	420	200	880

Now, if we agree to preserve the relative location of each entry in Table 1, we can summarize the set of data further, as follows:

$$\begin{bmatrix} 320 & 280 & 460 & 280 \\ 480 & 360 & 580 & 0 \\ 540 & 420 & 200 & 880 \end{bmatrix}$$

A matrix summarizing the data in Table 1

The array of numbers displayed here is an example of a matrix. Observe that the numbers in row 1 give the output of models A, B, C, and D of Acrosonic loudspeaker systems manufactured in location I; similarly, the numbers in rows 2 and 3 give the respective outputs of these loudspeaker systems in locations II and III. The numbers in each column of the matrix give the outputs of a particular model of loudspeaker system manufactured in each of the company's three manufacturing locations.

More generally, a matrix is an ordered rectangular array of real numbers. For example, each of the following arrays is a matrix:

$$A = \begin{bmatrix} 3 & 0 & -1 \\ 2 & 1 & 4 \end{bmatrix} \qquad B = \begin{bmatrix} 3 & 2 \\ 0 & 1 \\ -1 & 4 \end{bmatrix} \qquad C = \begin{bmatrix} 1 \\ 2 \\ 4 \\ 0 \end{bmatrix} \qquad D = \begin{bmatrix} 1 & 3 & 0 & 1 \end{bmatrix}$$

The real numbers that make up the array are called the **entries,** or *elements,* of the matrix. The entries in a row in the array are referred to as a **row** of the matrix, whereas the entries in a column in the array are referred to as a **column** of the matrix. Matrix *A*, for example, has two rows and three columns, which may be identified as follows:

$$
\begin{array}{ccc}
\text{Column 1} & \text{Column 2} & \text{Column 3}
\end{array}
$$

$$
\begin{array}{l}
\text{Row 1} \\
\text{Row 2}
\end{array}
\begin{bmatrix}
3 & 0 & -1 \\
2 & 1 & 4
\end{bmatrix}
$$

A 2 × 3 matrix

The **size,** or *dimension,* **of a matrix** is described in terms of the number of rows and columns of the matrix. For example, matrix *A* has two rows and three columns and is said to have size 2 by 3, denoted 2 × 3. In general, a matrix having *m* rows and *n* columns is said to have size *m* × *n*.

> ## Matrix
> A **matrix** is an ordered rectangular array of numbers. A matrix with *m* rows and *n* columns has size *m* × *n*. The entry in the *i*th row and *j*th column is denoted by a_{ij}.

A matrix of size 1 × *n*—a matrix having one row and *n* columns—is referred to as a **row matrix,** or *row vector,* of dimension *n*. For example, the matrix *D* is a row vector of dimension 4. Similarly, a matrix having *m* rows and one column is referred to as a **column matrix,** or *column vector,* of dimension *m*. The matrix *C* is a column vector of dimension 4. Finally, an *n* × *n* matrix—that is, a matrix having the same number of rows as columns—is called a **square matrix.** For example, the matrix

$$
\begin{bmatrix}
-3 & 8 & 6 \\
2 & \frac{1}{4} & 4 \\
1 & 3 & 2
\end{bmatrix}
$$

A 3 × 3 square matrix

is a square matrix of size 3 × 3, or simply of size 3.

APPLIED EXAMPLE 1   Organizing Production Data   Consider the matrix

$$
P = \begin{bmatrix}
320 & 280 & 460 & 280 \\
480 & 360 & 580 & 0 \\
540 & 420 & 200 & 880
\end{bmatrix}
$$

representing the output of loudspeaker systems of the Acrosonic Company discussed earlier (see Table 1).

a. What is the size of the matrix *P*?
b. Find a_{24} (the entry in row 2 and column 4 of the matrix *P*) and give an interpretation of this number.
c. Find the sum of the entries that make up row 1 of *P* and interpret the result.
d. Find the sum of the entries that make up column 4 of *P* and interpret the result.

Solution

a. The matrix *P* has three rows and four columns and hence has size 3 × 4.
b. The required entry lies in row 2 and column 4 and is the number 0. This means that no model D loudspeaker system was manufactured in location II in May.

c. The required sum is given by

$$320 + 280 + 460 + 280 = 1340$$

which gives the total number of loudspeaker systems manufactured in location I in May as 1340 units.

d. The required sum is given by

$$280 + 0 + 880 = 1160$$

giving the output of model D loudspeaker systems in all locations of the company in May as 1160 units.

Equality of Matrices

Two matrices are said to be *equal* if they have the same size and their corresponding entries are equal. For example,

$$\begin{bmatrix} 2 & 3 & 1 \\ 4 & 6 & 2 \end{bmatrix} = \begin{bmatrix} (3-1) & 3 & 1 \\ 4 & (4+2) & 2 \end{bmatrix}$$

Also,

$$\begin{bmatrix} 1 & 3 & 5 \\ 2 & 4 & 3 \end{bmatrix} \neq \begin{bmatrix} 1 & 2 \\ 3 & 4 \\ 5 & 3 \end{bmatrix}$$

since the matrix on the left has size 2×3, whereas the matrix on the right has size 3×2, and

$$\begin{bmatrix} 2 & 3 \\ 4 & 6 \end{bmatrix} \neq \begin{bmatrix} 2 & 3 \\ 4 & 7 \end{bmatrix}$$

since the corresponding elements in row 2 and column 2 of the two matrices are not equal.

> **Equality of Matrices**
> Two matrices are equal if they have the same size and their corresponding entries are equal.

EXAMPLE 2   Solve the following matrix equation for x, y, and z:

$$\begin{bmatrix} 1 & x & 3 \\ 2 & y-1 & 2 \end{bmatrix} = \begin{bmatrix} 1 & 4 & z \\ 2 & 1 & 2 \end{bmatrix}$$

Solution   Since the corresponding elements of the two matrices must be equal, we find that $x = 4$, $z = 3$, and $y - 1 = 1$, or $y = 2$.

Addition and Subtraction

Two matrices A and B of the *same size* can be added or subtracted to produce a matrix of the same size. This is done by adding or subtracting the corresponding entries in the two matrices. For example,

$$\begin{bmatrix} 1 & 3 & 4 \\ -1 & 2 & 0 \end{bmatrix} + \begin{bmatrix} 1 & 4 & 3 \\ 6 & 1 & -2 \end{bmatrix} = \begin{bmatrix} 1+1 & 3+4 & 4+3 \\ -1+6 & 2+1 & 0+(-2) \end{bmatrix} = \begin{bmatrix} 2 & 7 & 7 \\ 5 & 3 & -2 \end{bmatrix}$$

Adding two matrices of the same size

and

$$\begin{bmatrix} 1 & 2 \\ -1 & 3 \\ 4 & 0 \end{bmatrix} - \begin{bmatrix} 2 & -1 \\ 3 & 2 \\ -1 & 0 \end{bmatrix} = \begin{bmatrix} 1-2 & 2-(-1) \\ -1-3 & 3-2 \\ 4-(-1) & 0-0 \end{bmatrix} = \begin{bmatrix} -1 & 3 \\ -4 & 1 \\ 5 & 0 \end{bmatrix}$$

Subtracting two matrices of the same size

Addition and Subtraction of Matrices

If A and B are two matrices of the same size,

1. The *sum* $A + B$ is the matrix obtained by adding the corresponding entries in the two matrices.
2. The *difference* $A - B$ is the matrix obtained by subtracting the corresponding entries in B from A.

 APPLIED EXAMPLE 3   Organizing Production Data   The total output of Acrosonic for June is shown in Table 2.

TABLE 2

	Model A	Model B	Model C	Model D
Location I	210	180	330	180
Location II	400	300	450	40
Location III	420	280	180	740

The output for May was given earlier in Table 1. Find the total output of the company for May and June.

Solution   As we saw earlier, the production matrix for Acrosonic in May is given by

$$A = \begin{bmatrix} 320 & 280 & 460 & 280 \\ 480 & 360 & 580 & 0 \\ 540 & 420 & 200 & 880 \end{bmatrix}$$

Next, from Table 2, we see that the production matrix for June is given by

$$B = \begin{bmatrix} 210 & 180 & 330 & 180 \\ 400 & 300 & 450 & 40 \\ 420 & 280 & 180 & 740 \end{bmatrix}$$

Finally, the total output of Acrosonic for May and June is given by the matrix

$$A + B = \begin{bmatrix} 320 & 280 & 460 & 280 \\ 480 & 360 & 580 & 0 \\ 540 & 420 & 200 & 880 \end{bmatrix} + \begin{bmatrix} 210 & 180 & 330 & 180 \\ 400 & 300 & 450 & 40 \\ 420 & 280 & 180 & 740 \end{bmatrix}$$

$$= \begin{bmatrix} 530 & 460 & 790 & 460 \\ 880 & 660 & 1030 & 40 \\ 960 & 700 & 380 & 1620 \end{bmatrix}$$

The following laws hold for matrix addition.

Laws for Matrix Addition

If A, B, and C are matrices of the same size, then

1. $A + B = B + A$ Commutative law
2. $(A + B) + C = A + (B + C)$ Associative law

The *commutative law* for matrix addition states that the order in which matrix addition is performed is immaterial. The *associative law* states that when adding three matrices together, we may first add A and B and then add the resulting sum to C. Equivalently, we can add A to the sum of B and C.

A *zero matrix* is one in which all entries are zero. The zero matrix O has the property that

$$A + O = O + A = A$$

for any matrix A having the same size as that of O. For example, the zero matrix of size 3×2 is

$$O = \begin{bmatrix} 0 & 0 \\ 0 & 0 \\ 0 & 0 \end{bmatrix}$$

If A is any 3×2 matrix, then

$$A + O = \begin{bmatrix} a_{11} & a_{12} \\ a_{21} & a_{22} \\ a_{31} & a_{32} \end{bmatrix} + \begin{bmatrix} 0 & 0 \\ 0 & 0 \\ 0 & 0 \end{bmatrix} = \begin{bmatrix} a_{11} & a_{12} \\ a_{21} & a_{22} \\ a_{31} & a_{32} \end{bmatrix} = A$$

where a_{ij} denotes the entry in the ith row and jth column of the matrix A.

The matrix obtained by interchanging the rows and columns of a given matrix A is called the transpose of A and is denoted A^T. For example, if

$$A = \begin{bmatrix} 1 & 2 & 3 \\ 4 & 5 & 6 \\ 7 & 8 & 9 \end{bmatrix}$$

then

$$A^T = \begin{bmatrix} 1 & 4 & 7 \\ 2 & 5 & 8 \\ 3 & 6 & 9 \end{bmatrix}$$

Transpose of a Matrix

If A is an $m \times n$ matrix with elements a_{ij}, then the **transpose** of A is the $n \times m$ matrix A^T with elements a_{ji}.

Scalar Multiplication

A matrix A may be multiplied by a real number, called a **scalar** in the context of matrix algebra. The scalar product, denoted by cA, is a matrix obtained by multiplying each entry of A by c. For example, the scalar product of the matrix

$$A = \begin{bmatrix} 3 & -1 & 2 \\ 0 & 1 & 4 \end{bmatrix}$$

and the scalar 3 is the matrix

$$3A = 3 \begin{bmatrix} 3 & -1 & 2 \\ 0 & 1 & 4 \end{bmatrix} = \begin{bmatrix} 9 & -3 & 6 \\ 0 & 3 & 12 \end{bmatrix}$$

Scalar Product

If A is a matrix and c is a real number, then the **scalar product** cA is the matrix obtained by multiplying each entry of A by c.

EXAMPLE 4 Given

$$A = \begin{bmatrix} 3 & 4 \\ -1 & 2 \end{bmatrix} \quad \text{and} \quad B = \begin{bmatrix} 3 & 2 \\ -1 & 2 \end{bmatrix}$$

find the matrix X satisfying the *matrix equation* $2X + B = 3A$.

Solution From the given equation $2X + B = 3A$, we find that

$$2X = 3A - B$$

$$= 3 \begin{bmatrix} 3 & 4 \\ -1 & 2 \end{bmatrix} - \begin{bmatrix} 3 & 2 \\ -1 & 2 \end{bmatrix}$$

$$= \begin{bmatrix} 9 & 12 \\ -3 & 6 \end{bmatrix} - \begin{bmatrix} 3 & 2 \\ -1 & 2 \end{bmatrix} = \begin{bmatrix} 6 & 10 \\ -2 & 4 \end{bmatrix}$$

$$X = \frac{1}{2} \begin{bmatrix} 6 & 10 \\ -2 & 4 \end{bmatrix} = \begin{bmatrix} 3 & 5 \\ -1 & 2 \end{bmatrix}$$

APPLIED EXAMPLE 5 Production Planning The management of Acrosonic has decided to increase its July production of loudspeaker systems by 10% (over its June output). Find a matrix giving the targeted production for July.

Solution From the results of Example 3, we see that Acrosonic's total output for June may be represented by the matrix

$$B = \begin{bmatrix} 210 & 180 & 330 & 180 \\ 400 & 300 & 450 & 40 \\ 420 & 280 & 180 & 740 \end{bmatrix}$$

The required matrix is given by

$$(1.1)B = 1.1 \begin{bmatrix} 210 & 180 & 330 & 180 \\ 400 & 300 & 450 & 40 \\ 420 & 280 & 180 & 740 \end{bmatrix}$$

$$= \begin{bmatrix} 231 & 198 & 363 & 198 \\ 440 & 330 & 495 & 44 \\ 462 & 308 & 198 & 814 \end{bmatrix}$$

and is interpreted in the usual manner.

2.4   Self-Check Exercises

1. Perform the indicated operations:

$$\begin{bmatrix} 1 & 3 & 2 \\ -1 & 4 & 7 \end{bmatrix} - 3 \begin{bmatrix} 2 & 1 & 0 \\ 1 & 3 & 4 \end{bmatrix}$$

2. Solve the following matrix equation for x, y, and z:

$$\begin{bmatrix} x & 3 \\ z & 2 \end{bmatrix} + \begin{bmatrix} 2 - y & z \\ 2 - z & -x \end{bmatrix} = \begin{bmatrix} 3 & 7 \\ 2 & 0 \end{bmatrix}$$

3. Jack owns two gas stations, one downtown and the other in the Wilshire district. Over 2 consecutive days his gas stations recorded gasoline sales represented by the following matrices:

$$A = \begin{array}{r} \\ \text{Downtown} \\ \text{Wilshire} \end{array} \begin{array}{ccc} \text{Regular} & \begin{array}{c}\text{Regular}\\\text{plus}\end{array} & \text{Premium} \\ \begin{bmatrix} 1200 & 750 & 650 \\ 1100 & 850 & 600 \end{bmatrix} \end{array}$$

and

$$B = \begin{array}{r} \\ \text{Downtown} \\ \text{Wilshire} \end{array} \begin{array}{ccc} \text{Regular} & \begin{array}{c}\text{Regular}\\\text{plus}\end{array} & \text{Premium} \\ \begin{bmatrix} 1250 & 825 & 550 \\ 1150 & 750 & 750 \end{bmatrix} \end{array}$$

Find a matrix representing the total sales of the two gas stations over the 2-day period.

Solutions to Self-Check Exercises 2.4 can be found on page 117.

2.4   Concept Questions

1. Define (a) a *matrix*, (b) the *size of a matrix*, (c) a *row matrix*, (d) a *column matrix*, and (e) a *square matrix*.

2. When are two matrices equal? Give an example of two matrices that are equal.

3. Construct a 3 × 3 matrix A having the property that $A = A^T$. What special characteristic does A have?

2.4 Exercises

In Exercises 1–6, refer to the following matrices:

$$A = \begin{bmatrix} 2 & -3 & 9 & -4 \\ -11 & 2 & 6 & 7 \\ 6 & 0 & 2 & 9 \\ 5 & 1 & 5 & -8 \end{bmatrix}$$

$$B = \begin{bmatrix} 3 & -1 & 2 \\ 0 & 1 & 4 \\ 3 & 2 & 1 \\ -1 & 0 & 8 \end{bmatrix}$$

$$C = \begin{bmatrix} 1 & 0 & 3 & 4 & 5 \end{bmatrix}$$

$$D = \begin{bmatrix} 1 \\ 3 \\ -2 \\ 0 \end{bmatrix}$$

1. What is the size of A? Of B? Of C? Of D?

2. Find a_{14}, a_{21}, a_{31}, and a_{43}.   **3.** Find b_{13}, b_{31}, and b_{43}.

4. Identify the row matrix. What is its transpose?

5. Identify the column matrix. What is its transpose?

6. Identify the square matrix. What is its transpose?

In Exercises 7–12, refer to the following matrices:

$$A = \begin{bmatrix} -1 & 2 \\ 3 & -2 \\ 4 & 0 \end{bmatrix} \quad B = \begin{bmatrix} 2 & 4 \\ 3 & 1 \\ -2 & 2 \end{bmatrix}$$

$$C = \begin{bmatrix} 3 & -1 & 0 \\ 2 & -2 & 3 \\ 4 & 6 & 2 \end{bmatrix} \quad D = \begin{bmatrix} 2 & -2 & 4 \\ 3 & 6 & 2 \\ -2 & 3 & 1 \end{bmatrix}$$

7. What is the size of A? Of B? Of C? Of D?

8. Explain why the matrix $A + C$ does *not* exist.

9. Compute $A + B$.   **10.** Compute $2A - 3B$.

11. Compute $C - D$.   **12.** Compute $4D - 2C$.

In Exercises 13–20, perform the indicated operations.

13. $\begin{bmatrix} 6 & 3 & 8 \\ 4 & 5 & 6 \end{bmatrix} - \begin{bmatrix} 3 & -2 & -1 \\ 0 & -5 & -7 \end{bmatrix}$

14. $\begin{bmatrix} 2 & -3 & 4 & -1 \\ 3 & 1 & 0 & 0 \end{bmatrix} + \begin{bmatrix} 4 & 3 & -2 & -4 \\ 6 & 2 & 0 & -3 \end{bmatrix}$

15. $\begin{bmatrix} 1 & 4 & -5 \\ 3 & -8 & 6 \end{bmatrix} + \begin{bmatrix} 4 & 0 & -2 \\ 3 & 6 & 5 \end{bmatrix} - \begin{bmatrix} 2 & 8 & 9 \\ -11 & 2 & -5 \end{bmatrix}$

16. $3\begin{bmatrix} 1 & 1 & -3 \\ 3 & 2 & 3 \\ 7 & -1 & 6 \end{bmatrix} + 4\begin{bmatrix} -2 & -1 & 8 \\ 4 & 2 & 2 \\ 3 & 6 & 3 \end{bmatrix}$

17. $\begin{bmatrix} 1.2 & 4.5 & -4.2 \\ 8.2 & 6.3 & -3.2 \end{bmatrix} - \begin{bmatrix} 3.1 & 1.5 & -3.6 \\ 2.2 & -3.3 & -4.4 \end{bmatrix}$

18. $\begin{bmatrix} 0.06 & 0.12 \\ 0.43 & 1.11 \\ 1.55 & -0.43 \end{bmatrix} - \begin{bmatrix} 0.77 & -0.75 \\ 0.22 & -0.65 \\ 1.09 & -0.57 \end{bmatrix}$

19. $\dfrac{1}{2}\begin{bmatrix} 1 & 0 & 0 & -4 \\ 3 & 0 & -1 & 6 \\ -2 & 1 & -4 & 2 \end{bmatrix} + \dfrac{4}{3}\begin{bmatrix} 3 & 0 & -1 & 4 \\ -2 & 1 & -6 & 2 \\ 8 & 2 & 0 & -2 \end{bmatrix}$

$$-\dfrac{1}{3}\begin{bmatrix} 3 & -9 & -1 & 0 \\ 6 & 2 & 0 & -6 \\ 0 & 1 & -3 & 1 \end{bmatrix}$$

20. $0.5\begin{bmatrix} 1 & 3 & 5 \\ 5 & 2 & -1 \\ -2 & 0 & 1 \end{bmatrix} - 0.2\begin{bmatrix} 2 & 3 & 4 \\ -1 & 1 & -4 \\ 3 & 5 & -5 \end{bmatrix}$

$$+ 0.6\begin{bmatrix} 3 & 4 & -1 \\ 4 & 5 & 1 \\ 1 & 0 & 0 \end{bmatrix}$$

In Exercises 21–24, solve for u, x, y, and z in the matrix equation.

21. $\begin{bmatrix} 2x - 2 & 3 & 2 \\ 2 & 4 & y - 2 \\ 2z & -3 & 2 \end{bmatrix} = \begin{bmatrix} 3 & u & 2 \\ 2 & 4 & 5 \\ 4 & -3 & 2 \end{bmatrix}$

22. $\begin{bmatrix} x & -2 \\ 3 & y \end{bmatrix} + \begin{bmatrix} -2 & z \\ -1 & 2 \end{bmatrix} = \begin{bmatrix} 4 & -2 \\ 2u & 4 \end{bmatrix}$

23. $\begin{bmatrix} 1 & x \\ 2y & -3 \end{bmatrix} - 4\begin{bmatrix} 2 & -2 \\ 0 & 3 \end{bmatrix} = \begin{bmatrix} 3z & 10 \\ 4 & -u \end{bmatrix}$

24. $\begin{bmatrix} 1 & 2 \\ 3 & 4 \\ x & -1 \end{bmatrix} - 3\begin{bmatrix} y - 1 & 2 \\ 1 & 2 \\ 4 & 2z + 1 \end{bmatrix} = 2\begin{bmatrix} -4 & -u \\ 0 & -1 \\ 4 & 4 \end{bmatrix}$

$1 + 3 - 3y = -8$ $-1 \; -6z - 3 = 8$

$-3y = -12$ $-6z = 12$

$y = 4$

In Exercises 25 and 26, let

$$A = \begin{bmatrix} 2 & -4 & 3 \\ 4 & 2 & 1 \end{bmatrix} \quad B = \begin{bmatrix} 4 & -3 & 2 \\ 1 & 0 & 4 \end{bmatrix}$$

$$C = \begin{bmatrix} 1 & 0 & 2 \\ 3 & -2 & 1 \end{bmatrix}$$

25. Verify by direct computation the validity of the commutative law for matrix addition.

26. Verify by direct computation the validity of the associative law for matrix addition.

In Exercises 27–30, let

$$A = \begin{bmatrix} 3 & 1 \\ 2 & 4 \\ -4 & 0 \end{bmatrix} \quad \text{and} \quad B = \begin{bmatrix} 1 & 2 \\ -1 & 0 \\ 3 & 2 \end{bmatrix}$$

Verify each equation by direct computation.

27. $(3 + 5)A = 3A + 5A$　　**28.** $2(4A) = (2 \cdot 4)A = 8A$

29. $4(A + B) = 4A + 4B$　　**30.** $2(A - 3B) = 2A - 6B$

In Exercises 31–34, find the transpose of each matrix.

31. $\begin{bmatrix} 3 & 2 & -1 & 5 \end{bmatrix}$　　**32.** $\begin{bmatrix} 4 & 2 & 0 & -1 \\ 3 & 4 & -1 & 5 \end{bmatrix}$

33. $\begin{bmatrix} 1 & -1 & 2 \\ 3 & 4 & 2 \\ 0 & 1 & 0 \end{bmatrix}$　　**34.** $\begin{bmatrix} 1 & 2 & 6 & 4 \\ 2 & 3 & 2 & 5 \\ 6 & 2 & 3 & 0 \\ 4 & 5 & 0 & 2 \end{bmatrix}$

35. CHOLESTEROL LEVELS　Mr. Cross, Mr. Jones, and Mr. Smith each suffer from coronary heart disease. As part of their treatment, they were put on special low-cholesterol diets: Cross on diet I, Jones on diet II, and Smith on diet III. Progressive records of each patient's cholesterol level were kept. At the beginning of the first, second, third, and fourth months, the cholesterol levels of the three patients were:

Cross:　220, 215, 210, and 205
Jones:　220, 210, 200, and 195
Smith:　215, 205, 195, and 190

Represent this information in a 3×4 matrix.

36. INVESTMENT PORTFOLIOS　The following table gives the number of shares of certain corporations held by Leslie and Tom in their respective IRA accounts at the beginning of the year:

	IBM	GE	Ford	Wal-Mart
Leslie	500	350	200	400
Tom	400	450	300	200

Over the year, they added more shares to their accounts, as shown in the following table:

	IBM	GE	Ford	Wal-Mart
Leslie	50	50	0	100
Tom	0	80	100	50

a. Write a matrix A giving the holdings of Leslie and Tom at the beginning of the year and a matrix B giving the shares they have added to their portfolios.

b. Find a matrix C giving their total holdings at the end of the year.

37. BANKING　The numbers of three types of bank accounts on January 1 in the Central Bank and its branches are represented by matrix A:

	Checking accounts	Savings accounts	Fixed-deposit accounts
Main office	2820	1470	1120
$A =$ Westside branch	1030	520	480
Eastside branch	1170	540	460

The number and types of accounts opened during the first quarter are represented by matrix B, and the number and types of accounts closed during the same period are represented by matrix C. Thus,

$$B = \begin{bmatrix} 260 & 120 & 110 \\ 140 & 60 & 50 \\ 120 & 70 & 50 \end{bmatrix} \quad \text{and} \quad C = \begin{bmatrix} 120 & 80 & 80 \\ 70 & 30 & 40 \\ 60 & 20 & 40 \end{bmatrix}$$

a. Find matrix D, which represents the number of each type of account at the end of the first quarter at each location.

b. Because a new manufacturing plant is opening in the immediate area, it is anticipated that there will be a 10% increase in the number of accounts at each location during the second quarter. Write a matrix E to reflect this anticipated increase.

38. BOOKSTORE INVENTORIES　The Campus Bookstore's inventory of books is:

Hardcover: textbooks, 5280; fiction, 1680; nonfiction, 2320; reference, 1890

Paperback: fiction, 2810; nonfiction, 1490; reference, 2070; textbooks, 1940

The College Bookstore's inventory of books is:

Hardcover: textbooks, 6340; fiction, 2220; nonfiction, 1790; reference, 1980

Paperback: fiction, 3100; nonfiction, 1720; reference, 2710; textbooks, 2050

a. Represent Campus's inventory as a matrix A.
b. Represent College's inventory as a matrix B.
c. The two companies decide to merge, so now write a matrix C that represents the total inventory of the newly amalgamated company.

39. **INSURANCE CLAIMS** The property damage–claim frequencies per 100 cars in Massachusetts in the years 2000, 2001, and 2002 are 6.88, 7.05, and 7.18, respectively. The corresponding claim frequencies in the United States are 4.13, 4.09, and 4.06, respectively. Express this information using a 2×3 matrix.

Sources: Registry of Motor Vehicles; Federal Highway Administration

40. **MORTALITY RATES** Mortality actuarial tables in the United States have been revised in 2001, the fourth time since 1858. Based on the new life insurance mortality rates, 1% of 60-year-old men, 2.6% of 70-year-old men, 7% of 80-year-old men, 18.8% of 90-year-old men, and 36.3% of 100-year-old men would die within a year. The corresponding rates for women are 0.8%, 1.8%, 4.4%, 12.2%, and 27.6%, respectively. Express this information using a 2×5 matrix.

Source: Society of Actuaries

41. **LIFE EXPECTANCY** Figures for life expectancy at birth of Massachusetts residents in 2002 are 81, 76.1, and 82.2 years for white, black, and Hispanic women, respectively, and 76, 69.9, and 75.9 years for white, black, and Hispanic men,

respectively. Express this information using a 2×3 matrix and a 3×2 matrix.

Source: Massachusetts Department of Public Health

42. **MARKET SHARE OF MOTORCYCLES** The market share of motorcycles in the United States in 2001 follows: Honda 27.9%, Harley-Davidson 21.9%, Yamaha 19.2%, Suzuki 11%, Kawasaki 9.1%, and others 10.9%. The corresponding figures for 2002 are 27.6%, 23.3%, 18.2%, 10.5%, 8.8%, and 11.6%, respectively. Express this information using a 2×6 matrix. What is the sum of all the elements in the first row? In the second row? Is this expected? Which company gained the most market share between 2001 and 2002?

Source: Motorcycle Industry Council

In Exercises 43–46, determine whether the statement is true or false. If it is true, explain why it is true. If it is false, give an example to show why it is false.

43. If A and B are matrices of the same order and c is a scalar, then $c(A + B) = cA + cB$.

44. If A and B are matrices of the same order, $A - B = A + (-1)B$.

45. If A is a matrix and c is a nonzero scalar, then $(cA)^T = (1/c)A^T$.

46. If A is a matrix, then $(A^T)^T = A$.

2.4 Solutions to Self-Check Exercises

1. $\begin{bmatrix} 1 & 3 & 2 \\ -1 & 4 & 7 \end{bmatrix} - 3\begin{bmatrix} 2 & 1 & 0 \\ 1 & 3 & 4 \end{bmatrix} = \begin{bmatrix} 1 & 3 & 2 \\ -1 & 4 & 7 \end{bmatrix} - \begin{bmatrix} 6 & 3 & 0 \\ 3 & 9 & 12 \end{bmatrix}$

$= \begin{bmatrix} -5 & 0 & 2 \\ -4 & -5 & -5 \end{bmatrix}$

2. We are given

$\begin{bmatrix} x & 3 \\ z & 2 \end{bmatrix} + \begin{bmatrix} 2 - y & z \\ 2 - z & -x \end{bmatrix} = \begin{bmatrix} 3 & 7 \\ 2 & 0 \end{bmatrix}$

Performing the indicated operation on the left-hand side, we obtain

$\begin{bmatrix} 2 + x - y & 3 + z \\ 2 & 2 - x \end{bmatrix} = \begin{bmatrix} 3 & 7 \\ 2 & 0 \end{bmatrix}$

By the equality of matrices, we have

$$2 + x - y = 3$$
$$3 + z = 7$$
$$2 - x = 0$$

from which we deduce that $x = 2$, $y = 1$, and $z = 4$.

3. The required matrix is

$A + B = \begin{bmatrix} 1200 & 750 & 650 \\ 1100 & 850 & 600 \end{bmatrix} + \begin{bmatrix} 1250 & 825 & 550 \\ 1150 & 750 & 750 \end{bmatrix}$

$= \begin{bmatrix} 2450 & 1575 & 1200 \\ 2250 & 1600 & 1350 \end{bmatrix}$

USING TECHNOLOGY

■ Matrix Operations

Graphing Utility

A graphing utility can be used to perform matrix addition, matrix subtraction, and scalar multiplication. It can also be used to find the transpose of a matrix.

EXAMPLE 1 Let

$$A = \begin{bmatrix} 1.2 & 3.1 \\ -2.1 & 4.2 \\ 3.1 & 4.8 \end{bmatrix} \quad \text{and} \quad B = \begin{bmatrix} 4.1 & 3.2 \\ 1.3 & 6.4 \\ 1.7 & 0.8 \end{bmatrix}$$

Find (a) $A + B$, (b) $2.1A - 3.2B$, and (c) $(2.1A + 3.2B)^T$.

Solution We first enter the matrices A and B into the calculator.

a. Using matrix operations, we enter the expression $A + B$ and obtain

$$A + B = \begin{bmatrix} 5.3 & 6.3 \\ -0.8 & 10.6 \\ 4.8 & 5.6 \end{bmatrix}$$

b. Using matrix operations, we enter the expression $2.1A - 3.2B$ and obtain

$$2.1A - 3.2B = \begin{bmatrix} -10.6 & -3.73 \\ -8.57 & -11.66 \\ 1.07 & 7.52 \end{bmatrix}$$

c. Using matrix operations, we enter the expression $(2.1A + 3.2B)^T$ and obtain

$$(2.1A + 3.2B)^T = \begin{bmatrix} 15.64 & -.25 & 11.95 \\ 16.75 & 29.3 & 12.64 \end{bmatrix}$$

APPLIED EXAMPLE 2 John operates three gas stations in three locations, I, II, and III. Over 2 consecutive days, his gas stations recorded the following fuel sales (in gallons):

| | **Day 1** | | | |
	Regular	Regular Plus	Premium	Diesel
Location I	1400	1200	1100	200
Location II	1600	900	1200	300
Location III	1200	1500	800	500

| | **Day 2** | | | |
	Regular	Regular Plus	Premium	Diesel
Location I	1000	900	800	150
Location II	1800	1200	1100	250
Location III	800	1000	700	400

Find a matrix representing the total fuel sales at John's gas stations.

Solution The sales can be represented by the matrix A (day 1) and matrix B (day 2):

$$A = \begin{bmatrix} 1400 & 1200 & 1100 & 200 \\ 1600 & 900 & 1200 & 300 \\ 1200 & 1500 & 800 & 500 \end{bmatrix} \quad \text{and} \quad B = \begin{bmatrix} 1000 & 900 & 800 & 150 \\ 1800 & 1200 & 1100 & 250 \\ 800 & 1000 & 700 & 400 \end{bmatrix}$$

Next, we enter the matrices A and B into the calculator. Using matrix operations, we enter the expression $A + B$ and obtain

$$A + B = \begin{bmatrix} 2400 & 2100 & 1900 & 350 \\ 3400 & 2100 & 2300 & 550 \\ 2000 & 2500 & 1500 & 900 \end{bmatrix}$$

Excel

First, we show how basic operations on matrices can be carried out using Excel.

EXAMPLE 3 Given the following matrices,

$$A = \begin{bmatrix} 1.2 & 3.1 \\ -2.1 & 4.2 \\ 3.1 & 4.8 \end{bmatrix} \quad \text{and} \quad B = \begin{bmatrix} 4.1 & 3.2 \\ 1.3 & 6.4 \\ 1.7 & 0.8 \end{bmatrix}$$

a. Compute $A + B$.   **b.** Compute $2.1A - 3.2B$.

Solution

a. *First, represent the matrices A and B in a spreadsheet.* Enter the elements of each matrix in a block of cells as shown in Figure T1.

	A	B	C	D	E
1		A			B
2	1.2	3.1		4.1	3.2
3	-2.1	4.2		1.3	6.4
4	3.1	4.8		1.7	0.8

FIGURE T1
The elements of matrix A and matrix B in the spreadsheet

Second, compute the sum of matrix A and matrix B. Highlight the cells that will contain matrix $A + B$, type =, highlight the cells in matrix A, type +, highlight the cells in matrix B, and press $\boxed{\textbf{Ctrl-Shift-Enter}}$. The resulting matrix $A + B$ is shown in Figure T2.

	A	B
8		A + B
9	5.3	6.3
10	-0.8	10.6
11	4.8	5.6

FIGURE T2
The matrix $A + B$

Note: Boldfaced words/characters enclosed in a box (for example, $\boxed{\textbf{Enter}}$) indicate that an action (click, select, or press) is required. Words/characters printed blue (for example, Chart sub-type:) indicate words/characters that appear on the screen. Words/characters printed in a typewriter font (for example, = (-2/3)*A2+2) indicate words/characters that need to be typed and entered.

(continued)

b. *Highlight the cells that will contain matrix* (2.1A − 3.2B). Type = 2.1*, *highlight matrix* A, *type* −3.2*, *highlight the cells in matrix* B, *and press* Ctrl-Shift-Enter . The resulting matrix (2.1A − 3.2B) is shown in Figure T3.

	A	B
		2.1A - 3.2B
13		
14	−10.6	−3.73
15	−8.57	−11.66
16	1.07	7.52

FIGURE T3
The matrix (2.1A − 3.2B)

APPLIED EXAMPLE 4   John operates three gas stations in three locations I, II, and III. Over 2 consecutive days, his gas stations recorded the following fuel sales (in gallons):

	Day 1			
	Regular	Regular Plus	Premium	Diesel
Location I	1400	1200	1100	200
Location II	1600	900	1200	300
Location III	1200	1500	800	500

	Day 2			
	Regular	Regular Plus	Premium	Diesel
Location I	1000	900	800	150
Location II	1800	1200	1100	250
Location III	800	1000	700	400

Find a matrix representing the total fuel sales at John's gas stations.

Solution   The sales can be represented by the matrices A (day 1) and B (day 2):

$$A = \begin{bmatrix} 1400 & 1200 & 1100 & 200 \\ 1600 & 900 & 1200 & 300 \\ 1200 & 1500 & 800 & 500 \end{bmatrix} \quad \text{and} \quad B = \begin{bmatrix} 1000 & 900 & 800 & 150 \\ 1800 & 1200 & 1100 & 250 \\ 800 & 1000 & 700 & 400 \end{bmatrix}$$

We first enter the elements of the matrices A and B onto a spreadsheet. Next, we highlight the cells that will contain the matrix A + B, type =, highlight A, type +, highlight B, and then press Ctrl-Shift-Enter . The resulting matrix A + B is shown in Figure T4.

	A	B	C	D
		A + B		
23				
24	2400	2100	1900	350
25	3400	2100	2300	550
26	2000	2500	1500	900

FIGURE T4
The matrix A + B

TECHNOLOGY EXERCISES

Refer to the following matrices and perform the indicated operations.

$$A = \begin{bmatrix} 1.2 & 3.1 & -5.4 & 2.7 \\ 4.1 & 3.2 & 4.2 & -3.1 \\ 1.7 & 2.8 & -5.2 & 8.4 \end{bmatrix}$$

$$B = \begin{bmatrix} 6.2 & -3.2 & 1.4 & -1.2 \\ 3.1 & 2.7 & -1.2 & 1.7 \\ 1.2 & -1.4 & -1.7 & 2.8 \end{bmatrix}$$

1. $12.5A$

2. $-8.4B$

3. $A - B$

4. $B - A$

5. $1.3A + 2.4B$

6. $2.1A - 1.7B$

7. $3(A + B)$

8. $1.3(4.1A - 2.3B)$

2.5 Multiplication of Matrices

■ Matrix Product

In Section 2.4, we saw how matrices of the same size may be added or subtracted and how a matrix may be multiplied by a scalar (real number), an operation referred to as scalar multiplication. In this section, we see how, with certain restrictions, one matrix may be multiplied by another matrix.

To define matrix multiplication, let's consider the following problem: On a certain day, Al's Service Station sold 1600 gallons of regular, 1000 gallons of regular plus, and 800 gallons of premium gasoline. If the price of gasoline on this day was $1.69 for regular, $1.79 for regular plus, and $1.89 for premium gasoline, find the total revenue realized by Al's for that day.

The day's sale of gasoline may be represented by the matrix

$$A = [1600 \quad 1000 \quad 800] \qquad \text{Row matrix } (1 \times 3)$$

Next, we let the unit selling price of regular, regular plus, and premium gasoline be the entries in the matrix

$$B = \begin{bmatrix} 1.69 \\ 1.79 \\ 1.89 \end{bmatrix} \qquad \text{Column matrix } (3 \times 1)$$

The first entry in matrix A gives the number of gallons of regular gasoline sold, and the first entry in matrix B gives the selling price for each gallon of regular gasoline, so their product $(1600)(1.69)$ gives the revenue realized from the sale of regular gasoline for the day. A similar interpretation of the second and third entries in the two matrices suggests that we multiply the corresponding entries to obtain the respective revenues realized from the sale of regular, regular-plus, and premium gasoline. Finally, the total revenue realized by Al's from the sale of gasoline is given by adding these products to obtain

$$(1600)(1.69) + (1000)(1.79) + (800)(1.89) = 6006$$

or $6006.00.

This example suggests that if we have a row matrix of size $1 \times n$,

$$A = [a_1 \quad a_2 \quad a_3 \cdots a_n]$$

and a column matrix of size $n \times 1$,

$$B = \begin{bmatrix} b_1 \\ b_2 \\ b_3 \\ \vdots \\ b_n \end{bmatrix}$$

we may define the **matrix product** of A and B, written AB, by

$$AB = [a_1 \quad a_2 \quad a_3 \quad \cdots \quad a_n] \begin{bmatrix} b_1 \\ b_2 \\ b_3 \\ \vdots \\ b_n \end{bmatrix} = a_1 b_1 + a_2 b_2 + a_3 b_3 + \cdots + a_n b_n \qquad \textbf{(11)}$$

EXAMPLE 1   Let

$$A = [1 \quad -2 \quad 3 \quad 5] \quad \text{and} \quad B = \begin{bmatrix} 2 \\ 3 \\ 0 \\ -1 \end{bmatrix}$$

Then,

$$AB = [1 \quad -2 \quad 3 \quad 5] \begin{bmatrix} 2 \\ 3 \\ 0 \\ -1 \end{bmatrix} = (1)(2) + (-2)(3) + (3)(0) + (5)(-1) = -9$$

APPLIED EXAMPLE 2   **Stock Transactions**   Judy's stock holdings are given by the matrix

$$\begin{matrix} \quad\quad \text{GM} \quad\quad \text{IBM} \quad\quad \text{BAC} \\ A = \begin{bmatrix} 700 & 400 & 200 \end{bmatrix} \end{matrix}$$

At the close of trading on a certain day, the prices (in dollars per share) of these stocks are

$$B = \begin{bmatrix} 50 \\ 120 \\ 42 \end{bmatrix} \begin{matrix} \text{GM} \\ \text{IBM} \\ \text{BAC} \end{matrix}$$

What is the total value of Judy's holdings as of that day?

Solution Judy's holdings are worth

$$AB = [700 \quad 400 \quad 200] \begin{bmatrix} 50 \\ 120 \\ 42 \end{bmatrix} = (700)(50) + (400)(120) + (200)(42)$$

or $91,400. ∎

Returning once again to the matrix product AB in Equation (11), observe that the number of columns of the row matrix A is *equal* to the number of rows of the column matrix B. Observe further that the product matrix AB has size 1×1 (a real number may be thought of as a 1×1 matrix). Schematically,

Size of A Size of B

$(1 \times n)$ $(n \times 1)$

(1×1)

Size of AB

More generally, if A is a matrix of size $m \times n$ and B is a matrix of size $n \times p$ (the number of columns of A equals the numbers of rows of B), then the *matrix product* of A and B, AB, is defined and is a matrix of size $m \times p$. Schematically,

Size of A Size of B

$(m \times n)$ $(n \times p)$

$(m \times p)$

Size of AB

Next, let's illustrate the mechanics of matrix multiplication by computing the product of a 2×3 matrix A and a 3×4 matrix B. Suppose

$$A = \begin{bmatrix} a_{11} & a_{12} & a_{13} \\ a_{21} & a_{22} & a_{23} \end{bmatrix}$$

$$B = \begin{bmatrix} b_{11} & b_{12} & b_{13} & b_{14} \\ b_{21} & b_{22} & b_{23} & b_{24} \\ b_{31} & b_{32} & b_{33} & b_{34} \end{bmatrix}$$

From the schematic

Same

Size of A (2×3) (3×4) Size of B

(2×4)

Size of AB

we see that the matrix product $C = AB$ is defined (the number of columns of A equals the number of rows of B) and has size 2×4. Thus,

$$C = \begin{bmatrix} c_{11} & c_{12} & c_{13} & c_{14} \\ c_{21} & c_{22} & c_{23} & c_{24} \end{bmatrix}$$

The entries of C are computed as follows: The entry c_{11} (the entry in the *first* row, *first* column of C) is the product of the row matrix composed of the entries from the *first* row of A and the column matrix composed of the *first* column of B. Thus,

$$c_{11} = [a_{11} \quad a_{12} \quad a_{13}] \begin{bmatrix} b_{11} \\ b_{21} \\ b_{31} \end{bmatrix} = a_{11}b_{11} + a_{12}b_{21} + a_{13}b_{31}$$

The entry c_{12} (the entry in the *first* row, *second* column of C) is the product of the row matrix composed of the *first* row of A and the column matrix composed of the *second* column of B. Thus,

$$c_{12} = [a_{11} \quad a_{12} \quad a_{13}] \begin{bmatrix} b_{12} \\ b_{22} \\ b_{32} \end{bmatrix} = a_{11}b_{12} + a_{12}b_{22} + a_{13}b_{32}$$

The other entries in C are computed in a similar manner.

EXAMPLE 3   Let

$$A = \begin{bmatrix} 3 & 1 & 4 \\ -1 & 2 & 3 \end{bmatrix} \quad \text{and} \quad B = \begin{bmatrix} 1 & 3 & -3 \\ 4 & -1 & 2 \\ 2 & 4 & 1 \end{bmatrix}$$

Compute AB.

Solution   The size of matrix A is 2×3, and the size of matrix B is 3×3. Since the number of columns of matrix A is equal to the number of rows of matrix B, the matrix product $C = AB$ is defined. Furthermore, the size of matrix C is 2×3. Thus,

$$\begin{bmatrix} 3 & 1 & 4 \\ -1 & 2 & 3 \end{bmatrix} \begin{bmatrix} 1 & 3 & -3 \\ 4 & -1 & 2 \\ 2 & 4 & 1 \end{bmatrix} = \begin{bmatrix} c_{11} & c_{12} & c_{13} \\ c_{21} & c_{22} & c_{23} \end{bmatrix}$$

It remains now to determine the entries c_{11}, c_{12}, c_{13}, c_{21}, c_{22}, and c_{23}. We have

$$c_{11} = [3 \quad 1 \quad 4] \begin{bmatrix} 1 \\ 4 \\ 2 \end{bmatrix} = (3)(1) + (1)(4) + (4)(2) = 15$$

$$c_{12} = [3 \quad 1 \quad 4] \begin{bmatrix} 3 \\ -1 \\ 4 \end{bmatrix} = (3)(3) + (1)(-1) + (4)(4) = 24$$

$$c_{13} = [3 \quad 1 \quad 4] \begin{bmatrix} -3 \\ 2 \\ 1 \end{bmatrix} = (3)(-3) + (1)(2) + (4)(1) = -3$$

$$c_{21} = [-1 \quad 2 \quad 3] \begin{bmatrix} 1 \\ 4 \\ 2 \end{bmatrix} = (-1)(1) + (2)(4) + (3)(2) = 13$$

$$c_{22} = \begin{bmatrix} -1 & 2 & 3 \end{bmatrix} \begin{bmatrix} 3 \\ -1 \\ 4 \end{bmatrix} = (-1)(3) + (2)(-1) + (3)(4) = 7$$

$$c_{23} = \begin{bmatrix} -1 & 2 & 3 \end{bmatrix} \begin{bmatrix} -3 \\ 2 \\ 1 \end{bmatrix} = (-1)(-3) + (2)(2) + (3)(1) = 10$$

so the required product AB is given by

$$AB = \begin{bmatrix} 15 & 24 & -3 \\ 13 & 7 & 10 \end{bmatrix}$$

EXAMPLE 4 Let

$$A = \begin{bmatrix} 3 & 2 & 1 \\ -1 & 2 & 3 \\ 3 & 1 & 4 \end{bmatrix} \quad \text{and} \quad B = \begin{bmatrix} 1 & 3 & 4 \\ 2 & 4 & 1 \\ -1 & 2 & 3 \end{bmatrix}$$

Then,

$$AB = \begin{bmatrix} 3 \cdot 1 + 2 \cdot 2 + 1 \cdot (-1) & 3 \cdot 3 + 2 \cdot 4 + 1 \cdot 2 & 3 \cdot 4 + 2 \cdot 1 + 1 \cdot 3 \\ (-1) \cdot 1 + 2 \cdot 2 + 3 \cdot (-1) & (-1) \cdot 3 + 2 \cdot 4 + 3 \cdot 2 & (-1) \cdot 4 + 2 \cdot 1 + 3 \cdot 3 \\ 3 \cdot 1 + 1 \cdot 2 + 4 \cdot (-1) & 3 \cdot 3 + 1 \cdot 4 + 4 \cdot 2 & 3 \cdot 4 + 1 \cdot 1 + 4 \cdot 3 \end{bmatrix}$$

$$= \begin{bmatrix} 6 & 19 & 17 \\ 0 & 11 & 7 \\ 1 & 21 & 25 \end{bmatrix}$$

$$BA = \begin{bmatrix} 1 \cdot 3 + 3 \cdot (-1) + 4 \cdot 3 & 1 \cdot 2 + 3 \cdot 2 + 4 \cdot 1 & 1 \cdot 1 + 3 \cdot 3 + 4 \cdot 4 \\ 2 \cdot 3 + 4 \cdot (-1) + 1 \cdot 3 & 2 \cdot 2 + 4 \cdot 2 + 1 \cdot 1 & 2 \cdot 1 + 4 \cdot 3 + 1 \cdot 4 \\ (-1) \cdot 3 + 2 \cdot (-1) + 3 \cdot 3 & (-1) \cdot 2 + 2 \cdot 2 + 3 \cdot 1 & (-1) \cdot 1 + 2 \cdot 3 + 3 \cdot 4 \end{bmatrix}$$

$$= \begin{bmatrix} 12 & 12 & 26 \\ 5 & 13 & 18 \\ 4 & 5 & 17 \end{bmatrix}$$

As the last example shows, in general, $AB \neq BA$ for any two square matrices A and B. However, the following laws are valid for matrix multiplication.

Laws for Matrix Multiplication

If the products and sums are defined for the matrices A, B, and C, then

1. $(AB)C = A(BC)$ Associative law
2. $A(B + C) = AB + AC$ Distributive law

The square matrix of size n having 1s along the main diagonal and zeros elsewhere is called the identity matrix of size n.

Identity Matrix
The **identity matrix** of size n is given by

$$I_n = \begin{bmatrix} 1 & 0 & . & . & . & 0 \\ 0 & 1 & . & . & . & 0 \\ . & & . & & & . \\ . & & & . & & . \\ . & & & & . & . \\ 0 & 0 & . & . & . & 1 \end{bmatrix} \quad n \text{ rows}$$

n columns

The identity matrix has the property that $I_nA = A$ for any $n \times r$ matrix A, and $BI_n = B$ for any $s \times n$ matrix B. In particular, if A is a square matrix of size n, then

$$I_nA = AI_n = A$$

EXAMPLE 5   Let

$$A = \begin{bmatrix} 1 & 3 & 1 \\ -4 & 3 & 2 \\ 1 & 0 & 1 \end{bmatrix}$$

Then,

$$I_3A = \begin{bmatrix} 1 & 0 & 0 \\ 0 & 1 & 0 \\ 0 & 0 & 1 \end{bmatrix}\begin{bmatrix} 1 & 3 & 1 \\ -4 & 3 & 2 \\ 1 & 0 & 1 \end{bmatrix} = \begin{bmatrix} 1 & 3 & 1 \\ -4 & 3 & 2 \\ 1 & 0 & 1 \end{bmatrix} = A$$

$$AI_3 = \begin{bmatrix} 1 & 3 & 1 \\ -4 & 3 & 2 \\ 1 & 0 & 1 \end{bmatrix}\begin{bmatrix} 1 & 0 & 0 \\ 0 & 1 & 0 \\ 0 & 0 & 1 \end{bmatrix} = \begin{bmatrix} 1 & 3 & 1 \\ -4 & 3 & 2 \\ 1 & 0 & 1 \end{bmatrix} = A$$

so $I_3A = AI_3$, confirming our result for this special case.   ■

APPLIED EXAMPLE 6   Production Planning   Ace Novelty received an order from Magic World Amusement Park for 900 "Giant Pandas," 1200 "Saint Bernards," and 2000 "Big Birds." Ace's management decided that 500 Giant Pandas, 800 Saint Bernards, and 1300 Big Birds could be manufactured in their Los Angeles plant, and the balance of the order could be filled by their Seattle plant. Each Panda requires 1.5 square yards of plush, 30 cubic feet of stuffing, and 5 pieces of trim; each Saint Bernard requires 2 square yards of plush, 35 cubic feet of stuffing, and 8 pieces of trim; and each Big Bird requires 2.5 square yards of plush, 25 cubic feet of stuffing, and 15 pieces of trim. The plush costs $4.50 per square yard, the stuffing costs 10 cents per cubic foot, and the trim costs 25 cents per unit.

a. Find how much of each type of material must be purchased for each plant.
b. What is the total cost of materials incurred by each plant and the total cost of materials incurred by Ace Novelty in filling the order?

Solution   The quantities of each type of stuffed animal to be produced at each plant location may be expressed as a 2×3 *production matrix P*. Thus,

$$P = \begin{array}{c} \text{L.A.} \\ \text{Seattle} \end{array} \begin{bmatrix} \overset{\text{Pandas}}{500} & \overset{\text{St. Bernards}}{800} & \overset{\text{Birds}}{1300} \\ 400 & 400 & 700 \end{bmatrix}$$

Similarly, we may represent the amount and type of material required to manufacture each type of animal by a 3×3 *activity matrix A*. Thus,

$$A = \begin{array}{c} \text{Pandas} \\ \text{St. Bernards} \\ \text{Birds} \end{array} \begin{bmatrix} \overset{\text{Plush}}{1.5} & \overset{\text{Stuffing}}{30} & \overset{\text{Trim}}{5} \\ 2 & 35 & 8 \\ 2.5 & 25 & 15 \end{bmatrix}$$

Finally, the unit cost for each type of material may be represented by the 3×1 *cost matrix C*.

$$C = \begin{array}{c} \text{Plush} \\ \text{Stuffing} \\ \text{Trim} \end{array} \begin{bmatrix} 4.50 \\ 0.10 \\ 0.25 \end{bmatrix}$$

a. The amount of each type of material required for each plant is given by the matrix *PA*. Thus,

$$PA = \begin{bmatrix} 500 & 800 & 1300 \\ 400 & 400 & 700 \end{bmatrix} \begin{bmatrix} 1.5 & 30 & 5 \\ 2 & 35 & 8 \\ 2.5 & 25 & 15 \end{bmatrix}$$

$$= \begin{array}{c} \text{L.A.} \\ \text{Seattle} \end{array} \begin{bmatrix} \overset{\text{Plush}}{5600} & \overset{\text{Stuffing}}{75{,}500} & \overset{\text{Trim}}{28{,}400} \\ 3150 & 43{,}500 & 15{,}700 \end{bmatrix}$$

b. The total cost of materials for each plant is given by the matrix *PAC*:

$$PAC = \begin{bmatrix} 5600 & 75{,}500 & 28{,}400 \\ 3150 & 43{,}500 & 15{,}700 \end{bmatrix} \begin{bmatrix} 4.50 \\ 0.10 \\ 0.25 \end{bmatrix}$$

$$= \begin{array}{c} \text{L.A.} \\ \text{Seattle} \end{array} \begin{bmatrix} 39{,}850 \\ 22{,}450 \end{bmatrix}$$

or \$39,850 for the L.A. plant and \$22,450 for the Seattle plant. Thus, the total cost of materials incurred by Ace Novelty is \$62,300.   ∎

Matrix Representation

Example 7 shows how a system of linear equations may be written in a compact form with the help of matrices. (We will use this matrix equation representation in Section 2.6.)

EXAMPLE 7   Write the following system of linear equations in matrix form.

$$2x - 4y + z = 6$$
$$-3x + 6y - 5z = -1$$
$$x - 3y + 7z = 0$$

Solution   Let's write

$$A = \begin{bmatrix} 2 & -4 & 1 \\ -3 & 6 & -5 \\ 1 & -3 & 7 \end{bmatrix} \qquad X = \begin{bmatrix} x \\ y \\ z \end{bmatrix} \qquad B = \begin{bmatrix} 6 \\ -1 \\ 0 \end{bmatrix}$$

Note that A is just the 3×3 matrix of coefficients of the system, X is the 3×1 column matrix of unknowns (variables), and B is the 3×1 column matrix of constants. We now show that the required matrix representation of the system of linear equations is

$$AX = B$$

To see this, observe that

$$AX = \begin{bmatrix} 2 & -4 & 1 \\ -3 & 6 & -5 \\ 1 & -3 & 7 \end{bmatrix} \begin{bmatrix} x \\ y \\ z \end{bmatrix} = \begin{bmatrix} 2x - 4y + z \\ -3x + 6y - 5z \\ x - 3y + 7z \end{bmatrix}$$

Equating this 3×1 matrix with matrix B now gives

$$\begin{bmatrix} 2x - 4y + z \\ -3x + 6y - 5z \\ x - 3y + 7z \end{bmatrix} = \begin{bmatrix} 6 \\ -1 \\ 0 \end{bmatrix}$$

which, by matrix equality, is easily seen to be equivalent to the given system of linear equations. ▪

2.5 Self-Check Exercises

1. Compute

$$\begin{bmatrix} 1 & 3 & 0 \\ 2 & 4 & -1 \end{bmatrix} \begin{bmatrix} 3 & 1 & 4 \\ 2 & 0 & 3 \\ 1 & 2 & -1 \end{bmatrix}$$

2. Write the following system of linear equations in matrix form:

$$y - 2z = 1$$
$$2x - y + 3z = 0$$
$$x \qquad + 4z = 7$$

3. On June 1, the stock holdings of Ash and Joan Robinson were given by the matrix

$$A = \begin{array}{c} \\ \text{Ash} \\ \text{Joan} \end{array} \begin{array}{c} \begin{array}{cccc} \text{AT\&T} & \text{AOL} & \text{IBM} & \text{GM} \end{array} \\ \begin{bmatrix} 2000 & 1000 & 500 & 5000 \\ 1000 & 2500 & 2000 & 0 \end{bmatrix} \end{array}$$

and the closing prices of AT&T, AOL, IBM, and GM were $54, $113, $112, and $70/share, respectively. Use matrix multiplication to determine the separate values of Ash's and Joan's stock holdings as of that date.

Solutions to Self-Check Exercises 2.5 can be found on page 133.

2.5 Concept Questions

1. What is the difference between *scalar multiplication* and *matrix multiplication*? Give examples of each operation.

2 **a.** Suppose A and B are matrices whose products AB and BA are both defined. What can you say about the sizes of A and B?

b. If A, B, and C are matrices such that $A(B + C)$ is defined, what can you say about the relationship between the number of columns of A and the number of rows of C? Explain.

2.5 Exercises

In Exercises 1–4, the sizes of matrices *A* and *B* are given. Find the size of *AB* and *BA* whenever they are defined.

1. A is of size 2×3, and B is of size 3×5.

2. A is of size 3×4, and B is of size 4×3.

3. A is of size 1×7, and B is of size 7×1.

4. A is of size 4×4, and B is of size 4×4.

5. Let A be a matrix of size $m \times n$ and B be a matrix of size $s \times t$. Find conditions on m, n, s, and t so that both matrix products AB and BA are defined.

6. Find condition(s) on the size of a matrix A so that A^2 (that is, AA) is defined.

In Exercises 7–24, compute the indicated products.

7. $\begin{bmatrix} 1 & 2 \\ 3 & 0 \end{bmatrix} \begin{bmatrix} 1 \\ -1 \end{bmatrix}$

8. $\begin{bmatrix} -1 & 3 \\ 5 & 0 \end{bmatrix} \begin{bmatrix} 7 \\ 2 \end{bmatrix}$

9. $\begin{bmatrix} 3 & 1 & 2 \\ -1 & 2 & 4 \end{bmatrix} \begin{bmatrix} 4 \\ 1 \\ -2 \end{bmatrix}$

10. $\begin{bmatrix} 3 & 2 & -1 \\ 4 & -1 & 0 \\ -5 & 2 & 1 \end{bmatrix} \begin{bmatrix} 3 \\ -2 \\ 0 \end{bmatrix}$

11. $\begin{bmatrix} -1 & 2 \\ 3 & 1 \end{bmatrix} \begin{bmatrix} 2 & 4 \\ 3 & 1 \end{bmatrix}$

12. $\begin{bmatrix} 1 & 3 \\ -1 & 2 \end{bmatrix} \begin{bmatrix} 1 & 3 & 0 \\ 3 & 0 & 2 \end{bmatrix}$

13. $\begin{bmatrix} 2 & 1 & 2 \\ 3 & 2 & 4 \end{bmatrix} \begin{bmatrix} -1 & 2 \\ 4 & 3 \\ 0 & 1 \end{bmatrix}$

14. $\begin{bmatrix} -1 & 2 \\ 4 & 3 \\ 0 & 1 \end{bmatrix} \begin{bmatrix} 2 & 1 & 2 \\ 3 & 2 & 4 \end{bmatrix}$

15. $\begin{bmatrix} 0.1 & 0.9 \\ 0.2 & 0.8 \end{bmatrix} \begin{bmatrix} 1.2 & 0.4 \\ 0.5 & 2.1 \end{bmatrix}$

16. $\begin{bmatrix} 1.2 & 0.3 \\ 0.4 & 0.5 \end{bmatrix} \begin{bmatrix} 0.2 & 0.6 \\ 0.4 & -0.5 \end{bmatrix}$

17. $\begin{bmatrix} 6 & -3 & 0 \\ -2 & 1 & -8 \\ 4 & -4 & 9 \end{bmatrix} \begin{bmatrix} 1 & 0 & 0 \\ 0 & 1 & 0 \\ 0 & 0 & 1 \end{bmatrix}$

18. $\begin{bmatrix} 2 & 4 \\ -1 & -5 \\ 3 & -1 \end{bmatrix} \begin{bmatrix} 2 & -2 & 4 \\ 1 & 3 & -1 \end{bmatrix}$

19. $\begin{bmatrix} 3 & 0 & -2 & 1 \\ 1 & 2 & 0 & -1 \end{bmatrix} \begin{bmatrix} 2 & 1 & -1 \\ -1 & 2 & 0 \\ 0 & 0 & 1 \\ -1 & -2 & 2 \end{bmatrix}$

20. $\begin{bmatrix} 2 & 1 & -3 & 0 \\ 4 & -2 & -1 & 1 \\ -1 & 2 & 0 & 1 \end{bmatrix} \begin{bmatrix} 2 & -1 \\ 1 & 4 \\ 3 & -3 \\ 0 & -5 \end{bmatrix}$

21. $4 \begin{bmatrix} 1 & -2 & 0 \\ 2 & -1 & 1 \\ 3 & 0 & -1 \end{bmatrix} \begin{bmatrix} 1 & 3 & 1 \\ 1 & 4 & 0 \\ 0 & 1 & -2 \end{bmatrix}$

22. $3 \begin{bmatrix} 2 & -1 & 0 \\ 2 & 1 & 2 \\ 1 & 0 & -1 \end{bmatrix} \begin{bmatrix} 2 & 3 & 1 \\ 3 & -3 & 0 \\ 0 & 1 & -1 \end{bmatrix}$

23. $\begin{bmatrix} 1 & 0 \\ 0 & 1 \end{bmatrix} \begin{bmatrix} 4 & -3 & 2 \\ 7 & 1 & -5 \end{bmatrix} \begin{bmatrix} 1 & 0 & 0 \\ 0 & 1 & 0 \\ 0 & 0 & 1 \end{bmatrix}$

24. $2 \begin{bmatrix} 3 & 2 & -1 \\ 0 & 1 & 3 \\ 2 & 0 & 3 \end{bmatrix} \begin{bmatrix} 1 & 0 & 0 \\ 0 & 1 & 0 \\ 0 & 0 & 1 \end{bmatrix} \begin{bmatrix} 1 & 2 & 0 \\ 0 & -1 & -2 \\ 1 & 3 & 1 \end{bmatrix}$

In Exercises 25 and 26, let

$$A = \begin{bmatrix} 1 & 0 & -2 \\ 1 & -3 & 2 \\ -2 & 1 & 1 \end{bmatrix} \quad B = \begin{bmatrix} 3 & 1 & 0 \\ 2 & 2 & 0 \\ 1 & -3 & -1 \end{bmatrix}$$

$$C = \begin{bmatrix} 2 & -1 & 0 \\ 1 & -1 & 2 \\ 3 & -2 & 1 \end{bmatrix}$$

25. Verify the validity of the associative law for matrix multiplication.

26. Verify the validity of the distributive law for matrix multiplication.

27. Let

$$A = \begin{bmatrix} 1 & 2 \\ 3 & 4 \end{bmatrix} \quad \text{and} \quad B = \begin{bmatrix} 2 & 1 \\ 4 & 3 \end{bmatrix}$$

Compute AB and BA and hence deduce that matrix multiplication is, in general, not commutative.

28. Let

$$A = \begin{bmatrix} 0 & 3 & 0 \\ 1 & 0 & 1 \\ 0 & 2 & 0 \end{bmatrix} \quad B = \begin{bmatrix} 2 & 4 & 5 \\ 3 & -1 & -6 \\ 4 & 3 & 4 \end{bmatrix}$$

$$C = \begin{bmatrix} 4 & 5 & 6 \\ 3 & -1 & -6 \\ 2 & 2 & 3 \end{bmatrix}$$

a. Compute AB.
b. Compute AC.
c. Using the results of parts (a) and (b), conclude that $AB = AC$ does *not* imply that $B = C$.

29. Let

$$A = \begin{bmatrix} 3 & 0 \\ 8 & 0 \end{bmatrix} \quad \text{and} \quad B = \begin{bmatrix} 0 & 0 \\ 4 & 5 \end{bmatrix}$$

Show that $AB = 0$, thereby demonstrating that for matrix multiplication the equation $AB = 0$ does not imply that one or both of the matrices A and B must be the zero matrix.

30. Let

$$A = \begin{bmatrix} 2 & 2 \\ -2 & -2 \end{bmatrix}$$

Show that $A^2 = 0$. Compare this with the equation $a^2 = 0$, where a is a real number.

31. Find the matrix A such that

$$A\begin{bmatrix} 1 & 0 \\ -1 & 3 \end{bmatrix} = \begin{bmatrix} -1 & -3 \\ 3 & 6 \end{bmatrix}$$

Hint: Let $A = \begin{bmatrix} a & b \\ c & d \end{bmatrix}$.

32. Let

$$A = \begin{bmatrix} 3 & 1 \\ 0 & 2 \end{bmatrix} \quad \text{and} \quad B = \begin{bmatrix} 4 & -2 \\ 2 & 1 \end{bmatrix}$$

a. Compute $(A + B)^2$.
b. Compute $A^2 + 2AB + B^2$.
c. From the results of parts (a) and (b), show that in general $(A + B)^2 \neq A^2 + 2AB + B^2$.

33. Let

$$A = \begin{bmatrix} 2 & 4 \\ 5 & -6 \end{bmatrix} \quad \text{and} \quad B = \begin{bmatrix} 4 & 8 \\ -7 & 3 \end{bmatrix}$$

a. Find A^T and show that $(A^T)^T = A$.
b. Show that $(A + B)^T = A^T + B^T$.
c. Show that $(AB)^T = B^T A^T$.

34. Let

$$A = \begin{bmatrix} 1 & 3 \\ -2 & -1 \end{bmatrix} \quad \text{and} \quad B = \begin{bmatrix} 3 & -4 \\ 2 & -2 \end{bmatrix}$$

a. Find A^T and show that $(A^T)^T = A$.
b. Show that $(A + B)^T = A^T + B^T$.
c. Show that $(AB)^T = B^T A^T$.

In Exercises 35–40, write the given system of linear equations in matrix form.

35. $\begin{aligned} 2x - 3y &= 7 \\ 3x - 4y &= 8 \end{aligned}$

36. $\begin{aligned} 2x \phantom{{}- 2y} &= 7 \\ 3x - 2y &= 12 \end{aligned}$

37. $\begin{aligned} 2x - 3y + 4z &= 6 \\ 2y - 3z &= 7 \\ x - y + 2z &= 4 \end{aligned}$

38. $\begin{aligned} x - 2y + 3z &= -1 \\ 3x + 4y - 2z &= 1 \\ 2x - 3y + 7z &= 6 \end{aligned}$

39. $\begin{aligned} -x_1 + x_2 + x_3 &= 0 \\ 2x_1 - x_2 - x_3 &= 2 \\ -3x_1 + 2x_2 + 4x_3 &= 4 \end{aligned}$

40. $\begin{aligned} 3x_1 - 5x_2 + 4x_3 &= 10 \\ 4x_1 + 2x_2 - 3x_3 &= -12 \\ -x_1 \phantom{{}+ 2x_2} + x_3 &= -2 \end{aligned}$

41. **INVESTMENTS** William and Michael's stock holdings are given by the matrix

$$A = \begin{array}{c} \\ \text{William} \\ \text{Michael} \end{array} \begin{bmatrix} \overset{\text{BAC}}{200} & \overset{\text{GM}}{300} & \overset{\text{IBM}}{100} & \overset{\text{TRW}}{200} \\ 100 & 200 & 400 & 0 \end{bmatrix}$$

At the close of trading on a certain day, the prices (in dollars per share) of the stocks are given by the matrix

$$B = \begin{array}{c} \text{BAC} \\ \text{GM} \\ \text{IBM} \\ \text{TRW} \end{array} \begin{bmatrix} 54 \\ 48 \\ 98 \\ 82 \end{bmatrix}$$

a. Find AB.

b. Explain the meaning of the entries in the matrix AB.

42. FOREIGN EXCHANGE Kaitlyn just returned to London from a European trip and wishes to exchange the various currencies she has accumulated for Euros. She finds that she has 80 Austrian schillings, 26 French francs, 18 Dutch guilders, and 20 German marks. Suppose the foreign exchange rates are €0.0727 for one Austrian schilling, €0.1524 for one French franc, €0.4538 for one Dutch guilder, and €0.5113 for one German mark.

a. Write a row matrix A giving the values of the various currencies that Kaitlyn holds.

b. Write a column matrix B giving the exchange rates for the various currencies.

c. If Kaitlyn exchanges all her foreign currencies for Euros, how much will she have?

43. REAL ESTATE Bond Brothers, a real estate developer, builds houses in three states. The projected number of units of each model to be built in each state is given by the matrix

$$\begin{array}{cccc} & & \text{Model} & \\ & \text{I} & \text{II} & \text{III} & \text{IV} \end{array}$$
$$A = \begin{array}{c} \text{N.Y.} \\ \text{Conn.} \\ \text{Mass.} \end{array} \begin{bmatrix} 60 & 80 & 120 & 40 \\ 20 & 30 & 60 & 10 \\ 10 & 15 & 30 & 5 \end{bmatrix}$$

The profits to be realized are $20,000, $22,000, $25,000, and $30,000, respectively, for each model I, II, III, and IV house sold.

a. Write a column matrix B representing the profit for each type of house.

b. Find the total profit Bond Brothers expects to earn in each state if all the houses are sold.

44. BOX-OFFICE RECEIPTS Four theaters comprise the Cinema Center: cinemas I, II, III, and IV. The admission price for one feature at the Center is $2 for children, $3 for students, and $4 for adults. The attendance for the Sunday matinee is given by the matrix

$$\begin{array}{cccc} & \text{Children} & \text{Students} & \text{Adults} \end{array}$$
$$A = \begin{array}{c} \text{Cinema I} \\ \text{Cinema II} \\ \text{Cinema III} \\ \text{Cinema IV} \end{array} \begin{bmatrix} 225 & 110 & 50 \\ 75 & 180 & 225 \\ 280 & 85 & 110 \\ 0 & 250 & 225 \end{bmatrix}$$

Write a column vector B representing the admission prices. Then compute AB, the column vector showing the gross

receipts for each theater. Finally, find the total revenue collected at the Cinema Center for admission that Sunday afternoon.

45. POLITICS: VOTER AFFILIATION Matrix A gives the percentage of eligible voters in the city of Newton, classified according to party affiliation and age group.

$$\begin{array}{cccc} & \text{Dem.} & \text{Rep.} & \text{Ind.} \end{array}$$
$$A = \begin{array}{c} \text{Under 30} \\ \text{30 to 50} \\ \text{Over 50} \end{array} \begin{bmatrix} 0.50 & 0.30 & 0.20 \\ 0.45 & 0.40 & 0.15 \\ 0.40 & 0.50 & 0.10 \end{bmatrix}$$

The population of eligible voters in the city by age group is given by the matrix B:

$$\begin{array}{ccc} & \text{Under 30} & \text{30 to 50} & \text{Over 50} \end{array}$$
$$B = \begin{bmatrix} 30{,}000 & 40{,}000 & 20{,}000 \end{bmatrix}$$

Find a matrix giving the total number of eligible voters in the city who will vote Democratic, Republican, and Independent.

46. 401(K) RETIREMENT PLANS Three network consultants, Alan, Maria, and Steven, each received a year-end bonus of $10,000, which they decided to invest in a 401(K) retirement plan sponsored by their employer. Under this plan, each employee is allowed to place their investments in three funds—an equity index fund (I), a growth fund (II), and a global equity fund (III). The allocations of the investments (in dollars) of the three employees at the beginning of the year are summarized in the matrix

$$\begin{array}{cccc} & \text{I} & \text{II} & \text{III} \end{array}$$
$$A = \begin{array}{c} \text{Alan} \\ \text{Maria} \\ \text{Steven} \end{array} \begin{bmatrix} 4000 & 3000 & 3000 \\ 2000 & 5000 & 3000 \\ 2000 & 3000 & 5000 \end{bmatrix}$$

The returns of the three funds after 1 yr are given in the matrix

$$B = \begin{array}{c} \text{I} \\ \text{II} \\ \text{III} \end{array} \begin{bmatrix} 0.18 \\ 0.24 \\ 0.12 \end{bmatrix}$$

Which employee realized the best returns on his or her investment for the year in question? The worst return?

47. COLLEGE ADMISSIONS A university admissions committee anticipates an enrollment of 8000 students in its freshman class next year. To satisfy admission quotas, incoming students have been categorized according to their sex and place of residence. The number of students in each category is given by the matrix

		Male	Female
	In-state	2700	3000
$A =$	Out-of-state	800	700
	Foreign	500	300

By using data accumulated in previous years, the admissions committee has determined that these students will elect to enter the College of Letters and Science, the College of Fine Arts, the School of Business Administration, and the School of Engineering according to the percentages that appear in the following matrix:

		L. & S.	Fine Arts	Bus. Ad.	Eng.
$B =$	Male	0.25	0.20	0.30	0.25
	Female	0.30	0.35	0.25	0.10

Find the matrix AB that shows the number of in-state, out-of-state, and foreign students expected to enter each discipline.

48. **PRODUCTION PLANNING** Refer to Example 6 in this section. Suppose Ace Novelty received an order from another amusement park for 1200 Pink Panthers, 1800 Giant Pandas, and 1400 Big Birds. The quantity of each type of stuffed animal to be produced at each plant is shown in the following production matrix:

		Panthers	Pandas	Birds
$P =$	L.A.	700	1000	800
	Seattle	500	800	600

Each Panther requires 1.3 yd^2 of plush, 20 ft^3 of stuffing, and 12 pieces of trim. Assume the materials required to produce the other two stuffed animals and the unit cost for each type of material are the same as those given in Example 6.
a. How much of each type of material must be purchased for each plant?
b. What is the total cost of materials that will be incurred at each plant?
c. What is the total cost of materials incurred by Ace Novelty in filling the order?

49. **COMPUTING PHONE BILLS** Cindy regularly makes long distance phone calls to three foreign cities—London, Tokyo, and Hong Kong. The matrices A and B give the lengths (in minutes) of her calls during peak and nonpeak hours, respectively, to each of these three cities during the month of June.

	London	Tokyo	Hong Kong
$A = [$	80	60	40 $]$

and

	London	Tokyo	Hong Kong
$B = [$	300	150	250 $]$

The costs for the calls (in dollars per minute) for the peak and nonpeak periods in the month in question are given, respectively, by the matrices

	London		.34
$C =$	Tokyo		.42
	Hong Kong		.48

and

	London		.24
$D =$	Tokyo		.31
	Hong Kong		.35

Compute the matrix $AC + BD$ and explain what it represents.

50. **PRODUCTION PLANNING** The total output of loudspeaker systems of the Acrosonic Company in their three production facilities for May and June is given by the matrices A and B, respectively, where

		Model A	Model B	Model C	Model D
	Location I	320	280	460	280
$A =$	Location II	480	360	580	0
	Location III	540	420	200	880

		Model A	Model B	Model C	Model D
	Location I	210	180	330	180
$B =$	Location II	400	300	450	40
	Location III	420	280	180	740

The unit production costs and selling prices for these loudspeakers are given by matrices C and D, respectively, where

Model A	120
Model B	180
Model C	260
Model D	500

$C =$ (the above) and $D =$

Model A	160
Model B	250
Model C	350
Model D	700

Compute the following matrices and explain the meaning of the entries in each matrix.
a. AC b. AD c. BC d. BD e. $(A + B)C$
f. $(A + B)D$ g. $A(D - C)$
h. $B(D - C)$ i. $(A + B)(D - C)$

51. **DIET PLANNING** A dietitian plans a meal around three foods. The number of units of vitamin A, vitamin C, and calcium in each ounce of these foods is represented by the matrix M, where

		Food I	Food II	Food III
	Vitamin A	400	1200	800
$M =$	Vitamin C	110	570	340
	Calcium	90	30	60

The matrices A and B represent the amount of each food (in ounces) consumed by a girl at two different meals, where

	Food I	Food II	Food III
$A =$	[7	1	6]

	Food I	Food II	Food III
$B =$	[9	3	2]

Calculate the following matrices and explain the meaning of the entries in each matrix:
a. MA^T b. MB^T c. $M(A + B)^T$

52. **PRODUCTION PLANNING** Hartman Lumber Company has two branches in the city. The sales of four of its products for the last year (in thousands of dollars) are represented by the matrix

$$
B = \begin{matrix} \text{Branch I} \\ \text{Branch II} \end{matrix}
\overset{\begin{matrix} \text{A} & \text{B} & \text{C} & \text{D} \end{matrix}}
{\begin{bmatrix} 5 & 2 & 8 & 10 \\ 3 & 4 & 6 & 8 \end{bmatrix}}
$$

For the present year, management has projected that the sales of the four products in branch I will be 10% over the corresponding sales for last year and the sales of the four products in branch II will be 15% over the corresponding sales for last year.

a. Show that the sales of the four products in the two branches for the current year are given by the matrix AB where

$$
A = \begin{bmatrix} 1.1 & 0 \\ 0 & 1.15 \end{bmatrix}
$$

Compute AB.

b. Hartman has m branches nationwide, and the sales of n of its products (in thousands of dollars) last year are represented by the matrix

$$
B = \begin{matrix} \text{Branch 1} \\ \text{Branch 2} \\ \vdots \\ \text{Branch } m \end{matrix}
\overset{\begin{matrix} 1 & 2 & 3 & \cdots & n \end{matrix}}
{\begin{bmatrix} a_{11} & a_{12} & a_{13} & \cdots & a_{1n} \\ a_{21} & a_{22} & a_{23} & \cdots & a_{2n} \\ \vdots & \vdots & \vdots & & \vdots \\ a_{m1} & a_{m2} & a_{m3} & \cdots & a_{mn} \end{bmatrix}}
$$

Also, management has projected that the sales of the n products in branch 1, branch 2, . . . , branch m will be $r_1\%$, $r_2\%$, . . . , $r_m\%$, respectively, over the corresponding sales for last year. Write the matrix A such that AB gives the sales of the n products in the m branches for the current year.

In Exercises 53–56, determine whether the statement is true or false. If it is true, explain why it is true. If it is false, give an example to show why it is false.

53. If A and B are matrices such that AB and BA are both defined, then A and B must be square matrices of the same order.

54. If A and B are matrices such that AB is defined and if c is a scalar, then $(cA)B = A(cB) = cAB$.

55. If A, B, and C are matrices and $A(B + C)$ is defined, then B must have the same size as C and the number of columns of A must be equal to the number of rows of B.

56. If A is a 2 × 4 matrix and B is a matrix such that ABA is defined, then the size of B must be 4 × 2.

2.5 Solutions to Self-Check Exercises

1. We compute

$$
\begin{bmatrix} 1 & 3 & 0 \\ 2 & 4 & -1 \end{bmatrix}
\begin{bmatrix} 3 & 1 & 4 \\ 2 & 0 & 3 \\ 1 & 2 & -1 \end{bmatrix}
$$

$$
= \begin{bmatrix} 1(3) + 3(2) + 0(1) & 1(1) + 3(0) + 0(2) & 1(4) + 3(3) + 0(-1) \\ 2(3) + 4(2) - 1(1) & 2(1) + 4(0) - 1(2) & 2(4) + 4(3) - 1(-1) \end{bmatrix}
$$

$$
= \begin{bmatrix} 9 & 1 & 13 \\ 13 & 0 & 21 \end{bmatrix}
$$

2. Let

$$
A = \begin{bmatrix} 0 & 1 & -2 \\ 2 & -1 & 3 \\ 1 & 0 & 4 \end{bmatrix} \quad X = \begin{bmatrix} x \\ y \\ z \end{bmatrix} \quad B = \begin{bmatrix} 1 \\ 0 \\ 7 \end{bmatrix}
$$

Then, the given system may be written as the matrix equation

$$
AX = B
$$

3. Write

$$
B = \begin{bmatrix} 54 \\ 113 \\ 112 \\ 70 \end{bmatrix} \begin{matrix} \text{AT\&T} \\ \text{AOL} \\ \text{IBM} \\ \text{GM} \end{matrix}
$$

and compute

$$
AB = \begin{matrix} \text{Ash} \\ \text{Joan} \end{matrix} \begin{bmatrix} 2000 & 1000 & 500 & 5000 \\ 1000 & 2500 & 2000 & 0 \end{bmatrix} \begin{bmatrix} 54 \\ 113 \\ 112 \\ 70 \end{bmatrix}
$$

$$
= \begin{bmatrix} 627,000 \\ 560,500 \end{bmatrix} \begin{matrix} \text{Ash} \\ \text{Joan} \end{matrix}
$$

We conclude that Ash's stock holdings were worth \$627,000 and Joan's stock holdings were worth \$560,500 on June 1.

USING TECHNOLOGY

▬ Matrix Multiplication

Graphing Utility

A graphing utility can be used to perform matrix multiplication.

EXAMPLE 1 Let

$$A = \begin{bmatrix} 1.2 & 3.1 & -1.4 \\ 2.7 & 4.2 & 3.4 \end{bmatrix}$$

$$B = \begin{bmatrix} 0.8 & 1.2 & 3.7 \\ 6.2 & -0.4 & 3.3 \end{bmatrix}$$

$$C = \begin{bmatrix} 1.2 & 2.1 & 1.3 \\ 4.2 & -1.2 & 0.6 \\ 1.4 & 3.2 & 0.7 \end{bmatrix}$$

Find (a) AC and (b) $(1.1A + 2.3B)C$.

Solution First, we enter the matrices A, B, and C into the calculator.

a. Using matrix operations, we enter the expression $A*C$. We obtain the matrix

$$\begin{bmatrix} 12.5 & -5.68 & 2.44 \\ 25.64 & 11.51 & 8.41 \end{bmatrix}$$

(You need to scroll the display on the screen to obtain the complete matrix.)

b. Using matrix operations, we enter the expression $(1.1A + 2.3B)C$. We obtain the matrix

$$\begin{bmatrix} 39.464 & 21.536 & 12.689 \\ 52.078 & 67.999 & 32.55 \end{bmatrix}$$

Excel

We use the MMULT function in Excel to perform matrix multiplication.

EXAMPLE 2 Let

$$A = \begin{bmatrix} 1.2 & 3.1 & -1.4 \\ 2.7 & 4.2 & 3.4 \end{bmatrix} \quad B = \begin{bmatrix} 0.8 & 1.2 & 3.7 \\ 6.2 & -0.4 & 3.3 \end{bmatrix} \quad C = \begin{bmatrix} 1.2 & 2.1 & 1.3 \\ 4.2 & -1.2 & 0.6 \\ 1.4 & 3.2 & 0.7 \end{bmatrix}$$

Find (a) AC and (b) $(1.1A + 2.3B)C$.

Note: Boldfaced words/characters in a box (for example, [**Enter**]) indicate that an action (click, select, or press) is required. Words/characters printed blue (for example, Chart sub-type:) indicate words/characters that appear on the screen. Words/characters printed in a typewriter font (for example, =(-2/3)*A2+2) indicate words/characters that need to be typed and entered.

Solution

a. *First, enter the matrices A, B, and C onto a spreadsheet* (Figure T1).

	A	B	C	D	E	F	G
1		A					B	
2	1.2	3.1	−1.4		0.8	1.2	3.7
3	2.7	4.2	3.4		6.2	−0.4	3.3
4							
5		C					
6	1.2	2.1	1.3				
7	4.2	−1.2	0.6				
8	1.4	3.2	0.7				

FIGURE T1
Spreadsheet showing the matrices *A*, *B*, and *C*

Second, compute AC. Highlight the cells that will contain the matrix product *AC*, which has order 2 × 3. Type =MMULT(, highlight the cells in matrix *A*, type ,, highlight the cells in matrix *C*, type), and press **Ctrl-Shift-Enter** . The matrix product *AC* shown in Figure T2 will appear on your spreadsheet.

	A	B	C
10		AC	
11	12.5	−5.68	2.44
12	25.64	11.51	8.41

FIGURE T2
The matrix product *AC*

b. *Compute (1.1A + 2.3B)C.* Highlight the cells that will contain the matrix product (1.1*A* + 2.3*B*)*C*. Next, type =MMULT(1.1*, highlight the cells in matrix *A*, type +2.3*, highlight the cells in matrix *B*, type ,, highlight the cells in matrix *C*, type), and then press **Ctrl-Shift-Enter** . The matrix product shown in Figure T3 will appear on your spreadsheet.

	A	B	C
13		(1.1A+2.3B)C	
14	39.464	21.536	12.689
15	52.078	67.999	32.55

FIGURE T3
The matrix product (1.1*A* + 2.3*B*)*C*

(*continued*)

TECHNOLOGY EXERCISES

In Exercises 1–8, refer to the following matrices and perform the indicated operations. Express your answers accurate to two decimal places.

$$A = \begin{bmatrix} 1.2 & 3.1 & -1.2 & 4.3 \\ 7.2 & 6.3 & 1.8 & -2.1 \\ 0.8 & 3.2 & -1.3 & 2.8 \end{bmatrix}$$

$$B = \begin{bmatrix} 0.7 & 0.3 & 1.2 & -0.8 \\ 1.2 & 1.7 & 3.5 & 4.2 \\ -3.3 & -1.2 & 4.2 & 3.2 \end{bmatrix}$$

$$C = \begin{bmatrix} 0.8 & 7.1 & 6.2 \\ 3.3 & -1.2 & 4.8 \\ 1.3 & 2.8 & -1.5 \\ 2.1 & 3.2 & -8.4 \end{bmatrix}$$

1. AC

2. CB

3. $(A + B)C$

4. $(2A + 3B)C$

5. $(2A - 3.1B)C$

6. $C(2.1A + 3.2B)$

7. $(4.1A + 2.7B)1.6C$

8. $2.5C(1.8A - 4.3B)$

In Exercises 9–12, refer to the following matrices and perform the indicated operations. Express your answers accurate to two decimal places.

$$A = \begin{bmatrix} 2 & 5 & -4 & 2 & 8 \\ 6 & 7 & 2 & 9 & 6 \\ 4 & 5 & 4 & 4 & 4 \\ 9 & 6 & 8 & 3 & 2 \end{bmatrix}$$

$$B = \begin{bmatrix} 2 & 6 & 7 & 5 \\ 3 & 4 & 6 & 2 \\ -5 & 8 & 4 & 3 \\ 8 & 6 & 9 & 5 \\ 4 & 7 & 8 & 8 \end{bmatrix}$$

$$C = \begin{bmatrix} 6.2 & 7.3 & -4.0 & 7.1 & 9.3 \\ 4.8 & 6.5 & 8.4 & -6.3 & 8.4 \\ 5.4 & 3.2 & 6.3 & 9.1 & -2.8 \\ 8.2 & 7.3 & 6.5 & 4.1 & 9.8 \\ 10.3 & 6.8 & 4.8 & -9.1 & 20.4 \end{bmatrix}$$

$$D = \begin{bmatrix} 4.6 & 3.9 & 8.4 & 6.1 & 9.8 \\ 2.4 & -6.8 & 7.9 & 11.4 & 2.9 \\ 7.1 & 9.4 & 6.3 & 5.7 & 4.2 \\ 3.4 & 6.1 & 5.3 & 8.4 & 6.3 \\ 7.1 & -4.2 & 3.9 & -6.4 & 7.1 \end{bmatrix}$$

9. Find AB and BA.

10. Find CD and DC. Is $CD = DC$?

11. Find $AC + AD$.

12. Find
  a. AC
  b. AD
  c. $A(C + D)$
  d. Is $A(C + D) = AC + AD$?

2.6   The Inverse of a Square Matrix

▬ The Inverse of a Square Matrix

In this section, we discuss a procedure for finding the inverse of a matrix and show how the inverse can be used to help us solve a system of linear equations. The inverse of a matrix also plays a central role in the Leontief input–output model, which we will discuss in Section 2.7.

Recall that if a is a nonzero real number, then there exists a unique real number a^{-1} $\left(\text{that is, } \frac{1}{a}\right)$ such that

$$a^{-1}a = \left(\frac{1}{a}\right)(a) = 1$$

The use of the (multiplicative) inverse of a real number enables us to solve algebraic equations of the form

$$ax = b \tag{12}$$

For if $a \neq 0$, then $a^{-1} = \frac{1}{a}$. Upon multiplying both sides of (12) by a^{-1}, we have

$$a^{-1}(ax) = a^{-1}b$$
$$\left(\frac{1}{a}\right)(ax) = \frac{1}{a}(b)$$
$$x = \frac{b}{a}$$

For example, since the inverse of 2 is $2^{-1} = \frac{1}{2}$, we can solve the equation

$$2x = 5$$

by multiplying both sides of the equation by $2^{-1} = \frac{1}{2}$, giving

$$2^{-1}(2x) = 2^{-1} \cdot 5$$
$$x = \frac{5}{2}$$

We can use a similar procedure to solve the matrix equation

$$AX = B$$

where A, X, and B are matrices of the proper sizes. To do this we need the matrix equivalent of the inverse of a real number. Such a matrix, whenever it exists, is called the **inverse of a matrix.**

Inverse of a Matrix

Let A be a square matrix of size n. A square matrix A^{-1} of size n such that

$$A^{-1}A = AA^{-1} = I_n$$

is called the inverse of A.

Let's show that the matrix

$$A = \begin{bmatrix} 1 & 2 \\ 3 & 4 \end{bmatrix}$$

has as its inverse

$$A^{-1} = \begin{bmatrix} -2 & 1 \\ \frac{3}{2} & -\frac{1}{2} \end{bmatrix}$$

EXPLORE & DISCUSS

In defining the inverse of a matrix A, why is it necessary to require that A be a square matrix?

Since

$$AA^{-1} = \begin{bmatrix} 1 & 2 \\ 3 & 4 \end{bmatrix} \begin{bmatrix} -2 & 1 \\ \frac{3}{2} & -\frac{1}{2} \end{bmatrix} = \begin{bmatrix} 1 & 0 \\ 0 & 1 \end{bmatrix} = I$$

$$A^{-1}A = \begin{bmatrix} -2 & 1 \\ \frac{3}{2} & -\frac{1}{2} \end{bmatrix} \begin{bmatrix} 1 & 2 \\ 3 & 4 \end{bmatrix} = \begin{bmatrix} 1 & 0 \\ 0 & 1 \end{bmatrix} = I$$

we see that A^{-1} is the inverse of A, as asserted.

Not every square matrix has an inverse. A square matrix that has an inverse is said to be **nonsingular.** A matrix that does not have an inverse is said to be **singular.** An example of a singular matrix is given by

$$B = \begin{bmatrix} 0 & 1 \\ 0 & 0 \end{bmatrix}$$

If B had an inverse given by

$$B^{-1} = \begin{bmatrix} a & b \\ c & d \end{bmatrix}$$

where a, b, c, and d are some appropriate numbers, then, by the definition of an inverse, we would have $BB^{-1} = I$; that is,

$$\begin{bmatrix} 0 & 1 \\ 0 & 0 \end{bmatrix} \begin{bmatrix} a & b \\ c & d \end{bmatrix} = \begin{bmatrix} 1 & 0 \\ 0 & 1 \end{bmatrix}$$

$$\begin{bmatrix} c & d \\ 0 & 0 \end{bmatrix} = \begin{bmatrix} 1 & 0 \\ 0 & 1 \end{bmatrix}$$

which implies that $0 = 1$—an impossibility! This contradiction shows that B does not have an inverse.

▬ A Method for Finding the Inverse of a Square Matrix

The methods of Section 2.5 can be used to find the inverse of a nonsingular matrix. To discover such an algorithm, let's find the inverse of the matrix A, given by

$$A = \begin{bmatrix} 1 & 2 \\ -1 & 3 \end{bmatrix}$$

Suppose A^{-1} exists and is given by

$$A^{-1} = \begin{bmatrix} a & b \\ c & d \end{bmatrix}$$

where a, b, c, and d are to be determined. By the definition of an inverse, we have $AA^{-1} = I$; that is,

$$\begin{bmatrix} 1 & 2 \\ -1 & 3 \end{bmatrix} \begin{bmatrix} a & b \\ c & d \end{bmatrix} = \begin{bmatrix} 1 & 0 \\ 0 & 1 \end{bmatrix}$$

which simplifies to

$$\begin{bmatrix} a + 2c & b + 2d \\ -a + 3c & -b + 3d \end{bmatrix} = \begin{bmatrix} 1 & 0 \\ 0 & 1 \end{bmatrix}$$

But this matrix equation is equivalent to the two systems of linear equations

$$\left. \begin{aligned} a + 2c &= 1 \\ -a + 3c &= 0 \end{aligned} \right\} \quad \text{and} \quad \left. \begin{aligned} b + 2d &= 0 \\ -b + 3d &= 1 \end{aligned} \right\}$$

with augmented matrices given by

$$\left[\begin{array}{cc|c} 1 & 2 & 1 \\ -1 & 3 & 0 \end{array} \right] \quad \text{and} \quad \left[\begin{array}{cc|c} 1 & 2 & 0 \\ -1 & 3 & 1 \end{array} \right]$$

Note that the matrices of coefficients of the two systems are identical. This suggests that we solve the two systems of simultaneous linear equations by writing the following augmented matrix, which we obtain by joining the coefficient matrix and the two columns of constants:

$$\left[\begin{array}{cc|cc} 1 & 2 & 1 & 0 \\ -1 & 3 & 0 & 1 \end{array} \right]$$

Using the Gauss–Jordan elimination method, we obtain the following sequence of equivalent matrices:

$$\left[\begin{array}{cc|cc} 1 & 2 & 1 & 0 \\ -1 & 3 & 0 & 1 \end{array} \right] \xrightarrow{R_2 + R_1} \left[\begin{array}{cc|cc} 1 & 2 & 1 & 0 \\ 0 & 5 & 1 & 1 \end{array} \right] \xrightarrow{\frac{1}{5}R_2}$$

$$\left[\begin{array}{cc|cc} 1 & 2 & 1 & 0 \\ 0 & 1 & \frac{1}{5} & \frac{1}{5} \end{array} \right] \xrightarrow{R_1 - 2R_2} \left[\begin{array}{cc|cc} 1 & 0 & \frac{3}{5} & -\frac{2}{5} \\ 0 & 1 & \frac{1}{5} & \frac{1}{5} \end{array} \right]$$

Thus, $a = \frac{3}{5}$, $c = \frac{1}{5}$, $b = -\frac{2}{5}$, and $d = \frac{1}{5}$, giving

$$A^{-1} = \begin{bmatrix} \frac{3}{5} & -\frac{2}{5} \\ \frac{1}{5} & \frac{1}{5} \end{bmatrix}$$

The following computations verify that A^{-1} is indeed the inverse of A:

$$\begin{bmatrix} 1 & 2 \\ -1 & 3 \end{bmatrix} \begin{bmatrix} \frac{3}{5} & -\frac{2}{5} \\ \frac{1}{5} & \frac{1}{5} \end{bmatrix} = \begin{bmatrix} 1 & 0 \\ 0 & 1 \end{bmatrix} = \begin{bmatrix} \frac{3}{5} & -\frac{2}{5} \\ \frac{1}{5} & \frac{1}{5} \end{bmatrix} \begin{bmatrix} 1 & 2 \\ -1 & 3 \end{bmatrix}$$

The preceding example suggests the general algorithm for computing the inverse of a square matrix of size n when it exists.

Finding the Inverse of a Matrix
Given the $n \times n$ matrix A:

1. Adjoin the $n \times n$ identity matrix I to obtain the augmented matrix

$$[A \mid I]$$

2. Use a sequence of row operations to reduce $[A \mid I]$ to the form

$$[I \mid B]$$

  if possible.

The matrix B is the inverse of A.

Note Although matrix multiplication is not generally commutative, it is possible to prove that if A has an inverse and $AB = I$, then $BA = I$ also. Hence, to verify that B is the inverse of A, it suffices to show that $AB = I$. ■

EXAMPLE 1 Find the inverse of the matrix

$$A = \begin{bmatrix} 2 & 1 & 1 \\ 3 & 2 & 1 \\ 2 & 1 & 2 \end{bmatrix}$$

Solution We form the augmented matrix

$$\begin{bmatrix} 2 & 1 & 1 & | & 1 & 0 & 0 \\ 3 & 2 & 1 & | & 0 & 1 & 0 \\ 2 & 1 & 2 & | & 0 & 0 & 1 \end{bmatrix}$$

and use the Gauss–Jordan elimination method to reduce it to the form $[I \mid B]$:

$$\begin{bmatrix} 2 & 1 & 1 & | & 1 & 0 & 0 \\ 3 & 2 & 1 & | & 0 & 1 & 0 \\ 2 & 1 & 2 & | & 0 & 0 & 1 \end{bmatrix} \xrightarrow{R_1 - R_2} \begin{bmatrix} -1 & -1 & 0 & | & 1 & -1 & 0 \\ 3 & 2 & 1 & | & 0 & 1 & 0 \\ 2 & 1 & 2 & | & 0 & 0 & 1 \end{bmatrix}$$

$$\xrightarrow[\substack{-R_1 \\ R_2 + 3R_1 \\ R_3 + 2R_1}]{} \begin{bmatrix} 1 & 1 & 0 & | & -1 & 1 & 0 \\ 0 & -1 & 1 & | & 3 & -2 & 0 \\ 0 & -1 & 2 & | & 2 & -2 & 1 \end{bmatrix}$$

$$\xrightarrow[\substack{R_1 + R_2 \\ -R_2 \\ R_3 - R_2}]{} \begin{bmatrix} 1 & 0 & 1 & | & 2 & -1 & 0 \\ 0 & 1 & -1 & | & -3 & 2 & 0 \\ 0 & 0 & 1 & | & -1 & 0 & 1 \end{bmatrix}$$

$$\xrightarrow[\substack{R_1 - R_3 \\ R_2 + R_3}]{} \begin{bmatrix} 1 & 0 & 0 & | & 3 & -1 & -1 \\ 0 & 1 & 0 & | & -4 & 2 & 1 \\ 0 & 0 & 1 & | & -1 & 0 & 1 \end{bmatrix}$$

The inverse of A is the matrix

$$A^{-1} = \begin{bmatrix} 3 & -1 & -1 \\ -4 & 2 & 1 \\ -1 & 0 & 1 \end{bmatrix}$$

We leave it to you to verify these results. ■

Example 2 illustrates what happens to the reduction process when a matrix A does *not* have an inverse.

EXAMPLE 2 Find the inverse of the matrix

$$A = \begin{bmatrix} 1 & 2 & 3 \\ 2 & 1 & 2 \\ 3 & 3 & 5 \end{bmatrix}$$

Solution We form the augmented matrix

$$\left[\begin{array}{ccc|ccc} 1 & 2 & 3 & 1 & 0 & 0 \\ 2 & 1 & 2 & 0 & 1 & 0 \\ 3 & 3 & 5 & 0 & 0 & 1 \end{array}\right]$$

and use the Gauss–Jordan elimination method:

$$\left[\begin{array}{ccc|ccc} 1 & 2 & 3 & 1 & 0 & 0 \\ 2 & 1 & 2 & 0 & 1 & 0 \\ 3 & 3 & 5 & 0 & 0 & 1 \end{array}\right] \xrightarrow[R_3 - 3R_1]{R_2 - 2R_1} \left[\begin{array}{ccc|ccc} 1 & 2 & 3 & 1 & 0 & 0 \\ 0 & -3 & -4 & -2 & 1 & 0 \\ 0 & -3 & -4 & -3 & 0 & 1 \end{array}\right]$$

$$\xrightarrow[R_3 - R_2]{-R_2} \left[\begin{array}{ccc|ccc} 1 & 2 & 3 & 1 & 0 & 0 \\ 0 & 3 & 4 & 2 & -1 & 0 \\ 0 & 0 & 0 & -1 & -1 & 1 \end{array}\right]$$

Since all entries in the last row of the 3 × 3 submatrix that comprises the left-hand side of the augmented matrix just obtained are all equal to zero, the latter cannot be reduced to the form $[I \mid B]$. Accordingly, we draw the conclusion that A is singular—that is, does not have an inverse. ∎

More generally, we have the following criterion for determining when the inverse of a matrix does not exist.

EXPLORE & DISCUSS

Explain in terms of solutions to systems of linear equations why the final augmented matrix in Example 2 implies that A has no inverse. *Hint:* See the discussion on pages 138–139.

Matrices That Have No Inverses
If there is a row to the left of the vertical line in the augmented matrix containing all zeros, then the matrix does not have an inverse.

A Formula for the Inverse of a 2 × 2 Matrix

Before turning to some applications, we show an alternative method that employs a formula for finding the inverse of a 2 × 2 matrix. This method will prove useful in many situations—we will see an application in Example 5. The derivation of this formula is left as an exercise (Exercise 48).

Formula for the Inverse of a 2 × 2 Matrix
Let

$$A = \begin{bmatrix} a & b \\ c & d \end{bmatrix}$$

Suppose $D = ad - bc$ is not equal to zero. Then, A^{-1} exists and is given by

$$A^{-1} = \frac{1}{D}\begin{bmatrix} d & -b \\ -c & a \end{bmatrix} \qquad\qquad (13)$$

Note   As an aid to memorizing the formula, note that D is the product of the elements along the main diagonal minus the product of the elements along the other diagonal:

$$\begin{bmatrix} a & b \\ c & d \end{bmatrix} \qquad\qquad D = ad - bc$$

Main diagonal

Next, the matrix

$$\begin{bmatrix} d & -b \\ -c & a \end{bmatrix}$$

is obtained by interchanging a and d and reversing the signs of b and c. Finally, A^{-1} is obtained by dividing this matrix by D. ∎

EXAMPLE 3   Find the inverse of

$$A = \begin{bmatrix} 1 & 2 \\ 3 & 4 \end{bmatrix}$$

Solution   We first compute $D = (1)(4) - (2)(3) = 4 - 6 = -2$. Next, we write the matrix

$$\begin{bmatrix} 4 & -2 \\ -3 & 1 \end{bmatrix}$$

Finally, dividing this matrix by D, we obtain

$$A^{-1} = \frac{1}{-2}\begin{bmatrix} 4 & -2 \\ -3 & 1 \end{bmatrix} = \begin{bmatrix} -2 & 1 \\ \frac{3}{2} & -\frac{1}{2} \end{bmatrix}$$

∎

Solving Systems of Equations with Inverses

We now show how the inverse of a matrix may be used to solve certain systems of linear equations in which the number of equations in the system is equal to the number of variables. For simplicity, let's illustrate the process for a system of three linear equations in three variables:

$$\begin{aligned} a_{11}x_1 + a_{12}x_2 + a_{13}x_3 &= c_1 \\ a_{21}x_1 + a_{22}x_2 + a_{23}x_3 &= c_2 \\ a_{31}x_1 + a_{32}x_2 + a_{33}x_3 &= c_3 \end{aligned} \qquad\qquad (14)$$

Let's write

$$A = \begin{bmatrix} a_{11} & a_{12} & a_{13} \\ a_{21} & a_{22} & a_{23} \\ a_{31} & a_{32} & a_{33} \end{bmatrix} \qquad X = \begin{bmatrix} x_1 \\ x_2 \\ x_3 \end{bmatrix} \qquad B = \begin{bmatrix} b_1 \\ b_2 \\ b_3 \end{bmatrix}$$

You should verify that System (14) of linear equations may be written in the form of the matrix equation

$$AX = B \qquad\qquad (15)$$

If A is nonsingular, then the method of this section may be used to compute A^{-1}. Next, multiplying both sides of Equation (15) by A^{-1} (on the left), we obtain

$$A^{-1}AX = A^{-1}B \quad \text{or} \quad IX = A^{-1}B \quad \text{or} \quad X = A^{-1}B$$

the desired solution to the problem.

In the case of a system of n equations with n unknowns, we have the following, more general result.

Using Inverses to Solve Systems of Equations

If $AX = B$ is a linear system of n equations in n unknowns and if A^{-1} exists, then

$$X = A^{-1}B$$

is the unique solution of the system.

The use of inverses to solve systems of equations is particularly advantageous when we are required to solve more than one system of equations, $AX = B$, involving the same coefficient matrix, A, and different matrices of constants, B. As you will see in Examples 4 and 5, we need to compute A^{-1} just once in each case.

EXAMPLE 4 Solve the following systems of linear equations:

a. $\begin{aligned} 2x + y + z &= 1 \\ 3x + 2y + z &= 2 \\ 2x + y + 2z &= -1 \end{aligned}$ **b.** $\begin{aligned} 2x + y + z &= 2 \\ 3x + 2y + z &= -3 \\ 2x + y + 2z &= 1 \end{aligned}$

Solution We may write the given systems of equations in the form

$$AX = B \quad \text{and} \quad AX = C$$

respectively, where

$$A = \begin{bmatrix} 2 & 1 & 1 \\ 3 & 2 & 1 \\ 2 & 1 & 2 \end{bmatrix} \quad X = \begin{bmatrix} x \\ y \\ z \end{bmatrix} \quad B = \begin{bmatrix} 1 \\ 2 \\ -1 \end{bmatrix} \quad C = \begin{bmatrix} 2 \\ -3 \\ 1 \end{bmatrix}$$

The inverse of the matrix A,

$$A^{-1} = \begin{bmatrix} 3 & -1 & -1 \\ -4 & 2 & 1 \\ -1 & 0 & 1 \end{bmatrix}$$

was found in Example 1. Using this result, we find that the solution of the first system (a) is

$$X = A^{-1}B = \begin{bmatrix} 3 & -1 & -1 \\ -4 & 2 & 1 \\ -1 & 0 & 1 \end{bmatrix} \begin{bmatrix} 1 \\ 2 \\ -1 \end{bmatrix}$$

$$= \begin{bmatrix} (3)(1) + (-1)(2) + (-1)(-1) \\ (-4)(1) + (2)(2) + (1)(-1) \\ (-1)(1) + (0)(2) + (1)(-1) \end{bmatrix} = \begin{bmatrix} 2 \\ -1 \\ -2 \end{bmatrix}$$

or $x = 2$, $y = -1$, and $z = -2$.

The solution of the second system (b) is

$$X = A^{-1}C = \begin{bmatrix} 3 & -1 & -1 \\ -4 & 2 & 1 \\ -1 & 0 & 1 \end{bmatrix} \begin{bmatrix} 2 \\ -3 \\ 1 \end{bmatrix} = \begin{bmatrix} 8 \\ -13 \\ -1 \end{bmatrix}$$

or $x = 8$, $y = -13$, and $z = -1$. ▪

APPLIED EXAMPLE 5   Capital Expenditure Planning   The management of Checkers Rent-A-Car plans to expand its fleet of rental cars for the
next quarter by purchasing compact and full-size cars. The average cost of a compact car is $10,000, and the average cost of a full-size car is $24,000.

a. If a total of 800 cars is to be purchased with a budget of $12 million, how
many cars of each size will be acquired?
b. If the predicted demand calls for a total purchase of 1000 cars with a budget
of $14 million, how many cars of each type will be acquired?

Solution   Let x and y denote the number of compact and full-size cars to be purchased. Furthermore, let n denote the total number of cars to be acquired and b
the amount of money budgeted for the purchase of these cars. Then,

$$x + \qquad y = n$$
$$10,000x + 24,000y = b$$

This system of two equations in two variables may be written in the matrix form

$$AX = B$$

where

$$A = \begin{bmatrix} 1 & 1 \\ 10,000 & 24,000 \end{bmatrix} \qquad X = \begin{bmatrix} x \\ y \end{bmatrix} \qquad B = \begin{bmatrix} n \\ b \end{bmatrix}$$

Therefore,

$$X = A^{-1}B$$

Since A is a 2×2 matrix, its inverse may be found by using Formula (13).
We find $D = (1)(24,000) - (1)(10,000) = 14,000$, so

$$A^{-1} = \frac{1}{14,000} \begin{bmatrix} 24,000 & -1 \\ -10,000 & 1 \end{bmatrix} = \begin{bmatrix} \frac{24,000}{14,000} & -\frac{1}{14,000} \\ -\frac{10,000}{14,000} & \frac{1}{14,000} \end{bmatrix}$$

Thus,

$$X = \begin{bmatrix} \frac{12}{7} & -\frac{1}{14,000} \\ -\frac{5}{7} & \frac{1}{14,000} \end{bmatrix} \begin{bmatrix} n \\ b \end{bmatrix}$$

a. Here, $n = 800$ and $b = 12,000,000$, so

$$X = A^{-1}B = \begin{bmatrix} \frac{12}{7} & -\frac{1}{14,000} \\ -\frac{5}{7} & \frac{1}{14,000} \end{bmatrix} \begin{bmatrix} 800 \\ 12,000,000 \end{bmatrix} = \begin{bmatrix} 514.3 \\ 285.7 \end{bmatrix}$$

Therefore, 514 compact cars and 286 full-size cars will be acquired in this
case.

b. Here, $n = 1000$ and $b = 14,000,000$, so

$$X = A^{-1}B = \begin{bmatrix} \frac{12}{7} & -\frac{1}{14,000} \\ -\frac{5}{7} & \frac{1}{14,000} \end{bmatrix} \begin{bmatrix} 1000 \\ 14,000,000 \end{bmatrix} = \begin{bmatrix} 714.3 \\ 285.7 \end{bmatrix}$$

Therefore, 714 compact cars and 286 full-size cars will be purchased in this case. ∎

2.6 Self-Check Exercises

1. Find the inverse of the matrix

$$A = \begin{bmatrix} 2 & 1 & -1 \\ 1 & 1 & -1 \\ -1 & -2 & 3 \end{bmatrix}$$

if it exists.

2. Solve the system of linear equations

$$
\begin{aligned}
2x + y - z &= b_1 \\
x + y - z &= b_2 \\
-x - 2y + 3z &= b_3
\end{aligned}
$$

where (a) $b_1 = 5$, $b_2 = 4$, $b_3 = -8$ and (b) $b_1 = 2$, $b_2 = 0$, $b_3 = 5$, by finding the inverse of the coefficient matrix.

3. Grand Canyon Tours offers air and ground scenic tours of the Grand Canyon. Tickets for the $7\frac{1}{2}$-hr tour cost $169 for an adult and $129 for a child, and each tour group is limited to 19 people. On three recent fully booked tours, total receipts were $2931 for the first tour, $3011 for the second tour, and $2771 for the third tour. Determine how many adults and how many children were in each tour.

Solutions to Self-Check Exercises 2.6 can be found on page 148.

2.6 Concept Questions

1. What is the inverse of a matrix A?

2. Explain how you would find the inverse of a nonsingular matrix.

3. Give the formula for the inverse of the 2×2 matrix

$$A = \begin{bmatrix} a & b \\ c & d \end{bmatrix}$$

4. Explain how the inverse of a matrix can be used to solve a system of n linear equations in n unknowns. Can the method work for a system of m linear equations in n unknowns with $m \neq n$? Explain.

2.6 Exercises

In Exercises 1–4, show that the matrices are inverses of each other by showing that their product is the identity matrix *I*.

1. $\begin{bmatrix} 1 & -3 \\ 1 & -2 \end{bmatrix}$ and $\begin{bmatrix} -2 & 3 \\ -1 & 1 \end{bmatrix}$

2. $\begin{bmatrix} 4 & 5 \\ 2 & 3 \end{bmatrix}$ and $\begin{bmatrix} \frac{3}{2} & -\frac{5}{2} \\ -1 & 2 \end{bmatrix}$

3. $\begin{bmatrix} 3 & 2 & 3 \\ 2 & 2 & 1 \\ 2 & 1 & 1 \end{bmatrix}$ and $\begin{bmatrix} -\frac{1}{3} & -\frac{1}{3} & \frac{4}{3} \\ 0 & 1 & -1 \\ \frac{2}{3} & -\frac{1}{3} & -\frac{2}{3} \end{bmatrix}$

4. $\begin{bmatrix} 2 & 4 & -2 \\ -4 & -6 & 1 \\ 3 & 5 & -1 \end{bmatrix}$ and $\begin{bmatrix} \frac{1}{2} & -3 & -4 \\ -\frac{1}{2} & 2 & 3 \\ -1 & 1 & 2 \end{bmatrix}$

In Exercises 5–16, find the inverse of the matrix, if it exists. Verify your answer.

5. $\begin{bmatrix} 2 & 5 \\ 1 & 3 \end{bmatrix}$

6. $\begin{bmatrix} 2 & 3 \\ 3 & 5 \end{bmatrix}$

7. $\begin{bmatrix} 3 & -3 \\ -2 & 2 \end{bmatrix}$

8. $\begin{bmatrix} 4 & 2 \\ 6 & 3 \end{bmatrix}$

9. $\begin{bmatrix} 2 & -3 & -4 \\ 0 & 0 & -1 \\ 1 & -2 & 1 \end{bmatrix}$

10. $\begin{bmatrix} 1 & -1 & 3 \\ 2 & 1 & 2 \\ -2 & -2 & 1 \end{bmatrix}$

11. $\begin{bmatrix} 4 & 2 & 2 \\ -1 & -3 & 4 \\ 3 & -1 & 6 \end{bmatrix}$

12. $\begin{bmatrix} 1 & 2 & 0 \\ -3 & 4 & -2 \\ -5 & 0 & -2 \end{bmatrix}$

13. $\begin{bmatrix} 1 & 4 & -1 \\ 2 & 3 & -2 \\ -1 & 2 & 3 \end{bmatrix}$

14. $\begin{bmatrix} 3 & -2 & 7 \\ -2 & 1 & 4 \\ 6 & -5 & 8 \end{bmatrix}$

15. $\begin{bmatrix} 1 & 1 & -1 & 1 \\ 2 & 1 & 1 & 0 \\ 2 & 1 & 0 & 1 \\ 2 & -1 & -1 & 3 \end{bmatrix}$

16. $\begin{bmatrix} 1 & 1 & 2 & 3 \\ 2 & 3 & 0 & -1 \\ 0 & 2 & -1 & 1 \\ 1 & 2 & 1 & 1 \end{bmatrix}$

In Exercises 17–24, (a) write a matrix equation that is equivalent to the system of linear equations and (b) solve the system using the inverses found in Exercises 5–16.

17. $2x + 5y = 3$
$\quad\ x + 3y = 2$
(See Exercise 5.)

18. $2x + 3y = 5$
$\quad 3x + 5y = 8$
(See Exercise 6.)

19. $2x - 3y - 4z = 4$
$\qquad\qquad -z = 3$
$\quad x - 2y + z = -8$
(See Exercise 9.)

20. $\quad x_1 - x_2 + 3x_3 = 2$
$\quad 2x_1 + x_2 + 2x_3 = 2$
$-2x_1 - 2x_2 + x_3 = 3$
(See Exercise 10.)

21. $\quad x + 4y - z = 3$
$\quad 2x + 3y - 2z = 1$
$-x + 2y + 3z = 7$
(See Exercise 13.)

22. $3x_1 - 2x_2 + 7x_3 = 6$
$-2x_1 + x_2 + 4x_3 = 4$
$6x_1 - 5x_2 + 8x_3 = 4$
(See Exercise 14.)

23. $x_1 + x_2 - x_3 + x_4 = 6$
$2x_1 + x_2 + x_3 \qquad = 4$
$2x_1 + x_2 \qquad + x_4 = 7$
$2x_1 - x_2 - x_3 + 3x_4 = 9$
(See Exercise 15.)

24. $x_1 + x_2 + 2x_3 + 3x_4 = 4$
$2x_1 + 3x_2 \qquad - x_4 = 11$
$\qquad 2x_2 - x_3 + x_4 = 7$
$x_1 + 2x_2 + x_3 + x_4 = 6$
(See Exercise 16.)

In Exercises 25–32, (a) write each system of equations as a matrix equation and (b) solve the system of equations by using the inverse of the coefficient matrix.

25. $\qquad\qquad\qquad x + 2y = b_1$
$\qquad\qquad\qquad 2x - y = b_2$
where (i) $b_1 = 14, b_2 = 5$
and (ii) $b_1 = 4, b_2 = -1$

26. $\qquad\qquad\qquad 3x - 2y = b_1$
$\qquad\qquad\qquad 4x + 3y = b_2$
where (i) $b_1 = -6, b_2 = 10$
and (ii) $b_1 = 3, b_2 = -2$

27. $\qquad\qquad x + 2y + z = b_1$
$\qquad\qquad x + y + z = b_2$
$\qquad\qquad 3x + y + z = b_3$
where (i) $b_1 = 7, b_2 = 4, b_3 = 2$
and (ii) $b_1 = 5, b_2 = -3, b_3 = -1$

28. $\qquad\qquad x_1 + x_2 + x_3 = b_1$
$\qquad\qquad x_1 - x_2 + x_3 = b_2$
$\qquad\qquad x_1 - 2x_2 - x_3 = b_3$
where (i) $b_1 = 5, b_2 = -3, b_3 = -1$
and (ii) $b_1 = 1, b_2 = 4, b_3 = -2$

29. $\qquad\qquad 3x + 2y - z = b_1$
$\qquad\qquad 2x - 3y + z = b_2$
$\qquad\qquad x - y - z = b_3$
where (i) $b_1 = 2, b_2 = -2, b_3 = 4$
and (ii) $b_1 = 8, b_2 = -3, b_3 = 6$

30. $\qquad\qquad 2x_1 + x_2 + x_3 = b_1$
$\qquad\qquad x_1 - 3x_2 + 4x_3 = b_2$
$\qquad\qquad -x_1 \qquad + x_3 = b_3$
where (i) $b_1 = 1, b_2 = 4, b_3 = -3$
and (ii) $b_1 = 2, b_2 = -5, b_3 = 0$

31. $\qquad\qquad x_1 + x_2 + x_3 + x_4 = b_1$
$\qquad\qquad x_1 - x_2 - x_3 + x_4 = b_2$
$\qquad\qquad x_2 + 2x_3 + 2x_4 = b_3$
$\qquad\qquad x_1 + 2x_2 + x_3 - 2x_4 = b_4$
where (i) $b_1 = 1, b_2 = -1, b_3 = 4, b_4 = 0$
and (ii) $b_1 = 2, b_2 = 8, b_3 = 4, b_4 = -1$

32. $\qquad\qquad x_1 + x_2 + 2x_3 + x_4 = b_1$
$\qquad\qquad 4x_1 + 5x_2 + 9x_3 + x_4 = b_2$
$\qquad\qquad 3x_1 + 4x_2 + 7x_3 + x_4 = b_3$
$\qquad\qquad 2x_1 + 3x_2 + 4x_3 + 2x_4 = b_4$
where (i) $b_1 = 3, b_2 = 6, b_3 = 5, b_4 = 7$
and (ii) $b_1 = 1, b_2 = -1, b_3 = 0, b_4 = -4$

33. Let

$$A = \begin{bmatrix} 2 & 3 \\ -4 & -5 \end{bmatrix}$$

a. Find A^{-1}.

b. Show that $(A^{-1})^{-1} = A$.

34. Let

$$A = \begin{bmatrix} 6 & -4 \\ -4 & 3 \end{bmatrix} \quad \text{and} \quad B = \begin{bmatrix} 3 & -5 \\ 4 & -7 \end{bmatrix}$$

a. Find AB, A^{-1}, and B^{-1}.
b. Show that $(AB)^{-1} = B^{-1}A^{-1}$.

35. Let

$$A = \begin{bmatrix} 2 & -5 \\ 1 & -3 \end{bmatrix} \quad B = \begin{bmatrix} 4 & 3 \\ 1 & 1 \end{bmatrix} \quad C = \begin{bmatrix} 2 & 3 \\ -2 & 1 \end{bmatrix}$$

a. Find ABC, A^{-1}, B^{-1}, and C^{-1}.
b. Show that $(ABC)^{-1} = C^{-1}B^{-1}A^{-1}$.

36. TICKET REVENUES Rainbow Harbor Cruises charges $8/adult and $4/child for a round-trip ticket. The records show that, on a certain weekend, 1000 people took the cruise on Saturday and 800 people took the cruise on Sunday. The total receipts for Saturday were $6400, and the total receipts for Sunday were $4800. Determine how many adults and children took the cruise on Saturday and on Sunday.

37. PRICING BelAir Publishing publishes a deluxe leather edition and a standard edition of its Daily Organizer. The company's marketing department estimates that x copies of the deluxe edition and y copies of the standard edition will be demanded per month when the unit prices are p dollars and q dollars, respectively, where x, y, p, and q are related by the following system of linear equations:

$$5x + y = 1000(70 - p)$$
$$x + 3y = 1000(40 - q)$$

Find the monthly demand for the deluxe edition and the standard edition when the unit prices are set according to the following schedules:
a. $p = 50$ and $q = 25$ **b.** $p = 45$ and $q = 25$
c. $p = 45$ and $q = 20$

38. NUTRITION/DIET PLANNING Bob, a nutritionist attached to the University Medical Center, has been asked to prepare special diets for two patients, Susan and Tom. Bob has decided that Susan's meals should contain at least 400 mg of calcium, 20 mg of iron, and 50 mg of vitamin C, whereas Tom's meals should contain at least 350 mg of calcium, 15 mg of iron, and 40 mg of vitamin C. Bob has also decided that the meals are to be prepared from three basic foods: food A, food B, and food C. The special nutritional contents of these foods are summarized in the accompanying table. Find how many ounces of each type of food should be used in a meal so that the minimum requirements of calcium, iron, and vitamin C are met for each patient's meals.

| | Contents (mg/oz) | | |
	Calcium	Iron	Vitamin C
Food A	30	1	2
Food B	25	1	5
Food C	20	2	4

39. AGRICULTURE Jackson Farms have allotted a certain amount of land for cultivating soybeans, corn, and wheat. Cultivating 1 acre of soybeans requires 2 labor-hours, and cultivating 1 acre of corn or wheat requires 6 labor-hours. The cost of seeds for 1 acre of soybeans is $12, for 1 acre of corn is $20, and for 1 acre of wheat is $8. If all resources are to be used, how many acres of each crop should be cultivated if the following hold?
a. 1000 acres of land are allotted, 4400 labor-hours are available, and $13,200 is available for seeds.
b. 1200 acres of land are allotted, 5200 labor-hours are available, and $16,400 is available for seeds.

40. MIXTURE PROBLEM—FERTILIZER Lawnco produces three grades of commercial fertilizers. A 100-lb bag of grade A fertilizer contains 18 lb of nitrogen, 4 lb of phosphate, and 5 lb of potassium. A 100-lb bag of grade B fertilizer contains 20 lb of nitrogen and 4 lb each of phosphate and potassium. A 100-lb bag of grade C fertilizer contains 24 lb of nitrogen, 3 lb of phosphate, and 6 lb of potassium. How many 100-lb bags of each of the three grades of fertilizers should Lawnco produce if
a. 26,400 lb of nitrogen, 4900 lb of phosphate, and 6200 lb of potassium are available and all the nutrients are used?
b. 21,800 lb of nitrogen, 4200 lb of phosphate, and 5300 lb of potassium are available and all the nutrients are used?

41. INVESTMENT CLUB A private investment club has a certain amount of money earmarked for investment in stocks. To arrive at an acceptable overall level of risk, the stocks that management is considering have been classified into three categories: high-risk, medium-risk, and low-risk. Management estimates that high-risk stocks will have a rate of return of 15%/year; medium-risk stocks, 10%/year; and low-risk stocks, 6%/year. The members have decided that the investment in low-risk stocks should be equal to the sum of the investments in the stocks of the other two categories. Determine how much the club should invest in each type of stock in each of the following scenarios. (In all cases, assume that the entire sum available for investment is invested.)
a. The club has $200,000 to invest, and the investment goal is to have a return of $20,000/year on the total investment.
b. The club has $220,000 to invest, and the investment goal is to have a return of $22,000/year in the total investment.
c. The club has $240,000 to invest, and the investment goal is to have a return of $22,000/year on the total investment.

42. RESEARCH FUNDING The Carver Foundation funds three nonprofit organizations engaged in alternate-energy research activities. From past data, the proportion of funds spent by each organization in research on solar energy, energy from harnessing the wind, and energy from the motion of ocean tides is given in the accompanying table.

	Proportion of Money Spent		
	Solar	**Wind**	**Tides**
Organization I	0.6	0.3	0.1
Organization II	0.4	0.3	0.3
Organization III	0.2	0.6	0.2

Find the amount awarded to each organization if the total amount spent by all three organizations on solar, wind, and tidal research is

a. $9.2 million, $9.6 million, and $5.2 million, respectively.
b. $8.2 million, $7.2 million, and $3.6 million, respectively.

43. Find the value(s) of k so that

$$A = \begin{bmatrix} 1 & 2 \\ k & 3 \end{bmatrix}$$

has an inverse. What is the inverse of A?
Hint: Use Formula 13.

44. Find the value(s) of k so that

$$A = \begin{bmatrix} 1 & 0 & 1 \\ -2 & 1 & k \\ -1 & 2 & k^2 \end{bmatrix}$$

has an inverse.
Hint: Find the value(s) of k such that the augmented matrix $[A \mid I]$ can be reduced to the form $[I \mid B]$.

In Exercises 45–47, determine whether the statement is true or false. If it is true, explain why it is true. If it is false, give an example to show why it is false.

45. If A is a square matrix with inverse A^{-1} and c is a nonzero real number, then

$$(cA)^{-1} = \left(\frac{1}{c}\right)A^{-1}$$

46. The matrix

$$A = \begin{bmatrix} a & b \\ c & d \end{bmatrix}$$

has an inverse if and only if $ad - bc = 0$.

47. If A^{-1} does not exist, then the system $AX = B$ of n linear equations in n unknowns does not have a unique solution.

48. Let

$$A = \begin{bmatrix} a & b \\ c & d \end{bmatrix}$$

a. Find A^{-1}.
b. Find the necessary condition for A to be nonsingular.
c. Verify that $AA^{-1} = A^{-1}A = I$.

2.6   Solutions to Self-Check Exercises

1. We form the augmented matrix

$$\begin{bmatrix} 2 & 1 & -1 & | & 1 & 0 & 0 \\ 1 & 1 & -1 & | & 0 & 1 & 0 \\ -1 & -2 & 3 & | & 0 & 0 & 1 \end{bmatrix}$$

and row-reduce as follows:

$$\begin{bmatrix} 2 & 1 & -1 & | & 1 & 0 & 0 \\ 1 & 1 & -1 & | & 0 & 1 & 0 \\ -1 & -2 & 3 & | & 0 & 0 & 1 \end{bmatrix} \xrightarrow{R_1 \leftrightarrow R_2}$$

$$\begin{bmatrix} 1 & 1 & -1 & | & 0 & 1 & 0 \\ 2 & 1 & -1 & | & 1 & 0 & 0 \\ -1 & -2 & 3 & | & 0 & 0 & 1 \end{bmatrix} \begin{matrix} \\ \xrightarrow{R_2 - 2R_1} \\ \xrightarrow{R_3 + R_1} \end{matrix}$$

$$\begin{bmatrix} 1 & 1 & -1 & | & 0 & 1 & 0 \\ 0 & -1 & 1 & | & 1 & -2 & 0 \\ 0 & -1 & 2 & | & 0 & 1 & 1 \end{bmatrix} \begin{matrix} \xrightarrow{R_1 + R_2} \\ \xrightarrow{-R_2} \\ \xrightarrow{R_3 - R_2} \end{matrix}$$

$$\begin{bmatrix} 1 & 0 & 0 & | & 1 & -1 & 0 \\ 0 & 1 & -1 & | & -1 & 2 & 0 \\ 0 & 0 & 1 & | & -1 & 3 & 1 \end{bmatrix} \xrightarrow{R_2 + R_3}$$

$$\begin{bmatrix} 1 & 0 & 0 & | & 1 & -1 & 0 \\ 0 & 1 & 0 & | & -2 & 5 & 1 \\ 0 & 0 & 1 & | & -1 & 3 & 1 \end{bmatrix}$$

From the preceding results, we see that

$$A^{-1} = \begin{bmatrix} 1 & -1 & 0 \\ -2 & 5 & 1 \\ -1 & 3 & 1 \end{bmatrix}$$

2. a. We write the systems of linear equations in the matrix form

$$AX = B_1$$

where

$$A = \begin{bmatrix} 2 & 1 & -1 \\ 1 & 1 & -1 \\ -1 & -2 & 3 \end{bmatrix} \quad X = \begin{bmatrix} x \\ y \\ z \end{bmatrix} \quad B_1 = \begin{bmatrix} 5 \\ 4 \\ -8 \end{bmatrix}$$

Now, using the results of Exercise 1, we have

$$X = \begin{bmatrix} x \\ y \\ z \end{bmatrix} = A^{-1}B_1 = \begin{bmatrix} 1 & -1 & 0 \\ -2 & 5 & 1 \\ -1 & 3 & 1 \end{bmatrix} \begin{bmatrix} 5 \\ 4 \\ -8 \end{bmatrix} = \begin{bmatrix} 1 \\ 2 \\ -1 \end{bmatrix}$$

Therefore, $x = 1$, $y = 2$, and $z = -1$.

b. Here A and X are as in part (a), but

$$B_2 = \begin{bmatrix} 2 \\ 0 \\ 5 \end{bmatrix}$$

Therefore,

$$X = \begin{bmatrix} x \\ y \\ z \end{bmatrix} = A^{-1}B_2 = \begin{bmatrix} 1 & -1 & 0 \\ -2 & 5 & 1 \\ -1 & 3 & 1 \end{bmatrix} \begin{bmatrix} 2 \\ 0 \\ 5 \end{bmatrix} = \begin{bmatrix} 2 \\ 1 \\ 3 \end{bmatrix}$$

or $x = 2$, $y = 1$, and $z = 3$.

3. Let x denote the number of adults and y the number of children in a tour. Since the tours are filled to capacity, we have

$$x + y = 19$$

Next, using the fact that the total receipts for the first tour were $2931 leads to the equation

$$169x + 129y = 2931$$

Therefore, the number of adults and the number of children in the first tour is found by solving the system of linear equations

$$\begin{align} x + y &= 19 \\ 169x + 129y &= 2931 \end{align} \qquad \textbf{(a)}$$

Similarly, we see that the number of adults and the number of children in the second and third tours are found by solving the systems

$$\begin{align} x + y &= 19 \\ 169x + 129y &= 3011 \end{align} \qquad \textbf{(b)}$$

$$\begin{align} x + y &= 19 \\ 169x + 129y &= 2771 \end{align} \qquad \textbf{(c)}$$

These systems may be written in the form

$$AX = B_1 \quad AX = B_2 \quad AX = B_3$$

where

$$A = \begin{bmatrix} 1 & 1 \\ 169 & 129 \end{bmatrix} \quad X = \begin{bmatrix} x \\ y \end{bmatrix}$$

$$B_1 = \begin{bmatrix} 19 \\ 2931 \end{bmatrix} \quad B_2 = \begin{bmatrix} 19 \\ 3011 \end{bmatrix} \quad B_3 = \begin{bmatrix} 19 \\ 2771 \end{bmatrix}$$

To solve these systems, we first find A^{-1}. Using Formula (13), we obtain

$$A^{-1} = \begin{bmatrix} -\frac{129}{40} & \frac{1}{40} \\ \frac{169}{40} & -\frac{1}{40} \end{bmatrix}$$

Then, solving each system, we find

$$X = \begin{bmatrix} x \\ y \end{bmatrix} = A^{-1}B_1$$

$$= \begin{bmatrix} -\frac{129}{40} & \frac{1}{40} \\ \frac{169}{40} & -\frac{1}{40} \end{bmatrix} \begin{bmatrix} 19 \\ 2931 \end{bmatrix} = \begin{bmatrix} 12 \\ 7 \end{bmatrix} \qquad \textbf{(a)}$$

$$X = \begin{bmatrix} x \\ y \end{bmatrix} = A^{-1}B_2$$

$$= \begin{bmatrix} -\frac{129}{40} & \frac{1}{40} \\ \frac{169}{40} & -\frac{1}{40} \end{bmatrix} \begin{bmatrix} 19 \\ 3011 \end{bmatrix}$$

$$= \begin{bmatrix} 14 \\ 5 \end{bmatrix} \qquad \textbf{(b)}$$

$$X = \begin{bmatrix} x \\ y \end{bmatrix} = A^{-1}B_3$$

$$= \begin{bmatrix} -\frac{129}{40} & \frac{1}{40} \\ \frac{169}{40} & -\frac{1}{40} \end{bmatrix} \begin{bmatrix} 19 \\ 2771 \end{bmatrix} = \begin{bmatrix} 8 \\ 11 \end{bmatrix} \qquad \textbf{(c)}$$

We conclude that there were
a. 12 adults and 7 children on the first tour.
b. 14 adults and 5 children on the second tour.
c. 8 adults and 11 children on the third tour.

USING TECHNOLOGY

■ **Finding the Inverse of a Square Matrix**

Graphing Utility

A graphing utility can be used to find the inverse of a square matrix.

EXAMPLE 1   Use a graphing utility to find the inverse of

$$\begin{bmatrix} 1 & 3 & 5 \\ -2 & 2 & 4 \\ 5 & 1 & 3 \end{bmatrix}$$

Solution   We first enter the given matrix as

$$A = \begin{bmatrix} 1 & 3 & 5 \\ -2 & 2 & 4 \\ 5 & 1 & 3 \end{bmatrix}$$

Then, recalling the matrix A and using the $\boxed{x^{-1}}$ key, we find

$$A^{-1} = \begin{bmatrix} 0.1 & -0.2 & 0.1 \\ 1.3 & -1.1 & -0.7 \\ -0.6 & 0.7 & 0.4 \end{bmatrix}$$

EXAMPLE 2   Use a graphing utility to solve the system

$$\begin{aligned} x + 3y + 5z &= 4 \\ -2x + 2y + 4z &= 3 \\ 5x + \ \ y + 3z &= 2 \end{aligned}$$

by using the inverse of the coefficient matrix.

Solution   The given system can be written in the matrix form $AX = B$, where

$$A = \begin{bmatrix} 1 & 3 & 5 \\ -2 & 2 & 4 \\ 5 & 1 & 3 \end{bmatrix} \quad X = \begin{bmatrix} x \\ y \\ z \end{bmatrix} \quad B = \begin{bmatrix} 4 \\ 3 \\ 2 \end{bmatrix}$$

The solution is $X = A^{-1}B$. Entering the matrices A and B in the graphing utility and using the matrix multiplication capability of the utility gives the output shown in Figure T1—that is, $x = 0$, $y = 0.5$, and $z = 0.5$.

```
[A]⁻¹ [B]
                        [ [0 ]
                          [.5]
                          [.5] ]
 Ans →
```

FIGURE T1
The TI-83 screen showing $A^{-1}B$

Excel

We use the function **MINVERSE** to find the inverse of a square matrix using Excel.

EXAMPLE 3   Find the inverse of

$$A = \begin{bmatrix} 1 & 3 & 5 \\ -2 & 2 & 4 \\ 5 & 1 & 3 \end{bmatrix}$$

Solution

1. Enter the elements of matrix A onto a spreadsheet (Figure T2).
2. Compute the inverse of the matrix A: Highlight the cells that will contain the inverse matrix A^{-1}, type = MINVERSE (, highlight the cells containing matrix A, type), and press **Ctrl-Shift-Enter**. The desired matrix will appear in your spreadsheet (Figure T2).

FIGURE T2
Matrix A and its inverse, matrix A^{-1}

	A	B	C
1		Matrix A	
2	1	3	5
3	−2	2	4
5	5	1	3
6			
7		Matrix A⁻¹	
8	0.1	−0.2	0.1
9	1.3	−1.1	−0.7
10	−0.6	0.7	0.4

EXAMPLE 4   Solve the system

$$x + 3y + 5z = 4$$
$$-2x + 2y + 4z = 3$$
$$5x + y + 3z = 2$$

by using the inverse of the coefficient matrix.

Solution   The given system can be written in the matrix form $AX = B$, where

$$A = \begin{bmatrix} 1 & 3 & 5 \\ -2 & 2 & 4 \\ 5 & 1 & 3 \end{bmatrix} \quad X = \begin{bmatrix} x \\ y \\ z \end{bmatrix} \quad B = \begin{bmatrix} 4 \\ 3 \\ 2 \end{bmatrix}$$

The solution is $X = A^{-1}B$.

Note: Boldfaced words/characters enclosed in a box (for example, **Enter**) indicate that an action (click, select, or press) is required. Words/characters printed blue (for example, Chart sub-type:) indicate words/characters on the screen. Words/characters printed in a typewriter font (for example, = (−2/3) *A2+2) indicate words/characters that need to be typed and entered.

(continued)

	A
12	Matrix X
13	5.55112E−17
14	0.5
15	0.5

FIGURE T3
Matrix X gives the solution to the problem

1. *Enter the matrix B on a spreadsheet.*
2. *Compute $A^{-1}B$.* Highlight the cells that will contain the matrix X, and then type =MMULT (, highlight the cells in the matrix A^{-1}, type , , highlight the cells in the matrix B, type), and press ⃞ **Ctrl-Shift-Enter** ⃞. (*Note:* The matrix A^{-1} was found in Example 3.) The matrix X shown in Figure T3 will appear on your spreadsheet. Thus, $x = 0$, $y = 0.5$, and $z = 0.5$. ∎

TECHNOLOGY EXERCISES

In Exercises 1–6, find the inverse of the matrix. Express your answers accurate to two decimal places.

1.
$$\begin{bmatrix} 1.2 & 3.1 & -2.1 \\ 3.4 & 2.6 & 7.3 \\ -1.2 & 3.4 & -1.3 \end{bmatrix}$$

2.
$$\begin{bmatrix} 4.2 & 3.7 & 4.6 \\ 2.1 & -1.3 & -2.3 \\ 1.8 & 7.6 & -2.3 \end{bmatrix}$$

3.
$$\begin{bmatrix} 1.1 & 2.3 & 3.1 & 4.2 \\ 1.6 & 3.2 & 1.8 & 2.9 \\ 4.2 & 1.6 & 1.4 & 3.2 \\ 1.6 & 2.1 & 2.8 & 7.2 \end{bmatrix}$$

4.
$$\begin{bmatrix} 2.1 & 3.2 & -1.4 & -3.2 \\ 6.2 & 7.3 & 8.4 & 1.6 \\ 2.3 & 7.1 & 2.4 & -1.3 \\ -2.1 & 3.1 & 4.6 & 3.7 \end{bmatrix}$$

5.
$$\begin{bmatrix} 2 & -1 & 3 & 2 & 4 \\ 3 & 2 & -1 & 4 & 1 \\ 3 & 2 & 6 & 4 & -1 \\ 2 & 1 & -1 & 4 & 2 \\ 3 & 4 & 2 & 5 & 6 \end{bmatrix}$$

6.
$$\begin{bmatrix} 1 & 4 & 2 & 3 & 1.4 \\ 6 & 2.4 & 5 & 1.2 & 3 \\ 4 & 1 & 2 & 3 & 1.2 \\ -1 & 2 & -3 & 4 & 2 \\ 1.1 & 2.2 & 3 & 5.1 & 4 \end{bmatrix}$$

In Exercises 7–10, solve the system of linear equations by first writing the system in the form $AX = B$ and then solving the resulting system by using A^{-1}. Express your answers accurate to two decimal places.

7.
$$\begin{aligned} 2x - 3y + 4z &= 2.4 \\ 3x + 2y - 7z &= -8.1 \\ x + 4y - 2z &= 10.2 \end{aligned}$$

8.
$$\begin{aligned} 3.2x - 4.7y + 3.2z &= 7.1 \\ 2.1x + 2.6y + 6.2z &= 8.2 \\ 5.1x - 3.1y - 2.6z &= -6.5 \end{aligned}$$

9.
$$\begin{aligned} 3x_1 - 2x_2 + 4x_3 - 8x_4 &= 8 \\ 2x_1 + 3x_2 - 2x_3 + 6x_4 &= 4 \\ 3x_1 + 2x_2 - 6x_3 - 7x_4 &= -2 \\ 4x_1 - 7x_2 + 4x_3 + 6x_4 &= 22 \end{aligned}$$

10.
$$\begin{aligned} 1.2x_1 + 2.1x_2 - 3.2x_3 + 4.6x_4 &= 6.2 \\ 3.1x_1 - 1.2x_2 + 4.1x_3 - 3.6x_4 &= -2.2 \\ 1.8x_1 + 3.1x_2 - 2.4x_3 + 8.1x_4 &= 6.2 \\ 2.6x_1 - 2.4x_2 + 3.6x_3 - 4.6x_4 &= 3.6 \end{aligned}$$

2.7 Leontief Input–Output Model (Optional)

■ Input–Output Analysis

One of the many important applications of matrix theory to the field of economics is the study of the relationship between industrial production and consumer demand. At the heart of this analysis is the Leontief input–output model, pioneered by Wassily Leontief, who was awarded a Nobel Prize in economics in 1973 for his contributions to the field.

To illustrate this concept, let's consider an oversimplified economy consisting of three sectors: agriculture (*A*), manufacturing (*M*), and service (*S*). In general, part of the output of one sector is absorbed by another sector through interindustry purchases, with the excess available to fulfill consumer demands. The relationship governing both intraindustrial and interindustrial sales and purchases is conveniently represented by means of an **input–output matrix:**

$$
\begin{array}{c}
\text{Input} \\
\text{(amount used in production)}
\end{array}
\quad
\begin{array}{cc}
 & \text{Output (amount produced)} \\
 & \begin{array}{ccc} A & M & S \end{array} \\
\begin{array}{c} A \\ M \\ S \end{array} &
\left[
\begin{array}{ccc}
0.2 & 0.2 & 0.1 \\
0.2 & 0.4 & 0.1 \\
0.1 & 0.2 & 0.3
\end{array}
\right]
\end{array}
\qquad \textbf{(16)}
$$

The first column (read from top to bottom) tells us that the production of 1 unit of agricultural products requires the consumption of 0.2 unit of agricultural products, 0.2 unit of manufactured goods, and 0.1 unit of services. The second column tells us that the production of 1 unit of manufactured products requires the consumption of 0.2 unit of agricultural products, 0.4 unit of manufactured products, and 0.2 unit of services. Finally, the third column tells us that the production of 1 unit of services requires the consumption of 0.1 unit each of agricultural goods and manufactured products, and 0.3 unit of services.

  APPLIED EXAMPLE 1 Input–Output Analysis Refer to the input–output matrix (16).

a. If the units are measured in millions of dollars, determine the amount of agricultural products consumed in the production of $100 million worth of manufactured goods.

b. Determine the dollar amount of manufactured products required to produce $200 million worth of all goods and services in the economy.

Solution

a. The production of 1 unit—that is, $1 million worth of manufactured goods—requires the consumption of 0.2 unit of agricultural products. Thus, the amount of agricultural products consumed in the production of $100 million worth of manufactured goods is given by (100)(0.2), or $20 million.

b. The amount of manufactured goods required to produce 1 unit of all goods and services in the economy is given by adding the numbers of the second row of the input–output matrix—that is, 0.2 + 0.4 + 0.1, or 0.7 unit. Therefore, the production of $200 million worth of all goods and services in the economy requires 200(0.7), or $140 million worth, of manufactured products. ■

Next, suppose the total output of goods of the agriculture and manufacturing sectors and the total output from the service sector of the economy are given by x, y, and z units, respectively. What is the value of agricultural products consumed in the internal process of producing this total output of various goods and services?

To answer this question, we first note, by examining the input–output matrix

$$
\begin{array}{c}
 & \begin{array}{ccc} & \text{Output} & \\ A & M & S \end{array} \\
\text{Input} \quad \begin{array}{c} A \\ M \\ S \end{array} &
\begin{bmatrix}
0.2 & 0.2 & 0.1 \\
0.2 & 0.4 & 0.1 \\
0.1 & 0.2 & 0.3
\end{bmatrix}
\end{array}
$$

that 0.2 unit of agricultural products is required to produce 1 unit of agricultural products, so the amount of agricultural goods required to produce x units of agricultural products is given by $0.2x$ unit. Next, again referring to the input–output matrix, we see that 0.2 unit of agricultural products is required to produce 1 unit of manufactured products, so the requirement for producing y units of the latter is $0.2y$ unit of agricultural products. Finally, we see that 0.1 unit of agricultural goods is required to produce 1 unit of services, so the value of agricultural products required to produce z units of services is $0.1z$ unit. Thus, the total amount of agricultural products required to produce the total output of goods and services in the economy is

$$0.2x + 0.2y + 0.1z$$

units. In a similar manner, we see that the total amount of manufactured products and the total value of services to produce the total output of goods and services in the economy are given by

$$0.2x + 0.4y + 0.1z$$
$$0.1x + 0.2y + 0.3z$$

respectively.

These results could also be obtained using matrix multiplication. To see this, write the total output of goods and services x, y, and z as a 3×1 matrix

$$X = \begin{bmatrix} x \\ y \\ z \end{bmatrix} \qquad \text{Gross production matrix}$$

The matrix X is called the **gross production matrix**. Letting A denote the input–output matrix, we have

$$A = \begin{bmatrix} 0.2 & 0.2 & 0.1 \\ 0.2 & 0.4 & 0.1 \\ 0.1 & 0.2 & 0.3 \end{bmatrix} \qquad \text{Input–output matrix}$$

Then, the product

$$
\begin{aligned}
AX &= \begin{bmatrix} 0.2 & 0.2 & 0.1 \\ 0.2 & 0.4 & 0.1 \\ 0.1 & 0.2 & 0.3 \end{bmatrix} \begin{bmatrix} x \\ y \\ z \end{bmatrix} \\
&= \begin{bmatrix} 0.2x + 0.2y + 0.1z \\ 0.2x + 0.4y + 0.1z \\ 0.1x + 0.2y + 0.3z \end{bmatrix} \qquad \text{Internal consumption matrix}
\end{aligned}
$$

is a 3 × 1 matrix whose entries represent the respective values of the agricultural products, manufactured products, and services consumed in the internal process of production. The matrix AX is referred to as the **internal consumption matrix.**

Now, since X gives the total production of goods and services in the economy, and AX, as we have just seen, gives the amount of products and services consumed in the production of these goods and services, the 3 × 1 matrix $X - AX$ gives the net output of goods and services that is exactly enough to satisfy consumer demands. Letting matrix D represent these consumer demands, we are led to the following matrix equation:

$$X - AX = D$$
$$(I - A)X = D$$

where I is the 3 × 3 identity matrix.

Assuming that the inverse of $(I - A)$ exists, multiplying both sides of the last equation by $(I - A)^{-1}$ yields

$$X = (I - A)^{-1}D$$

Leontief Input–Output Model

In a **Leontief input–output model,** the matrix equation giving the net output of goods and services needed to satisfy consumer demand is

$$\underset{\substack{\text{Total}\\\text{output}}}{X} \; - \; \underset{\substack{\text{Internal}\\\text{consumption}}}{AX} \; = \; \underset{\substack{\text{Consumer}\\\text{demand}}}{D}$$

where X is the total output matrix, A is the input–output matrix, and D is the matrix representing consumer demand.

The solution to this equation is

$$X = (I - A)^{-1}D \qquad \text{Assuming that } (I - A)^{-1} \text{ exists} \qquad \textbf{(17)}$$

which gives the amount of goods and services that must be produced to satisfy consumer demand.

Equation (17) gives us a means of finding the amount of goods and services to be produced in order to satisfy a given level of consumer demand, as illustrated by the following example.

APPLIED EXAMPLE 2   Input–Output Model for a Three-Sector Economy   For the three-sector economy with input–output matrix given by (16), which is reproduced here,

$$\begin{bmatrix} 0.2 & 0.2 & 0.1 \\ 0.2 & 0.4 & 0.1 \\ 0.1 & 0.2 & 0.3 \end{bmatrix} \qquad \text{Each unit equals \$1 million.}$$

a. Find the gross output of goods and services needed to satisfy a consumer demand of $100 million worth of agricultural products, $80 million worth of manufactured products, and $50 million worth of services.

b. Find the value of the goods and services consumed in the internal process of production in order to meet this gross output.

Solution

a. We are required to determine the gross production matrix

$$X = \begin{bmatrix} x \\ y \\ z \end{bmatrix}$$

where x, y, and z denote the value of the agricultural products, the manufactured products, and services. The matrix representing the consumer demand is given by

$$D = \begin{bmatrix} 100 \\ 80 \\ 50 \end{bmatrix}$$

Next, we compute

$$I - A = \begin{bmatrix} 1 & 0 & 0 \\ 0 & 1 & 0 \\ 0 & 0 & 1 \end{bmatrix} - \begin{bmatrix} 0.2 & 0.2 & 0.1 \\ 0.2 & 0.4 & 0.1 \\ 0.1 & 0.2 & 0.3 \end{bmatrix} = \begin{bmatrix} 0.8 & -0.2 & -0.1 \\ -0.2 & 0.6 & -0.1 \\ -0.1 & -0.2 & 0.7 \end{bmatrix}$$

Using the method of Section 2.6, we find (to two decimal places)

$$(I - A)^{-1} = \begin{bmatrix} 1.43 & 0.57 & 0.29 \\ 0.54 & 1.96 & 0.36 \\ 0.36 & 0.64 & 1.57 \end{bmatrix}$$

Finally, using Equation (17), we find

$$X = (I - A)^{-1} D = \begin{bmatrix} 1.43 & 0.57 & 0.29 \\ 0.54 & 1.96 & 0.36 \\ 0.36 & 0.64 & 1.57 \end{bmatrix} \begin{bmatrix} 100 \\ 80 \\ 50 \end{bmatrix} = \begin{bmatrix} 203.1 \\ 228.8 \\ 165.7 \end{bmatrix}$$

To fulfill consumer demand, $203 million worth of agricultural products, $229 million worth of manufactured products, and $166 million worth of services should be produced.

b. The amount of goods and services consumed in the internal process of production is given by AX, or equivalently by $X - D$. In this case it is more convenient to use the latter, which gives the required result of

$$\begin{bmatrix} 203.1 \\ 228.8 \\ 165.7 \end{bmatrix} - \begin{bmatrix} 100 \\ 80 \\ 50 \end{bmatrix} = \begin{bmatrix} 103.1 \\ 148.8 \\ 115.7 \end{bmatrix}$$

or $103 million worth of agricultural products, $149 million worth of manufactured products, and $116 million worth of services.

APPLIED EXAMPLE 3   An Input–Output Model for a Three-Product Company   TKK Corporation, a large conglomerate, has three subsidiaries engaged in producing raw rubber, manufacturing tires, and manufacturing other rubber-based goods. The production of 1 unit of raw rubber requires the consumption of 0.08 unit of rubber, 0.04 unit of tires, and 0.02 unit of other rubber-based goods. To produce 1 unit of tires requires 0.6 unit of raw rubber, 0.02 unit of tires, and 0 units of other rubber-based goods. To produce 1 unit of other rubber-based goods requires 0.3 unit of raw rubber, 0.01 unit of tires, and 0.06 unit of other rubber-based goods. Market research indicates that the demand for the following year will be $200 million for raw rubber, $800 million for tires, and $120 million for other rubber-based products. Find the level of production for each subsidiary in order to satisfy this demand.

Solution   View the corporation as an economy having three sectors, with an input–output matrix given by

$$A = \begin{array}{c} \\ \text{Raw rubber} \\ \text{Tires} \\ \text{Goods} \end{array} \begin{array}{c} \overset{\text{Raw}}{\underset{\text{rubber}}{}} \quad \text{Tires} \quad \text{Goods} \\ \begin{bmatrix} 0.08 & 0.60 & 0.30 \\ 0.04 & 0.02 & 0.01 \\ 0.02 & 0 & 0.06 \end{bmatrix} \end{array}$$

Using Equation (17), we find that the required level of production is given by

$$X = \begin{bmatrix} x \\ y \\ z \end{bmatrix} = (I - A)^{-1}D$$

where x, y, and z denote the outputs of raw rubber, tires, and other rubber-based goods, and

$$D = \begin{bmatrix} 200 \\ 800 \\ 120 \end{bmatrix}$$

Now,

$$I - A = \begin{bmatrix} 0.92 & -0.60 & -0.30 \\ -0.04 & 0.98 & -0.01 \\ -0.02 & 0 & 0.94 \end{bmatrix}$$

You are asked to verify that

$$(I - A)^{-1} = \begin{bmatrix} 1.13 & 0.69 & 0.37 \\ 0.05 & 1.05 & 0.03 \\ 0.02 & 0.02 & 1.07 \end{bmatrix} \qquad \text{See Exercise 7.}$$

Therefore,

$$X = (I - A)^{-1}D = \begin{bmatrix} 1.13 & 0.69 & 0.37 \\ 0.05 & 1.05 & 0.03 \\ 0.02 & 0.02 & 1.07 \end{bmatrix} \begin{bmatrix} 200 \\ 800 \\ 120 \end{bmatrix} = \begin{bmatrix} 822.4 \\ 853.6 \\ 148.4 \end{bmatrix}$$

To fulfill the predicted demand, $822 million worth of raw rubber, $854 million worth of tires, and $148 million worth of other rubber-based goods should be produced.

2.7 Self-Check Exercises

1. Solve the matrix equation $(I - A)X = D$ for x and y given that

$$A = \begin{bmatrix} 0.4 & 0.1 \\ 0.2 & 0.2 \end{bmatrix} \quad X = \begin{bmatrix} x \\ y \end{bmatrix} \quad D = \begin{bmatrix} 50 \\ 10 \end{bmatrix}$$

2. A simple economy consists of two sectors: agriculture (A) and transportation (T). The input–output matrix for this economy is given by

$$A = \begin{array}{cc} & \begin{array}{cc} A & T \end{array} \\ \begin{array}{c} A \\ T \end{array} & \begin{bmatrix} 0.4 & 0.1 \\ 0.2 & 0.2 \end{bmatrix} \end{array}$$

a. Find the gross output of agricultural products needed to satisfy a consumer demand for $50 million worth of agricultural products and $10 million worth of transportation.

b. Find the value of agricultural products and transportation consumed in the internal process of production in order to meet the gross output.

Solutions to Self-Check Exercises 2.7 can be found on page 160.

2.7 Concept Questions

1. What do the quantities X, AX, and D represent in the matrix equation $X - AX = D$ for a Leontief input–output model?

2. What is the solution to the matrix equation $X - AX = D$? Does the solution to this equation always exist? Why or why not?

2.7 Exercises

1. **AN INPUT–OUTPUT MATRIX FOR A THREE-SECTOR ECONOMY** A simple economy consists of three sectors: agriculture (A), manufacturing (M), and transportation (T). The input–output matrix for this economy is given by

$$\begin{array}{cc} & \begin{array}{ccc} A & M & T \end{array} \\ \begin{array}{c} A \\ M \\ T \end{array} & \begin{bmatrix} 0.4 & 0.1 & 0.1 \\ 0.1 & 0.4 & 0.3 \\ 0.2 & 0.2 & 0.2 \end{bmatrix} \end{array}$$

a. If the units are measured in millions of dollars, determine the amount of agricultural products consumed in the production of $100 million worth of manufactured goods.

b. Determine the dollar amount of manufactured products required to produce $200 million worth of all goods in the economy.

c. Which sector consumes the greatest amount of agricultural products in the production of a unit of goods in that sector? The least?

2. **AN INPUT–OUTPUT MATRIX FOR A FOUR-SECTOR ECONOMY** The relationship governing the intraindustrial and interindustrial sales and purchases of four basic industries—agriculture (A), manufacturing (M), transportation (T), and energy (E)—of a certain economy is given by the following input–output matrix.

$$\begin{array}{cc} & \begin{array}{cccc} A & M & T & E \end{array} \\ \begin{array}{c} A \\ M \\ T \\ E \end{array} & \begin{bmatrix} 0.3 & 0.2 & 0 & 0.1 \\ 0.2 & 0.3 & 0.2 & 0.1 \\ 0.2 & 0.2 & 0.1 & 0.3 \\ 0.1 & 0.2 & 0.3 & 0.2 \end{bmatrix} \end{array}$$

a. How many units of energy are required to produce 1 unit of manufactured goods?

b. How many units of energy are required to produce 3 units of all goods in the economy?

c. Which sector of the economy is least dependent on the cost of energy?

d. Which sector of the economy has the smallest intraindustry purchases (sales)?

In Exercises 3–6, solve the matrix equation $(I - A) X = D$ for the matrices A and D.

3. $A = \begin{bmatrix} 0.4 & 0.2 \\ 0.3 & 0.1 \end{bmatrix}$ and $D = \begin{bmatrix} 10 \\ 12 \end{bmatrix}$

4. $A = \begin{bmatrix} 0.2 & 0.3 \\ 0.5 & 0.2 \end{bmatrix}$ and $D = \begin{bmatrix} 4 \\ 8 \end{bmatrix}$

5. $A = \begin{bmatrix} 0.5 & 0.2 \\ 0.2 & 0.5 \end{bmatrix}$ and $D = \begin{bmatrix} 10 \\ 20 \end{bmatrix}$

6. $A = \begin{bmatrix} 0.6 & 0.2 \\ 0.1 & 0.4 \end{bmatrix}$ and $D = \begin{bmatrix} 8 \\ 12 \end{bmatrix}$

7. Let

$$A = \begin{bmatrix} 0.08 & 0.60 & 0.30 \\ 0.04 & 0.02 & 0.01 \\ 0.02 & 0 & 0.06 \end{bmatrix}$$

Show that

$$(I - A)^{-1} = \begin{bmatrix} 1.13 & 0.69 & 0.37 \\ 0.05 & 1.05 & 0.03 \\ 0.02 & 0.02 & 1.07 \end{bmatrix}$$

8. AN INPUT–OUTPUT MODEL FOR A TWO-SECTOR ECONOMY A simple economy consists of two industries: agriculture and manufacturing. The production of 1 unit of agricultural products requires the consumption of 0.2 unit of agricultural products and 0.3 unit of manufactured goods. The production of 1 unit of manufactured products requires the consumption of 0.4 unit of agricultural products and 0.3 unit of manufactured goods.

   a. Find the gross output of goods needed to satisfy a consumer demand for $100 million worth of agricultural products and $150 million worth of manufactured products.

   b. Find the value of the goods consumed in the internal process of production in order to meet the gross output.

9. Rework Exercise 8 if the consumer demand for the output of agricultural goods and the consumer demand for manufactured products are $120 million and $140 million, respectively.

10. Refer to Example 3. Suppose the demand for raw rubber increases by 10%, the demand for tires increases by 20%, and the demand for other rubber-based products decreases by 10%. Find the level of production for each subsidiary in order to meet this demand.

11. AN INPUT–OUTPUT MODEL FOR A THREE-SECTOR ECONOMY Consider the economy of Exercise 1, consisting of three sectors: agriculture (A), manufacturing (M), and transportation (T), with an input–output matrix given by

$$\begin{array}{c} \\ A \\ M \\ T \end{array} \begin{array}{ccc} A & M & T \\ \begin{bmatrix} 0.4 & 0.1 & 0.1 \\ 0.1 & 0.4 & 0.3 \\ 0.2 & 0.2 & 0.2 \end{bmatrix} \end{array}$$

   a. Find the gross output of goods needed to satisfy a consumer demand for $200 million worth of agricultural products, $100 million worth of manufactured products, and $60 million worth of transportation.

   b. Find the value of goods and transportation consumed in the internal process of production in order to meet this gross output.

12. AN INPUT–OUTPUT MODEL FOR A THREE-SECTOR ECONOMY Consider a simple economy consisting of three sectors: food, clothing, and shelter. The production of 1 unit of food requires the consumption of 0.4 unit of food, 0.2 unit of clothing, and 0.2 unit of shelter. The production of 1 unit of clothing requires the consumption of 0.1 unit of food, 0.2 unit of clothing, and 0.3 unit of shelter. The production of 1 unit of shelter requires the consumption of 0.3 unit of food, 0.1 unit of clothing, and 0.1 unit of shelter. Find the level of production for each sector in order to satisfy the demand for $100 million worth of food, $30 million worth of clothing, and $250 million worth of shelter.

In Exercises 13–16, matrix *A* is an input–output matrix associated with an economy, and matrix *D* (units in millions of dollars) is a demand vector. In each problem, find the final outputs of each industry so that the demands of both industry and the open sector are met.

13. $A = \begin{bmatrix} 0.4 & 0.2 \\ 0.3 & 0.5 \end{bmatrix}$ and $D = \begin{bmatrix} 12 \\ 24 \end{bmatrix}$

14. $A = \begin{bmatrix} 0.1 & 0.4 \\ 0.3 & 0.2 \end{bmatrix}$ and $D = \begin{bmatrix} 5 \\ 10 \end{bmatrix}$

15. $A = \begin{bmatrix} \frac{1}{5} & \frac{2}{5} & \frac{1}{5} \\ \frac{1}{2} & 0 & \frac{1}{2} \\ 0 & \frac{1}{5} & 0 \end{bmatrix}$ and $D = \begin{bmatrix} 10 \\ 5 \\ 15 \end{bmatrix}$

16. $A = \begin{bmatrix} 0.2 & 0.4 & 0.1 \\ 0.3 & 0.2 & 0.1 \\ 0.1 & 0.2 & 0.2 \end{bmatrix}$ and $D = \begin{bmatrix} 6 \\ 8 \\ 10 \end{bmatrix}$

2.7   Solutions to Self-Check Exercises

1. Multiplying both sides of the given equation on the left by $(I - A)^{-1}$, we see that

$$X = (I - A)^{-1}D$$

Now,

$$I - A = \begin{bmatrix} 1 & 0 \\ 0 & 1 \end{bmatrix} - \begin{bmatrix} 0.4 & 0.1 \\ 0.2 & 0.2 \end{bmatrix} = \begin{bmatrix} 0.6 & -0.1 \\ -0.2 & 0.8 \end{bmatrix}$$

Next, we use the Gauss–Jordan procedure to compute $(I - A)^{-1}$ (to two decimal places):

$$\left[\begin{array}{cc|cc} 0.6 & -0.1 & 1 & 0 \\ -0.2 & 0.8 & 0 & 1 \end{array} \right] \xrightarrow{\frac{1}{0.6} R_1}$$

$$\left[\begin{array}{cc|cc} 1 & -0.17 & 1.67 & 0 \\ -0.2 & 0.8 & 0 & 1 \end{array} \right] \xrightarrow{R_2 + 0.2R_1}$$

$$\left[\begin{array}{cc|cc} 1 & -0.17 & 1.67 & 0 \\ 0 & 0.77 & 0.33 & 1 \end{array} \right] \xrightarrow{\frac{1}{0.77} R_2}$$

$$\left[\begin{array}{cc|cc} 1 & -0.17 & 1.67 & 0 \\ 0 & 1 & 0.43 & 1.30 \end{array} \right] \xrightarrow{R_1 + 0.17R_2}$$

$$\left[\begin{array}{cc|cc} 1 & 0 & 1.74 & 0.22 \\ 0 & 1 & 0.43 & 1.30 \end{array} \right]$$

giving

$$(I - A)^{-1} = \begin{bmatrix} 1.74 & 0.22 \\ 0.43 & 1.30 \end{bmatrix}$$

Therefore,

$$X = \begin{bmatrix} x \\ y \end{bmatrix} = (I - A)^{-1}D = \begin{bmatrix} 1.74 & 0.22 \\ 0.43 & 1.30 \end{bmatrix} \begin{bmatrix} 50 \\ 10 \end{bmatrix} = \begin{bmatrix} 89.2 \\ 34.5 \end{bmatrix}$$

or $x = 89.2$ and $y = 34.5$.

2. **a.** Let

$$X = \begin{bmatrix} x \\ y \end{bmatrix}$$

denote the gross production matrix, where x denotes the value of the agricultural products and y the value of transportation. Also, let

$$D = \begin{bmatrix} 50 \\ 10 \end{bmatrix}$$

denote the consumer demand. Then,

$$(I - A) X = D$$

or equivalently,

$$X = (I - A)^{-1}D$$

Using the results of Exercise 1, we find that $x = 89.2$ and $y = 34.5$. That is, to fulfill consumer demands, $89.2 million worth of agricultural products must be produced and $34.5 million worth of transportation services must be used.

b. The amount of agricultural products consumed and transportation services used is given by

$$X - D = \begin{bmatrix} 89.2 \\ 34.5 \end{bmatrix} - \begin{bmatrix} 50 \\ 10 \end{bmatrix} = \begin{bmatrix} 39.2 \\ 24.5 \end{bmatrix}$$

or $39.2 million worth of agricultural products and $24.5 million worth of transportation services.

USING TECHNOLOGY

■ The Leontief Input–Output Model

Graphing Utility

Since the solution to a problem involving a Leontief input–output model often involves several matrix operations, a graphing utility can be used to facilitate the necessary computations.

APPLIED EXAMPLE 1 Suppose the input–output matrix associated with an economy is given by A and the matrix D is a demand vector, where

$$A = \begin{bmatrix} 0.2 & 0.4 & 0.15 \\ 0.3 & 0.1 & 0.4 \\ 0.25 & 0.4 & 0.2 \end{bmatrix} \quad \text{and} \quad D = \begin{bmatrix} 20 \\ 15 \\ 40 \end{bmatrix}$$

Find the final outputs of each industry so that the demands of both industry and the open sector are met.

Solution First, we enter the matrices I (the identity matrix), A, and D. We are required to compute the output matrix $X = (I - A)^{-1}D$. Using the matrix operations of the graphing utility, we find

$$X = (I - A)^{-1} * D = \begin{bmatrix} 110.28 \\ 116.95 \\ 142.94 \end{bmatrix}$$

So, the final outputs of the first, second, and third industries are 110.28, 116.95, and 142.94 units, respectively. ∎

Excel

Here we show how to solve a problem involving a Leontief input–output model using matrix operations on a spreadsheet.

APPLIED EXAMPLE 2 Suppose the input–output matrix associated with an economy is given by matrix A and the matrix D is a demand vector, where

$$A = \begin{bmatrix} 0.2 & 0.4 & 0.15 \\ 0.3 & 0.1 & 0.4 \\ 0.25 & 0.4 & 0.2 \end{bmatrix} \quad \text{and} \quad D = \begin{bmatrix} 20 \\ 15 \\ 40 \end{bmatrix}$$

Find the final outputs of each industry so that the demands of both industry and the open sector are met.

Solution

1. *Enter the elements of the matrix A and D onto a spreadsheet* (Figure T1).

	A	B	C	D	E
1		Matrix A			Matrix D
2	0.2	0.4	0.15		20
3	0.3	0.1	0.4		15
4	0.25	0.4	0.2		40

FIGURE T1
Spreadsheet showing matrix A and matrix D

2. *Find* $(I - A)^{-1}$. Enter the elements of the 3×3 identity matrix I onto a spreadsheet. Highlight the cells that will contain the matrix $(I - A)^{-1}$. Type =MINVERSE(, highlight the cells containing the matrix I; type -, highlight the
(*continued*)

cells containing the matrix A; type), and press $\boxed{\textbf{Ctrl-Shift-Enter}}$. These results are shown in Figure T2.

	A	B	C
6		Matrix I	
7	1	0	0
8	0	1	0
9	0	0	1
10			
11		Matrix $(I-A)^{-1}$	
12	2.151777	1.460134	1.133525
13	1.306436	2.315082	1.402498
14	1.325648	1.613833	2.305476

FIGURE T2
Matrix I and matrix $(I-A)^{-1}$

3. *Compute* $(I - A)^{-1} * D$. Highlight the cells that will contain the matrix $(I - A)^{-1} * D$. Type `=MMULT(`, highlight the cells containing the matrix $(I - A)^{-1}$, type , , highlight the cells containing the matrix D, type), and press $\boxed{\textbf{Ctrl-Shift-Enter}}$. The resulting matrix is shown in Figure T3. So, the final outputs of the first, second, and third industries are 110.28, 116.95, and 142.94, respectively.

	A
16	Matrix $(I-A)^{-1}$ *D
17	110.2786
18	116.9549
19	142.9395

FIGURE T3
Matrix $(I-A)^{-1} * D$

TECHNOLOGY EXERCISES

In Exercises 1–4, A is an input–output matrix associated with an economy, and D (in units of dollars) is a demand vector. Find the final outputs of each industry so that the demands of both industry and the open sector are met.

1.
$$A = \begin{bmatrix} 0.3 & 0.2 & 0.4 & 0.1 \\ 0.2 & 0.1 & 0.2 & 0.3 \\ 0.3 & 0.1 & 0.2 & 0.3 \\ 0.4 & 0.2 & 0.1 & 0.2 \end{bmatrix} \text{ and } D = \begin{bmatrix} 40 \\ 60 \\ 70 \\ 20 \end{bmatrix}$$

2.
$$A = \begin{bmatrix} 0.12 & 0.31 & 0.40 & 0.05 \\ 0.31 & 0.22 & 0.12 & 0.20 \\ 0.18 & 0.32 & 0.05 & 0.15 \\ 0.32 & 0.14 & 0.22 & 0.05 \end{bmatrix} \text{ and } D = \begin{bmatrix} 50 \\ 20 \\ 40 \\ 60 \end{bmatrix}$$

3.
$$A = \begin{bmatrix} 0.2 & 0.2 & 0.3 & 0.05 \\ 0.1 & 0.1 & 0.2 & 0.3 \\ 0.3 & 0.2 & 0.1 & 0.4 \\ 0.2 & 0.05 & 0.2 & 0.1 \end{bmatrix} \text{ and } D = \begin{bmatrix} 25 \\ 30 \\ 50 \\ 40 \end{bmatrix}$$

4.
$$A = \begin{bmatrix} 0.2 & 0.4 & 0.3 & 0.1 \\ 0.1 & 0.2 & 0.1 & 0.3 \\ 0.2 & 0.1 & 0.4 & 0.05 \\ 0.3 & 0.1 & 0.2 & 0.05 \end{bmatrix} \text{ and } D = \begin{bmatrix} 40 \\ 20 \\ 30 \\ 60 \end{bmatrix}$$

Note: Boldfaced words/characters in a box (for example, $\boxed{\textbf{Enter}}$) indicate that an action (click, select, or press) is required. Words/characters printed blue (for example, Chart sub-type:) indicate words/characters on the screen. Words/characters printed in a typewriter font (for example, `=(-2/3)*A2+2`) indicate words/characters that need to be typed and entered.

| CHAPTER 2 | **Summary of Principal Formulas and Terms** |

FORMULAS

1. Laws for matrix addition	
a. Commutative law	$A + B = B + A$
b. Associative law	$(A + B) + C = A + (B + C)$
2. Laws for matrix multiplication	
a. Associative law	$(AB)C = A(BC)$
b. Distributive law	$A(B + C) = AB + AC$
3. Inverse of a 2×2 matrix	If $\quad A = \begin{bmatrix} a & b \\ c & d \end{bmatrix}$ and $\quad D = ad - bc \neq 0$ then $\quad A^{-1} = \dfrac{1}{D}\begin{bmatrix} d & -b \\ -c & a \end{bmatrix}$
4. Solution of system $AX = B$ (A, nonsingular)	$X = A^{-1}B$

TERMS

system of linear equations (72)

solution of a system of linear equations (72)

parameter (74)

dependent system (74)

inconsistent system (74)

Gauss–Jordan elimination method (80)

equivalent system (80)

coefficient matrix (83)

augmented matrix (83)

row-reduced form of a matrix (84)

row operations (85)

unit column (85)

pivoting (85)

size of a matrix (109)

matrix (109)

row matrix (109)

column matrix (109)

square matrix (109)

transpose of a matrix (113)

scalar (113)

scalar product (113)

matrix product (122)

identity matrix (126)

inverse of a matrix (137)

nonsingular matrix (138)

singular matrix (138)

input–output matrix (153)

gross production matrix (154)

internal consumption matrix (155)

Leontief input–output model (155)

| CHAPTER 2 | **Concept Review Questions** |

Fill in the blanks.

1. a. Two lines in the plane can intersect at (a) exactly _____ point, (b) infinitely _____ points, or (c) at _____ point.

   b. A system of two linear equations in two variables can have (a) exactly _____ solution, (b) infinitely _____ solutions, or (c) _____ solution.

2. To find the point(s) of intersection of two lines, we solve the system of _____ describing the two lines.

3. The row operations used in the Gauss–Jordan elimination method are denoted by _____, _____, and _____. The use of each of these operations does not alter the _____ of the system of linear equations.

4. a. A system of linear equations with fewer equations than variables cannot have a/an _____ solution.

  b. A system of linear equations with at least as many equations as variables may have _____ solution, _____ _____ solutions, or a _____ solution.

5. Two matrices are equal provided they have the same _____ and their corresponding _____ are equal.

6. Two matrices may be added (subtracted) together if they both have the same _____. To add or subtract two matrices, we add or subtract their _____ entries.

7. The transpose of a/an _____ matrix with elements a_{ij} is the matrix of size _____ with entries _____.

8. The scalar product of a matrix A by the scalar c is the matrix _____ obtained by multiplying each entry of A by _____.

9. a. For the product AB of two matrices A and B to be defined, the number of _____ of A must be equal to the number of _____ of B.

  b. If A is an $m \times n$ matrix and B is an $n \times p$ matrix, then the size of AB is _____.

10. a. If the products and sums are defined for the matrices A, B, and C, then the associative law states that $(AB)C = $ _____; the distributive law states that $A(B + C) = $ _____.

  b. If I is an identity matrix of order n, then $IA = A$ if A is any matrix of order _____.

11. A matrix A is nonsingular if there exists a matrix A^{-1} such that _____ $= $ _____ $= I$. If A^{-1} does not exist, then A is said to be _____.

12. A system of n linear equations in n variables written in the form $AX = B$ has a unique solution given by $X = $ _____ if A has an inverse.

CHAPTER 2 Review Exercises

In Exercises 1–4, perform the operations, if possible.

1. $\begin{bmatrix} 1 & 2 \\ -1 & 3 \\ 2 & 1 \end{bmatrix} + \begin{bmatrix} 1 & 0 \\ 0 & 1 \\ 1 & 2 \end{bmatrix}$

2. $\begin{bmatrix} -1 & 2 \\ 3 & 4 \end{bmatrix} - \begin{bmatrix} 1 & 2 \\ 5 & -2 \end{bmatrix}$

3. $\begin{bmatrix} -3 & 2 & 1 \end{bmatrix} \begin{bmatrix} 2 & 1 \\ -1 & 0 \\ 2 & 1 \end{bmatrix}$

4. $\begin{bmatrix} 1 & 3 & 2 \\ -1 & 2 & 3 \end{bmatrix} \begin{bmatrix} 1 \\ 4 \\ 2 \end{bmatrix}$

In Exercises 5–8, find the values of the variables.

5. $\begin{bmatrix} 1 & x \\ y & 3 \end{bmatrix} = \begin{bmatrix} z & 2 \\ 3 & w \end{bmatrix}$

6. $\begin{bmatrix} 3 & x \\ y & 3 \end{bmatrix} \begin{bmatrix} 1 \\ 2 \end{bmatrix} = \begin{bmatrix} 7 \\ 4 \end{bmatrix}$

7. $\begin{bmatrix} 3 & a+3 \\ -1 & b \\ c+1 & d \end{bmatrix} = \begin{bmatrix} 3 & 6 \\ e+2 & 4 \\ -1 & 2 \end{bmatrix}$

8. $\begin{bmatrix} x & 3 & 1 \\ 0 & y & 2 \end{bmatrix} \begin{bmatrix} 1 & 1 \\ 3 & z \\ 4 & 2 \end{bmatrix} = \begin{bmatrix} 12 & 4 \\ 2 & 2 \end{bmatrix}$

In Exercises 9–16, compute the expressions, if possible, given that

$$A = \begin{bmatrix} 1 & 3 & 1 \\ -2 & 1 & 3 \\ 4 & 0 & 2 \end{bmatrix}$$

$$B = \begin{bmatrix} 2 & 1 & 3 \\ -2 & -1 & -1 \\ 1 & 4 & 2 \end{bmatrix}$$

$$C = \begin{bmatrix} 3 & -1 & 2 \\ 1 & 6 & 4 \\ 2 & 1 & 3 \end{bmatrix}$$

9. $2A + 3B$

10. $3A - 2B$

11. $2(3A)$

12. $2(3A - 4B)$

13. $A(B - C)$

14. $AB + AC$

15. $A(BC)$

16. $\left(\dfrac{1}{2}\right)(CA - CB)$

In Exercises 17–24, solve the system of linear equations using the Gauss–Jordan elimination method.

17. $2x - 3y = 5$
$3x + 4y = -1$

18. $3x + 2y = 3$
$2x - 4y = -14$

19. $x - y + 2z = 5$
$3x + 2y + z = 10$
$2x - 3y - 2z = -10$

20. $3x - 2y + 4z = 16$
$2x + y - 2z = -1$
$x + 4y - 8z = -18$

21. $3x - 2y + 4z = 11$
$2x - 4y + 5z = 4$
$x + 2y - z = 10$

22. $x - 2y + 3z + 4w = 17$
$2x + y - 2z - 3w = -9$
$3x - y + 2z - 4w = 0$
$4x + 2y - 3z + w = -2$

23. $3x - 2y + z = 4$
$x + 3y - 4z = -3$
$2x - 3y + 5z = 7$
$x - 8y + 9z = 10$

24. $2x - 3y + z = 10$
$3x + 2y - 2z = -2$
$x - 3y - 4z = -7$
$4x + y - z = 4$

In Exercises 25–32, find the inverse of the matrix (if it exists).

25. $A = \begin{bmatrix} 3 & 1 \\ 1 & 2 \end{bmatrix}$

26. $A = \begin{bmatrix} 2 & 4 \\ 1 & 6 \end{bmatrix}$

27. $A = \begin{bmatrix} 3 & 4 \\ 2 & 2 \end{bmatrix}$

28. $A = \begin{bmatrix} 2 & 4 \\ 1 & -2 \end{bmatrix}$

29. $A = \begin{bmatrix} 2 & 3 & 1 \\ 1 & -1 & 2 \\ 1 & 2 & 1 \end{bmatrix}$

30. $A = \begin{bmatrix} 1 & 2 & 4 \\ 2 & 1 & 3 \\ -1 & 0 & 2 \end{bmatrix}$

31. $A = \begin{bmatrix} 1 & 2 & 4 \\ 3 & 1 & 2 \\ 1 & 0 & -6 \end{bmatrix}$

32. $A = \begin{bmatrix} 2 & 1 & -3 \\ 1 & 2 & -4 \\ 3 & 1 & -2 \end{bmatrix}$

In Exercises 33–36, compute the value of the expressions, if possible, given that

$$A = \begin{bmatrix} 1 & 2 \\ -1 & 2 \end{bmatrix} \quad B = \begin{bmatrix} 3 & 1 \\ 4 & 2 \end{bmatrix} \quad C = \begin{bmatrix} 1 & 1 \\ -1 & 2 \end{bmatrix}$$

33. $(A^{-1}B)^{-1}$

34. $(ABC)^{-1}$

35. $(2A - C)^{-1}$

36. $(A + B)^{-1}$

In Exercises 37–40, write each system of linear equations in the form $AX = C$. Find A^{-1} and use the result to solve the system.

37. $2x + 3y = -8$
$x - 2y = 3$

38. $x - 3y = -1$
$2x + 4y = 8$

39. $x - 2y + 4z = 13$
$2x + 3y - 2z = 0$
$x + 4y - 6z = -15$

40. $2x - 3y + 4z = 17$
$x + 2y - 4z = -7$
$3x - y + 2z = 14$

41. GASOLINE SALES Gloria Newburg operates three self-service gasoline stations in different parts of town. On a certain day, station A sold 600 gal of premium, 800 gal of super, 1000 gal of regular gasoline, and 700 gal of diesel fuel; station B sold 700 gal of premium, 600 gal of super, 1200 gal of regular gasoline, and 400 gal of diesel fuel; station C sold 900 gal of premium, 700 gal of super, 1400 gal of regular gasoline, and 800 gal of diesel fuel. Assume that the price of gasoline was $1.60/gal for premium, $1.40/gal for super, and $1.20/gal for regular, and diesel fuel sold for $1.50/gal. Use matrix algebra to find the total revenue at each station.

42. COMMON STOCK TRANSACTIONS Jack Spaulding bought 10,000 shares of stock X, 20,000 shares of stock Y, and 30,000 shares of stock Z at a unit price of $20, $30, and $50/share, respectively. Six months later, the closing prices of stocks X, Y, and Z were $22, $35, and $51/share, respectively. Jack made no other stock transactions during the period in question. Compare the value of Jack's stock holdings at the time of purchase and 6 mo later.

43. MACHINE SCHEDULING Desmond Jewelry wishes to produce three types of pendants: type A, type B, and type C. To manufacture a type-A pendant requires 2 min on machines I and II and 3 min on machine III. A type-B pendant requires 2 min on machine I, 3 min on machine II, and 4 min on machine III. A type-C pendant requires 3 min on machine I, 4 min on machine II, and 3 min on machine III. There are $3\frac{1}{2}$ hr available on machine I, $4\frac{1}{2}$ hr available on machine II, and 5 hr available on machine III. How many pendants of each type should Desmond make in order to use all the available time?

44. PETROLEUM PRODUCTION Wildcat Oil Company has two refineries—one located in Houston and the other in Tulsa. The Houston refinery ships 60% of its petroleum to a Chicago distributor and 40% of its petroleum to a Los Angeles distributor. The Tulsa refinery ships 30% of its petroleum to the Chicago distributor and 70% of its petroleum to the Los Angeles distributor. Assume that, over the year, the Chicago distributor received 240,000 gal of petroleum and the Los Angeles distributor received 460,000 gal of petroleum. Find the amount of petroleum produced at each of Wildcat's refineries.

1. Solve the following system of linear equations, using the Gauss–Jordan elimination method:

$$2x + y - z = -1$$
$$x + 3y + 2z = 2$$
$$3x + 3y - 3z = -5$$

2. Find the solution(s), if it exists, of the system of linear equations whose augmented matrix in reduced form follows.

a. $\begin{bmatrix} 1 & 0 & 0 & 2 \\ 0 & 1 & 0 & -3 \\ 0 & 0 & 1 & 1 \end{bmatrix}$
  b. $\begin{bmatrix} 1 & 0 & 0 & 3 \\ 0 & 1 & 0 & 0 \\ 0 & 0 & 0 & 1 \end{bmatrix}$

c. $\begin{bmatrix} 1 & 0 & 0 & 2 \\ 0 & 1 & 3 & 1 \\ 0 & 0 & 0 & 0 \end{bmatrix}$
  d. $\begin{bmatrix} 1 & 0 & 0 & 0 & 0 \\ 0 & 1 & 0 & 0 & 0 \\ 0 & 0 & 1 & 0 & 0 \\ 0 & 0 & 0 & 1 & 0 \end{bmatrix}$

e. $\begin{bmatrix} 1 & 0 & -1 & 2 \\ 0 & 1 & 2 & 3 \end{bmatrix}$

3. Solve each system of linear equations, using the Gauss–Jordan elimination method.

a. $x + 2y = 3$
    $3x - y = -5$
    $4x + y = -2$

b. $x - 2y + 4z = 2$
    $3x + y - 2z = 1$

4. Let

$$A = \begin{bmatrix} 1 & -2 & 4 \\ 3 & 0 & 1 \end{bmatrix} \qquad B = \begin{bmatrix} 1 & -1 & 2 \\ 3 & 1 & -1 \\ 2 & 1 & 0 \end{bmatrix}$$

$$C = \begin{bmatrix} 2 & -2 \\ 1 & 1 \\ 3 & 4 \end{bmatrix}$$

Find (a) AB, (b) $(A + C^T)B$, and (c) $C^TB - AB^T$.

5. Find A^{-1} if

$$A = \begin{bmatrix} 2 & 1 & 2 \\ 0 & -1 & 3 \\ 1 & 1 & 0 \end{bmatrix}$$

6. Solve the system

$$2x \qquad + z = 4$$
$$2x + y - z = -1$$
$$3x + y - z = 0$$

by first writing it in the matrix form $AX = B$ and then finding A^{-1}.

3 Linear Programming: A Geometric Approach

How should the aircraft engines be shipped? Curtis-Roe Aviation Industries manufactures jet engines in two different locations. These engines are to be shipped to the company's two main assembly plants. In Example 3, page 178, we will show how many engines should be produced and shipped from each manufacturing plant to each assembly plant in order to minimize shipping costs.

© Michael Melford/The Image Bank/Getty Images

MANY PRACTICAL PROBLEMS involve maximizing or minimizing a function subject to certain constraints. For example, we may wish to maximize a profit function subject to certain limitations on the amount of material and labor available. Maximization or minimization problems that can be formulated in terms of a *linear* objective function and constraints in the form of linear inequalities are called *linear programming problems*. In this chapter we look at linear programming problems involving two variables. These problems are amenable to geometric analysis, and the method of solution introduced here will shed much light on the basic nature of a linear programming problem.

3.1   Graphing Systems of Linear Inequalities in Two Variables

Graphing Linear Inequalities

In Chapter 1, we saw that a linear equation in two variables x and y

$$ax + by + c = 0 \qquad \text{a, b not both equal to zero}$$

has a *solution set* that may be exhibited graphically as points on a straight line in the xy-plane. We now show that there is also a simple graphical representation for **linear inequalities** in two variables:

$$ax + by + c < 0 \qquad ax + by + c \le 0$$
$$ax + by + c > 0 \qquad ax + by + c \ge 0$$

Before turning to a general procedure for graphing such inequalities, let's consider a specific example. Suppose we wish to graph

$$2x + 3y < 6 \tag{1}$$

We first graph the equation $2x + 3y = 6$, which is obtained by replacing the given inequality "$<$" with an equality "$=$" (Figure 1).

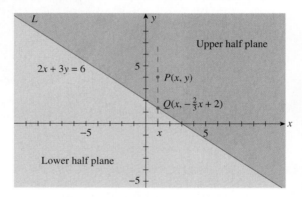

FIGURE 1
A straight line divides the xy-plane into two half planes.

Observe that this line divides the xy-plane into two half planes: an upper half plane and a lower half plane. Let's show that the upper half plane is the graph of the linear inequality

$$2x + 3y > 6 \tag{2}$$

whereas the lower half plane is the graph of the linear inequality

$$2x + 3y < 6 \tag{3}$$

To see this, let's write Equations (2) and (3) in the equivalent forms

$$y > -\frac{2}{3}x + 2 \tag{4}$$

and

$$y < -\frac{2}{3}x + 2 \tag{5}$$

The equation of the line itself is

$$y = -\frac{2}{3}x + 2 \tag{6}$$

Now pick any point $P(x, y)$ lying above the line L. Let Q be the point lying on L and directly below P (see Figure 1). Since Q lies on L, its coordinates must satisfy Equation (6). In other words, Q has representation $Q(x, -\frac{2}{3}x + 2)$. Comparing the y-coordinates of P and Q and recalling that P lies above Q so that its y-coordinate must be larger than that of Q, we have

$$y > -\frac{2}{3}x + 2$$

But this inequality is just Inequality (4) or, equivalently, Inequality (2). Similarly, we can show that any point lying below L must satisfy Inequality (5) and therefore (3).

This analysis shows that the lower half plane provides a solution to our problem (Figure 2). (The dashed line shows that the points on L do not belong to the solution set.) Observe that the two half planes in question are mutually exclusive; that is, they do not have any points in common. Because of this, there is an alternative and easier method of determining the solution to the problem.

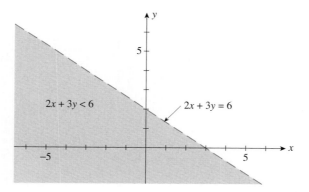

FIGURE 2
The set of points lying below the dashed line satisfies the given inequality.

To determine the required half plane, let's pick *any* point lying in one of the half planes. For simplicity, pick the origin $(0, 0)$, which lies in the lower half plane. Substituting $x = 0$ and $y = 0$ (the coordinates of this point) into the given Inequality (1), we find

$$2(0) + 3(0) < 6$$

or $0 < 6$, which is certainly true. This tells us that the required half plane is the half plane containing the test point—namely, the lower half plane.

Next, let's see what happens if we choose the point $(2, 3)$, which lies in the upper half plane. Substituting $x = 2$ and $y = 3$ into the given inequality, we find

$$2(2) + 3(3) < 6$$

or $13 < 6$, which is false. This tells us that the upper half plane is *not* the required half plane, as expected. Note, too, that no point $P(x, y)$ lying on the line constitutes a solution to our problem, because of the *strict* inequality $<$.

This discussion suggests the following procedure for graphing a linear inequality in two variables.

Procedure for Graphing Linear Inequalities
1. Draw the graph of the equation obtained for the given inequality by replacing the inequality sign with an equal sign. Use a dashed or dotted line if the problem involves a strict inequality, $<$ or $>$. Otherwise, use a solid line to indicate that the line itself constitutes part of the solution.
2. Pick a test point lying in one of the half planes determined by the line sketched in step 1 and substitute the values of x and y into the given inequality. Use the origin whenever possible.
3. If the inequality is satisfied, the graph of the inequality includes the half plane containing the test point. Otherwise, the solution includes the half plane not containing the test point.

EXAMPLE 1   Determine the solution set for the inequality $2x + 3y \geq 6$.

Solution   Replacing the inequality $\geq$ with an equality $=$, we obtain the equation $2x + 3y = 6$, whose graph is the straight line shown in Figure 3. Instead of a

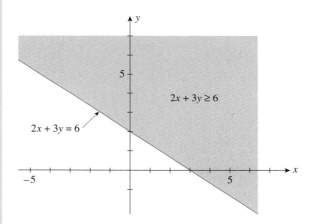

FIGURE 3
The set of points lying on the line and in the upper half plane satisfies the given inequality.

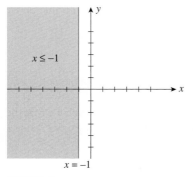

FIGURE 4
The set of points lying on the line $x = -1$ and in the left half plane satisfies the given inequality.

dashed line as before, we use a solid line to show that all points on the line are also solutions to the problem. Picking the origin as our test point, we find $2(0) + 3(0) \geq 6$, or $0 \geq 6$, which is impossible. So we conclude that the solution set is made up of the half plane not containing the origin, including, in this case, the line given by $2x + 3y = 6$.   ■

EXAMPLE 2   Graph $x \leq -1$.

Solution   The graph of $x = -1$ is the vertical line shown in Figure 4. Picking the origin $(0, 0)$ as a test point, we find $0 \leq -1$, which is false. Therefore, the required solution is the *left* half plane, which does not contain the origin.   ■

EXAMPLE 3   Graph $x - 2y > 0$.

Solution   We first graph the equation $x - 2y = 0$, or $y = (\frac{1}{2})x$ (Figure 5). Since the origin lies on the line, we may not use it as a test point. (Why?) Let's pick $(1, 2)$ as a test point. Substituting $x = 1$ and $y = 2$ into the given inequality, we find $1 - 2(2) > 0$, or $-3 > 0$, which is false. Therefore, the required solution is the half plane that does not contain the test point—namely, the lower half plane.

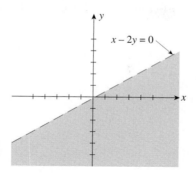

FIGURE 5
The set of points in the lower half plane satisfies $x - 2y > 0$.

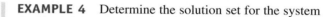

Graphing Systems of Linear Inequalities

By the **solution set of a system of linear inequalities** in the two variables x and y, we mean the set of all points (x, y) satisfying each inequality of the system. The graphical solution of such a system may be obtained by graphing the solution set for each inequality independently and then determining the region in common with each solution set.

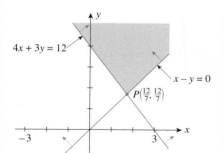

FIGURE 6
The set of points in the shaded area satisfies the system

$$4x + 3y \geq 12$$
$$x - y \leq 0$$

EXAMPLE 4 Determine the solution set for the system

$$4x + 3y \geq 12$$
$$x - y \leq 0$$

Solution Proceeding as in the previous examples, you should have no difficulty locating the half planes determined by each of the linear inequalities that make up the system. These half planes are shown in Figure 6. The intersection of the two half planes is the shaded region. A point in this region is an element of the solution set for the given system. The point P, the intersection of the two straight lines determined by the equations, is found by solving the simultaneous equations

$$4x + 3y = 12$$
$$x - y = 0$$

EXAMPLE 5 Sketch the solution set for the system

$$x \geq 0$$
$$y \geq 0$$
$$x + y - 6 \leq 0$$
$$2x + y - 8 \leq 0$$

Solution The first inequality in the system defines the right half plane—all points to the right of the y-axis plus all points lying on the y-axis itself. The second inequality in the system defines the upper half plane, including the x-axis. The half planes defined by the third and fourth inequalities are indicated by arrows in Figure 7. Thus, the required region, the intersection of the four half planes defined by the four inequalities in the given system of linear inequalities, is the shaded region. The point P is found by solving the simultaneous equations $x + y - 6 = 0$ and $2x + y - 8 = 0$.

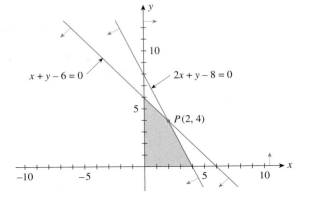

FIGURE 7
The set of points in the shaded region, including the x- and y-axes, satisfies the given inequalities.

The solution set found in Example 5 is an example of a bounded set. Observe that the set can be enclosed by a circle. For example, if you draw a circle of radius 10 with center at the origin, you will see that the set lies entirely inside the circle. On the other hand, the solution set of Example 4 cannot be enclosed by a circle and is said to be unbounded.

Bounded and Unbounded Solution Sets
The solution set of a system of linear inequalities is **bounded** if it can be enclosed by a circle. Otherwise, it is **unbounded**.

EXAMPLE 6   Determine the graphical solution set for the following system of linear inequalities:

$$2x + y \geq 50$$
$$x + 2y \geq 40$$
$$x \geq 0$$
$$y \geq 0$$

Solution   The required solution set is the unbounded region shown in Figure 8.

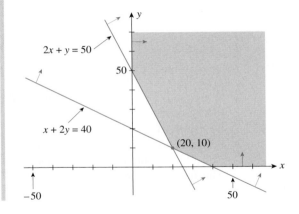

FIGURE 8
The solution set is an unbounded region.

3.1   Self-Check Exercises

1. Determine graphically the solution set for the following system of inequalities:

$$x + 2y \leq 10$$
$$5x + 3y \leq 30$$
$$x \geq 0, y \geq 0$$

2. Determine graphically the solution set for the following system of inequalities:

$$5x + 3y \geq 30$$
$$x - 3y \leq 0$$
$$x \geq 2$$

Solutions to Self-Check Exercises 3.1 can be found on page 175.

3.1   Concept Questions

1. **a.** What is the difference, geometrically, between the solution set of $ax + by < c$ and the solution set of $ax + by \leq c$?
   b. Describe the set that is obtained by intersecting the solution set of $ax + by \leq c$ with the solution set of $ax + by \geq c$.

2. **a.** What is the solution set of a system of linear inequalities?
   b. How do you find the graphical solution of a system of linear inequalities?

3.1   Exercises

In Exercises 1–10, find the graphical solution of each inequality.

1. $4x - 8 < 0$

2. $3y + 2 > 0$

3. $x - y \leq 0$

4. $3x + 4y \leq -2$

5. $x \leq -3$

6. $y \geq -1$

7. $2x + y \leq 4$

8. $-3x + 6y \geq 12$

9. $4x - 3y \leq -24$

10. $5x - 3y \geq 15$

In Exercises 11–18, write a system of linear inequalities that describes the shaded region.

11.

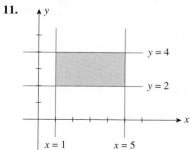

12.

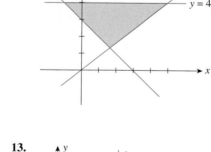

13.

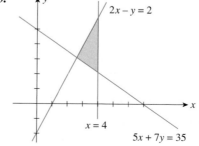

14.

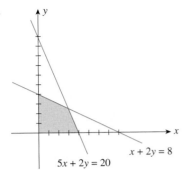

5x + 2y = 20

15.

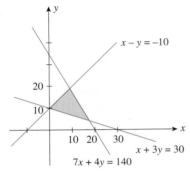

16.

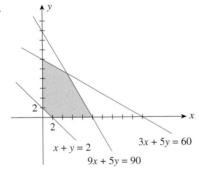

17.

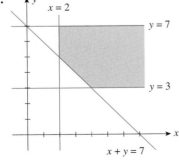

18.

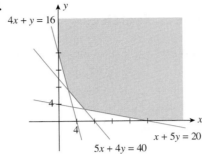

In Exercises 19–36, determine graphically the solution set for each system of inequalities and indicate whether the solution set is bounded or unbounded.

19. $2x + 4y > 16$
$-x + 3y \geq 7$

20. $3x - 2y > -13$
$-x + 2y > 5$

21. $x - y \leq 0$
$2x + 3y \geq 10$

22. $x + y \geq -2$
$3x - y \leq 6$

23. $x + 2y \geq 3$
$2x + 4y \leq -2$

24. $2x - y \geq 4$
$4x - 2y < -2$

25. $x + y \leq 6$
$0 \leq x \leq 3$
$y \geq 0$

26. $4x - 3y \leq 12$
$5x + 2y \leq 10$
$x \geq 0, y \geq 0$

27. $3x - 6y \leq 12$
$-x + 2y \leq 4$
$x \geq 0, y \geq 0$

28. $x + y \geq 20$
$x + 2y \geq 40$
$x \geq 0, y \geq 0$

29. $3x - 7y \geq -24$
$x + 3y \geq 8$
$x \geq 0, y \geq 0$

30. $3x + 4y \geq 12$
$2x - y \geq -2$
$0 \leq y \leq 3$
$x \geq 0$

31. $x + 2y \geq 3$
$5x - 4y \leq 16$
$0 \leq y \leq 2$
$x \geq 0$

32. $x + y \leq 4$
$2x + y \leq 6$
$2x - y \geq -1$
$x \geq 0, y \geq 0$

33. $6x + 5y \leq 30$
$3x + y \geq 6$
$x + y \geq 4$
$x \geq 0, y \geq 0$

34. $6x + 7y \leq 84$
$12x - 11y \leq 18$
$6x - 7y \leq 28$
$x \geq 0, y \geq 0$

35. $x - y \geq -6$
$x - 2y \leq -2$
$x + 2y \geq 6$
$x - 2y \geq -14$
$x \geq 0, y \geq 0$

36. $x - 3y \geq -18$
$3x - 2y \geq 2$
$x - 3y \leq -4$
$3x - 2y \leq 16$
$x \geq 0, y \geq 0$

In Exercises 37–40, determine whether the statement is true or false. If it is true, explain why it is true. If it is false, give an example to show why it is false.

37. The solution set of a linear inequality involving two variables is either a half plane or a straight line.

38. The solution set of the inequality $ax + by + c \leq 0$ is either a left half plane or a lower half plane.

39. The solution set of a system of linear inequalities in two variables is bounded if it can be enclosed by a rectangle.

40. The solution set of the system

$$ax + by \leq e$$
$$cx + dy \leq f$$
$$x \geq 0, y \geq 0$$

where a, b, c, d, e, and f are positive real numbers is a bounded set.

3.1 Solutions to Self-Check Exercises

1. The required solution set is shown in the following figure:

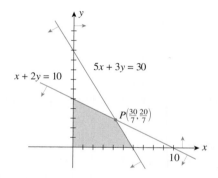

The point P is found by solving the system of equations

$$x + 2y = 10$$
$$5x + 3y = 30$$

Solving the first equation for x in terms of y gives

$$x = 10 - 2y$$

Substituting this value of x into the second equation of the system gives

$$5(10 - 2y) + 3y = 30$$
$$50 - 10y + 3y = 30$$
$$-7y = -20$$

so $y = \frac{20}{7}$. Substituting this value of y into the expression for x found earlier, we obtain

$$x = 10 - 2\left(\frac{20}{7}\right) = \frac{30}{7}$$

giving the point of intersection as $\left(\frac{30}{7}, \frac{20}{7}\right)$.

2. The required solution set is shown in the following figure:

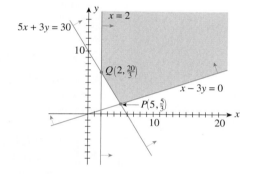

To find the coordinates of P, we solve the system

$$5x + 3y = 30$$
$$x - 3y = 0$$

Solving the second equation for x in terms of y and substituting this value of x in the first equation gives

$$5(3y) + 3y = 30$$

or $y = \frac{5}{3}$. Substituting this value of y into the second equation gives $x = 5$. Next, the coordinates of Q are found by solving the system

$$5x + 3y = 30$$
$$x = 2$$

obtaining $x = 2$ and $y = \frac{20}{3}$.

3.2   Linear Programming Problems

In many business and economic problems we are asked to optimize (maximize or minimize) a function subject to a system of equalities or inequalities. The function to be optimized is called the **objective function.** Profit functions and cost functions are examples of objective functions. The system of equalities or inequalities to which the objective function is subjected reflects the constraints (for example, limitations on resources such as materials and labor) imposed on the solution(s) to the problem. Problems of this nature are called **mathematical programming problems.** In particular, problems in which both the objective function and the constraints are expressed as linear equations or inequalities are called linear programming problems.

> ### Linear Programming Problem
> A **linear programming problem** consists of a linear objective function to be maximized or minimized subject to certain constraints in the form of linear equations or inequalities.

▬ A Maximization Problem

As an example of a linear programming problem in which the objective function is to be maximized, let's consider the following simplified version of a production problem involving two variables.

APPLIED EXAMPLE 1   A Production Problem   Ace Novelty wishes to produce two types of souvenirs: type A and type B. Each type-A souvenir will result in a profit of $1, and each type-B souvenir will result in a profit of $1.20. To manufacture a type-A souvenir requires 2 minutes on machine I and 1 minute on machine II. A type-B souvenir requires 1 minute on machine I and 3 minutes on machine II. There are 3 hours available on machine I and 5 hours available on machine II for processing the order. How many souvenirs of each type should Ace make in order to maximize its profit?

Solution   As a first step toward the mathematical formulation of this problem, we tabulate the given information, as shown in Table 1.

TABLE 1			
	Type A	**Type B**	**Time Available**
Machine I	2 min	1 min	180 min
Machine II	1 min	3 min	300 min
Profit/Unit	$1	$1.20	

Let x be the number of type-A souvenirs and y be the number of type-B souvenirs to be made. Then, the total profit P (in dollars) is given by

$$P = x + 1.2y$$

which is the objective function to be maximized.

The total amount of time that machine I is used is given by $2x + y$ minutes and must not exceed 180 minutes. Thus, we have the inequality

$$2x + y \le 180$$

Similarly, the total amount of time that machine II is used is $x + 3y$ minutes, which cannot exceed 300 minutes, so we are led to the inequality

$$x + 3y \le 300$$

Finally, neither x nor y can be negative, so

$$x \ge 0$$
$$y \ge 0$$

To summarize, the problem at hand is one of maximizing the objective function $P = x + 1.2y$ subject to the system of inequalities

$$2x + y \le 180$$
$$x + 3y \le 300$$
$$x \ge 0$$
$$y \ge 0$$

The solution to this problem will be completed in Example 1, Section 3.3. ∎

Minimization Problems

In the following example of a linear programming problem, the objective function is to be minimized.

 APPLIED EXAMPLE 2   A Nutrition Problem   A nutritionist advises an individual who is suffering from iron and vitamin-B deficiency to take at least 2400 milligrams (mg) of iron, 2100 mg of vitamin B_1 (thiamine), and 1500 mg of vitamin B_2 (riboflavin) over a period of time. Two vitamin pills are suitable, brand A and brand B. Each brand A pill contains 40 mg of iron, 10 mg of vitamin B_1, and 5 mg of vitamin B_2, and costs 6 cents. Each brand B pill contains 10 mg of iron and 15 mg each of vitamins B_1 and B_2, and costs 8 cents (Table 2). What combination of pills should the individual purchase in order to meet the minimum iron and vitamin requirements at the lowest cost?

TABLE 2

	Brand A	Brand B	Minimum Requirement
Iron	40 mg	10 mg	2400 mg
Vitamin B_1	10 mg	15 mg	2100 mg
Vitamin B_2	5 mg	15 mg	1500 mg
Cost/Pill	6¢	8¢	

Solution   Let x be the number of brand A pills and y be the number of brand B pills to be purchased. The cost C (in cents) is given by

$$C = 6x + 8y$$

and is the objective function to be minimized.

The amount of iron contained in x brand A pills and y brand B pills is given by $40x + 10y$ mg, and this must be greater than or equal to 2400 mg. This translates into the inequality

$$40x + 10y \geq 2400$$

Similar considerations involving the minimum requirements of vitamins B_1 and B_2 lead to the inequalities

$$10x + 15y \geq 2100$$
$$5x + 15y \geq 1500$$

respectively. Thus, the problem here is to minimize $C = 6x + 8y$ subject to

$$40x + 10y \geq 2400$$
$$10x + 15y \geq 2100$$
$$5x + 15y \geq 1500$$
$$x \geq 0, y \geq 0$$

The solution to this problem will be completed in Example 2, Section 3.3.   ■

APPLIED EXAMPLE 3 A Transportation Problem Curtis-Roe Aviation Industries has two plants, I and II, that produce the Zephyr jet engines used in their light commercial airplanes. The maximum production capacities of these two plants are 100 units and 110 units per month, respectively. The engines are shipped to two of Curtis-Roe's main assembly plants, A and B. The shipping costs (in dollars) per engine from plants I and II to the main assembly plants A and B are as follows:

From	To Assembly Plant A	B
Plant I	100	60
Plant II	120	70

In a certain month, assembly plant A needs 80 engines, whereas assembly plant B needs 70 engines. Find how many engines should be shipped from each plant to each main assembly plant if shipping costs are to be kept to a minimum.

Solution   Let x denote the number of engines shipped from plant I to assembly plant A and let y denote the number of engines shipped from plant I to assembly plant B. Since the requirements of assembly plants A and B are 80 and 70 engines, respectively, the number of engines shipped from plant II to assembly plants A and B are $(80 - x)$ and $(70 - y)$, respectively. These numbers may be displayed in a schematic. With the aid of the accompanying schematic and the shipping cost schedule, we find that the total shipping cost incurred by Curtis-Roe is given by

$$C = 100x + 60y + 120(80 - x) + 70(70 - y)$$
$$= 14,500 - 20x - 10y$$

Next, the production constraints on plants I and II lead to the inequalities

$$x + y \leq 100$$
$$(80 - x) + (70 - y) \leq 110$$

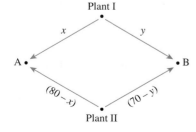

The last inequality simplifies to

$$x + y \geq 40$$

Also, the requirements of the two main assembly plants lead to the inequalities

$$x \geq 0 \qquad y \geq 0 \qquad 80 - x \geq 0 \qquad 70 - y \geq 0$$

The last two may be written as $x \leq 80$ and $y \leq 70$.

Summarizing, we have the following linear programming problem: Minimize the objective (cost) function $C = 14,500 - 20x - 10y$ subject to the constraints

$$x + y \geq 40$$
$$x + y \leq 100$$
$$x \leq 80$$
$$y \leq 70$$

where $x \geq 0$ and $y \geq 0$.

You will be asked to complete the solution to this problem in Exercise 43, Section 3.3.

APPLIED EXAMPLE 4   A Warehouse Problem Acrosonic manufactures its model F loudspeaker systems in two separate locations, plant I and plant II. The output at plant I is at most 400 per month, whereas the output at plant II is at most 600 per month. These loudspeaker systems are shipped to three warehouses that serve as distribution centers for the company. For the warehouses to meet their orders, the minimum monthly requirements of warehouses A, B, and C are 200, 300, and 400, respectively. Shipping costs from plant I to warehouses A, B, and C are $20, $8, and $10 per loudspeaker system, respectively, and shipping costs from plant II to each of these warehouses are $12, $22, and $18, respectively. What should the shipping schedule be if Acrosonic wishes to meet the requirements of the distribution centers and at the same time keep its shipping costs to a minimum?

Solution The respective shipping costs (in dollars) per loudspeaker system may be tabulated as in Table 3. Letting x_1 denote the number of loudspeaker systems shipped from plant I to warehouse A, x_2 the number shipped from plant I to warehouse B, and so on leads to Table 4.

TABLE 3

Plant	Warehouse A	B	C
I	20	8	10
II	12	22	18

TABLE 4

Plant	Warehouse A	B	C	Max. Prod.
I	x_1	x_2	x_3	400
II	x_4	x_5	x_6	600
Min. Req.	200	300	400	

From Tables 3 and 4 we see that the cost of shipping x_1 loudspeaker systems from plant I to warehouse A is $20x_1$, the cost of shipping x_2 loudspeaker systems from plant I to warehouse B is $\$8x_2$, and so on. Thus, the total monthly shipping cost incurred by Acrosonic is given by

$$C = 20x_1 + 8x_2 + 10x_3 + 12x_4 + 22x_5 + 18x_6$$

Next, the production constraints on plants I and II lead to the inequalities

$$x_1 + x_2 + x_3 \leq 400$$
$$x_4 + x_5 + x_6 \leq 600$$

(see Table 4). Also, the minimum requirements of each of the three warehouses lead to the three inequalities

$$x_1 + x_4 \geq 200$$
$$x_2 + x_5 \geq 300$$
$$x_3 + x_6 \geq 400$$

Summarizing, we have the following linear programming problem:

Minimize $C = 20x_1 + 8x_2 + 10x_3 + 12x_4 + 22x_5 + 18x_6$
subject to $x_1 + x_2 + x_3 \leq 400$
$x_4 + x_5 + x_6 \leq 600$
$x_1 + x_4 \geq 200$
$x_2 + x_5 \geq 300$
$x_3 + x_6 \geq 400$
$x_1 \geq 0, x_2 \geq 0, \ldots, x_6 \geq 0$

The solution to this problem will be completed in Section 4.2, Example 5.   ■

3.2   Self-Check Exercise

Gino Balduzzi, proprietor of Luigi's Pizza Palace, allocates $\$9000$ a month for advertising in two newspapers, the *City Tribune* and the *Daily News*. The *City Tribune* charges $\$300$ for a certain advertisement, whereas the *Daily News* charges $\$100$ for the same ad. Gino has stipulated that the ad is to appear in at least 15 but no more than 30 editions of the *Daily News* per month. The *City Tribune* has a daily circulation of 50,000, and the *Daily News* has a circulation of 20,000. Under these condi-

tions, determine how many ads Gino should place in each newspaper in order to reach the largest number of readers. Formulate but do not solve the problem. (The solution to this problem can be found in Exercise 3 of Solutions to Self-Check Exercises 3.3.)

The solution to Self-Check Exercise 3.2 can be found on page 184.

3.2   Concept Questions

1. What is a *linear programming problem*?

2. Suppose you are asked to formulate a linear programming problem in two variables x and y. How would you express the fact that x and y are nonnegative? Why are these conditions often required in practical problems?

3. What is the difference between a *maximization linear programming problem* and a *minimization linear programming problem*?

3.2 Exercises

Formulate but do not solve each of the following exercises as a linear programming problem. You will be asked to solve these problems later.

1. MANUFACTURING—PRODUCTION SCHEDULING A company manufactures two products, A and B, on two machines, I and II. It has been determined that the company will realize a profit of $3 on each unit of product A and a profit of $4 on each unit of product B. To manufacture a unit of product A requires 6 min on machine I and 5 min on machine II. To manufacture a unit of product B requires 9 min on machine I and 4 min on machine II. There are 5 hr of machine time available on machine I and 3 hr of machine time available on machine II in each work shift. How many units of each product should be produced in each shift to maximize the company's profit?

2. MANUFACTURING—PRODUCTION SCHEDULING National Business Machines manufactures two models of fax machines: A and B. Each model A costs $100 to make, and each model B costs $150. The profits are $30 for each model A and $40 for each model B fax machine. If the total number of fax machines demanded per month does not exceed 2500 and the company has earmarked no more than $600,000/month for manufacturing costs, how many units of each model should National make each month in order to maximize its monthly profits?

3. MANUFACTURING—PRODUCTION SCHEDULING Kane Manufacturing has a division that produces two models of fireplace grates, model A and model B. To produce each model A grate requires 3 lb of cast iron and 6 min of labor. To produce each model B grate requires 4 lb of cast iron and 3 min of labor. The profit for each model A grate is $2.00, and the profit for each model B grate is $1.50. If 1000 lb of cast iron and 20 hr of labor are available for the production of grates per day, how many grates of each model should the division produce per day in order to maximize Kane's profits?

4. MANUFACTURING—PRODUCTION SCHEDULING Refer to Exercise 3. Because of a backlog of orders on model A grates, the manager of Kane Manufacturing has decided to produce at least 150 of these models a day. Operating under this additional constraint, how many grates of each model should Kane produce to maximize profit?

5. FINANCE—ALLOCATION OF FUNDS Madison Finance has a total of $20 million earmarked for homeowner and auto loans. On the average, homeowner loans have a 10% annual rate of return, whereas auto loans yield a 12% annual rate of return. Management has also stipulated that the total amount of homeowner loans should be greater than or equal to four times the total amount of automobile loans. Determine the total amount of loans of each type Madison should extend to each category in order to maximize its returns.

6. INVESTMENTS—ASSET ALLOCATION A financier plans to invest up to $500,000 in two projects. Project A yields a return of 10% on the investment, whereas project B yields a return of 15% on the investment. Because the investment in project B is riskier than the investment in project A, she has decided that the investment in project B should not exceed 40% of the total investment. How much should the financier invest in each project in order to maximize the return on her investment?

7. MANUFACTURING—PRODUCTION SCHEDULING Acoustical Company manufactures a CD storage cabinet that can be bought fully assembled or as a kit. Each cabinet is processed in the fabrications department and the assembly department. If the fabrication department only manufactures fully assembled cabinets, then it can produce 200 units/day; but if it only manufactures kits, it can produce 200 units/day. If the assembly department only produces fully assembled cabinets, then it can produce 100 units/day; but if it only produces kits, then it can produce 300 units/day. Each fully assembled cabinet contributes $50 to the profits of the company, whereas each kit contributes $40 to its profits. How many fully assembled units and how many kits should the company produce per day in order to maximize its profits?

8. AGRICULTURE—CROP PLANNING A farmer plans to plant two crops, A and B. The cost of cultivating crop A is $40/acre, whereas that of crop B is $60/acre. The farmer has a maximum of $7400 available for land cultivation. Each acre of crop A requires 20 labor-hours, and each acre of crop B requires 25 labor-hours. The farmer has a maximum of 3300 labor-hours available. If she expects to make a profit of $150/acre on crop A and $200/acre on crop B, how many acres of each crop should she plant in order to maximize her profit?

9. MINING—PRODUCTION Perth Mining Company operates two mines for the purpose of extracting gold and silver. The Saddle Mine costs $14,000/day to operate, and it yields 50 oz of gold and 3000 oz of silver each day. The Horseshoe Mine costs $16,000/day to operate, and it yields 75 oz of gold and 1000 oz of silver each day. Company management has set a target of at least 650 oz of gold and 18,000 oz of silver. How many days should each mine be operated so that the target can be met at a minimum cost?

10. TRANSPORTATION Deluxe River Cruises operates a fleet of river vessels. The fleet has two types of vessels: A type-A vessel has 60 deluxe cabins and 160 standard cabins, whereas a type-B vessel has 80 deluxe cabins and 120 standard cabins. Under a charter agreement with Odyssey Travel Agency, Deluxe River Cruises is to provide Odyssey with a minimum of 360 deluxe and 680 standard cabins for their 15-day cruise in May. It costs $44,000 to operate a type-A

vessel and $54,000 to operate a type-B vessel for that period. How many of each type vessel should be used in order to keep the operating costs to a minimum?

11. NUTRITION—DIET PLANNING   A nutritionist at the Medical Center has been asked to prepare a special diet for certain patients. She has decided that the meals should contain a minimum of 400 mg of calcium, 10 mg of iron, and 40 mg of vitamin C. She has further decided that the meals are to be prepared from foods A and B. Each ounce of food A contains 30 mg of calcium, 1 mg of iron, 2 mg of vitamin C, and 2 mg of cholesterol. Each ounce of food B contains 25 mg of calcium, 0.5 mg of iron, 5 mg of vitamin C, and 5 mg of cholesterol. Find how many ounces of each type of food should be used in a meal so that the cholesterol content is minimized and the minimum requirements of calcium, iron, and vitamin C are met.

12. SOCIAL PROGRAMS PLANNING   AntiFam, a hunger-relief organization, has earmarked between $2 and $2.5 million, inclusive, for aid to two African countries, country A and country B. Country A is to receive between $1 and $1.5 million, inclusive, in aid, and country B is to receive at least $0.75 million in aid. It has been estimated that each dollar spent in country A will yield an effective return of $.60, whereas a dollar spent in country B will yield an effective return of $.80. How should the aid be allocated if the money is to be utilized most effectively according to these criteria? Hint: If x and y denote the amount of money to be given to country A and country B, respectively, then the objective function to be maximized is $P = 0.6x + 0.8y$.

13. ADVERTISING   Everest Deluxe World Travel has decided to advertise in the Sunday editions of two major newspapers in town. These advertisements are directed at three groups of potential customers. Each advertisement in newspaper I is seen by 70,000 group A customers, 40,000 group B customers, and 20,000 group C customers. Each advertisement in newspaper II is seen by 10,000 group A, 20,000 group B, and 40,000 group C customers. Each advertisement in newspaper I costs $1000, and each advertisement in newspaper II costs $800. Everest would like their advertisements to be read by at least 2 million people from group A, 1.4 million people from group B, and 1 million people from group C. How many advertisements should Everest place in each newspaper to achieve its advertisement goals at a minimum cost?

14. MANUFACTURING—SHIPPING COSTS   TMA manufactures 19-in. color television picture tubes in two separate locations, location I and location II. The output at location I is at most 6000 tubes/month, whereas the output at location II is at most 5000/month. TMA is the main supplier of picture tubes to Pulsar Corporation, its holding company, which has priority in having all its requirements met. In a certain month, Pulsar placed orders for 3000 and 4000 picture tubes to be shipped to two of its factories located in city A and city B, respectively. The shipping costs (in dollars) per picture tube from the two TMA plants to the two Pulsar factories are as follows:

From TMA	To Pulsar Factories	
	City A	City B
Location I	$3	$2
Location II	$4	$5

Find a shipping schedule that meets the requirements of both companies while keeping costs to a minimum.

15. INVESTMENTS—ASSET ALLOCATION   A financier plans to invest up to $2 million in three projects. She estimates that project A will yield a return of 10% on her investment, project B will yield a return of 15% on her investment, and project C will yield a return of 20% on her investment. Because of the risks associated with the investments, she decided to put not more than 20% of her total investment in project C. She also decided that her investments in projects B and C should not exceed 60% of her total investment. Finally, she decided that her investment in project A should be at least 60% of her investments in projects B and C. How much should the financier invest in each project if she wishes to maximize the total returns on her investments?

16. INVESTMENTS—ASSET ALLOCATION   Ashley has earmarked at most $250,000 for investment in three mutual funds: a money market fund, an international equity fund, and a growth-and-income fund. The money market fund has a rate of return of 6%/year, the international equity fund has a rate of return of 10%/year, and the growth-and-income fund has a rate of return of 15%/year. Ashley has stipulated that no more than 25% of her total portfolio should be in the growth-and-income fund and that no more than 50% of her total portfolio should be in the international equity fund. To maximize the return on her investment, how much should Ashley invest in each type of fund?

17. MANUFACTURING—PRODUCTION SCHEDULING   A company manufactures products A, B, and C. Each product is processed in three departments: I, II, and III. The total available labor-hours per week for departments I, II, and III are 900, 1080, and 840, respectively. The time requirements (in hours per unit) and profit per unit for each product are as follows:

	Product A	Product B	Product C
Dept. I	2	1	2
Dept. II	3	1	2
Dept. III	2	2	1
Profit	$18	$12	$15

How many units of each product should the company produce in order to maximize its profit?

18. **ADVERTISING** As part of a campaign to promote its annual clearance sale, the Excelsior Company decided to buy television advertising time on Station KAOS. Excelsior's advertising budget is $102,000. Morning time costs $3000/minute, afternoon time costs $1000/minute, and evening (prime) time costs $12,000/minute. Because of previous commitments, KAOS cannot offer Excelsior more than 6 min of prime time or more than a total of 25 min of advertising time over the 2 wk in which the commercials are to be run. KAOS estimates that morning commercials are seen by 200,000 people, afternoon commercials are seen by 100,000 people, and evening commercials are seen by 600,000 people. How much morning, afternoon, and evening advertising time should Excelsior buy to maximize exposure of its commercials?

19. **MANUFACTURING—PRODUCTION SCHEDULING** Custom Office Furniture Company is introducing a new line of executive desks made from a specially selected grade of walnut. Initially, three different models—A, B, and C—are to be marketed. Each model A desk requires $1\frac{1}{4}$ hr for fabrication, 1 hr for assembly, and 1 hr for finishing; each model B desk requires $1\frac{1}{2}$ hr for fabrication, 1 hr for assembly, and 1 hr for finishing; each model C desk requires $1\frac{1}{2}$ hr, $\frac{3}{4}$ hr, and $\frac{1}{2}$ hr for fabrication, assembly, and finishing, respectively. The profit on each model A desk is $26, the profit on each model B desk is $28, and the profit on each model C desk is $24. The total time available in the fabrication department, the assembly department, and the finishing department in the first month of production is 310 hr, 205 hr, and 190 hr, respectively. To maximize Custom's profit, how many desks of each model should be made in the month?

20. **MANUFACTURING—SHIPPING COSTS** Acrosonic of Example 4 also manufactures a model G loudspeaker system in plants I and II. The output at plant I is at most 800 systems/month, whereas the output at plant II is at most 600/month. These loudspeaker systems are also shipped to the three warehouses—A, B, and C—whose minimum monthly requirements are 500, 400, and 400, respectively. Shipping costs from plant I to warehouse A, warehouse B, and warehouse C are $16, $20, and $22/loudspeaker system, respectively, and shipping costs from plant II to each of these warehouses are $18, $16, and $14, respectively. What shipping schedule will enable Acrosonic to meet the warehouses' requirements and at the same time keep its shipping costs to a minimum?

21. **MANUFACTURING—SHIPPING COSTS** Steinwelt Piano manufactures uprights and consoles in two plants, plant I and plant II. The output of plant I is at most 300/month, whereas the output of plant II is at most 250/month. These pianos are shipped to three warehouses that serve as distribution centers for the company. To fill current and projected future orders, warehouse A requires a minimum of 200 pianos/month, warehouse B requires at least 150 pianos/month, and warehouse C requires at least 200 pianos/month. The shipping cost of each piano from plant I to warehouse A, warehouse

B, and warehouse C is $60, $60, and $80, respectively, and the shipping cost of each piano from plant II to warehouse A, warehouse B, and warehouse C is $80, $70, and $50, respectively. What shipping schedule will enable Steinwelt to meet the warehouses' requirements while keeping shipping costs to a minimum?

22. **MANUFACTURING—PREFABRICATED HOUSING PRODUCTION** Boise Lumber has decided to enter the lucrative prefabricated housing business. Initially, it plans to offer three models: standard, deluxe, and luxury. Each house is prefabricated and partially assembled in the factory, and the final assembly is completed on site. The dollar amount of building material required, the amount of labor required in the factory for prefabrication and partial assembly, the amount of on-site labor required, and the profit per unit are as follows:

	Standard Model	Deluxe Model	Luxury Model
Material	$6,000	$8,000	$10,000
Factory Labor (hr)	240	220	200
On-site Labor (hr)	180	210	300
Profit	$3,400	$4,000	$5,000

For the first year's production, a sum of $8.2 million is budgeted for the building material; the number of labor-hours available for work in the factory (for prefabrication and partial assembly) is not to exceed 218,000 hr; and the amount of labor for on-site work is to be less than or equal to 237,000 labor-hours. Determine how many houses of each type Boise should produce (market research has confirmed that there should be no problems with sales) to maximize its profit from this new venture.

23. **PRODUCTION—JUICE PRODUCTS** CalJuice Company has decided to introduce three fruit juices made from blending two or more concentrates. These juices will be packaged in 2-qt (64 fluid-oz) cartons. One carton of pineapple–orange juice requires 8 oz each of pineapple and orange juice concentrates. One carton of orange–banana juice requires 12 oz of orange juice concentrate and 4 oz of banana pulp concentrate. Finally, one carton of pineapple–orange–banana juice requires 4 oz of pineapple juice concentrate, 8 oz of orange juice concentrate, and 4 oz of banana pulp. The company has decided to allot 16,000 oz of pineapple juice concentrate, 24,000 oz of orange juice concentrate, and 5000 oz of banana pulp concentrate for the initial production run. The company also stipulated that the production of pineapple–orange–banana juice should not exceed 800 cartons. Its profit on one carton of pineapple–orange juice is $1.00; its profit on one carton of orange–banana juice is $.80, and its profit on one carton of pineapple–orange–banana juice is $.90. To realize a maximum profit, how many cartons of each blend should the company produce?

24. **MANUFACTURING—COLD FORMULA PRODUCTION** Bayer Pharmaceutical produces three kinds of cold formulas: formula I, formula II, and formula III. It takes 2.5 hr to produce 1000 bottles of formula I, 3 hr to produce 1000 bottles of formula II, and 4 hr to produce 1000 bottles of formula III. The profits for each 1000 bottles of formula I, formula II, and formula III are $180, $200, and $300, respectively. For a certain production run, there are enough ingredients on hand to make at most 9000 bottles of formula I, 12,000 bottles of formula II, and 6000 bottles of formula III. Furthermore, the time for the production run is limited to a maximum of 70 hr. How many bottles of each formula should be produced in this production run so that the profit is maximized?

In Exercises 25 and 26, determine whether the statement is true or false. If it is true, explain why it is true. If it is false, give an example to show why it is false.

25. The problem

$$\text{Maximize} \quad P = xy$$
$$\text{subject to} \quad 2x + 3y \le 12$$
$$2x + y \le 8$$
$$x \ge 0, y \ge 0$$

is a linear programming problem.

26. The problem

$$\text{Minimize} \quad C = 2x + 3y$$
$$\text{subject to} \quad 2x + 3y \le 6$$
$$x - y = 0$$
$$x \ge 0, y \ge 0$$

is a linear programming problem.

3.2 Solution to Self-Check Exercise

Let x denote the number of ads to be placed in the *City Tribune* and y the number to be placed in the *Daily News*. The total cost for placing x ads in the *City Tribune* and y ads in the *Daily News* is $300x + 100y$ dollars, and since the monthly budget is $9000, we must have

$$300x + 100y \le 9000$$

Next, the condition that the ad must appear in at least 15 but no more than 30 editions of the *Daily News* translates into the inequalities

$$y \ge 15$$
$$y \le 30$$

Finally, the objective function to be maximized is

$$P = 50{,}000x + 20{,}000y$$

To summarize, we have the following linear programming problem:

$$\text{Maximize} \quad P = 50{,}000x + 20{,}000y$$
$$\text{subject to} \quad 300x + 100y \le 9000$$
$$y \ge 15$$
$$y \le 30$$
$$x \ge 0, y \ge 0$$

3.3 Graphical Solution of Linear Programming Problems

▬ The Graphical Method

Linear programming problems in two variables have relatively simple geometric interpretations. For example, the system of linear constraints associated with a two-dimensional linear programming problem, unless it is inconsistent, defines a planar region whose boundary is composed of straight-line segments and/or half lines. Such problems are therefore amenable to graphical analysis.

Consider the following two-dimensional linear programming problem:

$$\text{Maximize} \quad P = 3x + 2y$$
$$\text{subject to} \quad 2x + 3y \le 12$$
$$2x + y \le 8 \tag{7}$$
$$x \ge 0, y \ge 0$$

The system of linear inequalities (7) defines the planar region S shown in Figure 9. Each point in S is a candidate for the solution of the problem at hand and is referred to as a **feasible solution.** The set S itself is referred to as a **feasible set.** Our goal is to find, from among all the points in the set S, the point(s) that optimizes the objective function P. Such a feasible solution is called an **optimal solution** and constitutes the solution to the linear programming problem under consideration.

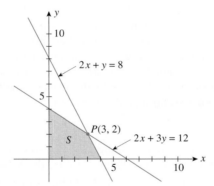

FIGURE 9
Each point in the feasible set S is a candidate for the optimal solution.

As noted earlier, each point $P(x, y)$ in S is a candidate for the optimal solution to the problem at hand. For example, the point $(1, 3)$ is easily seen to lie in S and is therefore in the running. The value of the objective function P at the point $(1, 3)$ is given by $P = 3(1) + 2(3) = 9$. Now, if we could compute the value of P corresponding to each point in S, then the point(s) in S that gave the largest value to P would constitute the solution set sought. Unfortunately, in most problems, the number of candidates is either too large or, as in this problem, the number is infinite. Thus, this method is at best unwieldy and at worst impractical.

Let's turn the question around. Instead of asking for the value of the objective function P at a feasible point, let's assign a value to the objective function P and ask whether there are feasible points that would correspond to the given value of P. To this end, suppose we assign a value of 6 to P. Then, the objective function P becomes $3x + 2y = 6$, a linear equation in x and y, and therefore has a graph that is a straight line L_1 in the plane. In Figure 10 we have drawn the graph of this straight line superimposed on the feasible set S.

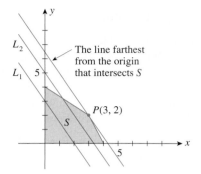

FIGURE 10
A family of parallel lines that intersect the feasible set S

It is clear that each point on the straight-line segment given by the intersection of the straight line L_1 and the feasible set S corresponds to the given value 6, of P. For this reason the line L_1 is called an **isoprofit line**. Let's repeat the process, this time assigning a value of 10 to P. We obtain the equation $3x + 2y = 10$ and the line L_2 (see Figure 10), which suggests that there are feasible points that correspond to a larger value of P. Observe that the line L_2 is parallel to the line L_1 since both lines have slope equal to $-\frac{3}{2}$, which is easily seen by casting the corresponding equations in the slope-intercept form.

In general, by assigning different values to the objective function, we obtain a family of parallel lines, each with slope equal to $-\frac{3}{2}$. Furthermore, a line corresponding to a larger value of P lies farther away from the origin than one with a smaller value of P. The implication is clear. To obtain the optimal solution(s) to the problem at hand, find the straight line, from this family of straight lines, that is farthest from the origin and still intersects the feasible set S. The required line is the one that passes through the point $P(3, 2)$ (see Figure 10), so the solution to the problem is given by $x = 3$, $y = 2$, resulting in a maximum value of $P = 3(3) + 2(2) = 13$.

That the optimal solution to this problem was found to occur at a vertex of the feasible set S is no accident. In fact, the result is a consequence of the following basic theorem on linear programming, which we state without proof.

THEOREM 1

Linear Programming

If a linear programming problem has a solution, then it must occur at a vertex, or corner point, of the feasible set S associated with the problem. Furthermore, if the objective function P is optimized at two adjacent vertices of S, then it is optimized at every point on the line segment joining these vertices, in which case there are infinitely many solutions to the problem.

Theorem 1 tells us that our search for the solution(s) to a linear programming problem may be restricted to the examination of the set of vertices of the feasible set S associated with the problem. Since a feasible set S has finitely many vertices, the theorem suggests that the solution(s) to the linear programming problem may be found by inspecting the values of the objective function P at these vertices.

Although Theorem 1 sheds some light on the nature of the solution of a linear programming problem, it does not tell us when a linear programming problem has a solution. The following theorem states some conditions that will guarantee when a linear programming problem has a solution.

THEOREM 2

Existence of a Solution

Suppose we are given a linear programming problem with a feasible set S and an objective function $P = ax + by$.

a. If S is bounded, then P has both a maximum and a minimum value on S.

b. If S is unbounded and both a and b are nonnegative, then P has a minimum value on S provided that the constraints defining S include the inequalities $x \geq 0$ and $y \geq 0$.

c. If S is the empty set, then the linear programming problem has no solution; that is, P has neither a maximum nor a minimum value.

The **method of corners,** a simple procedure for solving linear programming problems based on Theorem 1, follows.

The Method of Corners

1. Graph the feasible set.
2. Find the coordinates of all corner points (vertices) of the feasible set.
3. Evaluate the objective function at each corner point.
4. Find the vertex that renders the objective function a maximum (minimum). If there is only one such vertex, then this vertex constitutes a unique solution to the problem. If the objective function is maximized (minimized) at two adjacent corner points of S, there are infinitely many optimal solutions given by the points on the line segment determined by these two vertices.

APPLIED EXAMPLE 1   Maximizing Profits   We are now in a position to complete the solution to the production problem posed in Example 1, Section 3.2. Recall that the mathematical formulation led to the following linear programming problem:

$$
\begin{aligned}
\text{Maximize} \quad & P = x + 1.2y \\
\text{subject to} \quad & 2x + y \leq 180 \\
& x + 3y \leq 300 \\
& x \geq 0, \, y \geq 0
\end{aligned}
$$

Solution   The feasible set S for the problem is shown in Figure 11.

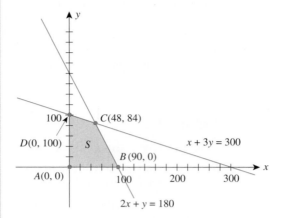

FIGURE 11
The corner point that yields the maximum profit is $C(48, 84)$.

The vertices of the feasible set are $A(0, 0)$, $B(90, 0)$, $C(48, 84)$, and $D(0, 100)$. The values of P at these vertices may be tabulated as follows:

Vertex	$P = x + 1.2y$
$A(0, 0)$	0
$B(90, 0)$	90
$C(48, 84)$	148.8
$D(0, 100)$	120

From the table, we see that the maximum of $P = x + 1.2y$ occurs at the vertex $(48, 84)$ and has a value of 148.8. Recalling what the symbols x, y, and P represent, we conclude that Ace Novelty would maximize its profit (a figure of $148.80) by producing 48 type-A souvenirs and 84 type-B souvenirs.   ■

EXPLORE & DISCUSS

Consider the linear programming problem

$$\text{Maximize}\quad P = 4x + 3y$$
$$\text{subject to}\quad 2x + \ \ y \le 10$$
$$2x + 3y \le 18$$
$$x \ge 0, y \ge 0$$

1. Sketch the feasible set S for the linear programming problem.
2. Draw the isoprofit lines superimposed on S corresponding to $P = 12, 16, 20,$ and 24 and show that these lines are parallel to each other.
3. Show that the solution to the linear programming problem is $x = 3$ and $y = 4$. Is this result the same as that found using the method of corners?

  APPLIED EXAMPLE 2   A Nutrition Problem   Complete the solution of the nutrition problem posed in Example 2, Section 3.2.

Solution   Recall that the mathematical formulation of the problem led to the following linear programming problem in two variables:

$$\text{Minimize} \quad C = 6x + 8y$$
$$\text{subject to} \quad 40x + 10y \geq 2400$$
$$10x + 15y \geq 2100$$
$$5x + 15y \geq 1500$$
$$x \geq 0, y \geq 0$$

The feasible set S defined by the system of constraints is shown in Figure 12.

FIGURE 12
The corner point that yields the minimum cost is $B(30, 120)$.

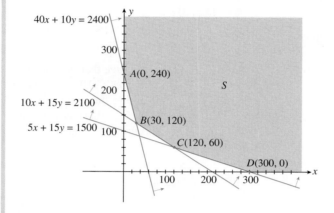

The vertices of the feasible set S are $A(0, 240)$, $B(30, 120)$, $C(120, 60)$, and $D(300, 0)$. The values of the objective function C at these vertices are given in the following table:

Vertex	$C = 6x + 8y$
$A(0, 240)$	1920
$B(30, 120)$	1140
$C(120, 60)$	1200
$D(300, 0)$	1800

From the table, we can see that the minimum for the objective function $C = 6x + 8y$ occurs at the vertex $B(30, 120)$ and has a value of 1140. Thus, the individual should purchase 30 brand A pills and 120 brand B pills at a minimum cost of $11.40. ■

EXAMPLE 3   A Linear Programming Problem with Multiple Solutions
Find the maximum and minimum of $P = 2x + 3y$ subject to the following system of linear inequalities:

$$2x + 3y \leq 30$$
$$y - x \leq 5$$
$$x + y \geq 5$$
$$x \leq 10$$
$$x \geq 0, y \geq 0$$

Solution   The feasible set S is shown in Figure 13. The vertices of the feasible set S are $A(5, 0)$, $B(10, 0)$, $C(10, \frac{10}{3})$, $D(3, 8)$, and $E(0, 5)$. The values of the objective function P at these vertices are given in the following table:

Vertex	$P = 2x + 3y$
$A(5, 0)$	10
$B(10, 0)$	20
$C(10, \frac{10}{3})$	30
$D(3, 8)$	30
$E(0, 5)$	15

From the table, we see that the maximum for the objective function $P = 2x + 3y$ occurs at the vertices $C(10, \frac{10}{3})$ and $D(3, 8)$. This tells us that every point on the line segment joining the points $C(10, \frac{10}{3})$ and $D(3, 8)$ maximizes P, giving it a value of 30 at each of these points. From the table, it is also clear that P is minimized at the point $(5, 0)$, where it attains a value of 10.

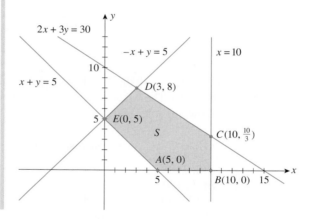

FIGURE 13
Every point lying on the line segment joining C and D maximizes P.

EXPLORE & DISCUSS

Consider the linear programming problem

$$\text{Maximize} \quad P = 2x + 3y$$
$$\text{subject to} \quad 2x + y \le 10$$
$$2x + 3y \le 18$$
$$x \ge 0, y \ge 0$$

1. Sketch the feasible set S for the linear programming problem.
2. Draw the isoprofit lines superimposed on S corresponding to $P = 6, 8, 12,$ and 18 and show that these lines are parallel to each other.
3. Show that there are infinitely many solutions to the problem. Is this result as predicted by the method of corners?

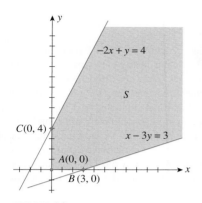

FIGURE 14
The maximization problem has no solution, because the feasible set is unbounded.

We close this section by examining two situations in which a linear programming problem may have no solution.

EXAMPLE 4 An Unbounded Linear Programming Problem with No Solution Solve the following linear programming problem:

$$\text{Maximize} \quad P = x + 2y$$
$$\text{subject to} \quad -2x + y \le 4$$
$$x - 3y \le 3$$
$$x \ge 0, y \ge 0$$

Solution The feasible set S for this problem is shown in Figure 14. Since the set S is unbounded (both x and y can take on arbitrarily large positive values), we see that we can make P as large as we please by making x and y large enough. This problem has no solution. The problem is said to be unbounded. ∎

EXAMPLE 5 An Infeasible Linear Programming Problem Solve the following linear programming problem:

$$\text{Maximize} \quad P = x + 2y$$
$$\text{subject to} \quad x + 2y \le 4$$
$$2x + 3y \ge 12$$
$$x \ge 0, y \ge 0$$

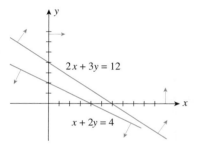

FIGURE 15
The problem is inconsistent because there is no point that satisfies all given inequalities.

Solution The half planes described by the constraints (inequalities) have no points in common (Figure 15). Therefore, there are no feasible points, and the problem has no solution. In this situation, we say that this problem is **infeasible,** or **inconsistent.** (These situations are unlikely to occur in well-posed problems arising from practical applications of linear programming.) ∎

The method of corners is particularly effective in solving two-variable linear programming problems with a small number of constraints, as the preceding examples have amply demonstrated. Its effectiveness, however, decreases rapidly as the number of variables and/or constraints increases. For example, it may be shown that a linear programming problem in three variables and five constraints may have up to 10 feasible corner points. The determination of the feasible corner points calls for the solution of ten 3×3 systems of linear equations and then the verification, by the substitution of each of these solutions into the system of constraints to see if it is in fact a feasible point. When the number of variables and constraints goes up to five and ten, respectively (still a very small system from the standpoint of applications in economics), the number of vertices to be found and checked for feasible corner points increases dramatically to 252, and each of these vertices is found by solving a 5×5 linear system! For this reason, the method of corners is seldom used to solve linear programming problems; its redeeming value lies in the fact that much insight is gained into the nature of the solutions of linear programming problems through its use in solving two-variable problems.

3.3　Self-Check Exercises

1. Use the method of corners to solve the following linear programming problem:

$$\text{Maximize} \quad P = 4x + 5y$$
$$\text{subject to} \quad x + 2y \le 10$$
$$5x + 3y \le 30$$
$$x \ge 0, y \ge 0$$

2. Use the method of corners to solve the following linear programming problem:

$$\text{Minimize} \quad C = 5x + 3y$$
$$\text{subject to} \quad 5x + 3y \ge 30$$
$$x - 3y \le 0$$
$$x \ge 2$$

3. Gino Balduzzi, proprietor of Luigi's Pizza Palace, allocates $9000 a month for advertising in two newspapers, the *City Tribune* and the *Daily News*. The *City Tribune* charges $300 for a certain advertisement, whereas the *Daily News* charges $100 for the same ad. Gino has stipulated that the ad is to appear in at least 15 but no more than 30 editions of the *Daily News* per month. The *City Tribune* has a daily circulation of 50,000, and the *Daily News* has a circulation of 20,000. Under these conditions, determine how many ads Gino should place in each newspaper in order to reach the largest number of readers.

Solutions to Self-Check Exercises 3.3 can be found on page 197.

3.3　Concept Questions

1. **a.** What is the *feasible set* associated with a linear programming problem?
   b. What is a *feasible solution* of a linear programming problem?
   c. What is an *optimal solution* of a linear programming problem?

2. Describe the method of corners.

3.3　Exercises

In Exercises 1–6, find the optimal (maximum and/or minimum) value(s) of the objective function on the feasible set S.

1. $Z = 2x + 3y$

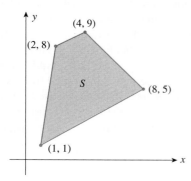

2. $Z = 3x - y$

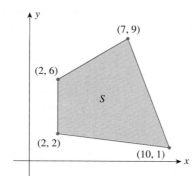

3. $Z = 3x + 4y$

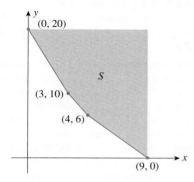

4. $Z = 7x + 9y$

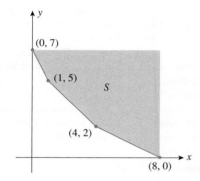

5. $Z = x + 4y$

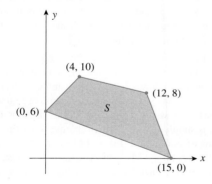

6. $Z = 3x + 2y$

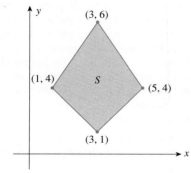

In Exercises 7–28, solve each linear programming problem by the method of corners.

7. Maximize $P = 2x + 3y$
subject to $\quad x + y \leq 6$
$\quad\quad\quad\quad x \leq 3$
$\quad\quad\quad\quad x \geq 0, y \geq 0$

8. Maximize $P = x + 2y$
subject to $\quad x + y \leq 4$
$\quad\quad\quad 2x + y \leq 5$
$\quad\quad\quad\quad x \geq 0, y \geq 0$

9. Maximize $P = 2x + y$ subject to the constraints of Exercise 8.

10. Maximize $P = 4x + 2y$
subject to $\quad x + y \leq 8$
$\quad\quad\quad 2x + y \leq 10$
$\quad\quad\quad\quad x \geq 0, y \geq 0$

11. Maximize $P = x + 8y$ subject to the constraints of Exercise 10.

12. Maximize $P = 3x - 4y$
subject to $\quad x + 3y \leq 15$
$\quad\quad\quad 4x + y \leq 16$
$\quad\quad\quad\quad x \geq 0, y \geq 0$

13. Maximize $P = x + 3y$
subject to $\quad 2x + y \leq 6$
$\quad\quad\quad\quad x + y \leq 4$
$\quad\quad\quad\quad\quad x \leq 1$
$\quad\quad\quad\quad x \geq 0, y \geq 0$

14. Maximize $P = 2x + 5y$
subject to $\quad 2x + y \leq 16$
$\quad\quad\quad 2x + 3y \leq 24$
$\quad\quad\quad\quad\quad y \leq 6$
$\quad\quad\quad\quad x \geq 0, y \geq 0$

15. Minimize $C = 3x + 4y$
subject to $\quad x + y \geq 3$
$\quad\quad\quad x + 2y \geq 4$
$\quad\quad\quad x \geq 0, y \geq 0$

16. Minimize $C = 2x + 4y$ subject to the constraints of Exercise 15.

17. Minimize $C = 3x + 6y$
subject to $\quad x + 2y \geq 40$
$\quad\quad\quad x + y \geq 30$
$\quad\quad\quad x \geq 0, y \geq 0$

18. Minimize $C = 3x + y$ subject to the constraints of Exercise 17.

19. Minimize $C = 2x + 10y$
subject to $\quad 5x + 2y \geq 40$
$\quad\quad\quad x + 2y \geq 20$
$\quad\quad\quad y \geq 3, x \geq 0$

20. Minimize $C = 2x + 5y$
subject to $4x + y \geq 40$
$2x + y \geq 30$
$x + 3y \geq 30$
$x \geq 0, y \geq 0$

21. Minimize $C = 10x + 15y$
subject to $x + y \leq 10$
$3x + y \geq 12$
$-2x + 3y \geq 3$
$x \geq 0, y \geq 0$

22. Maximize $P = 2x + 5y$ subject to the constraints of Exercise 21.

23. Maximize $P = 3x + 4y$
subject to $x + 2y \leq 50$
$5x + 4y \leq 145$
$2x + y \geq 25$
$y \geq 5, x \geq 0$

24. Maximize $P = 4x - 3y$ subject to the constraints of Exercise 23.

25. Maximize $P = 2x + 3y$
subject to $x + y \leq 48$
$x + 3y \geq 60$
$9x + 5y \leq 320$
$x \geq 10, y \geq 0$

26. Minimize $C = 5x + 3y$ subject to the constraints of Exercise 25.

27. Find the maximum and minimum of $P = 10x + 12y$ subject to
$$12.6, 31.5$$
$$5x + 2y \geq 63$$
$$x + y \geq 18$$
$$3x + 2y \leq 51 \quad 17, 25.5$$
$$x \geq 0, y \geq 0$$

28. Find the maximum and minimum of $P = 4x + 3y$ subject to
$$3x + 5y \geq 20$$
$$3x + y \leq 16$$
$$-2x + y \leq 1$$
$$x \geq 0, y \geq 0$$

The problems in Exercises 29–42 correspond to those in Exercises 1–14, Section 3.2. Use the results of your previous work to help you solve these problems.

29. MANUFACTURING—PRODUCTION SCHEDULING   A company manufactures two products, A and B, on two machines, I and II. It has been determined that the company will realize a profit of $3/unit of product A and a profit of $4/unit of product B. To manufacture a unit of product A requires 6 min on machine I and 5 min on machine II. To manufacture a unit of product B requires 9 min on machine I and 4 min on machine II.

There are 5 hr of machine time available on machine I and 3 hr of machine time available on machine II in each work shift. How many units of each product should be produced in each shift to maximize the company's profits? What is the optimal profit?

30. MANUFACTURING—PRODUCTION SCHEDULING   National Business Machines manufactures two models of fax machines: A and B. Each model A costs $100 to make, and each model B costs $150. The profits are $30 for each model A and $40 for each model B fax machine. If the total number of fax machines demanded per month does not exceed 2500 and the company has earmarked no more than $600,000/month for manufacturing costs, how many units of each model should National make each month in order to maximize its monthly profits? What is the optimal profit?

31. MANUFACTURING—PRODUCTION SCHEDULING   Kane Manufacturing has a division that produces two models of fireplace grates, model A and model B. To produce each model A grate requires 3 lb of cast iron and 6 min of labor. To produce each model B grate requires 4 lb of cast iron and 3 min of labor. The profit for each model A grate is $2.00, and the profit for each model B grate is $1.50. If 1000 lb of cast iron and 20 labor-hours are available for the production of fireplace grates per day, how many grates of each model should the division produce in order to help maximize Kane's profits? What is the optimal profit?

32. MANUFACTURING—PRODUCTION SCHEDULING   Refer to Exercise 31. Because of a backlog of orders for model A grates, Kane's manager had decided to produce at least 150 of these models a day. Operating under this additional constraint, how many grates of each model should Kane produce to maximize profit? What is the optimal profit?

33. FINANCE—ALLOCATION OF FUNDS   Madison Finance has a total of $20 million earmarked for homeowner and auto loans. On the average, homeowner loans have a 10% annual rate of return, whereas auto loans yield a 12% annual rate of return. Management has also stipulated that the total amount of homeowner loans should be greater than or equal to 4 times the total amount of automobile loans. Determine the total amount of loans of each type Madison should extend to each category in order to maximize its returns. What are the optimal returns?

34. INVESTMENTS—ASSET ALLOCATION   A financier plans to invest up to $500,000 in two projects. Project A yields a return of 10% on the investment, whereas project B yields a return of 15% on the investment. Because the investment in project B is riskier than the investment in project A, she has decided that the investment in project B should not exceed 40% of the total investment. How much should the financier invest in each project in order to maximize the return on her investment? What is the maximum return?

35. MANUFACTURING—PRODUCTION SCHEDULING   Acoustical manufactures a CD storage cabinet that can be bought fully assembled or as a kit. Each cabinet is processed in the fabrications department and the assembly department. If the fabrication department only manufactures fully assembled cabinets, then it can produce 200 units/day; but if it only manufactures kits, it can produce 200 units/day. If the assembly department produces only fully assembled cabinets, then it can produce 100 units/day; but if it produces only kits, then it can produce 300 units/day. Each fully assembled cabinet contributes $50 to the profits of the company, whereas each kit contributes $40 to its profits. How many fully assembled units and how many kits should the company produce per day in order to maximize its profits? What is the optimal profit?

36. AGRICULTURE—CROP PLANNING   A farmer plans to plant two crops, A and B. The cost of cultivating crop A is $40/acre, whereas that of crop B is $60/acre. The farmer has a maximum of $7400 available for land cultivation. Each acre of crop A requires 20 labor-hours, and each acre of crop B requires 25 labor-hours. The farmer has a maximum of 3300 labor-hours available. If she expects to make a profit of $150/acre on crop A and $200/acre on crop B, how many acres of each crop should she plant in order to maximize her profit? What is the optimal profit?

37. MINING—PRODUCTION   Perth Mining Company operates two mines for the purpose of extracting gold and silver. The Saddle Mine costs $14,000/day to operate, and it yields 50 oz of gold and 3000 oz of silver each day. The Horseshoe Mine costs $16,000/day to operate, and it yields 75 oz of gold and 1000 oz of silver each day. Company management has set a target of at least 650 oz of gold and 18,000 oz of silver. How many days should each mine be operated so that the target can be met at a minimum cost? What is the minimum cost?

38. TRANSPORTATION   Deluxe River Cruises operates a fleet of river vessels. The fleet has two types of vessels: A type-A vessel has 60 deluxe cabins and 160 standard cabins, whereas a type-B vessel has 80 deluxe cabins and 120 standard cabins. Under a charter agreement with Odyssey Travel Agency, Deluxe River Cruises is to provide Odyssey with a minimum of 360 deluxe and 680 standard cabins for their 15-day cruise in May. It costs $44,000 to operate a type-A vessel and $54,000 to operate a type-B vessel for that period. How many of each type vessel should be used in order to keep the operating costs to a minimum? What is the minimum cost?

39. NUTRITION—DIET PLANNING   A nutritionist at the Medical Center has been asked to prepare a special diet for certain patients. She has decided that the meals should contain a minimum of 400 mg of calcium, 10 mg of iron, and 40 mg of vitamin C. She has further decided that the meals are to be prepared from foods A and B. Each ounce of food A contains 30 mg of calcium, 1 mg of iron, 2 mg of vitamin C, and 2 mg of cholesterol. Each ounce of food B contains 25 mg of calcium, 0.5 mg of iron, 5 mg of vitamin C, and 5 mg of cholesterol. Find how many ounces of each type of food should be used in a meal so that the cholesterol content is minimized

and the minimum requirements of calcium, iron, and vitamin C are met.

40. SOCIAL PROGRAMS PLANNING   AntiFam, a hunger-relief organization, has earmarked between $2 and $2.5 million, inclusive, for aid to two African countries, country A and country B. Country A is to receive between $1 and $1.5 million, inclusive, in aid, and country B is to receive at least $0.75 million in aid. It has been estimated that each dollar spent in country A will yield an effective return of $.60, whereas a dollar spent in country B will yield an effective return of $.80. How should the aid be allocated if the money is to be utilized most effectively according to these criteria?
Hint: If x and y denote the amount of money to be given to country A and country B, respectively, then the objective function to be maximized is $P = 0.6x + 0.8y$.

41. ADVERTISING   Everest Deluxe World Travel has decided to advertise in the Sunday editions of two major newspapers in town. These advertisements are directed at three groups of potential customers. Each advertisement in newspaper I is seen by 70,000 group A customers, 40,000 group B customers, and 20,000 group C customers. Each advertisement in newspaper II is seen by 10,000 group A, 20,000 group B, and 40,000 group C customers. Each advertisement in newspaper I costs $1000, and each advertisement in newspaper II costs $800. Everest would like their advertisements to be read by at least 2 million people from group A, 1.4 million people from group B, and 1 million people from group C. How many advertisements should Everest place in each newspaper to achieve its advertising goals at a minimum cost? What is the minimum cost?
Hint: Use different scales for drawing the feasible set.

42. MANUFACTURING—SHIPPING COSTS   TMA manufactures 19-in. color television picture tubes in two separate locations, locations I and II. The output at location I is at most 6000 tubes/month, whereas the output at location II is at most 5000/month. TMA is the main supplier of picture tubes to the Pulsar Corporation, its holding company, which has priority in having all its requirements met. In a certain month, Pulsar placed orders for 3000 and 4000 picture tubes to be shipped to two of its factories located in city A and city B, respectively. The shipping costs (in dollars) per picture tube from the two TMA plants to the two Pulsar factories are as follows:

| From | To Pulsar Factories | |
	City A	City B
TMA, Loc. I	$3	$2
TMA, Loc. II	$4	$5

Find a shipping schedule that meets the requirements of both companies while keeping costs to a minimum.

43. Complete the solution to Example 3, Section 3.2.

44. MANUFACTURING—PRODUCTION SCHEDULING   Bata Aerobics manufactures two models of steppers used for aerobic exercises.

To manufacture each luxury model requires 10 lb of plastic and 10 min of labor. To manufacture each standard model requires 16 lb of plastic and 8 min of labor. The profit for each luxury model is $40, and the profit for each standard model is $30. If 6000 lb of plastic and 60 labor-hours are available for the production of the steppers per day, how many steppers of each model should Bata produce each day in order to maximize its profits? What is the optimal profit?

45. INVESTMENT PLANNING   Patricia has at most $30,000 to invest in securities in the form of corporate stocks. She has narrowed her choices to two groups of stocks: growth stocks that she assumes will yield a 15% return (dividends and capital appreciation) within a year and speculative stocks that she assumes will yield a 25% return (mainly in capital appreciation) within a year. Determine how much she should invest in each group of stocks in order to maximize the return on her investments within a year if she has decided to invest at least 3 times as much in growth stocks as in speculative stocks.

46. VETERINARY SCIENCE   A veterinarian has been asked to prepare a diet for a group of dogs to be used in a nutrition study at the School of Animal Science. It has been stipulated that each serving should be no larger than 8 oz and must contain at least 29 units of nutrient I and 20 units of nutrient II. The vet has decided that the diet may be prepared from two brands of dog food: brand A and brand B. Each ounce of brand A contains 3 units of nutrient I and 4 units of nutrient II. Each ounce of brand B contains 5 units of nutrient I and 2 units of nutrient II. Brand A costs 3 cents/ounce and brand B costs 4 cents/ounce. Determine how many ounces of each brand of dog food should be used per serving to meet the given requirements at a minimum cost.

47. MARKET RESEARCH   Trendex, a telephone survey company, has been hired to conduct a television-viewing poll among urban and suburban families in the Los Angeles area. The client has stipulated that a maximum of 1500 families is to be interviewed. At least 500 urban families must be interviewed, and at least half of the total number of families interviewed must be from the suburban area. For this service, Trendex will be paid $3000 plus $4 for each completed interview. From previous experience, Trendex has determined that it will incur an expense of $2.20 for each successful interview with an urban family and $2.50 for each successful interview with a suburban family. How many urban and suburban families should Trendex interview in order to maximize its profit?

In Exercises 48–51, determine whether the statement is true or false. If it is true, explain why it is true. If it is false, give an example to show why it is false.

48. An optimal solution of a linear programming problem is a feasible solution, but a feasible solution of a linear programming problem need not be an optimal solution.

49. A linear programming problem can have exactly three (optimal) solutions.

50. If a maximization problem has no solution, then the feasible set associated with the linear programming problem must be unbounded.

51. Suppose you are given the following linear programming problem: Maximize $P = ax + by$ on the unbounded feasible set S shown in the accompanying figure.

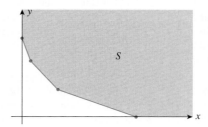

   a. If $a > 0$ or $b > 0$, then the linear programming problem has no optimal solution.
   b. If $a \leq 0$ and $b \leq 0$, then the linear programming problem has at least one optimal solution.

52. Suppose you are given the following linear programming problem: Maximize $P = ax + by$, where $a > 0$ and $b > 0$, on the feasible set S shown in the accompanying figure.

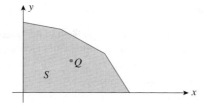

Explain, without using Theorem 1, why the optimal solution of the linear programming problem cannot occur at the point Q.

53. Suppose you are given the following linear programming problem: Maximize $P = ax + by$, where $a > 0$ and $b > 0$, on the feasible set S shown in the accompanying figure.

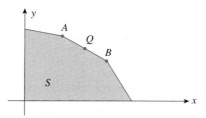

Explain, without using Theorem 1, why the optimal solution of the linear programming problem cannot occur at the point

Q unless the problem has infinitely many solutions lying along the line segment joining the vertices A and B.
Hint: Let $A(x_1, y_1)$ and $B(x_2, y_2)$. Then $Q(\bar{x}, \bar{y})$, where $\bar{x} = x_1 + (x_2 - x_1)t$ and $\bar{y} = y_2 + (y_2 - y_1)t$ with $0 < t < 1$. Study the value of P at and near Q.

54. Consider the linear programming problem

$$\text{Maximize} \quad P = 2x + 7y$$
$$\text{subject to} \quad 2x + y \geq 8$$
$$x + y \geq 6$$
$$x \geq 0, y \geq 0$$

a. Sketch the feasible set S.
b. Find the corner points of S.
c. Find the values of P at the corner points of S found in part (b).
d. Show that the linear programming problem has no (optimal) solution. Does this contradict Theorem 1?

55. Consider the linear programming problem

$$\text{Minimize} \quad C = -2x + 5y$$
$$\text{subject to} \quad x + y \leq 3$$
$$2x + y \leq 4$$
$$5x + 8y \geq 40$$
$$x \geq 0, y \geq 0$$

a. Sketch the feasible set.
b. Find the solution(s) of the linear programming problem, if it exists.

3.3 Solutions to Self-Check Exercises

1. The feasible set S for the problem was graphed in the solution to Exercise 1, Self-Check Exercises 3.1. It is reproduced in the accompanying figure.

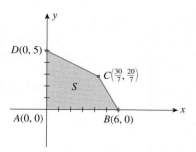

The values of the objective function P at the vertices of S are summarized in the accompanying table.

Vertex	$P = 4x + 5y$
$A(0, 0)$	0
$B(6, 0)$	24
$C\left(\frac{30}{7}, \frac{20}{7}\right)$	$\frac{220}{7} = 31\frac{3}{7}$
$D(0, 5)$	25

From the table, we see that the maximum for the objective function P is attained at the vertex $C\left(\frac{30}{7}, \frac{20}{7}\right)$. Therefore, the solution to the problem is $x = \frac{30}{7}$, $y = \frac{20}{7}$, and $P = 31\frac{3}{7}$.

2. The feasible set S for the problem was graphed in the solution to Exercise 2, Self-Check Exercises 3.1. It is reproduced in the accompanying figure.

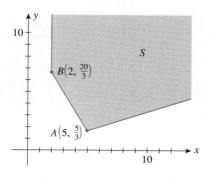

Evaluating the objective function $C = 5x + 3y$ at each corner point, we obtain the table

Vertex	$C = 5x + 3y$
$A\left(5, \frac{5}{3}\right)$	30
$B\left(2, \frac{20}{3}\right)$	30

We conclude that the objective function is minimized at every point on the line segment joining the points $\left(5, \frac{5}{3}\right)$ and $\left(2, \frac{20}{3}\right)$ and the minimum value of C is 30.

3. Refer to Self-Check Exercise 3.2. The problem is to maximize $P = 50,000x + 20,000y$ subject to

$$300x + 100y \leq 9000$$
$$y \geq 15$$
$$y \leq 30$$
$$x \geq 0, y \geq 0$$

The feasible set S for the problem is shown in the accompanying figure.

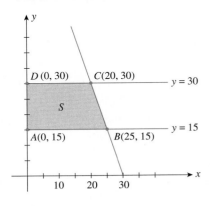

Evaluating the objective function $P = 50,000x + 20,000y$ at each vertex of S, we obtain

Vertex	$P = 50{,}000x + 20{,}000y$
$A(0, 15)$	300,000
$B(25, 15)$	1,550,000
$C(20, 30)$	1,600,000
$D(0, 30)$	600,000

From the table, we see that P is maximized when $x = 20$ and $y = 30$. Therefore, Gino should place 20 ads in the *City Tribune* and 30 with the *Daily News*.

3.4   Sensitivity Analysis (Optional)

In this section we investigate how changes in the parameters of a linear programming problem affect its optimal solution. This type of analysis is called **sensitivity analysis.** As in the previous sections, we restrict our analysis to the two-variable case, which is amenable to graphical analysis.

Recall the production problem posed in Example 1, Section 3.2, and solved in Example 1, Section 3.3:

$$\text{Maximize} \quad P = x + 1.2y \qquad \text{Objective function}$$
$$\text{subject to} \quad 2x + y \leq 180 \qquad \text{Constraint 1}$$
$$x + 3y \leq 300 \qquad \text{Constraint 2}$$
$$x \geq 0, y \geq 0$$

where x denotes the number of type-A souvenirs and y denotes the number of type-B souvenirs to be made. The optimal solution of this problem is $x = 48$, $y = 84$ (corresponding to the point C), and $P = 148.8$ (Figure 16).

The following questions arise in connection with this production problem:

1. How do changes made to the coefficients of the objective function affect the optimal solution?

2. How do changes made to the constants on the right-hand side of the constraints affect the optimal solution?

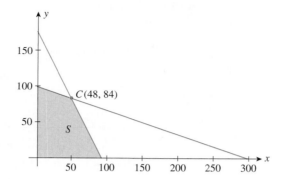

FIGURE 16
The optimal solution occurs at the point $C(48, 84)$.

Changes in the Coefficients of the Objective Function

In the production problem under consideration, the objective function is $P = x + 1.2y$. The coefficient of x, a 1, tells us that the contribution to the profit for each type-A souvenir is $1.00. The coefficient of y, 1.2, tells us that the contribution to the profit for each type-B souvenir is $1.20. Now suppose the contribution to the profit for each type-B souvenir remains fixed at $1.20 per souvenir. By how much can the contribution to the profit for each type-A souvenir vary without affecting the current optimal solution?

To answer this question, suppose the contribution to the profit of each type-A souvenir is c so that

$$P = cx + 1.2y \tag{8}$$

We need to determine the range of values of c so that the solution remains optimal.

We begin by rewriting Equation (8) for the isoprofit line in the slope-intercept form. Thus,

$$y = -\frac{c}{1.2}x + \frac{P}{1.2} \tag{9}$$

The slope of the isoprofit line is $-c/1.2$. If the slope of the isoprofit line exceeds that of the line associated with constraint 2, then the optimal solution shifts from point C to point D (see Figure 17 on page 200).

On the other hand, if the slope of the isoprofit line is *less than or equal to* the slope of the line associated with constraint 2, then the optimal solution remains unaffected. (You may verify that $-\frac{1}{3}$ is the slope of the line associated with constraint 2, by writing the equation $x + 3y = 300$ in the slope-intercept form.) In other words, we must have

$$-\frac{c}{1.2} \leq -\frac{1}{3}$$

$$\frac{c}{1.2} \geq \frac{1}{3} \qquad \text{Multiplying each side by } -1 \text{ reverses the inequality sign.}$$

$$c \geq \frac{1.2}{3} = 0.4$$

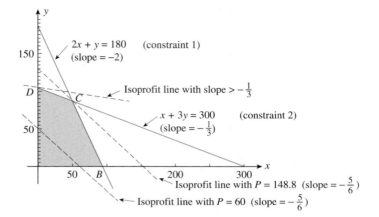

FIGURE 17
Increasing the slope of the isoprofit line $P = cx + 1.2y$ beyond $-\frac{1}{3}$ shifts the optimal solution from point C to point D.

A similar analysis shows that if the slope of the isoprofit line is less than that of the line associated with constraint 1, then the optimal solution shifts from point C to point B. Since the slope of the line associated with constraint 1 is -2, we see that point C will remain optimal, provided that the slope of the isoprofit line is *greater than or equal to* -2; that is, if

$$-\frac{c}{1.2} \ge -2$$

$$\frac{c}{1.2} \le 2$$

$$c \le 2.4$$

Thus, we have shown that if $0.4 \le c \le 2.4$, then the optimal solution obtained previously remains unaffected.

This result tells us that if the contribution to the profit of each type-A souvenir lies between \$.40 and \$2.40, then Ace Novelty should still make 48 type-A souvenirs and 84 type-B souvenirs. Of course, the profit of the company will change with a change in the value of c—it's the product mix that stays the same. For example, if the contribution to the profit of a type-A souvenir is \$1.50, then the profit of the company will be \$172.80. (See Exercise 1.) Incidentally, our analysis shows that the parameter c is not a sensitive parameter.

We leave it as an exercise for you to show that, with the contribution to the profit of type-A souvenirs held constant at \$1.00 per souvenir, the contribution to each type-B souvenir can vary between \$.50 and \$3.00 without affecting the product mix for the optimal solution (see Exercise 1).

APPLIED EXAMPLE 1 Profit Function Analysis Kane Manufacturing has a division that produces two models of grates, model A and model B. To produce each model A grate requires 3 pounds of cast iron and 6 minutes of labor. To produce each model B grate requires 4 pounds of cast iron and 3 minutes of labor. The profit for each model A grate is \$2.00, and the profit for each model B grate is \$1.50. Available for grate production each day are 1000 pounds of cast iron and 20 labor-hours. Because of an excess inventory of model A grates, management has decided to limit the production of model A grates to no more than 180 grates per day.

a. Use the method of corners to determine the number of grates of each model Kane should produce in order to maximize its profits.

b. Find the range of values that the contribution to the profit of a model A grate can assume without changing the optimal solution.

c. Find the range of values that the contribution to the profit of a model B grate can assume without changing the optimal solution.

Solution

a. Let x denote the number of model A grates and y the number of model B grates produced. Then verify that we are led to the following linear programming problem:

$$\text{Maximize} \quad P = 2x + 1.5y$$
$$\text{subject to} \quad 3x + 4y \le 1000 \quad \text{Constraint 1}$$
$$6x + 3y \le 1200 \quad \text{Constraint 2}$$
$$x \le 180 \quad \text{Constraint 3}$$
$$x \ge 0, y \ge 0$$

The graph of the feasible set S is shown in Figure 18.

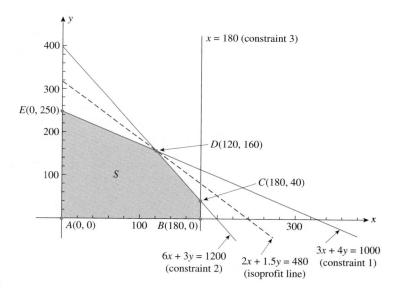

FIGURE 18
The shaded region is the feasible set S.
Also shown are the lines of the equations associated with the constraints.

From the following table of values,

Vertex	$P = 2x + 1.5y$
$A(0, 0)$	0
$B(180, 0)$	360
$C(180, 40)$	420
$D(120, 160)$	480
$E(0, 250)$	375

we see that the maximum of $P = 2x + 1.5y$ occurs at the vertex $D(120, 160)$ with a value of 480. Thus, Kane realizes a maximum profit of \$480 per day by producing 120 model A grates and 160 model B grates each day.

b. Let c (in dollars) denote the contribution to the profit of a model A grate. Then $P = cx + 1.5y$ or, upon solving for y,

$$y = -\frac{c}{1.5}x + \frac{P}{1.5}$$

$$= \left(-\frac{2}{3}c\right)x + \frac{2}{3}P$$

Referring to Figure 18 on page 201, you can see that if the slope of the isoprofit line is greater than the slope of the line associated with constraint 1, then the optimal solution will shift from point D to point E. Thus, for the optimal solution to remain unaffected, the slope of the isoprofit line must be less than or equal to the slope of the line associated with constraint 1. But the slope of the line associated with constraint 1 is $-\frac{3}{4}$, which you can see by rewriting the equation $3x + 4y = 1000$ in the slope-intercept form $y = -\frac{3}{4}x + 250$. Since the slope of the isoprofit line is $-2c/3$, we must have

$$-\frac{2c}{3} \le -\frac{3}{4}$$

$$\frac{2c}{3} \ge \frac{3}{4}$$

$$c \ge \left(\frac{3}{4}\right)\left(\frac{3}{2}\right) = \frac{9}{8} = 1.125$$

Again, referring to Figure 18, you can see that if the slope of the isoprofit line is less than that of the line associated with constraint 2, then the optimal solution shifts from point D to point C. Since the slope of the line associated with constraint 2 is -2 (rewrite the equation $6x + 3y = 1200$ in the slope-intercept form $y = -2x + 400$), we see that the optimal solution remains at point D, provided that the slope of the isoprofit line is greater than or equal to -2; that is,

$$-\frac{2c}{3} \ge -2$$

$$\frac{2c}{3} \le 2$$

$$c \le (2)\left(\frac{3}{2}\right) = 3$$

We conclude that the contribution to the profit of a model A grate can assume values between \$1.125 and \$3.00 without changing the optimal solution.

c. Let c (in dollars) denote the contribution to the profit of a model B grate. Then

$$P = 2x + cy$$

or, upon solving for y,

$$y = -\frac{2}{c}x + \frac{P}{c}$$

An analysis similar to that performed in part (b) with respect to constraint 1 shows that the optimal solution will remain in effect, provided that

$$-\frac{2}{c} \leq -\frac{3}{4}$$

$$\frac{2}{c} \geq \frac{3}{4}$$

$$c \leq 2\left(\frac{4}{3}\right) = \frac{8}{3} = 2\tfrac{2}{3}$$

Performing an analysis with respect to constraint 2 shows that the optimal solution will remain in effect, provided that

$$-\frac{2}{c} \geq -2$$

$$\frac{2}{c} \leq 2$$

$$c \geq 1$$

Thus, the contribution to the profit of a model B grate can assume values between \$1.00 and \$2.67 without changing the optimal solution. ∎

Changes to the Constants on the Right-Hand Side of the Constraint Inequalities

Let's return to the production problem posed at the beginning of this section:

$$\text{Maximize}\quad P = x + 1.2y$$
$$\text{subject to}\quad 2x + y \leq 180 \qquad \text{Constraint 1}$$
$$x + 3y \leq 300 \qquad \text{Constraint 2}$$
$$x \geq 0,\, y \geq 0$$

Now suppose the time available on machine I is changed from 180 minutes to $(180 + h)$ minutes, where h is a real number. Then the constraint on machine I is changed to

$$2x + y \leq 180 + h$$

But the line with equation $2x + y = 180 + h$ is parallel to the line $2x + y = 180$ associated with the original constraint 1.

As you can see from Figure 19 on page 204, the result of adding the constant h to the right-hand side of constraint 1 is to shift the current optimal solution from the point C to the new optimal solution occurring at the point C'. To find the coordinates of C', we observe that C' is the point of intersection of the lines with equations

$$2x + y = 180 + h \quad \text{and} \quad x + 3y = 300$$

Thus, the coordinates of the point are found by solving the system of linear equations

$$2x + y = 180 + h$$
$$x + 3y = 300$$

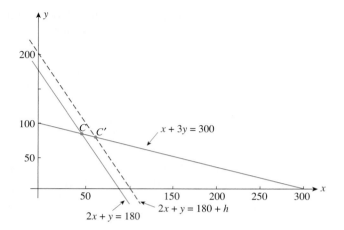

FIGURE 19
The lines with equations $2x + y = 180$ and $2x + y = 180 + h$ are parallel to each other.

The solutions are

$$x = \frac{3}{5}(80 + h) \qquad \text{and} \qquad y = \frac{1}{5}(420 - h) \tag{10}$$

The nonnegativity of x implies that

$$\frac{3}{5}(80 + h) \geq 0$$

$$80 + h \geq 0$$

$$h \geq -80$$

Next, the nonnegativity of y implies that

$$\frac{1}{5}(420 - h) \geq 0$$

$$420 - h \geq 0$$

$$h \leq 420$$

Thus, h must satisfy the inequalities $-80 \leq h \leq 420$. Our computations reveal that for a meaningful solution the time available for machine I must range between $(180 - 80)$ and $(180 + 420)$ minutes—that is, between 100 and 600 minutes. Under these conditions, Ace Novelty should produce $\frac{3}{5}(80 + h)$ type-A souvenirs and $\frac{1}{5}(420 - h)$ type-B souvenirs.

For example, if Ace Novelty can manage to increase the time available on machine I by 10 minutes, then it should produce $\frac{3}{5}(80 + 10)$, or 54, type-A souvenirs and $\frac{1}{5}(420 - 10)$, or 82, type-B souvenirs, with a resulting profit of

$$P = x + 1.2y = 54 + (1.2)(82) = 152.4$$

or $152.40.

We leave it as an exercise for you to show that if the time available on machine II is changed from 300 minutes to $(300 + k)$ minutes with no change in the maximum capacity for machine I, then k must satisfy the inequalities $-210 \leq k \leq 240$. Thus, for a meaningful solution to the problem, the time available on machine II must lie between 90 and 540 min (see Exercise 2). Furthermore, in this case, Ace Novelty should produce $\frac{1}{5}(240 - k)$ type-A souvenirs and $\frac{1}{5}(420 + 2k)$ type-B souvenirs (see Exercise 2).

Shadow Prices

We have just seen that if Ace Novelty could increase the maximum available time on machine I by 10 minutes, then the profit would increase from the original optimal value of $148.80 to $152.40. In this case, finding the extra time on machine I proved beneficial to the company. More generally, to study the economic benefits derived from increasing its resources, a company looks at the shadow prices associated with the respective resources. More specifically, we define the *shadow price* for the *i*th resource (associated with the *i*th constraint of the linear programming problem) to be the amount by which the value of the objective function is improved—increased in a maximization problem and decreased in a minimization problem—if the right-hand side of the *i*th constraint is increased by 1 unit.

In the Ace Novelty example discussed earlier, we showed that if the right-hand side of constraint 1 is increased by *h* units, then the optimal solution is given by (10):

$$x = \frac{3}{5}(80 + h) \quad \text{and} \quad y = \frac{1}{5}(420 - h)$$

The resulting profit is calculated as follows:

$$
\begin{aligned}
P &= x + 1.2y \\
&= x + \frac{6}{5}y \\
&= \frac{3}{5}(80 + h) + \left(\frac{6}{5}\right)\left(\frac{1}{5}\right)(420 - h) \\
&= \frac{3}{25}(1240 + 3h)
\end{aligned}
$$

Upon setting $h = 1$, we find

$$
\begin{aligned}
P &= \frac{3}{25}(1240 + 3) \\
&= 149.16
\end{aligned}
$$

Since the optimal profit for the original problem is $148.80, we see that the shadow price for the first resource is $149.16 - 148.80$, or $.36. To summarize, Ace Novelty's profit increases at the rate of $.36 per 1-minute increase in the time available on machine I.

We leave it as an exercise for you to show that the shadow price for resource 2 (associated with constraint 2) is $.28 (see Exercise 2).

APPLIED EXAMPLE 2 Shadow Prices Consider the problem posed in Example 1:

$$
\begin{array}{lll}
\text{Maximize} & P = 2x + 1.5y & \\
\text{subject to} & 3x + 4y \le 1000 & \text{Constraint 1} \\
& 6x + 3y \le 1200 & \text{Constraint 2} \\
& x \le 180 & \text{Constraint 3} \\
& x \ge 0, y \ge 0 &
\end{array}
$$

a. Find the range of values that resource 1 (the constant on the right-hand side of constraint 1) can assume.

b. Find the shadow price for resource 1.

Solution

a. Suppose the right-hand side of constraint 1 is replaced by $1000 + h$, where h is a real number. Then the new optimal solution occurs at the point D' (Figure 20).

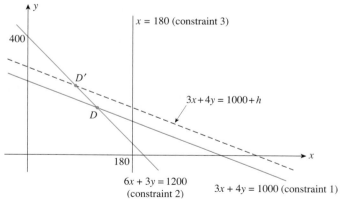

FIGURE 20
As the amount of resource 1 changes, the point at which the optimal solution occurs shifts from D to D'.

To find the coordinates of D', we solve the system

$$3x + 4y = 1000 + h$$
$$6x + 3y = 1200$$

206

Multiplying the first equation by -2 and adding the resulting equation to the second equation gives

$$-5y = -800 - 2h$$

$$y = \frac{2}{5}(400 + h)$$

Substituting this value of y into the second equation in the system gives

$$6x + \frac{6}{5}(400 + h) = 1200$$

$$x + \frac{1}{5}(400 + h) = 200$$

$$x = \frac{1}{5}(600 - h)$$

The nonnegativity of y implies that $h \geq -400$, and the nonnegativity of x implies that $h \leq 600$. But constraint 3 dictates that x must also satisfy

$$x = \frac{1}{5}(600 - h) \leq 180$$

$$600 - h \leq 900$$

$$-h \leq 300$$

$$h \geq -300$$

Therefore, h must satisfy $-300 \leq h \leq 600$. This tells us the resource 1 must lie between $1000 - 300$, or 700, and $1000 + 600$, or 1600—that is, between 700 and 1600 pounds.

b. If we set $h = 1$ in part (a), we obtain

$$x = \frac{1}{5}(600 - 1) = \frac{599}{5}$$

$$y = \frac{2}{5}(400 + 1) = \frac{802}{5}$$

Therefore, the profit realized at this level of production is

$$P = 2x + \frac{3}{2}y = 2\left(\frac{599}{5}\right) + \frac{3}{2}\left(\frac{802}{5}\right)$$

$$= \frac{2401}{5} = 480.2$$

Since the original optimal profit is \$480 (see Example 1), we see that the shadow price for resource 1 is \$.20. ■

If you examine Figure 20, you can see that increasing resource 3 (the constant on the right-hand side of constraint 3) has no effect on the optimal solution $D(120, 160)$ of the problem at hand. In other words, an increase in the resource associated with constraint 3 has no economic benefit for Kane Manufacturing. The shadow price for this resource is *zero*. There is a *surplus* of this resource. We say that the constraint $x \leq 180$ is not binding on the optimal solution $D(120, 160)$.

On the other hand, constraints 1 and 2, which *hold with equality* at the optimal solution $D(120, 160)$, are said to be **binding constraints.** The objective function cannot be increased without increasing these resources. They have *positive* shadow prices.

Importance of Sensitivity Analysis

We conclude this section by pointing out the importance of sensitivity analysis in solving real-world problems. The values of the parameters in these problems may change. For example, the management of Ace Novelty might wish to increase the price of a type-A souvenir because of increased demand for the product, or they might want to see how a change in the time available on machine I affects the (optimal) profit of the company.

If a parameter of a linear programming problem is changed, it is true that one needs only to re-solve the problem to obtain a new solution to the problem. But since a real-world linear programming problem often involves thousands of parameters, the amount of work involved in finding a new solution is prohibitive. Another disadvantage in using this approach is that it often takes many trials with different values of a parameter in order to see their effect on the optimal solution of the problem. Thus, a more analytical approach such as that discussed earlier is desirable.

Returning to the discussion of Ace Novelty, our analysis of the changes in the coefficients of the objective (profit) function suggests that if management decides to raise the price of a type-A souvenir, it can do so with the assurance that the optimal solution holds as long as the new price leaves the contribution to the profit of a type-A souvenir between $.40 and $2.40. There is no need to re-solve the linear programming problem for each new price being considered. Also, our analysis of the changes of the parameters on the right-hand side of the constraints suggests, for example, that for a meaningful solution to the problem the time available for machine I must lie in the range between 100 and 600 minutes. Furthermore, the analysis tells us how to compute the increase (decrease) in the optimal profit when the resource is adjusted, through the use of the shadow price associated with that constraint. Again, there is no need to re-solve the linear programming problem each time a change in the resource available is anticipated.

Using Technology examples and exercises that are solved using Excel's Solver can be found on pages 239–242 and 257–260.

3.4   Self-Check Exercises

Consider the linear programming problem:

Maximize $P = 2x + 4y$
subject to $2x + 5y \leq 19$ Constraint 1
$3x + 2y \leq 12$ Constraint 2
$x \geq 0, y \geq 0$

1. Use the method of corners to solve the problem.

2. Find the range of values that the coefficient of x can assume without changing the optimal solution.

3. Find the range of values that resource 1 (the constant on the right-hand side of constraint 1) can assume without changing the optimal solution.

4. Find the shadow price for resource 1.

5. Identify the binding and nonbinding constraints.

Solutions to Self-Check Exercises can be found on page 208.

3.4 Concept Questions

1. Suppose $P = 3x + 4y$ is the objective function in a linear programming (maximization) problem, where x denotes the number of units of product A and y denotes the number of units of product B to be made. What does the coefficient of x represent? The coefficient of y?

2. Given the linear programming problem

$$\text{Maximize} \quad P = 3x + 4y$$
$$\text{subject to} \quad x + y \le 4 \quad \text{Resource 1}$$
$$2x + y \le 5 \quad \text{Resource 2}$$

a. Write the inequality that represents an increase of h units in resource 1.
b. Write the inequality that represents an increase of k units in resource 2.

3. Explain the meaning of (a) a *shadow price* and (b) a *binding constraint*.

3.4 Exercises

1. Refer to the production problem discussed on pages 198–200.
   a. Show that the optimal solution holds if the contribution to the profit of a type-B souvenir lies between $.50 and $3.00.
   b. Show that if the contribution to the profit of a type-A souvenir is $1.50 (with the contribution to the profit of a type-B souvenir held at $1.20), then the optimal profit of the company will be $172.80.
   c. What will be the optimal profit of the company if the contribution to the profit of a type-B souvenir is $2.00 (with the contribution to the profit of a type-A souvenir held at $1.00)?

2. Refer to the production problem discussed on pages 203–205.
   a. Show that for a meaningful solution, the time available on machine II must lie between 90 and 540 min.
   b. Show that if the time available on machine II is changed from 300 min to $(300 + k)$ min, with no change in the maximum capacity for machine I, then Ace Novelty's profit is maximized by producing $\frac{1}{5}(240 - k)$ type-A souvenirs and $\frac{1}{5}(420 + 2k)$ type-B souvenirs, where $-210 \le k \le 240$.
   c. Show that the shadow price for resource 2 (associated with constraint 2) is $.28.

3. Refer to Example 2.
   a. Find the range of values that resource 2 can assume.
   b. By how much can the right-hand side of constraint 3 be changed so that the current optimal solution still holds?

4. Refer to Example 2.
   a. Find the shadow price for resource 2.
   b. Identify the binding and nonbinding constraints.

In Exercises 5–10, you are given a linear programming problem.
a. **Use the method of corners to solve the problem.**
b. **Find the range of values that the coefficient of x can assume without changing the optimal solution.**
c. **Find the range of values that resource 1 (requirement 1) can assume.**
d. **Find the shadow price for resource 1 (requirement 1).**
e. **Identify the binding and nonbinding constraints.**

5. Maximize $P = 3x + 4y$
   subject to $2x + 3y \le 12$ Resource 1
   $2x + y \le 8$ Resource 2
   $x \ge 0, y \ge 0$

6. Maximize $P = 2x + 5y$
   subject to $x + 3y \le 15$ Resource 1
   $4x + y \le 16$ Resource 2
   $x \ge 0, y \ge 0$

7. Minimize $C = 2x + 5y$
   subject to $x + 2y \ge 4$ Requirement 1
   $x + y \ge 3$ Requirement 2
   $x \ge 0, y \ge 0$

8. Minimize $C = 3x + 4y$
   subject to $x + 3y \ge 8$ Requirement 1
   $x + y \ge 4$ Requirement 2
   $x \ge 0, y \ge 0$

9. Maximize $P = 4x + 3y$
subject to $5x + 3y \le 30$ Resource 1
$2x + 3y \le 21$ Resource 2
$x \le 4$ Resource 3
$x \ge 0, y \ge 0$

10. Maximize $P = 4x + 5y$
subject to $x + y \le 30$ Resource 1
$x + 2y \le 40$ Resource 2
$x \le 25$ Resource 3
$x \ge 0, y \ge 0$

11. MANUFACTURING—PRODUCTION SCHEDULING   A company manufactures two products, A and B, on machines I and II. The company will realize a profit of $3/unit of product A and a profit of $4/unit of product B. Manufacturing 1 unit of product A requires 6 min on machine I and 5 min on machine II. Manufacturing 1 unit of product B requires 9 min on machine I and 4 min on machine II. There are 5 hr of time available on machine I and 3 hr of time available on machine II in each work shift.
a. How many units of each product should be produced in each shift to maximize the company's profit?
b. Find the range of values that the contribution to the profit of 1 unit of product A can assume without changing the optimal solution.
c. Find the range of values that the resource associated with the time constraint on machine I can assume.
d. Find the shadow price for the resource associated with the time constraint on machine I.

12. AGRICULTURE—CROP PLANNING   A farmer plans to plant two crops, A and B. The cost of cultivating crop A is $40/acre, whereas that of crop B is $60/acre. The farmer has a maximum of $7400 available for land cultivation. Each acre of crop A requires 20 labor-hours, and each acre of crop B requires 25 labor-hours. The farmer has a maximum of 3300 labor-hours available. If she expects to make a profit of $150/acre on crop A and $200/acre on crop B, how many acres of each crop should she plant in order to maximize her profit?
a. Find the range of values that the contribution to the profit of an acre of crop A can assume without changing the optimal solution.
b. Find the range of values that the resource associated with the constraint on the land available can assume.
c. Find the shadow price for the resource associated with the constraint on the land available.

13. MINING—PRODUCTION   Perth Mining Company operates two mines for the purpose of extracting gold and silver. The Saddle Mine costs $14,000/day to operate, and it yields 50 oz of gold and 3000 oz of silver per day. The Horseshoe Mine costs $16,000/day to operate, and it yields 75 oz of gold and 1000 ounces of silver each day. Company management has set a target of at least 650 oz of gold and 18,000 oz of silver.

a. How many days should each mine be operated so that the target can be met at a minimum cost?
b. Find the range of values that the cost of operating the Saddle Mine per day can assume without changing the optimal solution.
c. Find the range of values that the requirement for gold can assume.
d. Find the shadow price for the requirement for gold.

14. TRANSPORTATION   Deluxe River Cruises operates a fleet of river vessels. The fleet has two types of vessels: a type-A vessel has 60 deluxe cabins and 160 standard cabins, whereas a type-B vessel has 80 deluxe cabins and 120 standard cabins. Under a charter agreement with the Odyssey Travel Agency, Deluxe River Cruises is to provide Odyssey with a minimum of 360 deluxe and 680 standard cabins for their 15-day cruise in May. It costs $44,000 to operate a type-A vessel and $54,000 to operate a type-B vessel for that period.
a. How many of each type of vessel should be used in order to keep the operating costs to a minimum?
b. Find the range of values that the cost of operating a type-A vessel can assume without changing the optimal solution.
c. Find the range of values that the requirement for deluxe cabins can assume.
d. Find the shadow price for the requirement for deluxe cabins.

15. MANUFACTURING—PRODUCTION SCHEDULING   Soundex produces two models of clock radios. Model A requires 15 min of work on assembly line I and 10 min of work on assembly line II. Model B requires 10 min of work on assembly line I and 12 min of work on assembly line II. At most 25 hr of assembly time on line I and 22 hr of assembly time on line II are available each day. Soundex anticipates a profit of $12 on model A and $10 on model B. Because of previous overproduction, management decides to limit the production of model A clock radios to no more than 80/day.
a. To maximize Soundex's profit, how many clock radios of each model should be produced each day?
b. Find the range of values that the contribution to the profit of a model A clock radio can assume without changing the optimal solution.
c. Find the range of values that the resource associated with the time constraint on machine I can assume.
d. Find the shadow price for the resource associated with the time constraint on machine I.
e. Identify the binding and nonbinding constraints.

16. MANUFACTURING   Refer to Exercise 15.
a. If the contribution to the profit of a model A clock radio is changed to $8.50/radio, will the original optimal solution still hold? What will be the optimal profit?
b. If the contribution to the profit of a model A clock radio is changed to $14.00/radio, will the original optimal solution still hold? What will be the optimal profit?

17. **MANUFACTURING—PRODUCTION SCHEDULING** Kane Manufacturing has a division that produces two models of fireplace grates, model A and model B. To produce each model A grate requires 3 lb of cast iron and 6 min of labor. To produce each model B grate requires 4 lb of cast iron and 3 min of labor. The profit for each model A grate is $2, and the profit for each model B grate is $1.50. 1000 lb of cast iron and 20 labor-hours are available for the production of grates each day. Because of an excess inventory of model B grates, management has decided to limit the production of model B grates to no more than 200 grates per day. How many grates of each model should the division produce daily to maximize Kane's profits?
    a. Use the method of corners to solve the problem.
    b. Find the range of values that the coefficient of x can assume without changing the optimal solution.

c. Find the range of values that the resource for cast iron can assume without changing the optimal solution.
d. Find the shadow price for the resource for cast iron.
e. Identify the binding and nonbinding constraints.

18. **MANUFACTURING** Refer to Exercise 17.
    a. If the contribution to the profit of a model A grate is changed to $1.75/grate, will the original optimal solution still hold? What will be the new optimal solution?
    b. If the contribution to the profit of a model A grate is changed to $2.50/grate, will the original optimal solution still hold? What will be the new optimal solution?

3.4 Exercises

1. The feasible set for the problem is shown in the accompanying figure.

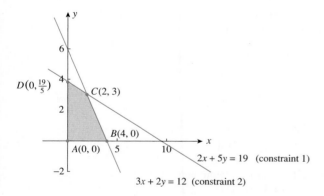

Evaluating the objective function $P = 2x + 4y$ at each feasible corner point, we obtain the following table:

Vertex	$P = 2x + 4y$
$A(0, 0)$	0
$B(4, 0)$	8
$C(2, 3)$	16
$D(0, \frac{19}{5})$	$15\frac{1}{5}$

We conclude that the maximum value of P is 16 attained at the point $(2, 3)$.

2. Assume that $P = cx + 4y$. Then

$$y = -\frac{c}{4}x + \frac{P}{4}$$

The slope of the isoprofit line is $-\frac{c}{4}$ and must be less than or equal to the slope of the line associated with constraint 1; that is,

$$-\frac{c}{4} \leq -\frac{2}{5}$$

Solving, we find $c \geq \frac{8}{5}$. A similar analysis shows that the slope of the isoprofit line must be greater than or equal to the slope of the line associated with constraint 2; that is,

$$-\frac{c}{4} \geq -\frac{3}{2}$$

Solving, we find $c \leq 6$. Thus, we have shown that if $1.6 \leq c \leq 6$, then the optimal solution obtained previously remains unaffected.

3. Suppose the right-hand side of constraint 1 is replaced by $19 + h$, where h is a real number. Then the new optimal solution occurs at the point whose coordinates are found by solving the system

$$2x + 5y = 19 + h$$
$$3x + 2y = 12$$

Multiplying the second equation by -5 and adding the resulting equation to 2 times the first equation gives

$$-11x = -60 + 2(19 + h) = -22 + 2h$$

$$x = 2 - \frac{2}{11}h$$

Substituting this value of x into the second equation in the system gives

$$3\left(2 - \frac{2}{11}h\right) + 2y = 12$$

$$2y = 12 - 6 + \frac{6}{11}h$$

$$y = 3 + \frac{3}{11}h$$

The nonnegativity of x implies $2 - \frac{2}{11}h \geq 0$, or $h \leq 11$. The nonnegativity of y implies $3 + \frac{3}{11}h \geq 0$, or $h \geq -11$. Therefore, h must satisfy $-11 \leq h \leq 11$. This tells us that resource 1 must lie between $19 - 11$ and $19 + 11$—that is, between 8 and 30.

4. If we set $h = 1$ in Exercise 3, we find that $x = \frac{20}{11}$ and $y = \frac{36}{11}$. Therefore, for these values of x and y,

$$P = 2\left(\frac{20}{11}\right) + 4\left(\frac{36}{11}\right) = \frac{184}{11} = 16\frac{8}{11}$$

Since the original optimal value of P is 16, we see that the shadow price for resource 1 is $\frac{8}{11}$.

5. Since both constraints hold with equality of the optimal solution $C(2, 3)$, they are binding constraints.

CHAPTER 3 **Summary of Principal Terms**

TERMS

solution set of a system of linear inequalities (171)	feasible solution (185)	sensitivity analysis (198)
bounded solution set (172)	feasible set (185)	shadow price (205)
unbounded solution set (172)	optimal solution (185)	binding constraint (207)
objective function (176)	isoprofit line (186)	
linear programming problem (176)	method of corners (187)	

CHAPTER 3 **Concept Review Questions**

Fill in the blanks.

1. **a.** The solution set of the inequality $ax + by < c$ is a/an _____ _____ that does not include the _____ with equation $ax + by = c$.
   b. If $ax + by < c$ describes the lower half plane, then the inequality _____ describes the lower half plane together with the line having equation _____.

2. **a.** The solution set of a system of linear inequalities in the two variables x and y is the set of all _____ satisfying _____ inequality of the system.
   b. The solution set of a system of linear inequalities is _____ if it can be _____ by a circle.

3. A linear programming problem consists of a linear function, called a/an _____ _____ to be _____ or _____ subject to constraints in the form of _____ equations or _____.

4. **a.** If a linear programming problem has a solution, then it must occur at a/an _____ _____ of the feasible set.
   b. If the objective function of a linear programming problem is optimized at two adjacent vertices of the feasible set, then it is optimized at every point on the _____ segment joining these vertices.

5. In sensitivity analysis, we investigate how changes in the _____ of a linear programming problem affect the _____ solution.

6. The shadow price for the ith _____ is the _____ by which the _____ of the objective function is _____ if the right-hand side of the ith constraint is _____ by 1 unit.

| CHAPTER 3 | **Review Exercises** |

In Exercises 1 and 2, find the optimal value(s) of the objective function on the feasible set S.

1. $Z = 2x + 3y$

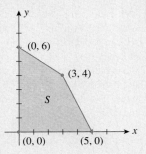

2. $Z = 4x + 3y$

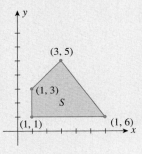

In Exercises 3–12, use the method of corners to solve the linear programming problem.

3. Maximize $P = 3x + 5y$
subject to $2x + 3y \leq 12$
$x + y \leq 5$
$x \geq 0, y \geq 0$

4. Maximize $P = 2x + 3y$
subject to $2x + y \leq 12$
$x - 2y \leq 1$
$x \geq 0, y \geq 0$

5. Minimize $C = 2x + 5y$
subject to $x + 3y \geq 15$
$4x + y \geq 16$
$x \geq 0, y \geq 0$

6. Minimize $C = 3x + 4y$
subject to $2x + y \geq 4$
$2x + 5y \geq 10$
$x \geq 0, y \geq 0$

7. Maximize $P = 3x + 2y$
subject to $2x + y \leq 16$
$2x + 3y \leq 36$
$4x + 5y \geq 28$
$x \geq 0, y \geq 0$

8. Maximize $P = 6x + 2y$
subject to $x + 2y \leq 12$
$x + y \leq 8$
$2x - 3y \geq 6$
$x \geq 0, y \geq 0$

9. Minimize $C = 2x + 7y$
subject to $3x + 5y \geq 45$
$3x + 10y \geq 60$
$x \geq 0, y \geq 0$

10. Minimize $C = 4x + y$
subject to $6x + y \geq 18$
$2x + y \geq 10$
$x + 4y \geq 12$
$x \geq 0, y \geq 0$

11. Find the maximum and minimum of $Q = x + y$ subject to

$$5x + 2y \geq 20$$
$$x + 2y \geq 8$$
$$x + 4y \leq 22$$
$$x \geq 0, y \geq 0$$

12. Find the maximum and minimum of $Q = 2x + 5y$ subject to

$$x + y \geq 4$$
$$-x + y \leq 6$$
$$x + 3y \leq 30$$
$$x \leq 12$$
$$x \geq 0, y \geq 0$$

13. FINANCIAL ANALYSIS An investor has decided to commit no more than $80,000 to the purchase of the common stocks of two companies, company A and company B. He has also estimated that there is a chance of at most a 1% capital loss on his investment in company A and a chance of at most a 4% loss on his investment in company B, and he has decided that these losses should not exceed $2000. On the other hand, he expects to make a 14% profit from his investment in company A and a 20% profit from his investment in company B. Determine how much he should invest in the stock of each company in order to maximize his investment returns.

14. MANUFACTURING—PRODUCTION SCHEDULING Soundex produces two models of clock radios. Model A requires 15 min of work on assembly line I and 10 min of work on assembly line II. Model B requires 10 min of work on assembly line I and 12 min of work on assembly line II. At most, 25 labor-hours of assembly time on line I and 22 labor-hours of assembly time on line II are available each day. It is anticipated that Soundex will realize a profit of $12 on model A and $10 on model B. How many clock radios of each model should be produced each day in order to maximize Soundex's profit?

15. **MANUFACTURING—PRODUCTION SCHEDULING** Kane Manufacturing has a division that produces two models of grates, model A and model B. To produce each model A grate requires 3 lb of cast iron and 6 min of labor. To produce each model B grate requires 4 lb of cast iron and 3 min of labor. The profit for each model A grate is $2.00, and the profit for

each model B grate is $1.50. Available for grate production each day are 1000 lb of cast iron and 20 labor-hours. Because of a backlog of orders for model B grates, Kane's manager has decided to produce at least 180 model B grates/day. How many grates of each model should Kane produce to maximize its profits?

16. **MINIMIZING SHIPPING COSTS** A manufacturer of projection TVs must ship a total of at least 1000 TVs to its two central warehouses. Each warehouse can hold a maximum of 750 TVs. The first warehouse already has 150 TVs on hand, whereas the second has 50 TVs on hand. It costs $40 to ship a TV to the first warehouse, and it costs $80 to ship a TV to the second warehouse. How many TVs should be shipped to each warehouse to minimize the cost?

CHAPTER 3   Before Moving On . . .

1. Determine graphically the solution set for the following systems of inequalities.

   a. $2x + y \le 10$
   $x + 3y \le 15$
   $x \le 4$
   $x \ge 0, y \ge 0$

   b. $2x + y \ge 8$
   $2x + 3y \ge 15$
   $x \ge 0$
   $y \ge 2$

2. Find the maximum and minimum values of $Z = 3x - y$ on the following feasible set.

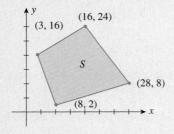

3. Maximize $P = x + 3y$
   subject to $2x + 3y \le 11$
   $3x + 7y \le 24$
   $x \ge 0, y \ge 0$

4. Minimize $C = 4x + y$
   subject to $2x + y \ge 10$
   $2x + 3y \ge 24$
   $x + 3y \ge 15$
   $x \ge 0, y \ge 0$

5. **Sensitivity Analysis (Optional)** Consider the following linear programming problem:

$$\text{Maximize} \quad P = 2x + 3y$$
$$\text{subject to} \quad x + 2y \le 16$$
$$3x + 2y \le 24$$
$$x \ge 0, y \ge 0$$

   a. Solve the problem.
   b. Find the range of values that the coefficient of x can assume without changing the optimal solution.
   c. Find the range of values that resource 1 (requirement 1) can assume.
   d. Find the shadow price for resource 1.
   e. Identify the binding and nonbinding constraints.

Linear Programming:
An Algebraic Approach

How much profit? The Ace Novelty Company produces three types of souvenirs. Each type requires a certain amount of time on each of three different machines. Each machine may be operated for a certain amount of time per day. In Example 5, page 229, we will determine how many souvenirs of each type Ace Novelty should make per day in order to maximize its daily profits.

© Roy Gumpel/Stone/Getty Images

THE GEOMETRIC APPROACH introduced in the previous chapter may be used to solve linear programming problems involving two or even three variables. But for linear programming problems involving more than two variables, an algebraic approach is preferred. One such technique, the *simplex method*, was developed by George Dantzig in the late 1940s and remains in wide use to this day.

We begin Chapter 4 by developing the simplex method for solving *standard maximization problems*. We then see how, thanks to the principle of duality discovered by the great mathematician John von Neumann, this method can be used to solve a restricted class of *standard minimization problems*. Finally, we see how the simplex method can be adapted to solve *nonstandard problems*—problems that do not belong to the other aforementioned categories.

4.1   The Simplex Method: Standard Maximization Problems

■ The Simplex Method

As mentioned in Chapter 3, the method of corners is not suitable for solving linear programming problems when the number of variables or constraints is large. Its major shortcoming is that a knowledge of all the corner points of the feasible set S associated with the problem is required. What we need is a method of solution that is based on a judicious selection of the corner points of the feasible set S, thereby reducing the number of points to be inspected. One such technique, called the *simplex method*, was developed in the late 1940s by George Dantzig and is based on the Gauss–Jordan elimination method. The simplex method is readily adaptable to the computer, which makes it ideally suitable for solving linear programming problems involving large numbers of variables and constraints.

Basically, the simplex method is an iterative procedure; that is, it is repeated over and over again. Beginning at some initial feasible solution (a corner point of the feasible set S, usually the origin), each iteration brings us to another corner point of S with an improved (but certainly no worse) value of the objective function. The iteration is terminated when the optimal solution is reached (if it exists).

In this section we describe the simplex method for solving a large class of problems that are referred to as standard maximization problems.

Before stating a formal procedure for solving standard linear programming problems based on the simplex method, let's consider the following analysis of a two-variable problem. The ensuing discussion will clarify the general procedure and at the same time enhance our understanding of the simplex method by examining the motivation that led to the steps of the procedure.

A Standard Linear Programming Problem

A standard maximization problem is one in which

1. The objective function is to be maximized.
2. All the variables involved in the problem are nonnegative.
3. Each linear constraint may be written so that the expression involving the variables is less than or equal to a nonnegative constant.

Consider the linear programming problem presented at the beginning of Section 3.3:

$$\text{Maximize} \quad P = 3x + 2y \tag{1}$$
$$\text{subject to} \quad 2x + 3y \le 12$$
$$2x + y \le 8 \tag{2}$$
$$x \ge 0, y \ge 0$$

You can easily verify that this is a standard maximization problem. The feasible set S associated with this problem is reproduced in Figure 1, where we have labeled the four feasible corner points $A(0, 0)$, $B(4, 0)$, $C(3, 2)$, and $D(0, 4)$. Recall that the optimal solution to the problem occurs at the corner point $C(3, 2)$.

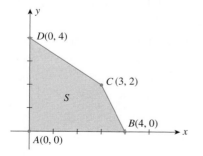

FIGURE 1
The optimal solution occurs at $C(3, 2)$.

As a first step in the solution using the simplex method, we replace the system of inequality constraints (2) with a system of equality constraints. This may be accomplished by using nonnegative variables called **slack variables.** Let's begin by considering the inequality

$$2x + 3y \leq 12$$

Observe that the left-hand side of this equation is always less than or equal to the right-hand side. Therefore, by adding a nonnegative variable u to the left-hand side to compensate for this difference, we obtain the equality

$$2x + 3y + u = 12$$

For example, if $x = 1$ and $y = 1$ [you can see by referring to Figure 1 that the point $(1, 1)$ is a feasible point of S], then $u = 7$. Thus,

$$2(1) + 3(1) + 7 = 12$$

If $x = 2$ and $y = 1$ [the point $(2, 1)$ is also a feasible point of S], then $u = 5$. Thus,

$$2(2) + 3(1) + 5 = 12$$

The variable u is a slack variable.

Similarly, the inequality $2x + y \leq 8$ is converted into the equation $2x + y + v = 8$ through the introduction of the slack variable v. System (2) of linear inequalities may now be viewed as the system of linear equations

$$
\begin{aligned}
2x + 3y + u \quad\; &= 12 \\
2x + \;\; y \quad\;\; + v &= 8
\end{aligned}
$$

where x, y, u, and v are all nonnegative.

Finally, rewriting the objective function (1) in the form $-3x - 2y + P = 0$, where the coefficient of P is $+1$, we are led to the following system of linear equations:

$$
\begin{aligned}
2x + 3y + u \qquad\qquad\; &= 12 \\
2x + \;\; y \quad\;\; + v \qquad &= 8 \qquad\qquad \textbf{(3)} \\
-3x - 2y \qquad\quad\; + P &= 0
\end{aligned}
$$

Since System (3) consists of three linear equations in the five variables x, y, u, v, and P, we may solve for three of the variables in terms of the other two. Thus, there are infinitely many solutions to this system expressible in terms of two parameters. Our linear programming problem is now seen to be equivalent to the following: From among all the solutions of System (3) for which x, y, u, and v are nonnegative (such solutions are called **feasible solutions**), determine the solution(s) that renders P a maximum.

The augmented matrix associated with System (3) is

Nonbasic variables ————⌐ ⌐——— Basic variables
⌐— Column of constants

$$
\begin{array}{ccccc}
x & y & u & v & P \\
\end{array}
$$

$$
\left[
\begin{array}{ccccc|c}
2 & 3 & 1 & 0 & 0 & 12 \\
2 & 1 & 0 & 1 & 0 & 8 \\
-3 & -2 & 0 & 0 & 1 & 0
\end{array}
\right] \qquad\qquad \textbf{(4)}
$$

Observe that each of the u-, v-, and P-columns of the augmented matrix (4) is a unit column (see page 85). The variables associated with unit columns are called **basic variables;** all other variables are called **nonbasic variables.**

Now, the configuration of the augmented matrix (4) suggests that we solve for the basic variables u, v, and P in terms of the nonbasic variables x and y, obtaining

$$\begin{aligned} u &= 12 - 2x - 3y \\ v &= 8 - 2x - y \\ P &= 3x + 2y \end{aligned} \qquad \textbf{(5)}$$

Of the infinitely many feasible solutions obtainable by assigning arbitrary nonnegative values to the parameters x and y, a particular solution is obtained by letting $x = 0$ and $y = 0$. In fact, this solution is given by

$$x = 0 \qquad y = 0 \qquad u = 12 \qquad v = 8 \qquad P = 0$$

Such a solution, obtained by setting the nonbasic variables equal to zero, is called a **basic solution** of the system. This particular solution corresponds to the corner point $A(0, 0)$ of the feasible set associated with the linear programming problem (see Figure 1). Observe that $P = 0$ at this point.

Now, if the value of P cannot be increased, we have found the optimal solution to the problem at hand. To determine whether the value of P can in fact be improved, let's turn our attention to the objective function in (1). Since both the coefficients of x and y are positive, the value of P can be improved by increasing x and/or y—that is, by moving away from the origin. Note that we arrive at the same conclusion by observing that the last row of the augmented matrix (4) contains entries that are *negative*. (Compare the original objective function, $P = 3x + 2y$, with the rewritten objective function, $-3x - 2y + P = 0$.)

Continuing our quest for an optimal solution, our next task is to determine whether it is more profitable to increase the value of x or that of y (increasing x and y simultaneously is more difficult). Since the coefficient of x is greater than that of y, a unit increase in the x-direction will result in a greater increase in the value of the objective function P than a unit increase in the y-direction. Thus, we should increase the value of x while holding y constant. How much can x be increased while holding $y = 0$? Upon setting $y = 0$ in the first two equations of (5), we see that

$$\begin{aligned} u &= 12 - 2x \\ v &= 8 - 2x \end{aligned} \qquad \textbf{(6)}$$

Since u must be nonnegative, the first equation of (6) implies that x cannot exceed $\frac{12}{2}$, or 6. The second equation of (6) and the nonnegativity of v implies that x cannot exceed $\frac{8}{2}$, or 4. Thus, we conclude that x can be increased by at most 4.

Now, if we set $y = 0$ and $x = 4$ in System (5), we obtain the solution

$$x = 4 \qquad y = 0 \qquad u = 4 \qquad v = 0 \qquad P = 12$$

which is a basic solution to System (3), this time with y and v as nonbasic variables. (Recall that the nonbasic variables are precisely the variables that are set equal to zero.)

Let's see how this basic solution may be found by working with the augmented matrix of the system. Since x is to replace v as a basic variable, our aim is to find an augmented matrix that is equivalent to the matrix (4) and has a configuration in which the x-column is in the unit form

$$\begin{bmatrix} 0 \\ 1 \\ 0 \end{bmatrix}$$

replacing what is presently the form of the v-column in (4). Now, this may be accomplished by pivoting about the circled number 2.

$$
\begin{array}{ccccc}
x & y & u & v & P \quad \text{Const.}
\end{array}
\left[\begin{array}{ccccc|c}
2 & 3 & 1 & 0 & 0 & 12 \\
\hline
② & 1 & 0 & 1 & 0 & 8 \\
\hline
-3 & -2 & 0 & 0 & 1 & 0
\end{array}\right]
\xrightarrow{\frac{1}{2}R_2}
\begin{array}{ccccc}
x & y & u & v & P \quad \text{Const.}
\end{array}
\left[\begin{array}{ccccc|c}
2 & 3 & 1 & 0 & 0 & 12 \\
① & \frac{1}{2} & 0 & \frac{1}{2} & 0 & 4 \\
-3 & -2 & 0 & 0 & 1 & 0
\end{array}\right]
\quad (7)
$$

$$
\xrightarrow[R_3 + 3R_2]{R_1 - 2R_2}
\begin{array}{ccccc}
x & y & u & v & P \quad \text{Const.}
\end{array}
\left[\begin{array}{ccccc|c}
0 & 2 & 1 & -1 & 0 & 4 \\
1 & \frac{1}{2} & 0 & \frac{1}{2} & 0 & 4 \\
0 & -\frac{1}{2} & 0 & \frac{3}{2} & 1 & 12
\end{array}\right]
\quad (8)
$$

Using (8), we now solve for the basic variables x, u, and P in terms of the nonbasic variables y and v, obtaining

$$x = 4 - \frac{1}{2}y - \frac{1}{2}v$$

$$u = 4 - 2y + v$$

$$P = 12 + \frac{1}{2}y - \frac{3}{2}v$$

Setting the nonbasic variables y and v equal to zero gives

$$x = 4 \qquad y = 0 \qquad u = 4 \qquad v = 0 \qquad P = 12$$

as before.

We have now completed one iteration of the simplex procedure, and our search has brought us from the feasible corner point $A(0, 0)$, where $P = 0$, to the feasible corner point $B(4, 0)$, where P attained a value of 12, which is certainly an improvement! (See Figure 2.)

Before going on, let's introduce the following terminology. The circled element 2 in the first augmented matrix of (7), which was to be converted into a 1, is called a *pivot element*. The column containing the pivot element is called the *pivot column*. The pivot column is associated with a nonbasic variable that is to be converted to a basic variable. Note that *the last entry in the pivot column is the negative number with the largest absolute value to the left of the vertical line in the last row—* precisely the criterion for choosing the direction of maximum increase in P.

The row containing the pivot element is called the *pivot row*. The pivot row can also be found by dividing each positive number in the pivot column into the corresponding number in the last column (the column of constants). *The pivot row is the one with the smallest ratio.* In the augmented matrix (7), the pivot row is the second row since the ratio $\frac{8}{2}$, or 4, is less than the ratio $\frac{12}{2}$, or 6. (Compare this with the earlier

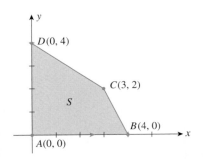

FIGURE 2
One iteration has taken us from $A(0, 0)$, where $P = 0$, to $B(4, 0)$, where $P = 12$.

analysis pertaining to the determination of the largest permissible increase in the value of x.)

The following is a summary of the procedure for selecting the pivot element.

> **Selecting the Pivot Element**
> 1. *Select the pivot column:* Locate the most negative entry to the left of the vertical line in the last row. The column containing this entry is the **pivot column.** (If there is more than one such column, choose any one.)
> 2. *Select the pivot row:* Divide each positive entry in the pivot column into its corresponding entry in the column of constants. The **pivot row** is the row corresponding to the smallest ratio thus obtained. (If there is more than one such entry, choose any one.)
> 3. The **pivot element** is the element common to both the pivot column and the pivot row.

Continuing with the solution to our problem, we observe that the last row of the augmented matrix (8) contains a negative number—namely, $-\frac{1}{2}$. This indicates that P is not maximized at the feasible corner point $B(4, 0)$, and another iteration is required. Without once again going into a detailed analysis, we proceed immediately to the selection of a pivot element. In accordance with the rules, we perform the necessary row operations as follows:

$$
\begin{array}{c}
\text{Pivot} \\ \text{row}
\end{array} \rightarrow
\begin{array}{cccccc}
& x & y & u & v & P & \\
& \left[\begin{array}{c} 0 \\ 1 \\ 0 \end{array}\right. & \begin{array}{c} ② \\ \frac{1}{2} \\ -\frac{1}{2} \end{array} & \begin{array}{c} 1 \\ 0 \\ 0 \end{array} & \begin{array}{c} -1 \\ \frac{1}{2} \\ \frac{3}{2} \end{array} & \begin{array}{c} 0 \\ 0 \\ 1 \end{array} & \left.\begin{array}{c} 4 \\ 4 \\ 12 \end{array}\right]
\end{array}
\quad
\begin{array}{l}
\text{Ratio} \\
\frac{4}{2} = 2 \\
\frac{4}{1/2} = 8
\end{array}
$$

$\uparrow$ Pivot column

$$
\xrightarrow{\frac{1}{2}R_1}
\begin{array}{cccccc}
& x & y & u & v & P \\
& \left[\begin{array}{c} 0 \\ 1 \\ 0 \end{array}\right. & \begin{array}{c} ① \\ \frac{1}{2} \\ -\frac{1}{2} \end{array} & \begin{array}{c} \frac{1}{2} \\ 0 \\ 0 \end{array} & \begin{array}{c} -\frac{1}{2} \\ \frac{1}{2} \\ \frac{3}{2} \end{array} & \begin{array}{c} 0 \\ 0 \\ 1 \end{array} & \left.\begin{array}{c} 2 \\ 4 \\ 12 \end{array}\right]
\end{array}
$$

$$
\begin{array}{c} R_2 - \frac{1}{2}R_1 \\ \xrightarrow{\hspace{1cm}} \\ R_3 + \frac{1}{2}R_1 \end{array}
\begin{array}{cccccc}
& x & y & u & v & P \\
& \left[\begin{array}{c} 0 \\ 1 \\ 0 \end{array}\right. & \begin{array}{c} 1 \\ 0 \\ 0 \end{array} & \begin{array}{c} \frac{1}{2} \\ -\frac{1}{4} \\ \frac{1}{4} \end{array} & \begin{array}{c} -\frac{1}{2} \\ \frac{3}{4} \\ \frac{5}{4} \end{array} & \begin{array}{c} 0 \\ 0 \\ 1 \end{array} & \left.\begin{array}{c} 2 \\ 3 \\ 13 \end{array}\right]
\end{array}
$$

Interpreting the last augmented matrix in the usual fashion, we find the basic solution $x = 3$, $y = 2$, and $P = 13$. Since there are no negative entries in the last row, the solution is optimal and P cannot be increased further. The optimal solution is the feasible corner point $C(3, 2)$ (Figure 3). Observe that this agrees with the solution we found using the method of corners in Section 3.3.

Having seen how the simplex method works, let's list the steps involved in the procedure. The first step is to set up the initial **simplex tableau.**

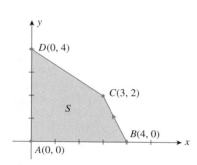

FIGURE 3
The next iteration has taken us from $B(4, 0)$, where $P = 12$, to $C(3, 2)$, where $P = 13$.

Setting Up the Initial Simplex Tableau

1. Transform the system of linear inequalities into a system of linear equations by introducing slack variables.
2. Rewrite the objective function

$$P = c_1 x_1 + c_2 x_2 + \cdots + c_n x_n$$

in the form

$$-c_1 x_1 - c_2 x_2 - \cdots - c_n x_n + P = 0$$

where all the variables are on the left and the coefficient of P is $+1$. Write this equation below the equations of step 1.
3. Write the augmented matrix associated with this system of linear equations.

EXAMPLE 1 Set up the initial simplex tableau for the linear programming problem posed in Example 1, Section 3.2.

Solution The problem at hand is to maximize

$$P = x + 1.2y$$

or, equivalently,

$$P = x + \frac{6}{5}y$$

subject to

$$
\begin{aligned}
2x + \ \ y &\le 180 \\
x + 3y &\le 300 \\
x \ge 0, y &\ge \ \ \ 0
\end{aligned}
\tag{9}
$$

This is a standard maximization problem and may be solved by the simplex method. Since System (9) has two linear inequalities (other than $x \ge 0$, $y \ge 0$), we introduce the two slack variables u and v to convert it to a system of linear equations:

$$
\begin{aligned}
2x + \ \ y + u \ \ \ \ \ &= 180 \\
x + 3y \ \ \ \ \ \ + v &= 300
\end{aligned}
$$

Next, by rewriting the objective function in the form

$$-x - \frac{6}{5}y + P = 0$$

where the coefficient of P is $+1$, and placing it below the system of equations, we obtain the system of linear equations

$$
\begin{aligned}
2x + \ \ y + u \ \ \ \ \ \ \ \ \ \ \ \ &= 180 \\
x + \ 3y \ \ \ \ \ + v \ \ \ \ \ \ &= 300 \\
-x - \frac{6}{5}y \ \ \ \ \ \ \ \ \ \ + P &= \ \ \ 0
\end{aligned}
$$

The initial simplex tableau associated with this system is

x	y	u	v	P	Constant
2	1	1	0	0	180
1	3	0	1	0	300
-1	$-\frac{6}{5}$	0	0	1	0

Before completing the solution to the problem posed in Example 1, let's summarize the main steps of the **simplex method.**

The Simplex Method
1. *Set up the initial simplex tableau.*
2. *Determine whether the optimal solution has been reached by examining all entries in the last row to the left of the vertical line.*
   a. If all the entries are nonnegative, the optimal solution has been reached. Proceed to step 4.
   b. If there are one or more negative entries, the optimal solution has not been reached. Proceed to step 3.
3. *Perform the pivot operation.* Locate the pivot element and convert it to a 1 by dividing all the elements in the pivot row by the pivot element. Using row operations, convert the pivot column into a unit column by adding suitable multiples of the pivot row to each of the other rows as required. Return to step 2.
4. *Determine the optimal solution(s).* The value of the variable heading each unit column is given by the entry lying in the column of constants in the row containing the 1. The variables heading columns not in unit form are assigned the value zero.

EXAMPLE 2　Complete the solution to the problem discussed in Example 1.

Solution　The first step in our procedure, setting up the initial simplex tableau, was completed in Example 1. We continue with Step 2.

Step 2　*Determine whether the optimal solution has been reached.* First, refer to the initial simplex tableau:

x	y	u	v	P	Constant
2	1	1	0	0	180
1	3	0	1	0	300
-1	$-\frac{6}{5}$	0	0	1	0

(10)

Since there are negative entries in the last row of the initial simplex tableau, the initial solution is not optimal. We proceed to Step 3.

Step 3　*Perform the following iterations.* First, locate the pivot element:
a. Since the entry $-\frac{6}{5}$ is the most negative entry to the left of the vertical line in the last row of the initial simplex tableau, the second column in the tableau is the pivot column.

b. Divide each positive number of the pivot column into the corresponding entry in the column of constants and compare the ratios thus obtained. We see that the ratio $\frac{300}{3}$ is less than the ratio $\frac{180}{1}$, so row 2 in the tableau is the pivot row.

c. The entry 3 lying in the pivot column and the pivot row is the pivot element.

Next, we convert this pivot element into a 1 by multiplying all the entries in the pivot row by $\frac{1}{3}$. Then, using elementary row operations, we complete the conversion of the pivot column into a unit column. The details of the iteration are recorded as follows:

	x	y	u	v	P	Constant		Ratio
	2	1	1	0	0	180		$\frac{180}{1} = 180$
Pivot $\rightarrow$ row	1	③	0	1	0	300		$\frac{300}{3} = 100$
	-1	$-\frac{6}{5}$	0	0	1	0		

↑
Pivot
column

	x	y	u	v	P	Constant
	2	1	1	0	0	180
$\frac{1}{3}R_2$ $\longrightarrow$	$\frac{1}{3}$	①	0	$\frac{1}{3}$	0	100
	-1	$-\frac{6}{5}$	0	0	1	0

	x	y	u	v	P	Constant
$R_1 - R_2$ $\longrightarrow$	$\frac{5}{3}$	0	1	$-\frac{1}{3}$	0	80
$R_3 + \frac{6}{5}R_2$	$\frac{1}{3}$	1	0	$\frac{1}{3}$	0	100
	$-\frac{3}{5}$	0	0	$\frac{2}{5}$	1	120

(11)

This completes one iteration. The last row of the simplex tableau contains a negative number, so an optimal solution has not been reached. Therefore, we repeat the iterative step once again, as follows:

	x	y	u	v	P	Constant		Ratio
Pivot $\rightarrow$ row	⑤⁄₃	0	1	$-\frac{1}{3}$	0	80		$\frac{80}{5/3} = 48$
	$\frac{1}{3}$	1	0	$\frac{1}{3}$	0	100		$\frac{100}{1/3} = 300$
	$-\frac{3}{5}$	0	0	$\frac{2}{5}$	1	120		

↑
Pivot
column

	x	y	u	v	P	Constant
$\frac{3}{5}R_1$ $\longrightarrow$	①	0	$\frac{3}{5}$	$-\frac{1}{5}$	0	48
	$\frac{1}{3}$	1	0	$\frac{1}{3}$	0	100
	$-\frac{3}{5}$	0	0	$\frac{2}{5}$	1	120

	x	y	u	v	P	Constant
$R_2 - \frac{1}{3}R_1$	1	0	$\frac{3}{5}$	$-\frac{1}{5}$	0	48
$R_3 + \frac{3}{5}R_1$	0	1	$-\frac{1}{5}$	$\frac{2}{5}$	0	84
	0	0	$\frac{9}{25}$	$\frac{7}{25}$	1	$148\frac{4}{5}$

(12)

The last row of the simplex tableau (12) contains no negative numbers, and we therefore conclude that the optimal solution has been reached.

Step 4 *Determine the optimal solution.* Locate the basic variables in the final tableau. In this case the basic variables (those heading unit columns) are x, y, and P. The value assigned to the basic variable x is the number 48, which is the entry lying in the column of constants and in row 1 (the row that contains the 1).

x	y	u	v	P	Constant
①	0	$\frac{3}{5}$	$-\frac{1}{5}$	0	48
0	①	$-\frac{1}{5}$	$\frac{2}{5}$	0	84
0	0	$\frac{9}{25}$	$\frac{7}{25}$	①	$148\frac{4}{5}$

Similarly, we conclude that $y = 84$ and $P = 148.8$. Next, we note that the variables u and v are nonbasic and are accordingly assigned the values $u = 0$ and $v = 0$. These results agree with those obtained in Example 1, Section 3.3. ■

EXAMPLE 3

$$\text{Maximize} \quad P = 2x + 2y + z$$
$$\text{subject to} \quad 2x + y + 2z \le 14$$
$$2x + 4y + z \le 26$$
$$x + 2y + 3z \le 28$$
$$x \ge 0, y \ge 0, z \ge 0$$

Solution Introducing the slack variables u, v, and w and rewriting the objective function in the standard form gives the system of linear equations

$$2x + y + 2z + u = 14$$
$$2x + 4y + z + v = 26$$
$$x + 2y + 3z + w = 28$$
$$-2x - 2y - z + P = 0$$

The initial simplex tableau is given by

x	y	z	u	v	w	P	Constant
2	1	2	1	0	0	0	14
2	4	1	0	1	0	0	26
1	2	3	0	0	1	0	28
−2	−2	−1	0	0	0	1	0

Since the most negative entry in the last row (-2) occurs twice, we may choose either the x- or the y-column as the pivot column. Choosing the x-column as the pivot column and proceeding with the first iteration, we obtain the following sequence of tableaus:

	x	y	z	u	v	w	P	Constant		Ratio
Pivot row →	②	1	2	1	0	0	0	14		$\frac{14}{2} = 7$
	2	4	1	0	1	0	0	26		$\frac{26}{2} = 13$
	1	2	3	0	0	1	0	28		$\frac{28}{1} = 28$
	-2	-2	-1	0	0	0	1	0		

Pivot column (↑)

	x	y	z	u	v	w	P	Constant
$\frac{1}{2}R_1$ →	①	$\frac{1}{2}$	1	$\frac{1}{2}$	0	0	0	7
	2	4	1	0	1	0	0	26
	1	2	3	0	0	1	0	28
	-2	-2	-1	0	0	0	1	0

	x	y	z	u	v	w	P	Constant
	1	$\frac{1}{2}$	1	$\frac{1}{2}$	0	0	0	7
$R_2 - 2R_1$	0	3	-1	-1	1	0	0	12
$R_3 - R_1$ $R_4 + 2R_1$	0	$\frac{3}{2}$	2	$-\frac{1}{2}$	0	1	0	21
	0	-1	1	1	0	0	1	14

Since there is a negative number in the last row of the simplex tableau, we perform another iteration, as follows:

	x	y	z	u	v	w	P	Constant		Ratio
	1	$\frac{1}{2}$	1	$\frac{1}{2}$	0	0	0	7		$\frac{7}{1/2} = 14$
Pivot row →	0	③	-1	-1	1	0	0	12		$\frac{12}{3} = 4$
	0	$\frac{3}{2}$	2	$-\frac{1}{2}$	0	1	0	21		$\frac{21}{3/2} = 14$
	0	-1	1	1	0	0	1	14		

Pivot column (↑)

	x	y	z	u	v	w	P	Constant
	1	$\frac{1}{2}$	1	$\frac{1}{2}$	0	0	0	7
$\frac{1}{3}R_2$ →	0	①	$-\frac{1}{3}$	$-\frac{1}{3}$	$\frac{1}{3}$	0	0	4
	0	$\frac{3}{2}$	2	$-\frac{1}{2}$	0	1	0	21
	0	-1	1	1	0	0	1	14

	x	y	z	u	v	w	P	Constant
	1	0	$\frac{7}{6}$	$\frac{2}{3}$	$-\frac{1}{6}$	0	0	5
$R_1 - \frac{1}{2}R_2$	0	1	$-\frac{1}{3}$	$-\frac{1}{3}$	$\frac{1}{3}$	0	0	4
$R_3 - \frac{3}{2}R_2$ $R_4 + R_2$	0	0	$\frac{5}{2}$	0	$-\frac{1}{2}$	1	0	15
	0	0	$\frac{2}{3}$	$\frac{2}{3}$	$\frac{1}{3}$	0	1	18

All entries in the last row are nonnegative, so we have reached the optimal solution. We conclude that $x = 5$, $y = 4$, $z = 0$, $u = 0$, $v = 0$, $w = 15$, and $P = 18$. ∎

EXPLORE & DISCUSS

Consider the linear programming problem

$$\text{Maximize} \quad P = x + 2y$$
$$\text{subject to} \quad -2x + y \leq 4$$
$$x - 3y \leq 3$$
$$x \geq 0, y \geq 0$$

1. Sketch the feasible set S for the linear programming problem and explain why the problem has an unbounded solution.

2. Use the simplex method to solve the problem as follows:
   a. Perform one iteration on the initial simplex tableau. Interpret your result. Indicate the point on S corresponding to this (nonoptimal) solution.
   b. Show that the simplex procedure breaks down when you attempt to perform another iteration by demonstrating that there is no pivot element.
   c. Describe what happens if you violate the rule for finding the pivot element by allowing the ratios to be negative and proceeding with the iteration.

The following example is constructed to illustrate the geometry associated with the simplex method when used to solve a problem in three-dimensional space. We sketch the feasible set for the problem and show the path dictated by the simplex method in arriving at the optimal solution for the problem. The use of a calculator will help in the arithmetic operations if you wish to verify the steps.

EXAMPLE 4

$$\text{Maximize} \quad P = 20x + 12y + 18z$$
$$\text{subject to} \quad 3x + y + 2z \leq 9$$
$$2x + 3y + z \leq 8$$
$$x + 2y + 3z \leq 7$$
$$x \geq 0, y \geq 0, z \geq 0$$

Solution Introducing the slack variables u, v, and w and rewriting the objective function in standard form gives the system of linear equations:

$$3x + y + 2z + u \qquad\qquad = 9$$
$$2x + 3y + z \qquad + v \qquad\quad = 8$$
$$x + 2y + 3z \qquad\qquad + w \quad = 7$$
$$-20x - 12y - 18z \qquad\qquad\qquad + P = 0$$

The initial simplex tableau is given by

x	y	z	u	v	w	P	Constant
3	1	2	1	0	0	0	9
2	3	1	0	1	0	0	8
1	2	3	0	0	1	0	7
−20	−12	−18	0	0	0	1	0

Since the most negative entry in the last row (-20) occurs in the x-column, we choose the x-column as the pivot column. Proceeding with the first iteration, we obtain the following sequence of tableaus:

	x	y	z	u	v	w	P	Constant	Ratio
Pivot row →	③	1	2	1	0	0	0	9	$\frac{9}{3} = 3$
	2	3	1	0	1	0	0	8	$\frac{8}{2} = 4$
	1	2	3	0	0	1	0	7	$\frac{7}{1} = 7$
	-20	-12	-18	0	0	0	1	0	

Pivot column ↑

	x	y	z	u	v	w	P	Constant
$\frac{1}{3}R_1$ →	①	$\frac{1}{3}$	$\frac{2}{3}$	$\frac{1}{3}$	0	0	0	3
	2	3	1	0	1	0	0	8
	1	2	3	0	0	1	0	7
	-20	-12	-18	0	0	0	1	0

$R_2 - 2R_1$
$R_3 - R_1$
$R_4 + 20R_1$ →

	x	y	z	u	v	w	P	Constant	Ratio
	1	$\frac{1}{3}$	$\frac{2}{3}$	$\frac{1}{3}$	0	0	0	3	9
Pivot row	0	⑦/₃	$-\frac{1}{3}$	$-\frac{2}{3}$	1	0	0	2	$\frac{6}{7}$
	0	$\frac{5}{3}$	$\frac{7}{3}$	$-\frac{1}{3}$	0	1	0	4	$\frac{12}{5}$
	0	$-\frac{16}{3}$	$-\frac{14}{3}$	$\frac{20}{3}$	0	0	1	60	

Pivot column ↑

After one iteration we are at the point $(3, 0, 0)$ with $P = 60$. (See Figure 4 on page 228.) Since the most negative entry in the last row is $-\frac{16}{3}$, we choose the y-column as the pivot column. Proceeding with this iteration, we obtain

	x	y	z	u	v	w	P	Constant
	1	$\frac{1}{3}$	$\frac{2}{3}$	$\frac{1}{3}$	0	0	0	3
$\frac{3}{7}R_2$ →	0	①	$-\frac{1}{7}$	$-\frac{2}{7}$	$\frac{3}{7}$	0	0	$\frac{6}{7}$
	0	$\frac{5}{3}$	$\frac{7}{3}$	$-\frac{1}{3}$	0	1	0	4
	0	$-\frac{16}{3}$	$-\frac{14}{3}$	$\frac{20}{3}$	0	0	1	60

$R_1 - \frac{1}{3}R_2$
$R_3 - \frac{5}{3}R_2$
$R_4 + \frac{16}{3}R_2$ →

	x	y	z	u	v	w	P	Constant	Ratio
	1	0	$\frac{5}{7}$	$\frac{3}{7}$	$-\frac{1}{7}$	0	0	$\frac{19}{7}$	$\frac{19}{5}$
	0	1	$-\frac{1}{7}$	$-\frac{2}{7}$	$\frac{3}{7}$	0	0	$\frac{6}{7}$	—
	0	0	⑱/₇	$\frac{1}{7}$	$-\frac{5}{7}$	1	0	$\frac{18}{7}$	1
	0	0	$-\frac{38}{7}$	$\frac{36}{7}$	$\frac{16}{7}$	0	1	$64\frac{4}{7}$	

Pivot column ↑

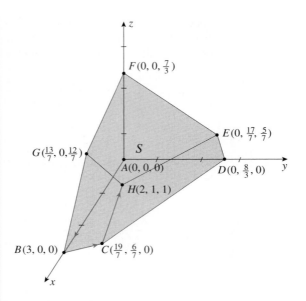

FIGURE 4
The simplex method brings us from the point A to the point H, at which the objective function is maximized.

The second iteration brings us to the point $\left(\frac{19}{7}, \frac{6}{7}, 0\right)$ with $P = 64\frac{4}{7}$. (See Figure 4.) Since there is a negative number in the last row of the simplex tableau, we perform another iteration, as follows:

$$\xrightarrow{\frac{7}{18}R_3}$$

	x	y	z	u	v	w	P	Constant
	1	0	$\frac{5}{7}$	$\frac{3}{7}$	$-\frac{1}{7}$	0	0	$\frac{19}{7}$
	0	1	$-\frac{1}{7}$	$-\frac{2}{7}$	$\frac{3}{7}$	0	0	$\frac{6}{7}$
	0	0	①	$\frac{1}{18}$	$-\frac{5}{18}$	$\frac{7}{18}$	0	1
	0	0	$-\frac{38}{7}$	$\frac{36}{7}$	$\frac{16}{7}$	0	1	$64\frac{4}{7}$

$$\begin{array}{c} R_1 - \frac{5}{7}R_3 \\ R_2 + \frac{1}{7}R_3 \\ \hline R_4 + \frac{38}{7}R_3 \end{array}$$

	x	y	z	u	v	w	P	Constant
	1	0	0	$\frac{7}{18}$	$\frac{1}{18}$	$-\frac{5}{18}$	0	2
	0	1	0	$-\frac{5}{18}$	$\frac{7}{18}$	$\frac{1}{18}$	0	1
	0	0	1	$\frac{1}{18}$	$-\frac{5}{18}$	$\frac{7}{18}$	0	1
	0	0	0	$\frac{49}{9}$	$\frac{7}{9}$	$\frac{19}{9}$	1	70

All entries in the last row are nonnegative, so we have reached the optimal solution. We conclude that $x = 2$, $y = 1$, $z = 1$, $u = 0$, $v = 0$, $w = 0$, and $P = 70$.

The feasible set S for the problem is the hexahedron shown in Figure 5. It is the intersection of the half spaces determined by the planes P_1, P_2, and P_3 with equations $3x + y + 2z = 9$, $2x + 3y + z = 8$, $x + 2y + 3z = 7$, respectively, and the coordinate planes $x = 0$, $y = 0$, and $z = 0$. That portion of the figure showing the feasible set S is shown in Figure 4. Observe that the first iteration of the simplex method brings us from $A(0, 0, 0)$ with $P = 0$ to $B(3, 0, 0)$ with $P = 60$. The second iteration brings us from $B(3, 0, 0)$ to $C\left(\frac{19}{7}, \frac{6}{7}, 0\right)$ with $P = 64\frac{4}{7}$, and the third iteration brings us from $C\left(\frac{19}{7}, \frac{6}{7}, 0\right)$ to the point $H(2, 1, 1)$ with an optimal value of 70 for P.

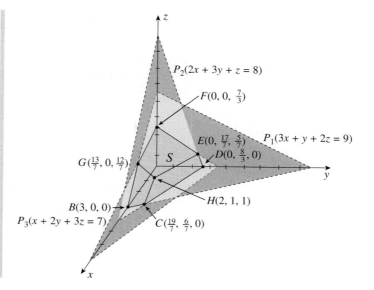

FIGURE 5
The feasible set S is obtained from the intersection of the half spaces determined by P_1, P_2, and P_3 with the coordinate planes $x = 0$, $y = 0$, and $z = 0$.

APPLIED EXAMPLE 5   Production Planning   Ace Novelty Company has determined that the profits are $6, $5, and $4 for each type-A, type-B, and type-C souvenir that it plans to produce. To manufacture a type-A souvenir requires 2 minutes on machine I, 1 minute on machine II, and 2 minutes on machine III. A type-B souvenir requires 1 minute on machine I, 3 minutes on machine II, and 1 minute on machine III. A type-C souvenir requires 1 minute on machine I and 2 minutes on each of machines II and III. Each day there are 3 hours available on machine I, 5 hours available on machine II, and 4 hours available on machine III for manufacturing these souvenirs. How many souvenirs of each type should Ace Novelty make per day in order to maximize its profit? (Compare with Example 1, Section 2.1.)

Solution   The given information is tabulated as follows:

	Type A	Type B	Type C	Time Available (min)
Machine I	2	1	1	180
Machine II	1	3	2	300
Machine III	2	1	2	240
Profit per Unit ($)	6	5	4	

Let x, y, and z denote the respective numbers of type-A, type-B, and type-C souvenirs to be made. The total amount of time that machine I is used is given by $2x + y + z$ minutes and must not exceed 180 minutes. Thus, we have the inequality

$$2x + y + z \le 180$$

Similar considerations on the use of machines II and III lead to the inequalities

$$x + 3y + 2z \le 300$$
$$2x + y + 2z \le 240$$

The profit resulting from the sale of the souvenirs produced is given by

$$P = 6x + 5y + 4z$$

The mathematical formulation of this problem leads to the following standard linear programming problem: Maximize the objective (profit) function $P = 6x + 5y + 4z$ subject to

$$
\begin{aligned}
2x + y + z &\le 180 \\
x + 3y + 2z &\le 300 \\
2x + y + 2z &\le 240 \\
x \ge 0, y \ge 0, z &\ge 0
\end{aligned}
$$

Introducing the slack variables u, v, and w gives the system of linear equations

$$
\begin{aligned}
2x + y + z + u &= 180 \\
x + 3y + 2z + v &= 300 \\
2x + y + 2z + w &= 240 \\
-6x - 5y - 4z + P &= 0
\end{aligned}
$$

The tableaus resulting from the use of the simplex algorithm are

	x	y	z	u	v	w	P	Constant		Ratio
Pivot row →	②	1	1	1	0	0	0	180		$\frac{180}{2} = 90$
	1	3	2	0	1	0	0	300		$\frac{300}{1} = 300$
	2	1	2	0	0	1	0	240		$\frac{240}{2} = 120$
	−6	−5	−4	0	0	0	1	0		

↑ Pivot column

$\xrightarrow{\frac{1}{2}R_1}$

	x	y	z	u	v	w	P	Constant
	①	$\frac{1}{2}$	$\frac{1}{2}$	$\frac{1}{2}$	0	0	0	90
	1	3	2	0	1	0	0	300
	2	1	2	0	0	1	0	240
	−6	−5	−4	0	0	0	1	0

$\begin{array}{c} R_2 - R_1 \\ \xrightarrow{\hspace{1cm}} \\ R_3 - 2R_1 \\ R_4 + 6R_1 \end{array}$ Pivot row

	x	y	z	u	v	w	P	Constant		Ratio
	1	$\frac{1}{2}$	$\frac{1}{2}$	$\frac{1}{2}$	0	0	0	90		$\frac{90}{1/2} = 180$
	0	$\frac{5}{2}$	$\frac{3}{2}$	$-\frac{1}{2}$	1	0	0	210		$\frac{210}{5/2} = 84$
	0	0	1	−1	0	1	0	60		
	0	−2	−1	3	0	0	1	540		

↑ Pivot column

$\xrightarrow{\frac{2}{5}R_2}$

	x	y	z	u	v	w	P	Constant
	1	$\frac{1}{2}$	$\frac{1}{2}$	$\frac{1}{2}$	0	0	0	90
	0	①	$\frac{3}{5}$	$-\frac{1}{5}$	$\frac{2}{5}$	0	0	84
	0	0	1	−1	0	1	0	60
	0	−2	−1	3	0	0	1	540

	x	y	z	u	v	w	P	Constant
	1	0	$\frac{1}{5}$	$\frac{3}{5}$	$-\frac{1}{5}$	0	0	48
$\xrightarrow{\begin{array}{c} R_1 - \frac{1}{2}R_2 \\ R_4 + 2R_2 \end{array}}$	0	1	$\frac{3}{5}$	$-\frac{1}{5}$	$\frac{2}{5}$	0	0	84
	0	0	1	-1	0	1	0	60
	0	0	$\frac{1}{5}$	$\frac{13}{5}$	$\frac{4}{5}$	0	1	708

From the final simplex tableau, we read off the solution

$$x = 48 \qquad y = 84 \qquad z = 0 \qquad u = 0 \qquad v = 0 \qquad w = 60 \qquad P = 708$$

Thus, in order to maximize its profit, Ace Novelty should produce 48 type-A souvenirs, 84 type-B souvenirs, and no type-C souvenirs. The resulting profit is $708 per day. The value of the slack variable $w = 60$ tells us that 1 hour of the available time on machine III is left unused.   ■

Interpreting Our Results   It is instructive to compare the results obtained here with those obtained in Example 7, Section 2.2. Recall that, to use all available machine time on each of the three machines, Ace Novelty had to produce 36 type-A, 48 type-B, and 60 type-C souvenirs. This would have resulted in a profit of $696. Example 5 shows how, through the optimal use of equipment, a company can boost its profit while reducing machine wear!

▬ Problems with Multiple Solutions and Problems with No Solutions

As we saw in Section 3.3, a linear programming problem may have infinitely many solutions. We also saw that a linear programming problem may have no solution. How do we spot each of these phenomena when using the simplex method to solve a problem?

   A linear programming problem may have infinitely many solutions if and only if the last row to the left of the vertical line of the final simplex tableau has a zero in a column that is not a unit column. Next, a linear programming problem will have no solution if the simplex method breaks down at some stage. For example, if at some stage there are no nonnegative ratios in our computation, then the linear programming problem has no solution (see Exercise 42).

EXPLORE & DISCUSS

Consider the linear programming problem

$$\text{Maximize} \quad P = 4x + 6y$$
$$\text{subject to} \quad 2x + \ y \le 10$$
$$2x + 3y \le 18$$
$$x \ge 0, y \ge 0$$

1. Sketch the feasible set for the linear programming problem.

2. Use the method of corners to show that there are infinitely many optimal solutions. What are they?

3. Use the simplex method to solve the problem as follows:
   a. Perform one iteration on the initial simplex tableau and conclude that you have arrived at an optimal solution. What is the value of P, and where is it attained? Compare this result with that obtained in step 2.
   b. Observe that the tableau obtained in part (a) indicates that there are infinitely many solutions (see the comment on multiple solutions on this page). Now perform another iteration on the simplex tableau using the x-column as the pivot column. Interpret the final tableau.

4.1 Self-Check Exercises

1. Solve the following linear programming problem by the simplex method:

$$\text{Maximize} \quad P = 2x + 3y + 6z$$
$$\text{subject to} \quad 2x + 3y + \ z \le 10$$
$$x + \ y + 2z \le 8$$
$$2y + 3z \le 6$$
$$x \ge 0, y \ge 0, z \ge 0$$

2. The LaCrosse Iron Works makes two models of cast-iron fireplace grates, model A and model B. Producing one model A grate requires 20 lb of cast iron and 20 min of labor, whereas producing one model B grate requires 30 lb of cast iron and 15 min of labor. The profit for a model A grate is \$6, and the profit for a model B grate is \$8. There are 7200 lb of cast iron and 100 labor-hours available each week. Because of a surplus from the previous week, the proprietor has decided that he should make no more than 150 units of model A grates this week. Determine how many of each model he should make in order to maximize his profits.

Solutions to Self-Check Exercises 4.1 can be found on page 237.

4.1 Concept Questions

1. Give the three characteristics of a standard maximization linear programming problem.

2. a. When the initial simplex tableau is set up, how is the system of linear inequalities transformed into a system of linear equations? How is the objective function $P = c_1x_1 + c_2x_2 + \cdots + c_nx_n$ rewritten?

   b. If you are given a simplex tableau, how do you determine whether the optimal solution has been reached?

3. In the simplex method, how is a pivot column selected? A pivot row? A pivot element?

4.1 Exercises

In Exercises 1–10, determine whether the given simplex tableau is in final form. If so, find the solution to the associated regular linear programming problem. If not, find the pivot element to be used in the next iteration of the simplex method.

1.

x	y	u	v	P	Constant
0	1	$\frac{5}{7}$	$-\frac{1}{7}$	0	$\frac{20}{7}$
1	0	$-\frac{3}{7}$	$\frac{2}{7}$	0	$\frac{30}{7}$
0	0	$\frac{13}{7}$	$\frac{3}{7}$	1	$\frac{220}{7}$

2.

x	y	u	v	P	Constant
1	1	1	0	0	6
1	0	-1	1	0	2
3	0	5	0	1	30

3.

x	y	u	v	P	Constant
0	$\frac{1}{2}$	1	$-\frac{1}{2}$	0	2
1	$\frac{1}{2}$	0	$\frac{1}{2}$	0	4
0	$-\frac{1}{2}$	0	$\frac{3}{2}$	1	12

4.

x	y	z	u	v	w	P	Constant
3	0	5	1	1	0	0	28
2	1	3	0	1	0	0	16
2	0	8	0	3	0	1	48

5.

x	y	z	u	v	w	P	Constant
1	$-\frac{1}{3}$	0	$\frac{1}{3}$	0	$-\frac{2}{3}$	0	$\frac{1}{3}$
0	2	0	0	1	1	0	6
0	$\frac{2}{3}$	1	$\frac{1}{3}$	0	$\frac{1}{3}$	0	$\frac{13}{3}$
0	4	0	1	0	2	1	17

6.

x	y	z	u	v	w	P	Constant
$\frac{1}{2}$	0	$\frac{1}{4}$	1	$-\frac{1}{4}$	0	0	$\frac{19}{2}$
$\frac{1}{2}$	1	$\frac{3}{4}$	0	$\frac{1}{4}$	0	0	$\frac{21}{2}$
2	0	3	0	0	1	0	30
-1	0	$-\frac{1}{2}$	6	$\frac{3}{2}$	0	1	63

7.

x	y	z	s	t	u	v	P	Constant
$\frac{5}{2}$	3	0	1	0	0	-4	0	46
1	0	0	0	1	0	0	0	9
0	1	0	0	0	1	0	0	12
0	0	1	0	0	0	1	0	6
-180	-200	0	0	0	0	300	1	1800

8.

x	y	z	s	t	u	v	P	Constant
1	0	0	$\frac{2}{5}$	0	$-\frac{6}{5}$	$-\frac{8}{5}$	0	4
0	0	0	$-\frac{2}{5}$	1	$\frac{6}{5}$	$\frac{8}{5}$	0	5
0	1	0	0	0	1	0	0	12
0	0	1	0	0	0	1	0	6
0	0	0	72	0	-16	12	1	4920

9.

x	y	z	u	v	P	Constant
1	0	$\frac{3}{5}$	0	$\frac{1}{5}$	0	30
0	1	$-\frac{19}{5}$	1	$-\frac{3}{5}$	0	10
0	0	$\frac{26}{5}$	0	$\frac{2}{5}$	1	60

10.

x	y	z	u	v	w	P	Constant
0	$\frac{1}{2}$	0	1	$-\frac{1}{2}$	0	0	2
1	$\frac{1}{2}$	1	0	$\frac{1}{2}$	0	0	13
2	$\frac{1}{2}$	0	0	$-\frac{3}{2}$	1	0	4
-1	3	0	0	1	0	1	26

In Exercises 11–25, solve each linear programming problem by the simplex method.

11. Maximize $P = 3x + 4y$
subject to $x + y \le 4$
$2x + y \le 5$
$x \ge 0, y \ge 0$

12. Maximize $P = 5x + 3y$
subject to $x + y \le 80$
$3x \le 90$
$x \ge 0, y \ge 0$

13. Maximize $P = 10x + 12y$
subject to $x + 2y \le 12$
$3x + 2y \le 24$
$x \ge 0, y \ge 0$

14. Maximize $P = 5x + 4y$
subject to $3x + 5y \le 78$
$4x + y \le 36$
$x \ge 0, y \ge 0$

15. Maximize $P = 4x + 6y$
subject to $3x + y \le 24$
$2x + y \le 18$
$x + 3y \le 24$
$x \ge 0, y \ge 0$

16. Maximize $P = 15x + 12y$
subject to $x + y \leq 12$
$3x + y \leq 30$
$10x + 7y \leq 70$
$x \geq 0, y \geq 0$

17. Maximize $P = 3x + 4y + 5z$
subject to $x + y + z \leq 8$
$3x + 2y + 4z \leq 24$
$x \geq 0, y \geq 0, z \geq 0$

18. Maximize $P = 3x + 3y + 4z$
subject to $x + y + 3z \leq 15$
$4x + 4y + 3z \leq 65$
$x \geq 0, y \geq 0, z \geq 0$

19. Maximize $P = 3x + 4y + z$
subject to $3x + 10y + 5z \leq 120$
$5x + 2y + 8z \leq 6$
$8x + 10y + 3z \leq 105$
$x \geq 0, y \geq 0, z \geq 0$

20. Maximize $P = x + 2y - z$
subject to $2x + y + z \leq 14$
$4x + 2y + 3z \leq 28$
$2x + 5y + 5z \leq 30$
$x \geq 0, y \geq 0, z \geq 0$

21. Maximize $P = 4x + 6y + 5z$
subject to $x + y + z \leq 20$
$2x + 4y + 3z \leq 42$
$2x + 3z \leq 30$
$x \geq 0, y \geq 0, z \geq 0$

22. Maximize $P = x + 4y - 2z$
subject to $3x + y - z \leq 80$
$2x + y - z \leq 40$
$-x + y + z \leq 80$
$x \geq 0, y \geq 0, z \geq 0$

23. Maximize $P = 12x + 10y + 5z$
subject to $2x + y + z \leq 10$
$3x + 5y + z \leq 45$
$2x + 5y + z \leq 40$
$x \geq 0, y \geq 0, z \geq 0$

24. Maximize $P = 2x + 6y + 6z$
subject to $2x + y + 3z \leq 10$
$4x + y + 2z \leq 56$
$6x + 4y + 3z \leq 126$
$2x + y + z \leq 32$
$x \geq 0, y \geq 0, z \geq 0$

25. Maximize $P = 24x + 16y + 23z$
subject to $2x + y + 2z \leq 7$
$2x + 3y + z \leq 8$
$x + 2y + 3z \leq 7$
$x \geq 0, y \geq 0, z \geq 0$

26. Rework Example 3 using the y-column as the pivot column in the first iteration of the simplex method.

27. Show that the following linear programming problem

Maximize $P = 2x + 2y - 4z$
subject to $3x + 3y - 2z \leq 100$
$5x + 5y + 3z \leq 150$
$x \geq 0, y \geq 0, z \geq 0$

has optimal solutions $x = 30$, $y = 0$, $z = 0$, $P = 60$ and $x = 0$, $y = 30$, $z = 0$, $P = 60$.

28. MANUFACTURING—PRODUCTION SCHEDULING   A company manufactures two products, A and B, on two machines, I and II. It has been determined that the company will realize a profit of \$3/unit of product A and a profit of \$4/unit of product B. To manufacture 1 unit of product A requires 6 min on machine I and 5 min on machine II. To manufacture 1 unit of product B requires 9 min on machine I and 4 min on machine II. There are 5 hr of machine time available on machine I and 3 hr of machine time available on machine II in each work shift. How many units of each product should be produced in each shift to maximize the company's profit? What is the largest profit the company can realize? Is there any time left unused on the machines?

29. MANUFACTURING—PRODUCTION SCHEDULING   National Business Machines Corporation manufactures two models of fax machines: A and B. Each model A costs \$100 to make, and each model B costs \$150. The profits are \$30 for each model A and \$40 for each model B fax machine. If the total number of fax machines demanded each month does not exceed 2500 and the company has earmarked no more than \$600,000/month for manufacturing costs, find how many units of each model National should make each month in order to maximize its monthly profits. What is the largest monthly profit the company can make?

30. MANUFACTURING—PRODUCTION SCHEDULING   Kane Manufacturing has a division that produces two models of hibachis, model A and model B. To produce each model A hibachi requires 3 lb of cast iron and 6 min of labor. To produce each model B hibachi requires 4 lb of cast iron and 3 min of labor. The profit for each model A hibachi is \$2, and the profit for each model B hibachi is \$1.50. If 1000 lb of cast iron and 20 labor-hours are available for the production of hibachis each day, how many hibachis of each model should the division produce to maximize Kane's profits? What is the largest profit the company can realize? Is there any raw material left over?

31. AGRICULTURE—CROP PLANNING A farmer has 150 acres of land suitable for cultivating crops A and B. The cost of cultivating crop A is $40/acre, whereas that of crop B is $60/acre. The farmer has a maximum of $7400 available for land cultivation. Each acre of crop A requires 20 labor-hours, and each acre of crop B requires 25 labor-hours. The farmer has a maximum of 3300 labor-hours available. If he expects to make a profit of $150/acre on crop A and $200/acre on crop B, how many acres of each crop should he plant in order to maximize his profit? What is the largest profit the farmer can realize? Are there any resources left over?

32. INVESTMENTS—ASSET ALLOCATION A financier plans to invest up to $500,000 in two projects. Project A yields a return of 10% on the investment, whereas project B yields a return of 15% on the investment. Because the investment in project B is riskier than the investment in project A, she has decided that the investment in project B should not exceed 40% of the total investment. How much should the financier invest in each project in order to maximize the return on her investment? What is the maximum return?

33. INVESTMENTS—ASSET ALLOCATION Ashley has earmarked at most $250,000 for investment in three mutual funds: a money market fund, an international equity fund, and a growth-and-income fund. The money market fund has a rate of return of 6%/year, the international equity fund has a rate of return of 10%/year, and the growth-and-income fund has a rate of return of 15%/year. Ashley has stipulated that no more than 25% of her total portfolio should be in the growth-and-income fund and that no more than 50% of her total portfolio should be in the international equity fund. To maximize the return on her investment, how much should Ashley invest in each type of fund? What is the maximum return?

34. MANUFACTURING—PRODUCTION SCHEDULING A company manufactures products A, B, and C. Each product is processed in three departments: I, II, and III. The total available labor-hours per week for departments I, II, and III are 900, 1080, and 840, respectively. The time requirements (in hours per unit) and profit per unit for each product are as follows:

	Product A	Product B	Product C
Dept. I	2	1	2
Dept. II	3	1	2
Dept. III	2	2	1
Profit ($)	18	12	15

How many units of each product should the company produce in order to maximize its profit? What is the largest profit the company can realize? Are there any resources left over?

35. ADVERTISING—TELEVISION COMMERCIALS As part of a campaign to promote its annual clearance sale, Excelsior Company decided to buy television-advertising time on Station KAOS.

Excelsior's television-advertising budget is $102,000. Morning time costs $3000/minute, afternoon time costs $1000/minute, and evening (prime) time costs $12,000/minute. Because of previous commitments, KAOS cannot offer Excelsior more than 6 min of prime time or more than a total of 25 min of advertising time over the 2 wk in which the commercials are to be run. KAOS estimates that morning commercials are seen by 200,000 people, afternoon commercials are seen by 100,000 people, and evening commercials are seen by 600,000 people. How much morning, afternoon, and evening advertising time should Excelsior buy to maximize exposure of its commercials?

36. INVESTMENTS—ASSET ALLOCATION Sharon has a total of $200,000 to invest in three types of mutual funds: growth, balanced, and income funds. Growth funds have a rate of return of 12%/year, balanced funds have a rate of return of 10%/year, and income funds have a return of 6%/year. The growth, balanced, and income mutual funds are assigned risk factors of 0.1, 0.06, and 0.02, respectively. Sharon has decided that at least 50% of her total portfolio is to be in income funds and at least 25% of it in balanced funds. She has also decided that the average risk factor for her investment should not exceed 0.05. How much should Sharon invest in each type of fund in order to realize a maximum return on her investment? What is the maximum return?
Hint: The average risk factor for the investment is given by $0.1x + 0.06y + 0.02z \leq 0.05(x + y + z)$.

37. MANUFACTURING—PRODUCTION CONTROL Custom Office Furniture is introducing a new line of executive desks made from a specially selected grade of walnut. Initially, three models—A, B, and C—are to be marketed. Each model A desk requires $1\frac{1}{4}$ hr for fabrication, 1 hr for assembly, and 1 hr for finishing; each model B desk requires $1\frac{1}{2}$ hr for fabrication, 1 hr for assembly, and 1 hr for finishing; each model C desk requires $1\frac{1}{2}$ hr, $\frac{3}{4}$ hr, and $\frac{1}{2}$ hr for fabrication, assembly, and finishing, respectively. The profit on each model A desk is $26, the profit on each model B desk is $28, and the profit on each model C desk is $24. The total time available in the fabrication department, the assembly department, and the finishing department in the first month of production is 310 hr, 205 hr, and 190 hr, respectively. To maximize Custom's profit, how many desks of each model should be made in the month? What is the largest profit the company can realize? Are there any resources left over?

38. MANUFACTURING—PREFABRICATED HOUSING PRODUCTION Boise Lumber has decided to enter the lucrative prefabricated housing business. Initially, it plans to offer three models: standard, deluxe, and luxury. Each house is prefabricated and partially assembled in the factory, and the final assembly is completed on site. The dollar amount of building material required, the amount of labor required in the factory for prefabrication and partial assembly, the amount of on-site labor required, and the profit per unit are as follows:

	Standard Model	Deluxe Model	Luxury Model
Material ($)	6,000	8,000	10,000
Factory Labor (hr)	240	220	200
On-Site Labor ($)	180	210	300
Profit	3,400	4,000	5,000

For the first year's production, a sum of $8,200,000 is budgeted for the building material; the number of labor-hours available for work in the factory (for prefabrication and partial assembly) is not to exceed 218,000 hr; and the amount of labor for on-site work is to be less than or equal to 237,000 labor-hours. Determine how many houses of each type Boise should produce (market research has confirmed that there should be no problems with sales) to maximize its profit from this new venture.

39. **MANUFACTURING—COLD FORMULA PRODUCTION** Bayer Pharmaceutical produces three kinds of cold formulas: I, II, and III. It takes 2.5 hr to produce 1000 bottles of formula I, 3 hr to produce 1000 bottles of formula II, and 4 hr to produce 1000 bottles of formula III. The profits for each 1000 bottles of formula I, formula II, and formula III are $180, $200, and $300, respectively. Suppose, for a certain production run, there are enough ingredients on hand to make at most 9000 bottles of formula I, 12,000 bottles of formula II, and 6000 bottles of formula III. Furthermore, suppose the time for the production run is limited to a maximum of 70 hr. How many bottles of each formula should be produced in this production run so that the profit is maximized? What is the maximum profit realizable by the company? Are there any resources left over?

40. **PRODUCTION—JUICE PRODUCTS** CalJuice Company has decided to introduce three fruit juices made from blending two or more concentrates. These juices will be packaged in 2-qt (64 fluid-oz) cartons. One carton of pineapple–orange juice requires 8 oz each of pineapple and orange juice concentrates. One carton of orange–banana juice requires 12 oz of orange juice concentrate and 4 oz of banana pulp concentrate. Finally, one carton of pineapple–orange–banana juice requires 4 oz of pineapple juice concentrate, 8 oz of orange juice concentrate, and 4 oz of banana pulp. The company has decided to allot 16,000 oz of pineapple juice concentrate, 24,000 oz of orange juice concentrate, and 5000 oz of banana pulp concentrate for the initial production run. The company also stipulated that the production of pineapple–orange–banana juice should not exceed 800 cartons. Its profit on one carton of pineapple– orange juice is $1.00; its profit on one carton of orange–banana juice is $.80, and its profit on one carton of pineapple–orange–banana juice is $.90. To realize a maximum profit, how many cartons of each blend should the company produce? What is the largest profit it can realize? Are there any concentrates left over?

41. **INVESTMENTS—ASSET ALLOCATION** A financier plans to invest up to $2 million in three projects. She estimates that project

A will yield a return of 10% on her investment, project B will yield a return of 15% on her investment, and project C will yield a return of 20% on her investment. Because of the risks associated with the investments, she decided to put not more than 20% of her total investment in project C. She also decided that her investments in projects B and C should not exceed 60% of her total investment. Finally, she decided that her investment in project A should be at least 60% of her investments in projects B and C. How much should the financier invest in each project if she wishes to maximize the total returns on her investments? What is the maximum amount she can expect to make from her investments?

42. Consider the linear programming problem

$$\text{Maximize} \quad P = 3x + 2y$$
$$\text{subject to} \quad x - y \leq 3$$
$$x \leq 2$$
$$x \geq 0, y \geq 0$$

a. Sketch the feasible set for the linear programming problem.

b. Show that the linear programming problem is unbounded.

c. Solve the linear programming problem using the simplex method. How does the method break down?

d. Explain why the result in part (c) implies that no solution exists for the linear programming problem.

In Exercises 43–46, determine whether the statement is true or false. If it is true, explain why it is true. If it is false, give an example to show why it is false.

43. If at least one of the coefficients $a_1, a_2, \ldots, a_n$ of the objective function $P = a_1x_1 + a_2x_2 + \cdots + a_nx_n$ is positive, then $(0, 0, \ldots, 0)$ cannot be the optimal solution of the standard (maximization) linear programming problem.

44. Choosing the pivot row by requiring that the ratio associated with that row be the smallest ensures that the iteration will not take us from a feasible point to a nonfeasible point.

45. Choosing the pivot column by requiring that it be the column associated with the most negative entry to the left of the vertical line in the last row of the simplex tableau ensures that the iteration will result in the greatest increase or, at worse, no decrease in the objective function.

46. If, at any stage of an iteration of the simplex method, it is not possible to compute the ratios (division by zero) or the ratios are negative, then we can conclude that the standard linear programming problem may have no solution.

4.1 Solutions to Self-Check Exercises

1. Introducing the slack variables u, v, and w, we obtain the system of linear equations

$$
\begin{aligned}
2x + 3y + z + u &= 10 \\
x + y + 2z + v &= 8 \\
2y + 3z + w &= 6 \\
-2x - 3y - 6z + P &= 0
\end{aligned}
$$

The initial simplex tableau and the successive tableaus resulting from the use of the simplex procedure follow:

	x	y	z	u	v	w	P	Constant	Ratio
	2	3	1	1	0	0	0	10	$\frac{10}{1} = 10$
	1	1	2	0	1	0	0	8	$\frac{8}{2} = 4$
Pivot row →	0	2	③	0	0	1	0	6	$\frac{6}{3} = 2$
	−2	−3	−6	0	0	0	1	0	

$\frac{1}{3}R_3 \longrightarrow$

Pivot column

x	y	z	u	v	w	P	Constant
2	3	1	1	0	0	0	10
1	1	2	0	1	0	0	8
0	$\frac{2}{3}$	①	0	0	$\frac{1}{3}$	0	2
−2	−3	−6	0	0	0	1	0

$\xrightarrow{\begin{array}{c} R_1 - R_3 \\ R_2 - 2R_3 \\ R_4 + 6R_3 \end{array}}$

	x	y	z	u	v	w	P	Constant	Ratio
	2	$\frac{7}{3}$	0	1	0	$-\frac{1}{3}$	0	8	$\frac{8}{2} = 4$
Pivot row →	①	$-\frac{1}{3}$	0	0	1	$-\frac{2}{3}$	0	4	$\frac{4}{1} = 4$
	0	$\frac{2}{3}$	1	0	0	$\frac{1}{3}$	0	2	—
	−2	1	0	0	0	2	1	12	

$\xrightarrow{\begin{array}{c} R_1 - 2R_2 \\ R_4 + 2R_2 \end{array}}$

Pivot column

x	y	z	u	v	w	P	Constant
0	3	0	1	−2	1	0	0
1	$-\frac{1}{3}$	0	0	1	$-\frac{2}{3}$	0	4
0	$\frac{2}{3}$	1	0	0	$\frac{1}{3}$	0	2
0	$\frac{1}{3}$	0	0	2	$\frac{2}{3}$	1	20

All entries in the last row are nonnegative, and the tableau is final. We conclude that $x = 4$, $y = 0$, $z = 2$, and $P = 20$.

2. Let x denote the number of model A grates and y the number of model B grates to be made this week. Then the profit function to be maximized is given by

$$P = 6x + 8y$$

The limitations on the availability of material and labor may be expressed by the linear inequalities

$$
\begin{array}{lll}
20x + 30y \le 7200 & \text{or} & 2x + 3y \le 720 \\
20x + 15y \le 6000 & \text{or} & 4x + 3y \le 1200
\end{array}
$$

Finally, the condition that no more than 150 units of model A grates be made each week may be expressed by the linear inequality

$$x \le 150$$

Thus, we are led to the following linear programming problem:

$$
\begin{aligned}
\text{Maximize} \quad & P = 6x + 8y \\
\text{subject to} \quad & 2x + 3y \le 720 \\
& 4x + 3y \le 1200 \\
& x \le 150 \\
& x \ge 0, y \ge 0
\end{aligned}
$$

To solve this problem, we introduce slack variables u, v, and w and use the simplex method, obtaining the following sequence of simplex tableaus:

	x	y	u	v	w	P	Constant	Ratio
Pivot row →	2	③	1	0	0	0	720	$\frac{720}{3} = 240$
	4	3	0	1	0	0	1200	$\frac{1200}{3} = 400$
	1	0	0	0	1	0	150	—
	−6	−8	0	0	0	1	0	

Pivot column

$\xrightarrow{\frac{1}{3}R_1}$

x	y	u	v	w	P	Constant
$\frac{2}{3}$	①	$\frac{1}{3}$	0	0	0	240
4	3	0	1	0	0	1200
1	0	0	0	1	0	150
−6	−8	0	0	0	1	0

$\xrightarrow{\begin{array}{c} R_2 - 3R_1 \\ R_4 + 8R_1 \end{array}}$

Pivot row →

	x	y	u	v	w	P	Constant	Ratio
	$\frac{2}{3}$	1	$\frac{1}{3}$	0	0	0	240	$\frac{240}{2/3} = 360$
	2	0	−1	1	0	0	480	$\frac{480}{2} = 240$
	①	0	0	0	1	0	150	$\frac{150}{1} = 150$
	$-\frac{2}{3}$	0	$\frac{8}{3}$	0	0	1	1920	

Pivot column

$\xrightarrow{\begin{array}{c} R_1 - \frac{2}{3}R_3 \\ R_2 - 2R_3 \\ R_4 + \frac{2}{3}R_3 \end{array}}$

x	y	u	v	w	P	Constant
0	1	$\frac{1}{3}$	0	$-\frac{2}{3}$	0	140
0	0	−1	1	−2	0	180
1	0	0	0	1	0	150
0	0	$\frac{8}{3}$	0	$\frac{2}{3}$	1	2020

The last tableau is final, and we see that $x = 150$, $y = 140$, and $P = 2020$. Therefore, LaCrosse should make 150 model A grates and 140 model B grates this week. The profit will be $2020.

USING TECHNOLOGY

■ The Simplex Method: Solving Maximization Problems

Graphing Utility

A graphing utility can be used to solve a linear programming problem by the simplex method as illustrated in Example 1.

EXAMPLE 1   (Refer to Example 5, Section 4.1.) The problem reduces to the following linear programming problem:

$$\text{Maximize} \quad P = 6x + 5y + 4z$$
$$\text{subject to} \quad 2x + y + z \le 180$$
$$x + 3y + 2z \le 300$$
$$2x + y + 2z \le 240$$
$$x \ge 0, y \ge 0, z \ge 0$$

With u, v, and w as slack variables, we are led to the following sequence of simplex tableaus, where the first tableau is entered as the matrix A:

	x	y	z	u	v	w	P	Constant
Pivot row →	②	1	1	1	0	0	0	180
	1	3	2	0	1	0	0	300
	2	1	2	0	0	1	0	240
	−6	−5	−4	0	0	0	1	0

Ratio:
$\frac{180}{2} = 90$
$\frac{300}{1} = 300$
$\frac{240}{2} = 120$

$\xrightarrow{\ *\mathbf{row}(\frac{1}{2}, A, 1) \blacktriangleright B\ }$

↑ Pivot column

x	y	z	u	v	w	P	Constant
①	0.5	0.5	0.5	0	0	0	90
1	3	2	0	1	0	0	300
2	1	2	0	0	1	0	240
−6	−5	−4	0	0	0	1	0

$\xrightarrow{\ *\mathbf{row+}(-1, B, 1, 2) \blacktriangleright C\ }$
$\xrightarrow{\ *\mathbf{row+}(-2, C, 1, 3) \blacktriangleright B\ }$
$\ *\mathbf{row+}(6, B, 1, 4) \ \blacktriangleright C$

	x	y	z	u	v	w	P	Constant
	1	0.5	0.5	0.5	0	0	0	90
Pivot row →	0	②.5	1.5	−0.5	1	0	0	210
	0	0	1	−1	0	1	0	60
	0	−2	−1	3	0	0	1	540

Ratio:
$\frac{90}{0.5} = 180$
$\frac{210}{2.5} = 84$

$\xrightarrow{\ *\mathbf{row}(\frac{1}{2.5}, C, 2) \blacktriangleright B\ }$

↑ Pivot column

x	y	z	u	v	w	P	Constant
1	0.5	0.5	0.5	0	0	0	90
0	①	0.6	−0.2	0.4	0	0	84
0	0	1	−1	0	1	0	60
0	−2	−1	3	0	0	1	540

$$*(\mathbf{row}+(-0.5, B, 2, 1) \blacktriangleright C$$
$$*\mathbf{row}+(2, C, 2, 4) \quad \blacktriangleright B$$

x	y	z	u	v	w	P	Constant
1	0	0.2	0.6	−0.2	0	0	48
0	1	0.6	−0.2	0.4	0	0	84
0	0	1	−1	0	1	0	60
0	0	0.2	2.6	0.8	0	1	708

The final simplex tableau is the same as the one obtained earlier. We see that $x = 48$, $y = 84$, $z = 0$, and $P = 708$. So, Ace Novelty should produce 48 type-A souvenirs, 84 type-B souvenirs, and no type-C souvenirs—resulting in a profit of $708 per day. ∎

Excel

Solver is an Excel add-in that is used to solve linear programming problems. When you start the Excel program, check the *Tools* menu for the *Solver* command. If it is not there, you will need to install it. (Check your manual for installation instructions.)

EXAMPLE 2   Solve the following linear programming problem:

$$\begin{aligned}
\text{Maximize} \quad & P = 6x + 5y + 4z \\
\text{subject to} \quad & 2x + y + z \le 180 \\
& x + 3y + 2z \le 300 \\
& 2x + y + 2z \le 240 \\
& x \ge 0, y \ge 0, z \ge 0
\end{aligned}$$

Solution

1. *Enter the data for the linear programming problem onto a spreadsheet.* Enter the labels shown in column A and the variables with which we are working under Decision Variables in cells B4:B6, as shown in Figure T1. This optional step will help us organize our work.

Note: Boldfaced words/characters enclosed in a box (for example, $\boxed{\textbf{Enter}}$) indicate that an action (click, select, or press) is required. Words/characters printed blue (for example, Chart sub-type:) indicate words/characters that appear on the screen. Words/characters printed in a typewriter font (for example, = (−2/3)*A2+2) indicate words/characters that need to be typed and entered.

(continued)

	A	B	C	D	E	F	G	H	I
1	Maximization Problem								
2						Formulas for indicated cells			
3	Decision Variables					C8: = 6∗C4 + 5∗C5 + 4∗C6			
4			x			C11: = 2∗C4 + C5 + C6			
5			y			C12: = C4 + 3∗C5 + 2∗C6			
6			z			C13: = 2∗C4 + C5 + 2∗C6			
7									
8	Objective Function		0						
9									
10	Constraints								
11			0	<=	180				
12			0	<=	300				
13			0	<=	240				

FIGURE T1
Setting up the spreadsheet for Solver

For the moment, the cells that will contain the values of the variables (C4:C6) are left blank. In C8 type the formula for the objective function: =6*C4+5*C5+4*C6. In C11 type the formula for the left-hand side of the first constraint: =2*C4+C5+C6. In C12 type the formula for the left-hand side of the second constraint: =C4+3*C5+2*C6. In C13 type the formula for the left-hand side of the third constraint: =2*C4+C5+2*C6. Zeros will then appear in cell B8 and cells C11:C13. In cells D11:D13, type <= to indicate that each constraint is of the form ≤. Finally, in cells E11:E13, type the right-hand value of each constraint—in this case, 180, 300, and 240, respectively. Note that we need not enter the nonnegativity constraints, $x \geq 0$, $y \geq 0$, and $z \geq 0$. The resulting spreadsheet is shown in Figure T1, where the formulas that were entered for the objective function and the constraints are shown in the comment box.

2. *Use Solver to solve the problem.* Click **Tools** on the menu bar and then click **Solver** . The Solver Parameters dialog box will appear.

   a. The pointer will be in the Set Target Cell: box (refer to Figure T2). Highlight the cell on your spreadsheet containing the formula for the objective function—in this case, C8.

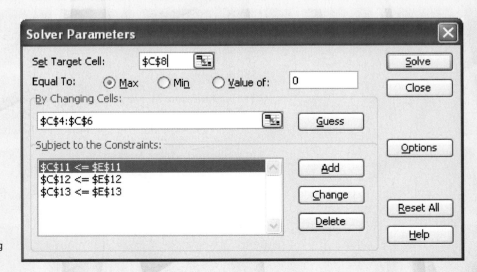

FIGURE T2
The completed Solver Parameters dialog box

Then, next to Equal To: select $\boxed{\text{Max}}$. Select the $\boxed{\textbf{By Changing Cells:}}$ box and highlight the cells in your spreadsheet that will contain the values of the variables—in this case, C4:C6. Select the $\boxed{\textbf{Subject to the Constraints:}}$ box and then click $\boxed{\textbf{Add}}$. The Add Constraint dialog box will appear (Figure T3).

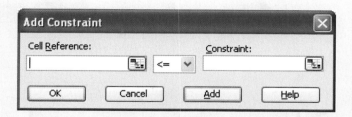

FIGURE T3
The Add Constraint dialog box

b. The pointer will appear in the Cell Reference: box. Highlight the cells on your spreadsheet that contain the formula for the left-hand side of the first constraint—in this case, C11. Next, select the symbol for the appropriate constraint—in this case, $\boxed{\text{<=}}$. Select the $\boxed{\textbf{Constraint:}}$ box and highlight the value of the right-hand side of the first constraint on your spreadsheet—in this case, 180. Click $\boxed{\textbf{Add}}$ and then follow the same procedure to enter the second and third constraint. Click $\boxed{\textbf{OK}}$. The resulting Solver Parameters dialog box shown in Figure T2 will appear.

c. In the Solver Parameters dialog box, click $\boxed{\textbf{Options}}$ (see Figure T2). In the Solver Options dialog box that appears, select $\boxed{\textbf{Assume Linear Model}}$ and $\boxed{\textbf{Assume Non-Negative}}$ constraints (Figure T4). Click $\boxed{\textbf{OK}}$.

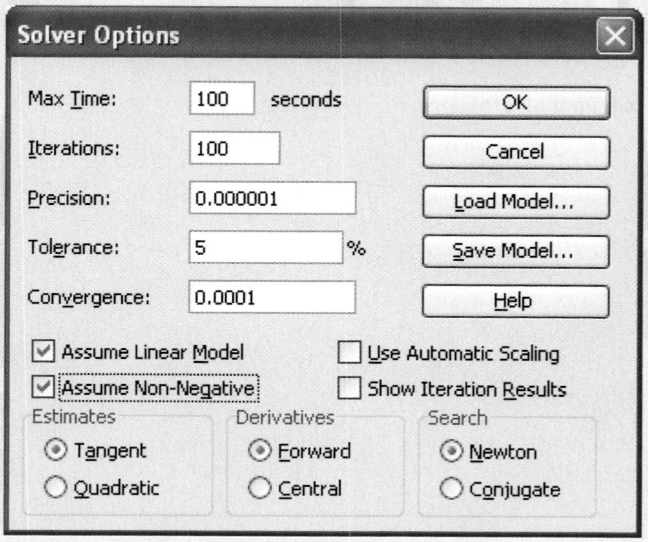

FIGURE T4
The Solver Options dialog box

(*continued*)

d. In the Solver Parameters dialog box that appears (see Figure T2), click Solve . A Solver Results dialog box will then appear and at the same time the answers will appear on your spreadsheet (Figure T5).

	A	B	C	D	E
1	Maximization Problem				
2					
3	Decision Variables				
4		x	48		
5		y	84		
6		z	0		
7					
8	Objective Function		708		
9					
10	Constraints				
11			180	<=	180
12			300	<=	300
13			240	<=	240

FIGURE T5
Completed spreadsheet after using Solver

3. *Read off your answers.* From the spreadsheet, we see that the objective function attains a maximum value of 708 (cell C8) when $x = 48$, $y = 84$, and $z = 0$ (cells C4:C6). ∎

TECHNOLOGY EXERCISES

Solve the linear programming problems.

1. Maximize $P = 2x + 3y + 4z + 2w$
  subject to $x + 2y + 3z + 2w \leq 6$
  $2x + 4y + z - w \leq 4$
  $3x + 2y - 2z + 3w \leq 12$
  $x \geq 0, y \geq 0, z \geq 0, w \geq 0$

2. Maximize $P = 3x + 2y + 2z + w$
  subject to $2x + y - z + 2w \leq 8$
  $2x - y + 2z + 3w \leq 20$
  $x + y + z + 2w \leq 8$
  $4x - 2y + z + 3w \leq 24$
  $x \geq 0, y \geq 0, z \geq 0, w \geq 0$

3. Maximize $P = x + y + 2z + 3w$
  subject to $3x + 6y + 4z + 2w \leq 12$
  $x + 4y + 8z + 4w \leq 16$
  $2x + y + 4z + w \leq 10$
  $x \geq 0, y \geq 0, z \geq 0, w \geq 0$

4. Maximize $P = 2x + 4y + 3z + 5w$
  subject to $x - 2y + 3z + 4w \leq 8$
  $2x + 2y + 4z + 6w \leq 12$
  $3x + 2y + z + 5w \leq 10$
  $2x + 8y - 2z + 6w \leq 24$
  $x \geq 0, y \geq 0, z \geq 0, w \geq 0$

4.2 The Simplex Method: Standard Minimization Problems

■ Minimization with ≤ Constraints

In the last section, we developed a procedure, called the simplex method, for solving standard linear programming problems. Recall that a standard maximization problem satisfies three conditions:

1. The objective function is to be maximized.
2. All the variables involved are nonnegative.
3. Each linear constraint may be written so that the expression involving the variables is less than or equal to a nonnegative constant.

In this section we see how the simplex method may be used to solve certain classes of problems that are not necessarily standard maximization problems. In particular, we see how a modified procedure may be used to solve problems involving the minimization of objective functions.

We begin by considering the class of linear programming problems that calls for the minimization of objective functions but otherwise satisfies Conditions 2 and 3 for standard maximization problems. The method used to solve these problems is illustrated in the following example.

EXAMPLE 1

$$\text{Minimize} \quad C = -2x - 3y$$
$$\text{subject to} \quad 5x + 4y \le 32$$
$$x + 2y \le 10$$
$$x \ge 0, y \ge 0$$

Solution This problem involves the minimization of the objective function and is accordingly *not* a standard maximization problem. Note, however, that all other conditions for a standard maximization problem hold true. To solve a problem of this type, we observe that minimizing the objective function C is equivalent to maximizing the objective function $P = -C$. Thus, the solution to this problem may be found by solving the following associated standard maximization problem: Maximize $P = 2x + 3y$ subject to the given constraints. Using the simplex method with u and v as slack variables, we obtain the following sequence of simplex tableaus:

	x	y	u	v	P	Constant	
	5	4	1	0	0	32	$\frac{32}{4} = 8$
Pivot row →	1	②	0	1	0	10	$\frac{10}{2} = 5$
	−2	−3	0	0	1	0	

Ratio

↑ Pivot column

	x	y	u	v	P	Constant
	5	4	1	0	0	32
$\xrightarrow{\frac{1}{2}R_2}$	$\frac{1}{2}$	①	0	$\frac{1}{2}$	0	5
	−2	−3	0	0	1	0

Pivot row	x	y	u	v	P	Constant		Ratio
$\xrightarrow{}$	③	0	1	-2	0	12		$\frac{12}{3} = 4$
$\begin{array}{c} R_1 - 4R_2 \\ \overrightarrow{R_3 + 3R_2} \end{array}$	$\frac{1}{2}$	1	0	$\frac{1}{2}$	0	5		$\frac{5}{1/2} = 10$
	$-\frac{1}{2}$	0	0	$\frac{3}{2}$	1	15		

Pivot column

	x	y	u	v	P	Constant
$\xrightarrow{\frac{1}{3}R_1}$	①	0	$\frac{1}{3}$	$-\frac{2}{3}$	0	4
	$\frac{1}{2}$	1	0	$\frac{1}{2}$	0	5
	$-\frac{1}{2}$	0	0	$\frac{3}{2}$	1	15

	x	y	u	v	P	Constant
$\begin{array}{c} R_2 - \frac{1}{2}R_1 \\ \overrightarrow{R_3 + \frac{1}{2}R_1} \end{array}$	1	0	$\frac{1}{3}$	$-\frac{2}{3}$	0	4
	0	1	$-\frac{1}{6}$	$\frac{5}{6}$	0	3
	0	0	$\frac{1}{6}$	$\frac{7}{6}$	1	17

The last tableau is in final form. The solution to the standard maximization problem associated with the given linear programming problem is $x = 4$, $y = 3$, and $P = 17$, so the required solution is given by $x = 4$, $y = 3$, and $C = -17$. You may verify that the solution is correct by using the method of corners. ∎

■ The Dual Problem

Another special class of linear programming problems we encounter in practical applications is characterized by the following conditions:

1. The objective function is to be *minimized*.
2. All the variables involved are nonnegative.
3. Each linear constraint may be written so that the expression involving the variables is *greater than or equal to* a constant.

Such problems are called **standard minimization problems.**

   A convenient method for solving this type of problem is based on the following observation. Each maximization linear programming problem is associated with a minimization problem, and vice versa. For the purpose of identification, the given problem is called the **primal problem;** the problem related to it is called the **dual problem.** The following example illustrates the technique for constructing the dual of a given linear programming problem.

EXAMPLE 2   Write the dual problem associated with the following problem:

Minimize the objective function $C = 6x + 8y$

subject to $40x + 10y \geq 2400$

$10x + 15y \geq 2100$

$5x + 15y \geq 1500$

$x \geq 0, y \geq \quad 0$

$\left.\begin{array}{c} \\ \\ \\ \\ \\ \end{array}\right\}$ Primal problem

Solution   We first write down the following tableau for the given primal problem:

x	y	Constant
40	10	2400
10	15	2100
5	15	1500
6	8	

Next, we interchange the columns and rows of the foregoing tableau and head the three columns of the resulting array with the three variables u, v, and w, obtaining the tableau

u	v	w	Constant
40	10	5	6
10	15	15	8
2400	2100	1500	

Interpreting the last tableau as if it were part of the initial simplex tableau for a standard maximization problem, with the exception that the signs of the coefficients pertaining to the objective function are not reversed, we construct the required dual problem as follows:

Maximize the objective function $P = 2400u + 2100v + 1500w$
subject to $40u + 10v + 5w \le 6$
              $10u + 15v + 15w \le 8$
where $u \ge 0$, $v \ge 0$, and $w \ge 0$.

> Dual problem

The connection between the solution of the primal problem and that of the dual problem is given by the following theorem. The theorem, attributed to John von Neumann (1903–1957), is stated without proof.

THEOREM 1

The Fundamental Theorem of Duality

A primal problem has a solution if and only if the corresponding dual problem has a solution. Furthermore, if a solution exists, then

a. The objective functions of both the primal and the dual problem attain the same optimal value.

b. The optimal solution to the primal problem appears under the slack variables in the last row of the final simplex tableau associated with the dual problem.

Armed with this theorem, we will solve the problem posed in Example 2.

EXAMPLE 3   Complete the solution to the problem posed in Example 2.

Solution   Observe that the dual problem associated with the given (primal) problem is a standard maximization problem. The solution may thus be found using the simplex algorithm. Introducing the slack variables x and y, we obtain the system of linear equations

$$
\begin{aligned}
40u + 10v + 5w + x &= 6 \\
10u + 15v + 15w + y &= 8 \\
-2400u - 2100v - 1500w + P &= 0
\end{aligned}
$$

Continuing with the simplex algorithm, we obtain the sequence of simplex tableaus

	u	v	w	x	y	P	Constant
Pivot row $\rightarrow$	④⓪	10	5	1	0	0	6
	10	15	15	0	1	0	8
	-2400	-2100	-1500	0	0	1	0

Ratio: $\frac{6}{40} = \frac{3}{20}$; $\frac{8}{10} = \frac{4}{5}$

↑ Pivot column

	u	v	w	x	y	P	Constant
$\frac{1}{40}R_1 \rightarrow$	①	$\frac{1}{4}$	$\frac{1}{8}$	$\frac{1}{40}$	0	0	$\frac{3}{20}$
	10	15	15	0	1	0	8
	-2400	-2100	-1500	0	0	1	0

	u	v	w	x	y	P	Constant
$R_2 - 10R_1$	1	$\frac{1}{4}$	$\frac{1}{8}$	$\frac{1}{40}$	0	0	$\frac{3}{20}$
$R_3 + 2400R_1 \rightarrow$	0	$\frac{25}{2}$	$\frac{55}{4}$	$-\frac{1}{4}$	1	0	$\frac{13}{2}$
	0	-1500	-1200	60	0	1	360

Ratio: $\frac{3/20}{1/4} = \frac{3}{5}$; $\frac{13/2}{25/2} = \frac{13}{25}$

	u	v	w	x	y	P	Constant
$\frac{2}{25}R_2 \rightarrow$	1	$\frac{1}{4}$	$\frac{1}{8}$	$\frac{1}{40}$	0	0	$\frac{3}{20}$
	0	①	$\frac{11}{10}$	$-\frac{1}{50}$	$\frac{2}{25}$	0	$\frac{13}{25}$
	0	-1500	-1200	60	0	1	360

	u	v	w	x	y	P	Constant
$R_1 - \frac{1}{4}R_2$	1	0	$-\frac{3}{20}$	$\frac{3}{100}$	$-\frac{1}{50}$	0	$\frac{1}{50}$
$R_3 + 1500R_2 \rightarrow$	0	1	$\frac{11}{10}$	$-\frac{1}{50}$	$\frac{2}{25}$	0	$\frac{13}{25}$
	0	0	450	30	120	1	1140

Solution for the primal problem (under x and y columns)

The last tableau is final. The fundamental theorem of duality tells us that the solution to the primal problem is $x = 30$ and $y = 120$ with a minimum value for C of 1140. Observe that the solution to the dual (maximization) problem may be read from the simplex tableau in the usual manner: $u = \frac{1}{50}$, $v = \frac{13}{25}$, $w = 0$, and $P = 1140$. Note that the maximum value of P is equal to the minimum value of

C as guaranteed by the fundamental theorem of duality. The solution to the primal problem agrees with the solution of the same problem solved in Section 3.3, Example 2, using the method of corners. ∎

Notes

1. We leave it to you to demonstrate that the dual of a standard minimization problem is always a standard maximization problem provided that the coefficients of the objective function in the primal problem are all nonnegative. Such problems can always be solved by applying the simplex method to solve the dual problem.
2. Standard minimization problems in which the coefficients of the objective function are not all nonnegative do not necessarily have a dual problem that is a standard maximization problem. ∎

EXAMPLE 4

$$\text{Minimize} \quad C = 3x + 2y$$
$$\text{subject to} \quad 8x + y \geq 80$$
$$8x + 5y \geq 240$$
$$x + 5y \geq 100$$
$$x \geq 0, y \geq 0$$

Solution   We begin by writing the dual problem associated with the given primal problem. First, we write down the following tableau for the primal problem:

x	y	Constant
8	1	80
8	5	240
1	5	100
3	2	

Next, interchanging the columns and rows of this tableau and heading the three columns of the resulting array with the three variables, u, v, and w, we obtain the tableau

u	v	w	Constant
8	8	1	3
1	5	5	2
80	240	100	

Interpreting the last tableau as if it were part of the initial simplex tableau for a standard maximization problem, with the exception that the signs of the coefficients pertaining to the objective function are not reversed, we construct the dual problem as follows: Maximize the objective function $P = 80u + 240v + 100w$ subject to the constraints

$$8u + 8v + w \leq 3$$
$$u + 5v + 5w \leq 2$$

where $u \geq 0$, $v \geq 0$, and $w \geq 0$. Having constructed the dual problem, which is a standard maximization problem, we now solve it using the simplex method. Introducing the slack variables x and y, we obtain the system of linear equations

$$
\begin{aligned}
8u + 8v + w + x &= 3 \\
u + 5v + 5w + y &= 2 \\
-80u - 240v - 100w + P &= 0
\end{aligned}
$$

Continuing with the simplex algorithm, we obtain the sequence of simplex tableaus

	u	v	w	x	y	P	Constant		Ratio
Pivot row $\rightarrow$	8	⑧	1	1	0	0	3		$\frac{3}{8}$
	1	5	5	0	1	0	2		$\frac{2}{5}$
	-80	-240	-100	0	0	1	0		

↑
Pivot column

	u	v	w	x	y	P	Constant
$\frac{1}{8}R_1$	1	①	$\frac{1}{8}$	$\frac{1}{8}$	0	0	$\frac{3}{8}$
	1	5	5	0	1	0	2
	-80	-240	-100	0	0	1	0

	u	v	w	x	y	P	Constant		Ratio
	1	1	$\frac{1}{8}$	$\frac{1}{8}$	0	0	$\frac{3}{8}$		3
$\begin{aligned} R_2 - 5R_1 \\ R_3 + 240R_1 \end{aligned}$	-4	0	$\frac{35}{8}$	$-\frac{5}{8}$	1	0	$\frac{1}{8}$		$\frac{1}{35}$
Pivot row	160	0	-70	30	0	1	90		

↑
Pivot column

	u	v	w	x	y	P	Constant
$\frac{8}{35}R_2$	1	1	$\frac{1}{8}$	$\frac{1}{8}$	0	0	$\frac{3}{8}$
	$-\frac{32}{35}$	0	①	$-\frac{1}{7}$	$\frac{8}{35}$	0	$\frac{1}{35}$
	160	0	-70	30	0	1	90

	u	v	w	x	y	P	Constant
$R_1 - \frac{1}{8}R_2$	$\frac{39}{35}$	1	0	$\frac{1}{7}$	$-\frac{1}{35}$	0	$\frac{13}{35}$
$R_3 + 70R_2$	$-\frac{32}{35}$	0	1	$-\frac{1}{7}$	$\frac{8}{35}$	0	$\frac{1}{35}$
	96	0	0	20	16	1	92

Solution for the
primal problem

The last tableau is final. The fundamental theorem of duality tells us that the solution to the primal problem is $x = 20$ and $y = 16$ with a minimum value for C of 92.

Our last example illustrates how the warehouse problem posed in Section 3.2 may be solved by duality.

 APPLIED EXAMPLE 5   A Warehouse Problem   Complete the solution to the warehouse problem given in Section 3.2, Example 4 (page 179).

Minimize

$$C = 20x_1 + 8x_2 + 10x_3 + 12x_4 + 22x_5 + 18x_6 \tag{13}$$

subject to

$$
\begin{aligned}
x_1 + x_2 + x_3 \qquad\qquad\qquad &\le 400 \\
x_4 + x_5 + x_6 &\le 600 \\
x_1 \qquad\quad + x_4 \qquad\qquad &\ge 200 \\
x_2 \qquad\quad + x_5 \qquad &\ge 300 \\
x_3 \qquad\quad + x_6 &\ge 400 \\
x_1 \ge 0,\; x_2 \ge 0, \ldots, x_6 &\ge \quad 0
\end{aligned}
\tag{14}
$$

Solution   Upon multiplying each of the first two inequalities of (14) by -1, we obtain the following equivalent system of constraints in which each of the expressions involving the variables is greater than or equal to a constant:

$$
\begin{aligned}
-x_1 - x_2 - x_3 \qquad\qquad\qquad &\ge -400 \\
- x_4 - x_5 - x_6 &\ge -600 \\
x_1 \qquad\quad + x_4 \qquad\qquad &\ge \quad 200 \\
x_2 \qquad\quad + x_5 \qquad &\ge \quad 300 \\
x_3 \qquad\quad + x_6 &\ge \quad 400 \\
x_1 \ge 0,\; x_2 \ge 0, \ldots, x_6 &\ge \quad 0
\end{aligned}
$$

The problem may now be solved by duality. First, we write the array of numbers:

x_1	x_2	x_3	x_4	x_5	x_6	Constant
-1	-1	-1	0	0	0	-400
0	0	0	-1	-1	-1	-600
1	0	0	1	0	0	200
0	1	0	0	1	0	300
0	0	1	0	0	1	400
20	8	10	12	22	18	

Interchanging the rows and columns of this array of numbers and heading the five columns of the resulting array of numbers by the variables u_1, u_2, u_3, u_4, and u_5 leads to

u_1	u_2	u_3	u_4	u_5	Constant
-1	0	1	0	0	20
-1	0	0	1	0	8
-1	0	0	0	1	10
0	-1	1	0	0	12
0	-1	0	1	0	22
0	-1	0	0	1	18
-400	-600	200	300	400	

from which we construct the associated dual problem: Maximize $P = -400u_1 - 600u_2 + 200u_3 + 300u_4 + 400u_5$ subject to

$$
\begin{aligned}
-u_1 \quad\quad + u_3 \quad\quad\quad\quad &\le 20 \\
-u_1 \quad\quad\quad\quad + u_4 \quad\quad &\le 8 \\
-u_1 \quad\quad\quad\quad\quad\quad + u_5 &\le 10 \\
-u_2 + u_3 \quad\quad\quad\quad &\le 12 \\
-u_2 \quad\quad + u_4 \quad\quad &\le 22 \\
-u_2 \quad\quad\quad\quad + u_5 &\le 18 \\
u_1 \ge 0,\ u_2 \ge 0,\ \ldots,\ u_5 &\ge 0
\end{aligned}
$$

Solving the standard maximization problem by the simplex algorithm, we obtain the following sequence of tableaus ($x_1, x_2, \ldots, x_6$ are slack variables):

u_1	u_2	u_3	u_4	u_5	x_1	x_2	x_3	x_4	x_5	x_6	P	Constant	Ratio
-1	0	1	0	0	1	0	0	0	0	0	0	20	—
-1	0	0	1	0	0	1	0	0	0	0	0	8	—
-1	0	0	0	①	0	0	1	0	0	0	0	10	10
0	-1	1	0	0	0	0	0	1	0	0	0	12	—
0	-1	0	1	0	0	0	0	0	1	0	0	22	—
0	-1	0	0	1	0	0	0	0	0	1	0	18	18
400	600	-200	-300	-400	0	0	0	0	0	0	1	0	

(Pivot row → third row; Pivot column → u_5)

u_1	u_2	u_3	u_4	u_5	x_1	x_2	x_3	x_4	x_5	x_6	P	Constant	Ratio
-1	0	1	0	0	1	0	0	0	0	0	0	20	—
-1	0	0	①	0	0	1	0	0	0	0	0	8	8
-1	0	0	0	1	0	0	1	0	0	0	0	10	—
0	-1	1	0	0	0	0	0	1	0	0	0	12	—
0	-1	0	1	0	0	0	0	0	1	0	0	22	22
1	-1	0	0	0	0	0	-1	0	0	1	0	8	—
0	600	-200	-300	0	0	0	400	0	0	0	1	4000	

($R_6 - R_3$, $R_7 + 400R_3$; Pivot row → second row; Pivot column → u_4)

u_1	u_2	u_3	u_4	u_5	x_1	x_2	x_3	x_4	x_5	x_6	P	Constant	Ratio
-1	0	1	0	0	1	0	0	0	0	0	0	20	—
-1	0	0	1	0	0	1	0	0	0	0	0	8	—
-1	0	0	0	1	0	0	1	0	0	0	0	10	—
0	-1	1	0	0	0	0	0	1	0	0	0	12	—
1	-1	0	0	0	0	-1	0	0	1	0	0	14	14
①	-1	0	0	0	0	0	-1	0	0	1	0	8	8
-300	600	-200	0	0	0	300	400	0	0	0	1	6400	

($R_5 - R_2$, $R_7 + 300R_2$; Pivot row → sixth row; Pivot column → u_1)

	u_1	u_2	u_3	u_4	u_5	x_1	x_2	x_3	x_4	x_5	x_6	P	Constant	Ratio
	0	−1	1	0	0	1	0	−1	0	0	1	0	28	28
$R_1 + R_6$	0	−1	0	1	0	0	1	−1	0	0	1	0	16	—
$R_2 + R_6$	0	−1	0	0	1	0	0	0	0	0	1	0	18	—
$R_3 + R_6$	0	−1	①	0	0	0	0	0	1	0	0	0	12	12
$R_5 − R_6$	0	0	0	0	0	0	−1	1	0	1	−1	0	6	—
$R_7 + 300R_6$ Pivot row	1	−1	0	0	0	0	0	−1	0	0	1	0	8	—
	0	300	−200	0	0	0	300	100	0	0	300	1	8800	

Pivot column (↑ under u_3)

	u_1	u_2	u_3	u_4	u_5	x_1	x_2	x_3	x_4	x_5	x_6	P	Constant
	0	0	0	0	0	1	0	−1	−1	0	1	0	16
	0	−1	0	1	0	0	1	−1	0	0	1	0	16
$R_1 − R_4$	0	−1	0	0	1	0	0	0	0	0	1	0	18
$R_7 + 200R_4$	0	−1	1	0	0	0	0	0	1	0	0	0	12
	0	0	0	0	0	0	−1	1	0	1	−1	0	6
	1	−1	0	0	0	0	0	−1	0	0	1	0	8
	0	100	0	0	0	0	300	100	200	0	300	1	11,200

The last tableau is final, and we find that

$$x_1 = 0 \qquad x_2 = 300 \qquad x_3 = 100 \qquad x_4 = 200$$
$$x_5 = 0 \qquad x_6 = 300 \qquad P = 11,200$$

Thus, to minimize shipping costs, Acrosonic should ship 300 loudspeaker systems from plant I to warehouse B, 100 systems from plant I to warehouse C, 200 systems from plant II to warehouse A, and 300 systems from plant II to warehouse C, at a total cost of $11,200. ∎

4.2 Self-Check Exercises

1. Write the dual problem associated with the following problem:

$$
\begin{aligned}
\text{Minimize} \quad & C = 2x + 5y \\
\text{subject to} \quad & 4x + y \geq 40 \\
& 2x + y \geq 30 \\
& x + 3y \geq 30 \\
& x \geq 0, y \geq 0
\end{aligned}
$$

2. Solve the primal problem posed in Exercise 1.

Solutions to Self-Check Exercises 4.2 can be found on page 254.

4.2 Concept Questions

1. Suppose you are given the linear programming problem

$$\text{Minimize} \quad C = -3x - 5y$$
$$\text{subject to} \quad 5x + 2y \leq 30$$
$$x + 3y \leq 21$$
$$x \geq 0, y \geq 0$$

Give the associated standard maximization problem that you would use to solve this linear programming problem, using the simplex method.

2. Give three characteristics of a standard minimization linear programming problem.

3. What is the primal problem associated with a standard minimization linear programming problem? The dual problem?

4. **a.** What does the fundamental theorem of duality tell us about the existence of a solution to a primal problem?
   b. How are the optimal values of the primal and dual problems related?
   c. Given the final simplex tableau associated with a dual problem, how would you determine the optimal solution to the associated primal problem?

4.2 Exercises

In Exercises 1–6, use the technique developed in this section to solve the minimization problem.

1. Minimize $\quad C = -2x + y$
   subject to $\quad x + 2y \leq 6$
   $3x + 2y \leq 12$
   $x \geq 0, y \geq 0$

2. Minimize $\quad C = -2x - 3y$
   subject to $\quad 3x + 4y \leq 24$
   $7x - 4y \leq 16$
   $x \geq 0, y \geq 0$

3. Minimize $C = -3x - 2y$ subject to the constraints of Exercise 2.

4. Minimize $\quad C = x - 2y + z$
   subject to $\quad x - 2y + 3z \leq 10$
   $2x + y - 2z \leq 15$
   $2x + y + 3z \leq 20$
   $x \geq 0, y \geq 0, z \geq 0$

5. Minimize $\quad C = 2x - 3y - 4z$
   subject to $\quad -x + 2y - z \leq 8$
   $x - 2y + 2z \leq 10$
   $2x + 4y - 3z \leq 12$
   $x \geq 0, y \geq 0, z \geq 0$

6. Minimize $C = -3x - 2y - z$ subject to the constraints of Exercise 5.

In Exercises 7–10, you are given the final simplex tableau for the dual problem. Give the solution to the primal problem and to the associated dual problem.

7. Problem: Minimize $\quad C = 8x + 12y$
   subject to $\quad x + 3y \geq 2$
   $2x + 2y \geq 3$
   $x \geq 0, y \geq 0$

   Final tableau:

u	v	x	y	P	Constant
0	1	$\frac{3}{4}$	$-\frac{1}{4}$	0	3
1	0	$-\frac{1}{2}$	$\frac{1}{2}$	0	2
0	0	$\frac{5}{4}$	$\frac{1}{4}$	1	13

8. Problem: Minimize $\quad C = 3x + 2y$
   subject to $\quad 5x + y \geq 10$
   $2x + 2y \geq 12$
   $x + 4y \geq 12$
   $x \geq 0, y \geq 0$

   Final tableau:

u	v	w	x	y	P	Constant
1	0	$-\frac{3}{4}$	$\frac{1}{4}$	$-\frac{1}{4}$	0	$\frac{1}{4}$
0	1	$\frac{19}{8}$	$-\frac{1}{8}$	$\frac{5}{8}$	0	$\frac{7}{8}$
0	0	9	1	5	1	13

9. Problem: Minimize $C = 10x + 3y + 10z$

subject to $2x + y + 5z \geq 20$

$4x + y + z \geq 30$

$x \geq 0, y \geq 0, z \geq 0$

Final tableau:

u	v	x	y	z	P	Constant
0	1	$\frac{1}{2}$	-1	0	0	2
1	0	$-\frac{1}{2}$	2	0	0	1
0	0	2	-9	1	0	3
0	0	5	10	0	1	80

10. Problem: Minimize $C = 2x + 3y$

subject to $x + 4y \geq 8$

$x + y \geq 5$

$2x + y \geq 7$

$x \geq 0, y \geq 0$

Final tableau:

u	v	w	x	y	P	Constant
0	1	$\frac{7}{3}$	$\frac{4}{3}$	$-\frac{1}{3}$	0	$\frac{5}{3}$
1	0	$-\frac{1}{3}$	$-\frac{1}{3}$	$\frac{1}{3}$	0	$\frac{1}{3}$
0	0	2	4	1	1	11

In Exercises 11–20, construct the dual problem associated with the primal problem. Solve the primal problem.

11. Minimize $C = 2x + 5y$

subject to $x + 2y \geq 4$

$3x + 2y \geq 6$

$x \geq 0, y \geq 0$

12. Minimize $C = 3x + 2y$

subject to $2x + 3y \geq 90$

$3x + 2y \geq 120$

$x \geq 0, y \geq 0$

13. Minimize $C = 6x + 4y$

subject to $6x + y \geq 60$

$2x + y \geq 40$

$x + y \geq 30$

$x \geq 0, y \geq 0$

14. Minimize $C = 10x + y$

subject to $4x + y \geq 16$

$x + 2y \geq 12$

$x \geq 2$

$x \geq 0, y \geq 0$

15. Minimize $C = 200x + 150y + 120z$

subject to $20x + 10y + z \geq 10$

$x + y + 2z \geq 20$

$x \geq 0, y \geq 0, z \geq 0$

16. Minimize $C = 40x + 30y + 11z$

subject to $2x + y + z \geq 8$

$x + y - z \geq 6$

$x \geq 0, y \geq 0, z \geq 0$

17. Minimize $C = 6x + 8y + 4z$

subject to $x + 2y + 2z \geq 10$

$2x + y + z \geq 24$

$x + y + z \geq 16$

$x \geq 0, y \geq 0, z \geq 0$

18. Minimize $C = 12x + 4y + 8z$

subject to $2x + 4y + z \geq 6$

$3x + 2y + 2z \geq 2$

$4x + y + z \geq 2$

$x \geq 0, y \geq 0, z \geq 0$

19. Minimize $C = 30x + 12y + 20z$

subject to $2x + 4y + 3z \geq 6$

$6x + z \geq 2$

$6y + 2z \geq 4$

$x \geq 0, y \geq 0, z \geq 0$

20. Minimize $C = 8x + 6y + 4z$

subject to $2x + 3y + z \geq 6$

$x + 2y - 2z \geq 4$

$x + y + 2z \geq 2$

$x \geq 0, y \geq 0, z \geq 0$

21. TRANSPORTATION Deluxe River Cruises operates a fleet of river vessels. The fleet has two types of vessels: A type-A vessel has 60 deluxe cabins and 160 standard cabins, whereas a type-B vessel has 80 deluxe cabins and 120 standard cabins. Under a charter agreement with Odyssey Travel Agency, Deluxe River Cruises is to provide Odyssey with a minimum of 360 deluxe and 680 standard cabins for their 15-day cruise in May. It costs $44,000 to operate a type-A vessel and $54,000 to operate a type-B vessel for that period. How many of each type vessel should be used in order to keep the operating costs to a minimum? What is the minimum cost?

22. SHIPPING COSTS Acrosonic manufactures a model G loudspeaker system in plants I and II. The output at plant I is at most 800/month, and the output at plant II is at most 600/month. Model G loudspeaker systems are also shipped to the three warehouses—A, B, and C—whose minimum monthly requirements are 500, 400, and 400, respectively. Shipping costs from plant I to warehouse A, warehouse B, and warehouse C are $16, $20, and $22/loudspeaker system, respectively, and shipping costs from plant II to each of these warehouses are $18, $16, and $14, respectively. What shipping schedule will enable Acrosonic to meet the requirements of the warehouses while keeping its shipping costs to a minimum? What is the minimum cost?

23. **ADVERTISING** Everest Deluxe World Travel has decided to advertise in the Sunday editions of two major newspapers in town. These advertisements are directed at three groups of potential customers. Each advertisement in newspaper I is seen by 70,000 group A customers, 40,000 group B customers, and 20,000 group C customers. Each advertisement in newspaper II is seen by 10,000 group A, 20,000 group B, and 40,000 group C customers. Each advertisement in newspaper I costs $1000, and each advertisement in newspaper II costs $800. Everest would like their advertisements to be read by at least 2 million people from group A, 1.4 million people from group B, and 1 million people from group C. How many advertisements should Everest place in each newspaper to achieve its advertising goals at a minimum cost? What is the minimum cost?

24. **SHIPPING COSTS** Steinwelt Piano manufactures uprights and consoles in two plants, plant I and plant II. The output of plant I is at most 300/month, and the output of plant II is at most 250/month. These pianos are shipped to three warehouses that serve as distribution centers for Steinwelt. To fill current and projected future orders, warehouse A requires a minimum of 200 pianos/month, warehouse B requires at least 150 pianos/month, and warehouse C requires at least 200 pianos/month. The shipping cost of each piano from plant I to warehouse A, warehouse B, and warehouse C is $60, $60, and $80, respectively, and the shipping cost of each piano from plant II to warehouse A, warehouse B, and warehouse C is $80, $70, and $50, respectively. What shipping schedule will enable Steinwelt to meet the requirements of the warehouses while keeping the shipping costs to a minimum? What is the minimum cost?

25. **NUTRITION—DIET PLANNING** The owner of the Health JuiceBar wishes to prepare a low-calorie fruit juice with a high vitamin-A and -C content by blending orange juice and pink grapefruit juice. Each glass of the blended juice is to contain at least 1200 International Units (IU) of vitamin A and 200 IU of vitamin C. One ounce of orange juice contains 60 IU of vitamin A, 16 IU of vitamin C, and 14 calories; each ounce of pink grapefruit juice contains 120 IU of vitamin A, 12 IU of vitamin C, and 11 calories. How many ounces of each juice should a glass of the blend contain if it is to meet the minimum vitamin requirements while containing a minimum number of calories?

26. **PRODUCTION CONTROL** An oil company operates two refineries in a certain city. Refinery I has an output of 200, 100, and 100 barrels of low-, medium-, and high-grade oil per day, respectively. Refinery II has an output of 100, 200, and 600 barrels of low-, medium-, and high-grade oil per day, respectively. The company wishes to produce at least 1000, 1400, and 3000 barrels of low-, medium-, and high-grade oil to fill an order. If it costs $200/day to operate refinery I and $300/day to operate refinery II, determine how many days each refinery should be operated to meet the requirements of the order at minimum cost to the company. What is the minimum cost?

In Exercises 27 and 28, determine whether the statement is true or false. If it is true, explain why it is true. If it is false, give an example to show why it is false.

27. If a standard minimization linear programming problem has a unique solution, then so does the corresponding maximization problem with objective function $P = -C$, where $C = a_1x_1 + a_2x_2 + \cdots + a_nx_n$ is the objective function for the minimization problem.

28. The optimal value attained by the objective function of the primal problem may be different from that attained by the objective function of the dual problem.

4.2 Solutions to Self-Check Exercises

1. We first write down the following tableau for the given (primal) problem:

x	y	Constant
4	1	40
2	1	30
1	3	30
2	5	0

Next, we interchange the columns and rows of the tableau and head the three columns of the resulting array with the three variables, u, v, and w, obtaining the tableau

u	v	w	Constant
4	2	1	2
1	1	3	5
40	30	30	0

Interpreting the last tableau as if it were the initial tableau for a standard linear programming problem, with the exception that the signs of the coefficients pertaining to the objective function are not reversed, we construct the required dual problem as follows:

Maximize $P = 40u + 30v + 30w$
subject to $4u + 2v + w \le 2$
$u + v + 3w \le 5$
$u \ge 0, v \ge 0, w \ge 0$

2. We introduce slack variables x and y to obtain the system of linear equations

$$4u + 2v + w + x = 2$$
$$u + v + 3w + y = 5$$
$$-40u - 30v - 30w + P = 0$$

Using the simplex algorithm, we obtain the sequence of simplex tableaus

	u	v	w	x	y	P	Constant	Ratio	
Pivot row →	④	2	1	1	0	0	2	$\frac{2}{4} = \frac{1}{2}$	$\frac{1}{4}R_1$
	1	1	3	0	1	0	5	$\frac{5}{1} = 5$	
	-40	-30	-30	0	0	1	0		

↑ Pivot column

	u	v	w	x	y	P	Constant	
	①	$\frac{1}{2}$	$\frac{1}{4}$	$\frac{1}{4}$	0	0	$\frac{1}{2}$	
	1	1	3	0	1	0	5	$R_2 - R_1$
	-40	-30	-30	0	0	1	0	$R_3 + 40R_1$

	u	v	w	x	y	P	Constant	Ratio	
	1	$\frac{1}{2}$	$\frac{1}{4}$	$\frac{1}{4}$	0	0	$\frac{1}{2}$	$\frac{1/2}{1/4} = 2$	$\frac{4}{11}R_2$
Pivot row →	0	$\frac{1}{2}$	⑪⁄₄	$-\frac{1}{4}$	1	0	$\frac{9}{2}$	$\frac{9/2}{11/4} = \frac{18}{11}$	
	0	-10	-20	10	0	1	20		

↑ Pivot column

	u	v	w	x	y	P	Constant	
	1	$\frac{1}{2}$	$\frac{1}{4}$	$\frac{1}{4}$	0	0	$\frac{1}{2}$	$R_1 - \frac{1}{4}R_2$
	0	$\frac{2}{11}$	①	$-\frac{1}{11}$	$\frac{4}{11}$	0	$\frac{18}{11}$	$R_3 + 20R_2$
	0	-10	-20	10	0	1	20	

	u	v	w	x	y	P	Constant	Ratio	
Pivot row →	1	⑤⁄₁₁	0	$\frac{3}{11}$	$-\frac{1}{11}$	0	$\frac{1}{11}$	$\frac{1/11}{5/11} = \frac{1}{5}$	$\frac{11}{5}R_1$
	0	$\frac{2}{11}$	1	$-\frac{1}{11}$	$\frac{4}{11}$	0	$\frac{18}{11}$	$\frac{18/11}{2/11} = 9$	
	0	$-\frac{70}{11}$	0	$\frac{90}{11}$	$\frac{80}{11}$	1	$\frac{580}{11}$		

↑ Pivot column

	u	v	w	x	y	P	Constant	
	$\frac{11}{5}$	①	0	$\frac{3}{5}$	$-\frac{1}{5}$	0	$\frac{1}{5}$	$R_2 - \frac{2}{11}R_1$
	0	$\frac{2}{11}$	1	$-\frac{1}{11}$	$\frac{4}{11}$	0	$\frac{18}{11}$	$R_3 + \frac{70}{11}R_1$
	0	$-\frac{70}{11}$	0	$\frac{90}{11}$	$\frac{80}{11}$	1	$\frac{580}{11}$	

u	v	w	x	y	P	Constant
$\frac{11}{5}$	1	0	$\frac{3}{5}$	$-\frac{1}{5}$	0	$\frac{1}{5}$
$-\frac{2}{5}$	0	1	$-\frac{1}{5}$	$\frac{2}{5}$	0	$\frac{8}{5}$
14	0	0	12	6	1	54

Solution for the primal problem

The last tableau is final, and the solution to the primal problem is $x = 12$ and $y = 6$ with a minimum value for C of 54.

USING TECHNOLOGY

■ The Simplex Method: Solving Minimization Problems

Graphing Utility

A graphing utility can be used to solve minimization problems, using the simplex method.

EXAMPLE 1

Minimize $C = 2x + 3y$
subject to $8x + y \ge 80$
$3x + 2y \ge 100$
$x + 4y \ge 80$
$x \ge 0, y \ge 0$

(continued)

Solution   We begin by writing the dual problem associated with the given primal problem. From the tableau for the primal problem

x	y	Constant
8	1	80
3	2	100
1	4	80
2	3	

we find, upon changing the columns and rows of this tableau and heading the three columns of the resulting array with the variables u, v, and w, the tableau

u	v	w	Constant
8	3	1	2
1	2	4	3
80	100	80	

This tells us that the dual problem is

$$\text{Maximize} \quad P = 80u + 100v + 80w$$
$$\text{subject to} \quad 8u + 3v + w \leq 2$$
$$u + 2v + 4w \leq 3$$
$$u \geq 0, v \geq 0, w \geq 0$$

To solve this standard maximization problem, we proceed as follows:

	u	v	w	x	y	P	Constant		Ratio	
Pivot row →	8	③	1	1	0	0	2		$\frac{2}{3}$	$*\textbf{row}(\frac{1}{3}, A, 1) \blacktriangleright B$
	1	2	4	0	1	0	3		$\frac{3}{2}$	
	−80	−100	−80	0	0	1	0			

↑
Pivot column

	u	v	w	x	y	P	Constant	
	2.67	①	0.33	0.33	0	0	0.67	$*\textbf{row}+(-2, B, 1, 2) \blacktriangleright C$
	1	2	4	0	1	0	3	$*\textbf{row}+(100, C, 1, 3) \blacktriangleright B$
	−80	−100	−80	0	0	1	0	

	u	v	w	x	y	P	Constant		Ratio	
	2.67	1	0.33	0.33	0	0	0.67		2	$*\textbf{row}(\frac{1}{3.33}, B, 2) \blacktriangleright C$
Pivot row →	−4.33	0	③.33	−0.67	1	0	1.67		0.5	
	186.67	0	−46.67	33.33	0	1	66.67			

↑
Pivot column

	u	v	w	x	y	P	Constant	
	2.67	1	0.33	0.33	0	0	0.67	$*\textbf{row}+(-0.33, C, 2, 1) \blacktriangleright B$
	−1.30	0	1	−0.2	0.3	0	0.5	$*\textbf{row}+(46.67, B, 2, 3) \blacktriangleright C$
	186.67	0	−46.67	33.33	0	1	66.67	

u	v	w	x	y	P	Constant
3.1	1	0	0.4	−0.1	0	0.50
−1.3	0	1	−0.2	0.3	0	0.50
125.93	0	0.05	23.99	14.02	1	90.03

$$\underbrace{}_{\text{Solution for the primal problem}}$$

From the last tableau, we see that $x = 23.99$, $y = 14.02$, and the minimum value of C is 90.03.

Excel

EXAMPLE 2

$$\begin{aligned}
\text{Minimize} \quad & C = 2x + 3y \\
\text{subject to} \quad & 8x + y \geq 80 \\
& 3x + 2y \geq 100 \\
& x + 4y \geq 80 \\
& x \geq 0, y \geq 0
\end{aligned}$$

Solution   We use Solver as outlined in Example 2, pages 239–242, to obtain the spreadsheet shown in Figure T1. (In this case, select ⎡**Min**⎤ next to Equal to: instead of Max because this is a minimization problem. (Also select ⎡**>=**⎤ in the Add Constraint dialog box because the inequalities in the problem are of the form ≥). From the spreadsheet, we read off the solution: $x = 24$, $y = 14$, and $C = 90$.

	A	B	C	D	E	F	G	H	I
1	Minimization Problem								
2					Formulas for indicated cells				
3	Decision Variables				C8: = 2*C4 + 3*C5				
4		x	24		C11: = 8*C4 + C5				
5		y	14		C12: = 3*C4 + 2*C5				
6					C13: = C4 + 4*C5				
7									
8	Objective Function		90						
9									
10	Constraints								
11			206	>=	80				
12			100	>=	100				
13			80	>=	80				

FIGURE T1
Completed spreadsheet after using Solver

EXAMPLE 3   Sensitivity Analysis*   Solve the following linear problem:

$$\begin{aligned}
\text{Maximize} \quad & P = 2x + 4y \\
\text{subject to} \quad & 2x + 5y \leq 19 \qquad \text{Constraint 1} \\
& 3x + 2y \leq 12 \qquad \text{Constraint 2} \\
& x \geq 0, y \geq 0
\end{aligned}$$

Note: Boldfaced words/characters enclosed in a box (for example, ⎡**Enter**⎤) indicate that an action (click, select, or press) is required. Words/characters printed blue (for example, Chart sub-type) indicate words/characters that appear on the screen. Words/characters printed in a typewriter font (for example, =(−2/3)*A2+2) indicate words/characters that need to be typed and entered.

*For those who have completed Section 3.4.

(*continued*)

a. Use Solver to solve the problem.
b. Find the range of values that the coefficient of x can assume without changing the optimal solution.
c. Find the range of values that resource 1 (the constant on the right-hand side of constraint 1) can assume without changing the optimal solution.
d. Find the shadow price for resource 1.
e. Identify the binding and nonbinding constraints.

Solution

a. Use Solver as outlined in Example 2, pages 239–242, to obtain the spreadsheet shown in Figure T2.

	A	B	C	D	E	F	G	H	I
1	Maximization Problem								
2					Formulas for indicated cells				
3	Decision Variables				C8: = 2∗C4 + 4∗C5				
4		x	2		C11: = 2∗C4 + 5∗C5				
5		y	3		C12: = 3∗C4 + 2∗C5				
6									
7									
8	Objective Function		16						
9									
10	Constraints								
11			19	<=	19				
12			12	<=	12				

FIGURE T2
Completed spreadsheet for a maximization problem

b. In the Solver Results dialog box, hold down Ctrl while selecting Answer and Sensitivity under Reports, and then click OK . By clicking the Answer Report 1 and Sensitivity Report 1 tabs that appear at the bottom of your worksheet, you can obtain the reports shown in Figure T3a and b.

Target Cell (Max)

Cell	Name	Original Value	Final Value
$C$8	Objective Function	0	16

Adjustable Cells

Cell	Name	Original Value	Final Value
$C$4	x	0	2
$C$5	y	0	3

Constraints

Cell	Name	Cell Value	Formula	Status	Slack
$C$11		19	$C$11<=$E$11	Binding	0
$C$12		12	$C$12<=$E$12	Binding	0

(a) The Answer Report

Adjustable Cells

Cell	Name	Final Value	Reduced Cost	Objective Coefficient	Allowable Increase	Allowable Decrease
$C$4	x	2	0	2	4	0.4
$C$5	y	3	0	4	1	2.666666667

Constraints

Cell	Name	Final Value	Shadow Price	Constraint R.H. Side	Allowable Increase	Allowable Decrease
$C$11		19	0.727272727	19	11	11
$C$12		12	0.181818182	12	16.5	4.4

FIGURE T3
The Solver reports

(b) The Sensitivity Report

From the sensitivity report, we see that the value of the Objective Coefficient for x is 2, the Allowable Increase for the coefficient is 4, and the Allowable Decrease for the coefficient is 0.4. Thus, the coefficient can vary from 1.6 to 6 without affecting the optimal solution.

c. From the sensitivity report, we see that the Final Value of the constraint for resource 1 is 19 and the Allowable Increase and Allowable Decrease for this value is 11. Thus, the value of the constraint must lie between $19 - 11$ and $19 + 11$—that is, between 8 and 30.

d. From the sensitivity report, we see that the Shadow Price for resource 1 is 0.727272727.

e. From the answer report, we conclude that both Constraints are binding. ∎

TECHNOLOGY EXERCISES

In Exercises 1–4, solve the linear programming problem by the simplex method.

1. Minimize $C = x + y + 3z$

subject to
$$2x + y + 3z \geq 6$$
$$x + 2y + 4z \geq 8$$
$$3x + y - 2z \geq 4$$
$$x \geq 0, y \geq 0, z \geq 0$$

2. Minimize $C = 2x + 4y + z$

subject to
$$x + 2y + 4z \geq 7$$
$$3x + y - z \geq 6$$
$$x + 4y + 2z \geq 24$$
$$x \geq 0, y \geq 0, z \geq 0$$

3. Minimize $C = x + 1.2y + 3.5z$

subject to
$$2x + 3y + 5z \geq 12$$
$$3x + 1.2y - 2.2z \geq 8$$
$$1.2x + 3y + 1.8z \geq 14$$
$$x \geq 0, y \geq 0, z \geq 0$$

4. Minimize $C = 2.1x + 1.2y + z$

subject to
$$x + y - z \geq 5.2$$
$$x - 2.1y + 4.2z \geq 8.4$$
$$x \geq 0, y \geq 0, z \geq 0$$

Exercises 5–8 are for Excel users only. You are given a linear programming problem:

a. Use Solver to solve the problem.

b. Find the range of values that the coefficient of x can assume without changing the optimal solution and the range of values that the coefficient of y can assume without changing the optimal solution.

c. Find the range of values that each resource (requirement) can assume.

d. Find the shadow price for each resource (requirement).

e. Identify the binding and nonbinding constraints.

5. Maximize $P = 3x + 4y$

subject to
$$2x + 3y \leq 12 \quad \text{Resource 1}$$
$$2x + y \leq 8 \quad \text{Resource 2}$$
$$x \geq 0, y \geq 0$$

(continued)

6. Maximize $P = 2x + 5y$
  subject to $x + 3y \le 15$ Resource 1
                 $4x + y \le 16$ Resource 2
                 $x \ge 0, y \ge 0$

7. Minimize $C = 2x + 5y$
  subject to $x + 2y \ge 4$ Requirement 1
               $x + y \ge 3$ Requirement 2
               $x \ge 0, \ y \ge 0$

8. Minimize $C = 3x + 4y$
  subject to $x + 3y \ge 8$ Requirement 1
               $x + y \ge 4$ Requirement 2
               $x \ge 0, \ y \ge 0$

4.3   The Simplex Method: Nonstandard Problems (Optional)

Section 4.1 showed how we can use the simplex method to solve standard maximization problems; Section 4.2 showed how, thanks to duality, we can use it to solve standard minimization problems, provided that the coefficients in the objective function are all nonnegative.

In this section, we see how the simplex method can be incorporated into a method for solving **nonstandard problems**—problems that do not fall into either of the two previous categories. We begin by recalling the characteristics of standard problems.

Standard Maximization Problem

1. The objective function is to be maximized.
2. All variables involved in the problem are nonnegative.
3. Each linear constraint may be written so that the expression involving the variables is less than or equal to a nonnegative constant.

Standard Minimization Problem (Restricted Version— See Condition 4)

1. The objective function is to be minimized.
2. All variables involved in the problem are nonnegative.
3. Each linear constraint may be written so that the expression involving the variables is greater than or equal to a constant.
4. *All coefficients in the objective function are nonnegative.*

Note   Recall that if all coefficients in the objective function are nonnegative, then a standard minimization problem can be solved by using the simplex method to solve the associated dual problem.   ∎

We now give some examples of linear programming problems that do not fit into these two categories of problems.

EXAMPLE 1   Explain why the following linear programming problem is *not* a standard maximization problem.

$$\begin{aligned} \text{Maximize} \quad & P = x + 2y \\ \text{subject to} \quad & 4x + 3y \leq 18 \\ & -x + 3y \geq 3 \\ & x \geq 0, y \geq 0 \end{aligned}$$

Solution   This is not a standard maximization problem because the second constraint,

$$-x + 3y \geq 3$$

violates condition 3. Observe that by multiplying both sides of this inequality by -1, we obtain

$$x - 3y \leq -3 \qquad \text{\small Recall that multiplying both sides of an inequality by} \\ \text{\small a negative number reverses the inequality sign.}$$

Now, the last equation still violates condition 3 because the constant on the right is *negative*.   ∎

Observe that the constraints in Example 1 involve both *less than or equal to constraints* ($\leq$) and *greater than or equal to constraints* ($\geq$). Such constraints are called **mixed constraints.** We will solve the problem posed in Example 1 later.

EXAMPLE 2   Explain why the following linear programming problem is *not* a restricted standard minimization problem.

$$\begin{aligned} \text{Minimize} \quad & C = 2x - 3y \\ \text{subject to} \quad & x + y \leq 5 \\ & x + 3y \geq 9 \\ & -2x + y \leq 2 \\ & x \geq 0, y \geq 0 \end{aligned}$$

Solution   Observe that the coefficients in the objective function C are not all nonnegative. Therefore, the problem is not a restricted standard minimization problem. By constructing the dual problem, you can convince yourself that the latter is not a standard maximization problem and thus cannot be solved using the methods described in Sections 4.1 and 4.2. Again, we will solve this problem later.   ∎

EXAMPLE 3   Explain why the following linear programming problem is *not* a standard maximization problem. Show that it cannot be rewritten as a restricted standard minimization problem.

$$\begin{aligned} \text{Maximize} \quad & P = x + 2y \\ \text{subject to} \quad & 2x + 3y \leq 12 \\ & -x + 3y = 3 \\ & x \geq 0, y \geq 0 \end{aligned}$$

Solution   The constraint equation $-x + 3y = 3$ is equivalent to the two inequalities

$$-x + 3y \leq 3 \quad \text{and} \quad -x + 3y \geq 3$$

By multiplying both sides of the second inequality by -1, it can be written in the form

$$x - 3y \leq -3$$

Therefore, the two given constraints are equivalent to the three constraints

$$\begin{aligned} 2x + 3y &\leq 12 \\ -x + 3y &\leq 3 \\ x - 3y &\leq -3 \end{aligned}$$

The third inequality violates Condition 3 for a standard maximization problem. Next, we see that the given problem is equivalent to the following:

$$\begin{aligned} \text{Minimize} \quad & C = -x - 2y \\ \text{subject to} \quad & -2x - 3y \geq -12 \\ & x - 3y \geq -3 \\ & -x + 3y \geq 3 \\ & x \geq 0, y \geq 0 \end{aligned}$$

Since the coefficients of the objective function are not all nonnegative, we conclude that the given problem cannot be rewritten as a restricted standard minimization problem. You will be asked to solve this problem in Exercise 11.   ∎

The Simplex Method for Solving Nonstandard Problems

To describe a technique for solving nonstandard problems, let's consider the problem of Example 1:

$$\begin{aligned} \text{Maximize} \quad & P = x + 2y \\ \text{subject to} \quad & 4x + 3y \leq 18 \\ & -x + 3y \geq 3 \\ & x \geq 0, y \geq 0 \end{aligned}$$

As a first step, we rewrite the constraints so that the second constraint involves a $\leq$ constraint. As in Example 1, we obtain

$$\begin{aligned} 4x + 3y &\leq 18 \\ x - 3y &\leq -3 \\ x \geq 0, y &\geq 0 \end{aligned}$$

Disregarding the fact that the constant on the right of the second inequality is negative, let's attempt to solve the problem using the simplex method for problems in standard form. Introducing the slack variables u and v gives the system of linear equations

$$\begin{aligned} 4x + 3y + u \qquad\qquad &= 18 \\ x - 3y \qquad + v \qquad &= -3 \\ -x - 2y \qquad\qquad + P &= 0 \end{aligned}$$

The initial simplex tableau is

x	y	u	v	P	Constant
4	3	1	0	0	18
1	−3	0	1	0	−3
−1	−2	0	0	1	0

Interpreting the tableau in the usual fashion, we see that

$$x = 0 \qquad y = 0 \qquad u = 18 \qquad v = -3$$

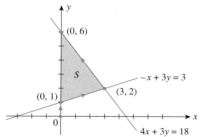

FIGURE 6
S is the feasible set for the problem.

Since the value of the slack variable v is negative, we see that this cannot be a feasible solution (remember, all variables must be nonnegative). In fact, you can see from Figure 6 that the point (0, 0) does not lie in the feasible set associated with the given problem. Since we must start from a feasible point when using the simplex method for problems in standard form, we see that this method is not applicable at this juncture.

Let's find a way to bring us from the nonfeasible point (0, 0) to *any* feasible point, after which we can switch to the simplex method for problems in standard form. This can be accomplished by pivoting as follows: Referring to the tableau,

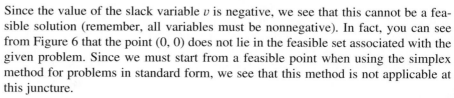

notice the negative number −3 lying in the column of constants, above the lower horizontal line. Locate any negative number to the left of this number (there must always be at least one such number if the problem has a solution). For the problem under consideration there is only one such number, the number −3 in the y-column. This column is designated as the pivot column. To find the pivot element, we form the *positive* ratios of the numbers in the column of constants to the corresponding numbers in the pivot column (above the last row). The pivot row is the row corresponding to the smallest ratio, and the pivot element is the element common to both the pivot row and the pivot column (the number circled in the foregoing tableau). Pivoting about this element, we have

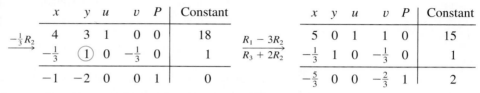

Interpreting the last tableau in the usual fashion, we see that

$$x = 0 \qquad y = 1 \qquad u = 15 \qquad v = 0 \qquad P = 2$$

Observe that the point (0, 1) is a feasible point (see Figure 6). Our iteration has brought us from a nonfeasible point to a feasible point in one iteration. Observe, too, that all the constants in the column of constants are now nonnegative, reflecting the fact that $(0, 1)$ is a feasible point, as we have just noted.

We can now use the simplex method for problems in standard form to complete the solution to our problem.

	x	y	u	v	P	Constant	Ratio
Pivot row $\rightarrow$	⑤	0	1	1	0	15	3
	$-\frac{1}{3}$	1	0	$-\frac{1}{3}$	0	1	—
	$-\frac{5}{3}$	0	0	$-\frac{2}{3}$	1	2	

Pivot column

	x	y	u	v	P	Constant
$\xrightarrow{\frac{1}{5}R_1}$	①	0	$\frac{1}{5}$	$\frac{1}{5}$	0	3
	$-\frac{1}{3}$	1	0	$-\frac{1}{3}$	0	1
	$-\frac{5}{3}$	0	0	$-\frac{2}{3}$	1	2

	x	y	u	v	P	Constant	Ratio
Pivot row $\rightarrow$ $\xrightarrow[R_3 + \frac{5}{3}R_1]{R_2 + \frac{1}{3}R_1}$	1	0	$\frac{1}{5}$	⑤$\frac{1}{5}$	0	3	15
	0	1	$\frac{1}{15}$	$-\frac{4}{15}$	0	2	—
	0	0	$\frac{1}{3}$	$-\frac{1}{3}$	1	7	

Pivot column

	x	y	u	v	P	Constant
$\xrightarrow{5R_1}$	5	0	1	①	0	15
	0	1	$\frac{1}{15}$	$-\frac{4}{15}$	0	2
	0	0	$\frac{1}{3}$	$-\frac{1}{3}$	1	7

	x	y	u	v	P	Constant
$\xrightarrow[R_3 + \frac{1}{3}R_1]{R_2 + \frac{4}{15}R_1}$	5	0	1	1	0	15
	$\frac{4}{3}$	1	$\frac{1}{3}$	0	0	6
	$\frac{5}{3}$	0	$\frac{2}{3}$	0	1	12

All entries in the last row are nonnegative, and the tableau is final. We see that the optimal solution is

$$x = 0 \qquad y = 6 \qquad u = 0 \qquad v = 15 \qquad P = 12$$

Observe that the maximum of P occurs at $(0, 6)$ (see Figure 6). The arrows indicate the path that our search for the maximum of P has taken us on.

Before looking at further examples, let's summarize the method for solving nonstandard problems.

The Simplex Method for Solving Nonstandard Problems

1. If necessary, rewrite the problem as a maximization problem (recall that minimizing C is equivalent to maximizing $-C$).
2. If necessary, rewrite all constraints (except $x \geq 0, y \geq 0, z \geq 0, \ldots$) using less than or equal to ($\leq$) inequalities.
3. Introduce slack variables and set up the initial simplex tableau.
4. Scan the upper part of the column of constants of the tableau for negative entries.
   a. If there are no negative entries, complete the solution using the simplex method for problems in standard form.
   b. If there are negative entries, proceed to step 5.
5. a. Pick any negative entry in the row in which a negative entry in the column of constants occurs. The column containing this entry is the pivot column.
   b. Compute the positive ratios of the numbers in the column of constants to the corresponding numbers in the pivot column (above the last row). The pivot row corresponds to the smallest ratio. The intersection of the pivot column and the pivot row determines the pivot element.
   c. Pivot the tableau about the pivot element. Then return to step 4.

We now apply the method to solve the nonstandard problem posed in Example 2.

EXAMPLE 4 Solve the problem of Example 2:

$$\text{Minimize} \quad C = 2x - 3y$$
$$\text{subject to} \quad x + y \leq 5$$
$$x + 3y \geq 9$$
$$-2x + y \leq 2$$
$$x \geq 0, y \geq 0$$

Solution We first rewrite the problem as a maximization problem with constraints using $\leq$, obtaining the following equivalent problem:

$$\text{Maximize} \quad P = -C = -2x + 3y$$
$$\text{subject to} \quad x + y \leq 5$$
$$-x - 3y \leq -9$$
$$-2x + y \leq 2$$
$$x \geq 0, y \geq 0$$

Introducing slack variables u, v, and w and following the procedure for solving nonstandard problems outlined earlier, we obtain the following sequence of tableaus:

	x	y	u	v	w	P	Constant	Ratio	
	1	1	1	0	0	0	5	$\frac{5}{1} = 5$	
	−1	−3	0	1	0	0	−9	$\frac{-9}{-3} = 3$	Column 1 could have been chosen
Pivot row →	−2	①	0	0	1	0	2	$\frac{2}{1} = 2$	as the pivot column as well.
	2	−3	0	0	0	1	0		

↑
Pivot column

	x	y	u	v	w	P	Constant
$R_1 - R_3$	3	0	1	0	-1	0	3
$R_2 + 3R_3$	$\boxed{-7}$	0	0	1	3	0	-3
$R_4 + 3R_3$	-2	1	0	0	1	0	2
	-4	0	0	0	3	1	6

$R_2 + 3R_3$ Pivot row

↑ Pivot column

Ratio
$\frac{3}{3} = 1$
$\frac{-3}{-7} = \frac{3}{7}$

	x	y	u	v	w	P	Constant
	3	0	1	0	-1	0	3
$\xrightarrow{-\frac{1}{7}R_2}$	$\boxed{1}$	0	0	$-\frac{1}{7}$	$-\frac{3}{7}$	0	$\frac{3}{7}$
	-2	1	0	0	1	0	2
	-4	0	0	0	3	1	6

	x	y	u	v	w	P	Constant
	0	0	1	$\boxed{\frac{3}{7}}$	$\frac{2}{7}$	0	$\frac{12}{7}$
$R_1 - 3R_2$	1	0	0	$-\frac{1}{7}$	$-\frac{3}{7}$	0	$\frac{3}{7}$
$R_3 + 2R_2$	0	1	0	$-\frac{2}{7}$	$\frac{1}{7}$	0	$\frac{20}{7}$
$R_4 + 4R_2$	0	0	0	$-\frac{4}{7}$	$\frac{9}{7}$	1	$\frac{54}{7}$

Pivot row

↑ Pivot column

Ratio
4
—
—

We now use the simplex method for problems in standard form to complete the problem.

	x	y	u	v	w	P	Constant
	0	0	$\frac{7}{3}$	$\boxed{1}$	$\frac{2}{3}$	0	4
$\xrightarrow{\frac{7}{3}R_1}$	1	0	0	$-\frac{1}{7}$	$-\frac{3}{7}$	0	$\frac{3}{7}$
	0	1	0	$-\frac{2}{7}$	$\frac{1}{7}$	0	$\frac{20}{7}$
	0	0	0	$-\frac{4}{7}$	$\frac{9}{7}$	1	$\frac{54}{7}$

	x	y	u	v	w	P	Constant
	0	0	$\frac{7}{3}$	1	$\frac{2}{3}$	0	4
$R_2 + \frac{1}{7}R_1$	1	0	$\frac{1}{3}$	0	$-\frac{1}{3}$	0	1
$R_3 + \frac{2}{7}R_1$	0	1	$\frac{2}{3}$	0	$\frac{1}{3}$	0	4
$R_4 + \frac{4}{7}R_1$	0	0	$\frac{4}{3}$	0	$\frac{5}{3}$	1	10

All the entries in the last row are nonnegative and the tableau is final. We see that the optimal solution is

$$x = 1 \qquad y = 4 \qquad u = 0 \qquad v = 4 \qquad w = 0 \qquad C = -P = -10$$

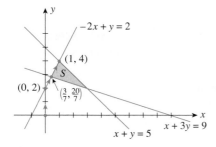

FIGURE 7
The feasible set S and the path leading from the initial nonfeasible point $(0, 0)$ to the optimal point $(1, 4)$

The feasible set S for this problem is shown in Figure 7. The path leading from the nonfeasible initial point $(0, 0)$ to the optimal point $(1, 4)$ goes through the nonfeasible point $(0, 2)$ and the feasible point $\left(\frac{3}{7}, \frac{20}{7}\right)$, in that order.

APPLIED EXAMPLE 5   Production Planning Rockford manufactures two models of exercise bicycles, a standard model and a deluxe model, in two separate plants—plant I and plant II. The maximum output at plant I is 1200 per month; the maximum output at plant II is 1000 per month. The profit per bike for standard and deluxe models manufactured at plant I is $40 and $60, respectively; the profit per bike for standard and deluxe models manufactured at plant II is $45 and $50, respectively.

For the month of May, Rockford received an order for 1000 standard models and 800 deluxe models. If prior commitments dictate that the number of deluxe models manufactured at plant I may not exceed the number of standard models manufactured there by more than 200, find how many of each model should be produced at each plant so as to satisfy the order and at the same time maximize Rockford's profit.

Solution   Let x and y denote the number of standard and deluxe models to be manufactured at plant I. Since the number of standard and deluxe models required are 1000 and 800, respectively, we see that the number of standard and deluxe models to be manufactured at plant II are $(1000 - x)$ and $(800 - y)$, respectively. Rockford's profit will then be

$$P = 40x + 60y + 45(1000 - x) + 50(800 - y)$$
$$= 85{,}000 - 5x + 10y$$

Since the maximum output of plant I is 1200, we have the constraint

$$x + y \leq 1200$$

Similarly, since the maximum output of plant II is 1000, we have

$$(1000 - x) + (800 - y) \leq 1000$$

or, equivalently,

$$-x - y \leq -800$$

Finally, the additional constraints placed on the production schedule at plants I and II translate into the inequalities

$$y - x \leq \ 200$$
$$x \leq 1000$$
$$y \leq \ 800$$

To summarize, the problem at hand is the following nonstandard problem:

$$
\begin{aligned}
\text{Maximize} \quad & P = 85{,}000 - 5x + 10y \\
\text{subject to} \quad & x + y \leq \ 1200 \\
& -x - y \leq -800 \\
& -x + y \leq \ 200 \\
& x \leq \ 1000 \\
& y \leq \ 800 \\
& x \geq 0, y \geq \ 0
\end{aligned}
$$

Let's introduce the slack variables, u v, w, r, and s. Using the simplex method for nonstandard problems, we obtain the following sequence of tableaus:

	x	y	u	v	w	r	s	P	Constant	Ratio
	1	1	1	0	0	0	0	0	1,200	$\frac{1200}{1}=1200$
Pivot row →	$\boxed{-1}$	−1	0	1	0	0	0	0	−800	$\frac{-800}{-1}=800$
	−1	1	0	0	1	0	0	0	200	—
	1	0	0	0	0	1	0	0	1,000	$\frac{1000}{1}=1000$
	0	1	0	0	0	0	1	0	800	—
	5	−10	0	0	0	0	0	1	85,000	

↑
Pivot column

$\xrightarrow{-R_2}$

	x	y	u	v	w	r	s	P	Constant
	1	1	1	0	0	0	0	0	1,200
	$\boxed{1}$	1	0	−1	0	0	0	0	800
	−1	1	0	0	1	0	0	0	200
	1	0	0	0	0	1	0	0	1,000
	0	1	0	0	0	0	1	0	800
	5	−10	0	0	0	0	0	1	85,000

$\begin{array}{l} R_1 - R_2 \\ R_3 + R_2 \\ \overline{R_4 - R_2} \\ R_6 - 5R_2 \end{array} \longrightarrow$

	x	y	u	v	w	r	s	P	Constant	Ratio	
	0	0	1	1	0	0	0	0	400	—	
	1	1	0	−1	0	0	0	0	800	$\frac{800}{1}=800$	We now use the
	0	$\boxed{2}$	0	−1	1	0	0	0	1,000	$\frac{1000}{2}=500$	simplex method for standard
	0	−1	0	1	0	1	0	0	200	—	problems.
Pivot row	0	1	0	0	0	0	1	0	800	$\frac{800}{1}=800$	
	0	−15	0	5	0	0	0	1	81,000		

↑
Pivot column

$\xrightarrow{\frac{1}{2}R_3}$

	x	y	u	v	w	r	s	P	Constant
	0	0	1	1	0	0	0	0	400
	1	1	0	−1	0	0	0	0	800
	0	$\boxed{1}$	0	$-\frac{1}{2}$	$\frac{1}{2}$	0	0	0	500
	0	−1	0	1	0	1	0	0	200
	0	1	0	0	0	0	1	0	800
	0	−15	0	5	0	0	0	1	81,000

$\begin{array}{l} R_2 - R_3 \\ R_4 + R_3 \\ \overline{R_5 - R_3} \\ R_6 + 15R_3 \end{array} \longrightarrow$

	x	y	u	v	w	r	s	P	Constant	Ratio
Pivot row →	0	0	1	$\boxed{1}$	0	0	0	0	400	$\frac{400}{1}=400$
	1	0	0	$-\frac{1}{2}$	$-\frac{1}{2}$	0	0	0	300	—
	0	1	0	$-\frac{1}{2}$	$\frac{1}{2}$	0	0	0	500	—
	0	0	0	$\frac{1}{2}$	$\frac{1}{2}$	1	0	0	700	$\frac{700}{1/2}=1400$
	0	0	0	$\frac{1}{2}$	$-\frac{1}{2}$	0	1	0	300	$\frac{300}{1/2}=600$
	0	0	0	$-\frac{5}{2}$	$\frac{15}{2}$	0	0	1	88,500	

↑
Pivot column

	x	y	u	v	w	r	s	P	Constant
	0	0	1	1	0	0	0	0	400
$R_2 + \frac{1}{2}R_1$	1	0	$\frac{1}{2}$	0	$-\frac{1}{2}$	0	0	0	500
$R_3 + \frac{1}{2}R_1$	0	1	$\frac{1}{2}$	0	$\frac{1}{2}$	0	0	0	700
$R_4 - \frac{1}{2}R_1$	0	0	$-\frac{1}{2}$	0	$\frac{1}{2}$	1	0	0	500
$R_5 - \frac{1}{2}R_1$									
$R_6 + \frac{5}{2}R_1$	0	0	$-\frac{1}{2}$	0	$-\frac{1}{2}$	0	1	0	100
	0	0	$\frac{5}{2}$	0	$\frac{15}{2}$	0	0	1	89,500

All entries in the last row are nonnegative, and the tableau is final. We see that $x = 500$, $y = 700$, and $P = 89,500$. This tells us that plant I should manufacture 500 standard and 700 deluxe exercise bicycles and that plant II should manufacture $(1000 - 500)$, or 500, standard and $(800 - 700)$, or 100, deluxe models. Rockford's profit will then be $89,500. ▪

EXPLORE & DISCUSS

Refer to Example 5.

1. Sketch the feasible set S for the linear programming problem.

2. Solve the problem using the method of corners.

3. Indicate on S the points (both nonfeasible and feasible) corresponding to each iteration of the simplex method and trace the path leading to the optimal solution.

4.3 Self-Check Exercises

1. Solve the following nonstandard problem using the method of this section:

$$\text{Maximize} \quad P = 2x + 3y$$
$$\text{subject to} \quad x + y \leq 40$$
$$-x + 2y \leq -10$$
$$x \geq 0, y \geq 0$$

2. A farmer has 150 acres of land suitable for cultivating crops A and B. The cost of cultivating crop A is $40/acre and that of crop B is $60/acre. The farmer has a maximum of $7400 available for land cultivation. Each acre of crop A requires 20 labor-hours, and each acre of crop B requires 25 labor-hours. The farmer has a maximum of 3300 labor-hours available. She has also decided that she will cultivate at least 70 acres of crop A. If she expects to make a profit of $150/acre on crop A and $200/acre on crop B, how many acres of each crop should she plant in order to maximize her profit?

Solutions to Self-Check Exercises 4.3 can be found on page 272.

4.3 Concept Questions

1. Explain why the following linear programming problem is *not* a standard maximization problem.

$$\text{Maximize} \quad P = 2x + 3y$$
$$\text{subject to} \quad 4x + 2y \leq 20$$
$$-2x + 3y \geq 5$$
$$x \geq 0, y \geq 0$$

2. Explain why the following linear programming problem is *not* a restricted standard minimization problem.

$$\text{Minimize} \quad C = 3x - y$$
$$\text{subject to} \quad x + y \geq 5$$
$$3x + 5y \geq 16$$
$$x \geq 0, y \geq 0$$

3. Explain why the following linear programming problem is *not* a standard maximization problem.

$$\begin{aligned}
\text{Maximize} \quad & P = x + 3y \\
\text{subject to} \quad & 3x + 4y \leq 16 \\
& 2x - 3y = 5 \\
& x \geq 0, y \geq 0
\end{aligned}$$

Can it be rewritten as a restricted standard minimization problem? Why or why not?

4.3   Exercises

In Exercises 1–4, rewrite each linear programming problem as a maximization problem with constraints involving inequalities of the form ≤ (with the exception of the inequalities $x \geq 0$, $y \geq 0$, and $z \geq 0$).

1. Minimize $C = 2x - 3y$
subject to $3x + 5y \geq 20$
$3x + y \leq 16$
$-2x + y \leq 1$
$x \geq 0, y \geq 0$

2. Minimize $C = 2x + 3y$
subject to $x + y \leq 10$
$x + 2y \geq 12$
$2x + y \geq 12$
$x \geq 0, y \geq 0$

3. Minimize $C = 5x + 10y + z$
subject to $2x + y + z \geq 4$
$x + 2y + 2z \geq 2$
$2x + 4y + 3z \leq 12$
$x \geq 0, y \geq 0, z \geq 0$

4. Maximize $P = 2x + y - 2z$
subject to $x + 2y + z \geq 10$
$3x + 4y + 2z \geq 5$
$2x + 5y + 12z \leq 20$
$x \geq 0, y \geq 0, z \geq 0$

In Exercises 5–20, use the method of this section to solve each linear programming problem.

5. Maximize $P = x + 2y$
subject to $2x + 5y \leq 20$
$x - 5y \leq -5$
$x \geq 0, y \geq 0$

6. Maximize $P = 2x + 3y$
subject to $x + 2y \leq 8$
$x - y \leq -2$
$x \geq 0, y \geq 0$

7. Minimize $C = -2x + y$
subject to $x + 2y \leq 6$
$3x + 2y \leq 12$
$x \geq 0, y \geq 0$

8. Minimize $C = -2x + 3y$
subject to $x + 3y \leq 60$
$2x + y \geq 45$
$x \leq 40$
$x \geq 0, y \geq 0$

9. Maximize $P = x + 4y$
subject to $x + 3y \leq 6$
$-2x + 3y \leq -6$
$x \geq 0, y \geq 0$

10. Maximize $P = 5x + y$
subject to $2x + y \leq 8$
$-x + y \geq 2$
$x \geq 0, y \geq 0$

11. Maximize $P = x + 2y$
subject to $2x + 3y \leq 12$
$-x + 3y = 3$
$x \geq 0, y \geq 0$

12. Minimize $C = x + 2y$
subject to $4x + 7y \leq 70$
$2x + y = 20$
$x \geq 0, y \geq 0$

13. Maximize $P = 5x + 4y + 2z$

subject to $x + 2y + 3z \leq 24$

$x - y + z \geq 6$

$x \geq 0, y \geq 0, z \geq 0$

14. Maximize $P = x - 2y + z$

subject to $2x + 3y + 2z \leq 12$

$x + 2y - 3z \geq 6$

$x \geq 0, y \geq 0, z \geq 0$

15. Minimize $C = x - 2y + z$

subject to $x - 2y + 3z \leq 10$

$2x + y - 2z \leq 15$

$2x + y + 3z \leq 20$

$x \geq 0, y \geq 0, z \geq 0$

16. Minimize $C = 2x - 3y + 4z$

subject to $-x + 2y - z \leq 8$

$x - 2y + 2z \leq 10$

$2x + 4y - 3z \leq 12$

$x \geq 0, y \geq 0, z \geq 0$

17. Maximize $P = 2x + y + z$

subject to $x + 2y + 3z \leq 28$

$2x + 3y - z \leq 6$

$x - 2y + z \geq 4$

$x \geq 0, y \geq 0, z \geq 0$

18. Minimize $C = 2x - y + 3z$

subject to $2x + y + z \geq 2$

$x + 3y + z \geq 6$

$2x + y + 2z \leq 12$

$x \geq 0, y \geq 0, z \geq 0$

19. Maximize $P = x + 2y + 3z$

subject to $x + 2y + z \leq 20$

$3x + y \leq 30$

$2x + y + z = 10$

$x \geq 0, y \geq 0, z \geq 0$

20. Minimize $C = 3x + 2y + z$

subject to $x + 2y + z \leq 20$

$3x + y \leq 30$

$2x + y + z = 10$

$x \geq 0, y \geq 0, z \geq 0$

21. AGRICULTURE—CROP PLANNING A farmer has 150 acres of land suitable for cultivating crops A and B. The cost of cultivating crop A is $40/acre and that of crop B is $60/acre. The farmer has a maximum of $7400 available for land cultivation. Each acre of crop A requires 20 labor-hours, and each

acre of crop B requires 25 labor-hours. The farmer has a maximum of 3300 labor-hours available. He has also decided that he will cultivate at least 80 acres of crop A. If he expects to make a profit of $150/acre on crop A and $200/acre on crop B, how many acres of each crop should he plant in order to maximize his profit?

22. VETERINARY SCIENCE A veterinarian has been asked to prepare a diet for a group of dogs to be used in a nutrition study at the School of Animal Science. It has been stipulated that each serving should be no larger than 8 oz and must contain at least 29 units of nutrient I and 20 units of nutrient II. The vet has decided that the diet may be prepared from two brands of dog food: brand A and brand B. Each ounce of brand A contains 3 units of nutrient I and 4 units of nutrient II. Each ounce of brand B contains 5 units of nutrient I and 2 units of nutrient II. Brand A costs 3 cents/ounce and brand B costs 4 cents/ounce. Determine how many ounces of each brand of dog food should be used per serving to meet the given requirements at a minimum cost.

23. FINANCE—ALLOCATION OF FUNDS The First Street branch of Capitol Bank has a sum of $60 million earmarked for home and commercial-development loans. The bank expects to realize an 8% annual rate of return on the home loans and a 6% annual rate of return on the commercial-development loans. Management has decided that the total amount of home loans is to be greater than or equal to 3 times the total amount of commercial-development loans. Owing to prior commitments, at least $10 million of the funds has been designated for commercial-development loans. Determine the amount of each type of loan the bank should extend in order to maximize its returns.

24. FINANCE—INVESTMENTS Natsano has at most $50,000 to invest in the common stocks of two companies. He estimates that an investment in company A will yield a return of 10%, whereas an investment in company B, which he feels is a riskier investment, will yield a return of 20%. If he decides that his investment in the stocks of company A is to exceed his investment in the stocks of company B by at least $20,000, determine how much he should invest in the stocks of each company in order to maximize the returns on his investment.

25. MANUFACTURING—PRODUCTION SCHEDULING A company manufactures products A, B, and C. Each product is processed in three departments: I, II, and III. The total available labor-hours per week for departments I, II, and III are 900, 1080, and 840, respectively. The time requirements (in hours/unit) and the profit per unit for each product are as follows:

	Product A	Product B	Product C
Dept. I	2	1	2
Dept. II	3	1	2
Dept. III	2	2	1
Profit	18	12	15

If management decides that the number of units of product B manufactured must equal or exceed the number of units of products A and C manufactured, how many units of each product should the company produce in order to maximize its profit?

26. MANUFACTURING—PRODUCTION SCHEDULING   Wayland Company manufactures two models of its twin-size futons, standard and deluxe, in two locations, I and II. The maximum output at location I is 600/week, whereas the maximum output at location II is 400/week. The profit per futon for standard and deluxe models manufactured at location I is $30 and $20, respectively; the profit per futon for standard and deluxe models manufactured at location II is $34 and $18, respectively. For a certain week, the company has received an order for 600 standard models and 300 deluxe models. If prior commitments dictate that the number of deluxe models manufactured at location II may not exceed the number of standard models manufactured there by more than 50, find how many of each model should be manufactured at each location so as to satisfy the order and at the same time maximize Wayland's profit.

27. NUTRITION—DIET PLANNING   A nutritionist at the Medical Center has been asked to prepare a special diet for certain patients. He has decided that the meals should contain a minimum of 400 mg of calcium, 10 mg of iron, and 40 mg of vitamin C. He has further decided that the meals are to be pre-

pared from foods A and B. Each ounce of food A contains 30 mg of calcium, 1 mg of iron, 2 mg of vitamin C, and 2 mg of cholesterol. Each ounce of food B contains 25 mg of calcium, 0.5 mg of iron, 5 mg of vitamin C, and 5 mg of cholesterol. Use the method of this section to determine how many ounces of each type of food the nutritionist should use in a meal so that the cholesterol content is minimized and the minimum requirements of calcium, iron, and vitamin C are met.

28. MANUFACTURING—SHIPPING COSTS   Steinwelt Piano manufactures uprights and consoles in two plants, plant I and plant II. The output of plant I is at most 300/month, whereas the output of plant II is at most 250/month. These pianos are shipped to three warehouses that serve as distribution centers for the company. To fill current and projected orders, warehouse A requires a minimum of 200 pianos/month, warehouse B requires at least 150 pianos/month, and warehouse C requires at least 200 pianos/month. The shipping cost of each piano from plant I to warehouse A, warehouse B, and warehouse C is $60, $60, and $80, respectively, and the shipping cost of each piano from plant II to warehouse A, warehouse B, and warehouse C is $80, $70, and $50, respectively. Use the method of this section to determine the shipping schedule that will enable Steinwelt to meet the warehouses' requirements while keeping the shipping costs to a minimum.

4.3   Solutions to Self-Check Exercises

1. We are given the problem

$$\text{Maximize} \quad P = 2x + 3y$$
$$\text{subject to} \quad x + y \le 40$$
$$-x + 2y \le -10$$
$$x \ge 0, y \ge 0$$

Using the method of this section and introducing the slack variables u and v, we obtain the following tableaus:

	x	y	u	v	P	Constant	Ratio
	1	1	1	0	0	40	$\frac{40}{1} = 40$
Pivot row →	⊝1	2	0	1	0	−10	$\frac{-10}{-1} = 10$
	−2	−3	0	0	1	0	

Pivot column

	x	y	u	v	P	Constant
	1	1	1	0	0	40
$-R_2 \longrightarrow$	①	−2	0	−1	0	10
	−2	−3	0	0	1	0

Pivot row →	x	y	u	v	P	Constant	Ratio
$\begin{array}{c} R_1 - R_2 \\ \longrightarrow \\ R_3 + 2R_2 \end{array}$	0	③	1	1	0	30	$\frac{30}{3} = 10$
	1	−2	0	−1	0	10	—
	0	−7	0	−2	1	20	

Pivot column

	x	y	u	v	P	Constant
$\frac{1}{3}R_1 \longrightarrow$	0	①	$\frac{1}{3}$	$\frac{1}{3}$	0	10
	1	−2	0	−1	0	10
	0	−7	0	−2	1	20

	x	y	u	v	P	Constant
$\begin{array}{c} R_2 + 2R_1 \\ \longrightarrow \\ R_3 + 7R_1 \end{array}$	0	1	$\frac{1}{3}$	$\frac{1}{3}$	0	10
	1	0	$\frac{2}{3}$	$-\frac{1}{3}$	0	30
	0	0	$\frac{7}{3}$	$\frac{1}{3}$	1	90

The last tableau is in final form, and we obtain the solution

$$x = 30 \qquad y = 10 \qquad u = 0 \qquad v = 0 \qquad P = 90$$

2. Let x denote the number of acres of crop A to be cultivated and y the number of acres of crop B to be cultivated. Since there is a total of 150 acres of land available for cultivation, we have $x + y \leq 150$. Next, the restriction on the amount of money available for land cultivation implies that $40x + 60y \leq 7400$. Similarly, the restriction on the amount of time available for labor implies that $20x + 25y \leq 3300$. Also, since she will cultivate at least 70 acres of crop A, $x \geq 70$. Since the profit on each acre of crop A is \$150 and the profit on each acre of crop B is \$200, we see that the profit realizable by the farmer is $P = 150x + 200y$. Summarizing, we have the following linear programming problem:

$$\text{Maximize} \quad P = 150x + 200y$$
$$\text{subject to} \quad x + y \leq 150$$
$$40x + 60y \leq 7400$$
$$20x + 25y \leq 3300$$
$$x \geq 70$$
$$x \geq 0, y \geq 0$$

To solve this nonstandard problem, we rewrite the fourth inequality in the form

$$-x \leq -70$$

Using the method of this section with u, v, w, and z as slack variables, we have the following tableaus:

	x	y	u	v	w	z	P	Constant	Ratio
	1	1	1	0	0	0	0	150	$\frac{150}{1} = 150$
	40	60	0	1	0	0	0	7,400	$\frac{7400}{40} = 185$
	20	25	0	0	1	0	0	3,300	$\frac{3300}{20} = 165$
Pivot row →	⓪-1	0	0	0	0	1	0	−70	$\frac{-70}{-1} = 70$
	−150	−200	0	0	0	0	1	0	

Pivot column ↑

	x	y	u	v	w	z	P	Constant
	1	1	1	0	0	0	0	150
	40	60	0	1	0	0	0	7,400
$-R_4$ →	20	25	0	0	1	0	0	3,300
	①	0	0	0	0	−1	0	70
	−150	−200	0	0	0	0	1	0

	x	y	u	v	w	z	P	Constant	Ratio
$R_1 - R_4$	0	1	1	0	0	1	0	80	$\frac{80}{1} = 80$
$R_2 - 40R_4$	0	60	0	1	0	40	0	4,600	$\frac{4600}{60} = 76\frac{2}{3}$
$R_3 - 20R_4$ →	0	㉕	0	0	1	20	0	1,900	$\frac{1900}{25} = 76$
$R_5 + 150R_4$	1	0	0	0	0	−1	0	70	—
Pivot row	0	−200	0	0	0	−150	1	10,500	

Pivot column ↑

	x	y	u	v	w	z	P	Constant
	0	1	1	0	0	1	0	80
$\frac{1}{25}R_3$ →	0	60	0	1	0	40	0	4,600
	0	①	0	0	$\frac{1}{25}$	$\frac{4}{5}$	0	76
	1	0	0	0	0	−1	0	70
	0	−200	0	0	0	−150	1	10,500

	x	y	u	v	w	z	P	Constant
	0	0	1	0	$-\frac{1}{25}$	$\frac{1}{5}$	0	4
$R_1 - R_3$ →	0	0	0	1	$-\frac{12}{5}$	−8	0	40
$R_2 - 60R_3$	0	1	0	0	$\frac{1}{25}$	$\frac{4}{5}$	0	76
$R_5 + 200R_3$	1	0	0	0	0	−1	0	70
	0	0	0	0	8	10	1	25,700

This last tableau is in final form, and the solution is

$$x = 70 \qquad y = 76 \qquad u = 4 \qquad v = 40$$
$$w = 0 \qquad z = 0 \qquad P = 25,700$$

Thus, by cultivating 70 acres of crop A and 76 acres of crop B, the farmer will attain a maximum profit of \$25,700.

CHAPTER 4   Summary of Principal Terms

TERMS

standard maximization problem (216)	pivot column (220)	standard minimization problem (244)
slack variable (217)	pivot row (220)	primal problem (244)
basic variable (218)	pivot element (220)	dual problem (244)
nonbasic variable (218)	simplex tableau (220)	nonstandard problem (260)
basic solution (218)	simplex method (222)	mixed constraints (261)

CHAPTER 4   Concept Review Questions

Fill in the blanks.

1. In a standard maximization problem the objective function is to be _____; all the variables involved in the problem are _____; each linear constraint may be written so that the expression involving the variables is _____ _____ or _____ _____ a nonnegative constant.

2. In setting up the initial simplex tableau, we first transform the system of linear inequalities into a system of linear _____, using_____ _____; the objective function is rewritten so that it has the form _____ and then is placed _____ the system of linear equations obtained earlier. Finally, the initial simplex tableau is the

_____ matrix associated with this system of linear equations.

3. In a standard minimization problem, the objective function is to be _____; all the variables involved in the problem are _____; each linear constraint may be written so that the expression involving the variables is _____ _____ or _____ _____ a constant.

4. The fundamental theorem of duality states that a primal problem has a solution if and only if the corresponding _____ problem has a solution. If a solution exists, then the _____ function of both the primal and dual problem attain the same _____ _____.

CHAPTER 4   Review Exercises

In Exercises 1–12, use the simplex method to solve each linear programming problem.

1. Maximize $P = 3x + 4y$
   subject to $x + 3y \le 15$
   $4x + y \le 16$
   $x \ge 0, y \ge 0$

2. Maximize $P = 2x + 5y$
   subject to $2x + y \le 16$
   $2x + 3y \le 24$
   $y \le 6$
   $x \ge 0, y \ge 0$

3. Maximize $P = 2x + 3y + 5z$
   subject to $x + 2y + 3z \le 12$
   $x - 3y + 2z \le 10$
   $x \ge 0, y \ge 0, z \ge 0$

4. Maximize $P = x + 2y + 3z$
   subject to $2x + y + z \le 14$
   $3x + 2y + 4z \le 24$
   $2x + 5y - 2z \le 10$
   $x \ge 0, y \ge 0, z \ge 0$

5. Minimize $C = 3x + 2y$
   subject to $2x + 3y \ge 6$
   $2x + y \ge 4$
   $x \ge 0, y \ge 0$

6. Minimize $C = x + 2y$
   subject to $3x + y \ge 12$
   $x + 4y \ge 16$
   $x \ge 0, y \ge 0$

7. Minimize $C = 24x + 18y + 24z$
   subject to $3x + 2y + z \ge 4$
   $x + y + 3z \ge 6$
   $x \ge 0, y \ge 0, z \ge 0$

8. Minimize $C = 4x + 2y + 6z$
   subject to $x + 2y + z \ge 4$
   $2x + y + 2z \ge 2$
   $3x + 2y + z \ge 3$
   $x \ge 0, y \ge 0, z \ge 0$

9. Maximize $P = 3x - 4y$
subject to $x + y \leq 45$
$x - 2y \geq 10$
$x \geq 0, y \geq 0$

10. Minimize $C = 2x + 3y$
subject to $x + y \leq 10$
$x + 2y \geq 12$
$2x + y \geq 12$
$x \geq 0, y \geq 0$

11. Maximize $P = 2x + 3y$
subject to $2x + 5y \leq 20$
$-x + 5y \geq 5$
$x \geq 0, y \geq 0$

12. Minimize $C = -3x - 4y$
subject to $x + y \leq 45$
$x + y \geq 15$
$x \leq 30$
$y \leq 25$
$x \geq 0, y \geq 0$

13. MINING—PRODUCTION Perth Mining Company operates two mines for the purpose of extracting gold and silver. The Saddle Mine costs $14,000/day to operate, and it yields 50 oz of gold and 3000 oz of silver each day. The Horseshoe Mine costs $16,000/day to operate and it yields 75 oz of gold and 1000 oz of silver each day. Company management has set a target of at least 650 oz of gold and 18,000 oz of silver. How many days should each mine be operated at so that the target can be met at a minimum cost to the company? What is the minimum cost?

14. INVESTMENT ANALYSIS Jorge has decided to invest at most $100,000 in securities in the form of corporate stocks. He has classified his options into three groups of stocks: blue-chip stocks that he assumes will yield a 10% return (dividends and capital appreciation) within a year, growth stocks that he assumes will yield a 15% return within a year, and speculative stocks that he assumes will yield a 20% return (mainly due to capital appreciation) within a year. Because of the relative risks involved in his investment, Jorge has further decided that no more than 30% of his investment should be in growth and speculative stocks and at least 50% of his investment should be in blue-chip and speculative stocks. Determine how much Jorge should invest in each group of stocks in the hope of maximizing the return on his investments.

15. MAXIMIZING PROFITS A company manufactures three products, A, B, and C, on two machines, I and II. It has been determined that the company will realize a profit of $4/unit of product A, $6/unit of product B, and $8/unit of product C. Manufacturing a unit of product A requires 9 min on machine I and 6 min on machine II; manufacturing a unit of product B requires 12 min on machine I and 6 min on machine II; manufacturing a unit of product C requires 18 min on machine I and 10 min on machine II. There are 6 hr of machine time available on machine I and 4 hr of machine time available on machine II in each work shift. How many units of each product should be produced in each shift in order to maximize the company's profit?

16. INVESTMENT ANALYSIS Sandra has at most $200,000 to invest in stocks, bonds, and money market funds. She expects annual yields of 15%, 10%, and 8%, respectively, on these investments. If Sandra wants at least $50,000 to be invested in money market funds and requires that the amount invested in bonds be greater than or equal to the sum of her investments in stocks and money market funds, determine how much she should invest in each vehicle in order to maximize the return on her investments.

CHAPTER 4 **Before Moving On . . .**

1. Consider the following linear programming problem:

Maximize $P = x + 2y - 3z$
subject to $2x + y - z \leq 3$
$x - 2y + 3z \leq 1$
$3x + 2y + 4z \leq 17$
$x \geq 0, y \geq 0, z \geq 0$

Write the initial simplex tableau for the problem and identify the pivot element to be used in the first iteration of the simplex method.

2. The following simplex tableau is in final form. Find the solution to the linear programming problem associated with this tableau.

x	y	z	u	v	w	P	Constant
0	$\frac{1}{2}$	0	1	$-\frac{1}{2}$	0	0	2
0	$\frac{1}{4}$	1	0	$\frac{5}{4}$	$-\frac{1}{2}$	0	11
1	$\frac{1}{4}$	0	0	$-\frac{3}{4}$	$\frac{1}{2}$	0	2
0	$\frac{13}{4}$	0	0	$\frac{1}{4}$	$\frac{1}{2}$	1	28

3. Using the simplex method, solve the following linear programming problem:

Maximize $P = 5x + 2y$
subject to $4x + 3y \leq 30$
$2x - 3y \leq 6$
$x \geq 0, y \geq 0$

4. Using the simplex method, solve the following linear programming problem:

$$\begin{aligned}
\text{Minimize} \quad & C = x + 2y \\
\text{subject to} \quad & x + y \geq 3 \\
& 2x + 3y \geq 6 \\
& x \geq 0, y \geq 0
\end{aligned}$$

5. **Mixed Constraints (Optional)** Using the simplex method, solve the following linear programming problem:

$$\begin{aligned}
\text{Maximize} \quad & P = 2x + y \\
\text{subject to} \quad & 2x + 5y \leq 20 \\
& 4x + 3y \geq 16 \\
& x \geq 0, y \geq 0
\end{aligned}$$

5 Mathematics of Finance

© Kim Kulish/Corbis

How much will the home mortgage payment be? The Blakelys received a bank loan to help finance the purchase of a house. They have agreed to repay the loan in equal monthly installments over a certain period of time. In Example 2, page 310, we will show how to determine the size of the monthly installment so that the loan is fully amortized at the end of the term.

INTEREST THAT IS periodically added to the principal and thereafter itself earns interest is called *compound interest*. We begin this chapter by deriving the *compound interest formula*, which gives the amount of money accumulated when an initial amount of money is invested in an account for a fixed term and earns compound interest.

An *annuity* is a sequence of payments made at regular intervals. We derive formulas giving the *future value of an annuity* (what you end up with) and the *present value of an annuity* (the lump sum that, when invested now, will yield the same future value as that of the annuity). Then, using these formulas, we answer questions involving the amortization of certain types of installment loans and questions involving *sinking funds* (funds that are set up to be used for a specific purpose at a future date).

5.1   Compound Interest

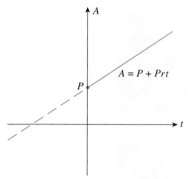

FIGURE 1
The accumulated amount is a linear
function of *t*.

■ Simple Interest

A natural application of linear functions to the business world is found in the computation of **simple interest**—interest that is computed on the original principal only. Thus, if *I* denotes the interest on a principal *P* (in dollars) at an interest rate of *r* per year for *t* years, then we have

$$I = Prt$$

The **accumulated amount** *A*, the sum of the principal and interest after *t* years, is given by

$$A = P + I = P + Prt$$
$$= P(1 + rt)$$

and is a linear function of *t* (see Exercise 40 at the end of this section). In business applications, we are normally interested only in the case where *t* is positive, so only that part of the line that lies in Quadrant I is of interest to us (Figure 1).

Simple Interest Formulas		
Interest:	$I = Prt$	**(1a)**
Accumulated amount:	$A = P(1 + rt)$	**(1b)**

EXAMPLE 1   A bank pays simple interest at the rate of 8% per year for certain deposits. If a customer deposits $1000 and makes no withdrawals for 3 years, what is the total amount on deposit at the end of 3 years? What is the interest earned in that period of time?

Solution   Using Equation (1b) with $P = 1000$, $r = 0.08$, and $t = 3$, we see that the total amount on deposit at the end of 3 years is given by

$$A = P(1 + rt)$$
$$= 1000[1 + (0.08)(3)] = 1240$$

or $1240.

The interest earned over the 3-year period is given by

$$I = Prt \qquad \text{Use (1a).}$$
$$= 1000(0.08)(3) = 240$$

or $240.

▨ EXPLORING WITH TECHNOLOGY

Refer to Example 1. Use a graphing utility to plot the graph of the function $A = 1000(1 + 0.08t)$, using the viewing window $[0, 10] \times [0, 2000]$.

1. What is the *A*-intercept of the straight line, and what does it represent?

2. What is the slope of the straight line, and what does it represent? (See Exercise 40.)

EXAMPLE 2   An amount of $2000 is invested in a 10-year trust fund that pays 6% annual simple interest. What is the total amount of the trust fund at the end of 10 years?

Solution   The total amount of the trust fund at the end of 10 years is given by

$$A = P(1 + rt)$$
$$= 2000[1 + (0.06)(10)] = 3200$$

or $3200. ■

▬ Compound Interest

In contrast to simple interest, earned interest that is periodically added to the principal and thereafter itself earns interest at the same rate is called **compound interest.** To find a formula for the accumulated amount, let's consider a numerical example. Suppose $1000 (the principal) is deposited in a bank for a term of 3 years, earning interest at the rate of 8% per year (called the **nominal,** or **stated, rate**) compounded annually. Then, using Equation (1b) with $P = 1000$, $r = 0.08$, and $t = 1$, we see that the accumulated amount at the end of the first year is

$$A_1 = P(1 + rt)$$
$$= 1000[1 + 0.08(1)] = 1000(1.08) = 1080$$

or $1080.

To find the accumulated amount A_2 at the end of the second year, we use (1b) once again, this time with $P = A_1$. (Remember, the principal *and* interest now earn interest over the second year.) We obtain

$$A_2 = P(1 + rt) = A_1(1 + rt)$$
$$= 1000[1 + 0.08(1)][1 + 0.08(1)]$$
$$= 1000[1 + 0.08]^2 = 1000(1.08)^2 = 1166.40$$

or $1166.40.

Finally, the accumulated amount A_3 at the end of the third year is found using (1b) with $P = A_2$, giving

$$A_3 = P(1 + rt) = A_2(1 + rt)$$
$$= 1000[1 + 0.08(1)]^2[1 + 0.08(1)]$$
$$= 1000[1 + 0.08]^3 = 1000(1.08)^3 \approx 1259.71$$

or approximately $1259.71.

If you reexamine our calculations, you will see that the accumulated amounts at the end of each year have the following form:

First year: $A_1 = 1000(1 + 0.08)$, or $A_1 = P(1 + r)$

Second year: $A_2 = 1000(1 + 0.08)^2$, or $A_2 = P(1 + r)^2$

Third year: $A_3 = 1000(1 + 0.08)^3$, or $A_3 = P(1 + r)^3$

These observations suggest the following general result: If P dollars is invested over a term of t years, earning interest at the rate of r per year compounded annually, then the accumulated amount is

$$A = P(1 + r)^t \tag{2}$$

Formula (2) was derived under the assumption that interest was compounded *annually*. In practice, however, interest is usually compounded more than once a year. The interval of time between successive interest calculations is called the **conversion period.**

If interest at a nominal rate of r per year is compounded m times a year on a principal of P dollars, then the simple interest rate per conversion period is

$$i = \frac{r}{m} \qquad \frac{\text{Annual interest rate}}{\text{Periods per year}}$$

For example, if the nominal interest rate is 8% per year ($r = 0.08$) and interest is compounded quarterly ($m = 4$), then

$$i = \frac{r}{m} = \frac{0.08}{4} = 0.02$$

or 2% per period.

To find a general formula for the accumulated amount when a principal of P dollars is deposited in a bank for a term of t years and earns interest at the (nominal) rate of r per year compounded m times per year, we proceed as before, using (1b) repeatedly with the interest rate $i = r/m$. We see that the accumulated amount at the end of each period is

First period: $A_1 = P(1 + i)$

Second period: $A_2 = A_1(1 + i)$ $= [P(1 + i)](1 + i) = P(1 + i)^2$

Third period: $A_3 = A_2(1 + i)$ $= [P(1 + i)^2](1 + i) = P(1 + i)^3$

$$\vdots \qquad\qquad\qquad \vdots$$

nth period: $A_n = A_{n-1}(1 + i) = [P(1 + i)^{n-1}](1 + i) = P(1 + i)^n$

But there are $n = mt$ periods in t years (number of conversion periods times the term). Therefore, the accumulated amount at the end of t years is given by

$$A = P(1 + i)^n$$

Compound Interest Formula (Accumulated Amount)

$$A = P(1 + i)^n \tag{3}$$

where $i = \dfrac{r}{m}$, $n = mt$, and

A = Accumulated amount at the end of n conversion periods
P = Principal
r = Nominal interest rate per year
m = Number of conversion periods per year
t = Term (number of years)

EXPLORING WITH TECHNOLOGY

Let $A_1(t)$ denote the accumulated amount of $100 earning simple interest at the rate of 10% per year over t years, and let $A_2(t)$ denote the accumulated amount of $100 earning interest at the rate of 10% per year compounded monthly over t years.

1. Find expressions for $A_1(t)$ and $A_2(t)$.

2. Use a graphing utility to plot the graphs of A_1 and A_2 on the same set of axes, using the viewing window $[0, 20] \times [0, 800]$.

3. Comment on the growth of $A_1(t)$ and $A_2(t)$ by referring to the graphs of A_1 and A_2.

EXAMPLE 3 Find the accumulated amount after 3 years if $1000 is invested at 8% per year compounded (a) annually, (b) semiannually, (c) quarterly, (d) monthly, and (e) daily.

Solution

a. Here, $P = 1000$, $r = 0.08$, and $m = 1$. Thus, $i = r = 0.08$ and $n = 3$, so Equation (3) gives

$$A = 1000(1 + 0.08)^3$$
$$\approx 1259.71$$

or $1259.71.

b. Here, $P = 1000$, $r = 0.08$, and $m = 2$. Thus, $i = \frac{0.08}{2}$ and $n = (3)(2) = 6$, so Equation (3) gives

$$A = 1000\left(1 + \frac{0.08}{2}\right)^6$$
$$\approx 1265.32$$

or $1265.32.

c. In this case, $P = 1000$, $r = 0.08$, and $m = 4$. Thus, $i = \frac{0.08}{4}$ and $n = (3)(4) = 12$, so Equation (3) gives

$$A = 1000\left(1 + \frac{0.08}{4}\right)^{12}$$
$$\approx 1268.24$$

or $1268.24

d. Here, $P = 1000$, $r = 0.08$, and $m = 12$. Thus, $i = \frac{0.08}{12}$ and $n = (3)(12) = 36$, so Equation (3) gives

$$A = 1000\left(1 + \frac{0.08}{12}\right)^{36}$$
$$\approx 1270.24$$

or $1270.24.

e. Here, $P = 1000$, $r = 0.08$, $m = 365$, and $t = 3$. Thus, $i = \frac{0.08}{365}$ and $n = (3)(365) = 1095$, so Equation (3) gives

$$A = 1000\left(1 + \frac{0.08}{365}\right)^{1095}$$
$$\approx 1271.22$$

or $1271.22. These results are summarized in Table 1.

TABLE 1

Nominal Rate, r	Conversion Period	Interest Rate/ Conversion Period	Initial Investment	Accumulated Amount
8%	Annual ($m = 1$)	8%	$1000	$1259.71
8	Semiannual ($m = 2$)	4	1000	1265.32
8	Quarterly ($m = 4$)	2	1000	1268.24
8	Monthly ($m = 12$)	2/3	1000	1270.24
8	Daily ($m = 365$)	8/365	1000	1271.22

EXPLORING WITH TECHNOLOGY

Investments allowed to grow over time can increase in value surprisingly fast. Consider the potential growth of $10,000 if earnings are reinvested. More specifically, suppose $A_1(t)$, $A_2(t)$, $A_3(t)$, $A_4(t)$, and $A_5(t)$ denote the accumulated values of an investment of $10,000 over a term of t years, and earning interest at the rate of 4%, 6%, 8%, 10%, and 12% per year compounded annually.

1. Find expressions for $A_1(t)$, $A_2(t)$, . . . , $A_5(t)$.

2. Use a graphing utility to plot the graphs of A_1, A_2, . . . , A_5 on the same set of axes, using the viewing window $[0, 20] \times [0, 100,000]$.

3. Use TRACE to find $A_1(20)$, $A_2(20)$, . . . , $A_5(20)$ and interpret your results.

Continuous Compounding of Interest

One question that arises naturally in the study of compound interest is, What happens to the accumulated amount over a fixed period of time if the interest is computed more and more frequently?

Intuition suggests that the more often interest is compounded, the larger the accumulated amount will be. This is confirmed by the results of Example 3, where we found that the accumulated amounts did in fact increase when we increased the number of conversion periods per year.

This leads us to another question: Does the accumulated amount keep growing, or does it approach a fixed number when the interest is computed more and more frequently over a fixed period of time?

To answer this question, let's look again at the compound interest formula:

$$A = P(1 + i)^n = P\left(1 + \frac{r}{m}\right)^{mt} \tag{4}$$

Recall that m is the number of conversion periods per year. So to find an answer to our question, we should let m get larger and larger in (4). If we let $u = \frac{m}{r}$ so that $m = ru$, then (4) becomes

$$A = P\left(1 + \frac{1}{u}\right)^{urt} \qquad \frac{r}{m} = \frac{1}{u}$$

$$= P\left[\left(1 + \frac{1}{u}\right)^u\right]^{rt} \qquad \text{Since } a^{xy} = (a^x)^y$$

TABLE 2

u	$\left(1 + \dfrac{1}{u}\right)^u$
10	2.59374
100	2.70481
1000	2.71692
10,000	2.71815
100,000	2.71827
1,000,000	2.71828

Now let's see what happens to the expression

$$\left(1 + \frac{1}{u}\right)^u$$

as u gets larger and larger. From Table 2, you can see that as u increases,

$$\left(1 + \frac{1}{u}\right)^u$$

seems to approach the number 2.71828 (we have rounded all our calculations to five decimal places).

It can be shown, although we will not do so here, that as u gets larger and larger, the value of the expression $\left(1 + \frac{1}{u}\right)^u$ approaches the irrational number 2.71828. . . , which we denote by e. (See the Exploring with Technology exercise that follows.)

 EXPLORING WITH TECHNOLOGY

To obtain a visual confirmation of the fact that the expression $\left(1 + \frac{1}{u}\right)^u$ approaches the number $e = 2.71828$ . . . as u gets larger and larger, plot the graph of $f(x) = \left(1 + \frac{1}{x}\right)^x$ in a suitable viewing window and observe that $f(x)$ approaches 2.71828. . . as x gets larger and larger. Use ZOOM and TRACE to find the value of $f(x)$ for large values of x.

Using this result, we can see that, as m gets larger and larger, A approaches $P(e)^{rt} = Pe^{rt}$. In this situation, we say that interest is *compounded continuously*. Let's summarize this important result.

Continuous Compound Interest Formula

$$A = Pe^{rt} \qquad\qquad (5)$$

where

$P = $ Principal

$r = $ Annual interest rate compounded continuously

$t = $ Time in years

$A = $ Accumulated amount at the end of t years

EXAMPLE 4 Find the accumulated amount after 3 years if $1000 is invested at 8% per year compounded (a) daily (assume a 365-day year) and (b) continuously.

Solution
a. Use Formula (3) with $P = 1000$, $r = 0.08$, $m = 365$, and $t = 3$. Thus, $i = 0.08/365$ and $n = (365)(3) = 1095$, so that

$$A = 1000\left(1 + \frac{0.08}{365}\right)^{(365)(3)} \approx 1271.22$$

or $1271.22.

b. Here we use Formula (5) with $P = 1000$, $r = 0.08$, and $t = 3$, obtaining

$$A = 1000e^{(0.08)(3)}$$
$$\approx 1271.25$$

or \$1271.25. ∎

Observe that the accumulated amounts corresponding to interest compounded daily and interest compounded continuously differ by very little. The continuous compound interest formula is a very important tool in theoretical work in financial analysis.

■ Effective Rate of Interest

Example 3 showed that the interest actually earned on an investment depends on the frequency with which the interest is compounded. Thus, the stated, or nominal, rate of 8% per year does not reflect the actual rate at which interest is earned. This suggests that we need to find a common basis for comparing interest rates. One such way of comparing interest rates is provided by the use of the *effective rate*. The **effective rate** is the *simple* interest rate that would produce the same accumulated amount in 1 year as the nominal rate compounded m times a year. The effective rate is also called the **effective annual yield.**

To derive a relationship between the nominal interest rate, r per year compounded m times, and its corresponding effective rate, R per year, let's assume an initial investment of P dollars. Then, the accumulated amount after 1 year at a simple interest rate of R per year is

$$A = P(1 + R)$$

Also, the accumulated amount after 1 year at an interest rate of r per year compounded m times a year is

$$A = P(1 + i)^n = P\left(1 + \frac{r}{m}\right)^m \qquad \text{Since } i = \frac{r}{m}$$

Equating the two expressions gives

$$P(1 + R) = P\left(1 + \frac{r}{m}\right)^m$$

$$1 + R = \left(1 + \frac{r}{m}\right)^m \qquad \text{Divide both sides by } P.$$

or, upon solving for R, we obtain the following formula for computing the effective rate of interest.

Effective Rate of Interest Formula

$$r_{\text{eff}} = \left(1 + \frac{r}{m}\right)^m - 1 \qquad\qquad \textbf{(6)}$$

where

r_{eff} = Effective rate of interest

r = Nominal interest rate per year

m = Number of conversion periods per year

VIDEO

EXAMPLE 5 Find the effective rate of interest corresponding to a nominal rate of 8% per year compounded (a) annually, (b) semiannually, (c) quarterly, (d) monthly, and (e) daily.

Solution

a. The effective rate of interest corresponding to a nominal rate of 8% per year compounded annually is of course given by 8% per year. This result is also confirmed by using Equation (4) with $r = 0.08$ and $m = 1$. Thus,

$$r_{\text{eff}} = (1 + 0.08) - 1 = 0.08$$

b. Let $r = 0.08$ and $m = 2$. Then, Equation (6) yields

$$r_{\text{eff}} = \left(1 + \frac{0.08}{2}\right)^2 - 1$$
$$= (1.04)^2 - 1$$
$$= 0.0816$$

so the required effective rate is 8.16% per year.

c. Let $r = 0.08$ and $m = 4$. Then, Equation (6) yields

$$r_{\text{eff}} = \left(1 + \frac{0.08}{4}\right)^4 - 1$$
$$= (1.02)^4 - 1$$
$$\approx 0.08243$$

so the corresponding effective rate in this case is 8.243% per year.

d. Let $r = 0.08$ and $m = 12$. Then, Equation (6) yields

$$r_{\text{eff}} = \left(1 + \frac{0.08}{12}\right)^{12} - 1$$
$$\approx 0.08300$$

so the corresponding effective rate in this case is 8.300% per year.

e. Let $r = 0.08$ and $m = 365$. Then Equation (6) yields

$$r_{\text{eff}} = \left(1 + \frac{0.08}{365}\right)^{365} - 1$$
$$\approx 0.08328$$

so the corresponding effective rate in this case is 8.328% per year. ◼

EXPLORE & DISCUSS

Recall the effective rate of interest formula:

$$r_{\text{eff}} = \left(1 + \frac{r}{m}\right)^m - 1$$

1. Show that
$$r = m[(1 + r_{\text{eff}})^{1/m} - 1]$$

2. A certificate of deposit (CD) is known to have an effective rate of 5.3%. If interest is compounded monthly, find the nominal rate of interest by using the result of part 1.

Now, if the effective rate of interest r_{eff} is known, the accumulated amount after t years on an investment of P dollars may be more readily computed by using the formula

$$A = P(1 + r_{\text{eff}})^t$$

The 1968 Truth in Lending Act passed by Congress requires that the effective rate of interest be disclosed in all contracts involving interest charges. The passage of this act has benefited consumers because they now have a common basis for comparing the various nominal rates quoted by different financial institutions. Furthermore, knowing the effective rate enables consumers to compute the actual charges involved in a transaction. Thus, if the effective rates of interest found in

Example 4 were known, the accumulated values of Example 3, shown in Table 3, could have been readily found.

TABLE 3

Nominal Rate, r	Frequency of Interest Payment	Effective Rate	Initial Investment	Accumulated Amount after 3 Years	
8%	Annually	8%	$1000	$1000(1 + 0.08)^3$	$\approx$ \$1259.71
8	Semiannually	8.16	1000	$1000(1 + 0.0816)^3$	$\approx$ 1265.32
8	Quarterly	8.243	1000	$1000(1 + 0.08243)^3$	$\approx$ 1268.23
8	Monthly	8.300	1000	$1000(1 + 0.08300)^3$	$\approx$ 1270.24
8	Daily	8.328	1000	$1000(1 + 0.08328)^3$	$\approx$ 1271.22

■ Present Value

Let's return to the compound interest Formula (3), which expresses the accumulated amount at the end of n periods when interest at the rate of r is compounded m times a year. The principal P in (3) is often referred to as the **present value,** and the accumulated value A is called the **future value,** since it is realized at a future date. In certain instances an investor may wish to determine how much money he should invest now, at a fixed rate of interest, so that he will realize a certain sum at some future date. This problem may be solved by expressing P in terms of A. Thus, from (3) we find

$$P = A(1 + i)^{-n}$$

Here, as before, $i = r/m$, where m is the number of conversion periods per year.

Present Value Formula for Compound Interest

$$P = A(1 + i)^{-n} \tag{7}$$

EXAMPLE 6 How much money should be deposited in a bank paying interest at the rate of 6% per year compounded monthly so that at the end of 3 years the accumulated amount will be \$20,000?

Solution Here, $r = 0.06$ and $m = 12$, so $i = \frac{0.06}{12}$ and $n = (3)(12) = 36$. Thus, the problem is to determine P, given that $A = 20,000$. Using Equation (7), we obtain

$$P = 20,000 \left(1 + \frac{0.06}{12}\right)^{-36}$$

$$\approx 16,713$$

or \$16,713. ■

EXAMPLE 7 Find the present value of \$49,158.60 due in 5 years at an interest rate of 10% per year compounded quarterly.

Solution Using Formula (7) with $r = 0.1$ and $m = 4$, so that $i = \frac{0.1}{4}$, $n = (4)(5) = 20$, and $A = 49{,}158.6$, we obtain

$$P = (49{,}158.6)\left(1 + \frac{0.1}{4}\right)^{-20} \approx 30{,}000.07$$

or approximately \$30,000. ▪

If we solve Formula (5) for P, we have

$$A = Pe^{rt}$$

and

$$P = Ae^{-rt} \qquad \textbf{(8)}$$

which gives the present value in terms of the future (accumulated) value for the case of continuous compounding.

APPLIED EXAMPLE 8   **Real Estate Investment** Blakely Investment Company owns an office building located in the commercial district of a city. As a result of the continued success of an urban renewal program, local business is enjoying a miniboom. The market value of Blakely's property is

$$V(t) = 300{,}000e^{\sqrt{t}/2}$$

where $V(t)$ is measured in dollars and t is the time in years from the present. If the expected rate of inflation is 9% compounded continuously for the next 10 years, find an expression for the present value $P(t)$ of the market price of the property valid for the next 10 years. Compute $P(7)$, $P(8)$, and $P(9)$ and interpret your results.

Solution   Using Formula (8) with $A = V(t)$ and $r = 0.09$, we find that the present value of the market price of the property t years from now is

$$\begin{aligned} P(t) &= V(t)e^{-0.09t} \\ &= 300{,}000e^{-0.09t+\sqrt{t}/2} \qquad (0 \le t \le 10) \end{aligned}$$

Letting $t = 7$, 8, and 9, respectively, we find that

$$P(7) = 300{,}000e^{-0.09(7)+\sqrt{7}/2} \approx 599{,}837, \text{ or } \$599{,}837$$
$$P(8) = 300{,}000e^{-0.09(8)+\sqrt{8}/2} \approx 600{,}640, \text{ or } \$600{,}640$$
$$P(9) = 300{,}000e^{-0.09(9)+\sqrt{9}/2} \approx 598{,}115, \text{ or } \$598{,}115$$

From the results of these computations, we see that the present value of the property's market price seems to decrease after a certain period of growth. This suggests there is an optimal time for the owners to sell. You can show that the highest present value of the property's market value is \$600,779 and it occurs at time $t \approx 7.72$ years by sketching the graph of the function P. ▪

The returns on certain investments such as zero coupon certificates of deposit (CDs) and zero coupon bonds are compared by quoting the time it takes for each investment to triple, or even quadruple. These calculations make use of the compound interest Formula (3).

APPLIED EXAMPLE 9   Investment Options   Jane has narrowed her investment options down to two:

1. Purchase a CD that matures in 12 years and pays interest upon maturity at the rate of 10% per year compounded daily (assume 365 days in a year).
2. Purchase a zero coupon CD that will triple her investment in the same period.

Which option will optimize her investment?

Solution   Let's compute the accumulated amount under option 1. Here,

$$r = 0.10 \qquad m = 365 \qquad t = 12$$

so $n = 12(365) = 4380$ and $i = 0.10/365$. The accumulated amount at the end of 12 years (after 4380 conversion periods) is

$$A = P\left(1 + \frac{0.10}{365}\right)^{4380} \approx 3.32P$$

or $3.32P$. If Jane chooses option 2, the accumulated amount of her investment after 12 years will be $3P. Therefore, she should choose option 1.   ∎

APPLIED EXAMPLE 10   IRAs   Moesha has an Individual Retirement Account (IRA) with a brokerage firm. Her money is invested in a money market mutual fund that pays interest on a daily basis. Over a 2-year period in which no deposits or withdrawals were made, her account grew from $4500 to $5268.24. Find the effective rate at which Moesha's account was earning interest over that period (assume 365 days in a year).

Solution   Let r_{eff} denote the required effective rate of interest. We have

$$5268.24 = 4500(1 + r_{eff})^2$$
$$(1 + r_{eff})^2 = 1.17072$$
$$1 + r_{eff} \approx 1.081998 \qquad \text{Take the square root on both sides.}$$

or $r_{eff} = 0.081998$. Therefore, the required effective rate is 8.20% per year.   ∎

5.1   Self-Check Exercises

1. Find the present value of $20,000 due in 3 yr at an interest rate of 12%/year compounded monthly.

2. Paul is a retiree living on Social Security and the income from his investment. Currently, his $100,000 investment in a 1-yr CD is yielding 10.6% interest compounded daily. If he rein-

vests the principal ($100,000) on the due date of the CD in another 1-yr CD paying 9.2% interest compounded daily, find the net decrease in his yearly income from his investment.

Solutions to Self-Check Exercises 5.1 can be found on page 292.

5.1   Concept Questions

1. Explain the difference between *simple interest* and *compound interest*.

2. What is the difference between the *accumulated amount* (future value) and the *present value* of an investment?

3. What is the *effective rate of interest*?

5.1 Exercises

1. Find the simple interest on a $500 investment made for 2 yr at an interest rate of 8%/year. What is the accumulated amount?

2. Find the simple interest on a $1000 investment made for 3 yr at an interest rate of 5%/year. What is the accumulated amount?

3. Find the accumulated amount at the end of 9 mo on an $800 deposit in a bank paying simple interest at a rate of 6%/year.

4. Find the accumulated amount at the end of 8 mo on a $1200 bank deposit paying simple interest at a rate of 7%/year.

5. If the accumulated amount is $1160 at the end of 2 yr and the simple rate of interest is 8%/year, what is the principal?

6. A bank deposit paying simple interest at the rate of 5%/year grew to a sum of $3100 in 10 mo. Find the principal.

7. How many days will it take for a sum of $1000 to earn $20 interest if it is deposited in a bank paying ordinary simple interest at the rate of 5%/year? (Use a 365-day year.)

8. How many days will it take for a sum of $1500 to earn $25 interest if it is deposited in a bank paying 5%/year? (Use a 365-day year.)

9. A bank deposit paying simple interest grew from an initial sum of $1000 to a sum of $1075 in 9 mo. Find the interest rate.

10. Determine the simple interest rate at which $1200 will grow to $1250 in 8 mo.

In Exercises 11–20, find the accumulated amount A if the principal P is invested at the interest rate of r per year for t years.

11. $P = \$1000$, $r = 7\%$, $t = 8$, compounded annually

12. $P = \$1000$, $r = 8\frac{1}{2}\%$, $t = 6$, compounded annually

13. $P = \$2500$, $r = 7\%$, $t = 10$, compounded semiannually

14. $P = \$2500$, $r = 9\%$, $t = 10.5$, compounded semiannually

15. $P = \$12,000$, $r = 8\%$, $t = 10.5$, compounded quarterly

16. $P = \$42,000$, $r = 7\frac{3}{4}\%$, $t = 8$, compounded quarterly

17. $P = \$150,000$, $r = 14\%$, $t = 4$, compounded monthly

18. $P = \$180,000$, $r = 9\%$, $t = 6\frac{1}{4}$, compounded monthly

19. $P = \$150,000$, $r = 12\%$, $t = 3$, compounded daily

20. $P = \$200,000$, $r = 8\%$, $t = 4$, compounded daily

In Exercises 21–24, find the effective rate corresponding to the given nominal rate.

21. 10%/year compounded semiannually

22. 9%/year compounded quarterly

23. 8%/year compounded monthly

24. 8%/year compounded daily

In Exercises 25–28, find the present value of $40,000 due in 4 years at the given rate of interest.

25. 8%/year compounded semiannually

26. 8%/year compounded quarterly

27. 7%/year compounded monthly

28. 9%/year compounded daily

29. Find the accumulated amount after 4 yr if $5000 is invested at 8%/year compounded continuously.

30. Find the accumulated amount after 6 yr if $6500 is invested at 7%/year compounded continuously.

31. Find the interest rate needed for an investment of $5000 to grow to an amount of $6000 in 3 yr if interest is compounded continuously.

32. Find the interest rate needed for an investment of $4000 to double in 5 yr if interest is compounded continuously.

33. How long will it take an investment of $6000 to grow to $7000 if the investment earns interest at the rate of $7\frac{1}{2}\%$ compounded continuously?

34. How long will it take an investment of $8000 to double if the investment earns interest at the rate of 8% compounded continuously?

35. CONSUMER DECISIONS Mitchell has been given the option of either paying his $300 bill now or settling it for $306 after 1 mo. If he chooses to pay after 1 mo, find the simple interest rate at which he would be charged.

36. COURT JUDGMENT Jennifer was awarded damages of $150,000 in a successful lawsuit she brought against her employer 5 yr ago. Interest (simple) on the judgment accrues at the rate of 12%/year from the date of filing. If the case were settled today, how much would Jennifer receive in the final judgment?

37. BRIDGE LOANS To help finance the purchase of a new house, the Abdullahs have decided to apply for a short-term loan (a bridge loan) in the amount of $120,000 for a term of 3 mo. If the bank charges simple interest at the rate of 12%/year, how much will the Abdullahs owe the bank at the end of the term?

38. CORPORATE BONDS David owns $20,000 worth of 10-yr bonds of Ace Corporation. These bonds pay interest every 6 mo at the rate of 7%/year (simple interest). How much income will David receive from this investment every 6 months? How much interest will David receive over the life of the bonds?

39. MUNICIPAL BONDS Maya paid $10,000 for a 7-yr bond issued by a city. She received interest amounting to $3500 over the life of the bonds. What rate of (simple) interest did the bond pay?

40. Write Equation (1b) in the slope-intercept form and interpret the meaning of the slope and the A-intercept in terms of r and P.
Hint: Refer to Figure 1.

41. HOSPITAL COSTS If the cost of a semiprivate room in a hospital was $480/day 5 yr ago and hospital costs have risen at the rate of 8%/year since that time, what rate would you expect to pay for a semiprivate room today?

42. FAMILY FOOD EXPENDITURE Today a typical family of four spends $600/month for food. If inflation occurs at the rate of 3%/year over the next 6 yr, how much should the typical family of four expect to spend for food 6 yr from now?

43. HOUSING APPRECIATION The Kwans are planning to buy a house 4 yr from now. Housing experts in their area have estimated that the cost of a home will increase at a rate of 5%/year during that period. If this economic prediction holds true, how much can the Kwans expect to pay for a house that currently costs $210,000?

44. ELECTRICITY CONSUMPTION A utility company in a western city of the United States expects the consumption of electricity to increase by 8%/year during the next decade, due mainly to the expected increase in population. If consumption does increase at this rate, find the amount by which the utility company will have to increase its generating capacity to meet the needs of the area at the end of the decade.

45. PENSION FUNDS The managers of a pension fund have invested $1.5 million in U.S. government certificates of deposit that pay interest at the rate of 5.5%/year compounded semiannually over a period of 10 yr. At the end of this period, how much will the investment be worth?

46. RETIREMENT FUNDS Five and a half years ago, Chris invested $10,000 in a retirement fund that grew at the rate of 10.82%/year compounded quarterly. What is his account worth today?

47. MUTUAL FUNDS Jodie invested $15,000 in a mutual fund 4 yr ago. If the fund grew at the rate of 9.8%/year compounded monthly, what would Jodie's account be worth today?

48. TRUST FUNDS A young man is the beneficiary of a trust fund established for him 21 yr ago at his birth. If the original amount placed in trust was $10,000, how much will he receive if the money has earned interest at the rate of

8%/year compounded annually? Compounded quarterly? Compounded monthly?

49. INVESTMENT PLANNING Find how much money should be deposited in a bank paying interest at the rate of 8.5%/year compounded quarterly so that at the end of 5 yr the accumulated amount will be $40,000.

50. PROMISSORY NOTES An individual purchased a 4-yr, $10,000 promissory note with an interest rate of 8.5%/year compounded semiannually. How much did the note cost?

51. FINANCING A COLLEGE EDUCATION The parents of a child have just come into a large inheritance and wish to establish a trust fund for her college education. If they estimate that they will need $100,000 in 13 yr, how much should they set aside in the trust now if they can invest the money at $8\frac{1}{2}$%/year compounded (a) annually, (b) semiannually, and (c) quarterly?

52. INVESTMENTS Anthony invested a sum of money 5 yr ago in a savings account that has since paid interest at the rate of 8%/year compounded quarterly. His investment is now worth $22,289.22. How much did he originally invest?

53. RATE COMPARISONS In the last 5 yr, Bendix Mutual Fund grew at the rate of 10.4%/year compounded quarterly. Over the same period, Acme Mutual Fund grew at the rate of 10.6%/year compounded semiannually. Which mutual fund has a better rate of return?

54. RATE COMPARISONS Fleet Street Savings Bank pays interest at the rate of 4.25%/year compounded weekly in a savings account, whereas Washington Bank pays interest at the rate of 4.125%/year compounded daily (assume a 365-day year). Which bank offers a better rate of interest?

55. LOAN CONSOLIDATION The proprietors of The Coachmen Inn secured two loans from Union Bank: one for $8000 due in 3 yr and one for $15,000 due in 6 yr, both at an interest rate of 10%/year compounded semiannually. The bank has agreed to allow the two loans to be consolidated into one loan payable in 5 yr at the same interest rate. What amount will the proprietors of the inn be required to pay the bank at the end of 5 yr?

56. EFFECTIVE RATE OF INTEREST Find the effective rate of interest corresponding to a nominal rate of 9%/year compounded annually, semiannually, quarterly, and monthly.

57. HOUSING APPRECIATION Georgia purchased a house in 1998 for $200,000. In 2004 she sold the house and made a net profit of $56,000. Find the effective annual rate of return on her investment over the 6-yr period.

58. COMMON STOCK TRANSACTION Steven purchased 1000 shares of a certain stock for $25,250 (including commissions). He sold the shares 2 yr later and received $32,100 after deducting commissions. Find the effective annual rate of return on his investment over the 2-yr period.

59. ZERO COUPON BONDS Nina purchased a zero coupon bond for $6724.53. The bond matures in 7 yr and has a face value of

$10,000. Find the effective rate of interest for the bond if interest is compounded semiannually.

Hint: Assume that the purchase price of the bond is the initial investment and that the face value of the bond is the accumulated amount.

60. ZERO COUPON BONDS Juan is contemplating buying a zero coupon bond that matures in 10 yr and has a face value of $10,000. If the bond yields a return of 5.25%/year, how much should Juan pay for the bond?

61. MONEY MARKET MUTUAL FUNDS Carlos invested $5000 in a money market mutual fund that pays interest on a daily basis. The balance in his account at the end of 8 mo (245 days) was $5170.42. Find the effective rate at which Carlos's account earned interest over this period (assume a 365-day year).

62. REVENUE GROWTH OF A HOME THEATER BUSINESS Maxwell started a home theater business in 2002. The revenue of his company for that year was $240,000. The revenue grew by 20% in 2003 and 30% in 2004. Maxwell projected that the revenue growth for his company in the next 3 yr will be at least 25%/year. How much does Maxwell expect his minimum revenue to be for 2007?

63. ONLINE RETAIL SALES Online retail sales stood at $23.5 billion for the year 2000. For the next 2 yr, they grew by 33.2% and 27.8% per year, respectively. For the next 6 yr, online retail sales are projected to grow at 30.5%, 19.9%, 24.3%, 14.0%, 17.6%, and 10.5% per year, respectively. What are the projected online sales for 2008?

Source: Jupiter Research

64. PURCHASING POWER The inflation rates in the U.S. economy in 2000 through 2003 are 3.4%, 2.8%, 1.6%, and 2.3%, respectively. What is the purchasing power of a dollar at the beginning of 2004 compared to that at the beginning of 2000?

Source: U.S. Census Bureau

65. INVESTMENT OPTIONS Investment A offers a 10% return compounded semiannually, and investment B offers a 9.75% return compounded continuously. Which investment has a higher rate of return over a 4-yr period?

66. EFFECT OF INFLATION ON SALARIES Leonard's current annual salary is $45,000. Ten years from now, how much will he need to earn in order to retain his present purchasing power if the rate of inflation over that period is 3%/year? Assume that inflation is continuously compounded.

67. SAVING FOR COLLEGE Having received a large inheritance, Jing-mei's parents wish to establish a trust for her college education. If 7 yr from now they need an estimated $70,000, how much should they set aside in trust now, if they invest the money at 10.5% compounded quarterly? Continuously?

68. PENSIONS Maria, who is now 50 years old, is employed by a firm that guarantees her a pension of $40,000/year at age 65. What is the present value of her first year's pension if infla-

tion over the next 15 yr is 6%? 8%? 12%? Assume that inflation is continuously compounded.

69. REAL ESTATE INVESTMENTS An investor purchased a piece of waterfront property. Because of the development of a marina in the vicinity, the market value of the property is expected to increase according to the rule

$$V(t) = 80,000e^{\sqrt{t/2}}$$

where $V(t)$ is measured in dollars and t is the time in years from the present. If the inflation rate is expected to be 9% compounded continuously for the next 8 yr, find an expression for the present value $P(t)$ of the property's market price valid for the next 8 yr. What is $P(t)$ expected to be in 4 yr?

70. The simple interest formula $A = P(1 + rt)$ [Formula (1b)] can be written in the form $A = Prt + P$, which is the slope-intercept form of a straight line with slope Pr and A-intercept P.
a. Describe the family of straight lines obtained by keeping the value of r fixed and allowing the value of P to vary. Interpret your results.
b. Describe the family of straight lines obtained by keeping the value of P fixed and allowing the value of r to vary. Interpret your results.

71. EFFECTIVE RATE OF INTEREST Suppose an initial investment of P grows to an accumulated amount of A in t yr. Show that the effective rate (annual effective yield) is

$$r_{eff} = (A/P)^{1/t} - 1$$

72. EFFECTIVE RATE OF INTEREST Martha invested $40,000 in a boutique 5 yr ago. Her investment is worth $70,000 today. What is the effective rate (annual effective yield) of her investment?

Hint: See Exercise 71.

In Exercises 73–76, determine whether the statement is true or false. If it is true, explain why it is true. If it is false, give an example to show why it is false.

73. When simple interest is used, the accumulated amount is a linear function of t.

74. If compound interest is converted annually, then the accumulated amount after t yr is the same as the accumulated amount under simple interest over t yr.

75. If interest is compounded annually, then the effective rate is the same as the nominal rate.

76. Susan's salary increased from $40,000/year to $50,000/year over a 5-yr period. Therefore, Susan got annual increases of 5% over that period.

5.1   Solutions to Self-Check Exercises

1. Using Equation (7) with $A = 20{,}000$, $r = 0.12$, and $m = 12$ so that $i = 0.12/12$ and $n = (3)(12) = 36$, we find the required present value to be

$$P = 20{,}000\left(1 + \frac{0.12}{12}\right)^{-36} \approx 13{,}978.50$$

or $13,978.50

2. The accumulated amount of Paul's current investment is found by using Equation (3) with $P = 100{,}000$, $r = 0.106$, and $m = 365$. Thus, $i = 0.106/365$ and $n = 365$, so the required accumulated amount is given by

$$A = 100{,}000\left(1 + \frac{0.106}{365}\right)^{365} \approx 111{,}180.48$$

or $111,180.48. Next, we compute the accumulated amount of Paul's reinvestment. Once again, using (3) with $P = 100{,}000$, $r = 0.092$, and $m = 365$ so that $i = 0.092/365$ and $n = 365$, we find the required accumulated amount in this case to be

$$\overline{A} = 100{,}000\left(1 + \frac{0.092}{365}\right)^{365}$$

or $109,635.21. Therefore, Paul can expect to experience a net decrease in yearly income of

$$111{,}180.48 - 109{,}635.21$$

or $1545.27.

USING TECHNOLOGY

■ **Finding the Accumulated Amount of an Investment, the Effective Rate of Interest, and the Present Value of an Investment**

Graphing Utility

Some graphing utilities have built-in routines for solving problems involving the mathematics of finance. For example, the TI-83 incorporates several functions that can be used to solve the problems that are encountered in Sections 5.1–5.3. Once again, the step-by-step procedures for using these functions are to be found on the Web site.

```
N     = 120
I%    = 10
PV    = -5000
PMT = 0
■ FV   = 13535.20745
P/Y   = 12
C/Y   = 12
PMT : END    BEGIN
```

FIGURE T1
The TI-83 screen showing the future value of an investment (FV)

EXAMPLE 1   Finding the Accumulated Amount of an Investment   Find the accumulated amount after 10 years if $5000 is invested at a rate of 10% per year compounded monthly.

Solution   Using the TI-83 TVM SOLVER with the following inputs,

$$N = 120 \qquad \text{(10)(12)}$$

$$I\% = 10$$

$$PV = -5000 \qquad \text{Recall that an investment is an outflow.}$$

$$PMT = 0$$

$$FV = 0$$

$$P/Y = 12 \qquad \text{The number of payments each year}$$

$$C/Y = 12 \qquad \text{The number of conversion periods each year}$$

$$\text{PMT:END BEGIN}$$

we obtain the display shown in Figure T1. We conclude that the required accumulated amount is $13,535.21.

EXAMPLE 2 Finding the Effective Rate of Interest Find the effective rate of interest corresponding to a nominal rate of 10% per year compounded quarterly.

Solution Here we use the **Eff** function of the TI-83 calculator to obtain the result shown in Figure T2. The required effective rate is approximately 10.38% per year.

```
► Eff (10, 4)
        10.38128906
```

FIGURE T2
The TI-83 screen showing the effective rate of interest (Eff)

EXAMPLE 3 Finding the Present Value of an Investment Find the present value of $20,000 due in 5 years if the interest rate is 7.5% per year compounded daily.

Solution Using the TI-83 **TVM SOLVER** with the following inputs,

$$N = 1825 \quad \text{(5)(365)}$$
$$I\% = 7.5$$
$$PV = 0$$
$$PMT = 0$$
$$FV = 20000$$
$$P/Y = 365 \quad \text{The number of payments each year}$$
$$C/Y = 365 \quad \text{The number of conversions each year}$$
$$PMT:END \ BEGIN$$

we obtain the display shown in Figure T3. We see that the required present value is approximately $13,746.32. Note that PV is negative because an investment is an outflow (money is paid out).

```
N    = 1825
I%   = 7.5
■ PV   = −13746.3151
PMT = 0
FV   = 20000
P/Y  = 365
C/Y  = 365
PMT : END    BEGIN
```

FIGURE T3
The TI-83 screen showing the present value of an investment (PV)

Excel

Excel has many built-in functions for solving problems involving the mathematics of finance. Here we illustrate the use of the FV (future value), EFFECT (effective rate), and the PV (present value) functions to solve problems of the type that we have encountered in Section 5.1.

Note: Boldfaced words/characters enclosed in a box (for example, ⌐Enter⌐) indicate that an action (click, select, or press) is required. Words/characters printed blue (for example, Chart sub-type:) indicate words/characters appearing on the screen. Words/characters in a typewriter font (for example, =(−2/3)*A2+2) indicate words/characters that need to be typed and entered.

(continued)

EXAMPLE 4 **Finding the Accumulated Amount of an Investment**   Find the accumulated amount after 10 years if $5000 is invested at a rate of 10% per year compounded monthly.

Solution   Here we are computing the future value of a lump-sum investment, so we use the FV (future value) function. Select $\boxed{f_x}$ from the toolbar to obtain the Insert Function dialog box. Then select $\boxed{\textbf{Financial}}$ from the Or select a category: list box. Next, select $\boxed{\textbf{FV}}$ under Select a function: and click $\boxed{\textbf{OK}}$. The Function Arguments dialog box will appear (see Figure T4). In our example, the mouse cursor is in the edit box headed by Type, so a definition of that term appears near the bottom of the box. Figure T4 shows the entries for each edit box in our example.

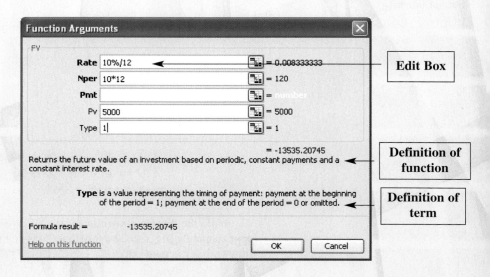

Note that the entry for Nper is given by the total number of periods for which the investment earns interest. The Pmt box is left blank since no money is added to the original investment. The PV entry is 5000. The entry for Type is a 1 because the lump-sum payment is made at the beginning of the investment period. The answer, $-\$13,535.21$, is shown at the bottom of the dialog box. It is negative because an investment is considered an outflow of money (money is paid out). (Click $\boxed{\textbf{OK}}$ and the answer will also appear on your spreadsheet.) ∎

EXAMPLE 5 **Finding the Effective Rate of Interest**   Find the effective rate of interest corresponding to a nominal rate of 10% per year compounded quarterly.

Solution   Here we use the EFFECT function to compute the effective rate of interest. Accessing this function from the Insert Function dialog box and making the required entries, we obtain the Function Arguments dialog box shown in Figure T5. The required effective rate is approximately 10.38% per year.

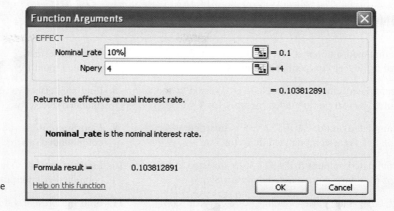

FIGURE T5
Excel's dialog box for the effective rate of interest function EFFECT

EXAMPLE 6   Finding the Present Value of an Investment   Find the present value of $20,000 due in 5 years if the interest rate is 7.5% per year compounded daily.

Solution   We use the PV function to compute the present value of a lump-sum investment. Accessing this function from the Insert Function dialog box and making the required entries, we obtain the PV dialog box shown in Figure T6. Once again, the Pmt edit box is left blank since no additional money is added to the original investment. The FV entry is 20000. The answer is negative because an investment is considered an outflow of money (money is paid out). We deduce that the required amount is $13,746.32.

Function Arguments   [X]

PV

Rate	7.5%/365	= 0.000205479
Nper	5*365	= 1825
Pmt		= number
Fv	20000	= 20000
Type	1	= 1

= -13746.3151

Returns the present value of an investment: the total amount that a series of future payments is worth now.

Type is a logical value: payment at the beginning of the period = 1; payment at the end of the period = 0 or omitted.

Formula result =   -13746.3151

Help on this function   [OK]   [Cancel]

FIGURE T6
Excel dialog box for the present value function (PV)

(*continued*)

TECHNOLOGY EXERCISES

1. Find the accumulated amount A if $5000 is invested at the interest rate of $5\frac{3}{8}\%$/year compounded monthly for 3 yr.

2. Find the accumulated amount A if $2850 is invested at the interest rate of $6\frac{5}{8}\%$/year compounded monthly for 4 yr.

3. Find the accumulated amount A if $327.35 is invested at the interest rate of $5\frac{1}{3}\%$/year compounded daily for 7 yr.

4. Find the accumulated amount A if $327.35 is invested at the interest rate of $6\frac{7}{8}\%$/year compounded daily for 8 yr.

5. Find the effective rate corresponding to $8\frac{2}{3}\%$/year compounded quarterly.

6. Find the effective rate corresponding to $10\frac{5}{8}\%$/year compounded monthly.

7. Find the effective rate corresponding to $9\frac{3}{4}\%$/year compounded monthly.

8. Find the effective rate corresponding to $4\frac{3}{8}\%$/year compounded quarterly.

9. Find the present value of $38,000 due in 3 yr at $8\frac{1}{4}\%$/year compounded quarterly.

10. Find the present value of $150,000 due in 5 yr at $9\frac{3}{8}\%$/year compounded monthly.

11. Find the present value of $67,456 due in 3 yr at $7\frac{7}{8}\%$/year compounded monthly.

12. Find the present value of $111,000 due in 5 yr at $11\frac{5}{8}\%$/year compounded monthly.

5.2   Annuities

Future Value of an Annuity

An **annuity** is a sequence of payments made at regular time intervals. The time period in which these payments are made is called the **term** of the annuity. Depending on whether the term is given by a *fixed time interval,* a time interval that begins at a definite date but extends indefinitely, or one that is not fixed in advance, an annuity is called an **annuity certain,** a *perpetuity,* or a *contingent annuity,* respectively. In general, the payments in an annuity need not be equal, but in many important applications they are equal. In this section we assume that annuity payments are equal. Examples of annuities are regular deposits to a savings account, monthly home mortgage payments, and monthly insurance payments.

Annuities are also classified by payment dates. An annuity in which the payments are made at the *end* of each payment period is called an **ordinary annuity,** whereas an annuity in which the payments are made at the beginning of each period is called an *annuity due.* Furthermore, an annuity in which the payment period coincides with the interest conversion period is called a **simple annuity,** whereas an annuity in which the payment period differs from the interest conversion period is called a *complex annuity.*

In this section, we consider ordinary annuities that are certain and simple, with periodic payments that are equal in size. In other words, we study annuities that are subject to the following conditions:

1. The terms are given by fixed time intervals.
2. The periodic payments are equal in size.
3. The payments are made at the *end* of the payment periods.
4. The payment periods coincide with the interest conversion periods.

To find a formula for the accumulated amount S of an annuity, suppose a sum of $100 is paid into an account at the end of each quarter over a period of 3 years. Furthermore, suppose the account earns interest on the deposit at the rate of 8% per year, compounded quarterly. Then, the first payment of $100 made at the end of the first quarter earns interest at the rate of 8% per year compounded four times a year (or 8/4 = 2% per quarter) over the remaining 11 quarters and therefore, by the compound interest formula, has an accumulated amount of

$$100\left(1 + \frac{0.08}{4}\right)^{11} \quad \text{or} \quad 100(1 + 0.02)^{11}$$

dollars at the end of the term of the annuity (Figure 2).

The second payment of $100 made at the end of the second quarter earns interest at the same rate over the remaining 10 quarters and therefore has an accumulated amount of

$$100(1 + 0.02)^{10}$$

dollars at the end of the term of the annuity, and so on. The last payment earns no interest since it is due at the end of the term. The amount of the annuity is obtained by adding all the terms in Figure 2. Thus,

$$S = 100 + 100(1 + 0.02) + 100(1 + 0.02)^2 + \cdots + 100(1 + 0.02)^{11}$$

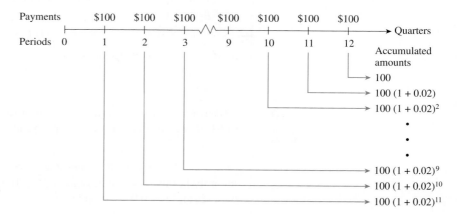

FIGURE 2
The sum of the accumulated amounts is the amount of the annuity.

The sum on the right is the sum of the first n terms of a *geometric progression* with first term R and common ratio $(1 + i)$. We will show in Section 5.4 that the sum S can be written in the more compact form

$$S = 100\left[\frac{(1 + 0.02)^{12} - 1}{0.02}\right]$$
$$\approx 1341.21$$

or approximately $1341.21.

To find a general formula for the accumulated amount S of an annuity, suppose a sum of R is paid into an account at the end of each period for n periods and the account earns interest at the rate of i per period. Then, proceeding as we did with the numerical example, we see that

$$S = R + R(1 + i) + R(1 + i)^2 + \cdots + R(1 + i)^{n-1}$$

$$= R\left[\frac{(1 + i)^n - 1}{i}\right] \tag{9}$$

The expression inside the brackets is commonly denoted by $s_{\overline{n}|i}$ (read "s angle n at i") and is called the **compound-amount factor.** Extensive tables have been constructed that give values of $s_{\overline{n}|i}$ for different values of i and n (such as Table 1, which is found on the Web site). In terms of the compound-amount factor,

$$S = Rs_{\overline{n}|i} \tag{10}$$

The quantity S in Equations (9) and (10) is realizable at some future date and is accordingly called the future value of an annuity.

Future Value of an Annuity

The future value S of an annuity of n payments of R dollars each, paid at the end of each investment period into an account that earns interest at the rate of i per period, is

$$S = R\left[\frac{(1 + i)^n - 1}{i}\right]$$

EXAMPLE 1   Find the amount of an ordinary annuity of 12 monthly payments of $100 that earn interest at 12% per year compounded monthly.

Solution   Since i is the interest rate per *period* and interest is compounded monthly in this case, we have $i = \frac{0.12}{12} = 0.01$. Using Equation (9) with $R = 100$, $n = 12$, and $i = 0.01$, we have

$$S = \frac{100[(1.01)^{12} - 1]}{0.01}$$

$$\approx 1268.25 \qquad \text{Use a calculator.}$$

or $1268.25. The same result is obtained by observing that

$$S = 100s_{\overline{12}|0.01}$$

$$= 100(12.6825)$$

$$= 1268.25 \qquad \text{Use Table 1 from the Web site.}$$

EXPLORE & DISCUSS

Future Value S of an Annuity Due

1. Consider an annuity satisfying Conditions 1, 2, and 4 on page 296, but with Condition 3 replaced by the condition that payments are made at the beginning of the payment periods. By using an argument similar to that used to establish Formula (9), show that the future value S of an annuity due of n payments of R dollars each, paid at the *beginning* of each investment into an account that earns interest at the rate of i per period, is

$$S = R(1 + i)\left[\frac{(1 + i)^n - 1}{i}\right]$$

2. Use the result of part 1 to see how much your nest egg will be at age 65 if you start saving $4000 annually at age 30, assuming a 10% average annual return; if you start saving at 35; if you start saving at 40. [Moral of the story: It is never too early to start saving!]

 EXPLORING WITH TECHNOLOGY

Refer to the Explore & Discuss problem on this page.

1. Show that if $R = 4000$ and $i = 0.1$, then $S = 44,000[(1.1)^n - 1]$. Using a graphing utility, plot the graph of $f(x) = 44,000[(1.1)^x - 1]$, using the viewing window $[0, 40] \times [0, 1,200,000]$.

2. Verify the results of part 1 by evaluating $f(35)$, $f(30)$, and $f(25)$ using the EVAL function.

Present Value of an Annuity

In certain instances, you may want to determine the current value P of a sequence of equal periodic payments that will be made over a certain period of time. After each payment is made, the new balance continues to earn interest at some nominal rate. The amount P is referred to as the present value of an annuity.

To derive a formula for determining the present value P of an annuity, we may argue as follows. The amount P invested now and earning interest at the rate of i per period will have an accumulated value of $P(1 + i)^n$ at the end of n periods. But this must be equal to the future value of the annuity S given by Formula (9). Therefore, equating the two expressions, we have

$$P(1 + i)^n = R\left[\frac{(1 + i)^n - 1}{i}\right]$$

Multiplying both sides of this equation by $(1 + i)^{-n}$ gives

$$P = R(1 + i)^{-n}\left[\frac{(1 + i)^n - 1}{i}\right]$$

$$= R\left[\frac{(1 + i)^n(1 + i)^{-n} - (1 + i)^{-n}}{i}\right] \qquad (1 + i)^n(1 + i)^{-n} = 1$$

$$= R\left[\frac{1 - (1 + i)^{-n}}{i}\right]$$

$$= Ra_{\overline{n}|i}$$

where the factor $a_{\overline{n}|i}$ (read "*a* angle *n* at *i*") represents the expression inside the brackets. Extensive tables have also been constructed giving values of $a_{\overline{n}|i}$ for different values of *i* and *n* (see Table 1, found on the Web site).

Present Value of an Annuity
The **present value P of an annuity** of *n* payments of *R* dollars each, paid at the end of each investment period into an account that earns interest at the rate of *i* per period, is

$$P = R\left[\frac{1 - (1 + i)^{-n}}{i}\right] \tag{11}$$

EXAMPLE 2 Find the present value of an ordinary annuity of 24 payments of $100 each made monthly and earning interest at 9% per year compounded monthly.

Solution Here, $R = 100$, $i = \frac{r}{m} = \frac{0.09}{12} = 0.0075$, and $n = 24$, so by Formula (11)

$$P = \frac{100[1 - (1.0075)^{-24}]}{0.0075}$$

$$\approx 2188.91$$

or $2188.91. The same result may be obtained by using Table 1 from the Web. Thus,

$$P = 100a_{\overline{24}|0.0075}$$

$$= 100(21.8891)$$

$$= 2188.91$$

APPLIED EXAMPLE 3 Saving for a College Education As a savings program toward Alberto's college education, his parents decide to deposit $100 at the end of every month into a bank account paying interest at the rate of 6% per year compounded monthly. If the savings program began when Alberto was 6 years old, how much money would have accumulated by the time he turns 18?

Solution By the time the child turns 18, the parents would have made 144 deposits into the account. Thus, $n = 144$. Furthermore, we have $R = 100$, $r = 0.06$, and $m = 12$, so $i = \frac{0.06}{12} = 0.005$. Using Equation (9), we find that the amount of money that would have accumulated is given by

$$S = \frac{100[(1.005)^{144} - 1]}{0.005}$$

$$\approx 21{,}015$$

or $21,015. ∎

APPLIED EXAMPLE 4   Financing a Car   After making a down payment of $2000 for an automobile, Murphy paid $200 per month for 36 months with interest charged at 12% per year compounded monthly on the unpaid balance. What was the original cost of the car? What portion of Murphy's total car payments went toward interest charges?

Solution The loan taken up by Murphy is given by the present value of the annuity

$$P = \frac{200[1 - (1.01)^{-36}]}{0.01} = 200a_{\overline{36}|0.01}$$

$$\approx 6021.50$$

or $6021.50. Therefore, the original cost of the automobile is $8021.50 ($6021.50 plus the $2000 down payment). The interest charges paid by Murphy are given by $(36)(200) - 6021.50$, or $1178.50. ∎

One important application of annuities is in the area of tax planning. During the 1980s, Congress created many tax-sheltered retirement savings plans, such as Individual Retirement Accounts (IRAs), Keogh plans, and Simplified Employee Pension (SEP) plans. These plans are examples of annuities in which the individual is allowed to make contributions (which are often tax deductible) to an investment account. The amount of the contribution is limited by congressional legislation. The taxes on the contributions and/or the interest accumulated in these accounts are deferred until the money is withdrawn, ideally during retirement, when tax brackets should be lower. In the interim period, the individual has the benefit of tax-free growth on his or her investment.

Suppose, for example, you are eligible to make a fully deductible contribution to an IRA and you are in a marginal tax bracket of 28%. Additionally, suppose you receive a year-end bonus of $2000 from your employer and have the option of depositing the $2000 into either an IRA or a regular savings account, both accounts earning interest at an effective annual rate of 8% per year. If you choose to invest your bonus in a regular savings account, you will first have to pay taxes on the $2000, leaving $1440 to invest. At the end of 1 year, you will also have to pay taxes on the interest earned, leaving you with

Accumulated amount	−	Tax on interest	=	Net amount
1555.20	−	32.26	=	1522.94

or $1522.94.

On the other hand, if you put the money into the IRA account, the entire sum will earn interest, and at the end of 1 year you will have (1.08)($2000), or $2160, in your account. Of course, you will still have to pay taxes on this money when you withdraw it, but you will have gained the advantage of tax-free growth of the larger principal over the years. The disadvantage of this option is that if you withdraw the money before you reach the age of $59\frac{1}{2}$, you will be liable for taxes on both your contributions and the interest earned and you will also have to pay a 10% penalty.

Note   In practice, the size of the contributions an individual might make to the various retirement plans might vary from year to year. Also, he or she might make the contributions at different payment periods. To simplify our discussion, we will consider examples in which fixed payments are made at regular intervals.

APPLIED EXAMPLE 5   IRA Accounts   Caroline is planning to make a contribution of $2000 on January 31 of each year into an IRA earning interest at an effective rate of 9% per year. After she makes her 25th payment on January 31 of the year following her retirement, how much will she have in her IRA?

Solution   The amount of money Caroline will have after her 25th payment into her account is found by using Equation (9) with $R = 2000$, $r = 0.09$, and $m = 1$, so $i = r/m = 0.09$ and $n = 25$. The required amount is given by

$$S = \frac{2000[(1.09)^{25} - 1]}{0.09}$$
$$\approx 169,401.79$$

or $169,401.79.

After-tax-deferred annuities are another type of investment vehicle that allows an individual to build assets for retirement, college funds, or other future needs. The advantage gained in this type of investment is that the tax on the accumulated interest is deferred to a later date. Note that in this type of investment the contributions themselves are not tax deductible. At first glance, the advantage thus gained may seem to be relatively inconsequential, but its true effect is illustrated by the next example.

APPLIED EXAMPLE 6   Investment Analysis   Both Clark and Colby are salaried individuals, 45 years of age, who are saving for their retirement 20 years from now. Both Clark and Colby are also in the 28% marginal tax bracket. Clark makes a $1000 contribution annually on December 31 into a savings account earning an effective rate of 8% per year. At the same time, Colby makes a $1000 annual payment to an insurance company for an after-tax-deferred annuity. The annuity also earns interest at an effective rate of 8% per year. (Assume that both men remain in the same tax bracket throughout this period and disregard state income taxes.)
  a. Calculate how much each man will have in his investment account at the end of 20 years.
  b. Compute the interest earned on each account.

c. Show that even if the interest on Colby's investment were subjected to a tax of 28% upon withdrawal of his investment at the end of 20 years, the net accumulated amount of his investment would still be greater than that of Clark's.

Solution

a. Because Clark is in the 28% marginal tax bracket, the net yield for his investment is $(0.72)(8)$, or 5.76%, per year.

Using Formula (9) with $R = 1000$, $r = 0.0576$, and $m = 1$, so that $i = 0.0576$ and $n = 20$, we see that Clark's investment will be worth

$$S = \frac{1000[(1 + 0.0576)^{20} - 1]}{0.0576}$$

$$\approx 35{,}850.49$$

or $35,850.49 at his retirement.

Colby has a tax-sheltered investment with an effective yield of 8% per year. Using Formula (9) with $R = 1000$, $r = 0.08$, and $m = 1$, so that $i = 0.08$ and $n = 20$, we see that Colby's investment will be worth

$$S = \frac{1000[(1 + 0.08)^{20} - 1]}{0.08}$$

$$\approx 45{,}761.96$$

or $45,761.96 at his retirement.

b. Each man will have paid $20(1000)$, or $20,000, into his account. Therefore, the total interest earned in Clark's account will be $(35{,}850.49 - 20{,}000)$, or $15,850.49, whereas the total interest earned in Colby's account will be $(45{,}761.96 - 20{,}000)$, or $25,761.96.

c. From part (b) we see that the total interest earned in Colby's account will be $25,761.96. If it were taxed at 28%, he would still end up with $(0.72)(25{,}761.96)$, or $18,548.61. This is larger than the total interest of $15,850.49 earned by Clark.   ■

5.2   Self-Check Exercises

1. Phyliss Fletcher opened an IRA on January 31, 1990, with a contribution of $2000. She plans to make a contribution of $2000 thereafter on January 31 of each year until her retirement in the year 2009 (20 payments). If the account earns interest at the rate of 8%/year compounded yearly, how much will Phyliss have in her account when she retires?

2. Denver Wildcatting Company has an immediate need for a loan. In an agreement worked out with its banker, Denver assigns its royalty income of $4800/month for the next 3 years from certain oil properties to the bank, with the first payment due at the end of the first month. If the bank charges interest at the rate of 9%/year compounded monthly, what is the amount of the loan negotiated between the parties?

Solutions to Self-Check Exercises 5.2 can be found on page 305.

5.2   Concept Questions

1. In an ordinary annuity, is the term fixed or variable? Are the periodic payments all of the same size, or do they vary in size? Are the payments made at the beginning or the end of the payment period? Do the payment periods coincide with the interest conversion periods?

2. What is the difference between an *ordinary annuity* and an *annuity due*?

3. What is the *future value of an annuity*? Give an example.

4. What is the *present value of an annuity*? Give an example.

5.2 Exercises

In Exercises 1–8, find the amount (future value) of each ordinary annuity.

1. $1000 a year for 10 yr at 10%/year compounded annually

2. $1500 per semiannual period for 8 yr at 9%/year compounded semiannually

3. $1800 per quarter for 6 yr at 8%/year compounded quarterly

4. $500 per semiannual period for 12 yr at 11%/year compounded semiannually

5. $600 per quarter for 9 yr at 12%/year compounded quarterly

6. $150 per month for 15 yr at 10%/year compounded monthly

7. $200/month for $20\frac{1}{4}$ yr at 9%/year compounded monthly

8. $100/week for $7\frac{1}{2}$ yr at 7.5%/year compounded weekly

In Exercises 9–14, find the present value of each ordinary annuity.

9. $5000 a year for 8 yr at 8%/year compounded annually

10. $1200 per semiannual period for 6 yr at 10%/year compounded semiannually

11. $4000 a year for 5 yr at 9%/year compounded yearly

12. $3000 per semiannual period for 6 yr at 11%/year compounded semiannually

13. $800 per quarter for 7 yr at 12%/year compounded quarterly

14. $150 per month for 10 yr at 8%/year compounded monthly

15. **IRAs** If a merchant deposits $1500 annually at the end of each tax year in an IRA account paying interest at the rate of 8%/year compounded annually, how much will she have in her account at the end of 25 yr?

16. **Savings Accounts** If Jackson deposits $100 at the end of each month in a savings account earning interest at the rate of 8%/year compounded monthly, how much will he have on deposit in his savings account at the end of 6 yr, assuming that he makes no withdrawals during that period?

17. **Savings Accounts** Linda has joined a "Christmas Fund Club" at her bank. At the end of every month, December through October inclusive, she will make a deposit of $40 in her fund. If the money earns interest at the rate of 7%/year compounded monthly, how much will she have in her account on December 1 of the following year?

18. **Keogh Accounts** Robin, who is self-employed, contributes $5000/year into a Keogh account. How much will he have in

the account after 25 yr if the account earns interest at the rate of 8.5%/year compounded yearly?

19. **Retirement Planning** As a fringe benefit for the past 12 yr, Colin's employer has contributed $100 at the end of each month into an employee retirement account for Colin that pays interest at the rate of 7%/year compounded monthly. Colin has also contributed $2000 at the end of each of the last 8 yr into an IRA that pays interest at the rate of 9%/year compounded yearly. How much does Colin have in his retirement fund at this time?

20. **Savings Accounts** The Pirerras are planning to go to Europe 3 yr from now and have agreed to set aside $150 each month for their trip. If they deposit this money at the end of each month into a savings account paying interest at the rate of 8%/year compounded monthly, how much money will be in their travel fund at the end of the third year?

21. **Investment Analysis** Karen has been depositing $150 at the end of each month in a tax-free retirement account since she was 25. Matt, who is the same age as Karen, started depositing $250 at the end of each month in a tax-free retirement account when he was 35. Assuming that both accounts have been and will be earning interest at the rate of 5%/year compounded monthly, who will end up with the larger retirement account at the age of 65?

22. **Investment Analysis** Luis has $150,000 in his retirement account at his present company. Because he is assuming a position with another company, Luis is planning to roll over his assets to a new account. Luis also plans to put $3000/quarter into the new account until his retirement 20 yr from now. If the account earns interest at the rate of 8%/year compounded quarterly, how much will Luis have in his account at the time of his retirement?
Hint: Use the compound interest formula and the annuity formula.

23. **Auto Leasing** The Betzes have leased an auto for 2 yr at $450/month. If money is worth 9%/year compounded monthly, what is the equivalent cash payment (present value) of this annuity?

24. **Auto Financing** Lupé made a down payment of $2000 toward the purchase of a new car. To pay the balance of the purchase price, she has secured a loan from her bank at the rate of 12%/year compounded monthly. Under the terms of her finance agreement, she is required to make payments of $210/month for 36 mo. What is the cash price of the car?

25. **Installment Plans** Pierce Publishing sells encyclopedias under two payment plans: cash or installment. Under the

installment plan, the customer pays $22/month over 3 yr with interest charged on the balance at a rate of 18%/year compounded monthly. Find the cash price for a set of encyclopedias if it is equivalent to the price paid by a customer using the installment plan.

26. **LOTTERY PAYOUTS** A state lottery commission pays the winner of the "Million Dollar" lottery 20 installments of $50,000/year. The commission makes the first payment of $50,000 immediately and the other $n = 19$ payments at the end of each of the next 19 yr. Determine how much money the commission should have in the bank initially to guarantee the payments, assuming that the balance on deposit with the bank earns interest at the rate of 8%/year compounded yearly.
Hint: Find the present value of an annuity.

27. **PURCHASING A HOME** The Johnsons have accumulated a nest egg of $25,000 that they intend to use as a down payment toward the purchase of a new house. Because their present gross income has placed them in a relatively high tax bracket, they have decided to invest a minimum of $1200/month in monthly payments (to take advantage of their tax deductions) toward the purchase of their house. However, because of other financial obligations, their monthly payments should not exceed $1500. If local mortgage rates are 9.5%/year compounded monthly for a conventional 30-yr mortgage, what is the price range of houses that they should consider?

28. **PURCHASING A HOME** Refer to Exercise 27. If local mortgage rates were increased to 10%, how would this affect the price range of houses the Johnsons should consider?

29. **PURCHASING A HOME** Refer to Exercise 27. If the Johnsons decide to secure a 15-yr mortgage instead of a 30-yr mortgage, what is the price range of houses they should consider when the local mortgage rate for this type of loan is 9%?

30. **SAVINGS PLAN** Lauren plans to deposit $5000 into a bank account at the beginning of next month and $200/month into

the same account at the end of that month and at the end of each subsequent month for the next 5 yr. If her bank pays interest at the rate of 6%/year compounded monthly, how much will Lauren have in her account at the end of 5 yr? (Assume she makes no withdrawals during the 5-yr period.)

31. **FINANCIAL PLANNING** Joe plans to deposit $200 at the end of each month into a bank account for a period of 2 yr, after which he plans to deposit $300 at the end of each month into the same account for another 3 yr. If the bank pays interest at the rate of 6%/year compounded monthly, how much will Joe have in his account by the end of 5 yr? (Assume no withdrawals are made during the 5-yr period.)

32. **INVESTMENT ANALYSIS** From age 25 to age 40, Jessica deposited $200 at the end of each month into a tax-free retirement account. She made no withdrawals or further contributions until age 65. Alex made deposits of $300 into his tax-free retirement account from age 40 to age 65. If both accounts earned interest at the rate of 5%/year compounded monthly, who ends up with a bigger nest egg upon reaching the age of 65?
Hint: Use both the annuity formula and the compound interest formula.

In Exercises 33 and 34, determine whether the statement is true or false. If it is true, explain why it is true. If it is false, give an example to show why it is false.

33. The future value of an annuity can be found by adding together all the payments that are paid into the account.

34. If the future value of an annuity of n payments of R dollars each, paid at the end of each investment period into an account that earns interest at the rate of i per period, is S dollars, then

$$R = \frac{iS}{(1 + i)^n - 1}$$

5.2 Solutions to Self-Check Exercises

1. The amount Phyliss will have in her account when she retires may be found by using Formula (9) with $R = 2000$, $r = 0.08$, $m = 1$ so that $i = r = 0.08$ and $n = 20$. Thus,

$$S = \frac{2000[(1.08)^{20} - 1]}{0.08}$$
$$\approx 91{,}523.93$$

or $91,523.93.

2. We want to find the present value of an ordinary annuity of 36 payments of $4800 each made monthly and earning interest at 9%/year compounded monthly. Using Formula (11) with $R = 4800$, $m = 12$, so that $i = r/m = 0.09/12 = 0.0075$ and $n = (12)(3) = 36$, we find

$$P = \frac{4800[1 - (1.0075)^{-36}]}{0.0075} \approx 150{,}944.67$$

or $150,944.67, the amount of the loan negotiated.

USING TECHNOLOGY

■ Finding the Amount of an Annuity

Graphing Utility

As mentioned in Using Technology, Section 5.1, the TI-83 can facilitate the solution of problems in finance. We continue to exploit its versatility in this section.

EXAMPLE 1　Finding the Future Value of an Annuity　Find the amount of an ordinary annuity of 36 quarterly payments of $220 each that earn interest at the rate of 10% per year compounded quarterly.

Solution　We use the TI-83 **TVM SOLVER** with the following inputs:

$$N = 36$$
$$I\% = 10$$
$$PV = 0$$
$$PMT = -220 \quad \text{Recall that a payment is an outflow.}$$
$$FV = 0$$
$$P/Y = 4 \quad \text{The number of payments each year}$$
$$C/Y = 4 \quad \text{The number of conversion periods each year}$$
$$\text{PMT:END BEGIN}$$

The result is displayed in Figure T1. We deduce that the desired amount is $12,606.31.

```
N    = 36
I%   = 10
PV   = 0
PMT  = −220
FV   = 12606.31078
P/Y  = 4
C/Y  = 4
PMT : END    BEGIN
```

FIGURE T1
The TI-83 screen showing the future value (FV) of an annuity

EXAMPLE 2　Finding the Present Value of an Annuity　Find the present value of an ordinary annuity of 48 payments of $300 each made monthly and earning interest at the rate of 9% per year compounded monthly.

Solution　We use the TI-83 **TVM SOLVER** with the following inputs:

$$N = 48$$
$$I\% = 9$$
$$PV = 0$$
$$PMT = -300 \quad \text{A payment is an outflow.}$$
$$FV = 0$$
$$P/Y = 12 \quad \text{The number of payments each year}$$
$$C/Y = 12 \quad \text{The number of conversion periods each year}$$
$$\text{PTM:END BEGIN}$$

The output is displayed in Figure T2. We see that the required present value of the annuity is $12,055.43.

```
N     = 48
I%    = 9
PV    = 12055.43457
PMT   = −300
FV    = 0
P/Y   = 12
C/Y   = 12
PMT : END    BEGIN
```

FIGURE T2
The TI-83 screen showing the present value (PV) of an ordinary annuity

Excel

Now we show how Excel can be used to solve financial problems involving annuities.

EXAMPLE 3 Finding the Future Value of an Annuity Find the amount of an ordinary annuity of 36 quarterly payments of $220 each that earn interest at the rate of 10% per year compounded quarterly.

Solution Here we are computing the future value of a series of equal payments, so we use the FV (future value) function. As before, we access the Insert Function dialog box to obtain the Function Arguments dialog box. After making each of the required entries, we obtain the dialog box shown in Figure T3.

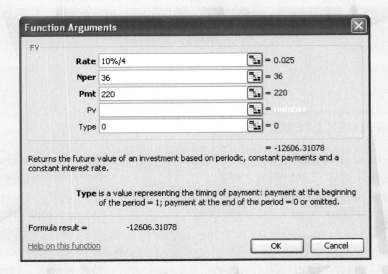

FIGURE T3
Excel's dialog box for the future value (FV) of an annuity

Note that a 0 is entered in the Type edit box because payments are made at the end of each payment period. Once again, the answer is negative because cash is paid out. We deduce that the desired amount is $12,606.31.

Note: Boldfaced words/characters enclosed in a box (for example, ⌐Enter⌐) indicate that an action (click, select, or press) is required. Words/characters printed blue (for example, Chart sub-type:) indicate words/characters that appear on the screen. Words/characters printed in a typewriter font (for example, = (−2/3) *A2+2) indicate words/characters that need to be typed and entered.

(continued)

EXAMPLE 4   Finding the Present Value of an Annuity   Find the present value of an ordinary annuity of 48 payments of $300 each made monthly and earning interest at the rate of 9% per year compounded monthly.

Solution   Here we use the PV function to compute the present value of an annuity. Accessing the PV (present value) function from the Insert Function dialog box and making the required entries, we obtain the PV dialog box shown in Figure T4. We see that the required present value of the annuity is $12,055.43.

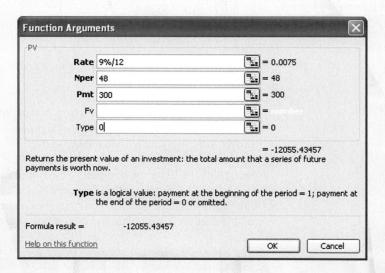

FIGURE T4
Excel's dialog box for computing the present value (PV) of an annuity

TECHNOLOGY EXERCISES

1. Find the amount of an ordinary annuity of 20 payments of $2500/quarter at $7\frac{1}{4}$%/year compounded quarterly.

2. Find the amount of an ordinary annuity of 24 payments of $1790/quarter at $8\frac{3}{4}$%/year compounded quarterly.

3. Find the amount of an ordinary annuity of $120/month for 5 yr at $6\frac{3}{8}$%/year compounded monthly.

4. Find the amount of an ordinary annuity of $225/month for 6 yr at $7\frac{5}{8}$%/year compounded monthly.

5. Find the present value of an ordinary annuity of $4500/semi-annual period for 5 yr earning interest at 9%/year compounded semiannually.

6. Find the present value of an ordinary annuity of $2100/quarter for 7 yr earning interest at $7\frac{1}{8}$%/year compounded quarterly.

7. Find the present value of an ordinary annuity of $245/month for 6 yr earning interest at $8\frac{3}{8}$%/year compounded monthly.

8. Find the present value of an ordinary annuity of $185/month for 12 yr earning interest at $6\frac{5}{8}$%/year compounded monthly.

5.3   Amortization and Sinking Funds

■ Amortization of Loans

The annuity formulas derived in Section 5.2 may be used to answer questions involving the amortization of certain types of installment loans. For example, in a typical housing loan, the mortgagor makes periodic payments toward reducing his indebtedness to the lender, who charges interest at a fixed rate on the unpaid portion of the debt. In practice, the borrower is required to repay the lender in periodic installments, usually of the same size and over a fixed term, so that the loan (principal plus interest charges) is amortized at the end of the term.

By thinking of the monthly loan repayments R as the payments in an annuity, we see that the original amount of the loan is given by P, the present value of the annuity. From Equation (11), Section 5.2, we have

$$P = R\left[\frac{1 - (1 + i)^{-n}}{i}\right] = Ra_{\overline{n}|i} \qquad (12)$$

A question a financier might ask is: How much should the monthly installment be so that a loan will be amortized at the end of the term of the loan? To answer this question, we simply solve (12) for R in terms of P, obtaining

$$R = \frac{Pi}{1 - (1 + i)^{-n}} = \frac{P}{a_{\overline{n}|i}}$$

Amortization Formula

The periodic payment R on a loan of P dollars to be amortized over n periods with interest charged at the rate of i per period is

$$R = \frac{Pi}{1 - (1 + i)^{-n}} \qquad (13)$$

EXAMPLE 1   A sum of \$50,000 is to be repaid over a 5-year period through equal installments made at the end of each year. If an interest rate of 8% per year is charged on the unpaid balance and interest calculations are made at the end of each year, determine the size of each installment so that the loan (principal plus interest charges) is amortized at the end of 5 years. Verify the result by displaying the amortization schedule.

Solution   Substituting $P = 50{,}000$, $i = r = 0.08$ (here, $m = 1$), and $n = 5$ into Formula (13), we obtain

$$R = \frac{(50{,}000)(0.08)}{1 - (1.08)^{-5}} \approx 12{,}522.82$$

giving the required yearly installment as \$12,522.82.

TABLE 4

An Amortization Schedule

End of Period	Interest Charged	Repayment Made	Payment Toward Principal	Outstanding Principal
0	—	—	—	$50,000.00
1	$4,000.00	$12,522.82	$ 8,522.82	41,477.18
2	3,318.17	12,522.82	9,204.65	32,272.53
3	2,581.80	12,522.82	9,941.02	22,331.51
4	1,786.52	12,522.82	10,736.30	11,595.21
5	927.62	12,522.82	11,595.20	0.01

The amortization schedule is presented in Table 4. The outstanding principal at the end of 5 years is of course zero. (The figure of $.01 in Table 4 is the result of round-off errors.) Observe that initially the larger portion of the repayment goes toward payment of interest charges, but as time goes by more and more of the payment goes toward repayment of the principal.   ◾

Financing a Home

APPLIED EXAMPLE 2   Home Mortgage Payments   The Blakelys borrowed $120,000 from a bank to help finance the purchase of a house. The bank charges interest at a rate of 9% per year on the unpaid balance, with interest computations made at the end of each month. The Blakelys have agreed to repay the loan in equal monthly installments over 30 years. How much should each payment be if the loan is to be amortized at the end of the term?

Solution   Here, $P = 120,000$, $i = \frac{r}{m} = \frac{0.09}{12} = 0.0075$, and $n = (30)(12) = 360$. Using Formula (13) we find that the size of each monthly installment required is given by

$$R = \frac{(120,000)(0.0075)}{1 - (1.0075)^{-360}}$$
$$\approx 965.55$$

or $965.55.   ◾

APPLIED EXAMPLE 3   Home Equity   Teresa and Raul purchased a house 10 years ago for $200,000. They made a down payment of 20% of the purchase price and secured a 30-year conventional home mortgage at 9% per year on the unpaid balance. The house is now worth $280,000. How much equity do Teresa and Raul have in their house now (after making 120 monthly payments)?

Solution   Since the down payment was 20%, we know that they secured a loan of 80% of $200,000, or $160,000. Furthermore, using Formula (13) with $P = 160,000$, $i = \frac{r}{m} = \frac{0.09}{12} = 0.0075$, and $n = (30)(12) = 360$, we determine that their monthly installment is

$$R = \frac{(160,000)(0.0075)}{1 - (1.0075)^{-360}}$$

$$\approx 1287.40$$

or $1287.40.

After 120 monthly payments have been made, the outstanding principal is given by the sum of the present values of the remaining installments (that is, $360 - 120 = 240$ installments). But this sum is just the present value of an annuity with $n = 240$, $R = 1287.40$, and $i = 0.0075$. Using Formula (11), we find

$$P = 1{,}287.40 \left[\frac{1 - (1 + 0.0075)^{-240}}{0.0075} \right]$$

$$\approx 143{,}088.01$$

or approximately $143,088. Therefore, Teresa and Raul have an equity of $280{,}000 - 143{,}088$, that is, $136,912.   ∎

EXPLORE & DISCUSS and EXPLORING WITH TECHNOLOGY

1. Consider the amortization Formula (13):

$$R = \frac{Pi}{1 - (1 + i)^{-n}}$$

Suppose you know the values of R, P, and n and you wish to determine i. Explain why you can accomplish this task by finding the point of intersection of the graphs of the functions

$$y_1 = R \quad \text{and} \quad y_2 = \frac{Pi}{1 - (1 + i)^{-n}}$$

2. Thalia knows that her monthly repayment on her 30-year conventional home loan of $150,000 is $1100.65 per month. Help Thalia determine the interest rate for her loan by verifying or executing the following steps:
   a. Plot the graphs of

$$y_1 = 1100.65 \quad \text{and} \quad y_2 = \frac{150{,}000x}{1 - (1 + x)^{-360}}$$

   using the viewing window $[0, 0.01] \times [0, 1200]$.
   b. Use the ISECT (intersection) function of the graphing utility to find the point of intersection of the graphs of part (a). Explain why this gives the value of i.
   c. Compute r from the relationship $r = 12i$.

EXPLORE & DISCUSS and EXPLORING WITH TECHNOLOGY

1. Suppose you secure a home mortgage loan of $\$P$ with an interest rate of r per year to be amortized over t years through monthly installments of $\$R$. Show that after N installments your outstanding principal is given by

$$B(N) = P\left[\frac{(1 + i)^n - (1 + i)^N}{(1 + i)^n - 1}\right] \qquad (0 \le N \le n)$$

Hint: $B(N) = R\left[\dfrac{1 - (1 + i)^{-n+N}}{i}\right]$. To see this, study Example 3, page 310. Replace R using Formula (13).

2. Refer to Example 3, page 310. Using the result of part 1 above, show that Teresa and Raul's outstanding balance after making N payments is

$$E(N) = \frac{160{,}000(1.0075^{360} - 1.0075^N)}{1.0075^{360} - 1} \qquad (0 \le N \le 360)$$

3. Using a graphing utility, plot the graph of

$$E(x) = \frac{160{,}000(1.0075^{360} - 1.0075^x)}{1.0075^{360} - 1}$$

using the viewing window $[0, 360] \times [0, 160{,}000]$.

4. Referring to the graph in part 3, observe that the outstanding principal drops off slowly in the early years and accelerates quickly to zero toward the end of the loan. Can you explain why?

5. How long does it take Teresa and Raul to repay one half of the loan of $\$160{,}000$?
   Hint: See the Explore & Discuss and Exploring with Technology box following Example 3.

APPLIED EXAMPLE 4   Home Affordability   The Jacksons have determined that after making a down payment they could afford at most $\$1000$ for a monthly house payment. The bank charges interest at the rate of 9.6% per year on the unpaid balance, with interest computations made at the end of each month. If the loan is to be amortized in equal monthly installments over 30 years, what is the maximum amount that the Jacksons can borrow from the bank?

Solution   Here, $i = \frac{r}{m} = \frac{0.096}{12} = 0.008$, $n = (30)(12) = 360$, $R = 1000$, and we are required to find P. From Equation (11), we have

$$P = \frac{R[1 - (1 + i)^{-n}]}{i}$$

Substituting the numerical values for R, n, and i into this expression for P, we obtain

$$P = \frac{1000[1 - (1.008)^{-360}]}{0.008} \approx 117{,}902$$

Therefore, the Jacksons can borrow at most $\$117{,}902$.   ∎

TITLE   Director of Finance
INSTITUTION   Thomson Higher Education

At Thomson Higher Education, we publish hundreds of college textbooks a year, and it is my job to oversee all the financial aspects of our division. One aspect of my position is that I must keep abreast of the rules and regulations that are set forth by the U.S. Securities and Exchange Commission (SEC). In an effort to apply newly designed rules of revenue recognition from the Sarbanes-Oxley Act of 2002, we review the investment costs for each program that we publish and amortize that investment to reflect the trend in the way each title sells. So, if we spend one million dollars on a program, we will take the *Sum of the Years' Digits* formula and determine the amount we will amortize when the book publishes. Since company policy requires that we amortize these costs over three years, we would amortize $500,000 in the first year, $333,000 in the second year and $167,000 in the final year.

The reason we use SOYD rather than straight lining the amortization is due to the pattern of sales. When a title is first published, the supply is limited exclusively to what the publisher sells. Therefore, no other supply exists and the publisher sells the highest quantity of titles within the life of the edition. In the second year, other avenues, such as used books, are available to the customer. Because the supply is higher but demand is the same, the number sold by the publisher is less. The further out these sales go (four or five years), the number of units the publisher sells decreases. So if the publisher straight-lined the development costs over three years, the amortization would be the same every year and would not reflect selling patterns.

Many titles also have an increasing number of media components. One of the most popular components is the online homework system. Since these are considered internally developed software systems, they must amortize under similar conditions but are considered a direct expense as opposed to a cost of goods sold (COGS).

Corporate accounting rules stipulate that these types of costs begin amortizing immediately after they go live rather than at the end of their copyright year, which is different from the rules of bookplate amortization.

Sinking Funds

Sinking funds are another important application of the annuity formulas. Simply stated, a **sinking fund** is an account that is set up for a specific purpose at some future date. For example, an individual might establish a sinking fund for the purpose of discharging a debt at a future date. A corporation might establish a sinking fund in order to accumulate sufficient capital to replace equipment that is expected to be obsolete at some future date.

By thinking of the amount to be accumulated by a specific date in the future as the future value of an annuity [Equation (9), Section 5.2], we can answer questions about a large class of sinking fund problems.

APPLIED EXAMPLE 5   Sinking Fund   The proprietor of Carson Hardware has decided to set up a sinking fund for the purpose of purchasing a truck in 2 years' time. It is expected that the truck will cost $30,000. If the fund earns 10% interest per year compounded quarterly, determine the size of each (equal) quarterly installment the proprietor should pay into the fund. Verify the result by displaying the schedule.

313

Solution The problem at hand is to find the size of each quarterly payment R of an annuity given that its future value is $S = 30,000$, the interest earned per conversion period is $i = \frac{r}{m} = \frac{0.1}{4} = 0.025$, and the number of payments is $n = (2)(4) = 8$. The formula for the annuity,

$$S = R\left[\frac{(1 + i)^n - 1}{i}\right]$$

when solved for R yields

$$R = \frac{iS}{(1 + i)^n - 1} \tag{14}$$

or, equivalently,

$$R = \frac{S}{s_{\overline{n}|i}}$$

Substituting the appropriate numerical values for i, S, and n into Equation (14), we obtain the desired quarterly payment

$$R = \frac{(0.025)(30,000)}{(1.025)^8 - 1} \approx 3434.02$$

or $3434.02. Table 5 shows the required schedule.

TABLE 5

A Sinking Fund Schedule

End of Period	Deposit Made	Interest Earned	Addition to Fund	Accumulated Amount in Fund
1	$3,434.02	0	$3,434.02	$ 3,434.02
2	3,434.02	$ 85.85	3,519.87	6,953.89
3	3,434.02	173.85	3,607.87	10,561.76
4	3,434.02	264.04	3,698.06	14,259.82
5	3,434.02	356.50	3,790.52	18,050.34
6	3,434.02	451.26	3,885.28	21,935.62
7	3,434.02	548.39	3,982.41	25,918.03
8	3,434.02	647.95	4,081.97	30,000.00

The formula derived in this last example is restated below.

Sinking Fund Payment

The periodic payment R required to accumulate a sum of S dollars over n periods with interest charged at the rate of i per period is

$$R = \frac{iS}{(1 + i)^n - 1}$$

5.3 Self-Check Exercises

1. The Mendozas wish to borrow $100,000 from a bank to help finance the purchase of a house. Their banker has offered the following plans for their consideration. In plan I, the Mendozas have 30 yr to repay the loan in monthly installments with interest on the unpaid balance charged at 10.5%/year compounded monthly. In plan II, the loan is to be repaid in monthly installments over 15 yr with interest on the unpaid balance charged at 9.75%/year compounded monthly.
   a. Find the monthly repayment for each plan.
   b. What is the difference in total payments made under each plan?

2. Harris, a self-employed individual who is 46 yr old, is setting up a defined-benefit retirement plan. If he wishes to have $250,000 in this retirement account by age 65, what is the size of each yearly installment he will be required to make into a savings account earning interest at $8\frac{1}{4}$%/year? (Assume that Harris is eligible to make each of the 20 required contributions.)

Solutions to Self-Check Exercises 5.3 can be found on page 318.

5.3 Concept Questions

1. Write the amortization formula.
   a. If P and i are fixed and n is allowed to increase, what will happen to R?
   b. Interpret the result of part (a).

2. Using the formula for computing a sinking fund payment, show that if the number of payments into a sinking fund increases, then the size of the periodic payment into the sinking fund decreases.

5.3 Exercises

In Exercises 1–8, find the periodic payment R required to amortize a loan of P dollars over t years with interest earned at the rate of r%/year compounded m times a year.

1. $P = 100,000, r = 8, t = 10, m = 1$

2. $P = 40,000, r = 3, t = 15, m = 2$

3. $P = 5000, r = 4, t = 3, m = 4$

4. $P = 16,000, r = 9, t = 4, m = 12$

5. $P = 25,000, r = 3, t = 12, m = 4$

6. $P = 80,000, r = 10.5, t = 15, m = 12$

7. $P = 80,000, r = 10.5, t = 30, m = 12$

8. $P = 100,000, r = 10.5, t = 25, m = 12$

In Exercises 9–14, find the periodic payment R required to accumulate a sum of S dollars over t years with interest earned at the rate of r%/year compounded m times a year.

9. $S = 20,000, r = 4, t = 6, m = 2$

10. $S = 40,000, r = 4, t = 9, m = 4$

11. $S = 100,000, r = 4.5, t = 20, m = 6$

12. $S = 120,000, r = 4.5, t = 30, m = 6$

13. $S = 250,000, r = 10.5, t = 25, m = 12$

14. $S = 350,000, r = 7.5, t = 10, m = 12$

15. Suppose payments were made at the end of each quarter into an ordinary annuity earning interest at the rate of 10%/year compounded quarterly. If the future value of the annuity after 5 yr is $50,000, what was the size of each payment?

16. Suppose payments were made at the end of each month into an ordinary annuity earning interest at the rate of 9%/year compounded monthly. If the future value of the annuity after 10 yr is $60,000, what was the size of each payment?

17. Suppose payments will be made for $6\frac{1}{2}$ yr at the end of each semiannual period into an ordinary annuity earning interest at the rate of 7.5%/year compounded semiannually. If the present value of the annuity is $35,000, what should be the size of each payment?

18. Suppose payments will be made for $9\frac{1}{4}$ yr at the end of each month into an ordinary annuity earning interest at the rate of 6.25%/year compounded monthly. If the present value of the annuity is $42,000, what should be the size of each payment?

19. **LOAN AMORTIZATION** A sum of $100,000 is to be repaid over a 10-yr period through equal installments made at the end of each year. If an interest rate of 10%/year is charged on the unpaid balance and interest calculations are made at the end of each year, determine the size of each installment so that the loan (principal plus interest charges) is amortized at the end of 10 yr.

20. **LOAN AMORTIZATION** What monthly payment is required to amortize a loan of $30,000 over 10 yr if interest at the rate of 12%/year is charged on the unpaid balance and interest calculations are made at the end of each month?

21. **HOME MORTGAGES** Complete the following table, which shows the monthly payments on a $100,000, 30-yr mortgage at the interest rates shown. Use this information to answer the following questions.

Amount of Mortgage, $	Interest Rate, %	Monthly Payment, $
$100,000	7	665.30
100,000	8	...
100,000	9	...
100,000	10	...
100,000	11	...
100,000	12	1028.61

a. What is the difference in monthly payments between a $100,000, 30-yr mortgage secured at 7%/year and one secured at 10%/year?

b. Use the table to calculate the monthly mortgage payments on a $150,000 mortgage at 10%/year over 30 yr and a $50,000 mortgage at 10%/year over 30 yr.

22. **FINANCING A HOME** The Flemings secured a bank loan of $96,000 to help finance the purchase of a house. The bank charges interest at a rate of 9%/year on the unpaid balance, and interest computations are made at the end of each month. The Flemings have agreed to repay the loan in equal monthly installments over 25 yr. What should be the size of each repayment if the loan is to be amortized at the end of the term?

23. **FINANCING A CAR** The price of a new car is $16,000. Assume an individual makes a down payment of 25% toward the purchase of the car and secures financing for the balance at the rate of 10%/year compounded monthly.

a. What monthly payment will she be required to make if the car is financed over a period of 36 mo? Over a period of 48 mo?

b. What will the interest charges be if she elects the 36-mo plan? The 48-mo plan?

24. **FINANCIAL ANALYSIS** A group of private investors purchased a condominium complex for $2 million. They made an initial down payment of 10% and obtained financing for the balance. If the loan is to be amortized over 15 yr at an interest

rate of 12%/year compounded quarterly, find the required quarterly payment.

25. **FINANCING A HOME** The Taylors have purchased a $270,000 house. They made an initial down payment of $30,000 and secured a mortgage with interest charged at the rate of 8%/year on the unpaid balance. Interest computations are made at the end of each month. If the loan is to be amortized over 30 yr, what monthly payment will the Taylors be required to make? What is their equity (disregarding appreciation) after 5 yr? After 10 yr? After 20 yr?

26. **FINANCIAL PLANNING** Jessica wants to accumulate $10,000 by the end of 5 yr in a special bank account, which she had opened for this purpose. To achieve this goal, Jessica plans to deposit a fixed sum of money into the account at the end of the month over the 5-yr period. If the bank pays interest at the rate of 5%/year compounded monthly, how much does she have to deposit each month into her account?

27. **SINKING FUNDS** A city has $2.5 million worth of school bonds that are due in 20 yr and has established a sinking fund to retire this debt. If the fund earns interest at the rate of 7%/year compounded annually, what amount must be deposited annually in this fund?

28. **TRUST FUNDS** Carl is the beneficiary of a $20,000 trust fund set up for him by his grandparents. Under the terms of the trust, he is to receive the money over a 5-yr period in equal installments at the end of each year. If the fund earns interest at the rate of 9%/year compounded annually, what amount will he receive each year?

29. **SINKING FUNDS** Lowell Corporation wishes to establish a sinking fund to retire a $200,000 debt that is due in 10 yr. If the investment will earn interest at the rate of 9%/year compounded quarterly, find the amount of the quarterly deposit that must be made in order to accumulate the required sum.

30. **SINKING FUNDS** The management of Gibraltar Brokerage Services anticipates a capital expenditure of $20,000 in 3 years' time for the purpose of purchasing new fax machines and has decided to set up a sinking fund to finance this purchase. If the fund earns interest at the rate of 10%/year compounded quarterly, determine the size of each (equal) quarterly installment that should be deposited in the fund.

31. **RETIREMENT ACCOUNTS** Andrea, a self-employed individual, wishes to accumulate a retirement fund of $250,000. How much should she deposit each month into her retirement account, which pays interest at the rate of 8.5%/year compounded monthly, to reach her goal upon retirement 25 yr from now?

32. **STUDENT LOANS** Joe secured a loan of $12,000 3 yr ago from a bank for use toward his college expenses. The bank charges interest at the rate of 4%/year compounded monthly on his loan. Now that he has graduated from college, Joe wishes to repay the loan by amortizing it through monthly payments over 10 yr at the same interest rate. Find the size of the monthly payments he will be required to make.

33. **RETIREMENT ACCOUNTS**   Robin wishes to accumulate a sum of $450,000 in a retirement account by the time of her retirement 30 yr from now. If she wishes to do this through monthly payments into the account that earn interest at the rate of 10%/year compounded monthly, what should be the size of each payment?

34. **FINANCING COLLEGE EXPENSES**   Yumi's grandparents presented her with a gift of $20,000 when she was 10 yr old to be used for her college education. Over the next 7 yr, until she turned 17, Yumi's parents had invested her money in a tax-free account that had yielded interest at the rate of 5.5%/year compounded monthly. Upon turning 17, Yumi now plans to withdraw her funds in equal annual installments over the next 4 yr, starting at age 18. If the college fund is expected to earn interest at the rate of 6%/year, compounded annually, what will be the size of each installment?

35. **IRAs**   Martin has deposited $375 in his IRA at the end of each quarter for the past 20 yr. His investment has earned interest at the rate of 8%/year compounded quarterly over this period. Now, at age 60, he is considering retirement. What quarterly payment will he receive over the next 15 yr? (Assume that the money is earning interest at the same rate and payments are made at the end of each quarter.) If he continues working and makes quarterly payments of the same amount in his IRA account until age 65, what quarterly payment will he receive from his fund upon retirement over the following 10 yr?

36. **FINANCING A CAR**   Darla purchased a new car during a special sales promotion by the manufacturer. She secured a loan from the manufacturer in the amount of $16,000 at a rate of 7.9%/year compounded monthly. Her bank is now charging 11.5%/year compounded monthly for new car loans. Assuming that each loan would be amortized by 36 equal monthly installments, determine the amount of interest she would have paid at the end of 3 yr for each loan. How much less will she have paid in interest payments over the life of the loan by borrowing from the manufacturer instead of her bank?

37. **FINANCING A HOME**   The Sandersons are planning to refinance their home. The outstanding principal on their original loan is $100,000 and was to be amortized in 240 equal monthly installments at an interest rate of 10%/year compounded monthly. The new loan they expect to secure is to be amortized over the same period at an interest rate of 7.8%/year compounded monthly. How much less can they expect to pay over the life of the loan in interest payments by refinancing the loan at this time?

38. **INVESTMENT ANALYSIS**   Since he was 22 yr old, Ben has been depositing $200 at the end of each month into a tax-free retirement account earning interest at the rate of 6.5%/year compounded monthly. Larry, who is the same age as Ben, decided to open a tax-free retirement account 5 yr after Ben opened his. If Larry's account earns interest at the same rate

as Ben's, determine how much Larry should deposit each month into his account so that both men will have the same amount of money in their accounts at age 65.

39. **FINANCING A HOME**   Eight years ago, Kim secured a bank loan of $180,000 to help finance the purchase of a house. The mortgage was for a term of 30 yr, with an interest rate of 9.5%/year compounded monthly on the unpaid balance to be amortized through monthly payments. What is the outstanding principal on Kim's house now?

40. **BALLOON PAYMENT MORTGAGE**   Olivia plans to secure a 5-yr balloon mortgage of $200,000 toward the purchase of a condominium. Her monthly payment for the 5 yr is calculated on the basis of a 30-yr conventional mortgage at the rate of 6%/year compounded monthly. At the end of the 5 yr, Olivia is required to pay the balance owed (the "balloon" payment). What will be her monthly payment, and what will be her balloon payment?

41. **FINANCING A HOME**   Sarah secured a bank loan of $200,000 for the purchase of a house. The mortgage is to be amortized through monthly payments for a term of 15 yr, with an interest rate of 6%/year compounded monthly on the unpaid balance. She plans to sell her house in 5 yr. How much will Sarah still owe on her house?

42. **HOME REFINANCING**   Five years ago, Diane secured a bank loan of $300,000 to help finance the purchase of a loft in the San Francisco Bay area. The term of the mortgage was 30 yr, and the interest rate was 9%/year compounded monthly on the unpaid balance. Because the interest rate for a conventional 30-yr home mortgage has now dropped to 7%/year compounded monthly, Diane is thinking of refinancing her property.
   a. What is Diane's current monthly mortgage payment?
   b. What is Diane's current outstanding principal?
   c. If Diane decides to refinance her property by securing a 30-yr home mortgage loan in the amount of the current outstanding principal at the prevailing interest rate of 7%/year compounded monthly, what will be her monthly mortgage payment?
   d. How much less would Diane's monthly mortgage payment be if she refinances?

43. **HOME REFINANCING**   Four years ago, Emily secured a bank loan of $200,000 to help finance the purchase of an apartment in Boston. The term of the mortgage is 30 yr, and the interest rate is 9.5%/year compounded monthly. Because the interest rate for a conventional 30-yr home mortgage has now dropped to 6.75%/year compounded monthly, Emily is thinking of refinancing her property.
   a. What is Emily's current monthly mortgage payment?
   b. What is Emily's current outstanding principal?
   c. If Emily decides to refinance her property by securing a 30-yr home mortgage loan in the amount of the current outstanding principal at the prevailing interest rate of

6.75%/year compounded monthly, what will be her monthly mortgage payment?

d. How much less would Emily's monthly mortgage payment be if she refinances?

44. ADJUSTABLE-RATE MORTGAGE Three years ago Samantha secured an adjustable-rate mortgage (ARM) loan to help finance the purchase of a house. The amount of the original loan was $150,000 for a term of 30 yr, with interest at the rate of 7.5%/year compounded monthly. Currently the interest rate is 7%/year compounded monthly, and Samantha's monthly payments are due to be recalculated. What would be her new monthly payment?

Hint: Calculate her current outstanding principal. Then, to amortize the loan in the next 27 yr, determine the monthly payment based on the current interest rate.

45. FINANCING A HOME After making a down payment of $25,000, the Meyers need to secure a loan of $280,000 to purchase a certain house. Their bank's current rate for 25-yr home loans is 11%/year compounded monthly. The owner has offered to finance the loan at 9.8%/year compounded monthly. Assuming that both loans would be amortized over

a 25-yr period by 300 equal monthly installments, determine the difference in the amount of interest the Meyers would pay by choosing the seller's financing rather than their bank's.

46. REFINANCING A HOME The Martinezes are planning to refinance their home. The outstanding balance on their original loan is $150,000. Their finance company has offered them two options:

Option A: A fixed-rate mortgage at an interest rate of 7.5%/year compounded monthly, payable over a 30-yr period in 360 equal monthly installments.

Option B: A fixed-rate mortgage at an interest rate of 7.25%/year compounded monthly, payable over a 15-yr period in 180 equal monthly installments. (Assume that there are no additional finance charges.)

a. Find the monthly payment required to amortize each of these loans over the life of the loan.

b. How much interest would the Martinezes save if they chose the 15-yr mortgage instead of the 30-yr mortgage?

5.3   Solutions to Self-Check Exercises

1. **a.** We use Equation (13) in each instance. Under plan I,

$$P = 100,000 \qquad i = \frac{r}{m} = \frac{0.105}{12} = 0.00875$$

$$n = (30)(12) = 360$$

Therefore, the size of each monthly repayment under plan I is

$$R = \frac{100,000(0.00875)}{1 - (1.00875)^{-360}}$$
$$\approx 914.74$$

or $914.74.

Under plan II,

$$P = 100,000 \qquad i = \frac{r}{m} = \frac{0.0975}{12} = 0.008125$$

$$n = (15)(12) = 180$$

Therefore, the size of each monthly repayment under plan II is

$$R = \frac{100,000(0.008125)}{1 - (1.008125)^{-180}}$$
$$\approx 1059.36$$

or $1059.36.

b. Under plan I, the total amount of repayments will be

$$(360)(914.74) = 329,306.40 \qquad \text{Number of payments} \\ \times \text{ the size of each installment}$$

or $329,306.40. Under plan II, the total amount of repayments will be

$$(180)(1059.36) = 190,684.80$$

or $190,684.80. Therefore, the difference in payments is

$$329,306.40 - 190,684.80 = 138,621.60$$

or $138,621.60.

2. We use Equation (14) with

$$S = 250,000$$
$$i = r = 0.0825 \qquad \text{Since } m = 1$$
$$n = 20$$

giving the required size of each installment as

$$R = \frac{(0.0825)(250,000)}{(1.0825)^{20} - 1}$$
$$\approx 5313.59$$

or $5313.59.

USING TECHNOLOGY

■ Amortizing a Loan

Graphing Utility

Here we use the TI-83 **TVM SOLVER** function to help us solve problems involving amortization and sinking funds.

EXAMPLE 1   Finding the Payment to Amortize a Loan   The Wongs are considering obtaining a preapproved 30-year loan of $120,000 to help finance the purchase of a house. The mortgage company charges interest at the rate of 8% per year on the unpaid balance, with interest computations made at the end of each month. What will be the monthly installments if the loan is amortized at the end of the term?

Solution   We use the TI-83 **TVM SOLVER** with the following inputs:

$$N = 360 \qquad (30)(12)$$
$$I\% = 8$$
$$PV = 120000$$
$$PMT = 0$$
$$FV = 0$$
$$P/Y = 12 \qquad \text{The number of payments each year}$$
$$C/Y = 12 \qquad \text{The number of conversion periods each year}$$
$$PMT\text{:END BEGIN}$$

From the output shown in Figure T1, we see that the required payment is $880.52.

```
N    = 360
I%   = 8
PV   = 120000
■ PMT = −880.51748...
FV   = 0
P/Y  = 12
C/Y  = 12
PMT : END    BEGIN
```

FIGURE T1
The TI-83 screen showing the monthly installment, PMT

EXAMPLE 2   Finding the Payment in a Sinking Fund   Heidi wishes to establish a retirement account that will be worth $500,000 in 20 years' time. She expects that the account will earn interest at the rate of 11% per year compounded monthly. What should be the monthly contribution into her account each month?

Solution   We use the TI-83 **TVM SOLVER** with the following inputs:

$$N = 240 \qquad (20)(12)$$
$$I\% = 11$$
$$PV = 0$$
$$PMT = 0$$
$$FV = 500000$$
$$P/Y = 12 \qquad \text{The number of payments each year}$$
$$C/Y = 12 \qquad \text{The number of conversion periods each year}$$
$$PMT\text{:END BEGIN}$$

```
N    = 240
I%   = 11
PV   = 0
■ PMT = −577.60862...
FV   = 500000
P/Y  = 12
C/Y  = 12
PMT : END    BEGIN
```

FIGURE T2
The TI-83 screen showing the monthly payment, PMT

The result is displayed in Figure T2. We see that Heidi's monthly contribution should be $577.61. (*Note:* The display for PMT is negative because it is an outflow.)

(continued)

Excel

Here we use Excel to help us solve problems involving amortization and sinking funds.

EXAMPLE 3 Finding the Payment to Amortize a Loan The Wongs are considering obtaining a preapproved 30-year loan of $120,000 to help finance the purchase of a house. The mortgage company charges interest at the rate of 8% per year on the unpaid balance, with interest computations made at the end of each month. What will be the monthly installments if the loan is amortized at the end of the term?

Solution We use the PMT function to solve this problem. Accessing this function from the Insert Function dialog box and making the required entries, we obtain the Function Arguments dialog box shown in Figure T3. We see that the desired result is $880.52. (Recall that cash you pay out is represented by a negative number.)

Function Arguments [×]

PMT

Rate	8%/12	= 0.006666667
Nper	30*12	= 360
Pv	120000	= 120000
Fv		= number
Type	0	= 0

= -880.5174887

Calculates the payment for a loan based on constant payments and a constant interest rate.

Type is a logical value: payment at the beginning of the period = 1; payment at the end of the period = 0 or omitted.

Formula result =   -880.5174887

Help on this function   [OK]   [Cancel]

FIGURE T3
Excel's dialog box giving the payment function, PMT

EXAMPLE 4 Finding the Payment in a Sinking Fund Heidi wishes to establish a retirement account that will be worth $500,000 in 20 years' time. She expects that the account will earn interest at the rate of 11% per year compounded monthly. What should be the monthly contribution into her account each month?

Solution As in Example 3, we use the PMT function, but this time we are given the future value of the investment. Accessing the PMT function from the Insert Function dialog box and making the required entries, we obtain the Function

---

Note: Words/characters printed blue (for example, Chart sub-type:) indicate words/characters on the screen.

Arguments dialog box shown in Figure T4. We see that Heidi's monthly contribution should be $577.61. (Note that the value for PMT is negative because it is an outflow.)

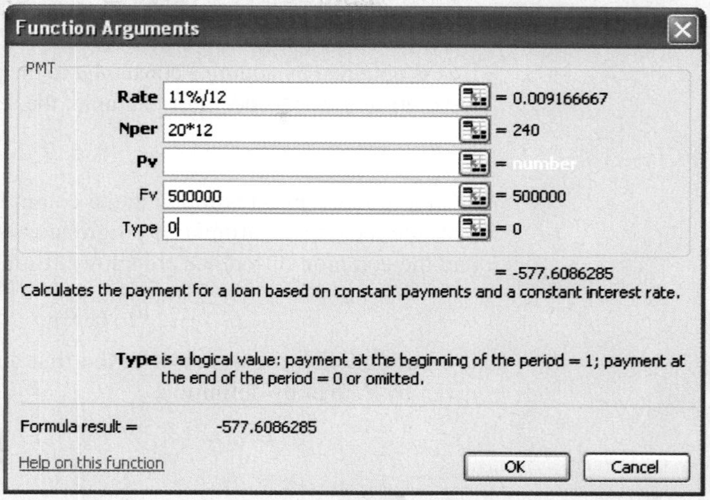

Function Arguments	⊠

PMT

Rate	11%/12	= 0.009166667
Nper	20*12	= 240
Pv		= number
Fv	500000	= 500000
Type	0	= 0

= -577.6086285

Calculates the payment for a loan based on constant payments and a constant interest rate.

Type is a logical value: payment at the beginning of the period = 1; payment at the end of the period = 0 or omitted.

Formula result =   -577.6086285

Help on this function   [OK]   [Cancel]

FIGURE T4
Excel's dialog box giving the payment function, PMT

TECHNOLOGY EXERCISES

1. Find the periodic payment required to amortize a loan of $55,000 over 120 periods with interest earned at the rate of $6\frac{5}{8}\%$/period.

2. Find the periodic payment required to amortize a loan of $178,000 over 180 periods with interest earned at the rate of $7\frac{1}{8}\%$/period.

3. Find the periodic payment required to amortize a loan of $227,000 over 360 periods with interest earned at the rate of $8\frac{1}{8}\%$/period.

4. Find the periodic payment required to amortize a loan of $150,000 over 360 periods with interest earned at the rate of $7\frac{3}{8}\%$/period.

5. Find the periodic payment required to accumulate $25,000 over 12 periods with interest earned at the rate of 2%/period.

6. Find the periodic payment required to accumulate $50,000 over 36 periods with interest earned at the rate of $2\frac{1}{4}\%$/ period.

7. Find the periodic payment required to accumulate $137,000 over 120 periods with interest earned at the rate of $\frac{3}{4}\%$/ period.

8. Find the periodic payment required to accumulate $144,000 over 120 periods with interest earned at the rate of $\frac{5}{8}\%$/ period.

9. A loan of $120,000 is to be repaid over a 10-yr period through equal installments made at the end of each year. If an interest rate of 8.5%/year is charged on the unpaid balance and interest calculations are made at the end of each year, determine the size of each installment so that the loan is amortized at the end of 10 yr. Verify the result by displaying the amortization schedule.

10. A loan of $265,000 is to be repaid over an 8-yr period through equal installments made at the end of each year. If an interest rate of 7.4%/year is charged on the unpaid balance and interest calculations are made at the end of each year, determine the size of each installment so that the loan is amortized at the end of 8 yr. Verify the result by displaying the amortization schedule.

5.4   Arithmetic and Geometric Progressions (Optional)

▬ Arithmetic Progressions

An **arithmetic progression** is a sequence of numbers in which each term after the first is obtained by adding a constant d to the preceding term. The constant d is called the **common difference.** For example, the sequence

$$3, 6, 9, 12, \ldots$$

is an arithmetic progression with the common difference equal to 3.

Observe that an arithmetic progression is completely determined if the first term and the common difference are known. In fact, if

$$a_1, a_2, a_3, \ldots, a_n, \ldots$$

is an arithmetic progression with the first term given by a and common difference given by d, then by definition,

$$a_1 = a$$
$$a_2 = a_1 + d = a + d$$
$$a_3 = a_2 + d = (a + d) + d = a + 2d$$
$$a_4 = a_3 + d = (a + 2d) + d = a + 3d$$
$$\vdots$$
$$a_n = a_{n-1} + d = a + (n - 2)d + d = a + (n - 1)d$$

Thus, we see that the nth term of an arithmetic progression with first term a and common difference d is given by

$$a_n = a + (n - 1)d \tag{15}$$

---

nth Term of an Arithmetic Progression
The nth term of an arithmetic progression with first term a and common difference d is given by

$$a_n = a + (n - 1)d$$

EXAMPLE 1   Find the 12th term of the arithmetic progression

$$2, 7, 12, 17, 22, \ldots$$

Solution   The first term of the arithmetic progression is $a_1 = a = 2$, and the common difference is $d = 5$, so upon setting $n = 12$ in Equation (15), we find

$$a_{12} = 2 + (12 - 1)5 = 57 \qquad \blacksquare$$

EXAMPLE 2   Write the first five terms of an arithmetic progression whose 3rd and 11th terms are 21 and 85, respectively.

Solution Using Equation (15), we obtain

$$a_3 = a + 2d = 21$$
$$a_{11} = a + 10d = 85$$

Subtracting the first equation from the second gives $8d = 64$, or $d = 8$. Substituting this value of d into the first equation yields $a + 16 = 21$, or $a = 5$. Thus, the required arithmetic progression is given by the sequence

$$5, 13, 21, 29, 37, \ldots \qquad \blacksquare$$

Let S_n denote the sum of the first n terms of an arithmetic progression with first term $a_1 = a$ and common difference d. Then

$$S_n = a + (a + d) + (a + 2d) + \cdots + [a + (n - 1)d] \qquad \textbf{(16)}$$

Rewriting the expression for S_n with the terms in reverse order gives

$$S_n = [a + (n - 1)d] + [a + (n - 2)d] + \cdots + (a + d) + a \qquad \textbf{(17)}$$

Adding Equations (16) and (17), we obtain

$$2S_n = [2a + (n - 1)d] + [2a + (n - 1)d]$$
$$+ \cdots + [2a + (n - 1)d]$$
$$= n[2a + (n - 1)d]$$
$$S_n = \frac{n}{2}[2a + (n - 1)d]$$

Sum of Terms in an Arithmetic Progression

The sum of the first n terms of an arithmetic progression with first term a and common difference d is given by

$$S_n = \frac{n}{2}[2a + (n - 1)d] \qquad \textbf{(18)}$$

EXAMPLE 3 Find the sum of the first 20 terms of the arithmetic progression of Example 1.

Solution Letting $a = 2$, $d = 5$, and $n = 20$ in Equation (18), we obtain

$$S_{20} = \frac{20}{2}[2 \cdot 2 + 19 \cdot 5] = 990 \qquad \blacksquare$$

APPLIED EXAMPLE 4 Company Sales Madison Electric Company had sales of $200,000 in its first year of operation. If the sales increased by $30,000 per year thereafter, find Madison's sales in the fifth year and its total sales over the first 5 years of operation.

Solution Madison's yearly sales follow an arithmetic progression, with the first term given by $a = 200,000$ and the common difference given by $d = 30,000$. The sales in the fifth year are found by using Equation (15) with $n = 5$. Thus,

$$a_5 = 200,000 + (5 - 1)30,000 = 320,000$$

or $320,000.

Madison's total sales over the first 5 years of operation are found by using (18) with $n = 5$. Thus,

$$S_5 = \frac{5}{2} [2(200{,}000) + (5 - 1)30{,}000]$$

$$= 1{,}300{,}000$$

or $1,300,000.　■

▰ Geometric Progressions

A **geometric progression** is a sequence of numbers in which each term after the first is obtained by multiplying the preceding term by a constant r. The constant r is called the **common ratio.**

A geometric progression is completely determined if the first term and the common ratio are known. Thus, if

$$a_1, a_2, a_3, \ldots, a_n, \ldots$$

is a geometric progression with the first term given by a and common ratio given by r, then by definition,

$$a_1 = a$$
$$a_2 = a_1 r = ar$$
$$a_3 = a_2 r = ar^2$$
$$a_4 = a_3 r = ar^3$$
$$\vdots$$
$$a_n = a_{n-1} r = ar^{n-1}$$

Thus, we see that the nth term of a geometric progression with first term a and common ratio r is given by

$$a_n = ar^{n-1} \tag{19}$$

> ### nth Term of a Geometric Progression
> The nth term of a geometric progression with first term a and common ratio r is given by
>
> $$a_n = ar^{n-1}$$

EXAMPLE 5　Find the eighth term of a geometric progression whose first five terms are 162, 54, 18, 6, and 2.

Solution　The common ratio is found by taking the ratio of any term other than the first to the preceding term. Taking the ratio of the fourth term to the third term, for example, gives $r = \frac{6}{18} = \frac{1}{3}$. To find the eighth term of the geometric progression, use Formula (19) with $a = 162$, $r = \frac{1}{3}$, and $n = 8$, obtaining

$$a_8 = 162\left(\frac{1}{3}\right)^7$$

$$= \frac{2}{27}$$

■

EXAMPLE 6   Find the tenth term of a geometric progression whose third term is 16 and whose seventh term is 1.

Solution   Using Equation (19) with $n = 3$ and $n = 7$, respectively, yields

$$a_3 = ar^2 = 16$$
$$a_7 = ar^6 = 1$$

Dividing a_7 by a_3 gives

$$\frac{ar^6}{ar^2} = \frac{1}{16}$$

from which we obtain $r^4 = \frac{1}{16}$, or $r = \frac{1}{2}$. Substituting this value of r into the expression for a_3, we obtain

$$a\left(\frac{1}{2}\right)^2 = 16 \quad \text{or} \quad a = 64$$

Finally, using (19) once again with $a = 64$, $r = \frac{1}{2}$, and $n = 10$ gives

$$a_{10} = 64\left(\frac{1}{2}\right)^9 = \frac{1}{8}$$

■

To find the sum of the first n terms of a geometric progression with the first term $a_1 = a$ and common ratio r, denote the required sum by S_n. Then,

$$S_n = a + ar + ar^2 + \cdots + ar^{n-2} + ar^{n-1} \tag{20}$$

Upon multiplying (20) by r, we obtain

$$rS_n = ar + ar^2 + ar^3 + \cdots + ar^{n-1} + ar^n \tag{21}$$

Subtracting (21) from (20) gives

$$S_n - rS_n = a - ar^n$$
$$(1 - r)S_n = a(1 - r^n)$$

If $r \neq 1$, we may divide both sides of the last equation by $(1 - r)$, obtaining

$$S_n = \frac{a(1 - r^n)}{(1 - r)}$$

If $r = 1$, then (20) gives

$$S_n = a + a + a + \cdots + a \qquad n \text{ terms}$$
$$= na$$

Thus,

$$S_n = \begin{cases} \dfrac{a(1 - r^n)}{1 - r} & \text{if } r \neq 1 \\ na & \text{if } r = 1 \end{cases}$$

Sum of Terms in a Geometric Progression

The sum of the first n terms of a geometric progression with first term a and common ratio r is given by

$$S_n = \begin{cases} \dfrac{a(1 - r^n)}{1 - r} & \text{if } r \neq 1 \\ na & \text{if } r = 1 \end{cases} \tag{22}$$

EXAMPLE 7 Find the sum of the first six terms of the following geometric progression

$$3, 6, 12, 24, \ldots$$

Solution Here, $a = 3$, $r = \frac{6}{3} = 2$, and $n = 6$, so Formula (22) gives

$$S_6 = \frac{3(1 - 2^6)}{1 - 2} = 189$$

■

APPLIED EXAMPLE 8   Company Sales Michaelson Land Development Company had sales of $1 million in its first year of operation. If sales increased by 10% per year thereafter, find Michaelson's sales in the fifth year and its total sales over the first 5 years of operation.

Solution Michaelson's yearly sales follow a geometric progression, with the first term given by $a = 1,000,000$ and the common ratio given by $r = 1.1$. The sales in the fifth year are found by using Formula (19) with $n = 5$. Thus,

$$a_5 = 1,000,000(1.1)^4 = 1,464,100$$

or $1,464,100.

Michaelson's total sales over the first 5 years of operation are found by using Equation (22) with $n = 5$. Thus,

$$S_5 = \frac{1,000,000[1 - (1.1)^5]}{1 - 1.1}$$
$$= 6,105,100$$

or $6,105,100.

■

Double Declining-Balance Method of Depreciation

In Section 1.3, we discussed the straight-line, or linear, method of depreciating an asset. Linear depreciation assumes that the asset depreciates at a constant rate. For certain assets, such as machines, whose market values drop rapidly in the early years of usage and thereafter less rapidly, another method of depreciation called the **double declining-balance method** is often used. In practice, a business firm normally employs the double declining-balance method for depreciating such assets for a certain number of years and then switches over to the linear method.

To derive an expression for the book value of an asset being depreciated by the double declining-balance method, let C (in dollars) denote the original cost of the asset and let the asset be depreciated over N years. Using this method, the amount depreciated each year is $\frac{2}{N}$ times the value of the asset at the beginning of that year. Thus, the amount by which the asset is depreciated in its first year of use is given by $\frac{2C}{N}$, so if $V(1)$ denotes the book value of the asset at the end of the first year, then

$$V(1) = C - \frac{2C}{N} = C\left(1 - \frac{2}{N}\right)$$

Next, if $V(2)$ denotes the book value of the asset at the end of the second year, then a similar argument leads to

$$V(2) = C\left(1 - \frac{2}{N}\right) - C\left(1 - \frac{2}{N}\right)\frac{2}{N}$$

$$= C\left(1 - \frac{2}{N}\right)\left(1 - \frac{2}{N}\right)$$

$$= C\left(1 - \frac{2}{N}\right)^2$$

Continuing, we find that if $V(n)$ denotes the book value of the asset at the end of n years, then the terms $C, V(1), V(2), \ldots, V(n)$ form a geometric progression with first term C and common ratio $\left(1 - \frac{2}{N}\right)$. Consequently, the nth term, $V(n)$, is given by

$$V(n) = C\left(1 - \frac{2}{N}\right)^n \qquad (1 \le n \le N) \qquad \textbf{(23)}$$

Also, if $D(n)$ denotes the amount by which the asset has been depreciated by the end of the nth year, then

$$D(n) = C - C\left(1 - \frac{2}{N}\right)^n$$

$$= C\left[1 - \left(1 - \frac{2}{N}\right)^n\right] \qquad \textbf{(24)}$$

APPLIED EXAMPLE 9 Depreciation of Equipment A tractor purchased at a cost of $60,000 is to be depreciated by the double declining-balance method over 10 years. What is the book value of the tractor at the end of 5 years? By what amount has the tractor been depreciated by the end of the fifth year?

Solution We have $C = 60,000$ and $N = 10$. Thus, using Formula (23) with $n = 5$ gives the book value of the tractor at the end of 5 years as

$$V(5) = 60,000\left(1 - \frac{2}{10}\right)^5$$

$$= 60,000\left(\frac{4}{5}\right)^5 = 19,660.80$$

or $19,660.80.

The amount by which the tractor has been depreciated by the end of the fifth year is given by

$$60{,}000 - 19{,}660.80 = 40{,}339.20$$

or \$40,339.20. You may verify the last result by using Equation (24) directly. ■

EXPLORING WITH TECHNOLOGY

A tractor purchased at a cost of \$60,000 is to be depreciated over 10 years with a residual value of \$0. Using the double declining-balance method, its value at the end of n years is $V_1(n) = 60{,}000(0.8)^n$ dollars. Using straight-line depreciation, its value at the end of n years is $V_2(n) = 60{,}000 - 6000n$. Use a graphing utility to sketch the graphs of V_1 and V_2 in the viewing window $[0, 10] \times [0, 70{,}000]$. Comment on the relative merits of each method of depreciation.

5.4 Self-Check Exercises

1. Find the sum of the first five terms of the geometric progression with first term -24 and common ratio $-\frac{1}{2}$.

2. Office equipment purchased for \$75,000 is to be depreciated by the double declining-balance method over 5 years. Find the book value at the end of 3 yr.

3. Derive the formula for the future value of an annuity [Equation (9), Section 5.2].

Solutions to Self-Check Exercises 5.4 can be found on page 330.

5.4 Concept Questions

1. Suppose an arithmetic progression has first term a and common difference d.
  a. What is the formula for the nth term of this progression?
  b. What is the formula for the sum of the first n terms of this progression?

2. Suppose a geometric progression has first term a and common ratio r.
  a. What is the formula for the nth term of this progression?
  b. What is the formula for the sum of the first n terms of this progression?

5.4 Exercises

In Exercises 1–4, find the nth term of the arithmetic progression that has the given values of a, d, and n.

1. $a = 6, d = 3, n = 9$

2. $a = -5, d = 3, n = 7$

3. $a = -15, d = \dfrac{3}{2}, n = 8$

4. $a = 1.2, d = 0.4, n = 98$

5. Find the first five terms of the arithmetic progression whose 4th and 11th terms are 30 and 107, respectively.

6. Find the first five terms of the arithmetic progression whose 7th and 23rd terms are -5 and -29, respectively.

7. Find the seventh term of the arithmetic progression: $x, x + y, x + 2y, \ldots$

8. Find the 11th term of the arithmetic progression: $a + b, 2a, 3a - b, \ldots$

9. Find the sum of the first 15 terms of the arithmetic progression: 4, 11, 18, . . .

10. Find the sum of the first 20 terms of the arithmetic progression: 5, -1, -7, . . .

11. Find the sum of the odd integers between 14 and 58.

12. Find the sum of the even integers between 21 and 99.

13. Find $f(1) + f(2) + f(3) + \cdots + f(20)$, given that $f(x) = 3x - 4$.

14. Find $g(1) + g(2) + g(3) + \cdots + g(50)$, given that $g(x) = 12 - 4x$.

15. Show that Equation (18) can be written as

$$S_n = \frac{n}{2}(a + a_n)$$

where a_n represents the last term of an arithmetic progression. Use this formula to find

a. The sum of the first 11 terms of the arithmetic progression whose 1st and 11th terms are 3 and 47, respectively.

b. The sum of the first 20 terms of the arithmetic progression whose 1st and 20th terms are 5 and -33, respectively.

16. SALES GROWTH   Moderne Furniture Company had sales of $1,500,000 during its first year of operation. If the sales increased by $160,000/year thereafter, find Moderne's sales in the fifth year and its total sales over the first 5 yr of operation.

17. EXERCISE PROGRAM   As part of her fitness program, Karen has taken up jogging. If she jogs 1 mi the first day and increases her daily run by 1/4 mi every week, how long will it take her to reach her goal of 10 mi/day?

18. COST OF DRILLING   A 100-ft oil well is to be drilled. The cost of drilling the first foot is $10.00, and the cost of drilling each additional foot is $4.50 more than that of the preceding foot. Find the cost of drilling the entire 100 ft.

19. CONSUMER DECISIONS   Kunwoo wishes to go from the airport to his hotel, which is 25 mi away. The taxi rate is $1.00 for the first mile and $.60 for each additional mile. The airport limousine also goes to his hotel and charges a flat rate of $7.50. How much money will the tourist save by taking the airport limousine?

20. SALARY COMPARISONS   Markeeta, a recent college graduate, received two job offers. Company A offered her an initial salary of $38,800 with guaranteed annual increases of $2000/year for the first 5 yr. Company B offered an initial salary of $40,400 with guaranteed annual increases of $1500 per year for the first 5 yr.

a. Which company is offering a higher salary for the fifth year of employment?

b. Which company is offering more money for the first 5 yr of employment?

21. SUM-OF-THE-YEARS'-DIGITS METHOD OF DEPRECIATION   One of the methods that the Internal Revenue Service allows for computing depreciation of certain business property is the sum-of-the-years'-digits method. If a property valued at C dollars

has an estimated useful life of N years and a salvage value of S dollars, then the amount of depreciation D_n allowed during the nth year is given by

$$D_n = (C - S)\frac{N - (n - 1)}{S_N} \qquad (0 \le n \le N)$$

where S_N is the sum of the first N positive integers representing the estimated useful life of the property. Thus,

$$S_N = 1 + 2 + \cdots + N = \frac{N(N + 1)}{2}$$

a. Verify that the sum of the arithmetic progression $S_N = 1 + 2 + \cdots + N$ is given by

$$\frac{N(N + 1)}{2}$$

b. If office furniture worth $6000 is to be depreciated by this method over $N = 10$ years and the salvage value of the furniture is $500, find the depreciation for the third year by computing D_3.

22. SUM-OF-THE-YEARS'-DIGITS METHOD OF DEPRECIATION   Refer to Example 2, Section 1.3, where the amount of depreciation allowed for a printing machine, which has an estimated useful life of 5 yr and an initial value of $100,000 (with no salvage value), was $20,000/year using the straight-line method of depreciation. Determine the amount of depreciation that would be allowed for the first year if the printing machine were depreciated using the sum-of-the-years'-digits method described in Exercise 21. Which method would result in a larger depreciation of the asset in its first year of use?

In Exercises 23–28, determine which of the sequences are geometric progressions. For each geometric progression, find the seventh term and the sum of the first seven terms.

23. $4, 8, 16, 32, \ldots$

24. $1, -\frac{1}{2}, \frac{1}{4}, -\frac{1}{8}, \ldots$

25. $\frac{1}{2}, -\frac{3}{8}, \frac{1}{4}, -\frac{9}{64}, \ldots$

26. $0.004, 0.04, 0.4, 4, \ldots$

27. $243, 81, 27, 9, \ldots$

28. $-1, 1, 3, 5, \ldots$

29. Find the 20th term and sum of the first 20 terms of the geometric progression $-3, 3, -3, 3, \ldots$

30. Find the 23rd term in a geometric progression having the first term $a = 0.1$ and ratio $r = 2$.

31. POPULATION GROWTH   It has been projected that the population of a certain city in the southwest will increase by 8% during each of the next 5 yr. If the current population is 200,000, what is the expected population in 5 yr?

32. SALES GROWTH   Metro Cable TV had sales of $2,500,000 in its first year of operation. If thereafter the sales increased by 12% of the previous year, find the sales of the company in the fifth year and the total sales over the first 5 yr of operation.

33. COLAs Suppose the cost-of-living index had increased by 3% during each of the past 6 yr and that a member of the EUW Union had been guaranteed an annual increase equal to 2% above the cost-of-living index over that period. What would be the present salary of a union member whose salary 6 yr ago was $32,000?

34. Savings Plans The parents of a 9-yr-old boy have agreed to deposit $10 in their son's bank account on his 10th birthday and to double the size of their deposit every year thereafter until his 18th birthday.
   a. How much will they have to deposit on his 18th birthday?
   b. How much will they have deposited by his 18th birthday?

35. Salary Comparisons An employee of Stenton Printing whose current annual salary is $48,000 has the option of taking an annual raise of 8%/year for the next 4 yr or a fixed annual raise of $4000/year. Which option would be more profitable to him considering his total earnings over the 4-yr period?

36. Bacteria Growth A culture of a certain bacteria is known to double in number every 3 hr. If the culture has an initial count of 20, what will be the population of the culture at the end of 24 hr?

37. Trust Funds Sarah is the recipient of a trust fund that she will receive over a period of 6 yr. Under the terms of the trust, she is to receive $10,000 the first year and each succeeding annual payment is to be increased by 15%.
   a. How much will she receive during the sixth year?
   b. What is the total amount of the six payments she will receive?

In Exercises 38–40, find the book value of office equipment purchased at a cost C at the end of the nth year if it is to be depreciated by the double declining-balance method over 10 years. Assume a salvage value of $0.

38. $C = \$20,000$, $n = 4$ **39.** $C = \$150,000$, $n = 8$

40. $C = \$80,000$, $n = 7$

41. Double Declining-Balance Method of Depreciation Restaurant equipment purchased at a cost of $150,000 is to be depreciated by the double declining-balance method over 10 yr. What is the book value of the equipment at the end of 6 yr? By what amount has the equipment been depreciated at the end of the sixth year?

42. Double Declining-Balance Method of Depreciation Refer to Exercise 22. Recall that a printing machine that had an estimated useful life of 5 yr and an initial value of $100,000 (with no salvage value) was to be depreciated. At the end of the first year, using the straight-line method of depreciation, the amount of depreciation allowed was $20,000, and when the sum-of-the-years'-digits method was used the depreciation was $33,333. Determine the amount of depreciation that would be allowed for the first year if the printing machine were depreciated by the double declining-balance method. Which of these three methods would result in the largest depreciation of the printing machine at the end of its first year of use?

In Exercises 43 and 44, determine whether the statement is true or false. If it is true, explain why it is true. If it is false, give an example to show why it is false.

43. If $a_1, a_2, a_3, \ldots, a_n$ and $b_1, b_2, b_3, \ldots, b_n$ are arithmetic progressions, then $a_1 + b_1, a_2 + b_2, a_3 + b_3, \ldots, a_n + b_n$ is also an arithmetic progression.

44. If $a_1, a_2, a_3, \ldots, a_n$ and $b_1, b_2, b_3, \ldots, b_n$ are geometric progressions, then $a_1b_1, a_2b_2, a_3b_3, \ldots, a_nb_n$ is also a geometric progression.

5.4 Solutions to Self-Check Exercises

1. Use Equation (22) with $a = -24$ and $r = -\frac{1}{2}$, obtaining

$$S_5 = \frac{-24\left[1 - (-\frac{1}{2})^5\right]}{1 - (-\frac{1}{2})}$$

$$= \frac{-24\left(1 + \frac{1}{32}\right)}{\frac{3}{2}} = -\frac{33}{2}$$

2. Use Equation (23) with $C = 75,000$, $N = 5$, and $n = 3$, giving the book value of the office equipment at the end of 3 yr as

$$V(3) = 75,000\left(1 - \frac{2}{5}\right)^3 = 16,200$$

   or $16,200.

3. We have

$$S = R + R(1 + i) + R(1 + i)^2 + \cdots + R(1 + i)^{n-1}$$

Now, the sum on the right is easily seen to be the sum of the first n terms of a geometric progression with first term R and common ratio $(1 + i)$, so by virtue of Formula (22),

$$S = \frac{R[1 - (1 + i)^n]}{1 - (1 + i)} = R\left[\frac{(1 + i)^n - 1}{i}\right]$$

CHAPTER 5 **Summary of Principal Formulas and Terms**

 FORMULAS

1. Simple interest (accumulated amount)	$A = P(1 + rt)$
2. Compound interest	
a. Accumulated amount	$A = P(1 + i)^n$
b. Present value	$P = A(1 + i)^{-n}$
c. Interest rate per conversion period	$i = \dfrac{r}{m}$
d. Number of conversion periods	$n = mt$
3. Continuous compound interest	
a. Accumulated amount	$A = Pe^{rt}$
b. Present value	$P = Ae^{-rt}$
4. Effective rate of interest	$r_{\text{eff}} = \left(1 + \dfrac{r}{m}\right)^m - 1$
5. Annuities	
a. Future value	$S = R\left[\dfrac{(1 + i)^n - 1}{i}\right]$
b. Present value	$P = R\left[\dfrac{1 - (1 + i)^{-n}}{i}\right]$
6. Amortization payment	$R = \dfrac{Pi}{1 - (1 + i)^{-n}}$
7. Sinking fund payment	$R = \dfrac{iS}{(1 + i)^n - 1}$

 TERMS

simple interest (278)	effective rate (284)	ordinary annuity (296)
accumulated amount (278)	present value (286)	simple annuity (296)
compound interest 279)	future value (286)	future value of an annuity (298)
nominal rate (stated rate) (279)	annuity (296)	present value of an annuity (300)
conversion period (280)	annuity certain (296)	sinking fund (313)

CHAPTER 5 **Concept Review Questions**

Fill in the blanks.

1. a. Simple interest is computed on the _____ principal only. The formula for simple interest is $A =$ _____.

b. In calculations using compound interest, earned interest is periodically added to the principal and thereafter itself earns _____. The formula for compound interest is $A =$ _____. Solving this equation for P gives the present value formula for compound interest $P =$ _____.

2. The effective rate of interest is the _____ interest rate that would produce the same accumulated amount in _____ year as the _____ rate compounded _____ times a year. The formula for calculating the effective rate is $r_{\text{eff}} = $ _____.

3. A sequence of payments made at regular time intervals is called a/an _____; if the payments are made at the end of each payment period, then it is called a/an _____; if the payment period coincides with the interest conversion period, it is called a/an _____.

4. **a.** The future value of an annuity is $S = $ _____.
   b. The present value of an annuity is $P = $ _____.

5. The periodic payment R on a loan of P dollars to be amortized over n periods with interest charged at the rate of i per period is $R = $ _____.

6. A sinking fund is an account that is set up for a specific purpose at some _____ date. The periodic payment R required to accumulate a sum of S dollars over n periods with interest charged at the rate of i per period is $R = $ _____.

7. An arithmetic progression is a sequence of numbers in which each term after the first is obtained by adding a/an _____ _____ to the preceding term. The nth term of an arithmetic progression is $a_n = $ _____. The sum of the first n terms of an arithmetic progression is $S_n = $ _____.

8. A geometric progression is a sequence of numbers in which each term after the first is obtained by multiplying the preceding term by a/an _____ _____. The nth term of a geometric progression is $a_n = $ _____. If $r \neq 1$, the sum of the first n terms of a geometric progression is $S_n = $ _____.

CHAPTER 5   Review Exercises

1. Find the accumulated amount after 4 yr if $5000 is invested at 10%/year compounded (a) annually, (b) semiannually, (c) quarterly, and (d) monthly.

2. Find the accumulated amount after 8 yr if $12,000 is invested at 6.5%/year compounded (a) annually, (b) semiannually, (c) quarterly, and (d) monthly.

3. Find the effective rate of interest corresponding to a nominal rate of 12%/year compounded (a) annually, (b) semiannually, (c) quarterly, and (d) monthly.

4. Find the effective rate of interest corresponding to a nominal rate of 11.5%/year compounded (a) annually, (b) semiannually, (c) quarterly, and (d) monthly.

5. Find the present value of $41,413 due in 5 yr at an interest rate of 6.5%/year compounded quarterly.

6. Find the present value of $64,540 due in 6 yr at an interest rate of 8%/year compounded monthly.

7. Find the amount (future value) of an ordinary annuity of $150/quarter for 7 yr at 8%/year compounded quarterly.

8. Find the future value of an ordinary annuity of $120/month for 10 yr at 9%/year compounded monthly.

9. Find the present value of an ordinary annuity of 36 payments of $250 each made monthly and earning interest at 9%/year compounded monthly.

10. Find the present value of an ordinary annuity of 60 payments of $5000 each made quarterly and earning interest at 8%/year compounded quarterly.

11. Find the payment R needed to amortize a loan of $22,000 at 8.5%/year compounded monthly with 36 monthly installments over a period of 3 yr.

12. Find the payment R needed to amortize a loan of $10,000 at 9.2%/year compounded monthly with 36 monthly installments over a period of 3 yr.

13. Find the payment R needed to accumulate $18,000 with 48 monthly installments over a period of 4 yr at an interest rate of 6%/year compounded monthly.

14. Find the payment R needed to accumulate $15,000 with 60 monthly installments over a period of 5 yr at an interest rate of 7.2%/year compounded monthly.

15. Find the rate of interest per year compounded on a daily basis that is equivalent to 7.2%/year compounded monthly.

16. Find the rate of interest per year compounded on a daily basis that is equivalent to 9.6%/year compounded monthly.

17. INVESTMENT RETURN   A hotel was purchased by a conglomerate for $4.5 million and sold 5 yr later for $8.2 million. Find the annual rate of return (compounded continuously).

18. Find the present value of $119,346 due in 4 yr at an interest rate of 10%/year compounded continuously.

19. COMPANY SALES   JCN Media Corporation had sales of $1,750,000 in the first year of operation. If the sales increased by 14%/year thereafter, find the company's sales in the fourth year and the total sales over the first 4 yr of operation.

20. **CDs** The manager of a money market fund has invested $4.2 million in certificates of deposit that pay interest at the rate of 5.4%/year compounded quarterly over a period of 5 yr. How much will the investment be worth at the end of 5 yr?

21. **SAVINGS ACCOUNTS** Emily deposited $2000 into a bank account 5 yr ago. The bank paid interest at the rate of 8%/year compounded weekly. What is Emily's account worth today?

22. **SAVINGS ACCOUNTS** Kim invested a sum of money 4 yr ago in a savings account that has since paid interest at the rate of 6.5%/year compounded monthly. Her investment is now worth $19,440.31. How much did she originally invest?

23. **SAVINGS ACCOUNTS** Andrew withdrew $5986.09 from a savings account, which he closed this morning. The account had earned interest at the rate of 6%/year compounded continuously during the 3-yr period that the money was on deposit. How much did Andrew originally deposit into the account?

24. **MUTUAL FUNDS** Juan invested $24,000 in a mutual fund 5 yr ago. Today his investment is worth $34,616. Find the effective annual rate of return on his investment over the 5-yr period.

25. **COLLEGE SAVINGS PROGRAM** The Blakes have decided to start a monthly savings program in order to provide for their son's college education. How much should they deposit at the end of each month in a savings account earning interest at the rate of 8%/year compounded monthly so that at the end of the tenth year the accumulated amount will be $40,000?

26. **RETIREMENT ACCOUNTS** Mai Lee has contributed $200 at the end of each month into her company's employee retirement account for the past 10 yr. Her employer has matched her contribution each month. If the account has earned interest at the rate of 8%/year compounded monthly over the 10-yr period, determine how much Mai Lee now has in her retirement account.

27. **AUTOMOBILE LEASING** Maria has leased an auto for 4 yr at $300/month. If money is worth 5%/year compounded monthly, what is the equivalent cash payment (present value) of this annuity? (Assume that the payments are made at the end of each month.)

28. **INSTALLMENT FINANCING** Peggy made a down payment of $400 toward the purchase of new furniture. To pay the balance of the purchase price, she has secured a loan from her bank at 12%/year compounded monthly. Under the terms of her finance agreement, she is required to make payments of $75.32 at the end of each month for 24 mo. What was the purchase price of the furniture?

29. **HOME FINANCING** The Turners have purchased a house for $150,000. They made an initial down payment of $30,000 and secured a mortgage with interest charged at the rate of 9%/year on the unpaid balance. (Interest computations are made at the end of each month.) Assume the loan is amortized over 30 yr.
   a. What monthly payment will the Turners be required to make?
   b. What will be their total interest payment?
   c. What will be their equity (disregard depreciation) after 10 yr?

30. **HOME FINANCING** Refer to Exercise 29. If the loan is amortized over 15 yr,
   a. What monthly payment will the Turners be required to make?
   b. What will be their total interest payment?
   c. What will be their equity (disregard depreciation) after 10 yr?

31. **SINKING FUNDS** The management of a corporation anticipates a capital expenditure of $500,000 in 5 yr for the purpose of purchasing replacement machinery. To finance this purchase, a sinking fund that earns interest at the rate of 10%/year compounded quarterly will be set up. Determine the amount of each (equal) quarterly installment that should be deposited in the fund. (Assume that the payments are made at the end of each quarter.)

32. **SINKING FUNDS** The management of a condominium association anticipates a capital expenditure of $120,000 in 2 yr for the purpose of painting the exterior of the condominium. To pay for this maintenance, a sinking fund will be set up that will earn interest at the rate of 5.8%/year compounded monthly. Determine the amount of each (equal) monthly installment the association will be required to deposit into the fund at the end of each month for the next 2 yr.

33. **CREDIT CARD PAYMENTS** The outstanding balance on Bill's credit card account is $3200. The bank issuing the credit card is charging 18.6%/year compounded monthly. If Bill decides to pay off this balance in equal monthly installments at the end of each month for the next 18 mo, how much will be his monthly payment? What is the effective rate of interest the bank is charging Bill?

34. **FINANCIAL PLANNING** Matt's parents have agreed to contribute $250/month toward the rent for his apartment in his junior year in college. The plan is for Matt's parents to deposit a lump sum in Matt's bank account on August 1 and have Matt withdraw $250 on the first of each month starting on September 1 and ending on May 1 the following year. If the bank pays interest on the balance at the rate of 5%/year compounded monthly, how much should Matt's parents deposit into his account?

CHAPTER 5　Before Moving On . . .

1. Find the accumulated amount at the end of 3 yr if $2000 is deposited in an account paying interest at the rate of 8%/year compounded monthly.

2. Find the effective rate of interest corresponding to a nominal rate of 6%/year compounded daily.

3. Find the future value of an ordinary annuity of $800/week for 10 yr at 6%/year compounded weekly.

4. Find the monthly payment required to amortize a loan of $100,000 over 10 yr with interest earned at the rate of 8%/year compounded monthly.

5. Find the weekly payment required to accumulate a sum of $15,000 over 6 yr with interest earned at the rate of 10%/year compounded weekly.

6. **A. P. and G. P. (Optional)**
   a. Find the sum of the first ten terms of the arithmetic progression 3, 7, 11, 15, 19, ...
   b. Find the sum of the first eight terms of the geometric progression $\frac{1}{2}$, 1, 2, 4, 8, ...

Sets and Counting

© Jason Homa/The Image Bank/Getty Images

What are the investment options? An investor has decided to purchase shares of stock from a recommended list of aerospace, energy development, and electronics companies. In Example 5, page 355, we will determine how many ways the investor may select a group of three companies from the list.

WE OFTEN DEAL with well-defined collections of objects called *sets*. In this chapter, we see how sets can be combined algebraically to yield other sets. We also look at some techniques for determining the number of elements in a set and for determining the number of ways the elements of a set can be arranged or combined. These techniques enable us to solve many practical problems, as you will see throughout the chapter.

6.1   Sets and Set Operations

■ Set Terminology and Notation

We often deal with collections of different kinds of objects. For example, in conducting a study of the distribution of the weights of newborn infants, we might consider the collection of all infants born in the Massachusetts General Hospital during 2004. In a study of the fuel consumption of compact cars, we might be interested in the collection of compact cars manufactured by General Motors in the 2004 model year. Such collections are examples of sets. More specifically, a **set** is a well-defined collection of objects. Thus, a set is not just any collection of objects, but it must be well defined in the sense that if we are given an object, then we should be able to determine whether or not it belongs to the collection.

The objects of a set are called the **elements,** or *members*, **of a set** and are usually denoted by lowercase letters $a, b, c, \ldots$; the sets themselves are usually denoted by uppercase letters $A, B, C, \ldots$. The elements of a set may be displayed by listing each element between braces. For example, using **roster notation,** the set A consisting of the first three letters of the English alphabet is written

$$A = \{a, b, c\}$$

The set B of all letters of the alphabet may be written

$$B = \{a, b, c, \ldots, z\}$$

Another notation commonly used is **set-builder notation.** Here, a rule is given that describes the definite property or properties an object x must satisfy to qualify for membership in the set. Using this notation, the set B is written as

$$B = \{x | x \text{ is a letter of the English alphabet}\}$$

and is read "B is the set of all elements x such that x is a letter of the English alphabet."

If a is an element of a set A, we write $a \in A$ and read "a belongs to A" or "a is an element of A." If, however, the element a does not belong to the set A, then we write $a \notin A$ and read "a does not belong to A." For example, if $A = \{1, 2, 3, 4, 5\}$, then $3 \in A$ but $6 \notin A$.

EXPLORE & DISCUSS

1. Let A denote the collection of all the days in August 2004 in which the average daily temperature in San Francisco was approximately 75°F. Is A a set? Explain your answer.

2. Let B denote the collection of all the days in August 2004 in which the average daily temperature in San Francisco was between 73.5°F and 81.2°F, inclusive. Is B a set? Explain your answer.

Set Equality

Two sets A and B are **equal,** written $A = B$, if and only if they have exactly the same elements.

EXAMPLE 1   Let A, B, and C be the sets

$$A = \{a, e, i, o, u\}$$
$$B = \{a, i, o, e, u\}$$
$$C = \{a, e, i, o\}$$

Then, $A = B$ since they both contain exactly the same elements. Note that the order in which the elements are displayed is immaterial. Also, $A \neq C$ since $u \in A$ but $u \notin C$. Similarly, we conclude that $B \neq C$. ■

Subset
If every element of a set A is also an element of a set B, then we say that A is a **subset** of B and write $A \subseteq B$.

By this definition, two sets A and B are equal if and only if (1) $A \subseteq B$ and (2) $B \subseteq A$. You may verify this (see Exercise 66).

EXAMPLE 2   Referring to Example 1, we find that $C \subseteq B$ since every element of C is also an element of B. Also, if D is the set

$$D = \{a, e, i, o, x\}$$

then D is not a subset of A, written $D \nsubseteq A$, since $x \in D$ but $x \notin A$. Observe that $A \nsubseteq D$ as well since $u \in A$ but $u \notin D$. ■

If A and B are sets such that $A \subseteq B$ but $A \neq B$, then we say that A is a **proper subset** of B. In other words, a set A is a proper subset of a set B, written $A \subset B$, if (1) $A \subseteq B$ and (2) there exists at least one element in B that is not in A. The latter condition states that the set A is properly "smaller" than the set B.

EXAMPLE 3   Let $A = \{1, 2, 3, 4, 5, 6\}$ and $B = \{2, 4, 6\}$. Then, B is a proper subset of A since (1) $B \subseteq A$, which is easily verified, and (2) there exists at least one element in A that is not in B—for example, the element 1. ■

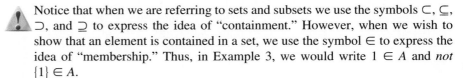 Notice that when we are referring to sets and subsets we use the symbols $\subset$, $\subseteq$, $\supset$, and $\supseteq$ to express the idea of "containment." However, when we wish to show that an element is contained in a set, we use the symbol $\in$ to express the idea of "membership." Thus, in Example 3, we would write $1 \in A$ and *not* $\{1\} \in A$.

Empty Set
The set that contains no elements is called the **empty set** and is denoted by $\varnothing$.

The empty set, $\varnothing$, is a subset of every set. To see this, observe that $\varnothing$ has no elements and, therefore, contains no element that is not also in A.

EXAMPLE 4   List all subsets of the set $A = \{a, b, c\}$.

Solution   There is one subset consisting of no elements—namely, the empty set, $\varnothing$. Next, observe that there are three subsets consisting of one element,

$$\{a\}, \quad \{b\}, \quad \{c\}$$

three subsets consisting of two elements,

$$\{a, b\}, \quad \{a, c\}, \quad \{b, c\}$$

and one subset consisting of three elements, the set A itself. Therefore, the subsets of A are

$$\varnothing, \quad \{a\}, \quad \{b\}, \quad \{c\}, \quad \{a, b\}, \quad \{a, c\}, \quad \{b, c\}, \quad \{a, b, c\} \quad \blacksquare$$

In contrast with the empty set, we have, on the other extreme, the notion of a largest, or universal, set. A **universal set** is the set of all elements of interest in a particular discussion. It is the largest in the sense that all sets considered in the discussion of the problem are subsets of the universal set. Of course, different universal sets are associated with different problems, as shown in Example 5.

EXAMPLE 5
a. If the problem at hand is to determine the ratio of female to male students in a college, then a logical choice of a universal set is the set consisting of the whole student body of the college.
b. If the problem is to determine the ratio of female to male students in the business department of the college in part (a), then the set of all students in the business department may be chosen as the universal set. $\blacksquare$

A visual representation of sets is realized through the use of **Venn diagrams,** which are of considerable help in understanding the concepts introduced earlier, as well as in solving problems involving sets. The universal set U is represented by a rectangle, and subsets of U are represented by regions lying inside the rectangle.

EXAMPLE 6   Use Venn diagrams to illustrate the following statements:
a. The sets A and B are equal.
b. The set A is a proper subset of the set B.
c. The sets A and B are not subsets of each other.

Solution   The respective Venn diagrams are shown in Figure 1a–c.

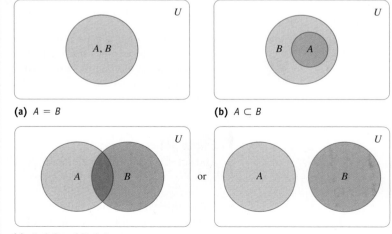

(a) $A = B$

(b) $A \subset B$

or

(c) $A \not\subset B$ and $B \not\subset A$ $\blacksquare$

FIGURE 1

■ Set Operations

Having introduced the concept of a set, our next task is to consider operations on sets—that is, to consider ways in which sets may be combined to yield other sets. These operations enable us to combine sets in much the same way the operations of addition and multiplication enable us to combine numbers to obtain other numbers. In what follows, all sets are assumed to be subsets of a given universal set U.

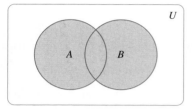

FIGURE 2
Set union $A \cup B$

> **Set Union**
>
> Let A and B be sets. The **union** of A and B, written $A \cup B$, is the set of all elements that belong to either A or B or both.
>
> $$A \cup B = \{x \mid x \in A \quad \text{or} \quad x \in B \quad \text{or} \quad \text{both}\}$$

The shaded portion of the Venn diagram (Figure 2) depicts the set $A \cup B$.

EXAMPLE 7   If $A = \{a, b, c\}$ and $B = \{a, c, d\}$, then $A \cup B = \{a, b, c, d\}$. ■

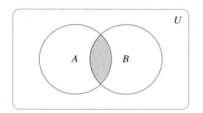

FIGURE 3
Set intersection $A \cap B$

> **Set Intersection**
>
> Let A and B be sets. The set of elements in common with the sets A and B, written $A \cap B$, is called the **intersection** of A and B.
>
> $$A \cap B = \{x \mid x \in A \quad \text{and} \quad x \in B\}$$

The shaded portion of the Venn diagram (Figure 3) depicts the set $A \cap B$.

EXAMPLE 8   Let $A = \{a, b, c\}$ and $B = \{a, c, d\}$. Then, $A \cap B = \{a, c\}$. (Compare this result with Example 7.) ■

EXAMPLE 9   Let $A = \{1, 3, 5, 7, 9\}$ and $B = \{2, 4, 6, 8, 10\}$. Then, $A \cap B = \varnothing$. ■

The two sets of Example 9 have null intersection. In general, the sets A and B are said to be **disjoint** if they have no elements in common—that is, if $A \cap B = \varnothing$.

EXAMPLE 10   If U is the set of all students in the classroom and $M = \{x \in U \mid x$ is male$\}$ and $F = \{x \in U \mid x$ is female$\}$, then $F \cap M = \varnothing$, and F and M are disjoint. ■

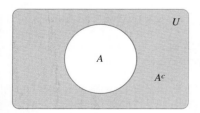

FIGURE 4
Set complementation

> **Complement of a Set**
>
> If U is a universal set and A is a subset of U, then the set of all elements in U that are not in A is called the **complement** of A and is denoted A^c.
>
> $$A^c = \{x \mid x \in U, x \notin A\}$$

The shaded portion of the Venn diagram (Figure 4) shows the set A^c.

EXPLORE & DISCUSS

Let A, B, and C be nonempty subsets of a set U.

1. Suppose $A \cap B \neq \varnothing$, $A \cap C \neq \varnothing$, and $B \cap C \neq \varnothing$. Can you conclude that $A \cap B \cap C \neq \varnothing$? Explain your answer with an example.

2. Suppose $A \cap B \cap C \neq \varnothing$. Can you conclude that $A \cap B \neq \varnothing$, $A \cap C \neq \varnothing$, and $B \cap C \neq \varnothing$ simultaneously? Explain your answer.

EXAMPLE 11   Let $U = \{1, 2, 3, 4, 5, 6, 7, 8, 9, 10\}$ and $A = \{2, 4, 6, 8, 10\}$. Then, $A^c = \{1, 3, 5, 7, 9\}$.

The following rules hold for the operation of complementation. See whether you can verify them.

Set Complementation

If U is a universal set and A is a subset of U, then

a. $U^c = \varnothing$ **b.** $\varnothing^c = U$ **c.** $(A^c)^c = A$
d. $A \cup A^c = U$ **e.** $A \cap A^c = \varnothing$

The following rules govern the operations on sets.

Set Operations

Let U be a universal set. If A, B, and C are arbitrary subsets of U, then

$A \cup B = B \cup A$	*Commutative law for union*
$A \cap B = B \cap A$	*Commutative law for intersection*
$A \cup (B \cup C) = (A \cup B) \cup C$	*Associative law for union*
$A \cap (B \cap C) = (A \cap B) \cap C$	*Associative law for intersection*
$A \cup (B \cap C)$	
$\quad = (A \cup B) \cap (A \cup C)$	*Distributive law for union*
$A \cap (B \cup C)$	
$\quad = (A \cap B) \cup (A \cap C)$	*Distributive law for intersection*

Two additional rules, referred to as De Morgan's laws, govern the operations on sets.

De Morgan's Laws

Let A and B be sets. Then,

$$(A \cup B)^c = A^c \cap B^c \qquad \textbf{(1)}$$
$$(A \cap B)^c = A^c \cup B^c \qquad \textbf{(2)}$$

Equation (1) states that the complement of the union of two sets is equal to the intersection of their complements. Equation (2) states that the complement of the intersection of two sets is equal to the union of their complements.

We will not prove De Morgan's laws here, but the plausibility of (2) is illustrated in the following example.

EXAMPLE 12   Using Venn diagrams, show that

$$(A \cap B)^c = A^c \cup B^c$$

Solution $(A \cap B)^c$ is the set of elements in U but not in $A \cap B$ and is thus the shaded region shown in Figure 5. Next, A^c and B^c are shown in Figure 6a–b. Their union, $A^c \cup B^c$, is easily seen to be equivalent to $(A \cap B)^c$ by referring once again to Figure 5.

FIGURE 5
$(A \cap B)^c$

FIGURE 6
$A^c \cup B^c$ is the set obtained by joining (a) and (b).

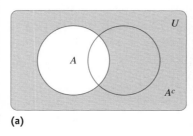

(a)

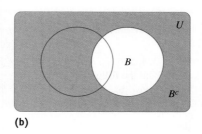

(b)

EXAMPLE 13   Let $U = \{1, 2, 3, 4, 5, 6, 7, 8, 9, 10\}$, $A = \{1, 2, 4, 8, 9\}$, and $B = \{3, 4, 5, 6, 8\}$. Verify by direct computation that $(A \cup B)^c = A^c \cap B^c$.

Solution $A \cup B = \{1, 2, 3, 4, 5, 6, 8, 9\}$, so $(A \cup B)^c = \{7, 10\}$. However, $A^c = \{3, 5, 6, 7, 10\}$ and $B^c = \{1, 2, 7, 9, 10\}$, so $A^c \cap B^c = \{7, 10\}$. The required result follows.

APPLIED EXAMPLE 14   Automobile Options   Let U denote the set of all cars in a dealer's lot and

$A = \{x \in U | x \text{ is equipped with automatic transmission}\}$

$B = \{x \in U | x \text{ is equipped with air conditioning}\}$

$C = \{x \in U | x \text{ is equipped with side air bags}\}$

Find an expression in terms of A, B, and C for each of the following sets:
a. The set of cars with at least one of the given options
b. The set of cars with exactly one of the given options
c. The set of cars with automatic transmission and side air bags but no air conditioning

Solution
a. The set of cars with at least one of the given options is $A \cup B \cup C$ (Figure 7a).
b. The set of cars with automatic transmission only is given by $A \cap B^c \cap C^c$. Similarly, we find that the set of cars with air conditioning only is given by $B \cap C^c \cap A^c$, whereas the set of cars with side air bags only is given by $C \cap A^c \cap B^c$. Thus, the set of cars with exactly one of the given options is $(A \cap B^c \cap C^c) \cup (B \cap C^c \cap A^c) \cup (C \cap A^c \cap B^c)$ (Figure 7b).

c. The set of cars with automatic transmission and side air bags but no air conditioning is given by $A \cap C \cap B^c$ (Figure 7c).

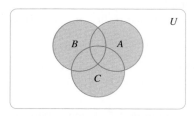

(a) The set of cars with at least one option

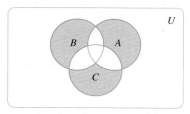

(b) The set of cars with exactly one option

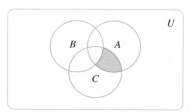

(c) The set of cars with automatic transmission and side air bags but no air conditioning

FIGURE 7

6.1 Self-Check Exercises

1. Let $U = \{1, 2, 3, 4, 5, 6, 7\}$, $A = \{1, 2, 3\}$, $B = \{3, 4, 5, 6\}$, and $C = \{2, 3, 4\}$. Find the following sets:

  a. A^c **b.** $A \cup B$ **c.** $B \cap C$
  d. $(A \cup B) \cap C$ **e.** $(A \cap B) \cup C$ **f.** $A^c \cap (B \cup C)^c$

2. Let U denote the set of all members of the House of Representatives. Let

$$D = \{x \in U | x \text{ is a Democrat}\}$$
$$R = \{x \in U | x \text{ is a Republican}\}$$

$$F = \{x \in U | x \text{ is a female}\}$$
$$L = \{x \in U | x \text{ is a lawyer by training}\}$$

Describe each of the following sets in words.

  a. $D \cap F$ **b.** $F^c \cap R$ **c.** $D \cap F \cap L^c$

Solutions to Self-Check Exercises 6.1 can be found on page 345.

6.1 Concept Questions

1. a. What is a *set*? Give an example.
  b. When are two sets equal? Give an example of two equal sets.
  c. What is the *empty set*?

2. What can you say about two sets A and B such that
  a. $A \cup B \subseteq A$ **b.** $A \cup B = \emptyset$
  c. $A \cap B = B$ **d.** $A \cap B = \emptyset$

3. a. If $A \subset B$, what can you say about the relationship between A^c and B^c?
  b. If $A^c = \emptyset$, what can you say about A?

6.1 Exercises

In Exercises 1–4, write the set in set-builder notation.

1. The set of gold medalists in the 2002 Winter Olympic Games

2. The set of football teams in the NFL

3. $\{3, 4, 5, 6, 7\}$

4. $\{1, 3, 5, 7, 9, 11, \ldots, 39\}$

In Exercises 5–8, list the elements of the set in roster notation.

5. $\{x | x \text{ is a digit in the number } 352{,}646\}$

6. $\{x | x \text{ is a letter in the word } HIPPOPOTAMUS\}$

7. $\{x | 2 - x = 4; x, \text{ an integer}\}$

8. $\{x | 2 - x = 4; x, \text{ a fraction}\}$

In Exercises 9–14, state whether the statements are true or false.

9. **a.** $\{a, b, c\} = \{c, a, b\}$ **b.** $A \in A$

10. **a.** $\emptyset \in A$ **b.** $A \subset A$

11. **a.** $0 \in \emptyset$ **b.** $0 = \emptyset$

12. **a.** $\{\emptyset\} = \emptyset$ **b.** $\{a, b\} \in \{a, b, c\}$

13. $\{$Chevrolet, Pontiac, Buick$\} \subset \{x | x$ is a division of General Motors$\}$

14. $\{x | x$ is a silver medalist in the 2002 Winter Olympic Games$\} = \emptyset$

In Exercises 15 and 16, let $A = \{1, 2, 3, 4, 5\}$. Determine whether the statements are true or false.

15. **a.** $2 \in A$ **b.** $A \subseteq \{2, 4, 6\}$

16. **a.** $0 \in A$ **b.** $\{1, 3, 5\} \in A$

17. Let $A = \{1, 2, 3\}$. Which of the following sets are equal to A?
 a. $\{2, 1, 3\}$ **b.** $\{3, 2, 1\}$
 c. $\{0, 1, 2, 3\}$

18. Let $A = \{a, e, l, t, r\}$. Which of the following sets are equal to A?
 a. $\{x | x$ is a letter of the word *later*$\}$
 b. $\{x | x$ is a letter of the word *latter*$\}$
 c. $\{x | x$ is a letter of the word *relate*$\}$

19. List all subsets of the following sets:
 a. $\{1, 2\}$ **b.** $\{1, 2, 3\}$ **c.** $\{1, 2, 3, 4\}$

20. List all subsets of the set $A = \{$IBM, U.S. Steel, Union Carbide, Boeing$\}$. Which of these are proper subsets of A?

In Exercises 21–24, find the smallest possible set (that is, the set with the least number of elements) that contains the sets as subsets.

21. $\{1, 2\}, \{1, 3, 4\}, \{4, 6, 8, 10\}$

22. $\{1, 2, 4\}, \{a, b\}$

23. $\{$Jill, John, Jack$\}, \{$Susan, Sharon$\}$

24. $\{$GM, Ford, Chrysler$\}, \{$Daimler-Benz, Volkswagen$\}, \{$Toyota, Nissan$\}$

25. Use Venn diagrams to represent the following relationships:
 a. $A \subset B$ and $B \subset C$
 b. $A \subset U$ and $B \subset U$, where A and B have no elements in common
 c. The sets $A, B,$ and C are equal.

26. Let U denote the set of all students who applied for admission to the freshman class at Faber College for the upcoming academic year and let

$A = \{x \in U | x$ is a successful applicant$\}$
$B = \{x \in U | x$ is a female student who enrolled in the freshman class$\}$
$C = \{x \in U | x$ is a male student who enrolled in the freshman class$\}$

 a. Use Venn diagrams to represent the sets $U, A, B,$ and C.
 b. Determine whether the following statements are true or false.
 i. $A \subseteq B$ **ii.** $B \subset A$ **iii.** $C \subset B$

In Exercises 27 and 28, shade the portion of the accompanying figure that represents each set.

27. **a.** $A \cap B^c$
 b. $A^c \cap B$

28. **a.** $A^c \cap B^c$
 b. $(A \cup B)^c$

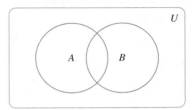

In Exercises 29–32, shade the portion of the accompanying figure that represents each set.

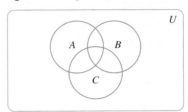

29. **a.** $A \cup B \cup C$ **b.** $A \cap B \cap C$

30. **a.** $A \cap B \cap C^c$ **b.** $A^c \cap B \cap C$

31. **a.** $A^c \cap B^c \cap C^c$ **b.** $(A \cup B)^c \cap C$

32. **a.** $A \cup (B \cap C)^c$ **b.** $(A \cup B \cup C)^c$

In Exercises 33–36, let $U = \{1, 2, 3, 4, 5, 6, 7, 8, 9, 10\}$, $A = \{1, 3, 5, 7, 9\}$, $B = \{2, 4, 6, 8, 10\}$, and $C = \{1, 2, 4, 5, 8, 9\}$. Find each set.

33. **a.** A^c **b.** $B \cup C$ **c.** $C \cup C^c$

34. **a.** $C \cap C^c$ **b.** $(A \cap C)^c$ **c.** $A \cup (B \cap C)$

35. **a.** $(A \cap B) \cup C$ **b.** $(A \cup B \cup C)^c$
 c. $(A \cap B \cap C)^c$

36. **a.** $A^c \cap (B \cap C^c)$ **b.** $(A \cup B^c) \cup (B \cap C^c)$
 c. $(A \cup B)^c \cap C^c$

In Exercises 37 and 38, determine whether the pairs of sets are disjoint.

37. a. $\{1, 2, 3, 4\}, \{4, 5, 6, 7\}$
   b. $\{a, c, e, g\}, \{b, d, f\}$

38. a. $\varnothing, \{1, 3, 5\}$
   b. $\{0, 1, 3, 4\}, \{0, 2, 5, 7\}$

In Exercises 39–42, let U denote the set of all employees at Universal Life Insurance Company. Let

$$T = \{x \in U | x \text{ drinks tea}\}$$
$$C = \{x \in U | x \text{ drinks coffee}\}$$

Describe each set in words.

39. a. T^c **b.** C^c

40. a. $T \cup C$ **b.** $T \cap C$

41. a. $T \cap C^c$ **b.** $T^c \cap C$

42. a. $T^c \cap C^c$ **b.** $(T \cup C)^c$

In Exercises 43–46, let U denote the set of all employees in a hospital. Let

$$N = \{x \in U | x \text{ is a nurse}\}$$
$$D = \{x \in U | x \text{ is a doctor}\}$$
$$A = \{x \in U | x \text{ is an administrator}\}$$
$$M = \{x \in U | x \text{ is a male}\}$$
$$F = \{x \in U | x \text{ is a female}\}$$

Describe each set in words.

43. a. D^c **b.** N^c

44. a. $N \cup D$ **b.** $N \cap M$

45. a. $D \cap M^c$ **b.** $D \cap A$

46. a. $N \cap F$ **b.** $(D \cup N)^c$

In Exercises 47 and 48, let U denote the set of all senators in Congress. Let

$$D = \{x \in U | x \text{ is a Democrat}\}$$
$$R = \{x \in U | x \text{ is a Republican}\}$$
$$F = \{x \in U | x \text{ is a female}\}$$
$$L = \{x \in U | x \text{ is a lawyer}\}$$

Write the set that represents each statement.

47. a. The set of all Democrats who are female
   b. The set of all Republicans who are male and are not lawyers

48. a. The set of all Democrats who are female or are lawyers
   b. The set of all senators who are not Democrats or are lawyers

In Exercises 49 and 50, let U denote the set of all students in the business college of a certain university. Let

$$A = \{x \in U | x \text{ had taken a course in Accounting}\}$$
$$B = \{x \in U | x \text{ had taken a course in Economics}\}$$
$$C = \{x \in U | x \text{ had taken a course in Marketing}\}$$

Write the set that represents each statement.

49. a. The set of students who have not had a course in Economics
   b. The set of students who have had courses in Accounting and Economics
   c. The set of students who have had courses in Accounting and Economics but not Marketing

50. a. The set of students who have had courses in Economics but not courses in Accounting or Marketing
   b. The set of students who have had at least one of the three courses
   c. The set of students who have had all three courses

In Exercises 51 and 52, refer to the following diagram where U is the set of all tourists surveyed over a 1-week period in London and

$$A = \{x \in U | x \text{ has taken the underground [subway]}\}$$
$$B = \{x \in U | x \text{ has taken a cab}\}$$
$$C = \{x \in U | x \text{ has taken a bus}\}$$

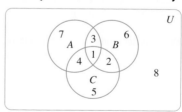

Express the indicated regions in set notation and in words.

51. a. Region 1
   b. Regions 1 and 4 together
   c. Regions 4, 5, 7, and 8 together

52. a. Region 3
   b. Regions 4 and 6 together
   c. Regions 5, 6, and 7 together

In Exercises 53–58, use Venn diagrams to illustrate each statement.

53. $A \subset A \cup B$; $B \subset A \cup B$ **54.** $A \cap B \subset A$; $A \cap B \subset B$

55. $A \cup (B \cup C) = (A \cup B) \cup C$

56. $A \cap (B \cap C) = (A \cap B) \cap C$

57. $A \cap (B \cup C) = (A \cap B) \cup (A \cap C)$

58. $(A \cup B)^c = A^c \cap B^c$

In Exercises 59 and 60, let

$$U = \{1, 2, 3, 4, 5, 6, 7, 8, 9, 10\}$$
$$A = \{1, 3, 5, 7, 9\}, B = \{1, 2, 4, 7, 8\}$$
$$C = \{2, 4, 6, 8\}$$

Verify by direct computation each equation.

59. a. $A \cup (B \cup C) = (A \cup B) \cup C$
   b. $A \cap (B \cap C) = (A \cap B) \cap C$

60. a. $A \cap (B \cup C) = (A \cap B) \cup (A \cap C)$
   b. $(A \cup B)^c = A^c \cap B^c$

In Exercises 61–64, refer to the accompanying figure and find the points that belong to each set.

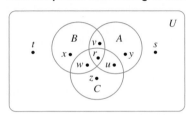

61. a. $A \cup B$ **b.** $A \cap B$

62. a. $A \cap (B \cup C)$ **b.** $(B \cap C)^c$

63. a. $(B \cup C)^c$ **b.** A^c

64. a. $(A \cap B) \cap C^c$ **b.** $(A \cup B \cup C)^c$

65. Suppose $A \subset B$ and $B \subset C$, where A and B are any two sets. What conclusion can be drawn regarding the sets A and C?

66. Verify the assertion that two sets A and B are equal if and only if (1) $A \subseteq B$ and (2) $B \subseteq A$.

In Exercises 67–72, determine whether the statement is true or false. If it is true, explain why it is true. If it is false, give an example to show why it is false.

67. A set is never a subset of itself.

68. A proper subset of a set is itself a subset of the set, but not vice versa.

69. If $A \cup B = \varnothing$, then $A = \varnothing$ and $B = \varnothing$.

70. If $A \cap B = \varnothing$, then $A = \varnothing$ and $B = \varnothing$ or both A and B are empty.

71. $(A \cup A^c)^c = \varnothing$

72. If $A \subseteq B$, then $A \cap B = A$.

6.1 Solutions to Self-Check Exercises

1. a. A^c is the set of all elements in U but not in A. Therefore,
$$A^c = \{4, 5, 6, 7\}$$

b. $A \cup B$ consists of all elements in A and/or B. So,
$$A \cup B = \{1, 2, 3, 4, 5, 6\}$$

c. $B \cap C$ is the set of all elements in both B and C. Therefore,
$$B \cap C = \{3, 4\}$$

d. Using the result from part (b), we find
$$(A \cup B) \cap C = \{1, 2, 3, 4, 5, 6\} \cap \{2, 3, 4\}$$
$$= \{2, 3, 4\}$$

e. First, we compute
$$A \cap B = \{3\}$$

Next, since $(A \cap B) \cup C$ is the set of all elements in $(A \cap B)$ and/or C, we conclude that
$$(A \cap B) \cup C = \{3\} \cup \{2, 3, 4\}$$
$$= \{2, 3, 4\}$$

f. From part (a), we have $A^c = \{4, 5, 6, 7\}$. Next, we compute
$$B \cup C = \{3, 4, 5, 6\} \cup \{2, 3, 4\}$$
$$= \{2, 3, 4, 5, 6\}$$

from which we deduce that
$$(B \cup C)^c = \{1, 7\} \qquad \text{The set of elements in } U \text{ but not in } B \cup C$$

Finally, using these results, we obtain
$$A^c \cap (B \cup C)^c = \{4, 5, 6, 7\} \cap \{1, 7\} = \{7\}$$

2. a. $D \cap F$ denotes the set of all elements in both D and F. Since an element in D is a Democrat and an element in F is a female representative, we see that $D \cap F$ is the set of all female Democrats in the House of Representatives.

b. Since F^c is the set of male representatives and R is the set of Republicans, we see that $F^c \cap R$ is the set of male Republicans in the House of Representatives.

c. L^c is the set of representatives who are not lawyers by training. Therefore, $D \cap F \cap L^c$ is the set of female Democratic representatives who are not lawyers by training.

6.2 The Number of Elements in a Finite Set

Counting the Elements in a Set

The solution to some problems in mathematics calls for finding the number of elements in a set. Such problems are called **counting problems** and constitute a field of study known as **combinatorics.** Our study of combinatorics is restricted to the results that will be required for our work in probability later on.

The number of elements in a finite set is determined by simply counting the elements in the set. If A is a set, then $n(A)$ denotes the number of elements in A. For example, if

$$A = \{1, 2, 3, \ldots, 20\} \qquad B = \{a, b\} \qquad C = \{8\}$$

then $n(A) = 20$, $n(B) = 2$, and $n(C) = 1$.

The empty set has no elements in it, so $n(\varnothing) = 0$. Another result that is easily seen to be true is the following: If A and B are disjoint sets, then

$$n(A \cup B) = n(A) + n(B) \tag{3}$$

EXAMPLE 1   If $A = \{a, c, d\}$ and $B = \{b, e, f, g\}$, then $n(A) = 3$ and $n(B) = 4$, so $n(A) + n(B) = 7$. However, $A \cup B = \{a, b, c, d, e, f, g\}$ and $n(A \cup B) = 7$. Thus, Equation (3) holds true in this case. Note that $A \cap B = \varnothing$.

In the general case, A and B need not be disjoint, which leads us to the formula

$$\boxed{n(A \cup B) = n(A) + n(B) - n(A \cap B)} \tag{4}$$

To see this, we observe that the set $A \cup B$ may be viewed as the union of three mutually disjoint sets with x, y, and z elements, respectively (Figure 8). This figure shows that

$$n(A \cup B) = x + y + z$$

Also,

$$n(A) = x + y \quad \text{and} \quad n(B) = y + z$$

so

$$n(A) + n(B) = (x + y) + (y + z)$$
$$= (x + y + z) + y$$
$$= n(A \cup B) + n(A \cap B) \qquad {\scriptstyle n(A \cap B) = y}$$

Solving for $n(A \cup B)$, we obtain

$$n(A \cup B) = n(A) + n(B) - n(A \cap B)$$

which is the desired result.

EXAMPLE 2   Let $A = \{a, b, c, d, e\}$ and $B = \{b, d, f, h\}$. Verify Equation (4) directly.

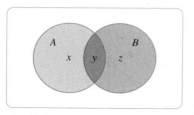

FIGURE 8
$n(A \cup B) = x + y + z$

Solution

$$A \cup B = \{a, b, c, d, e, f, h\} \quad \text{so} \quad n(A \cup B) = 7$$
$$A \cap B = \{b, d\} \quad \text{so} \quad n(A \cap B) = 2$$

Furthermore,

$$n(A) = 5 \quad \text{and} \quad n(B) = 4$$

so

$$n(A) + n(B) - n(A \cap B) = 5 + 4 - 2 = 7 = n(A \cup B) \qquad \blacksquare$$

APPLIED EXAMPLE 3   Consumer Surveys   In a survey of 100 coffee drinkers, it was found that 70 take sugar, 60 take cream, and 50 take both sugar and cream with their coffee. How many coffee drinkers take sugar or cream with their coffee?

Solution   Let U denote the set of 100 coffee drinkers surveyed and let

$$A = \{x \in U | x \text{ takes sugar}\}$$
$$B = \{x \in U | x \text{ takes cream}\}$$

Then, $n(A) = 70$, $n(B) = 60$, and $n(A \cap B) = 50$. The set of coffee drinkers who take sugar or cream with their coffee is given by $A \cup B$. Using (4), we find

$$n(A \cup B) = n(A) + n(B) - n(A \cap B)$$
$$= 70 + 60 - 50 = 80$$

Thus, 80 out of the 100 coffee drinkers surveyed take cream or sugar with their coffee. $\qquad \blacksquare$

EXPLORE & DISCUSS

Prove Formula (5), using an argument similar to that used to prove Formula (4). Another proof is outlined in Exercise 40 on page 352.

An equation similar to (4) may be derived for the case that involves any finite number of finite sets. For example, a relationship involving the number of elements in the sets A, B, and C is given by

$$n(A \cup B \cup C) = n(A) + n(B) + n(C) - n(A \cap B)$$
$$- n(A \cap C) - n(B \cap C) + n(A \cap B \cap C) \qquad (5)$$

As useful as equations such as (5) are, in practice it is often easier to attack a problem directly with the aid of Venn diagrams, as shown by the following example.

APPLIED EXAMPLE 4   Marketing Surveys   A leading cosmetics manufacturer advertises its products in three magazines: *Cosmopolitan*, *McCall's*, and the *Ladies Home Journal*. A survey of 500 customers by the manufacturer reveals the following information:

180 learned of its products from *Cosmopolitan*.

200 learned of its products from *McCall's*.

192 learned of its products from the *Ladies Home Journal*.

 84 learned of its products from *Cosmopolitan* and *McCall's*.

 52 learned of its products from *Cosmopolitan* and the *Ladies Home Journal*.

 64 learned of its products from *McCall's* and the *Ladies Home Journal*.

 38 learned of its products from all three magazines.

How many of the customers saw the manufacturer's advertisement in
a. At least one magazine?
b. Exactly one magazine?

Solution   Let U denote the set of all customers surveyed and let

$C = \{x \in U | x$ learned of the products from *Cosmopolitan*$\}$
$M = \{x \in U | x$ learned of the products from *McCall's*$\}$
$L = \{x \in U | x$ learned of the products from the *Ladies Home Journal*$\}$

The result that 38 customers learned of the products from all three magazines translates into $n(C \cap M \cap L) = 38$ (Figure 9a). Next, the result that 64 learned of the products from *McCall's* and the *Ladies Home Journal* translates into $n(M \cap L) = 64$. This leaves

$$64 - 38 = 26$$

who learned of the products from only *McCall's* and the *Ladies Home Journal* (Figure 9b). Similarly, $n(C \cap L) = 52$, so

$$52 - 38 = 14$$

learned of the products from only *Cosmopolitan* and the *Ladies Home Journal*, and $n(C \cap M) = 84$, so

$$84 - 38 = 46$$

learned of the products from only *Cosmopolitan* and *McCall's*. These numbers appear in the appropriate regions in Figure 9b.

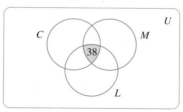

(a) All three magazines

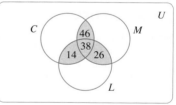

(b) Two or more magazines

FIGURE 9

Continuing, we have $n(L) = 192$, so the number who learned of the products from the *Ladies Home Journal* only is given by

$$192 - 14 - 38 - 26 = 114$$

(Figure 10). Similarly, $n(M) = 200$, so

$$200 - 46 - 38 - 26 = 90$$

learned of the products from only *McCall's*, and $n(C) = 180$, so

$$180 - 14 - 38 - 46 = 82$$

learned of the products from only *Cosmopolitan*. Finally,

$$500 - (90 + 26 + 114 + 14 + 82 + 46 + 38) = 90$$

learned of the products from other sources.

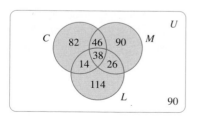

FIGURE 10
At least one magazine

We are now in a position to answer questions (a) and (b).

a. Referring to Figure 10, we see that the number of customers who learned of the products from at least one magazine is given by

$$n(C \cup M \cup L) = 90 + 26 + 114 + 14 + 82 + 46 + 38 = 410$$

b. The number of customers who learned of the products from exactly one magazine (Figure 11) is given by

$$n(L \cap C^c \cap M^c) + n(M \cap C^c \cap L^c) + n(C \cap L^c \cap M^c)$$
$$= 114 + 90 + 82 = 286$$

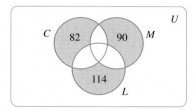

FIGURE 11
Exactly one magazine

6.2   Self-Check Exercises

1. Let A and B be subsets of a universal set U and suppose $n(U) = 100$, $n(A) = 60$, $n(B) = 40$, and $n(A \cap B) = 20$. Compute:
 a. $n(A \cup B)$ **b.** $n(A \cap B^c)$ **c.** $n(A^c \cap B)$

2. In a survey of 1000 readers of *Video Magazine*, it was found that 900 own at least one VCR in the VHS format, 240 own at

least one VCR in the S-VHS format, and 160 own VCRs in both formats. How many of the readers surveyed own VCRs in the VHS format only? How many of the readers surveyed do not own a VCR in either format?

Solutions to Self-Check Exercises 6.2 can be found on page 352.

6.2   Concept Questions

1. a. If A and B are sets with $A \cap B = \varnothing$, what can you say about $n(A) + n(B)$? Explain.
 b. If A and B are sets satisfying $n(A \cup B) \neq n(A) + n(B)$, what can you say about $A \cap B$? Explain.

2. Let A and B be subsets of U, the universal set, and suppose $A \cap B = \varnothing$. Is it true that $n(A) - n(B) = n(B^c) - n(A^c)$? Explain.

6.2   Exercises

In Exercises 1 and 2, verify the equation

$$n(A \cup B) = n(A) + n(B)$$

for the disjoint sets.

1. $A = \{a, e, i, o, u\}$ and $B = \{g, h, k, l, m\}$

2. $A = \{x \mid x$ is a whole number between 0 and 4$\}$
 $B = \{x \mid x$ is a negative integer greater than $-4\}$

3. Let $A = \{2, 4, 6, 8\}$ and $B = \{6, 7, 8, 9, 10\}$. Compute:
 a. $n(A)$ **b.** $n(B)$
 c. $n(A \cup B)$ **d.** $n(A \cap B)$

4. Let $U = \{1, 2, 3, 4, 5, 6, 7, a, b, c, d, e\}$. If $A = \{1, 2, a, e\}$ and $B = \{1, 2, 3, 4, a, b, c\}$, find
   a. $n(A^c)$ **b.** $n(A \cap B^c)$
   c. $n(A \cup B^c)$ **d.** $n(A^c \cap B^c)$

5. Verify directly that $n(A \cup B) = n(A) + n(B) - n(A \cap B)$ for the sets in Exercise 3.

6. Let $A = \{a, e, i, o, u\}$ and $B = \{b, d, e, o, u\}$. Verify by direct computation that $n(A \cup B) = n(A) + n(B) - n(A \cap B)$.

7. If $n(A) = 15$, $n(A \cap B) = 5$, and $n(A \cup B) = 30$, what is $n(B)$?

8. If $n(A) = 10$, $n(A \cup B) = 15$, and $n(B) = 8$, what is $n(A \cap B)$?

In Exercises 9 and 10, let A and B be subsets of a universal set U and suppose $n(U) = 200$, $n(A) = 100$, $n(B) = 80$, and $n(A \cap B) = 40$. Compute:

9. **a.** $n(A \cup B)$ **b.** $n(A^c)$ **c.** $n(A \cap B^c)$

10. **a.** $n(A^c \cap B)$ **b.** $n(B^c)$ **c.** $n(A^c \cap B^c)$

11. Find $n(A \cup B)$ given that $n(A) = 6$, $n(B) = 10$, and $n(A \cap B) = 3$.

12. If $n(B) = 6$, $n(A \cup B) = 14$, and $n(A \cap B) = 3$, find $n(A)$.

13. If $n(A) = 4$, $n(B) = 5$, and $n(A \cup B) = 9$, find $n(A \cap B)$.

14. If $n(A) = 16$, $n(B) = 16$, $n(C) = 14$, $n(A \cap B) = 6$, $n(A \cap C) = 5$, $n(B \cap C) = 6$, and $n(A \cup B \cup C) = 31$, find $n(A \cap B \cap C)$.

15. If $n(A) = 12$, $n(B) = 12$, $n(A \cap B) = 5$, $n(A \cap C) = 5$, $n(B \cap C) = 4$, $n(A \cap B \cap C) = 2$, and $n(A \cup B \cup C) = 25$, find $n(C)$.

16. A survey of 1000 subscribers to the *Los Angeles Times* revealed that 900 people subscribe to the daily morning edition and 500 subscribe to both the daily and the Sunday editions. How many subscribe to the Sunday edition? How many subscribe to the Sunday edition only?

17. On a certain day, the Wilton County Jail had 190 prisoners. Of these, 130 were accused of felonies, and 121 were accused of misdemeanors. How many prisoners were accused of both a felony and a misdemeanor?

18. Of 100 clock radios sold recently in a department store, 70 had FM circuitry, and 90 had AM circuitry. How many radios had both FM and AM circuitry? How many could receive FM transmission only? How many could receive AM transmission only?

19. CONSUMER SURVEYS In a survey of 120 consumers conducted in a shopping mall, 80 consumers indicated that they buy brand A of a certain product, 68 buy brand B, and 42 buy both brands. How many consumers participating in the survey buy
   a. At least one of these brands?
   b. Exactly one of these brands?
   c. Only brand A?
   d. None of these brands?

20. CONSUMER SURVEYS In a survey of 200 members of a local sports club, 100 members indicated that they plan to attend the next Summer Olympic Games, 60 indicated that they plan to attend the next Winter Olympic Games, and 40 indicated that they plan to attend both games. How many members of the club plan to attend
   a. At least one of the two games?
   b. Exactly one of the games?
   c. The Summer Olympic Games only?
   d. None of the games?

21. INVESTING In a poll conducted among 200 active investors, it was found that 120 use discount brokers, 126 use full-service brokers, and 64 use both discount and full-service brokers. How many investors
   a. Use at least one kind of broker?
   b. Use exactly one kind of broker?
   c. Use only discount brokers?
   d. Don't use a broker?

22. COMMUTER TRENDS Of 50 employees of a store located in downtown Boston, 18 people take the subway to work, 12 take the bus, and 7 take both the subway and the bus. How many employees
   a. Take the subway or the bus to work?
   b. Take only the bus to work?
   c. Take either the bus or the subway to work?
   d. Get to work by some other means?

23. CONSUMER SURVEYS In a survey of 200 households regarding the ownership of desktop and laptop computers, the following information was obtained:

   120 households own only desktop computers.

   10 households own only laptop computers.

   40 households own neither desktop nor laptop computers.

   How many households own both desktop and laptop computers?

24. CONSUMER SURVEYS In a survey of 400 households regarding the ownership of VCRs and DVD players, the following data was obtained:

   360 households own one or more VCRs.

   170 households own one or more VCRs and one or more DVD players.

   19 households do not own a VCR or a DVD player.

   How many households only own one or more DVD players?

In Exercises 25–28, let *A*, *B*, and *C* be subsets of a universal set *U* and suppose $n(U) = 100$, $n(A) = 28$, $n(B) = 30$, $n(C) = 34$, $n(A \cap B) = 8$, $n(A \cap C) = 10$, $n(B \cap C) = 15$, and $n(A \cap B \cap C) = 5$. Compute:

25. a. $n(A \cup B \cup C)$ **b.** $n(A^c \cap B \cap C)$

26. a. $n[A \cap (B \cup C)]$ **b.** $n[A \cap (B \cup C)^c]$

27. a. $n(A^c \cap B^c \cap C^c)$ **b.** $n[A^c \cap (B \cup C)]$

28. a. $n[A \cup (B \cap C)]$ **b.** $n(A^c \cap B^c \cap C^c)^c$

29. ECONOMIC SURVEYS   A survey of the opinions of 10 leading economists in a certain country showed that because oil prices were expected to drop in that country over the next 12 mo,

7 had lowered their estimate of the consumer inflation rate.

8 had raised their estimate of the gross national product (GNP) growth rate.

2 had lowered their estimate of the consumer inflation rate but had not raised their estimate of the GNP growth rate.

How many economists had both lowered their estimate of the consumer inflation rate and raised their estimate of the GNP growth rate for that period?

30. STUDENT DROPOUT RATE   Data released by the Department of Education regarding the rate (percentage) of ninth-grade students that don't graduate showed that out of 50 states,

12 states had an increase in the dropout rate during the past 2 yr.

15 states had a dropout rate of at least 30% during the past 2 yr.

21 states had an increase in the dropout rate and/or a dropout rate of at least 30% during the past 2 yr.

a. How many states had both a dropout rate of at least 30% and an increase in the dropout rate over the 2-yr period?
b. How many states had a dropout rate that was less than 30% but that had increased over the 2-yr period?

31. STUDENT READING HABITS   A survey of 100 college students who frequent the reading lounge of a university revealed the following results:

40 read *Time*.

30 read *Newsweek*.

25 read *U.S. News & World Report*.

15 read *Time* and *Newsweek*.

12 read *Time* and *U.S. News & World Report*.

10 read *Newsweek* and *U.S. News & World Report*.

4 read all three magazines.

How many of the students surveyed read
a. At least one magazine?
b. Exactly one magazine?
c. Exactly two magazines?
d. None of these magazines?

32. SAT SCORES   Results of a Department of Education survey of SAT test scores in 22 states showed that

10 states had an average composite test score of at least 1000 during the past 3 yr.

15 states had an increase of at least 10 points in the average composite score during the past 3 yr.

8 states had both an average composite SAT score of at least 1000 and an increase in the average composite score of at least 10 points during the past 3 yr.

a. How many of the 22 states had composite scores less than 1000 and showed an increase of at least 10 points over the 3-yr period?
b. How many of the 22 states had composite scores of at least 1000 and did not show an increase of at least 10 points over the 3-yr period?

33. CONSUMER SURVEYS   The 120 consumers of Exercise 19 were also asked about their buying preferences concerning another product that is sold in the market under three labels. The results were

12 buy only those sold under label A.

25 buy only those sold under label B.

26 buy only those sold under label C.

15 buy only those sold under labels A and B.

10 buy only those sold under labels A and C.

12 buy only those sold under labels B and C.

8 buy the product sold under all three labels.

How many of the consumers surveyed buy the product sold under
a. At least one of the three labels?
b. Labels A and B but not C?
c. Label A?
d. None of these labels?

34. STUDENT SURVEYS   To help plan the number of meals to be prepared in a college cafeteria, a survey was conducted, and the following data were obtained:

130 students ate breakfast.

180 students ate lunch.

275 students ate dinner.

68 students ate breakfast and lunch.

112 students ate breakfast and dinner.

90 students ate lunch and dinner.

58 students ate all three meals.

How many of the students ate
a. At least one meal in the cafeteria?
b. Exactly one meal in the cafeteria?
c. Only dinner in the cafeteria?
d. Exactly two meals in the cafeteria?

35. INVESTMENTS   In a survey of 200 employees of a company regarding their 401(K) investments, the following data were obtained:

141 had investments in stock funds.

91 had investments in bond funds.

60 had investments in money market funds.

47 had investments in stock funds and bond funds.

36 had investments in stock funds and money market funds.

36 had investments in bond funds and money market funds.

5 had investments in some other vehicle.

a. How many of the employees surveyed had investments in all three types of funds?
b. How many of the employees had investments in stock funds only?

36. NEWSPAPER SUBSCRIPTIONS   In a survey of 300 individual investors regarding subscriptions to the *New York Times* (*NYT*), *Wall Street Journal* (*WSJ*), and *USA Today* (*UST*), the following data were obtained:

122 subscribe to the *NYT*.

150 subscribe to the *WSJ*.

62 subscribe to the *UST*.

38 subscribe to the *NYT* and *WSJ*.

28 subscribe to the *WSJ* and *UST*.

20 subscribe to the *NYT* and *UST*.

36 do not subscribe to any of these newspapers.

a. How many of the individual investors surveyed subscribe to all three newspapers?
b. How many subscribe to only one of these newspapers?

In Exercises 37–39, determine whether the statement is true or false. If it is true, explain why it is true. If it is false, give an example to show why it is false.

37. If $A \cap B \neq \varnothing$, then $n(A \cup B) \neq n(A) + n(B)$.

38. If $A \subseteq B$, then $n(B) = n(A) + n(A^c \cap B)$.

39. If $n(A \cup B) = n(A) + n(B)$, then $A \cap B = \varnothing$.

40. Derive Equation (5).
    Hint: Equation (4) may be written as $n(D \cup E) = n(D) + n(E) - n(D \cap E)$. Now, put $D = A \cup B$ and $E = C$. Use (4) again if necessary.

6.2   Solutions to Self-Check Exercises

1. Refer to the following Venn diagram:

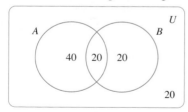

Using this result we see that
a. $n(A \cup B) = 40 + 20 + 20 = 80$
b. $n(A \cap B^c) = 40$
c. $n(A^c \cap B) = 20$

2. Let U denote the set of all readers surveyed and let

$A = \{x \in U | x$ owns at least one VCR in the VHS format$\}$
$B = \{x \in U | x$ owns at least one VCR in the S-VHS format$\}$

Then, the result that 160 of the readers own VCRs in both formats gives $n(A \cap B) = 160$. Also, $n(A) = 900$ and

$n(B) = 240$. Using this information, we obtain the following Venn diagram:

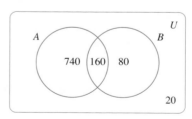

From the Venn diagram we see that the number of readers who own VCRs in the VHS format only is given by

$$n(A \cap B^c) = 740$$

The number of readers who do not own a VCR in either format is given by

$$n(A^c \cap B^c) = 20$$

6.3 The Multiplication Principle

▬ The Fundamental Principle of Counting

The solution of certain problems requires more sophisticated counting techniques than those developed in the previous section. We look at some such techniques in this and the following section. We begin by stating a fundamental principle of counting called the **multiplication principle.**

> **The Multiplication Principle**
>
> Suppose there are m ways of performing a task T_1 and n ways of performing a task T_2. Then, there are mn ways of performing the task T_1 followed by the task T_2.

EXAMPLE 1 Three trunk roads connect town A and town B, and two trunk roads connect town B and town C.
a. Use the multiplication principle to find the number of ways a journey from town A to town C via town B may be completed.
b. Verify part (a) directly by exhibiting all possible routes.

Solution
a. Since there are three ways of performing the first task (going from town A to town B) followed by two ways of performing the second task (going from town B to town C), the multiplication principle says that there are $3 \cdot 2$, or 6, ways to complete a journey from town A to town C via town B.
b. Label the trunk roads connecting town A and town B with the Roman numerals I, II, and III and the trunk roads connecting town B and town C with the lowercase letters a and b. A schematic of this is shown in Figure 12. Then the routes from town A to town C via town B may be exhibited

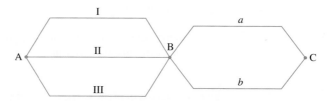

FIGURE 12
Roads from towns A to C

with the aid of a **tree diagram** (Figure 13). If we follow all of the branches from the initial point A to the right-hand edge of the tree, we obtain the six routes represented by the six ordered pairs

$$(\text{I}, a), \quad (\text{I}, b), \quad (\text{II}, a), \quad (\text{II}, b), \quad (\text{III}, a), \quad (\text{III}, b)$$

where (I, a) means that the journey from town A to town B is made on trunk road I with the rest of the journey from town B to town C to be completed on trunk road a, and so forth.

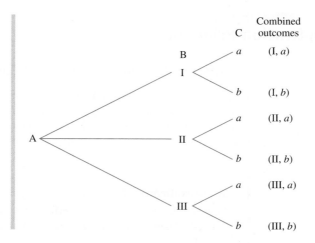

FIGURE 13
Tree diagram displaying the possible routes from town A to town C

EXPLORE & DISCUSS

One way of gauging the performance of an airline is to track the arrival times of its flights. Suppose we denote by E, O, and L, a flight that arrives early, on time, or late, respectively.

1. Use a tree diagram to exhibit the possible outcomes when you track two successive flights of the airline. How many outcomes are there?

2. How many outcomes are there if you track three successive flights? Justify your answer.

EXAMPLE 2   Diners at Angelo's Spaghetti Bar may select their entree from 6 varieties of pasta and 28 choices of sauce. How many such combinations are there that consist of 1 variety of pasta and 1 kind of sauce?

Solution   There are 6 ways of choosing a pasta followed by 28 ways of choosing a sauce, so by the multiplication principle, there are $6 \cdot 28$, or 168, combinations of this pasta dish.

The multiplication principle may be easily extended, which leads to the **generalized multiplication principle.**

Generalized Multiplication Principle

Suppose a task T_1 can be performed in N_1 ways, a task T_2 can be performed in N_2 ways, . . . , and, finally, a task T_n can be performed in N_n ways. Then, the number of ways of performing the tasks $T_1, T_2, \ldots, T_n$ in succession is given by the product

$$N_1 N_2 \cdots N_n$$

We now illustrate the application of the generalized multiplication principle to several diverse situations.

EXAMPLE 3 A coin is tossed 3 times, and the sequence of heads and tails is recorded.
a. Use the generalized multiplication principle to determine the number of outcomes of this activity.
b. Exhibit all the sequences by means of a tree diagram.

Solution
a. The coin may land in two ways. Therefore, in three tosses the number of outcomes (sequences) is given by $2 \cdot 2 \cdot 2$, or 8.
b. Let H and T denote the outcomes "a head" and "a tail," respectively. Then the required sequences may be obtained as shown in Figure 14, giving the sequence as HHH, HHT, HTH, HTT, THH, THT, TTH, and TTT.

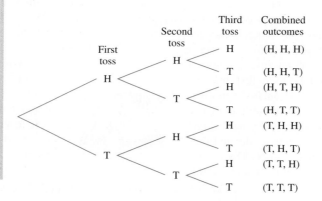

FIGURE 14
Tree diagram displaying possible outcomes of three consecutive coin tosses

APPLIED EXAMPLE 4 Combination Locks A combination lock is unlocked by dialing a sequence of numbers, first to the left, then to the right, and to the left again. If there are ten digits on the dial, determine the number of possible combinations.

Solution There are ten choices for the first number, followed by ten for the second and ten for the third, so by the generalized multiplication principle, there are $10 \cdot 10 \cdot 10$, or 1000, possible combinations.

APPLIED EXAMPLE 5 Investment Options An investor has decided to purchase shares in the stock of three companies: one engaged in aerospace activities, one involved in energy development, and one involved in electronics. After some research, the account executive of a brokerage firm has recommended that the investor consider stock from five aerospace companies, three energy development companies, and four electronics companies. In how many ways may the investor select the group of three companies from the executive's list?

PORTFOLIO   Stephanie Molina

TITLE   Computer Crimes Detective
INSTITUTION   Maricopa County Sheriff's Office

Working as a detective in the computer crimes division of the Maricopa County Sheriff's Office, I find applied mathematics techniques play a significant role in my job when I search for evidence contained on computer hard drives and other forms of media. To obtain evidence, I am required to have a working knowledge of certain applied mathematics skills so that I can effectively communicate with the computer forensic analyst who will be decoding the evidence. In order to conduct an effective investigation, I am also required to understand this data in a wide variety of formats. With this information, I can work with the analyst to reconstruct data that may play a significant roll in determining events that occurred pertaining to a crime.

During the course of an investigation, I have to look at the data not only in text, but also in code. Using this view, the analyst can decipher different file types and possible evidence in unallocated space throughout the hard drive. This unallocated space can contain deleted files that may contain potential evidence. The analyst also has to decode files by hand and, at this

point, recognizing patterns among the files becomes very important. From here, we can derive an algorithm to define those patterns. By producing an algorithm, it makes it possible to write a program that will decode the files for you.

For example, there was a case that involved a suspect that was receiving files through a mail server. This suspect was then opening the files and deleting the email. Members of my computer forensic laboratory and I viewed these files in their original code to try and discover any patterns or inconsistencies within the code to find a solution to the problem. We did find a clue buried within the code. We then derived an algorithm defining its pattern. By inputting the algorithm, we could then extract the files from the coded data.

While I do not have a solid background in computer science or, even, mathematics, my knowledge of applied mathematics helps me to understand the procedures involved in obtaining evidence. Best of all, I am able to clearly convey my needs to the forensic analysts in my department.

Solution   The investor has five choices for selecting an aerospace company, three choices for selecting an energy development company, and four choices for selecting an electronics company. Therefore, by the generalized multiplication principle, there are $5 \cdot 3 \cdot 4$, or 60, ways in which she can select a group of three companies, one from each industry group.   ∎

APPLIED EXAMPLE 6   Travel Options   Tom is planning to leave for New York City from Washington, D.C., on Monday morning and has decided that he will either fly or take the train. There are five flights and two trains departing for New York City from Washington that morning. When he returns on Sunday afternoon, Tom plans to either fly or hitch a ride with a friend. There are two flights departing from New York City to Washington that afternoon. In how many ways can Tom complete this round trip?

Solution   There are seven ways Tom can go from Washington, D.C., to New York City (five by plane and two by train). On the return trip, Tom can travel in three ways (two by plane and one by car). Therefore, by the multiplication principle, Tom can complete the round trip in $7 \cdot 3$, or 21, ways.   ∎

6.3   Self-Check Exercises

1. Encore Travel offers a "Theater Week in London" package originating from New York City. There is a choice of eight flights departing from New York City each week, a choice of five hotel accommodations, and a choice of one complimentary ticket to one of eight shows. How many such travel packages can one choose from?

2. The Café Napolean offers a dinner special on Wednesdays consisting of a choice of two entrées (Beef Bourguignon and

Chicken Basquaise); one dinner salad; one French roll; a choice of three vegetables; a choice of a carafe of Burgundy, Rosé, or Chablis wine; a choice of coffee or tea; and a choice of six French pastries for dessert. How many combinations of dinner specials are there?

Solutions to Self-Check Exercises 6.3 can be found on page 359.

6.3   Concept Questions

1. Explain the multiplication principle and illustrate it with a diagram.

2. Given the following tree diagram for an activity, what are the possible outcomes?

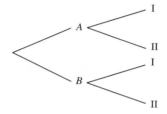

6.3   Exercises

1. **RENTAL RATES** Lynbrook West, an apartment complex financed by the State Housing Finance Agency, consists of one-, two-, three-, and four-bedroom units. The rental rate for each type of unit—low, moderate, or market—is determined by the income of the tenant. How many different rates are there?

2. **COMMUTER PASSES** Five different types of monthly commuter passes are offered by a city's local transit authority for three different groups of passengers: youths, adults, and senior citizens. How many different kinds of passes must be printed each month?

3. **BLACKJACK** In the game of blackjack, a 2-card hand consisting of an ace and either a face card or a 10 is called a "blackjack." If a standard 52-card deck is used, determine how many blackjack hands can be dealt.

4. **COIN TOSSES** A coin is tossed 4 times and the sequence of heads and tails is recorded.

a. Use the generalized multiplication principle to determine the number of outcomes of this activity.

b. Exhibit all the sequences by means of a tree diagram.

5. **WARDROBE SELECTION** A female executive selecting her wardrobe purchased two blazers, four blouses, and three skirts in coordinating colors. How many ensembles consisting of a blazer, a blouse, and a skirt can she create from this collection?

6. **COMMUTER OPTIONS** Four commuter trains and three express buses depart from city A to city B in the morning, and three commuter trains and three express buses operate on the return trip in the evening. In how many ways can a commuter from city A to city B complete a daily round trip via bus and/or train?

7. **PSYCHOLOGY EXPERIMENTS** A psychologist has constructed the following maze for use in an experiment. The maze is constructed so that a rat must pass through a series of one-way doors. How many different paths are there from start to finish?

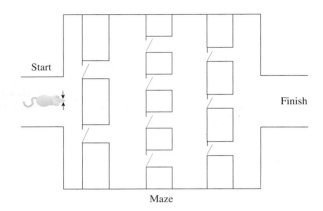

Start

Finish

Maze

8. UNION BARGAINING ISSUES   In a survey conducted by a union, members were asked to rate the importance of the following issues: (1) job security, (2) increased fringe benefits, and (3) improved working conditions. Five different responses were allowed for each issue. Among completed surveys, how many different responses to this survey were possible?

9. HEALTH-CARE PLAN OPTIONS   A new state employee is offered a choice of ten basic health plans, three dental plans, and two vision care plans. How many different health-care plans are there to choose from if one plan is selected from each category?

10. CODE WORDS   How many three-letter code words can be constructed from the first ten letters of the Greek alphabet if no repetitions are allowed?

11. SOCIAL SECURITY NUMBERS   A Social Security number has nine digits. How many Social Security numbers are possible?

12. SERIAL NUMBERS   Computers manufactured by a certain company have a serial number consisting of a letter of the alphabet followed by a four-digit number. If all the serial numbers of this type have been used, how many sets have already been manufactured?

13. COMPUTER DATING   A computer dating service uses the results of its compatibility survey for arranging dates. The survey consists of 50 questions, each having five possible answers. How many different responses are possible if every question is answered?

14. AUTOMOBILE SELECTION   An automobile manufacturer has three different subcompact cars in the line. Customers selecting one of these cars have a choice of three engine sizes, four body styles, and three color schemes. How many different selections can a customer make?

15. MENU SELECTIONS   Two soups, five entrées, and three desserts are listed on the "Special" menu at the Neptune Restaurant. How many different selections consisting of one soup, one entrée, and one dessert can a customer choose from this menu?

16. TELEVISION-VIEWING POLLS   An opinion poll is to be conducted among cable TV viewers. Six multiple-choice questions, each with four possible answers, will be asked. In how many different ways can a viewer complete the poll if exactly one response is given to each question?

17. ATM CARDS   To gain access to his account, a customer using an automatic teller machine (ATM) must enter a four-digit code. If repetition of the same four digits is not allowed (for example, 5555), how many possible combinations are there?

18. POLITICAL POLLS   An opinion poll was conducted by the Morris Polling Group. Respondents were classified according to their sex (M or F), political affiliation (D, I, R), and the region of the country in which they reside (NW, W, C, S, E, NE).
  a. Use the generalized multiplication principle to determine the number of possible classifications.
  b. Construct a tree diagram to exhibit all possible classifications of females.

19. LICENSE PLATE NUMBERS   Over the years, the state of California has used different combinations of letters of the alphabet and digits on its automobile license plates.
  a. At one time, license plates were issued that consisted of three letters followed by three digits. How many different license plates can be issued under this arrangement?
  b. Later on, license plates were issued that consisted of three digits followed by three letters. How many different license plates can be issued under this arrangement?

20. LICENSE PLATE NUMBERS   In recent years, the state of California issued license plates using a combination of one letter of the alphabet followed by three digits, followed by another three letters of the alphabet. How many different license plates can be issued using this configuration?

21. EXAMS   An exam consists of ten true-or-false questions. Assuming that every question is answered, in how many different ways can a student complete the exam? In how many ways may the exam be completed if a penalty is imposed for each incorrect answer, so that a student may leave some questions unanswered?

22. WARRANTY NUMBERS   A warranty identification number for a certain product consists of a letter of the alphabet followed by a five-digit number. How many possible identification numbers are there if the first digit of the five-digit number must be nonzero?

23. LOTTERIES   In a state lottery, there are 15 finalists eligible for the Big Money Draw. In how many ways can the first, second, and third prizes be awarded if no ticket holder may win more than one prize?

24. TELEPHONE NUMBERS
  a. How many seven-digit telephone numbers are possible if the first digit must be nonzero?

b. How many international direct-dialing numbers are possible if each number consists of a three-digit area code (the first digit of which must be nonzero) and a number of the type described in part (a)?

25. Slot Machines A "lucky dollar" is one of the nine symbols printed on each reel of a slot machine with three reels. A player receives one of various payouts whenever one or more "lucky dollars" appear in the window of the machine. Find the number of winning combinations for which the machine gives a payoff.

Hint: (a) Compute the number of ways in which the nine symbols on the first, second, and third wheels can appear in the window slot and (b) compute the number of ways in which the eight symbols other than the "lucky dollar" can appear in the window slot. The difference $(a - b)$ is the number of ways in which the "lucky dollar" can appear in the window slot. Why?

26. Staffing Student Painters, which specializes in painting the exterior of residential buildings, has five people available to be organized into two-person and three-person teams.

a. In how many ways can the two-person team be formed?
b. In how many ways can the three-person team be formed?
c. In how many ways can the company organize the available people into two- or three-person teams?

In Exercises 27 and 28, determine whether the statement is true or false. If it is true, explain why it is true. If it is false, give an example to show why it is false.

27. The number of three-digit numbers that can be formed from the digits 1, 2, 3, and 4 if the number must be odd is 32.

28. If there are six toppings available, then the number of different pizzas that can be made is 2^5, or 32, different pizzas.

6.3 Solutions to Self-Check Exercises

1. A tourist has a choice of eight flights, five hotel accommodations, and eight tickets. By the generalized multiplication principle, there are $8 \cdot 5 \cdot 8$, or 320, travel packages.

2. There is a choice of two entrées, one dinner salad, one French roll, a choice of three vegetables, a choice of three wines, a choice of two nonalcoholic beverages, and a choice of six pastries. Therefore, by the generalized multiplication principle, there are $2 \cdot 1 \cdot 1 \cdot 3 \cdot 3 \cdot 2 \cdot 6$, or 216, combinations of dinner specials.

6.4 Permutations and Combinations

■ Permutations

In this section, we apply the generalized multiplication principle to the solution of two types of counting problems. Both types involve determining the number of ways the elements of a set may be arranged, and both play an important role in the solution of problems in probability.

We begin by considering the permutations of a set. Specifically, given a set of distinct objects, a **permutation** of the set is an arrangement of these objects in a *definite order*. To see why the order in which objects are arranged is important in certain practical situations, suppose the winning number for the first prize in a raffle is 9237. Then, the number 2973, although it contains the same digits as the winning number, cannot be a first-prize winner (Figure 15). Here, the four objects—the numbers 9, 2, 3, and 7—are arranged in a different order; one arrangement is associated with the winning number for the first prize, and the other is not.

FIGURE 15
The same digits appear on each ticket, but the order of the digits is different.

EXAMPLE 1 Let $A = \{a, b, c\}$.
a. Find the number of permutations of A.
b. List all the permutations of A with the aid of a tree diagram.

Solution
a. Each permutation of A consists of a sequence of the three letters a, b, c. Therefore, we may think of such a sequence as being constructed by filling in each of the three blanks

$$\underline{}\quad\underline{}\quad\underline{}$$

with one of the three letters. Now, there are three ways in which we may fill the first blank—we may choose a, b, or c. Having selected a letter for the first blank, there are two letters left for the second blank. Finally, there is but one way left to fill the third blank. Schematically, we have

$$\underline{3}\quad\underline{2}\quad\underline{1}$$

Invoking the generalized multiplication principle, we conclude that there are $3 \cdot 2 \cdot 1$, or 6, permutations of the set A.
b. The tree diagram associated with this problem appears in Figure 16, and the six permutations of A are abc, acb, bac, bca, cab, and cba.

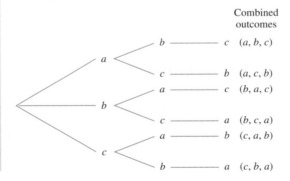

FIGURE 16
Permutations of three objects

Note Notice that, when the possible outcomes are listed in the tree diagram in Example 1, order is taken into account. Thus, (a, b, c) and (a, c, b) are two different arrangements.

EXAMPLE 2 Find the number of ways a baseball team consisting of nine people can arrange themselves in a line for a group picture.

Solution We want to determine the number of permutations of the nine members of the baseball team. Each permutation in this situation consists of an arrangement of the nine team members in a line. The nine positions can be represented by nine blanks. Thus,

Position $\underline{\quad}$ $\underline{\quad}$ $\underline{\quad}$ $\underline{\quad}$ $\underline{\quad}$ $\underline{\quad}$ $\underline{\quad}$ $\underline{\quad}$ $\underline{\quad}$
                  1   2   3   4   5   6   7   8   9

There are nine ways to choose from among the nine players to fill the first position. When that position is filled, eight players are left, which gives us eight ways to fill the second position. Proceeding in a similar manner, we find that there are seven ways to fill the third position, and so on. Schematically, we have

Number of ways to $\underline{9}$ $\underline{8}$ $\underline{7}$ $\underline{6}$ $\underline{5}$ $\underline{4}$ $\underline{3}$ $\underline{2}$ $\underline{1}$
fill each position

Invoking the generalized multiplication principle, we conclude that there are $9 \cdot 8 \cdot 7 \cdot 6 \cdot 5 \cdot 4 \cdot 3 \cdot 2 \cdot 1$, or 362,880, ways the baseball team can be arranged for the picture.   ■

 Whenever we are asked to determine the number of ways the objects of a set can be arranged in a line, order is important. For example, if we take a picture of two baseball players, A and B, then the two players can line up for the picture in two ways, AB or BA, and the two pictures will be different.

Pursuing the same line of argument used in solving the problems in the last two examples, we can derive an expression for the number of ways of permuting a set A of n distinct objects taken n at a time. In fact, each permutation may be viewed as being obtained by filling each of n blanks with one and only one element from the set. There are n ways of filling the first blank, followed by $(n - 1)$ ways of filling the second blank, and so on, so by the generalized multiplication principle there are

$$n(n - 1)(n - 2) \cdot \cdots \cdot 3 \cdot 2 \cdot 1$$

ways of permuting the elements of the set A.

Before stating this result formally, let's introduce a notation that will enable us to write in a compact form many of the expressions that follow. We use the symbol $n!$ (read *"n-factorial"*) to denote the product of the first n natural numbers.

---

n-Factorial

For any natural number n,

$$n! = n(n - 1)(n - 2) \cdot \cdots \cdot 3 \cdot 2 \cdot 1$$
$$0! = 1$$

For example,

$$1! = 1$$
$$2! = 2 \cdot 1 = 2$$
$$3! = 3 \cdot 2 \cdot 1 = 6$$
$$4! = 4 \cdot 3 \cdot 2 \cdot 1 = 24$$
$$5! = 5 \cdot 4 \cdot 3 \cdot 2 \cdot 1 = 120$$
$$\vdots$$
$$10! = 10 \cdot 9 \cdot 8 \cdot 7 \cdot 6 \cdot 5 \cdot 4 \cdot 3 \cdot 2 \cdot 1 = 3,628,800$$

Using this notation, we may express *the number of permutations of n distinct objects taken n at a time, P(n, n),* as

$$P(n, n) = n!$$

In many situations, we are interested in determining the number of ways of permuting n distinct objects taken r at a time, where $r \leq n$. To derive a formula for computing the number of ways of permuting a set consisting of n distinct objects taken r at a time, we observe that each such permutation may be viewed as being obtained by filling each of r blanks with precisely one element from the set. Now there are n ways of filling the first blank, followed by $(n - 1)$ ways of filling the second blank, and so on. Finally, there are $(n - r + 1)$ ways of filling the rth blank. We may represent this argument schematically:

Number of ways	n	$n - 1$	$n - 2$	$\cdots$	$n - r + 1$
Position	1st	2nd	3rd		rth

Using the generalized multiplication principle, we conclude that *the number of ways of permuting n distinct objects taken r at a time, P(n, r),* is given by

$$P(n, r) = \underbrace{n(n - 1)(n - 2) \cdots (n - r + 1)}_{r \text{ factors}}$$

Since

$$n(n - 1)(n - 2) \cdots (n - r + 1)$$

$$= [n(n - 1)(n - 2) \cdots (n - r + 1)] \cdot \underbrace{\frac{[(n - r)(n - r - 1) \cdot \cdots \cdot 3 \cdot 2 \cdot 1]}{[(n - r)(n - r - 1) \cdot \cdots \cdot 3 \cdot 2 \cdot 1]}}_{\text{Here we are multiplying by 1.}}$$

$$= \frac{[n(n - 1)(n - 2) \cdots (n - r + 1)][(n - r)(n - r - 1) \cdot \cdots \cdot 3 \cdot 2 \cdot 1]}{(n - r)(n - r - 1) \cdot \cdots \cdot 3 \cdot 2 \cdot 1}$$

$$= \frac{n!}{(n - r)!}$$

we have the following formula:

> ### Permutations of *n* Distinct Objects
> The number of *permutations* of n distinct objects taken r at a time is
>
> $$P(n, r) = \frac{n!}{(n - r)!} \tag{6}$$

Note　When $r = n$, Equation (6) reduces to

$$P(n, n) = \frac{n!}{0!} = \frac{n!}{1} = n! \qquad \text{Note that } 0! = 1.$$

In other words, the number of permutations of a set of n distinct objects, taken all together, is $n!$.

EXAMPLE 3   Compute (a) $P(4, 4)$ and (b) $P(4, 2)$ and interpret your results.

Solution

a. $P(4, 4) = \dfrac{4!}{(4 - 4)!} = \dfrac{4!}{0!} = \dfrac{4!}{1} = \dfrac{4 \cdot 3 \cdot 2 \cdot 1}{1} = 24$ Recall that $0! = 1$.

This gives the number of permutations of four objects taken four at a time.

b. $P(4, 2) = \dfrac{4!}{(4 - 2)!} = \dfrac{4!}{2!} = \dfrac{4 \cdot 3 \cdot 2 \cdot 1}{2 \cdot 1} = 12$

This is the number of permutations of four objects taken two at a time. ◼

EXAMPLE 4   Let $A = \{a, b, c, d\}$.

a. Use Equation (6) to compute the number of permutations of the set A taken two at a time.

b. Display the permutations of part (a) with the aid of a tree diagram.

Solution

a. Here, $n = 4$ and $r = 2$, so the required number of permutations is given by

$$P(4, 2) = \frac{4!}{(4 - 2)!} = \frac{4!}{2!} = \frac{4 \cdot 3 \cdot 2 \cdot 1}{2 \cdot 1} = 4 \cdot 3$$
$$= 12$$

b. The tree diagram associated with the problem is shown in Figure 17, and the permutations of A taken two at a time are

$$ab, \quad ac, \quad ad, \quad ba, \quad bc, \quad bd, \quad ca, \quad cb, \quad cd, \quad da, \quad db, \quad dc \quad ◼$$

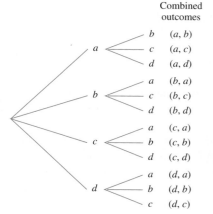

Combined outcomes

a — b — (a, b)
a — c — (a, c)
a — d — (a, d)
b — a — (b, a)
b — c — (b, c)
b — d — (b, d)
c — a — (c, a)
c — b — (c, b)
c — d — (c, d)
d — a — (d, a)
d — b — (d, b)
d — c — (d, c)

FIGURE 17
Permutations of four objects taken two at a time

EXAMPLE 5   Find the number of ways a chairman, a vice-chairman, a secretary, and a treasurer can be chosen from a committee of eight members.

Solution   The problem is equivalent to finding the number of permutations of eight distinct objects taken four at a time. Therefore, there are

$$P(8, 4) = \frac{8!}{(8 - 4)!} = \frac{8!}{4!} = 8 \cdot 7 \cdot 6 \cdot 5 = 1680$$

ways of choosing the four officials from the committee of eight members. ◼

The permutations considered thus far have been those involving sets of *distinct* objects. In many situations we are interested in finding the number of permutations of a set of objects in which not all of the objects are distinct.

Permutations of n Objects, Not All Distinct

Given a set of n objects in which n_1 objects are alike and of one kind, n_2 objects are alike and of another kind, . . . , and, finally, n_r objects are alike and of yet another kind so that

$$n_1 + n_2 + \cdots + n_r = n$$

then the number of permutations of these n objects taken n at a time is given by

$$\frac{n!}{n_1! n_2! \cdots n_r!} \tag{7}$$

To establish Equation (7), let's denote the number of such permutations by x. Now, if we *think* of the n_1 objects as being distinct, then they may be permuted in $n_1!$ ways. Similarly, if we *think* of the n_2 objects as being distinct, then they may be permuted in $n_2!$ ways, and so on. Therefore, if we *think* of the n objects as being distinct, then, by the generalized multiplication principle, there are $x \cdot n_1! \cdot n_2! \cdot \cdots \cdot n_r!$ permutations of these objects. But, the number of permutations of a set of n distinct objects taken n at a time is just equal to $n!$. Therefore, we have

$$x(n_1! \cdot n_2! \cdot \cdots \cdot n_r!) = n!$$

from which we deduce that

$$x = \frac{n!}{n_1! n_2! \cdots n_r!}$$

EXAMPLE 6   Find the number of permutations that can be formed from all the letters in the word *ATLANTA*.

Solution   There are seven objects (letters) involved, so $n = 7$. However, three of them are alike and of one kind (the three *A*s), two of them are alike and of another kind (the two *T*s), so that in this case, $n_1 = 3$, $n_2 = 2$, $n_3 = 1$, and $n_4 = 1$. Therefore, using Formula (7), there are

$$\frac{7!}{3!2!1!1!} = \frac{7 \cdot 6 \cdot 5 \cdot 4 \cdot 3 \cdot 2 \cdot 1}{3 \cdot 2 \cdot 1 \cdot 2 \cdot 1} = 420$$

required permutations. ■

APPLIED EXAMPLE 7   **Management Decisions**   Weaver and Kline, a stock brokerage firm, has received nine inquiries regarding new accounts. In how many ways can these inquiries be directed to three of the firm's account executives if each account executive is to handle three inquiries?

Solution   If we think of the nine inquiries as being slots arranged in a row with inquiry 1 on the left and inquiry 9 on the right, then the problem can be thought of as one of filling each slot with a business card from an account executive. Then nine business cards would be used, of which three are alike and of one kind, three are alike and of another kind, and three are alike and of yet another kind. Thus, using (7) with $n = 9$, $n_1 = n_2 = n_3 = 3$, there are

$$\frac{9!}{3!3!3!} = \frac{9 \cdot 8 \cdot 7 \cdot 6 \cdot 5 \cdot 4 \cdot 3 \cdot 2 \cdot 1}{3 \cdot 2 \cdot 1 \cdot 3 \cdot 2 \cdot 1 \cdot 3 \cdot 2 \cdot 1} = 1680$$

ways of assigning the inquiries. ■

Combinations

Up to now, we have dealt with permutations of a set—that is, with arrangements of the objects of the set in which the *order* of the elements is taken into consideration. In many situations one is interested in determining the number of ways of selecting r objects from a set of n objects without any regard to the order in which the objects are selected. Such a subset is called a **combination.**

For example, if one is interested in knowing the number of 5-card poker hands that can be dealt from a standard deck of 52 cards, then the order in which the poker hand is dealt is unimportant (Figure 18). In this situation, we are interested in determining the number of combinations of 5 cards (objects) selected from a deck (set) of 52 cards (objects). (We will solve this problem in Example 10.)

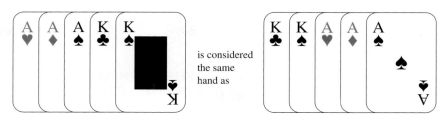

is considered the same hand as

FIGURE 18

To derive a formula for determining the number of combinations of n objects taken r at a time, written

$$C(n, r) \quad \text{or} \quad \binom{n}{r}$$

we observe that each of the $C(n, r)$ combinations of r objects can be permuted in $r!$ ways (Figure 19).

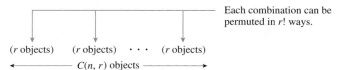

Each combination can be permuted in $r!$ ways.

(r objects) (r objects) $\cdots$ (r objects)

$\longleftarrow$ $C(n, r)$ objects $\longrightarrow$

FIGURE 19

Thus, by the multiplication principle, the product $r!C(n, r)$ gives the number of permutations of n objects taken r at a time; that is,

$$r!C(n, r) = P(n, r)$$

from which we find

$$C(n, r) = \frac{P(n, r)}{r!}$$

or, using Equation (6),

$$C(n, r) = \frac{n!}{r!(n - r)!}$$

Combinations of n Objects

The number of combinations of n distinct objects taken r at a time is given by

$$C(n, r) = \frac{n!}{r!(n - r)!} \qquad \text{(where } r \leq n\text{)} \tag{8}$$

EXAMPLE 8 Compute and interpret the results of (a) $C(4, 4)$ and (b) $C(4, 2)$.

Solution

a. $C(4, 4) = \dfrac{4!}{4!(4 - 4)!} = \dfrac{4!}{4!0!} = 1$ \qquad Recall that $0! = 1$.

This gives 1 as the number of combinations of four distinct objects taken four at a time.

b. $C(4, 2) = \dfrac{4!}{2!(4 - 2)!} = \dfrac{4!}{2!2!} = \dfrac{4 \cdot 3}{2} = 6$

This gives 6 as the number of combinations of four distinct objects taken two at a time.

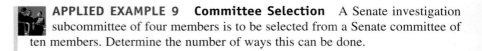

APPLIED EXAMPLE 9   Committee Selection   A Senate investigation subcommittee of four members is to be selected from a Senate committee of ten members. Determine the number of ways this can be done.

Solution   Since the order in which the members of the subcommittee are selected is unimportant, the number of ways of choosing the subcommittee is given by $C(10, 4)$, the number of combinations of ten objects taken four at a time. Therefore, there are

$$C(10, 4) = \dfrac{10!}{4!(10 - 4)!} = \dfrac{10!}{4!6!} = \dfrac{10 \cdot 9 \cdot 8 \cdot 7}{4 \cdot 3 \cdot 2 \cdot 1} = 210$$

ways of choosing such a subcommittee.

Note   Remember, a combination is a selection of objects *without* regard to order. Thus, in Example 9, we used a combination formula rather than a permutation formula to solve the problem because the order of selection was not important; that is, it did not matter whether a member of the subcommittee was selected first, second, third, or fourth.

APPLIED EXAMPLE 10   Poker   How many poker hands of 5 cards can be dealt from a standard deck of 52 cards?

Solution   The order in which the 5 cards are dealt is not important. The number of ways of dealing a poker hand of 5 cards from a standard deck of 52 cards is given by $C(52, 5)$, the number of combinations of 52 objects taken five at a time. Therefore, there are

$$\begin{aligned} C(52, 5) &= \dfrac{52!}{5!(52 - 5)!} = \dfrac{52!}{5!47!} \\ &= \dfrac{52 \cdot 51 \cdot 50 \cdot 49 \cdot 48}{5 \cdot 4 \cdot 3 \cdot 2 \cdot 1} \\ &= 2{,}598{,}960 \end{aligned}$$

ways of dealing such a poker hand.

The next several examples show that solving a counting problem often involves the repeated application of Equation (6) and/or (8), possibly in conjunction with the multiplication principle.

APPLIED EXAMPLE 11 Selecting Members of a Group The members of a string quartet composed of two violinists, a violist, and a cellist are to be selected from a group of six violinists, three violists, and two cellists, respectively.

a. In how many ways can the string quartet be formed?

b. In how many ways can the string quartet be formed if one of the violinists is to be designated as the first violinist and the other is to be designated as the second violinist?

Solution

a. Since the order in which each musician is selected is not important, we use combinations. The violinists may be selected in $C(6, 2)$, or 15, ways; the violist may be selected in $C(3, 1)$, or 3, ways; and the cellist may be selected in $C(2, 1)$, or 2, ways. By the multiplication principle, there are $15 \cdot 3 \cdot 2$, or 90, ways of forming the string quartet.

b. The order in which the violinists are selected is important here. Consequently, the number of ways of selecting the violinists is given by $P(6, 2)$, or 30, ways. The number of ways of selecting the violist and the cellist are, of course, 3 and 2, respectively. Therefore, the number of ways in which the string quartet may be formed is given by $30 \cdot 3 \cdot 2$, or 180, ways. ∎

Note The solution of Example 11 involves both a permutation and a combination. When we select two violinists from six violinists, order is not important, and we use a combination formula to solve the problem. However, when one of the violinists is designated as a first violinist, order is important, and we use a permutation formula to solve the problem. ∎

APPLIED EXAMPLE 12 Investment Options Refer to Example 5, page 355. Suppose the investor has decided to purchase shares in the stocks of two aerospace companies, two energy development companies, and two electronics companies. In how many ways may the investor select the group of six companies for the investment from the recommended list of five aerospace companies, three energy development companies, and four electronics companies?

Solution There are $C(5, 2)$ ways in which the investor may select the aerospace companies, $C(3, 2)$ ways in which she may select the companies involved in energy development, and $C(4, 2)$ ways in which she may select the electronics companies as investments. By the generalized multiplication principle, there are

$$C(5, 2)C(3, 2)C(4, 2) = \frac{5!}{2!3!} \cdot \frac{3!}{2!1!} \cdot \frac{4!}{2!2!}$$

$$= \frac{5 \cdot 4}{2} \cdot 3 \cdot \frac{4 \cdot 3}{2} = 180$$

ways of selecting the group of six companies for her investment. ∎

APPLIED EXAMPLE 13 Scheduling Performances The Futurists, a rock group, are planning a concert tour with performances to be given in five

cities: San Francisco, Los Angeles, San Diego, Denver, and Las Vegas. In how many ways can they arrange their itinerary if
a. There are no restrictions?
b. The three performances in California must be given consecutively?

Solution
a. The order is important here, and we see that there are

$$P(5, 5) = 5! = 120$$

ways of arranging their itinerary.
b. First, note that there are $P(3, 3)$ ways of choosing between performing in California and in the two cities outside that state. Next, there are $P(3, 3)$ ways of arranging their itinerary in the three cities in California. Therefore, by the multiplication principle, there are

$$P(3, 3)P(3, 3) = \frac{3!}{(3 - 3)!} \cdot \frac{3!}{(3 - 3)!} = (6)(6) = 36$$

ways of arranging their itinerary. ▪

APPLIED EXAMPLE 14   U.N. Security Council Voting The U.N. Security Council consists of 5 permanent members and 10 nonpermanent members. Decisions made by the council require nine votes for passage. However, any permanent member may veto a measure and thus block its passage. In how many ways can a measure be passed if all 15 members of the Council vote (no abstentions)?

Solution   If a measure is to be passed, then all 5 permanent members must vote for passage of that measure. This can be done in $C(5, 5)$, or 1, way.

Next, observe that since nine votes are required for passage of a measure, *at least* 4 of the 10 nonpermanent members must also vote for its passage. To determine the number of ways this can be done, notice that there are $C(10, 4)$ ways in which exactly 4 of the nonpermanent members can vote for passage of a measure, $C(10, 5)$ ways in which exactly 5 of them can vote for passage of a measure, and so on. Finally, there are $C(10, 10)$ ways in which all 10 nonpermanent members can vote for passage of a measure. Therefore, there are

$$C(10, 4) + C(10, 5) + \cdots + C(10, 10)$$

ways in which at least 4 of the 10 nonpermanent members can vote for a measure. So, by the multiplication principle, there are

$$C(5, 5)[C(10, 4) + C(10, 5) + \cdots + C(10, 10)]$$
$$= (1)\left[\frac{10!}{4!6!} + \frac{10!}{5!5!} + \cdots + \frac{10!}{10!0!}\right]$$
$$= (1)(210 + 252 + 210 + 120 + 45 + 10 + 1) = 848$$

ways a measure can be passed. ▪

6.4 Self-Check Exercises

1. Evaluate:
   a. 5!   **b.** $C(7, 4)$ **c.** $P(6, 2)$

2. A space shuttle crew consists of a shuttle commander, a pilot, three engineers, a scientist, and a civilian. The shuttle commander and pilot are to be chosen from 8 candidates, the three engineers from 12 candidates, the scientist from 5 candidates,

and the civilian from 2 candidates. How many such space shuttle crews can be formed?

Solutions to Self-Check Exercises 6.4 can be found on page 372.

6.4 Concept Questions

1. **a.** What is a *permutation* of a set of distinct objects?
   b. How many permutations of a set of five distinct objects taken three at a time are there?

2. Given a set of ten objects in which three are alike and of one kind, three are alike and of yet another kind, and four are alike and still yet of another kind, what is the formula for computing the permutation of these ten objects taken ten at a time?

3. **a.** What is a combination of a set of n distinct objects taken r at a time?
   b. How many combinations of six distinct objects taken three at a time are there?

6.4 Exercises

In Exercises 1–22, evaluate the expression.

1. $3(5!)$

2. $2(7!)$

3. $\dfrac{5!}{2!3!}$

4. $\dfrac{6!}{4!2!}$

5. $P(5, 5)$

6. $P(6, 6)$

7. $P(5, 2)$

8. $P(5, 3)$

9. $P(n, 1)$

10. $P(k, 2)$

11. $C(6, 6)$

12. $C(8, 8)$

13. $C(7, 4)$

14. $C(9, 3)$

15. $C(5, 0)$

16. $C(6, 5)$

17. $C(9, 6)$

18. $C(10, 3)$

19. $C(n, 2)$

20. $C(7, r)$

21. $P(n, n - 2)$

22. $C(n, n - 2)$

In Exercises 23–30, classify each problem according to whether it involves a permutation or a combination.

23. In how many ways can the letters of the word *GLACIER* be arranged?

24. A four-member executive committee is to be formed from a twelve-member board of directors. In how many ways can it be formed?

25. As part of a quality-control program, 3 cell phones are selected at random for testing from each of 100 cell phones produced by the manufacturer. In how many ways can this test batch be chosen?

26. How many three-digit numbers can be formed using the numerals in the set {3, 2, 7, 9} if repetition is not allowed?

27. In how many ways can nine different books be arranged on a shelf?

28. A member of a book club wishes to purchase two books from a selection of eight books recommended for a certain month. In how many ways can she choose them?

29. How many five-card poker hands can be dealt consisting of three queens and a pair?

30. In how many ways can a six-letter security password be formed from letters of the alphabet if no letter is repeated?

31. How many four-letter permutations can be formed from the first four letters of the alphabet?

32. How many three-letter permutations can be formed from the first five letters of the alphabet?

33. In how many ways can four students be seated in a row of four seats?

34. In how many ways can five people line up at a checkout counter in a supermarket?

35. How many different batting orders can be formed for a nine-member baseball team?

36. In how many ways can the names of six candidates for political office be listed on a ballot?

37. In how many ways can a member of a hiring committee select 3 of 12 job applicants for further consideration?

38. In how many ways can an investor select four mutual funds for his investment portfolio from a recommended list of eight mutual funds?

39. Find the number of distinguishable permutations that can be formed from the letters of the word *ANTARCTICA*.

40. Find the number of distinguishable permutations that can be formed from the letters of the word *PHILIPPINES*.

41. MANAGEMENT DECISIONS   In how many ways can a supermarket chain select 3 out of 12 possible sites for the construction of new supermarkets?

42. BOOK SELECTIONS   A student is given a reading list of ten books from which he must select two for an outside reading requirement. In how many ways can he make his selections?

43. QUALITY CONTROL   In how many ways can a quality-control engineer select a sample of 3 transistors for testing from a batch of 100 transistors?

44. STUDY GROUPS   A group of five students studying for a bar exam had formed a study group. Each member of the group will be responsible for preparing a study outline for one of five courses. In how many different ways can the five courses be assigned to the members of the group?

45. TELEVISION PROGRAMMING   In how many ways can a television-programming director schedule six different commercials in the six time slots allocated to commercials during a 1-hr program?

46. WAITING LINES   Seven people arrive at the ticket counter of a cinema at the same time. In how many ways can they line up to purchase their tickets?

47. MANAGEMENT DECISIONS   Weaver and Kline, a stock brokerage firm, has received six inquiries regarding new accounts. In how many ways can these inquiries be directed to its 12 account executives if each executive handles no more than one inquiry?

48. CAR POOLS   A company car that has a seating capacity of six is to be used by six employees who have formed a car pool. If only four of these employees can drive, how many possible seating arrangements are there for the group?

49. BOOK DISPLAYS   At a college library exhibition of faculty publications, three mathematics books, four social science books, and three biology books will be displayed on a shelf. (Assume that none of the books is alike.)

a. In how many ways can the ten books be arranged on the shelf?

b. In how many ways can the ten books be arranged on the shelf if books on the same subject matter are placed together?

50. SEATING   In how many ways can four married couples attending a concert be seated in a row of eight seats if
a. There are no restrictions?
b. Each married couple is seated together?
c. The members of each sex are seated together?

51. NEWSPAPER ADVERTISEMENTS   Four items from five different departments of Metro Department Store will be featured in a one-page newspaper advertisement as shown in the following diagram:

Advertisement

1	2	3	4
5	6	7	8
9	10	11	12
13	14	15	16
17	18	19	20

a. In how many different ways can the 20 featured items be arranged on the page?

b. If items from the same department must be in the same row, how many arrangements are possible?

52. MANAGEMENT DECISIONS   C & J Realty has received 12 inquiries from prospective home buyers. In how many ways can the inquiries be directed to four of the firm's real estate agents if each agent handles three inquiries?

53. SPORTS   In the women's tennis tournament at Wimbledon, two finalists, A and B, are competing for the title, which will be awarded to the first player to win two sets. In how many different ways can the match be completed?

54. SPORTS   In the men's tennis tournament at Wimbledon, two finalists, A and B, are competing for the title, which will be awarded to the first player to win three sets. In how many different ways can the match be completed?

55. U.N. VOTING   Refer to Example 14. In how many ways can a measure be passed if two particular permanent and two particular nonpermanent members of the Council abstain from voting?

56. JURY SELECTION   In how many different ways can a panel of 12 jurors and 2 alternate jurors be chosen from a group of 30 prospective jurors?

57. TEACHING ASSISTANTSHIPS   Twelve graduate students have applied for three available teaching assistantships. In how

many ways can the assistantships be awarded among these applicants if

a. No preference is given to any student?

b. One particular student must be awarded an assistantship?

c. The group of applicants includes seven men and five women and it is stipulated that at least one woman must be awarded an assistantship?

58. EXAMS A student taking an examination is required to answer 10 out of 15 questions.

a. In how many ways can the 10 questions be selected?

b. In how many ways can the 10 questions be selected if exactly 2 of the first 3 questions must be answered?

59. CONTRACT BIDDING UBS Television Company is considering bids submitted by seven different firms for three different contracts. In how many ways can the contracts be awarded among these firms if no firm is to receive more than two contracts?

60. SENATE COMMITTEES In how many ways can a subcommittee of four be chosen from a Senate committee of five Democrats and four Republicans if

a. All members are eligible?

b. The subcommittee must consist of two Republicans and two Democrats?

61. COURSE SELECTION A student planning her curriculum for the upcoming year must select one of five Business courses, one of three Mathematics courses, two of six elective courses, and either one of four History courses or one of three Social Science courses. How many different curricula are available for her consideration?

62. PERSONNEL SELECTION JCL Computers has five vacancies in its executive trainee program. In how many ways can the company select five trainees from a group of ten female and ten male applicants if the vacancies

a. May be filled by any combination of men and women?

b. Must be filled by two men and three women?

63. DRIVERS' TESTS A state Motor Vehicle Department requires learners to pass a written test on the motor vehicle laws of the state. The exam consists of ten true-or-false questions, of which eight must be answered correctly to qualify for a permit. In how many different ways can a learner who answers all the questions on the exam qualify for a permit?

64. QUALITY CONTROL Goodman Tire has 32 tires of a particular size and grade in stock, 2 of which are defective. If a set of 4 tires is to be selected,

a. How many different selections can be made?

b. How many different selections can be made that do not include any defective tires?

A list of poker hands ranked in order from the highest to the lowest is shown in the following table, along with a description and example of each hand. Use the table to answer Exercises 65–70.

Hand	Description	Example
Straight flush	5 cards in sequence in the same suit	A ♥ 2 ♥ 3 ♥ 4 ♥ 5 ♥
Four of a kind	4 cards of the same rank and any other card	K ♥ K ♦ K ♠ K ♣ 2 ♥
Full house	3 of a kind and a pair	3 ♥ 3 ♦ 3 ♣ 7 ♥ 7 ♦
Flush	5 cards of the same suit that are not all in sequence	5 ♥ 6 ♥ 9 ♥ J ♥ K ♥
Straight	5 cards in sequence but not all of the same suit	10 ♥ J ♦ Q ♣ K ♠ A ♥
Three of a kind	3 cards of the same rank and 2 unmatched cards	K ♥ K ♦ K ♠ 2 ♥ 4 ♦
Two pair	2 cards of the same rank and 2 cards of any other rank with an unmatched card	K ♥ K ♦ 2 ♥ 2 ♠ 4 ♣
One pair	2 cards of the same rank and 3 unmatched cards	K ♥ K ♦ 5 ♥ 2 ♠ 4 ♥

If a 5-card poker hand is dealt from a well-shuffled deck of 52 cards, how many different hands consist of the following:

65. A straight flush? (Note that an ace may be played as either a high or a low card in a straight sequence—that is, A, 2, 3, 4, 5 or 10, J, Q, K, A. Hence, there are ten possible sequences for a straight in one suit.)

66. A straight (but not a straight flush)?

67. A flush (but not a straight flush)?

68. Four of a kind?

69. A full house?

70. Two pairs?

71. BUS ROUTING The following is a schematic diagram of a city's street system between the points *A* and *B*. The City Transit Authority is in the process of selecting a route from *A* to *B* along which to provide bus service. If the company's intention is to keep the route as short as possible, how many routes must be considered?

Street system

72. SPORTS In the World Series, one National League team and one American League team compete for the coveted title that is awarded to the first team to win four games. In how many different ways can the series of seven games be completed?

73. VOTING QUORUMS A quorum (minimum) of 6 voting members is required at all meetings of the Curtis Townhomes Owners Association. If there is a total of 12 voting members in the group, find the number of ways this quorum can be formed.

74. CIRCULAR PERMUTATIONS Suppose n distinct objects are arranged in a circle. Show that the number of (different) circular arrangements of the n objects is $(n - 1)!$.
Hint: Consider the arrangement of the five letters A, B, C, D, and E in the accompanying figure. The permutations $ABCDE$, $BCDEA$, $CDEAB$, $DEABC$, and $EABCD$ are not distinguishable. Generalize this observation to the case of n objects.

75. Refer to Exercise 74. In how many ways can five TV commentators be seated at a round table for a discussion?

76. Refer to Exercise 74. In how many ways can four men and four women be seated at a round table at a dinner party if each guest is seated next to members of the opposite sex?

77. At the end of Section 3.3, we mentioned that in order to solve a linear programming problem in three variables and five constraints we have to solve 56 3×3 systems of linear equations. Verify this assertion.

78. Refer to Exercise 77. Show that in order to solve a linear programming problem in five variables and 10 constraints we have to solve 3003 5×5 systems of linear equations. This assertion was also made at the end of Section 3.3.

In Exercises 79–82, determine whether the statement is true or false. If it is true, explain why it is true. If it is false, give an example to show why it is false.

79. The number of permutations of n distinct objects taken all together is $n!$

80. $P(n, r) = r!C(n, r)$

81. The number of combinations of n objects taken $n - r$ at a time is the same as the number taken r at a time.

82. If a set of n objects consists of r elements of one kind and $n - r$ elements of another kind, then the number of permutations of the n objects taken all together is $P(n, r)$.

6.4 Solutions to Self-Check Exercises

1. a. $5! = 5 \cdot 4 \cdot 3 \cdot 2 \cdot 1 = 120$

  b. $C(7, 4) = \dfrac{7!}{4!3!} = \dfrac{7 \cdot 6 \cdot 5}{3 \cdot 2 \cdot 1} = 35$

  c. $P(6, 2) = \dfrac{6!}{4!} = 6 \cdot 5 = 30$

2. There are $P(8, 2)$ ways of picking the shuttle commander and pilot (the order *is* important here), $C(12, 3)$ ways of picking the engineers (the order is not important here), $C(5, 1)$ ways of picking the scientist, and $C(2, 1)$ ways of picking the civilian. By the multiplication principle, there are

$$P(8, 2) \cdot C(12, 3) \cdot C(5, 1) \cdot C(2, 1)$$
$$= \frac{8!}{6!} \cdot \frac{12!}{9!3!} \cdot \frac{5!}{4!1!} \cdot \frac{2!}{1!1!}$$
$$= \frac{(8)(7)(12)(11)(10)(5)(2)}{(3)(2)}$$
$$= 123,200$$

ways a crew can be selected.

USING TECHNOLOGY

■ Evaluating $n!$, $P(n, r)$, and $C(n, r)$

Graphing Utility

A graphing utility can be used to calculate factorials, permutations, and combinations with relative ease. A graphing utility is therefore an indispensable tool in solving counting problems involving large numbers of objects. Here we use the **nPr** (permutation) and **nCr** (combination) functions of a graphing utility.

EXAMPLE 1 Use a graphing utility to find (a) $12!$, (b) $P(52, 5)$, and (c) $C(38, 10)$.

Solution

a. Using the factorial function, we find that $12! = 479,001,600$.

b. Using the **nPr** function, we have

$$P(52, 5) = 52 \; \textbf{nPr} \; 5 = 311,875,200$$

c. Using the **nCr** function, we obtain

$$C(38, 10) = 38 \; \textbf{nCr} \; 10 = 472,733,756$$

■

Excel

Excel has built-in functions for calculating factorials, permutations, and combinations.

EXAMPLE 2 Use Excel to calculate

a. $12!$

b. $P(52, 5)$

c. $C(38, 10)$

Solution

a. In cell A1, enter =FACT(12) and press ⎡**Shift-Enter**⎤. The number 479001600 will appear.

b. In cell A2, enter =PERMUT(52,5) and press ⎡**Shift-Enter**⎤. The number 311875200 will appear.

c. In cell A3, enter =COMBIN(38,10) and press ⎡**Shift-Enter**⎤. The number 472733756 will appear.

■

Note: Boldfaced words/characters enclosed in a box (for example, ⎡**Enter**⎤) indicate that an action (click, select, or press) is required. Words/characters printed blue (for example, Chart sub-type:) indicate words/characters that appear on the screen. Words/characters printed in a typewriter font (for example, =(−2/3)*A2+2)) indicate words/characters that need to be typed and entered.

(continued)

TECHNOLOGY EXERCISES

In Exercises 1–10, evaluate the expression.

1. $15!$

2. $20!$

3. $4(18!)$

4. $\dfrac{30!}{18!}$

5. $P(52, 7)$

6. $P(24, 8)$

7. $C(52, 7)$

8. $C(26, 8)$

9. $P(10, 4)C(12, 6)$

10. $P(20, 5)C(9, 3)C(8, 4)$

11. A mathematics professor uses a computerized test bank to prepare her final exam. If 25 different problems are available for the first three exam questions, 40 different problems available for the next five questions, and 30 different problems available for the last two questions, how many different ten-question exams can she set? (Assume that the order of the questions within each group is not important.)

12. S & S Brokerage has received 100 inquiries from prospective clients. In how many ways can the inquiries be directed to five of the firm's brokers if each broker handles 20 inquiries?

 CHAPTER 6   **Summary of Principal Formulas and Terms**

 FORMULAS

1. Commutative laws	$A \cup B = B \cup A$ $A \cap B = B \cap A$
2. Associative laws	$A \cup (B \cup C) = (A \cup B) \cup C$ $A \cap (B \cap C) = (A \cap B) \cap C$
3. Distributive laws	$A \cup (B \cap C)$ $\quad = (A \cup B) \cap (A \cup C)$ $A \cap (B \cup C)$ $\quad = (A \cap B) \cup (A \cap C)$
4. De Morgan's laws	$(A \cup B)^c = A^c \cap B^c$ $(A \cap B)^c = A^c \cup B^c$
5. Number of elements in the union of two finite sets	$n(A \cup B) = n(A) + n(B)$ $\quad\quad - n(A \cap B)$
6. Permutation of n distinct objects, taken r at a time	$P(n, r) = \dfrac{n!}{(n - r)!}$
7. Permutation of n objects, not all distinct, taken n at a time	$\dfrac{n!}{n_1! n_2! \cdots n_r!}$
8. Combination of n distinct objects, taken r at a time	$C(n, r) = \dfrac{n!}{r!(n - r)!}$

 TERMS

set (336)	empty set (337)	multiplication principle (353)
element of a set (336)	universal set (338)	generalized multiplication principle (354)
roster notation (336)	Venn diagram (338)	permutation (359)
set-builder notation (336)	set union (339)	n-factorial (361)
set equality (336)	set intersection (339)	combination (364)
subset (337)	set complementation (339)	

CHAPTER 6 Concept Review Questions

Fill in the blanks.

1. A well-defined collection of objects is called a/an _____. These objects are also called _____ of the _____.

2. Two sets having exactly the same elements are said to be _____.

3. If every element of a set A is also an element of a set B, then A is a/an _____ of B.

4. **a.** The empty set $\varnothing$ is the set containing _____ elements.
   b. The universal set is the set containing _____ elements.

5. **a.** The set of all elements in A and/or B is called the _____ of A and B.
   b. The set of all elements in A and B is called the _____ of A and B.

6. The set of all elements in U that are not in A is called the _____ of A.

7. Applying De Morgan's law, we can write $(A \cup B \cup C)^c =$ _____.

8. An arrangement of a set of distinct objects in a definite order is called a/an _____; an arrangement in which the order is not important is a/an _____.

CHAPTER 6 Review Exercises

In Exercises 1–4, list the elements of each set in roster notation.

1. $\{x \mid 3x - 2 = 7; x, \text{ an integer}\}$

2. $\{x \mid x \text{ is a letter of the word } TALLAHASSEE\}$

3. The set whose elements are the even numbers between 3 and 11

4. $\{x \mid (x - 3)(x + 4) = 0; x, \text{ a negative integer}\}$

Let $A = \{a, c, e, r\}$. In Exercises 5–8, determine whether the set is equal to A.

5. $\{r, e, c, a\}$

6. $\{x \mid x \text{ is a letter of the word } career\}$

7. $\{x \mid x \text{ is a letter of the word } racer\}$

8. $\{x \mid x \text{ is a letter of the word } cares\}$

In Exercises 9–12, shade the portion of the accompanying figure that represents the set.

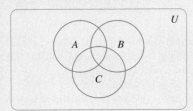

9. $A \cup (B \cap C)$

10. $(A \cap B \cap C)^c$

11. $A^c \cap B^c \cap C^c$

12. $A^c \cap (B^c \cup C^c)$

Let $U = \{a, b, c, d, e\}$, $A = \{a, b\}$, $B = \{b, c, d\}$, and $C = \{a, d, e\}$. In Exercises 13–16, verify the equation by direct computation.

13. $A \cup (B \cup C) = (A \cup B) \cup C$

14. $A \cap (B \cap C) = (A \cap B) \cap C$

15. $A \cap (B \cup C) = (A \cap B) \cup (A \cap C)$

16. $A \cup (B \cap C) = (A \cup B) \cap (A \cup C)$

Let $U = \{$all participants in a consumer-behavior survey conducted by a national polling group$\}$

$A = \{$consumers who avoided buying a product because it is not recyclable$\}$

$B = \{$consumers who used cloth rather than disposable diapers$\}$

$C = \{$consumers who boycotted a company's products because of their record on the environment$\}$

$D = \{$consumers who voluntarily recycled their garbage$\}$

In Exercises 17–20, describe each set in words.

17. $A \cap C$

18. $A \cup D$

19. $B^c \cap D$

20. $C^c \cup D^c$

Let *A* and *B* be subsets of a universal set *U* and suppose $n(U) = 350$, $n(A) = 120$, $n(B) = 80$, and $n(A \cap B) = 50$. In Exercises 21–26, find the number of elements in each set.

21. $n(A \cup B)$ **22.** $n(A^c)$

23. $n(B^c)$ **24.** $n(A^c \cap B)$

25. $n(A \cap B^c)$ **26.** $n(A^c \cap B^c)$

In Exercises 27–30, evaluate each quantity.

27. $C(20, 18)$ **28.** $P(9, 7)$

29. $C(5, 3) \cdot P(4, 2)$ **30.** $4 \cdot P(5, 3) \cdot C(7, 4)$

31. CREDIT CARD COMPARISONS   A comparison of five major credit cards showed that

3 offered cash advances.

3 offered extended payments for *all* goods and services purchased.

2 required an annual fee of less than $35.

2 offered both cash advances and extended payments.

1 offered extended payments and had an annual fee less than $35.

No card had an annual fee less than $35 and offered both cash advances and extended payments.

How many cards had an annual fee less than $35 and offered cash advances? (Assume that every card had at least one of the three mentioned features.)

32. STUDENT SURVEYS   The Department of Foreign Languages of a liberal arts college conducted a survey of its recent graduates to determine the foreign language courses they had taken while undergraduates at the college. Of the 480 graduates

200 had at least 1 yr of Spanish.

178 had at least 1 yr of French.

140 had at least 1 yr of German.

33 had at least 1 yr of Spanish and French.

24 had at least 1 yr of Spanish and German.

18 had at least 1 yr of French and German.

3 had at least 1 yr of all three languages.

How many of the graduates had
a. At least 1 yr of at least one of the three languages?
b. At least 1 yr of exactly one of the three languages?
c. Less than 1 yr of any of the three languages?

33. In how many ways can six different compact discs be arranged on a shelf?

34. In how many ways can three pictures be selected from a group of six different pictures?

35. Find the number of distinguishable permutations that can be formed from the letters of each word.
a. *CINCINNATI*   **b.** *HONOLULU*

36. How many three-digit numbers can be formed from the numerals in the set $\{1, 2, 3, 4, 5\}$ if
a. Repetition of digits is not allowed?
b. Repetition of digits is allowed?

37. INVESTMENTS   In a survey conducted by Helena, a financial consultant, it was revealed that of her 400 clients

300 own stocks.

180 own bonds.

160 own mutual funds.

110 own both stocks and bonds.

120 own both stocks and mutual funds.

90 own both bonds and mutual funds.

How many of Helena's clients own stocks, bonds, and mutual funds?

38. POKER   From a standard 52-card deck, how many 5-card poker hands can be dealt consisting of
a. Five clubs?   **b.** Three kings and one pair?

39. ELECTIONS   In an election being held by the Associated Students Organization, there are six candidates for president, four for vice president, five for secretary, and six for treasurer. How many different possible outcomes are there for this election?

40. TEAM SELECTION   There are eight seniors and six juniors in the Math Club at Jefferson High School. In how many ways can a math team consisting of four seniors and two juniors be selected from the members of the Math Club?

41. SEATING ARRANGEMENTS   In how many ways can seven students be assigned seats in a row containing seven desks if
a. There are no restrictions?
b. Two of the students must not be seated next to each other?

42. QUALITY CONTROL   From a shipment of 60 transistors, 5 of which are defective, a sample of 4 transistors is selected at random.
a. In how many different ways can the sample be selected?
b. How many samples contain 3 defective transistors?
c. How many samples do not contain any defective transistors?

43. RANDOM SAMPLES   A sample of 4 balls is to be selected at random from an urn containing 15 balls numbered 1 to 15. If 6 balls are green, 5 are white, and 4 are black,
a. How many different samples can be selected?
b. How many samples can be selected that contain at least 1 white ball?

CHAPTER 6   Before Moving On . . .

1. Let $U = \{a, b, c, d, e, f, g\}$, $A = \{a, d, f, g\}$, $B = \{d, f, g\}$, and $C = \{b, c, e, f\}$. Find
   a. $A \cap (B \cup C)$
   b. $(A \cap C) \cup (B \cup C)$
   c. A^c

2. Let A, B, and C be subsets of a universal set U and suppose $n(U) = 120$, $n(A) = 20$, $n(A \cap B) = 10$, $n(A \cap C) = 11$, $n(B \cap C) = 9$, and $n(A \cap B \cap C) = 4$. Find $n[A \cap (B \cup C)^c]$.

3. In how many ways can four compact discs be selected from six different compact discs?

4. From a standard 52-card deck, how many 5-card poker hands can be dealt consisting of 3 deuces and 2 face cards?

5. There are six seniors and five juniors in the Chess Club at Madison High School. In how many ways can a team consisting of three seniors and two juniors be selected from the members of the Chess Club?

7 Probability

Where did the defective picture tube come from? Picture tubes for the Pulsar 19-inch color television sets are manufactured in three locations and then shipped to the main plant of Vista Vision for final assembly. Each location produces a certain number of the picture tubes with different degrees of reliability. In Example 1, page 430, we will determine the likelihood that a defective picture tube is manufactured in a particular location.

THE SYSTEMATIC STUDY of probability began in the 17th century, when certain aristocrats wanted to discover superior strategies to use in the gaming rooms of Europe. Some of the best mathematicians of the period were engaged in this pursuit. Since then, probability has evolved in virtually every sphere of human endeavor in which an element of uncertainty is present.

We begin by introducing some of the basic terminology used in the study of the subject. Then, in Section 7.2, we give the technical meaning of the term *probability*. The rest of this chapter is devoted to the development of techniques for computing the probabilities of the occurrence of certain events.

7.1   Experiments, Sample Spaces, and Events

▬ Terminology

A number of specialized terms are used in the study of probability. We begin by defining the term *experiment*.

> **Experiment**
> An **experiment** is an activity with observable results.

The results of the experiment are called the **outcomes** of the experiment. Three examples of experiments are the following:

- Tossing a coin and observing whether it falls "heads" or "tails"
- Casting a die and observing which of the numbers 1, 2, 3, 4, 5, or 6 shows up
- Testing a spark plug from a batch of 100 spark plugs and observing whether or not it is defective

In our discussion of experiments, we use the following terms:

> **Sample Point, Sample Space, and Event**
> **Sample point:** An outcome of an experiment
> **Sample space:** The set consisting of all possible sample points of an experiment
> **Event:** A subset of a sample space of an experiment

The sample space of an experiment is a universal set whose elements are precisely the outcomes, or the sample points, of the experiment; the events of the experiment are the subsets of the universal set. A sample space associated with an experiment that has a finite number of possible outcomes (sample points) is called a **finite sample space.**

Since the events of an experiment are subsets of a universal set (the sample space of the experiment), we may use the results for set theory given in Chapter 6 to help us study probability. The event B is said to **occur** in a trial of an experiment whenever B contains the observed outcome. We begin by explaining the roles played by the empty set and a universal set when viewed as events associated with an experiment. The empty set, $\varnothing$, is called the *impossible event*; it cannot occur since the $\varnothing$ has no elements (outcomes). Next, the universal set S is referred to as the *certain event*; it must occur since S contains all the outcomes of the experiment.

This terminology is illustrated in the next several examples.

EXAMPLE 1   Describe the sample space associated with the experiment of tossing a coin and observing whether it falls "heads" or "tails." What are the events of this experiment?

Solution The two outcomes are "heads" and "tails," and the required sample space is given by $S = \{H, T\}$, where H denotes the outcome "heads" and T denotes the outcome "tails." The events of the experiment, the subsets of S, are

$$\emptyset, \quad \{H\}, \quad \{T\}, \quad S$$

Note that we have included the impossible event, $\emptyset$, and the certain event, S. ∎

Since the events of an experiment are subsets of the sample space of the experiment, we may talk about the union and intersection of any two events; we can also consider the complement of an event with respect to the sample space.

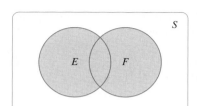

$E \cup F$

(a) The union of two events

> **Union of Two Events**
> The **union of** the **two events** E and F is the event $E \cup F$.

Thus, the event $E \cup F$ contains the set of outcomes of E and/or F.

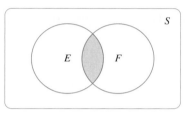

$E \cap F$

(b) The intersection of two events

> **Intersection of Two Events**
> The **intersection of** the **two events** E and F is the event $E \cap F$.

Thus, the event $E \cap F$ contains the set of outcomes of E and F.

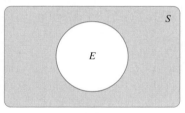

E^c

(c) The complement of the event E

FIGURE 1

> **Complement of an Event**
> The **complement of an event** E is the event E^c.

Thus, the event E^c is the set containing all the outcomes in the sample space S that are not in E.

Venn diagrams depicting the union, intersection, and complementation of events are shown in Figure 1.

These concepts are illustrated in the following example.

 VIDEO

EXAMPLE 2 Consider the experiment of casting a die and observing the number that falls uppermost. Let $S = \{1, 2, 3, 4, 5, 6\}$ denote the sample space of the experiment and $E = \{2, 4, 6\}$ and $F = \{1, 3\}$ be events of this experiment. Compute (a) $E \cup F$, (b) $E \cap F$, and (c) F^c. Interpret your results.

Solution
a. $E \cup F = \{1, 2, 3, 4, 6\}$ and is the event that the outcome of the experiment is a 1, a 2, a 3, a 4, or a 6.
b. $E \cap F = \emptyset$ is the impossible event; the number appearing uppermost when a die is cast cannot be both even and odd at the same time.
c. $F^c = \{2, 4, 5, 6\}$ is precisely the event that the event F does not occur. ∎

If two events cannot occur at the same time, they are said to be mutually exclusive. Using set notation, we have the following definition.

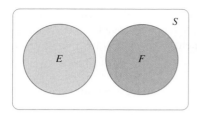

FIGURE 2
Mutually exclusive events

EXPLORE & DISCUSS

1. Suppose E and F are two complementary events. Must E and F be mutually exclusive? Explain your answer.

2. Suppose E and F are mutually exclusive events. Must E and F be complementary? Explain your answer.

Mutually Exclusive Events

E and F are **mutually exclusive** if $E \cap F = \emptyset$.

As before, we may use Venn diagrams to illustrate these events. In this case, the two mutually exclusive events are depicted as two nonintersecting circles (Figure 2).

EXAMPLE 3   An experiment consists of tossing a coin three times and observing the resulting sequence of "heads" and "tails."
  a. Describe the sample space S of the experiment.
  b. Determine the event E that exactly two heads appear.
  c. Determine the event F that at least one head appears.

Solution
  a. The sample points may be obtained with the aid of a tree diagram (Figure 3).

First toss	Second toss	Third toss	Sample points
		H	(H, H, H)
	H	T	(H, H, T)
H		H	(H, T, H)
	T	T	(H, T, T)
		H	(T, H, H)
	H	T	(T, H, T)
T		H	(T, T, H)
	T	T	(T, T, T)

FIGURE 3

The required sample space S is given by

$$S = \{HHH, HHT, HTH, HTT, THH, THT, TTH, TTT\}$$

  b. By scanning the sample space S obtained in part (a), we see that the outcomes in which exactly two heads appear are given by the event

$$E = \{HHT, HTH, THH\}$$

  c. Proceeding as in part (b), we find

$$F = \{HHH, HHT, HTH, HTT, THH, THT, TTH\}$$ ∎

EXAMPLE 4   An experiment consists of casting a pair of dice and observing the number that falls uppermost on each die.
  a. Describe an appropriate sample space S for this experiment.
  b. Determine the events $E_2, E_3, E_4, \ldots, E_{12}$ that the sum of the numbers falling uppermost is $2, 3, 4, \ldots, 12$, respectively.

Solution
  a. We may represent each outcome of the experiment by an ordered pair of numbers, the first representing the number that appears uppermost on the first die

and the second representing the number that appears uppermost on the second die. To distinguish between the two dice, think of the first die as being red and the second as being green. Since there are six possible outcomes for each die, the multiplication principle implies that there are $6 \cdot 6$, or 36, elements in the sample space:

$$S = \{(1, 1), (1, 2), (1, 3), (1, 4), (1, 5), (1, 6),$$
$$(2, 1), (2, 2), (2, 3), (2, 4), (2, 5), (2, 6),$$
$$(3, 1), (3, 2), (3, 3), (3, 4), (3, 5), (3, 6),$$
$$(4, 1), (4, 2), (4, 3), (4, 4), (4, 5), (4, 6),$$
$$(5, 1), (5, 2), (5, 3), (5, 4), (5, 5), (5, 6),$$
$$(6, 1), (6, 2), (6, 3), (6, 4), (6, 5), (6, 6)\}$$

b. With the aid of the results of part (a), we obtain the required list of events, shown in Table 1.

TABLE 1

Sum of Uppermost Numbers	Event
2	$E_2 = \{(1, 1)\}$
3	$E_3 = \{(1, 2), (2, 1)\}$
4	$E_4 = \{(1, 3), (2, 2), (3, 1)\}$
5	$E_5 = \{(1, 4), (2, 3), (3, 2), (4, 1)\}$
6	$E_6 = \{(1, 5), (2, 4), (3, 3), (4, 2), (5, 1)\}$
7	$E_7 = \{(1, 6), (2, 5), (3, 4), (4, 3), (5, 2), (6, 1)\}$
8	$E_8 = \{(2, 6), (3, 5), (4, 4), (5, 3), (6, 2)\}$
9	$E_9 = \{(3, 6), (4, 5), (5, 4), (6, 3)\}$
10	$E_{10} = \{(4, 6), (5, 5), (6, 4)\}$
11	$E_{11} = \{(5, 6), (6, 5)\}$
12	$E_{12} = \{(6, 6)\}$

APPLIED EXAMPLE 5 Movie Attendance   The manager of a local cinema records the number of patrons attending a first-run movie at the 1 p.m. screening. The theater has a seating capacity of 500.
a. What is an appropriate sample space for this experiment?
b. Describe the event E that fewer than 50 people attend the screening.
c. Describe the event F that the theater is more than half full at the screening.

Solution
a. The number of patrons at the screening (the outcome) could run from 0 to 500. Therefore, a sample space for this experiment is

$$S = \{0, 1, 2, 3, \ldots, 500\}$$

b. $E = \{0, 1, 2, 3, \ldots, 49\}$
c. $F = \{251, 252, 253, \ldots, 500\}$

APPLIED EXAMPLE 6   Family Composition   An experiment consists of studying the composition of a three-child family in which the children were born at different times.

a. Describe an appropriate sample space S for this experiment.
b. Describe the event E that there are two girls and a boy in the family.
c. Describe the event F that the oldest child is a girl.
d. Describe the event G that the oldest child is a girl and the youngest child is a boy.

Solution

a. The sample points of the experiment may be obtained with the aid of the tree diagram shown in Figure 4, where b denotes a boy and g denotes a girl.

First child	Second child	Third child	Sample points
		b	$b\,b\,b$
	b	g	$b\,b\,g$
b		b	$b\,g\,b$
	g	g	$b\,g\,g$
		b	$g\,b\,b$
	b	g	$g\,b\,g$
g		b	$g\,g\,b$
	g	g	$g\,g\,g$

FIGURE 4
Tree diagram for three-child families

We see from the tree diagram that the required sample space is given by

$$S = \{bbb,\ bbg,\ bgb,\ bgg,\ gbb,\ gbg,\ ggb,\ ggg\}$$

Using the tree diagram, we find that

b. $E = \{bgg,\ gbg,\ ggb\}$
c. $F = \{gbb,\ gbg,\ ggb,\ ggg\}$
d. $G = \{gbb,\ ggb\}$

The next example shows that sample spaces may be infinite.

APPLIED EXAMPLE 7   Testing New Products   EverBrite is developing a high-amperage, high-capacity battery as a source for powering electric cars. The battery is tested by installing it in a prototype electric car and running the car with a fully charged battery on a test track at a constant speed of 55 mph until the car runs out of power. The distance covered by the car is then observed.

a. What is the sample space for this experiment?
b. Describe the event E that the driving range under test conditions is less than 150 miles.
c. Describe the event F that the driving range is between 200 and 250 miles, inclusive.

Solution

a. Since the distance d covered by the car in any run may be given by any nonnegative number, the sample space S is given by

$$S = \{d\,|\,d \geq 0\}$$

b. The event E is given by

$$E = \{d \mid d < 150\}$$

c. The event F is given by

$$F = \{d \mid 200 \le d \le 250\}$$

∎

7.1 Self-Check Exercises

1. A sample of three apples taken from Cavallero's Fruit Stand are examined to determine whether they are good or rotten.
  a. What is an appropriate sample space for this experiment?
  b. Describe the event E that exactly one of the apples picked is rotten.
  c. Describe the event F that the first apple picked is rotten.

2. Refer to Self-Check Exercise 1.
  a. Find $E \cup F$.
  b. Find $E \cap F$.
  c. Find F^c.
  d. Are the events E and F mutually exclusive?

Solutions to Self-Check Exercises 7.1 can be found on page 388.

7.1 Concept Questions

1. Explain what is meant by an *experiment*. Give an example. For the example you have chosen, describe (a) a *sample point*, (b) the *sample space*, and (c) an *event*, of the experiment.

2. What does it mean for two events to be mutually exclusive? Give an example of two mutually exclusive events E and F. How can you prove that they are mutually exclusive?

7.1 Exercises

In Exercises 1–6, let $S = \{a, b, c, d, e, f\}$ be a sample space of an experiment and let $E = \{a, b\}$, $F = \{a, d, f\}$, and $G = \{b, c, e\}$ be events of this experiment.

1. Find the events $E \cup F$ and $E \cap F$.

2. Find the events $F \cup G$ and $F \cap G$.

3. Find the events F^c and $E \cap G^c$.

4. Find the events E^c and $F^c \cap G$.

5. Are the events E and F mutually exclusive?

6. Are the events $E \cup F$ and $E \cap F^c$ mutually exclusive?

In Exercises 7–14, let $S = \{1, 2, 3, 4, 5, 6\}$, $E = \{2, 4, 6\}$, $F = \{1, 3, 5\}$, and $G = \{5, 6\}$.

7. Find the event $E \cup F \cup G$.

8. Find the event $E \cap F \cap G$.

9. Find the event $(E \cup F \cup G)^c$.

10. Find the event $(E \cap F \cap G)^c$.

11. Are the events E and F mutually exclusive?

12. Are the events F and G mutually exclusive?

13. Are the events E and F complementary?

14. Are the events F and G complementary?

In Exercises 15–20, let S be any sample space and E, F, and G be any three events associated with the experiment. Describe the events, using the symbols $\cup$, $\cap$, and ^{c}.

15. The event that E and/or F occurs

16. The event that both E and F occur

17. The event that G does not occur

18. The event that E but not F occurs

19. The event that none of the events E, F, and G occurs

20. The event that E occurs but neither of the events F or G occurs

21. Consider the sample space S of Example 4, page 382.
   a. Determine the event that the number that falls uppermost on the first die is greater than the number that falls uppermost on the second die.
   b. Determine the event that the number that falls uppermost on the second die is double that of the number that falls on the first die.

22. Consider the sample space S of Example 4, page 382.
   a. Determine the event that the sum of the numbers falling uppermost is less than or equal to 7.
   b. Determine the event that the number falling uppermost on one die is a 4 and the number falling uppermost on the other die is greater than 4.

23. Let $S = \{a, b, c\}$ be a sample space of an experiment with outcomes a, b, and c. List all the events of this experiment.

24. Let $S = \{1, 2, 3\}$ be a sample space associated with an experiment.
   a. List all events of this experiment.
   b. How many subsets of S contain the number 3?
   c. How many subsets of S contain either the number 2 or the number 3?

25. An experiment consists of selecting a card from a standard deck of playing cards and noting whether it is black (B) or red (R).
   a. Describe an appropriate sample space for this experiment.
   b. What are the events of this experiment?

26. An experiment consists of selecting a letter at random from the letters in the word *MASSACHUSETTS* and observing the outcomes.
   a. What is an appropriate sample space for this experiment?
   b. Describe the event "the letter selected is a vowel."

27. An experiment consists of tossing a coin and casting a die and observing the outcomes.
   a. Describe an appropriate sample space for this experiment.
   b. Describe the event "a head is tossed and an even number is cast."

28. An experiment consists of spinning the hand of the numbered disc shown in the following figure and observing the region in which the pointer stops. (If the needle stops on a line, the result is discounted and the needle is spun again.)

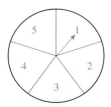

   a. What is an appropriate sample space for this experiment?
   b. Describe the event "the spinner points to the number 2."
   c. Describe the event "the spinner points to an odd number."

29. A die is cast and the number that falls uppermost is observed. Let E denote the event that the number shown is a 2 and let F denote the event that the number shown is an even number.
   a. Are the events E and F mutually exclusive?
   b. Are the events E and F complementary?

30. A die is cast and the number that falls uppermost is observed. Let E denote the event that the number shown is even and let F denote the event that the number is an odd number.
   a. Are the events E and F mutually exclusive?
   b. Are the events E and F complementary?

31. QUALITY CONTROL A sample of three transistors taken from a local electronics store was examined to determine whether the transistors were defective (d) or nondefective (n). What is an appropriate sample space for this experiment?

32. BLOOD TYPING Human blood is classified by the presence or absence of three main antigens (A, B, and Rh). When a blood specimen is typed, the presence of the A and/or B antigen is indicated by listing the letter A and/or the letter B. If neither the A nor B antigen is present, the letter O is used. The presence or absence of the Rh antigen is indicated by the symbols $+$ or $-$, respectively. Thus, if a blood specimen is classified as AB^+, it contains the A and the B antigens as well as the Rh antigen. Similarly, O^- blood contains none of the three antigens. Using this information, determine the sample space corresponding to the different blood groups.

33. GAME SHOWS In a television game show, the winner is asked to select three prizes from five different prizes, A, B, C, D, and E.
   a. Describe a sample space of possible outcomes (order is not important).
   b. How many points are there in the sample space corresponding to a selection that includes A?
   c. How many points are there in the sample space corresponding to a selection that includes A and B?
   d. How many points are there in the sample space corresponding to a selection that includes either A or B?

34. AUTOMATIC TELLERS The manager of a local bank observes how long it takes a customer to complete his transactions at the automatic bank teller.
   a. Describe an appropriate sample space for this experiment.
   b. Describe the event that it takes a customer between 2 and 3 min to complete his transactions at the automatic bank teller.

35. COMMON STOCKS Robin purchased shares of a machine tool company and shares of an airline company. Let E be the event that the shares of the machine tool company increase in value over the next 6 mo, and let F be the event that the shares of the airline company increase in value over the next 6 mo. Using the symbols $\cup$, $\cap$, and ^{c}, describe the following events.

a. The shares in the machine tool company do not increase in value.

b. The shares in both the machine tool company and the airline company do not increase in value.

c. The shares of at least one of the two companies increase in value.

d. The shares of only one of the two companies increase in value.

36. **CUSTOMER SERVICE SURVEYS** The customer service department of Universal Instruments, manufacturer of the Galaxy home computer, conducted a survey among customers who had returned their purchase registration cards. Purchasers of its deluxe model home computer were asked to report the length of time (t) in days before service was required.

a. Describe a sample space corresponding to this survey.

b. Describe the event E that a home computer required service before a period of 90 days had elapsed.

c. Describe the event F that a home computer did not require service before a period of 1 yr had elapsed.

37. **ASSEMBLY-TIME STUDIES** A time study was conducted by the production manager of Vista Vision to determine the length of time in minutes required by an assembly worker to complete a certain task during the assembly of its Pulsar color television sets.

a. Describe a sample space corresponding to this time study.

b. Describe the event E that an assembly worker took 2 min or less to complete the task.

c. Describe the event F that an assembly worker took more than 2 min to complete the task.

38. **POLITICAL POLLS** An opinion poll is conducted among a state's electorate to determine the relationship between their income levels and their stands on a proposition aimed at reducing state income taxes. Voters are classified as belonging to either the low-, middle-, or upper-income group. They are asked whether they favor, oppose, or are undecided about the proposition. Let the letters L, M, and U represent the low-, middle-, and upper-income groups, respectively, and let the letters f, o, and u represent the responses—favor, oppose, and undecided, respectively.

a. Describe a sample space corresponding to this poll.

b. Describe the event E_1 that a respondent favors the proposition.

c. Describe the event E_2 that a respondent opposes the proposition and does not belong to the low-income group.

d. Describe the event E_3 that a respondent does not favor the proposition and does not belong to the upper-income group.

39. **QUALITY CONTROL** As part of a quality-control procedure, an inspector at Bristol Farms randomly selects ten eggs from each consignment of eggs he receives and records the number of broken eggs.

a. What is an appropriate sample space for this experiment?

b. Describe the event E that at most three eggs are broken.

c. Describe the event F that at least five eggs are broken.

40. **POLITICAL POLLS** In the opinion poll of Exercise 38, the voters were also asked to indicate their political affiliations—Democrat, Republican, or Independent. As before, let the letters L, M, and U represent the low-, middle-, and upper-income groups, respectively, and let the letters D, R, and I represent Democrat, Republican, and Independent, respectively.

a. Describe a sample space corresponding to this poll.

b. Describe the event E_1 that a respondent is a Democrat.

c. Describe the event E_2 that a respondent belongs to the upper-income group and is a Republican.

d. Describe the event E_3 that a respondent belongs to the middle-income group and is not a Democrat.

41. **SHUTTLE BUS USAGE** A certain airport hotel operates a shuttle bus service between the hotel and the airport. The maximum capacity of a bus is 20 passengers. On alternate trips of the shuttle bus over a period of 1 wk, the hotel manager kept a record of the number of passengers arriving at the hotel in each bus.

a. What is an appropriate sample space for this experiment?

b. Describe the event E that a shuttle bus carried fewer than ten passengers.

c. Describe the event F that a shuttle bus arrived with a full capacity.

42. **SPORTS** Eight players, A, B, C, D, E, F, G, and H, are competing in a series of elimination matches of a tennis tournament in which the winner of each preliminary match will advance to the semifinals and the winner of the semifinals will advance to the finals. An outline of the scheduled matches follows. Describe a sample space listing the possible participants in the finals.

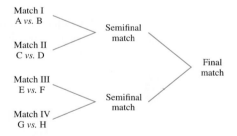

43. An experiment consists of selecting a card at random from a well-shuffled 52-card deck. Let E denote the event that an ace is drawn and let F denote the event that a spade is drawn. Show that $n(E \cup F) = n(E) + n(F) - n(E \cap F)$.

44. Let S be a sample space for an experiment. Show that if E is any event of an experiment, then E and E^c are mutually exclusive.

45. Let S be a sample space for an experiment and let E and F be events of this experiment. Show that the events $E \cup F$ and $E^c \cap F^c$ are mutually exclusive.
Hint: Use De Morgan's law.

46. Let S be a sample space of an experiment with n outcomes. Determine the number of events of this experiment.

In Exercises 47 and 48, determine whether the statement is true or false. If it is true, explain why it is true. If it is false, give an example to show why it is false.

47. If E and F are mutually exclusive and E and G are mutually exclusive, then F and G are mutually exclusive.

48. The numbers 1, 2, and 3 are written separately on three pieces of paper. These slips of paper are then placed in a bowl. If you draw two slips from the bowl, one at a time, without replacement, then the sample space for this experiment consists of six elements.

7.1   Solutions to Self-Check Exercises

1. a. Let g denote a good apple and r a rotten apple. Thus, the required sample points may be obtained with the aid of a tree diagram (compare with Example 3). The required sample space is given by

$$S = \{ggg, ggr, grg, grr, rgg, rgr, rrg, rrr\}$$

b. By scanning the sample space S obtained in part (a), we identify the outcomes in which exactly one apple is rotten. We find

$$E = \{ggr, grg, rgg\}$$

c. Proceeding as in part (b), we find

$$F = \{rgg, rgr, rrg, rrr\}$$

2. Using the results of Self-Check Exercise 1, we find
  a. $E \cup F = \{ggr, grg, rgg, rgr, rrg, rrr\}$
  b. $E \cap F = \{rgg\}$
  c. F^c is the set of outcomes in S but not in F. Thus,

$$F^c = \{ggg, ggr, grg, grr\}$$

  d. Since $E \cap F \neq \varnothing$, we conclude that E and F are not mutually exclusive.

7.2   Definition of Probability

■ Finding the Probability of an Event

Let's return to the coin-tossing experiment. The sample space of this experiment is given by $S = \{H, T\}$, where the sample points H and T correspond to the two possible outcomes, a *head* and a *tail*. If the coin is *unbiased*, then there is *one chance out of two* of obtaining a head (or a tail), and we say that the *probability* of tossing a head (tail) is $\frac{1}{2}$, abbreviated

$$P(H) = \frac{1}{2} \quad \text{and} \quad P(T) = \frac{1}{2}$$

An alternative method of obtaining the values of $P(H)$ and $P(T)$ is based on continued experimentation and does not depend on the assumption that the two outcomes are equally likely. Table 2 summarizes the results of such an exercise.

Observe that the relative frequencies (column 3) differ considerably when the number of trials is small, but as the number of trials becomes very large, the relative frequency approaches the number .5. This result suggests that we assign to $P(H)$ the value $\frac{1}{2}$, as before.

More generally, consider an experiment that may be repeated over and over again under independent and similar conditions. Suppose that in n trials an event E

TABLE 2

As the Number of Trials Increases, the Relative Frequency Approaches .5

Number of Tosses, n	Number of Heads, m	Relative Frequency of Heads, m/n
10	4	.4000
100	58	.5800
1,000	492	.4920
10,000	5,034	.5034
20,000	10,024	.5012
40,000	20,032	.5008

occurs m times. We call the ratio m/n the **relative frequency** of the event E after n repetitions. If this relative frequency approaches some value $P(E)$ as n becomes larger and larger, then $P(E)$ is called the **empirical probability** of E. Thus, the probability $P(E)$ of an event occurring is a measure of the proportion of the time that the event E will occur in the long run. Observe that this method of computing the probability of a head occurring is effective even in the case when a biased coin is used in the experiment. The relative frequency distribution is often referred to as an observed or **empirical probability distribution.**

The **probability of an event** is a number that lies between 0 and 1. In general, the larger the probability of an event, the more likely the event will occur. Thus, an event with a probability of .8 is more likely to occur than an event with a probability of .6. An event with a probability of $\frac{1}{2}$, or .5, has a "fifty-fifty," or equal, chance of occurring.

Now suppose we are given an experiment and wish to determine the probabilities associated with certain events of the experiment. This problem could be solved by computing $P(E)$ directly for each event E of interest. However, in practice, the number of events that we may be interested in is usually quite large, so this approach is not satisfactory.

The following approach is particularly suitable when the sample space of an experiment is finite.* Let S be a finite sample space with n outcomes; that is,

$$S = \{s_1, s_2, s_3, \ldots, s_n\}$$

Then, the events

$$\{s_1\}, \{s_2\}, \{s_3\}, \ldots, \{s_n\}$$

which consist of exactly one point, are called **elementary,** or **simple, events** of the experiment. They are elementary in the sense that any (nonempty) event of the experiment may be obtained by taking a finite union of suitable elementary events. The simple events of an experiment are also **mutually exclusive;** that is, given any two simple events of the experiment, only one can occur.

By assigning probabilities to each of the simple events, we obtain the results shown in Table 3. This table is called a **probability distribution** for the experiment. The function P, which assigns a probability to each of the simple events, is called a **probability function.**

TABLE 3

A Probability Distribution

Simple Event	Probability*
$\{s_1\}$	$P(s_1)$
$\{s_2\}$	$P(s_2)$
$\{s_3\}$	$P(s_3)$
.	.
.	.
.	.
$\{s_n\}$	$P(s_n)$

*For simplicity, we use the notation $P(s_i)$ instead of the technically more correct $P(\{s_i\})$.

---

*For the remainder of the chapter, we assume that all sample spaces are finite.

The numbers $P(s_1), P(s_2), \ldots, P(s_n)$ have the following properties:

1. $0 \le P(s_i) \le 1 \qquad i = 1, 2, \ldots, n$
2. $P(s_1) + P(s_2) + \cdots + P(s_n) = 1$
3. $P(\{s_i\} \cup \{s_j\}) = P(s_i) + P(s_j) \qquad (i \ne j) \qquad i = 1, 2, \ldots, n; j = 1, 2, \ldots, n$

The first property simply states that the probability of a simple event must be between 0 and 1 inclusive. The second property states that the sum of the probabilities of all simple events of the sample space is 1. This follows from the fact that the event S is certain to occur. The third property states that the probability of the union of two mutually exclusive events is given by the sum of their probabilities.

As we saw earlier, there is no unique method for assigning probabilities to the simple events of an experiment. In practice, the methods used to determine these probabilities may range from theoretical considerations of the problem on the one extreme to the reliance on "educated guesses" on the other.

Sample spaces in which the outcomes are equally likely are called **uniform sample spaces.** Assigning probabilities to the simple events in these spaces is relatively easy.

Probability of an Event in a Uniform Sample Space
If

$$S = \{s_1, s_2, \ldots, s_n\}$$

is the sample space for an experiment in which the outcomes are equally likely, then we assign the probabilities

$$P(s_1) = P(s_2) = \cdots = P(s_n) = \frac{1}{n}$$

to each of the simple events $s_1, s_2, \ldots, s_n$.

EXAMPLE 1   A fair die is cast, and the number that falls uppermost is observed. Determine the probability distribution for the experiment.

Solution   The sample space for the experiment is $S = \{1, 2, 3, 4, 5, 6\}$, and the simple events are accordingly given by the sets $\{1\}, \{2\}, \{3\}, \{4\}, \{5\}$, and $\{6\}$. Since the die is assumed to be fair, the six outcomes are equally likely. We therefore assign a probability of $\frac{1}{6}$ to each of the simple events and obtain the probability distribution shown in Table 4.   ▪

TABLE 4

A Probability Distribution

Simple Event	Probability
$\{1\}$	$\frac{1}{6}$
$\{2\}$	$\frac{1}{6}$
$\{3\}$	$\frac{1}{6}$
$\{4\}$	$\frac{1}{6}$
$\{5\}$	$\frac{1}{6}$
$\{6\}$	$\frac{1}{6}$

EXPLORE & DISCUSS

You suspect that a die is biased.

1. Describe a method you might use to prove your assertion.

2. How would you assign the probability to each outcome 1 through 6 of the experiment of casting the die and observing which number lands uppermost?

The next example shows how the *relative frequency* interpretation of probability lends itself to the computation of probabilities.

TABLE 5

Data Obtained During 200 Test Runs of an Electric Car

Distance Covered in Miles, x	Frequency of Occurrence
$0 < x \leq 50$	4
$50 < x \leq 100$	10
$100 < x \leq 150$	30
$150 < x \leq 200$	100
$200 < x \leq 250$	40
$250 < x$	16

TABLE 6

A Probability Distribution

Simple Event	Probability
$\{s_1\}$	.02
$\{s_2\}$	.05
$\{s_3\}$	.15
$\{s_4\}$	.50
$\{s_5\}$	.20
$\{s_6\}$	.08

 APPLIED EXAMPLE 2 **Testing New Products** Refer to Example 7, Section 7.1. The data shown in Table 5 were obtained in tests involving 200 test runs. Each run was made with a fully charged battery.

a. Describe an appropriate sample space for this experiment.
b. Find the empirical probability distribution for this experiment.

Solution

a. Let s_1 denote the outcome that the distance covered by the car does not exceed 50 miles; let s_2 denote the outcome that the distance covered by the car is greater than 50 miles but does not exceed 100 miles, and so on. Finally, let s_6 denote the outcome that the distance covered by the car is greater than 250 miles. Then, the required sample space is given by

$$S = \{s_1, s_2, s_3, s_4, s_5, s_6\}$$

b. To compute the empirical probability distribution for the experiment, we turn to the relative frequency interpretation of probability. Accepting the inaccuracies inherent in a relatively small number of trials (200 runs), we take the probability of s_1 occurring as

$$P(s_1) = \frac{\text{Number of trials in which } s_1 \text{ occurs}}{\text{Total number of trials}}$$

$$= \frac{4}{200} = .02$$

In a similar manner, we assign probabilities to the other simple events, obtaining the probability distribution shown in Table 6. ∎

We are now in a position to give a procedure for computing the probability $P(E)$ of an arbitrary event E of an experiment.

Finding the Probability of an Event E

1. Determine a sample space S associated with the experiment.
2. Assign probabilities to the simple events of S.
3. If $E = \{s_1, s_2, s_3, \ldots, s_n\}$ where $\{s_1\}, \{s_2\}, \{s_3\}, \ldots, \{s_n\}$ are simple events, then

$$P(E) = P(s_1) + P(s_2) + P(s_3) + \cdots + P(s_n)$$

If E is the empty set, $\varnothing$, then $P(E) = 0$.

The principle stated in step 3 is called the **addition principle** and is a consequence of Property 3 of the probability function (page 390). This principle allows us to find the probabilities of all other events once the probabilities of the simple events are known.

⚠ The addition rule applies *only* to the addition of probabilities of simple events.

  APPLIED EXAMPLE 3 **Casting Dice** A pair of fair dice is cast.
a. Calculate the probability that the two dice show the same number.
b. Calculate the probability that the sum of the numbers of the two dice is 6.

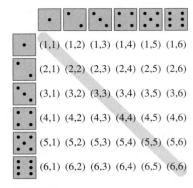

FIGURE 5
The event that the two dice show the same number

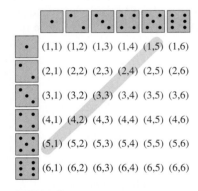

FIGURE 6
The event that the sum of the numbers on the two dice is 6

Solution   From the results of Example 4, page 382, we see that the sample space S of the experiment consists of 36 outcomes:

$$S = \{(1, 1), (1, 2), \ldots, (6, 5), (6, 6)\}$$

Since both dice are fair, each of the 36 outcomes is equally likely. Accordingly, we assign the probability of $\frac{1}{36}$ to each simple event. We are now in a position to answer the questions posed.

a. The event that the two dice show the same number is given by

$$E = \{(1, 1), (2, 2), (3, 3), (4, 4), (5, 5), (6, 6)\}$$

(Figure 5). Therefore, by the addition principle, the probability that the two dice show the same number is given by

$$P(E) = P[(1, 1)] + P[(2, 2)] + \cdots + P[(6, 6)]$$
$$= \frac{1}{36} + \frac{1}{36} + \cdots + \frac{1}{36} \qquad \text{Six terms}$$
$$= \frac{1}{6}$$

b. The event that the sum of the numbers of the two dice is 6 is given by

$$E_6 = \{(1, 5), (2, 4), (3, 3), (4, 2), (5, 1)\}$$

(Figure 6). Therefore, the probability that the sum of the numbers on the two dice is 6 is given by

$$P(E_6) = P[(1, 5)] + P[(2, 4)] + P[(3, 3)] + P[(4, 2)] + P[(5, 1)]$$
$$= \frac{1}{36} + \frac{1}{36} + \cdots + \frac{1}{36} \qquad \text{Five terms}$$
$$= \frac{5}{36}$$

 APPLIED EXAMPLE 4   Testing New Products   Consider the experiment by EverBrite in Example 2. What is the probability that the prototype car will travel more than 150 miles on a fully charged battery?

Solution   Using the results of Example 2, we see that the event that the car will travel more than 150 miles on a fully charged battery is given by $E = \{s_4, s_5, s_6\}$. Therefore, the probability that the car will travel more than 150 miles on one charge is given by

$$P(E) = P(s_4) + P(s_5) + P(s_6)$$

or, using the probability distribution for the experiment obtained in Example 2,

$$P(E) = .50 + .20 + .08 = .78$$

7.2   Self-Check Exercises

1. A biased die was cast repeatedly, and the results of the experiment are summarized in the following table:

Outcome	1	2	3	4	5	6
Frequency of Occurrence	142	173	158	175	162	190

Using the relative frequency interpretation of probability, find the empirical probability distribution for this experiment.

2. In an experiment conducted to study the effectiveness of an eye-level third brake light in the prevention of rear-end collisions, 250 of the 500 highway patrol cars of a certain state were equipped with such lights. At the end of the 1-yr trial period, the records revealed that of those equipped with a third brake light, there were 14 incidents of rear-end collision. There were 22 such incidents involving the cars not equipped with the accessory. Based on these data, what is the probability that a highway patrol car equipped with a third brake light

will be rear-ended within a 1-yr period? What is the probability that a car not so equipped will be rear-ended within a 1-yr period?

Solutions to Self-Check Exercises 7.2 can be found on page 397.

7.2 Concept Questions

1. Define (a) a *probability distribution* and (b) a *probability function*. Give examples of each.

2. If $S = \{s_1, s_2, \ldots s_n\}$ is the sample space for an experiment in which the outcomes are equally likely, what is the probability of each of the simple events $s_1, s_2, \ldots, s_n$? What do we call this type of sample space?

3. Suppose $E = \{s_1, s_2, s_3, \ldots s_n\}$, where E is an event of an experiment and $\{s_1\}, \{s_2\}, \{s_3\}, \ldots, \{s_n\}$ are simple events. If E is nonempty, what is $P(E)$? If E is empty, what is $P(E)$?

7.2 Exercises

In Exercises 1–8, list the simple events associated with each experiment.

1. A nickel and a dime are tossed, and the result of heads or tails is recorded for each coin.

2. A card is selected at random from a standard 52-card deck, and its suit—hearts (h), diamonds (d), spades (s), or clubs (c)—is recorded.

3. OPINION POLLS An opinion poll is conducted among a group of registered voters. Their political affiliation, Democrat (D), Republican (R), or Independent (I), and their sex, male (m) or female (f), are recorded.

4. QUALITY CONTROL As part of a quality-control procedure, eight circuit boards are checked, and the number of defectives is recorded.

5. MOVIE ATTENDANCE In a survey conducted to determine whether movie attendance is increasing (i), decreasing (d), or holding steady (s) among various sectors of the population, participants are classified as follows:

Group 1: Those aged 10–19

Group 2: Those aged 20–29

Group 3: Those aged 30–39

Group 4: Those aged 40–49

Group 5: Those 50 and over

The response of each participant and his or her age group are recorded.

6. DURABLE GOODS ORDERS Data concerning durable goods orders are obtained each month by an economist. A record is kept for a 1-yr period of any increase (i), decrease (d), or unchanged movement (u) in the number of durable goods orders for each month as compared with the number of such orders in the same month in the previous year.

7. BLOOD TYPES Blood tests are given as a part of the admission procedure at the Monterey Garden Community Hospital. The blood type of each patient (A, B, AB, or O) and the presence or absence of the Rh factor in each patient's blood (Rh^+ or Rh^-) are recorded.

8. METEOROLOGY A meteorologist preparing a weather map classifies the expected average temperature in each of five neighboring states for the upcoming week as follows:
   a. More than 10° below average
   b. Normal to 10° below average
   c. Higher than normal to 10° above average
   d. More than 10° above average
   Using each state's abbreviation and the categories—(a), (b), (c), and (d)—the meteorologist records these data.

9. GRADE DISTRIBUTIONS The grade distribution for a certain class is shown in the following table. Find the probability distribution associated with these data.

Grade	A	B	C	D	F
Frequency of Occurrence	4	10	18	6	2

10. **BLOOD TYPES** The percentage of the general population that has each blood type is shown in the following table. Determine the probability distribution associated with these data.

Blood Type	A	B	AB	O
Population, %	41	12	3	44

11. **TRAFFIC SURVEYS** The number of cars entering a tunnel leading to an airport in a major city over a period of 200 peak hours was observed and the following data were obtained:

Cars, x	Frequency of Occurrence
$0 < x \leq 200$	15
$200 < x \leq 400$	20
$400 < x \leq 600$	35
$600 < x \leq 800$	70
$800 < x \leq 1000$	45
$x > 1000$	15

a. Describe an appropriate sample space for this experiment.
b. Find the empirical probability distribution for this experiment.

12. **ARRIVAL TIMES** The arrival times of the 8 A.M. Boston-based commuter train as observed in the suburban town of Sharon over 120 weekdays is summarized below:

Arrival Time, x	Frequency of Occurrence
7:56 a.m. $< x \leq$ 7:58 a.m.	4
7:58 a.m. $< x \leq$ 8:00 a.m.	18
8:00 a.m. $< x \leq$ 8:02 a.m.	50
8:02 a.m. $< x \leq$ 8:04 a.m.	32
8:04 a.m. $< x \leq$ 8:06 a.m.	9
8:06 a.m. $< x \leq$ 8:08 a.m.	4
8:08 a.m. $< x \leq$ 8:10 a.m.	3

a. Describe an appropriate sample space for this experiment.
b. Find the empirical probability distribution for this experiment.

13. **SAME-SEX MARRIAGE** In a *Los Angeles Times* poll of 1936 California residents conducted in February 2004, the following question was asked: Do you favor or oppose an amendment to the U.S. Constitution barring same-sex marriage? The following results were obtained:

Opinion	Favor	Oppose	Don't know
Respondents	910	891	135

Determine the empirical probability distribution associated with these data.
Source: The Field Poll: Los Angeles Times

14. **EMAIL SERVICES** The number of subscribers to five leading email services is shown in the accompanying table:

Company	A	B	C
Subscribers	300,000	200,000	120,000

Company	D	E
Subscribers	80,000	60,000

Find the empirical probability distribution associated with these data.

15. **POLITICAL VIEWS** In a poll conducted among 2000 college freshmen to ascertain the political views of college students, the accompanying data were obtained. Determine the empirical probability distribution associated with these data.

Political Views	A	B	C	D	E
Respondents	52	398	1140	386	24

A: Far left

B: Liberal

C: Middle of the road

D: Conservative

E: Far right

16. **PRODUCT SURVEYS** The accompanying data were obtained from a survey of 1500 Americans who were asked: How safe are American-made consumer products? Determine the empirical probability distribution associated with these data.

Rating	A	B	C	D	E
Respondents	285	915	225	30	45

A: Very safe

B: Somewhat safe

C: Not too safe

D: Not safe at all

E: Don't know

17. **ASSEMBLY-TIME STUDIES** The results of a time study conducted by the production manager of Ace Novelty are shown in the accompanying table, where the number of space action-figures produced each quarter hour during an 8-hr workday has been tabulated. Find the empirical probability distribution associated with this experiment.

Figures Produced (in dozens)	Frequency of Occurrence
30	4
31	0
32	6
33	8
34	6
35	4
36	4

18. SERVICE-UTILIZATION STUDIES Metro Telephone Company compiled the accompanying information during a service-utilization study pertaining to the number of customers using their Dial-the-Time service from 7 a.m. to 9 a.m. on a certain weekday morning. Using these data, find the empirical probability distribution associated with the experiment.

Calls Received/Minute	Frequency of Occurrence
10	6
11	15
12	12
13	3
14	12
15	36
16	24
17	0
18	6
19	6

19. CORRECTIVE LENS USE According to Mediamark Research, 84 million out of 179 million adults in the United States correct their vision by using prescription eyeglasses, bifocals, or contact lenses. (Some respondents use more than one type.) What is the probability that an adult selected at random from the adult population uses corrective lenses?

Source: Mediamark Research

20. TRAFFIC DEATHS A study of deaths in car crashes from 1986 to 2002 revealed the following data on deaths in crashes by day of the week.

Day of the Week	Sunday	Monday	Tuesday	Wednesday
Average Number of Deaths	132	98	95	98

Day of the Week	Thursday	Friday	Saturday
Average Number of Deaths	105	133	158

Find the empirical probability distribution associated with these data.

Source: Insurance Institute for Highway Safety

21. FAMILY INCOME According to the 2000 U.S. Census Bureau, the income distribution of households and families was as follows:

Income, $	0–24,999	25,000–49,999	50,000–74,999
Households and Families:	30,261,220	30,965,514	20,540,604

Income $	75,000–99,999	100,000–124,999	125,000–149,999
Households and Families:	10,779,245	5,491,526	2,656,300

Income $	150,000–199,999	200,000 or more
Households and Families:	2,322,038	2,502,675

Find the empirical probability distribution associated with these data.

Source: U.S. Census Bureau

22. CORRECTIONAL SUPERVISION A study conducted by the Corrections Department of a certain state revealed that 163,605 people out of a total adult population of 1,778,314 were under correctional supervision (on probation, parole, or in jail). What is the probability that a person selected at random from the adult population in that state is under correctional supervision?

23. LIGHTNING DEATHS According to data obtained from the National Weather Service, 376 of the 439 people killed by lightning in the United States between 1985 and 1992 were men. (Job and recreational habits of men make them more vulnerable to lightning.) Assuming that this trend holds in the future, what is the probability that a person killed by lightning
   a. Is a male?   **b.** Is a female?

24. QUALITY CONTROL One light bulb is selected at random from a lot of 120 light bulbs, of which 5% are defective. What is the probability that the light bulb selected is defective?

25. EFFORTS TO STOP SHOPLIFTING According to a survey of 176 retailers, 46% of them use electronic tags as protection against shoplifting and employee theft. If one of these retailers is selected at random, what is the probability that he or she uses electronic tags as antitheft devices?

26. If a ball is selected at random from an urn containing three red balls, two white balls, and five blue balls, what is the probability that it will be a white ball?

27. If 1 card is drawn at random from a standard 52-card deck, what is the probability that the card drawn is
   a. A diamond?   **b.** A black card?
   c. An ace?

28. A pair of fair dice is cast. What is the probability that
   a. The sum of the numbers shown uppermost is less than 5?
   b. At least one 6 is cast?

29. TRAFFIC LIGHTS   What is the probability of arriving at a traffic light when it is red if the red signal is flashed for 30 sec, the yellow signal for 5 sec, and the green signal for 45 sec?

30. ROULETTE   What is the probability that a roulette ball will come to rest on an even number other than 0 or 00? (Assume that there are 38 equally likely outcomes consisting of the numbers 1–36, 0, and 00.)

31. Refer to Exercise 9. What is the probability that a student selected at random from this class received a passing grade (D or better)?

32. Refer to Exercise 11. What is the probability that more than 600 cars will enter the airport tunnel during a peak hour?

33. DISPOSITION OF CRIMINAL CASES   Of the 98 first-degree murder cases from 2002 through the first half of 2004 in the Suffolk superior court, 9 cases were thrown out of the system, 62 cases were plea-bargained, and 27 cases went to trial. What is the probability that a case selected at random
   a. Was settled through plea bargaining?
   b. Went to trial?
   Source: Boston Globe

34. SWEEPSTAKES   In a sweepstakes sponsored by Gemini Paper Products, 100,000 entries have been received. If 1 grand prize, 5 first prizes, 25 second prizes, and 500 third prizes are to be awarded, what is the probability that a person who has submitted one entry will win
   a. The grand prize?
   b. A prize?

35. A pair of fair dice is cast, and the sum of the two numbers falling uppermost observed. The probability of obtaining a sum of 2 is the same as that of obtaining a 7 since there is only one way of getting a 2—namely, by each die showing a 1; and there is only one way of obtaining a 7—namely, by one die showing a 3 and the other die showing a 4. What is wrong with this argument?

In Exercises 36–39, determine whether the given experiment has a sample space with equally likely outcomes.

36. A loaded die is cast, and the number appearing uppermost on the die is recorded.

37. Two fair dice are cast, and the sum of the numbers appearing uppermost is recorded.

38. A ball is selected at random from an urn containing six black balls and six red balls, and the color of the ball is recorded.

39. A weighted coin is thrown, and the outcome of heads or tails is recorded.

40. Let $S = \{s_1, s_2, s_3, s_4, s_5, s_6\}$ be the sample space associated with an experiment having the following probability distribution:

Outcome	s_1	s_2	s_3	s_4	s_5	s_6
Probability	$\frac{1}{12}$	$\frac{1}{4}$	$\frac{1}{12}$	$\frac{1}{6}$	$\frac{1}{3}$	$\frac{1}{12}$

Find the probability of the event:
   a. $A = \{s_1, s_3\}$
   b. $B = \{s_2, s_4, s_5, s_6\}$
   c. $C = S$

41. Consider the composition of a three-child family in which the children were born at different times. Assume that a girl is as likely as a boy at each birth. What is the probability that
   a. There are two girls and a boy in the family?
   b. The oldest child is a girl?
   c. The oldest child is a girl and the youngest child is a boy?

42. Let $S = \{s_1, s_2, s_3, s_4, s_5\}$ be the sample space associated with an experiment having the following probability distribution:

Outcome	s_1	s_2	s_3	s_4	s_5
Probability	$\frac{1}{14}$	$\frac{3}{14}$	$\frac{6}{14}$	$\frac{2}{14}$	$\frac{2}{14}$

Find the probability of the event:
   a. $A = \{s_1, s_2, s_4\}$ **b.** $B = \{s_1, s_5\}$
   c. $C = S$

43. AIRLINE SAFETY   In an attempt to study the leading causes of airline crashes, the following data were compiled from records of airline crashes from 1959 to 1994 (excluding sabotage and military action).

Primary Factor	Accidents
Flight crew	327
Airplane	49
Maintenance	14
Weather	22
Airport/air traffic control	19
Miscellaneous/other	15

Assume that you have just learned of an airline crash and that the data give a good indication of the causes of airline crashes, in general. Give an estimate of the probability that the primary cause of the crash was due to pilot error or bad weather.
   Source: National Transportation Safety Board

44. POLITICAL POLLS   An opinion poll was conducted among a group of registered voters in a certain state concerning a proposition aimed at limiting state and local taxes. Results of the poll indicated that 35% of the voters favored the proposition, 32% were against it, and the remaining group were undecided. If the results of the poll are assumed to be representative of the opinions of the state's electorate, what is the probability that a registered voter selected at random from the electorate
   a. Favors the proposition?
   b. Is undecided about the proposition?

In Exercises 45 and 46, determine whether the statement is true or false. If it is true, explain why it is true. If it is false, give an example to show why it is false.

45. If $S = \{s_1, s_2, \ldots, s_n\}$ is a uniform sample space with n outcomes, then $0 \le P(s_1) + P(s_2) + \cdots + P(s_n) \le 1$.

46. Let $S = \{s_1, s_2, \ldots, s_n\}$ be a uniform sample space for an experiment. If $n \ge 5$ and $E = \{s_1, s_2, s_5\}$, then $P(E) = 3/n$.

7.2 Solutions to Self-Check Exercises

1.

$$P(1) = \frac{\text{Number of trials in which a 1 appears uppermost}}{\text{Total number of trials}}$$

$$= \frac{142}{1000}$$

$$= .142$$

Similarly, we compute $P(2), \ldots, P(6)$, obtaining the following probability distribution:

Outcome	1	2	3	4	5	6
Probability	.142	.173	.158	.175	.162	.190

2. The probability that a highway patrol car equipped with a third brake light will be rear-ended within a 1-yr period is given by

$$\frac{\text{Number of rear-end collisions involving cars equipped with a third brake light}}{\text{Total number of such cars}} = \frac{14}{250} = .056$$

The probability that a highway patrol car not equipped with a third brake light will be rear-ended within a 1-yr period is given by

$$\frac{\text{Number of rear-end collisions involving cars not equipped with a third brake light}}{\text{Total number of such cars}} = \frac{22}{250} = .088$$

7.3 Rules of Probability

▬ Properties of the Probability Function and Their Applications

In this section, we examine some of the properties of the probability function and look at the role they play in solving certain problems. We begin by looking at the generalization of the three properties of the probability function, which were stated for simple events in the last section. Let S be a sample space of an experiment and suppose E and F are events of the experiment. We have:

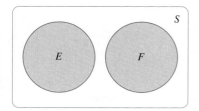

FIGURE 7
If E and F are mutually exclusive events, then $P(E \cup F) = P(E) + P(F)$.

Property 1: $P(E) \ge 0$ for any E
Property 2: $P(S) = 1$
Property 3: If E and F are mutually exclusive (that is, only one of them can occur, or equivalently, $E \cap F = \varnothing$), then

$$P(E \cup F) = P(E) + P(F)$$

(Figure 7).

Property 3 may be easily extended to the case involving any finite number of mutually exclusive events. Thus, if $E_1, E_2, \ldots, E_n$ are mutually exclusive events, then

$$P(E_1 \cup E_2 \cup \cdots \cup E_n) = P(E_1) + P(E_2) + \cdots + P(E_n)$$

TABLE 7

Probability Distribution

Score, x	Probability
$x > 700$	.01
$600 < x \leq 700$	.07
$500 < x \leq 600$	.19
$400 < x \leq 500$	.23
$300 < x \leq 400$	.31
$x \leq 300$	.19

 APPLIED EXAMPLE 1   SAT Verbal Scores   The superintendent of a metropolitan school district has estimated the probabilities associated with the SAT verbal scores of students from that district. The results are shown in Table 7. If a student is selected at random, what is the probability that his or her SAT verbal score will be

a. More than 400?
b. Less than or equal to 500?
c. Greater than 400 but less than or equal to 600?

Solution   Let A, B, C, D, E, and F denote, respectively, the event that the score is greater than 700, greater than 600 but less than or equal to 700, greater than 500 but less than or equal to 600, and so forth. Then, these events are mutually exclusive. Therefore,

a. The probability that the student's score will be more than 400 is given by

$$P(D \cup C \cup B \cup A) = P(D) + P(C) + P(B) + P(A)$$
$$= .23 + .19 + .07 + .01$$
$$= .5$$

b. The probability that the student's score will be less than or equal to 500 is given by

$$P(D \cup E \cup F) = P(D) + P(E) + P(F)$$
$$= .23 + .31 + .19 = .73$$

c. The probability that the student's score will be greater than 400 but less than or equal to 600 is given by

$$P(C \cup D) = P(C) + P(D)$$
$$= .19 + .23 = .42$$

Property 3 holds if and only if E and F are mutually exclusive. In the general case, we have the following rule:

Property 4: Addition Rule

If E and F are any two events of an experiment, then

$$P(E \cup F) = P(E) + P(F) - P(E \cap F)$$

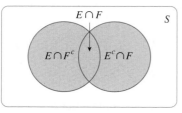

FIGURE 8
$E \cup F = (E \cap F^c) \cup (E \cap F) \cup (E^c \cap F)$

To derive this property, refer to Figure 8. Observe that we can write

$$E = (E \cap F^c) \cup (E \cap F) \quad \text{and} \quad F = (E^c \cap F) \cup (E \cap F)$$

as a union of disjoint sets. Therefore,

$$P(E) = P(E \cap F^c) + P(E \cap F) \quad \text{or} \quad P(E \cap F^c) = P(E) - P(E \cap F)$$

and

$$P(F) = P(E^c \cap F) + P(E \cap F) \quad \text{or} \quad P(E^c \cap F) = P(F) - P(E \cap F)$$

Finally, since $E \cup F = (E \cap F^c) \cup (E \cap F) \cup (E^c \cap F)$ is a union of disjoint sets, we have

$$
\begin{aligned}
P(E \cup F) &= P(E \cap F^c) + P(E \cap F) + P(E^c \cap F) \\
&= P(E) - P(E \cap F) + P(E \cap F) + P(F) - P(E \cap F) \qquad \text{Use the earlier results.} \\
&= P(E) + P(F) - P(E \cap F)
\end{aligned}
$$

Note Observe that when E and F are mutually exclusive—that is, when $E \cap F = \varnothing$—then the equation of Property 4 reduces to that of Property 3. In other words, if E and F are mutually exclusive events, then $P(E \cup F) = P(E) + P(F)$. If E and F are not mutually exclusive events, then $P(E \cup F) = P(E) + P(F) - P(E \cap F)$. ∎

EXAMPLE 2 A card is drawn from a well-shuffled deck of 52 playing cards. What is the probability that it is an ace or a spade?

Solution Let E denote the event that the card drawn is an ace and let F denote the event that the card drawn is a spade. Then,

$$
P(E) = \frac{4}{52} \quad \text{and} \quad P(F) = \frac{13}{52}
$$

Furthermore, E and F are not mutually exclusive events. In fact, $E \cap F$ is the event that the card drawn is an ace of spades. Consequently,

$$
P(E \cap F) = \frac{1}{52}
$$

The event that a card drawn is an ace or a spade is $E \cup F$, with probability given by

$$
\begin{aligned}
P(E \cup F) &= P(E) + P(F) - P(E \cap F) \\
&= \frac{4}{52} + \frac{13}{52} - \frac{1}{52} = \frac{16}{52} = \frac{4}{13}
\end{aligned}
$$

(Figure 9). This result, of course, can be obtained by arguing that 16 of the 52 cards are either spades or aces of other suits. ∎

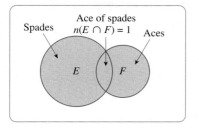

FIGURE 9
$P(E \cup F) = P(E) + P(F) - P(E \cap F)$

EXPLORE & DISCUSS

Let E, F, and G be any three events of an experiment. Use Formula (5) of Section 6.2 to show that

$$
\begin{aligned}
P(E \cup F \cup G) = {} & P(E) + P(F) + P(G) - P(E \cap F) - P(E \cap G) \\
& - P(F \cap G) + P(E \cap F \cap G)
\end{aligned}
$$

If E, F, and G are mutually exclusive, what is $P(E \cup F \cup G)$?

APPLIED EXAMPLE 3 Quality Control The quality-control department of Vista Vision, manufacturer of the Pulsar 19-inch color television set, has determined from records obtained from the company's service centers that 3% of the sets sold experience video problems, 1% experience audio problems, and 0.1% experience both video as well as audio problems before the expiration of the 90-day warranty. Find the probability that a set purchased by a consumer will experience video or audio problems before the warranty expires.

Solution   Let E denote the event that a set purchased will experience video problems within 90 days and let F denote the event that a set purchased will experience audio problems within 90 days. Then,

$$P(E) = .03 \qquad P(F) = .01 \qquad P(E \cap F) = .001$$

The event that a set purchased will experience video problems or audio problems before the warranty expires is $E \cup F$, and the probability of this event is given by

$$P(E \cup F) = P(E) + P(F) - P(E \cap F)$$
$$= .03 + .01 - .001$$
$$= .039$$

(Figure 10).

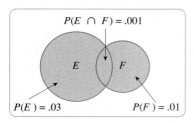

FIGURE 10
$P(E \cup F) =$
   $P(E) + P(F) - P(E \cap F)$

Another property of the probability function that is of considerable aid in computing probability follows.

Property 5: Rule of Complements
If E is an event of an experiment and E^c denotes the complement of E, then

$$P(E^c) = 1 - P(E)$$

Property 5 is an immediate consequence of Properties 2 and 3. Indeed, we have $E \cup E^c = S$ and $E \cap E^c = \varnothing$, so

$$1 = P(S) = P(E \cup E^c) = P(E) + P(E^c)$$

and, therefore,

$$P(E^c) = 1 - P(E)$$

APPLIED EXAMPLE 4   Refer to Example 3. What is the probability that a Pulsar 19-inch color television set bought by a consumer will not experience video or audio difficulties before the warranty expires?

Solution   Let E denote the event that a set bought by a consumer will experience video or audio difficulties before the warranty expires. Then, the event that the set will not experience either problem before the warranty expires is given by E^c, with probability

$$P(E^c) = 1 - P(E)$$
$$= 1 - .039$$
$$= .961$$

■ Computations Involving the Rules of Probability

We close this section by looking at two additional examples that illustrate the rules of probability.

EXAMPLE 5 Let E and F be two mutually exclusive events and suppose $P(E) = .1$ and $P(F) = .6$. Compute:

a. $P(E \cap F)$ **b.** $P(E \cup F)$ **c.** $P(E^c)$

d. $P(E^c \cap F^c)$ **e.** $P(E^c \cup F^c)$

Solution

a. Since the events E and F are mutually exclusive—that is, $E \cap F = \varnothing$—we have $P(E \cap F) = 0$.

b. $P(E \cup F) = P(E) + P(F)$ Since E and F are mutually exclusive

$= .1 + .6$

$= .7$

c. $P(E^c) = 1 - P(E)$ Property 5

$= 1 - .1$

$= .9$

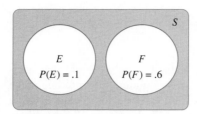

FIGURE 11
$P(E^c \cap F^c) = P[(E \cup F)^c]$

d. Observe that, by De Morgan's law, $E^c \cap F^c = (E \cup F)^c$, so

$$P(E^c \cap F^c) = P[(E \cup F)^c] \qquad \text{See Figure 11.}$$
$$= 1 - P(E \cup F) \qquad \text{Property 5}$$
$$= 1 - .7 \qquad \text{Use the result of part (b).}$$
$$= .3$$

e. Again, using De Morgan's law, we find

$$P(E^c \cup F^c) = P[(E \cap F)^c]$$
$$= 1 - P(E \cap F)$$
$$= 1 - 0 \qquad \text{Use the result of part (a).}$$
$$= 1$$

EXAMPLE 6   Let E and F be two events of an experiment with sample space S. Suppose $P(E) = .2$, $P(F) = .1$, and $P(E \cap F) = .05$. Compute:
a. $P(E \cup F)$
b. $P(E^c \cap F^c)$
c. $P(E^c \cap F)$ *Hint:* Draw a Venn diagram.

Solution
a. $P(E \cup F) = P(E) + P(F) - P(E \cap F)$ Property 4
$$= .2 + .1 - .05$$
$$= .25$$

b. Using De Morgan's law, we have

$$P(E^c \cap F^c) = P[(E \cup F)^c]$$
$$= 1 - P(E \cup F) \qquad \text{Property 5}$$
$$= 1 - .25 \qquad \text{Use the result of part (a).}$$
$$= .75$$

c. From the Venn diagram describing the relationship between E, F, and S (Figure 12), we have

$$P(E^c \cap F) = .05 \qquad \text{The shaded subset is the event } E^c \cap F.$$

This result may also be obtained by using the relationship

$$P(E^c \cap F) = P(F) - P(E \cap F)$$
$$= .1 - .05$$
$$= .05$$

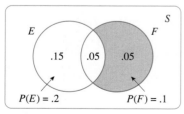

FIGURE 12
$P(E^c \cap F)$: the probability that the event F, but not the event E, will occur

as before.

7.3   Self-Check Exercises

1. Let E and F be events of an experiment with sample space S. Suppose $P(E) = .4$, $P(F) = .5$, and $P(E \cap F) = .1$. Compute:
   a. $P(E \cup F)$ **b.** $P(E \cap F^c)$

2. Susan Garcia wishes to sell or lease a condominium through a realty company. The realtor estimates that the probability of finding a buyer within a month of the date the property is listed for sale or lease is .3, the probability of finding a lessee

is .8, and the probability of finding both a buyer and a lessee is .1. Determine the probability that the property will be sold or leased within 1 mo from the date the property is listed for sale or lease.

Solutions to Self-Check Exercises 7.3 can be found on page 407.

7.3   Concept Questions

1. Suppose S is a sample space of an experiment, E and F are events of the experiment, and P is a probability function. Give the meaning of each of the following statements:
   a. $P(E) = 0$ **b.** $P(F) = 0.5$ **c.** $P(S) = 1$
   d. $P(E \cup F) = P(E) + P(F) - P(E \cap F)$

2. Give an example, based on a real-life situation, illustrating the property $P(E^c) = 1 - P(E)$, where E is an event and E^c is the complement of E.

7.3   Exercises

A pair of dice is cast, and the number that appears uppermost on each die is observed. In Exercises 1–6, refer to this experiment and find the probability of the event.

1. The sum of the numbers is an even number.

2. The sum of the numbers is either 7 or 11.

3. A pair of 1s is thrown.

4. A double is thrown.

5. One die shows a 6, and the other is a number less than 3.

6. The sum of the numbers is at least 4.

An experiment consists of selecting a card at random from a 52-card deck. In Exercises 7–12, refer to this experiment and find the probability of the event.

7. A king of diamonds is drawn.

8. A diamond or a king is drawn.

9. A face card is drawn.

10. A red face card is drawn.

11. An ace is not drawn.

12. A black face card is not drawn.

13. Five hundred people have purchased raffle tickets. What is the probability that a person holding one ticket will win the first prize? What is the probability that he or she will not win the first prize?

14. **TV HOUSEHOLDS**   The results of a recent television survey of American TV households revealed that 87 out of every 100 TV households have at least one remote control. What is the probability that a randomly selected TV household does not have at least one remote control?

In Exercises 15–24, explain why the statement is incorrect.

15. The sample space associated with an experiment is given by $S = \{a, b, c\}$, where $P(a) = .3$, $P(b) = .4$, and $P(c) = .4$.

16. The probability that a bus will arrive late at the Civic Center is .35, and the probability that it will be on time or early is .60.

17. A person participates in a weekly office pool in which he has one chance in ten of winning the purse. If he participates for 5 wk in succession, the probability of winning at least one purse is $\frac{5}{10}$.

18. The probability that a certain stock will increase in value over a period of 1 wk is .6. Therefore, the probability that the stock will decrease in value is .4.

19. A red die and a green die are tossed. The probability that a 6 will appear uppermost on the red die is $\frac{1}{6}$, and the probability that a 1 will appear uppermost on the green die is $\frac{1}{6}$. Hence, the probability that the red die will show a 6 or the green die will show a 1 is $\frac{1}{6} + \frac{1}{6}$.

20. Joanne, a high school senior, has applied for admission to four colleges, A, B, C, and D. She has estimated that the probability that she will be accepted for admission by A, B, C, and D is .5, .3, .1, and .08, respectively. Thus, the probability that she will be accepted for admission by at least one college is $P(A) + P(B) + P(C) + P(D) = .5 + .3 + .1 + .08 = .98$.

21. The sample space associated with an experiment is given by $S = \{a, b, c, d, e\}$. The events $E = \{a, b\}$ and $F = \{c, d\}$ are mutually exclusive. Hence, the events E^c and F^c are mutually exclusive.

22. A 5-card poker hand is dealt from a 52-card deck. Let A denote the event that a flush is dealt and let B be the event that a straight is dealt. Then the events A and B are mutually exclusive.

23. Mark Owens, an optician, estimates that the probability that a customer coming into his store will purchase one or more pairs of glasses but not contact lenses is .40, and the probability that he will purchase one or more pairs of contact lenses but not glasses is .25. Hence, Owens concludes that the probability that a customer coming into his store will purchase neither a pair of glasses nor a pair of contact lenses is .35.

24. There are eight grades in Garfield Elementary School. If a student is selected at random from the school, then the probability that the student is in the first grade is $\frac{1}{8}$.

25. Let E and F be two events that are mutually exclusive and suppose $P(E) = .2$ and $P(F) = .5$. Compute
a. $P(E \cap F)$　　　　　　**b.** $P(E \cup F)$
c. $P(E^c)$　　　　　　　　**d.** $P(E^c \cap F^c)$

26. Let E and F be two events of an experiment with sample space S. Suppose $P(E) = .6$, $P(F) = .4$, and $P(E \cap F) = .2$. Compute
a. $P(E \cup F)$　　　　　　**b.** $P(E^c)$
c. $P(F^c)$　　　　　　　　**d.** $P(E^c \cap F)$

27. Let $S = \{s_1, s_2, s_3, s_4\}$ be the sample space associated with an experiment having the probability distribution shown in the accompanying table. If $A = \{s_1, s_2\}$ and $B = \{s_1, s_3\}$, find
a. $P(A), P(B)$　　　　　　**b.** $P(A^c), P(B^c)$
c. $P(A \cap B)$　　　　　　**d.** $P(A \cup B)$

Outcome	Probability
s_1	$\frac{1}{8}$
s_2	$\frac{3}{8}$
s_3	$\frac{1}{4}$
s_4	$\frac{1}{4}$

28. Let $S = \{s_1, s_2, s_3, s_4, s_5, s_6\}$ be the sample space associated with an experiment having the probability distribution shown in the accompanying table. If $A = \{s_1, s_2\}$ and $B = \{s_1, s_5, s_6\}$, find
a. $P(A), P(B)$　　　　　　**b.** $P(A^c), P(B^c)$
c. $P(A \cap B)$　　　　　　**d.** $P(A \cup B)$
e. $P(A^c \cap B^c)$　　　　　**f.** $P(A^c \cup B^c)$

Outcome	Probability
s_1	$\frac{1}{3}$
s_2	$\frac{1}{8}$
s_3	$\frac{1}{6}$
s_4	$\frac{1}{6}$
s_5	$\frac{1}{12}$
s_6	$\frac{1}{8}$

29. Teacher Attitudes In a survey of 2140 teachers in a certain metropolitan area, conducted by a nonprofit organization regarding teacher attitudes, the following data were obtained:

900 said that lack of parental support is a problem.

890 said that abused or neglected children are problems.

680 said that malnutrition or students in poor health is a problem.

120 said that lack of parental support and abused or neglected children are problems.

110 said that lack of parental support and malnutrition or poor health are problems.

140 said that abused or neglected children and malnutrition or poor health are problems.

40 said that lack of parental support, abuse or neglect, and malnutrition or poor health are problems.

What is the probability that a teacher selected at random from this group said that lack of parental support is the only problem hampering a student's schooling?
Hint: Draw a Venn diagram.

30. Course Enrollments Among 500 freshmen pursuing a business degree at a university, 320 are enrolled in an Economics course, 225 are enrolled in a Mathematics course, and 140 are enrolled in both an Economics and a Mathematics course. What is the probability that a freshman selected at random from this group is enrolled in
a. An Economics and/or a Mathematics course?
b. Exactly one of these two courses?
c. Neither an Economics course nor a Mathematics course?

31. Consumer Surveys A leading manufacturer of kitchen appliances advertised its products in two magazines: *Good*

Housekeeping and the *Ladies Home Journal*. A survey of 500 customers revealed that 140 learned of its products from *Good Housekeeping*, 130 learned of its products from the *Ladies Home Journal*, and 80 learned of its products from both magazines. What is the probability that a person selected at random from this group saw the manufacturer's advertisement in

a. Both magazines?
b. At least one of the two magazines?
c. Exactly one magazine?

32. **ROLLOVER DEATHS**   The following table gives the number of people killed in rollover crashes in various types of vehicles in 2002:

Types of Vehicles	Cars	Pickups	SUVs	Vans
Deaths	4768	2742	2448	698

Find the empirical probability distribution associated with these data. If a fatality due to a rollover crash in 2002 is picked at random, what is the probability that the victim was in

a. A car?   **b.** An SUV?   **c.** A pickup or an SUV?

Source: National Highway Traffic Safety Administration

33. **INVESTMENT IN TECHNOLOGY**   One hundred sixty top regional executives were asked, "Do you plan to invest more or less in computers and information technology in the coming year?" The results of the poll follow:

Answer	Respondents
Less	27
The same	66
More	61
No answer	6

If one of the participants in the poll is selected at random, what is the probability that he or she said they would invest

a. More in computers and information technology in the coming year?
b. The same or less in computers and information technology in the coming year?

Source: Greater Boston Chamber of Commerce

34. **IN-FLIGHT SERVICE**   In a survey conducted in November 2002 of 1400 international business travelers concerning in-flight service over the past few years, the following information was obtained:

Comments on Quality of Service	Respondents
Has remained the same from two years ago.	630
Has diminished over that time frame.	406
Has improved over that time frame.	336
Weren't sure.	28

If a person in the survey is chosen at random, what is the probability that he or she has rated the in-flight service as

a. Remaining the same or improved over the time frame in question?
b. Remaining the same or diminished over the time frame in question?

Source: American Express

35. **SWITCHING JOBS**   Two hundred workers were asked, "Would a better economy lead you to switch jobs?" The results of the survey follow:

Answer	Very likely	Somewhat likely	Somewhat unlikely	Very unlikely	Don't know
Respondents	40	28	26	104	2

If a worker is chosen at random, what is the probability that he or she

a. Is very unlikely to switch jobs?
b. Is somewhat likely or very likely to switch jobs?

Source: Accountemps

36. **401(K) INVESTORS**   According to a study conducted in 2003 concerning the participation, by age, of 401(K) investors, the following data were obtained:

Age	20s	30s	40s	50s	60s
Percent	11	28	32	22	7

a. What percent of 401(K) investors are in their 20s or 60s?
b. What percent of 401(K) investors are under the age of 50?

Source: Investment Company Institute

37. **ASSET ALLOCATION**   When asked by a 30-yr-old bachelor for professional advice concerning the asset allocation of his $200,000 nest egg, the financial-planning firm of Sagemark Consulting suggested the following:

Asset A	Percent
U.S. mid/small cap stocks	22
U.S. large cap stocks	46
International stocks	17
Bonds/income	8
REITS	5
Cash	2

a. According to Sagemark Consulting, what percent of the bachelor's portfolio should be invested in U.S. large cap or U.S. mid/small cap stocks?
b. According to Sagemark Consulting, what percent of the bachelor's portfolio should be invested in vehicles other than bonds/incomes or REITS?

Source: Sagemark Consulting

38. **DOWNLOADING MUSIC**   The following table, compiled in 2004, gives the percent of music downloaded from the United States and other countries by U.S. users.

Country	U.S.	Germany	Canada	Italy	U.K.	France	Japan	Other
Percent	45.1	16.5	6.9	6.1	4.2	3.8	2.5	14.9

a. Verify that the table does give a probability distribution for the experiment.

b. What is the probability that a user who downloads music, selected at random, obtained it from either the United States or Canada?

c. What is the probability that a U.S. user who downloads music, selected at random, does not obtain it from Italy, the United Kingdom (U.K.), or France?

Source: Felix Oberholtzer-Gee and Koleman Strumpf

39. **ASSEMBLY-TIME STUDIES**   A time study was conducted by the production manager of Universal Instruments to determine how much time it took an assembly worker to complete a certain task during the assembly of its Galaxy home computers. Results of the study indicated that 20% of the workers were able to complete the task in less than 3 min, 60% of the workers were able to complete the task in 4 min or less, and 10% of the workers required more than 5 min to complete the task. If an assembly-line worker is selected at random from this group, what is the probability that:

a. He or she will be able to complete the task in 5 min or less?

b. He or she will not be able to complete the task within 4 min?

c. The time taken for the worker to complete the task will be between 3 and 4 min (inclusive)?

40. **PLANS TO KEEP CARS**   In a survey conducted to see how long Americans keep their cars, 2000 automobile owners were asked how long they plan to keep their present cars. The results of the survey follow:

Years Car Is Kept, x	Respondents
$0 \le x < 1$	60
$1 \le x < 3$	440
$3 \le x < 5$	360
$5 \le x < 7$	340
$7 \le x < 10$	240
$x \ge 10$	560

Find the probability distribution associated with these data. What is the probability that an automobile owner selected at random from those surveyed plans to keep his or her present car

a. Less than five years?

b. Three or more years?

41. **GUN-CONTROL LAWS**   A poll was conducted among 250 residents of a certain city regarding tougher gun-control laws. The results of the poll are shown in the table:

	Own Only a Handgun	Own Only a Rifle	Own a Handgun and a Rifle	Own Neither	Total
Favor Tougher Laws	0	12	0	138	150
Oppose Tougher Laws	58	5	25	0	88
No Opinion	0	0	0	12	12
Total	58	17	25	150	250

If one of the participants in this poll is selected at random, what is the probability that he or she

a. Favors tougher gun-control laws?

b. Owns a handgun?

c. Owns a handgun but not a rifle?

d. Favors tougher gun-control laws and does not own a handgun?

42. **RISK OF AN AIRPLANE CRASH**   According to a study of Western-built commercial jets involved in crashes from 1988 to 1998, the percent of airplane crashes that occur at each stage of flight are as follows:

Phase	Percent
On ground, taxiing	4
During takeoff	10
Climbing to cruise altitude	19
En route	5
Descent and approach	31
Landing	31

If one of the doomed flights in the period 1988–1998 is picked at random, what is the probability that it crashed

a. While taxiing on the ground or while en route?

b. During takeoff or landing?

If the study is indicative of airplane crashes in general, when is the risk of a plane crash the highest?

Source: National Transportation and Safety Board

43. Suppose the probability that Bill can solve a problem is p_1 and the probability that Mike can solve it is p_2. Show that the probability that Bill and Mike working independently can solve the problem is $p_1 + p_2 - p_1 p_2$.

44. Fifty raffle tickets are numbered 1 through 50, and one of them is drawn at random. What is the probability that the number is a multiple of 5 or 7? Consider the following "solution": Since 10 tickets bear numbers that are multiples of 5 and 7 tickets bear numbers that are multiples of 7, we conclude that the required probability is

$$\frac{10}{50} + \frac{7}{50} = \frac{17}{50}$$

What is wrong with this argument? What is the correct answer?

In Exercises 45–48, determine whether the statement is true or false. If it is true, explain why it is true. If it is false, give an example to show why it is false.

45. If A is a subset of B and $P(B) = 0$, then $P(A) = 0$.

46. If A is a subset of B, then $P(A) \le P(B)$.

47. If $E_1, E_2, \ldots, E_n$ are events of an experiment, then $P(E_1 \cup E_2 \cup \cdots \cup E_n) = P(E_1) + P(E_2) + \cdots + P(E_n)$.

48. If E is an event of an experiment, then $P(E) + P(E^c) = 1$.

7.3 Solutions to Self-Check Exercises

1. a. Using Property 4, we find

$$P(E \cup F) = P(E) + P(F) - P(E \cap F)$$
$$= .4 + .5 - .1$$
$$= .8$$

b. From the accompanying Venn diagram, in which the subset $E \cap F^c$ is shaded, we see that

$$P(E \cap F^c) = .3$$

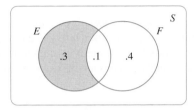

The result may also be obtained by using the relationship

$$P(E \cap F^c) = P(E) - P(E \cap F)$$
$$= .4 - .1 = .3$$

2. Let E denote the event that the property will be sold within 1 mo of the date it is listed for sale or lease and let F denote the event that the property will be leased within the same time period. Then,

$$P(E) = .3 \qquad P(F) = .8 \qquad P(E \cap F) = .1$$

The probability of the event that the property will be sold or leased within 1 mo of the date it is listed for sale or lease is given by

$$P(E \cup F) = P(E) + P(F) - P(E \cap F)$$
$$= .3 + .8 - .1 = 1$$

—that is, a certainty.

7.4 Use of Counting Techniques in Probability

▬ Further Applications of Counting Techniques

As we have seen many times before, a problem in which the underlying sample space has a small number of elements may be solved by first determining all such sample points. For problems involving sample spaces with a large number of sample points, however, this approach is neither practical nor desirable.

In this section, we see how the counting techniques studied in Chapter 6 may be employed to help us solve problems in which the associated sample spaces contain large numbers of sample points. In particular, we restrict our attention to the study of uniform sample spaces—that is, sample spaces in which the outcomes are equally likely. For such spaces we have the following result:

Computing the Probability of an Event in a Uniform Sample Space
Let S be a uniform sample space and let E be any event. Then,

$$P(E) = \frac{\text{Number of favorable outcomes in } E}{\text{Number of possible outcomes in } S} = \frac{n(E)}{n(S)} \qquad \text{(1)}$$

EXAMPLE 1   An unbiased coin is tossed six times. What is the probability that the coin will land heads
  a. Exactly three times?
  b. At most three times?
  c. On the first and the last toss?

Solution

a. Each outcome of the experiment may be represented as a sequence of heads and tails. Using the generalized multiplication principle, we see that the number of outcomes of this experiment is given by 2^6, or 64. Let E denote the event that the coin lands heads exactly three times. Since there are $C(6, 3)$ ways this can occur, we see that the required probability is

$$P(E) = \frac{n(E)}{n(S)} = \frac{C(6, 3)}{64} = \frac{\frac{6!}{3!3!}}{64} \qquad \text{\small S, sample space of the experiment}$$

$$= \frac{\frac{6 \cdot 5 \cdot 4}{3 \cdot 2}}{64} = \frac{20}{64} = \frac{5}{16} = .3125$$

b. Let F denote the event that the coin lands heads at most three times. Then $n(F)$ is given by the sum of the number of ways the coin lands heads zero times (no heads!), the number of ways it lands heads exactly once, the number of ways it lands heads exactly twice, and the number of ways it lands heads exactly three times. That is,

$$n(F) = C(6, 0) + C(6, 1) + C(6, 2) + C(6, 3)$$

$$= \frac{6!}{0!6!} + \frac{6!}{1!5!} + \frac{6!}{2!4!} + \frac{6!}{3!3!}$$

$$= 1 + 6 + \frac{(6)(5)}{2} + \frac{(6)(5)(4)}{(3)(2)} = 42$$

Therefore, the required probability is

$$P(F) = \frac{n(F)}{n(S)} = \frac{42}{64} = \frac{21}{32} \approx .66$$

c. Let F denote the event that the coin lands heads on the first and the last toss. Then $n(F) = 1 \cdot 2 \cdot 2 \cdot 2 \cdot 2 \cdot 1 = 2^4$, so the probability that this event occurs is

$$P(F) = \frac{2^4}{2^6}$$

$$= \frac{1}{2^2}$$

$$= \frac{1}{4}$$

EXAMPLE 2   Two cards are selected at random from a well-shuffled pack of 52 playing cards. What is the probability that
a. They are both aces?   **b.** Neither of them is an ace?

Solution

a. The experiment consists of selecting 2 cards from a pack of 52 playing cards. Since the order in which the cards are selected is immaterial, the sample points are combinations of 52 cards taken 2 at a time. Now, there are $C(52, 2)$ ways of selecting 52 cards taken 2 at a time, so the number of elements in the sample space S is given by $C(52, 2)$. Next, we observe that there are $C(4, 2)$ ways of selecting 2 aces from the 4 in the deck. Therefore, if E denotes the event that the cards selected are both aces, then

$$P(E) = \frac{n(E)}{n(S)}$$

$$= \frac{C(4, 2)}{C(52, 2)} = \frac{\dfrac{4!}{2!2!}}{\dfrac{52!}{2!50!}}$$

$$= \frac{1}{221}$$

b. Let F denote the event that neither of the two cards selected is an ace. Since there are $C(48, 2)$ ways of selecting two cards, neither of which is an ace, we find that

$$P(F) = \frac{n(F)}{n(S)} = \frac{C(48, 2)}{C(52, 2)} = \frac{\dfrac{48!}{2!46!}}{\dfrac{52!}{2!50!}} = \frac{48 \cdot 47}{2} \cdot \frac{2}{52 \cdot 51}$$

$$= \frac{188}{221}$$

APPLIED EXAMPLE 3   Quality Control   A bin in the hi-fi department of Building 20, a bargain outlet, contains 100 blank cassette tapes, of which 10 are known to be defective. If a customer selects 6 of these cassette tapes, determine the probability

a. That 2 of them are defective.

b. That at least 1 of them is defective.

Solution

a. There are $C(100, 6)$ ways of selecting a set of 6 cassette tapes from the 100, and this gives $n(S)$, the number of outcomes in the sample space associated with the experiment. Next, we observe that there are $C(10, 2)$ ways of selecting a set of 2 defective cassette tapes from the 10 defective cassette tapes and $C(90, 4)$ ways of selecting a set of 4 nondefective cassette tapes from the 90 nondefective cassette tapes (Figure 13). Thus, by the multiplication principle, there are $C(10, 2) \cdot C(90, 4)$ ways of selecting 2 defective and 4 nondefective cassette tapes. Therefore, the probability of selecting 6 cassette tapes, of which 2 are defective, is given by

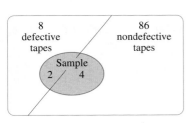

FIGURE 13
A sample of 6 tapes selected from 90 nondefective tapes and 10 defective tapes

$$\frac{C(10, 2) \cdot C(90, 4)}{C(100, 6)} = \frac{\dfrac{10!}{2!8!} \dfrac{90!}{4!86!}}{\dfrac{100!}{6!94!}}$$

$$= \frac{10 \cdot 9}{2} \cdot \frac{90 \cdot 89 \cdot 88 \cdot 87}{4 \cdot 3 \cdot 2 \cdot 1} \cdot \frac{6 \cdot 5 \cdot 4 \cdot 3 \cdot 2 \cdot 1}{100 \cdot 99 \cdot 98 \cdot 97 \cdot 96 \cdot 95}$$

$$\approx .096$$

b. Let E denote the event that none of the cassette tapes selected is defective. Then E^c gives the event that at least 1 of the cassette tapes is defective. But, by the rule of complements,

$$P(E^c) = 1 - P(E)$$

To compute $P(E)$, we observe that there are $C(90, 6)$ ways of selecting a set of 6 cassette tapes that are nondefective. Therefore,

$$P(E) = \frac{C(90, 6)}{C(100, 6)}$$

$$P(E^c) = 1 - \frac{C(90, 6)}{C(100, 6)}$$

$$= 1 - \frac{\dfrac{90!}{6!84!}}{\dfrac{100!}{6!94!}}$$

$$= 1 - \frac{90 \cdot 89 \cdot 88 \cdot 87 \cdot 86 \cdot 85}{6 \cdot 5 \cdot 4 \cdot 3 \cdot 2 \cdot 1} \cdot \frac{6 \cdot 5 \cdot 4 \cdot 3 \cdot 2 \cdot 1}{100 \cdot 99 \cdot 98 \cdot 97 \cdot 96 \cdot 95}$$

$$\approx .48$$

The Birthday Problem

EXAMPLE 4   A group of five people is selected at random. What is the probability that at least two of them have the same birthday?

Solution   For simplicity, we assume that none of the five people was born on February 29 of a leap year. Since the five people were selected at random, we may also assume that each of them is equally likely to have any of the 365 days of a year as his or her birthday. If we let A, B, C, D, and F represent the five people, then an outcome of the experiment may be represented by (a, b, c, d, f), where the numbers a, b, c, d, and f give the birthdays of A, B, C, D, and F, respectively.

We first observe that since there are 365 possibilities for each of the dates a, b, c, d, and f, the multiplication principle implies that there are

$$\boxed{365} \cdot \boxed{365} \cdot \boxed{365} \cdot \boxed{365} \cdot \boxed{365}$$
$$\quad a \qquad\quad b \qquad\quad c \qquad\quad d \qquad\quad f$$

or 365^5 outcomes of the experiment. Therefore,

$$n(S) = 365^5$$

where S denotes the sample space of the experiment.

Next, let E denote the event that two or more of the five people have the same birthday. It is now necessary to compute $P(E)$. However, a direct computation of $P(E)$ is relatively difficult. It is much easier to compute $P(E^c)$, where E^c is the event that no two of the five people have the same birthday, and then use the relation

$$P(E) = 1 - P(E^c)$$

To compute $P(E^c)$, observe that there are 365 ways (corresponding to the 365 dates) on which A's birthday can occur, followed by 364 ways on which B's birthday could occur if B were not to have the same birthday as A, and so on. Therefore, by the generalized multiplication principle,

$$n(E^c) = \underset{\substack{\text{A's}\\\text{birthday}}}{365} \cdot \underset{\substack{\text{B's}\\\text{birthday}}}{364} \cdot \underset{\substack{\text{C's}\\\text{birthday}}}{363} \cdot \underset{\substack{\text{D's}\\\text{birthday}}}{362} \cdot \underset{\substack{\text{F's}\\\text{birthday}}}{361}$$

Thus,

$$P(E^c) = \frac{n(E^c)}{n(S)}$$

$$= \frac{365 \cdot 364 \cdot 363 \cdot 362 \cdot 361}{365^5}$$

$$P(E) = 1 - P(E^c)$$

$$= 1 - \frac{365 \cdot 364 \cdot 363 \cdot 362 \cdot 361}{365^5}$$

$$\approx .027$$

We can extend the result obtained in Example 4 to the general case involving r people. In fact, if E denotes the event that at least two of the r people have the same birthday, an argument similar to that used in Example 4 leads to the result

$$P(E) = 1 - \frac{365 \cdot 364 \cdot 363 \cdot \cdots \cdot (365 - r + 1)}{365^r}$$

By letting r take on the values 5, 10, 15, 20, . . . , 50, in turn, we obtain the probabilities that at least 2 of 5, 10, 15, 20, . . . , 50 people, respectively, have the same birthday. These results are summarized in Table 8.

The results show that in a group of 23 randomly selected people the chances are greater than 50% that at least 2 of them will have the same birthday. In a group of 50 people, it is an excellent bet that at least 2 people in the group will have the same birthday.

TABLE 8

Probability That at Least Two People in a Randomly Selected Group of r People Have the Same Birthday

r	$P(E)$
5	.027
10	.117
15	.253
20	.411
22	.476
23	.507
25	.569
30	.706
40	.891
50	.970

EXPLORE & DISCUSS

During an episode of the *Tonight Show*, Johnny Carson related "The Birthday Problem" to the audience—noting that, in a group of 50 or more people, probabilists have calculated that the probability of at least 2 people having the same birthday is very high. To illustrate this point, he proceeded to conduct his own experiment. A person selected at random from the audience was asked to state his birthday. Carson then asked if anyone in the audience had the same birthday. The response was negative. He repeated the experiment. Once again, the response was negative. These results, observed Carson, were contrary to expectations. In a later episode of the show, Carson explained why this experiment had been improperly conducted. Explain why Carson failed to illustrate the point he was trying to make in the earlier episode.

7.4   Self-Check Exercises

1. Four balls are selected at random without replacement from an urn containing 10 white balls and 8 red balls. What is the probability that all the chosen balls are white?

2. A box contains 20 microchips, of which 4 are substandard. If 2 of the chips are taken from the box, what is the probability that they are both substandard?

Solutions to Self-Check Exercises 7.4 can be found on page 414.

7.4   Concept Questions

1. What is the probability of an event E in a uniform sample space S?

2. Suppose we want to find the probability that at least two people in a group of six randomly selected people have the same birthday.

   a. If S denotes the sample space of this experiment, what is $n(S)$?

   b. If E is the event that two or more of the six people in the group have the same birthday, explain how you would use $P(E^c)$ to determine $P(E)$.

7.4   Exercises

An unbiased coin is tossed five times. In Exercises 1–4, find the probability of the event.

1. The coin lands heads all five times.

2. The coin lands heads exactly once.

3. The coin lands heads at least once.

4. The coin lands heads more than once.

Two cards are selected at random without replacement from a well-shuffled deck of 52 playing cards. In Exercises 5–8, find the probability of the event.

5. A pair is drawn.   6. A pair is not drawn.

7. Two black cards are drawn.

8. Two cards of the same suit are drawn.

Four balls are selected at random without replacement from an urn containing three white balls and five blue balls. In Exercises 9–12, find the probability of the event.

9. Two of the balls are white, and two are blue.

10. All of the balls are blue.

11. Exactly three of the balls are blue.

12. Two or three of the balls are white.

Assume that the probability of a boy being born is the same as the probability of a girl being born. In Exercises 13–16, find the probability that a family with three children will have the given composition.

13. Two boys and one girl

14. At least one girl

15. No girls

16. The two oldest children are girls.

17. An exam consists of ten true-or-false questions. If a student guesses at every answer, what is the probability that he or she will answer exactly six questions correctly?

18. PERSONNEL SELECTION Jacobs & Johnson, an accounting firm, employs 14 accountants, of whom 8 are CPAs. If a delegation of 3 accountants is randomly selected from the firm to attend a conference, what is the probability that 3 CPAs will be selected?

19. QUALITY CONTROL Two light bulbs are selected at random from a lot of 24, of which 4 are defective. What is the probability that

    a. Both of the light bulbs are defective?

    b. At least 1 of the light bulbs is defective?

20. A customer at Cavallaro's Fruit Stand picks a sample of 3 oranges at random from a crate containing 60 oranges, of which 4 are rotten. What is the probability that the sample contains 1 or more rotten oranges?

21. QUALITY CONTROL A shelf in the Metro Department Store contains 80 colored ink cartridges for a popular ink-jet printer. Six of the cartridges are defective. If a customer selects 2 of these cartridges at random from the shelf, what is the probability that

    a. Both are defective?

    b. At least 1 is defective?

22. QUALITY CONTROL Electronic baseball games manufactured by Tempco Electronics are shipped in lots of 24. Before shipping, a quality-control inspector randomly selects a sample of 8 from each lot for testing. If the sample contains any defective games, the entire lot is rejected. What is the probability that a lot containing exactly 2 defective games will still be shipped?

23. PERSONNEL SELECTION The City Transit Authority plans to hire 12 new bus drivers. From a group of 100 qualified applicants, of which 60 are men and 40 are women, 12 names are to be selected by lot. Suppose that Mary and John Lewis are among the 100 qualified applicants.

a. What is the probability that Mary's name will be selected? That both Mary's and John's names will be selected?

b. If it is stipulated that an equal number of men and women are to be selected (6 men from the group of 60 men and 6 women from the group of 40 women), what is the probability that Mary's name will be selected? That Mary's and John's names will be selected?

24. PUBLIC HOUSING The City Housing Authority has received 50 applications from qualified applicants for eight low-income apartments. Three of the apartments are on the north side of town, and five are on the south side. If the apartments are to be assigned by means of a lottery, what is the probability that
    a. A specific qualified applicant will be selected for one of these apartments?
    b. Two specific qualified applicants will be selected for apartments on the same side of town?

25. A student studying for a vocabulary test knows the meanings of 12 words from a list of 20 words. If the test contains 10 words from the study list, what is the probability that at least 8 of the words on the test are words that the student knows?

26. DRIVERS' TESTS Four different written driving tests are administered by the Motor Vehicle Department. One of these four tests is selected at random for each applicant for a driver's license. If a group consisting of two women and three men apply for a license, what is the probability that
    a. Exactly two of the five will take the same test?
    b. The two women will take the same test?

27. BRAND SELECTION A druggist wishes to select three brands of aspirin to sell in his store. He has five major brands to choose from: A, B, C, D, and E. If he selects the three brands at random, what is the probability that he will select
    a. Brand B?
    b. Brands B and C?
    c. At least one of the two brands B and C?

28. BLACKJACK In the game of blackjack, a 2-card hand consisting of an ace and a face card or a 10 is called a blackjack.
    a. If a player is dealt 2 cards from a standard deck of 52 well-shuffled cards, what is the probability that the player will receive a blackjack?
    b. If a player is dealt 2 cards from 2 well-shuffled standard decks, what is the probability that the player will receive a blackjack?

29. SLOT MACHINES Refer to Exercise 25, Section 6.3, where the "lucky dollar" slot machine was described. What is the probability that the three "lucky dollar" symbols will appear in the window of the slot machine?

30. ROULETTE In 1959 a world record was set for the longest run on an ungaffed (fair) roulette wheel at the El San Juan Hotel in Puerto Rico. The number 10 appeared six times in a row.

What is the probability of the occurrence of this event? (Assume that there are 38 equally likely outcomes consisting of the numbers 1–36, 0, and 00.)

In "The Numbers Game," a state lottery, four numbers are drawn with replacement from an urn containing the digits 0–9, inclusive. In Exercises 31–34, find the probability of a ticket holder having the indicated winning ticket.

31. All four digits in exact order (the grand prize)

32. Two specified, consecutive digits in exact order (the first two digits, the middle two digits, or the last two digits)

33. One specified digit in exact order (the first, second, third, or fourth digit)

34. All four digits in any order (including the other winning tickets)

A list of poker hands ranked in order from the highest to the lowest is shown in the accompanying table along with a description and example of each hand. Use the table to answer Exercises 35–40.

Hand	Description	Example
Straight flush	5 cards in sequence in the same suit	A ♥ 2 ♥ 3 ♥ 4 ♥ 5 ♥
Four of a kind	4 cards of the same rank and any other card	K ♥ K ♦ K ♠ K ♣ 2 ♥
Full house	3 of a kind and a pair	3 ♥ 3 ♦ 3 ♣ 7 ♥ 7 ♦
Flush	5 cards of the same suit that are not all in sequence	5 ♥ 6 ♥ 9 ♥ J ♥ K ♥
Straight	5 cards in sequence but not all of the same suit	10 ♥ J ♦ Q ♣ K ♠ A ♥
Three of a kind	3 cards of the same rank and 2 unmatched cards	K ♥ K ♦ K ♠ 2 ♥ 4 ♦
Two pair	2 cards of the same rank and 2 cards of any other rank with an unmatched card	K ♥ K ♦ 2 ♥ 2 ♠ 4 ♣
One pair	2 cards of the same rank and 3 unmatched cards	K ♥ K ♦ 5 ♥ 2 ♠ 4 ♥

If a 5-card poker hand is dealt from a well-shuffled deck of 52 cards, what is the probability of being dealt the given hand?

35. A straight flush (Note that an ace may be played as either a high or low card in a straight sequence—that is, A, 2, 3, 4, 5 or 10, J, Q, K, A. Hence, there are ten possible sequences for a straight in one suit.)

36. A straight (but not a straight flush)

37. A flush (but not a straight flush)

38. Four of a kind

39. A full house

40. Two pairs

41. Zodiac Signs   There are 12 signs of the Zodiac: Aries, Taurus, Gemini, Cancer, Leo, Virgo, Libra, Scorpio, Sagittarius, Capricorn, Aquarius, and Pisces. Each sign corresponds to a different calendar period of approximately 1 mo. Assuming that a person is just as likely to be born under one sign as another, what is the probability that in a group of five people at least two of them

a. Have the same sign?

b. Were born under the sign of Aries?

42. Birthday Problem   What is the probability that at least two of the nine justices of the U.S. Supreme Court have the same birthday?

43. Birthday Problem   Fifty people are selected at random. What is the probability that none of the people in this group have the same birthday?

44. Birthday Problem   There were 42 different presidents of the United States from 1789 through 2000. What is the probability that at least two of them had the same birthday? Compare your calculation with the facts by checking an almanac or some other source.

7.4 Solutions to Self-Check Exercises

1. The probability that all 4 balls selected are white is given by

$$\frac{\text{The number of ways of selecting 4 white balls from the 10 in the urn}}{\text{The number of ways of selecting any 4 balls from the 18 balls in the urn}}$$

$$= \frac{C(10, 4)}{C(18, 4)}$$

$$= \frac{\dfrac{10!}{4!6!}}{\dfrac{18!}{4!14!}}$$

$$= \frac{10 \cdot 9 \cdot 8 \cdot 7}{4 \cdot 3 \cdot 2} \cdot \frac{4 \cdot 3 \cdot 2}{18 \cdot 17 \cdot 16 \cdot 15}$$

$$\approx .069$$

2. The probability that both chips are substandard is given by

$$\frac{\text{The number of ways of choosing any 2 of the 4 substandard chips}}{\text{The number of ways of choosing any 2 of the 20 chips}}$$

$$= \frac{C(4, 2)}{C(20, 2)}$$

$$= \frac{\dfrac{4!}{2!2!}}{\dfrac{20!}{2!18!}}$$

$$= \frac{4 \cdot 3}{2} \cdot \frac{2}{20 \cdot 19}$$

$$\approx .032$$

7.5 Conditional Probability and Independent Events

■ Conditional Probability

Three cities, A, B, and C, are vying to play host to the Summer Olympic Games in 2008. If each city has the same chance of winning the right to host the Games, then the probability of city A hosting the Games is $\frac{1}{3}$. Suppose city B then decides to pull out of contention because of fiscal problems. Then it would seem that city A's chances of playing host will increase. In fact, if each of the two remaining cities have equal chances of winning, then the probability of city A playing host to the Games is $\frac{1}{2}$.

In general, the probability of an event is affected by the occurrence of other events and/or by the knowledge of information relevant to the event. Basically, the injection of conditions into a problem modifies the underlying sample space of the original problem. This in turn leads to a change in the probability of the event.

> **EXAMPLE 1**   Two cards are drawn without replacement from a well-shuffled deck of 52 playing cards.
> **a.** What is the probability that the first card drawn is an ace?
> **b.** What is the probability that the second card drawn is an ace given that the first card drawn was not an ace?
> **c.** What is the probability that the second card drawn is an ace given that the first card drawn was an ace?

Solution
a. The sample space here consists of 52 equally likely outcomes, 4 of which are aces. Therefore, the probability that the first card drawn is an ace is $\frac{4}{52}$.
b. Having drawn the first card, there are 51 cards left in the deck. In other words, for the second phase of the experiment, we are working in a *reduced* sample space. If the first card drawn was not an ace, then this modified sample space of 51 points contains 4 "favorable" outcomes (the 4 aces), so the probability that the second card drawn is an ace is given by $\frac{4}{51}$.
c. If the first card drawn was an ace, then there are 3 aces left in the deck of 51 playing cards, so the probability that the second card drawn is an ace is given by $\frac{3}{51}$. ∎

Observe that in Example 1 the occurrence of the first event reduces the size of the original sample space. The information concerning the first card drawn also leads us to the consideration of modified sample spaces: In part (b) the deck contained 4 aces, and in part (c) the deck contained 3 aces.

The probability found in part (b) or (c) of Example 1 is known as a **conditional probability,** since it is the probability of an event occurring given that another event has already occurred. For example, in part (b) we computed the probability of the event that the second card drawn is an ace, given the event that the first card drawn was not an ace. In general, given two events A and B of an experiment, one may, under certain circumstances, compute the probability of the event B given that the event A has already occurred. This probability, denoted by $P(B|A)$, is called the **conditional probability of B given A.**

A formula for computing the conditional probability of B given A may be discovered with the aid of a Venn diagram. Consider an experiment with a uniform sample space S and suppose A and B are two events of the experiment (Figure 14).

The condition that the event A has occurred tells us that the possible outcomes of the experiment in the second phase are restricted to those outcomes (elements) in the set A. In other words, we may work with the reduced sample space A instead of the original sample space S in the experiment. Next, we observe that, with respect to the reduced sample space A, the outcomes favorable to the event B are precisely those elements in the set $A \cap B$. Consequently, the conditional probability of B given A is

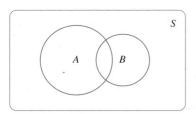

(a) Original sample space

(b) Reduced sample space A. The shaded area is $A \cap B$.

FIGURE 14

$$P(B|A) = \frac{\text{Number of elements in } A \cap B}{\text{Number of elements in } A}$$

$$= \frac{n(A \cap B)}{n(A)} \qquad n(A) \neq 0$$

Dividing the numerator and the denominator by $n(S)$, the number of elements in S, we have

$$P(B|A) = \frac{\dfrac{n(A \cap B)}{n(S)}}{\dfrac{n(A)}{n(S)}}$$

which is equivalent to the following formula:

> **Conditional Probability of an Event**
>
> If A and B are events in an experiment and $P(A) \neq 0$, then the conditional probability that the event B will occur given that the event A has already occurred is
>
> $$P(B|A) = \frac{P(A \cap B)}{P(A)} \qquad (2)$$

EXAMPLE 2   A pair of fair dice is cast. What is the probability that the sum of the numbers falling uppermost is 7 if it is known that one of the numbers is a 5?

Solution   Let A denote the event that the sum of the numbers falling uppermost is 7 and let B denote the event that one of the numbers is a 5. From the results of Example 4, Section 7.1, we find that

$$A = \{(6, 1), (5, 2), (4, 3), (3, 4), (2, 5), (1, 6)\}$$
$$B = \{(5, 1), (5, 2), (5, 3), (5, 4), (5, 5), (5, 6),$$
$$(1, 5), (2, 5), (3, 5), (4, 5), (6, 5)\}$$

so that

$$A \cap B = \{(5, 2), (2, 5)\}$$

(Figure 15). Since the dice are fair, each outcome of the experiment is equally likely; therefore,

$$P(A \cap B) = \frac{2}{36} \quad \text{and} \quad P(B) = \frac{11}{36} \qquad \text{Recall that } n(S) = 36.$$

Thus, the probability that the sum of the numbers falling uppermost is 7 given that one of the numbers is a 5 is, by virtue of Equation (2),

$$P(A|B) = \frac{\dfrac{2}{36}}{\dfrac{11}{36}} = \frac{2}{11}$$

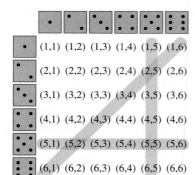

FIGURE 15
$A \cap B = \{(5, 2), (2, 5)\}$

APPLIED EXAMPLE 3   Color Blindness   In a test conducted by the U.S. Army, it was found that of 1000 new recruits, 600 men and 400 women, 50 of the men and 4 of the women were red-green color-blind. Given that a recruit selected at random from this group is red-green color-blind, what is the probability that the recruit is a male?

Solution Let C denote the event that a randomly selected subject is red-green color-blind and let M denote the event that the subject is a male recruit. Since 54 out of the 1000 subjects are color-blind, we may take

$$P(C) = \frac{54}{1000} = .054$$

Therefore, by Equation (2), the probability that a subject is male given that the subject is red-green color-blind is

$$P(M|C) = \frac{P(M \cap C)}{P(C)}$$

$$= \frac{.05}{.054} = \frac{25}{27}$$

EXPLORE & DISCUSS

Let A and B be events in an experiment, and $P(A) \neq 0$. In n trials the event A occurs m times, the event B occurs k times, and the events A and B occur together l times.

1. Explain why it makes good sense to call the ratio l/m the conditional relative frequency of the event B given the event A.

2. Show that the relative frequencies l/m, m/n, and l/n satisfy the equation

$$\frac{l}{m} = \frac{\dfrac{l}{n}}{\dfrac{m}{n}}$$

3. Explain why the result of part 2 suggests that Formula (2)

$$P(B|A) = \frac{P(A \cap B)}{P(A)} \qquad [P(A) \neq 0]$$

is plausible.

In certain problems, the probability of an event B occurring given that A has occurred, written $P(B|A)$, is known, and we wish to find the probability of A *and* B occurring. The solution to such a problem is facilitated by the use of the following formula:

Product Rule

$$P(A \cap B) = P(A) \cdot P(B|A) \tag{3}$$

This formula is obtained from (2) by multiplying both sides of the equation by $P(A)$. We illustrate the use of the product rule in the next several examples.

 APPLIED EXAMPLE 4 Seniors with Driver's Licenses There are 300 seniors in Jefferson High School, of which 140 are males. It is known that 80% of the males and 60% of the females have their driver's license. If a student is selected at random from this senior class, what is the probability that the student is

a. A male and has a driver's license?

b. A female who does not have a driver's license?

Solution

a. Let M denote the event that the student is a male and let D denote the event that the student has a driver's license. Then,

$$P(M) = \frac{140}{300} \quad \text{and} \quad P(D|M) = .8$$

Now, the event that the student selected at random is a male and has a driver's license is $M \cap D$, and, by the product rule, the probability of this event occurring is given by

$$P(M \cap D) = P(M) \cdot P(D|M)$$

$$= \left(\frac{140}{300}\right)(.8) = \frac{28}{75}$$

b. Let F denote the event that the student is a female. Then D^c is the event that the student does not have a driver's license. We have

$$P(F) = \frac{160}{300} \quad \text{and} \quad P(D^c|F) = 1 - .6 = .4$$

Note that we have used the rule of complements in the computation of $P(D^c|F)$. Now, the event that the student selected at random is a female and does not have a driver's license is $F \cap D^c$, so by the product rule, the probability of this event occurring is given by

$$P(F \cap D^c) = P(F) \cdot P(D^c|F)$$

$$= \left(\frac{160}{300}\right)(.4) = \frac{16}{75}$$
■

EXAMPLE 5 Two cards are drawn without replacement from a well-shuffled deck of 52 playing cards. What is the probability that the first card drawn is an ace and the second card drawn is a face card?

Solution Let A denote the event that the first card drawn is an ace and let F denote the event that the second card drawn is a face card. Then $P(A) = \frac{4}{52}$. After drawing the first card, there are 51 cards left in the deck, of which 12 are face cards. Therefore, the probability of drawing a face card given that the first card drawn was an ace is given by

$$P(F|A) = \frac{12}{51}$$

By the product rule, the probability that the first card drawn is an ace and the second card drawn is a face card is given by

$$P(A \cap F) = P(A) \cdot P(F|A)$$

$$= \left(\frac{4}{52}\right)\left(\frac{12}{51}\right) = \frac{4}{221}$$
■

EXPLORE & DISCUSS

The product rule can be extended to the case involving three or more events. For example, if A, B, and C are three events in an experiment, then it can be shown that

$$P(A \cap B \cap C) = P(A) \cdot P(B|A) \cdot P(C|A \cap B)$$

1. Explain the formula in words.

2. Suppose 3 cards are drawn without replacement from a well-shuffled deck of 52 playing cards. Use the given formula to find the probability that the 3 cards are aces.

The product rule may be generalized to the case involving any finite number of events. For example, in the case involving the three events E, F, and G, it may be shown that

$$P(E \cap F \cap G) = P(E) \cdot P(F|E) \cdot P(G|E \cap F) \qquad \textbf{(4)}$$

▰ More on Tree Diagrams

Formula (4) and its generalizations may be used to help us solve problems that involve finite stochastic processes. More specifically, a **finite stochastic process** is an experiment consisting of a finite number of stages in which the outcomes and associated probabilities of each stage depend on the outcomes and associated probabilities of the preceding stages.

We can use tree diagrams to help us solve problems involving finite stochastic processes. Consider, for example, the experiment consisting of drawing 2 cards without replacement from a well-shuffled deck of 52 playing cards. What is the probability that the second card drawn is a face card?

We may think of this experiment as a stochastic process with two stages. The events associated with the first stage are F, that the card drawn is a face card, and F^c, that the card drawn is not a face card. Since there are 12 face cards, we have

$$P(F) = \frac{12}{52} \quad \text{and} \quad P(F^c) = 1 - \frac{12}{52} = \frac{40}{52}$$

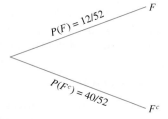

FIGURE 16
F is the probability that a face card is drawn.

The outcomes of this trial, together with the associated probabilities, may be represented along two branches of a tree diagram, as shown in Figure 16.

In the second trial, we again have two events: G, that the card drawn is a face card, and G^c, that the card drawn is not a face card. But the outcome of the second trial depends on the outcome of the first trial. For example, if the first card drawn was a face card, then the event G that the second card drawn is a face card has probability given by the *conditional probability* $P(G|F)$. Since the occurrence of a face card in the first draw leaves 11 face cards in a deck of 51 cards for the second draw, we see that

$$P(G|F) = \frac{11}{51} \quad \begin{array}{l} \text{The probability of drawing} \\ \text{a face card given that a face} \\ \text{card has already been drawn} \end{array}$$

Similarly, the occurrence of a face card in the first draw leaves 40 that are other than face cards in a deck of 51 cards for the second draw. Therefore, the probability of

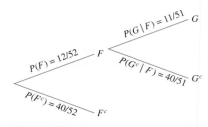

FIGURE 17
G is the probability that the second
card drawn is a face card.

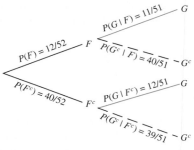

FIGURE 18
Tree diagram showing the two trials of
the experiment

drawing other than a face card in the second draw given that the first card drawn is
a face card is

$$P(G^c|F) = \frac{40}{51}$$

Using these results, we extend the tree diagram of Figure 16 by displaying
another two branches of the tree growing from its upper branch (Figure 17).

To complete the tree diagram, we compute $P(G|F^c)$ and $P(G^c|F^c)$, the condi-
tional probabilities that the second card drawn is a face card and other than a face
card, respectively, given that the first card drawn is not a face card. We find that

$$P(G|F^c) = \frac{12}{51} \quad \text{and} \quad P(G^c|F^c) = \frac{39}{51}$$

This leads to the completion of the tree diagram, shown in Figure 18, where the
branches of the tree that lead to the two outcomes of interest have been highlighted.

Having constructed the tree diagram associated with the problem, we are now
in a position to answer the question posed earlier—namely, "What is the probability
of the second card being a face card?" Observe that Figure 18 shows the two ways
in which a face card may result in the second draw—namely, the two *G*s on the
extreme right of the diagram.

Now, by the product rule, the probability that the second card drawn is a
face card and the first card drawn is a face card (this is represented by the upper
branch) is

$$P(G \cap F) = P(F) \cdot P(G|F)$$

Similarly, the probability that the second card drawn is a face card and the first
card drawn is other than a face card (this corresponds to the other branch) is

$$P(G \cap F^c) = P(F^c) \cdot P(G|F^c)$$

Observe that each of these probabilities is obtained by taking the *product of the
probabilities appearing on the respective branch.* Since $G \cap F$ and $G \cap F^c$ are
mutually exclusive events (why?), the probability that the second card drawn is a
face card is given by

$$P(G \cap F) + P(G \cap F^c) = P(F) \cdot P(G|F) + P(F^c) \cdot P(G|F^c)$$

or, upon replacing the probabilities on the right of the expression by their numerical
values,

$$P(G \cap F) + P(G \cap F^c) = \left(\frac{12}{52}\right)\left(\frac{11}{51}\right) + \left(\frac{40}{52}\right)\left(\frac{12}{51}\right)$$

$$= \frac{3}{13}$$

APPLIED EXAMPLE 6   Quality Control The picture tubes for the
Pulsar 19-inch color television sets are manufactured in three locations and
then shipped to the main plant of Vista Vision for final assembly. Plants A, B,
and C supply 50%, 30%, and 20%, respectively, of the picture tubes used by the
company. The quality-control department of the company has determined that
1% of the picture tubes produced by plant A are defective, whereas 2% of the picture tubes produced by plants B and C are defective. What is the probability that
a randomly selected Pulsar 19-inch color television set will have a defective picture tube?

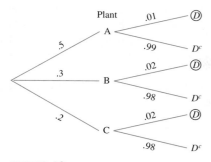

FIGURE 19
Tree diagram showing the probabilities of producing defective picture tubes at each plant

Solution Let A, B, and C denote the events that the set chosen has a picture tube manufactured in plant A, plant B, and plant C, respectively. Also, let D denote the event that a set has a defective picture tube. Using the given information, we draw the tree diagram shown in Figure 19. (The events that result in a set with a defective picture tube being selected are circled.) Taking the product of the probabilities along each branch leading to such an event and adding them yields the probability that a set chosen at random has a defective picture tube. Thus, the required probability is given by

$$(.5)(.01) + (.3)(.02) + (.2)(.02) = .005 + .006 + .004$$
$$= .015$$

APPLIED EXAMPLE 7 Quality Control A box contains eight 9-volt transistor batteries, of which two are known to be defective. The batteries are selected one at a time without replacement and tested until a nondefective one is found. What is the probability that the number of batteries tested is (a) one, (b) two, and (c) three?

Solution We may view this experiment as a multistage process with up to three stages. In the first stage, a battery is selected with a probability of $\frac{6}{8}$ of its being nondefective and a probability of $\frac{2}{8}$ of its being defective. If the battery selected is good, the experiment is terminated. Otherwise, a second battery is selected with probabilities of $\frac{6}{7}$ and $\frac{1}{7}$, respectively, of its being nondefective and defective. If the second battery selected is good, the experiment is terminated. Otherwise, a third battery is selected with probabilities of 1 and 0, respectively, of its being nondefective and defective. The tree diagram associated with this experiment is shown in Figure 20, where N denotes the event that the battery selected is nondefective and D denotes the event that the battery selected is defective.

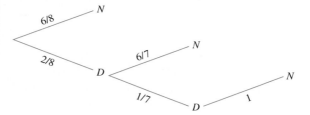

FIGURE 20
In this experiment, batteries are selected until a nondefective one is found.

With the aid of the tree diagram we see that (a) the probability that only one battery is selected is $\frac{6}{8}$, (b) the probability that two batteries are selected is $\left(\frac{2}{8}\right)\left(\frac{6}{7}\right)$, or $\frac{3}{14}$, and (c) the probability that three batteries are selected is $\left(\frac{2}{8}\right)\left(\frac{1}{7}\right)(1) = \frac{1}{28}$.

Independent Events

Let's return to the experiment of drawing 2 cards in succession without replacement from a well-shuffled deck of 52 playing cards considered in Example 5. Let E denote the event that the first card drawn is not a face card and let F denote the event that the second card drawn is a face card. It is intuitively clear that the events E and F are *not* independent of each other since whether or not the first card drawn is a face card affects the likelihood that the second card drawn is a face card.

Next, let's consider the experiment of tossing a coin twice and observing the outcomes: If H denotes the event that the first toss produces "heads" and T denotes the event that the second toss produces "tails," then it is intuitively clear that H and T *are* independent of each other since the outcome of the first toss does not affect the outcome of the second.

In general, two events A and B are independent if the outcome of one does not affect the outcome of the other. Thus, we have

Independent Events

If A and B are **independent events,** then

$$P(A|B) = P(A) \quad \text{and} \quad P(B|A) = P(B)$$

Using the product rule, we can find a simple test to determine the independence of two events. Suppose A and B are independent and $P(A) \neq 0$ and $P(B) \neq 0$. Then,

$$P(B|A) = P(B)$$

Thus, by the product rule, we have

$$P(A \cap B) = P(A) \cdot P(B|A) = P(A) \cdot P(B)$$

Conversely, if this equation holds, then it can be seen that $P(B|A) = P(B)$; that is, A and B are independent. Accordingly, we have the following test for the independence of two events.

Test for the Independence of Two Events

Two events A and B are independent if and only if

$$P(A \cap B) = P(A) \cdot P(B) \tag{5}$$

 Do not confuse *independent* events with *mutually exclusive* events. The former pertains to how the occurrence of one event affects the occurrence of another event, whereas the latter pertains to the question of whether the events can occur at the same time.

EXAMPLE 8   Consider the experiment consisting of tossing a fair coin twice and observing the outcomes. Show that the event of "heads" in the first toss and "tails" in the second toss are independent events.

Solution   Let A denote the event that the outcome of the first toss is a *head* and let B denote the event that the outcome of the second toss is a *tail*. The sample space of the experiment is

$$S = \{(H, H), (H, T), (T, H), (T, T)\}$$
$$A = \{(H, H), (H, T)\}$$
$$B = \{(H, T), (T, T)\}$$

so that

$$A \cap B = \{(H, T)\}$$

Next, we compute

$$P(A \cap B) = \frac{1}{4} \qquad P(A) = \frac{1}{2} \qquad P(B) = \frac{1}{2}$$

and observe that Equation (5) is satisfied in this case so that A and B are independent events, as we set out to show. ■

APPLIED EXAMPLE 9   Medical Surveys   A survey conducted by an independent agency for the National Lung Society found that of 2000 women, 680 were heavy smokers and 50 had emphysema. Of those who had emphysema, 42 were also heavy smokers. Using the data in this survey, determine whether the events "being a heavy smoker" and "having emphysema" are independent events.

Solution   Let A denote the event that a woman is a heavy smoker and let B denote the event that a woman has emphysema. Then, the probability that a woman is a heavy smoker and has emphysema is given by

$$P(A \cap B) = \frac{42}{2000} = .021$$

Next,

$$P(A) = \frac{680}{2000} = .34 \quad \text{and} \quad P(B) = \frac{50}{2000} = .025$$

so that

$$P(A) \cdot P(B) = (.34)(.025) = .0085$$

Since $P(A \cap B) \neq P(A) \cdot P(B)$, we conclude that A and B are not independent events. ■

The solution of many practical problems involves more than two independent events. In such cases we use the following result.

EXPLORE & DISCUSS

Let E and F be independent events in a sample space S. Are E^c and F^c independent?

Independence of More Than Two Events

If $E_1, E_2, \ldots, E_n$ are independent events, then

$$P(E_1 \cap E_2 \cap \cdots \cap E_n) = P(E_1) \cdot P(E_2) \cdot \cdots \cdot P(E_n) \qquad \textbf{(6)}$$

Formula (6) states that the probability of the simultaneous occurrence of n independent events is equal to the product of the probabilities of the n events.

 It is important to note that the mere requirement that the n events $E_1, E_2, \ldots, E_n$ satisfy (6) is not sufficient to guarantee that the n events are indeed independent. However, a criterion does exist for determining the independence of n events and may be found in more advanced texts on probability.

EXAMPLE 10   It is known that the three events A, B, and C are independent and $P(A) = .2$, $P(B) = .4$, and $P(C) = .5$. Compute:
a. $P(A \cap B)$ **b.** $P(A \cap B \cap C)$

Solution   Using Formulas (5) and (6), we find

a. $P(A \cap B) = P(A) \cdot P(B)$
$$= (.2)(.4) = .08$$

b. $P(A \cap B \cap C) = P(A) \cdot P(B) \cdot P(C)$
$$= (.2)(.4)(.5) = .04$$

APPLIED EXAMPLE 11   Quality Control   The Acrosonic model F loud-speaker system has four loudspeaker components: a woofer, a midrange, a tweeter, and an electrical crossover. The quality-control manager of Acrosonic has determined that on the average 1% of the woofers, 0.8% of the midranges, and 0.5% of the tweeters are defective, while 1.5% of the electrical crossovers are defective. Determine the probability that a loudspeaker system selected at random coming off the assembly line and before final inspection is not defective. Assume that the defects in the manufacturing of the components are unrelated.

Solution   Let A, B, C, and D denote, respectively, the events that the woofer, the midrange, the tweeter, and the electrical crossover are defective. Then,

$$P(A) = .01 \qquad P(B) = .008 \qquad P(C) = .005 \qquad P(D) = .015$$

and the probabilities of the corresponding complementary events are

$$P(A^c) = .99 \qquad P(B^c) = .992 \qquad P(C^c) = .995 \qquad P(D^c) = .985$$

The event that a loudspeaker system selected at random is not defective is given by $A^c \cap B^c \cap C^c \cap D^c$, and since the events A, B, C, and D (and therefore also A^c, B^c, C^c, and D^c) are assumed to be independent, we find that the required probability is given by

$$P(A^c \cap B^c \cap C^c \cap D^c) = P(A^c) \cdot P(B^c) \cdot P(C^c) \cdot P(D^c)$$
$$= (.99)(.992)(.995)(.985)$$
$$\approx .96$$

7.5   Self-Check Exercises

1. Let A and B be events in a sample space S such that $P(A) = .4$, $P(B) = .8$, and $P(A \cap B) = .3$. Find:
   a. $P(A|B)$ **b.** $P(B|A)$

2. Three friends—Alice, Betty, and Cathy—are unmarried and are 30, 35, and 40 years old, respectively. While they were having tea one afternoon, Alice recalled reading an article by Yale sociologist Neil Bennett in which he concluded that women who are still unmarried at age 30 have only a 20%

chance of marrying, those who are still unmarried at 35 have a 5.4% chance, and those who are still unmarried at 40 have a 1.3% chance. Betty wondered what the probability was that all three of them would eventually "tie the knot." Cathy, a statistician, pulled out her pocket calculator and answered Betty's question. What was her answer?

Solutions to Self-Check Exercises 7.5 can be found on page 429.

7.5   Concept Questions

1. What is *conditional probability*? Illustrate the concept with an example.

2. If A and B are events in an experiment and $P(A) \neq 0$, then what is the formula for computing $P(B|A)$?

3. If A and B are events in an experiment and the conditional probability $P(B|A)$ is known, give the formula that can be used to compute the event that both A and B will occur.

4. **a.** What is the test for determining the independence of two events?
   b. What is the difference between *mutually exclusive events* and *independent events*?

7.5   Exercises

1. Let A and B be events in a sample space S such that $P(A) = .6$, $P(B) = .5$, and $P(A \cap B) = .2$. Find:
   a. $P(A|B)$ **b.** $P(B|A)$

2. Let A and B be two events in a sample space S such that $P(A) = .4$, $P(B) = .6$, and $P(A \cap B) = .3$. Find:
   a. $P(A|B)$ **b.** $P(B|A)$

3. Let A and B be two events in a sample space S such that $P(A) = .6$ and $P(B|A) = .5$. Find $P(A \cap B)$.

4. Let A and B be the events described in Exercise 1. Find:
   a. $P(A|B^c)$ **b.** $P(B|A^c)$
   Hint: $(A \cap B^c) \cup (A \cap B) = A$.

In Exercises 5–8, determine whether the events A and B are independent.

5. $P(A) = .3$, $P(B) = .6$, $P(A \cap B) = .18$

6. $P(A) = .6$, $P(B) = .8$, $P(A \cap B) = .2$

7. $P(A) = .5$, $P(B) = .7$, $P(A \cup B) = .85$

8. $P(A^c) = .3$, $P(B^c) = .4$, $P(A \cap B) = .42$

9. If A and B are independent events, $P(A) = .4$, and $P(B) = .6$, find
   a. $P(A \cap B)$ **b.** $P(A \cup B)$

10. If A and B are independent events, $P(A) = .35$, and $P(B) = .45$, find
   a. $P(A \cap B)$ **b.** $P(A \cup B)$

11. The accompanying tree diagram represents an experiment consisting of two trials:

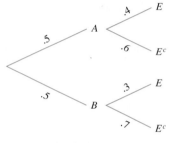

Use the diagram to find
a. $P(A)$ **b.** $P(E|A)$
c. $P(A \cap E)$ **d.** $P(E)$
e. Does $P(A \cap E) = P(A) \cdot P(E)$?
f. Are A and E independent events?

12. The accompanying tree diagram represents an experiment consisting of two trials. Use the diagram to find
a. $P(A)$ **b.** $P(E|A)$
c. $P(A \cap E)$ **d.** $P(E)$
e. Does $P(A \cap E) = P(A) \cdot P(E)$?
f. Are A and E independent events?

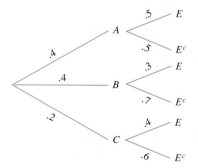

13. An experiment consists of two trials. The outcomes of the first trial are A and B with probabilities of occurring of .4 and .6. There are also two outcomes, C and D, in the second trial with probabilities of .3 and .7. Draw a tree diagram representing this experiment. Use this diagram to find
   a. $P(A)$ **b.** $P(C|A)$
   c. $P(A \cap C)$ **d.** $P(C)$
   e. Does $P(A \cap C) = P(A) \cdot P(C)$?
   f. Are A and C independent events?

14. An experiment consists of two trials. The outcomes of the first trial are A, B, and C, with probabilities of occurring of .2, .5, and .3, respectively. The outcomes of the second trial are E and F, with probabilities of occurring of .6 and .4. Draw a tree diagram representing this experiment. Use this diagram to find
   a. $P(B)$ **b.** $P(F|B)$
   c. $P(B \cap F)$ **d.** $P(F)$
   e. Does $P(B \cap F) = P(B) \cdot P(F)$?
   f. Are B and F independent events?

15. A pair of fair dice is cast. Let E denote the event that the number falling uppermost in the first die is 5 and let F denote the event that the sum of the numbers falling uppermost is 10.
   a. Compute $P(F)$.   **b.** Compute $P(E \cap F)$.
   c. Compute $P(F|E)$.   **d.** Compute $P(E)$.
   e. Are E and F independent events?

16. A pair of fair dice is cast. Let E denote the event that the number falling uppermost in the first die is 4 and let F denote the event that the sum of the numbers falling uppermost is 6.
   a. Compute $P(F)$.   **b.** Compute $P(E \cap F)$.
   c. Compute $P(F|E)$.   **d.** Compute $P(E)$.
   e. Are E and F independent events?

17. A pair of fair dice is cast. What is the probability that the sum of the numbers falling uppermost is less than 9, if it is known that one of the numbers is a 6?

18. A pair of fair dice is cast. What is the probability that the number landing uppermost on the first die is a 4, if it is known that the sum of the numbers falling uppermost is 7?

19. A pair of fair dice is cast. Let E denote the event that the number landing uppermost on the first die is a 3 and let F denote the event that the sum of the numbers falling uppermost is 7. Determine whether E and F are independent events.

20. A pair of fair dice is cast. Let E denote the event that the number landing uppermost on the first die is a 3 and let F denote the event that the sum of the numbers landing uppermost is 6. Determine whether E and F are independent events.

21. A card is drawn from a well-shuffled deck of 52 playing cards. Let E denote the event that the card drawn is black and let F denote the event that the card drawn is a spade. Determine whether E and F are independent events. Give an intuitive explanation for your answer.

22. A card is drawn from a well-shuffled deck of 52 playing cards. Let E denote the event that the card drawn is an ace and let F denote the event that the card drawn is a diamond. Determine whether E and F are independent events. Give an intuitive explanation for your answer.

23. **PRODUCT RELIABILITY**   The probability that a battery will last 10 hr or more is .80, and the probability that it will last 15 hr or more is .15. Given that a battery has lasted 10 hr, find the probability that it will last 15 hr or more.

24. Two cards are drawn without replacement from a well-shuffled deck of 52 playing cards.
   a. What is the probability that the first card drawn is a heart?
   b. What is the probability that the second card drawn is a heart given that the first card drawn was not a heart?
   c. What is the probability that the second card drawn is a heart given that the first card drawn was a heart?

25. Five black balls and four white balls are placed in an urn. Two balls are then drawn in succession. What is the probability that the second ball drawn is a white ball if
   a. The second ball is drawn without replacing the first?
   b. The first ball is replaced before the second is drawn?

26. **AUDITING TAX RETURNS**   A tax specialist has estimated the probability that a tax return selected at random will be audited is .02. Furthermore, he estimates the probability that an audited return will result in additional assessments being levied on the taxpayer is .60. What is the probability that a tax return selected at random will result in additional assessments being levied on the taxpayer?

27. **STUDENT ENROLLMENT**   At a certain medical school, $\frac{1}{7}$ of the students are from a minority group. Of those students who belong to a minority group, $\frac{1}{3}$ are black.
   a. What is the probability that a student selected at random from this medical school is black?
   b. What is the probability that a student selected at random from this medical school is black if it is known that the student is a member of a minority group?

28. **EDUCATIONAL LEVEL OF VOTERS**   In a survey of 1000 eligible voters selected at random, it was found that 80 had a college degree. Additionally, it was found that 80% of those who had a college degree voted in the last presidential election, whereas 55% of the people who did not have a college degree voted in the last presidential election. Assuming that the poll is representative of all eligible voters, find the probability that an eligible voter selected at random
   a. Had a college degree and voted in the last presidential election.

b. Did not have a college degree and did not vote in the last presidential election.

c. Voted in the last presidential election.

d. Did not vote in the last presidential election.

29. Three cards are drawn without replacement from a well-shuffled deck of 52 playing cards. What is the probability that the third card drawn is a diamond?

30. A coin is tossed 3 times. What is the probability that the coin will land heads

a. At least twice?

b. On the second toss given that heads were thrown on the first toss?

c. On the third toss given that tails were thrown on the first toss?

31. In a three-child family, what is the probability that all three children are girls given that one of the children is a girl? (Assume that the probability of a boy being born is the same as the probability of a girl being born.)

32. QUALITY CONTROL   An automobile manufacturer obtains the microprocessors used to regulate fuel consumption in its automobiles from three microelectronic firms: A, B, and C. The quality-control department of the company has determined that 1% of the microprocessors produced by firm A are defective, 2% of those produced by firm B are defective, and 1.5% of those produced by firm C are defective. Firms A, B, and C supply 45%, 25%, and 30%, respectively, of the microprocessors used by the company. What is the probability that a randomly selected automobile manufactured by the company will have a defective microprocessor?

33. CAR THEFT   Figures obtained from a city's police department seem to indicate that, of all motor vehicles reported as stolen, 64% were stolen by professionals whereas 36% were stolen by amateurs (primarily for joy rides). Of those vehicles presumed stolen by professionals, 24% were recovered within 48 hr, 16% were recovered after 48 hr, and 60% were never recovered. Of those vehicles presumed stolen by amateurs, 38% were recovered within 48 hr, 58% were recovered after 48 hr, and 4% were never recovered.

a. Draw a tree diagram representing these data.

b. What is the probability that a vehicle stolen by a professional in this city will be recovered within 48 hr?

c. What is the probability that a vehicle stolen in this city will never be recovered?

34. HOUSING LOANS   The chief loan officer of La Crosse Home Mortgage Company summarized the housing loans extended by the company in 2003 according to type and term of the loan. Her list shows that 70% of the loans were fixed-rate mortgages (F), 25% were adjustable-rate mortgages (A), and 5% belong to some other category (O) (mostly second trust-deed loans and loans extended under the graduated payment plan). Of the fixed-rate mortgages, 80% were 30-yr loans and 20% were 15-yr loans; of the adjustable-rate mortgages, 40% were 30-yr loans and 60% were 15-yr loans; finally, of

the other loans extended, 30% were 20-yr loans, 60% were 10-yr loans, and 10% were for a term of 5 yr or less.

a. Draw a tree diagram representing this experiment.

b. What is the probability that a home loan extended by La Crosse has an adjustable rate and is for a term of 15 yr?

c. What is the probability that a home loan extended by La Crosse is for a term of 15 yr?

35. COLLEGE ADMISSIONS   The admissions office of a private university released the following admission data for the preceding academic year: From a pool of 3900 male applicants, 40% were accepted by the university, and of these, 40% subsequently enrolled. Additionally, from a pool of 3600 female applicants, 45% were accepted by the university, and of these, 40% subsequently enrolled. What is the probability that

a. A male applicant will be accepted by and subsequently will enroll in the university?

b. A student who applies for admissions will be accepted by the university?

c. A student who applies for admission will be accepted by the university and subsequently will enroll?

36. QUALITY CONTROL   A box contains two defective Christmas tree lights that have been inadvertently mixed with eight nondefective lights. If the lights are selected one at a time without replacement and tested until both defective lights are found, what is the probability that both defective lights will be found after three trials?

37. QUALITY CONTROL   It is estimated that 0.80% of a large consignment of eggs in a certain supermarket is broken.

a. What is the probability that a customer who randomly selects a dozen of these eggs receives at least one broken egg?

b. What is the probability that a customer who selects these eggs at random will have to check three cartons before finding a carton without any broken eggs? (Each carton contains a dozen eggs.)

38. STUDENT FINANCIAL AID   The accompanying data were obtained from the financial aid office of a certain university:

	Receiving Financial Aid	Not Receiving Financial Aid	Total
Undergraduates	4,222	3,898	8,120
Graduates	1,879	731	2,610
Total	6,101	4,629	10,730

Let A be the event that a student selected at random from this university is an undergraduate student and let B be the event that a student selected at random is receiving financial aid.

a. Find each of the following probabilities: $P(A)$, $P(B)$, $P(A \cap B)$, $P(B|A)$, and $P(B|A^c)$.

b. Are the events A and B independent events?

39. EMPLOYEE EDUCATION AND INCOME   The personnel department of Franklin National Life Insurance Company compiled the accompanying data regarding the income and education of its employees:

	Income $50,000 or Below	Income Above $50,000
Noncollege Graduate	2040	840
College Graduate	400	720

Let A be the event that a randomly chosen employee has a college degree and B the event that the chosen employee's income is more than \$50,000.

a. Find each of the following probabilities: $P(A)$, $P(B)$, $P(A \cap B)$, $P(B|A)$, and $P(B|A^c)$.

b. Are the events A and B independent events?

40. Two cards are drawn without replacement from a well-shuffled deck of 52 cards. Let A be the event that the first card drawn is a heart and let B be the event that the second card drawn is a red card. Show that the events A and B are dependent events.

41. MEDICAL RESEARCH A nationwide survey conducted by the National Cancer Society revealed the following information: Of 10,000 people surveyed, 3200 were "heavy coffee drinkers" and 160 had cancer of the pancreas. Of those who had cancer of the pancreas, 132 were heavy coffee drinkers. Using the data in this survey, determine whether the events "being a heavy coffee drinker" and "having cancer of the pancreas" are independent events.

42. SWITCHING INTERNET SERVICE PROVIDERS (ISPs) According to a survey conducted in 2004 of 1000 American adults with Internet access, one in four households plan to switch ISPs in the next 6 mo. Of those who plan to switch, 1% of the households are likely to switch to a satellite connection, 27% to digital subscriber line (DSL), 28% to cable modem, 35% to dial-up modem, and 9% don't know which service provider they will switch to.

a. If a person participating in the survey is chosen at random, what is the probability that he or she will switch to a dial-up modem connection?

b. If a person in the survey has planned to switch ISPs, what is the probability that he or she will upgrade to high-speed service (satellite, DSL, or cable)?

Source: Ipsos-Insight

43. RELIABILITY OF SECURITY SYSTEMS Before being allowed to enter a maximum-security area at a military installation, a person must pass three identification tests: a voice-pattern test, a fingerprint test, and a handwriting test. If the reliability of the first test is 97%, the reliability of the second test is 98.5%, and that of the third is 98.5%, what is the probability that this security system will allow an improperly identified person to enter the maximum-security area?

44. RELIABILITY OF A HOME THEATER SYSTEM In a home theater system, the probability that the video component needs repair within 1 yr is .01, the probability that the electronic components need repair within 1 yr is .005, and the probability that the audio component needs repair within 1 yr is .001.

Assuming the probabilities are independent, find the probability that

a. At least one component will need repair within 1 yr.

b. Exactly one component will need repair within 1 yr.

45. PROBABILITY OF TRANSPLANT REJECTION The independent probabilities that the three patients who are scheduled to receive kidney transplants at General Hospital will suffer rejection are $\frac{1}{2}$, $\frac{1}{3}$, and $\frac{1}{10}$. Find the probabilities that

a. At least one patient will suffer rejection.

b. Exactly two patients will suffer rejection.

46. QUALITY CONTROL Copykwik has four photocopy machines A, B, C, and D. The probability that a given machine will break down on a particular day is

$$P(A) = \frac{1}{50} \qquad P(B) = \frac{1}{60} \qquad P(C) = \frac{1}{75} \qquad P(D) = \frac{1}{40}$$

Assuming independence, what is the probability on a particular day that

a. All four machines will break down?

b. None of the machines will break down?

c. Exactly one machine will break down?

47. PRODUCT RELIABILITY The proprietor of Cunningham's Hardware Store has decided to install floodlights on the premises as a measure against vandalism and theft. If the probability is .01 that a certain brand of floodlight will burn out within a year, find the minimum number of floodlights that must be installed to ensure that the probability that at least one of them will remain functional within the year is at least .99999. (Assume that the floodlights operate independently.)

48. Let E be any event in a sample space S.

a. Are E and S independent? Explain your answer.

b. Are E and $\varnothing$ independent? Explain your answer.

49. Suppose the probability that an event will occur in one trial is p. Show that the probability that the event will occur at least once in n independent trials is $1 - (1 - p)^n$.

50. Let E and F be mutually exclusive events and suppose $P(F) \neq 0$. Find $P(E|F)$ and interpret your result.

51. Let E and F be events such that $F \subset E$. Find $P(E|F)$ and interpret your result.

52. Suppose A and B are mutually exclusive events and $P(A \cup B) \neq 0$. What is $P(A|A \cup B)$?

In Exercises 53–56, determine whether the statement is true or false. If it is true, explain why it is true. If it is false, give an example to show why it is false.

53. If A and B are mutually exclusive and $P(B) \neq 0$, then $P(A|B) = 0$.

54. If A is an event of an experiment, then $P(A|A^c) \neq 0$.

55. If A and B are events of an experiment, then

$$P(A \cap B) = P(A|B) \cdot P(B) = P(B|A) \cdot P(A)$$

56. If A and B are independent events with $P(A) \neq 0$ and $P(B) \neq 0$, then $A \cap B \neq \varnothing$.

7.5 Solutions to Self-Check Exercises

1. a. $P(A|B) = \dfrac{P(A \cap B)}{P(B)}$

$= \dfrac{.3}{.8} = \dfrac{3}{8}$

b. $P(B|A) = \dfrac{P(A \cap B)}{P(A)}$

$= \dfrac{.3}{.4} = \dfrac{3}{4}$

2. Let A, B, and C denote the events that three women who are still unmarried at ages 30, 35, and 40 *will* marry eventually.

Then, it seems reasonable to assume that these events are independent, with $P(A) = .20$, $P(B) = .054$, and $P(C) = .013$. Based on these figures, the probability that all three women will be married eventually is given by

$$P(A) \cdot P(B) \cdot P(C) = (.2)(.054)(.013)$$
$$= .00014$$

7.6 Bayes' Theorem

■ A Posteriori Probabilities

Suppose three machines, A, B, and C, produce similar engine components. Machine A produces 45% of the total components, machine B produces 30%, and machine C, 25%. For the usual production schedule, 6% of the components produced by machine A do not meet established specifications; for machine B and machine C, the corresponding figures are 4% and 3%. One component is selected at random from the total output and is found to be defective. What is the probability that the component selected was produced by machine A?

The answer to this question is found by calculating the probability *after* the outcomes of the experiment have been observed. Such probabilities are called **a posteriori probabilities** as opposed to **a priori probabilities**—probabilities that give the likelihood that an event *will* occur, the subject of the last several sections.

Returning to the example under consideration, we need to determine the a posteriori probability for the event that the component selected was produced by machine A. To this end, let A, B, and C denote the event that a component is produced by machine A, machine B, and machine C, respectively. We may represent this experiment with a Venn diagram (Figure 21).

The three mutually exclusive events A, B, and C form a **partition** of the sample space S; that is, aside from being mutually exclusive, their union is precisely S. The event D that a component is defective is the shaded area. Again referring to Figure 21, we see that

1. The event D may be expressed as

$$D = (A \cap D) \cup (B \cap D) \cup (C \cap D)$$

2. The event that a component is defective and is produced by machine A is given by $A \cap D$.

Thus, the a posteriori probability that a defective component selected was produced by machine A is given by

$$P(A|D) = \frac{n(A \cap D)}{n(D)}$$

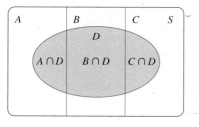

FIGURE 21
D is the event that a defective component is produced by machine A, machine B, or machine C.

Upon dividing both the numerator and the denominator by $n(S)$ and observing that the events $A \cap D$, $B \cap D$, and $C \cap D$ are mutually exclusive, we obtain

$$P(A|D) = \frac{P(A \cap D)}{P(D)}$$

$$= \frac{P(A \cap D)}{P(A \cap D) + P(B \cap D) + P(C \cap D)} \tag{7}$$

Next, using the product rule, we may express

$$P(A \cap D) = P(A) \cdot P(D|A)$$
$$P(B \cap D) = P(B) \cdot P(D|B)$$
$$P(C \cap D) = P(C) \cdot P(D|C)$$

so that Equation (7) may be expressed in the form

$$P(A|D) = \frac{P(A) \cdot P(D|A)}{P(A) \cdot P(D|A) + P(B) \cdot P(D|B) + P(C) \cdot P(D|C)} \tag{8}$$

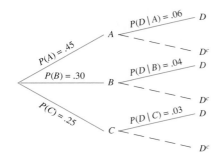

FIGURE 22
A tree diagram displaying the probabilities that a defective component is produced by machine A, machine B, or machine C

which is a special case of a result known as **Bayes' theorem.**

Observe that the expression on the right of (8) involves the probabilities $P(A)$, $P(B)$, and $P(C)$ and the conditional probabilities $P(D|A)$, $P(D|B)$, and $P(D|C)$, all of which may be calculated in the usual fashion. In fact, by displaying these quantities on a tree diagram, we obtain Figure 22. We may compute the required probability by substituting the relevant quantities into (8), or we may make use of the following device:

$$P(A|D) = \frac{\text{Product of probabilities along the limb through } A}{\text{Sum of products of the probabilities along each limb terminating at } D}$$

In either case, we obtain

$$P(A|D) = \frac{(.45)(.06)}{(.45)(.06) + (.3)(.04) + (.25)(.03)}$$
$$\approx .58$$

Before looking at any further examples, let's state the general form of Bayes' theorem.

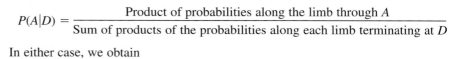

Bayes' Theorem

Let $A_1, A_2, \ldots, A_n$ be a partition of a sample space S and let E be an event of the experiment such that $P(E) \neq 0$. Then the a posteriori probability $P(A_i|E)$ $(1 \leq i \leq n)$ is given by

$$P(A_i|E) = \frac{P(A_i) \cdot P(E|A_i)}{P(A_1) \cdot P(E|A_1) + P(A_2) \cdot P(E|A_2) + \cdots + P(A_n) \cdot P(E|A_n)} \tag{9}$$

APPLIED EXAMPLE 1   Quality Control   The picture tubes for the Pulsar 19-inch color television sets are manufactured in three locations and then shipped to the main plant of Vista Vision for final assembly. Plants A, B, and C supply 50%, 30%, and 20%, respectively, of the picture tubes used by

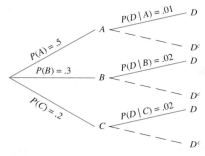

FIGURE 23
$P(C|D) =$

$$\frac{\text{Product of probabilities of branches to } D \text{ through } C}{\text{Sum of product of probabilities of branches leading to } D}$$

Vista Vision. The quality-control department of the company has determined that 1% of the picture tubes produced by plant A are defective, whereas 2% of the picture tubes produced by plants B and C are defective. If a Pulsar 19-inch color television set is selected at random and the picture tube is found to be defective, what is the probability that the picture tube was manufactured in plant C? (Compare with Example 6, page 420.)

Solution   Let A, B, and C denote the event that the set chosen has a picture tube manufactured in plant A, plant B, and plant C, respectively. Also, let D denote the event that a set has a defective picture tube. Using the given information, we may draw the tree diagram shown in Figure 23. Next, using Formula (9), we find that the required a posteriori probability is given by

$$P(C|D) = \frac{P(C) \cdot P(D|C)}{P(A) \cdot P(D|A) + P(B) \cdot P(D|B) + P(C) \cdot P(D|C)}$$

$$= \frac{(.2)(.02)}{(.5)(.01) + (.3)(.02) + (.2)(.02)}$$

$$\approx .27$$

APPLIED EXAMPLE 2   Income Distributions   A study was conducted in a large metropolitan area to determine the annual incomes of married couples in which the husbands were the sole providers and of those in which the husbands and wives were both employed. Table 9 gives the results of this study.

TABLE 9

Annual Family Income, $	Married Couples, %	Income Group with Both Spouses Working, %
125,000 and over	4	65
100,000–124,999	10	73
75,000–99,999	21	68
50,000–74,999	24	63
30,000–49,999	30	43
Under 30,000	11	28

a. What is the probability that a couple selected at random from this area has two incomes?

b. If a randomly chosen couple has two incomes, what is the probability that the annual income of this couple is over $125,000?

c. If a randomly chosen couple has two incomes, what is the probability that the annual income of this couple is greater than $49,999?

Solution   Let A denote the event that the annual income of the couple is $125,000 and over; let B denote the event that the annual income is between $100,000 and $124,999; let C denote the event that the annual income is between $75,000 and $99,999; and so on. Finally, let F denote the event that the annual income is less than $30,000 and let T denote the event that both spouses work. The probabilities of the occurrence of these events are displayed in Figure 24.

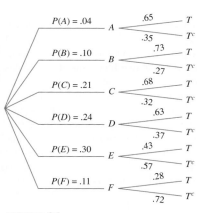

FIGURE 24

a. The probability that a couple selected at random from this group has two incomes is given by

$$P(T) = P(A) \cdot P(T|A) + P(B) \cdot P(T|B) + P(C) \cdot P(T|C)$$
$$+ P(D) \cdot P(T|D) + P(E) \cdot P(T|E) + P(F) \cdot P(T|F)$$
$$= (.04)(.65) + (.10)(.73) + (.21)(.68) + (.24)(.63)$$
$$+ (.30)(.43) + (.11)(.28)$$
$$= .5528$$

b. Using the results of part (a) and Bayes' theorem, we find that the probability that a randomly chosen couple has an annual income over \$125,000 given that both spouses are working is

$$P(A|T) = \frac{P(A) \cdot P(T|A)}{P(T)} = \frac{(.04)(.65)}{.5528}$$
$$\approx .047$$

c. The probability that a randomly chosen couple has an annual income greater than \$49,999 given that both spouses are working is

$$P(A|T) + P(B|T) + P(C|T) + P(D|T)$$
$$= \frac{P(A) \cdot P(T|A) + P(B) \cdot P(T|B) + P(C) \cdot P(T|C) + P(D) \cdot P(T|D)}{P(T)}$$
$$= \frac{(.04)(.65) + (.1)(.73) + (.21)(.68) + (.24)(.63)}{.5528}$$
$$\approx .711$$

7.6   Self-Check Exercises

1. The accompanying tree diagram represents a two-stage experiment. Use the diagram to find $P(B|D)$.

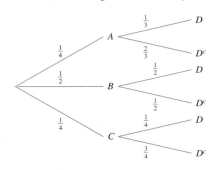

2. In a recent presidential election, it was estimated that the probability that the Republican candidate would be elected was $\frac{3}{5}$ and therefore the probability that the Democratic candidate would be elected was $\frac{2}{5}$ (the two Independent candidates were given little chance of being elected). It was also estimated that if the Republican candidate were elected, then the probability that research for a new manned bomber would continue was $\frac{4}{5}$. But if the Democratic candidate were successful, then the probability that the research would continue was $\frac{3}{10}$. Research was terminated shortly after the successful presidential candidate took office. What is the probability that the Republican candidate won that election?

Solutions to Self-Check Exercises 7.6 can be found on page 438.

7.6 Concept Questions

1. What are *a priori probabilities* and *a posteriori probabilities*? Give an example of each.

2. Suppose the events A, B, and C form the partition of a sample space S and suppose E is an event of an experiment such that $P(E) \neq 0$. Use Bayes' theorem to write the formula for the a posteriori probability $P(A|E)$.

3. Refer to Concept Question 2. If E is the event that a product was produced in factory A, factory B, or factory C, and $P(E) \neq 0$, what does $P(A|E)$ represent?

7.6 Exercises

In Exercises 1–3, refer to the accompanying Venn diagram. An experiment in which the three mutually exclusive events A, B, and C form a partition of the uniform sample space S is depicted in the diagram.

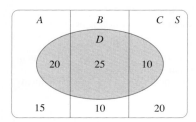

1. Draw a tree diagram using the information given in the Venn diagram illustrating the probabilities of the events A, B, C, and D.

2. Find: **a.** $P(D)$ **b.** $P(A|D)$

3. Find: **a.** $P(D^c)$ **b.** $P(B|D^c)$

In Exercises 4–6, refer to the accompanying Venn diagram. An experiment in which the three mutually exclusive events A, B, and C form a partition of the uniform sample space S is depicted in the diagram.

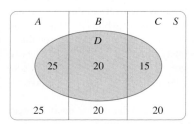

4. Draw a tree diagram using the information given in the Venn diagram illustrating the probabilities of the events A, B, C, and D.

5. Find: **a.** $P(D)$ **b.** $P(B|D)$

6. Find: **a.** $P(D^c)$ **b.** $P(B|D^c)$

7. The accompanying tree diagram represents a two-stage experiment. Use the diagram to find
   a. $P(A) \cdot P(D|A)$ **b.** $P(B) \cdot P(D|B)$
   c. $P(A|D)$

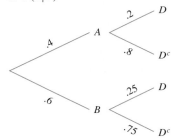

8. The accompanying tree diagram represents a two-stage experiment. Use the diagram to find
   a. $P(A) \cdot P(D|A)$ **b.** $P(B) \cdot P(D|B)$
   c. $P(A|D)$

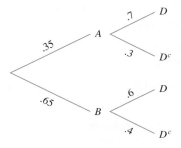

9. The accompanying tree diagram represents a two-stage experiment. Use the diagram to find
   a. $P(A) \cdot P(D|A)$ **b.** $P(B) \cdot P(D|B)$
   c. $P(C) \cdot P(D|C)$ **d.** $P(A|D)$

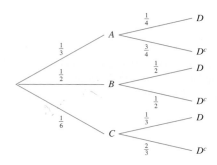

10. The accompanying tree diagram represents a two-stage experiment. Use this diagram to find
   a. $P(A \cap D)$ **b.** $P(B \cap D)$ **c.** $P(C \cap D)$ **d.** $P(D)$
   e. Verify:

$$P(A|D) = \frac{P(A \cap D)}{P(D)}$$

$$= \frac{P(A) \cdot P(D|A)}{P(A) \cdot P(D|A) + P(B) \cdot P(D|B) + P(C) \cdot P(D|C)}$$

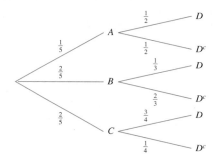

In Exercises 11–14, refer to the following experiment: Two cards are drawn in succession without replacement from a standard deck of 52 cards.

11. What is the probability that the first card is a heart given that the second card is a heart?

12. What is the probability that the first card is a heart given that the second card is a diamond?

13. What is the probability that the first card is a jack given that the second card is an ace?

14. What is the probability that the first card is a face card given that the second card is an ace?

In Exercises 15–18, refer to the following experiment: Urn A contains four white and six black balls. Urn B contains three white and five black balls. A ball is drawn from urn A and then transferred to urn B. A ball is then drawn from urn B.

15. Represent the probabilities associated with this two-stage experiment in the form of a tree diagram.

16. What is the probability that the transferred ball was white given that the second ball drawn was white?

17. What is the probability that the transferred ball was black given that the second ball drawn was white?

18. What is the probability that the transferred ball was black given that the second ball drawn was black?

19. Politics The 1992 U.S. Senate was composed of 57 Democrats and 43 Republicans. Of the Democrats, 38 served in the military, whereas 28 of the Republicans had seen military service. If a senator selected at random had served in the military, what is the probability that he was Republican? *Note:* No congresswoman had served in the military.

20. Quality Control Jansen Electronics has four machines that produce an identical component for use in its videocassette players. The proportion of the components produced by each machine and the probability of that component being defective are shown in the accompanying table. What is the probability that a component selected at random
   a. Is defective?
   b. Was produced by machine I, given it is defective?
   c. Was produced by machine II, given it is defective?

Machine	Proportion of Components Produced	Probability of Defective Component
I	.15	.04
II	.30	.02
III	.35	.02
IV	.20	.03

21. An experiment consists of randomly selecting one of three coins, tossing it, and observing the outcome—heads or tails. The first coin is a two-headed coin, the second is a biased coin such that $P(H) = .75$, and the third is a fair coin.
   a. What is the probability that the coin that is tossed will show heads?
   b. If the coin selected shows heads, what is the probability that this coin is the fair coin?

22. Reliability of Medical Tests A medical test has been designed to detect the presence of a certain disease. Among those who have the disease, the probability that the disease will be detected by the test is .95. However, the probability that the test will erroneously indicate the presence of the disease in those who do not actually have it is .04. It is estimated that 4% of the population who take this test have the disease.
   a. If the test administered to an individual is positive, what is the probability that the person actually has the disease?
   b. If an individual takes the test twice and both times the test is positive, what is the probability that the person actually has the disease?

23. **RELIABILITY OF MEDICAL TESTS** Refer to Exercise 22. Suppose 20% of the people who were referred to a clinic for the test did in fact have the disease. If the test administered to an individual from this group is positive, what is the probability that the person actually has the disease?

24. **QUALITY CONTROL** A desk lamp produced by Luminar was found to be defective. The company has three factories where the lamps are manufactured. The percentage of the total number of desk lamps produced by each factory and the probability that a lamp manufactured by that factory is defective are shown in the accompanying table. What is the probability that the defective lamp was manufactured in factory III?

Factory	Percentage of Total Production	Probability of Defective Component
I	.35	.015
II	.35	.01
III	.30	.02

25. **AUTO-ACCIDENT RATES** An insurance company has compiled the accompanying data relating the age of drivers and the accident rate (the probability of being involved in an accident during a 1-yr period) for drivers within that group:

Age Group	Percentage of Insured Drivers	Accident Rate
Under 25	.16	.055
25–44	.40	.025
45–64	.30	.02
65 and over	.14	.04

a. What is the probability that an insured driver will be involved in an accident during a particular 1-yr period?
b. What is the probability that an insured driver who is involved in an accident is under 25?

26. **SEAT-BELT COMPLIANCE** Data compiled by the Highway Patrol Department regarding the use of seat belts by drivers in a certain area after the passage of a compulsory seat-belt law are shown in the accompanying table:

Drivers	Percentage of Drivers in Group	Percentage of Group Stopped for Moving Violation
Group I (using seat belts)	.64	.002
Group II (not using seat belts)	.36	.005

If a driver in that area is stopped for a moving violation, what is the probability that he or she
a. Will have a seat belt on?
b. Will not have a seat belt on?

27. **MEDICAL RESEARCH** Based on data obtained from the National Institute of Dental Research, it has been determined that 42% of 12-yr-olds have never had a cavity, 34% of 13-yr-olds have never had a cavity, and 28% of 14-yr-olds have never had a cavity. If a child is selected at random from a group of 24 junior high school students comprising six 12-yr-olds, eight 13-yr-olds, and ten 14-yr-olds and this child does not have a cavity, what is the probability that this child is 14 yrs old?

28. **VOTING PATTERNS** In a recent senatorial election, 50% of the voters in a certain district were registered as Democrats, 35% were registered as Republicans, and 15% were registered as Independents. The incumbent Democratic senator was reelected over her Republican and Independent opponents. Exit polls indicated that she gained 75% of the Democratic vote, 25% of the Republican vote, and 30% of the Independent vote. Assuming that the exit poll is accurate, what is the probability that a vote for the incumbent was cast by a registered Republican?

29. **CRIME RATES** Data compiled by the Department of Justice on the number of people arrested for serious crimes (murder, forcible rape, robbery, and so on) in 1988 revealed that 89% were male and 11% were female. Of the males, 30% were under 18, whereas 27% of the females arrested were under 18.
a. What is the probability that a person arrested for a serious crime in 1988 was under 18?
b. If a person arrested for a serious crime in 1988 was known to be under 18, what is the probability that the person is female?

30. **OPINION POLLS** In a survey involving 600 Democrats, 400 Republicans, and 200 Independents, the question "Do you favor or oppose eliminating taxes on dividends paid to shareholders?" was asked. The following results were obtained:

Answer	Democrats, %	Republicans, %	Independents, %
Favor	29	66	48
Opposed	71	34	52

If a randomly chosen respondent in the survey answered "favor," what is the probability that he or she is an Independent?

Source: TechnoMetrica Market Intelligence

31. **OPINION POLLS** A poll was conducted among 500 registered voters in a certain area regarding their position on a national lottery to raise revenue for the government. The results of the poll are shown in the accompanying table:

Sex	Percentage of Voters Polled	Percentage Favoring Lottery	Percentage Not Favoring Lottery	Percentage Expressing No Opinion
Male	.51	.62	.32	.06
Female	.49	.68	.28	.04

What is the probability that a registered voter who
a. Favored a national lottery was a woman?
b. Expressed no opinion regarding the lottery was a woman?

32. SELECTION OF SUPREME COURT JUDGES   In a past presidential election, it was estimated that the probability that the Republican candidate would be elected was $\frac{3}{5}$, and therefore the probability that the Democratic candidate would be elected was $\frac{2}{5}$ (the two Independent candidates were given little chance of being elected). It was also estimated that if the Republican candidate were elected, the probabilities that a conservative, moderate, or liberal judge would be appointed to the Supreme Court (one retirement was expected during the presidential term) were $\frac{1}{2}$, $\frac{1}{3}$, and $\frac{1}{6}$, respectively. If the Democratic candidate were elected, the probabilities that a conservative, moderate, or liberal judge would be appointed to the Supreme Court would be $\frac{1}{8}$, $\frac{3}{8}$, and $\frac{1}{2}$, respectively. A conservative judge was appointed to the Supreme Court during the presidential term. What is the probability that the Democratic candidate was elected?

33. PERSONNEL SELECTION   Applicants for temporary office work at Carter Temporary Help Agency who have successfully completed a typing test are then placed in suitable positions by Nancy Dwyer and Darla Newberg. Employers who hire temporary help through the agency return a card indicating satisfaction or dissatisfaction with the work performance of those hired. From past experience it is known that 80% of the employees placed by Nancy are rated as satisfactory, whereas 70% of those placed by Darla are rated as satisfactory. Darla places 55% of the temporary office help at the agency and Nancy the remaining 45%. If a Carter office worker is rated unsatisfactory, what is the probability that he or she was placed by Darla?

34. COLLEGE MAJORS   The Office of Admissions and Records of a large western university released the accompanying information concerning the contemplated majors of its freshman class:

Major	Percentage of Freshmen Choosing This Major	Percentage of Females	Percentage of Males
Business	.24	.38	.62
Humanities	.08	.60	.40
Education	.08	.66	.34
Social Science	.07	.58	.42
Natural Sciences	.09	.52	.48
Other	.44	.48	.52

a. What is the probability that a student selected at random from the freshman class is a female?
b. What is the probability that a business student selected at random from the freshman class is a male?
c. What is the probability that a female student selected at random from the freshman class is majoring in business?

35. MEDICAL DIAGNOSES   A study was conducted among a certain group of union members whose health insurance policies required second opinions prior to surgery. Of those members whose doctors advised them to have surgery, 20% were informed by a second doctor that no surgery was needed. Of these, 70% took the second doctor's opinion and did not go through with the surgery. Of the members who were advised to have surgery by both doctors, 95% went through with the surgery. What is the probability that a union member who had surgery was advised to do so by a second doctor?

36. AGE DISTRIBUTION OF RENTERS   A study conducted by the Metro Housing Agency in a midwestern city revealed the accompanying information concerning the age distribution of renters within the city:

Age	Percentage of Adult Population	Percentage of Group Who Are Renters
21–44	.51	.58
45–64	.31	.45
65 and over	.18	.60

a. What is the probability that an adult selected at random from this population is a renter?
b. If a renter is selected at random, what is the probability that he or she is in the 21–44 age bracket?
c. If a renter is selected at random, what is the probability that he or she is 45 yr of age or older?

37. PERSONAL HABITS   There were 80 male guests at a party. The number of men in each of four age categories is given in the following table. The table also gives the probability that a man in the respective age category will keep his paper money in order of denomination.

Age	Number of Men	Percentage of Men Who Keep Paper Money in Order
21–34	25	.9
35–44	30	.61
45–54	15	.80
55 and over	10	.80

A man's wallet was retrieved and the paper money in it was kept in order of denomination. What is the probability that the wallet belonged to a male guest between the ages of 35 and 44?

38. THE SOCIAL LADDER   The following table summarizes the results of a poll conducted with 1154 adults:

Annual Household Income, $	Respondents within That Income Range, %	Percent of Respondents Who Call Themselves		
		Rich	Middle Class	Poor
Less than 15,000	11.2	0	24	76
15,000–29,999	18.6	3	60	37
30,000–49,999	24.5	0	86	14
50,000–74,999	21.9	2	90	8
75,000 and higher	23.8	5	91	4

a. What is the probability that a respondent chosen at random calls himself or herself middle class?

b. If a randomly chosen respondent calls himself or herself middle class, what is the probability that the annual household income of that individual is between $30,000 and $49,999, inclusive?

c. If a randomly chosen respondent calls himself or herself middle class, what is the probability that the individual's income is less than or equal to $29,999 or greater than or equal to $50,000?

Source: New York Times/CBS News; Wall Street Journal Almanac

39. OPINION POLLS   In a survey involving 400 likely Democratic voters and 300 likely Republican voters, the question "Do you support or oppose legislation that would require registration of all handguns?" was asked. The following results were obtained:

Answer	Democrats, %	Republicans, %
Support	77	59
Oppose	14	31
Don't know/refused	9	10

If a randomly chosen respondent in the survey answered "Oppose," what is the probability that he or she is a likely Democratic voter?

40. OPINION POLLS   In a survey involving 400 likely Democratic voters and 300 likely Republican voters, the question "Do you support or oppose legislation that would require trigger locks on guns, to prevent misuse by children?" was asked. The following results were obtained:

Answer	Democrats, %	Republicans, %
Support	88	71
Oppose	7	20
Don't know/refused	5	9

If a randomly chosen respondent in the survey answered "Support," what is the probability that he or she is a likely Republican voter?

41. VOTER TURNOUT BY INCOME   Voter turnout drops steadily as income level declines. The following table gives the percent of eligible voters in a certain city, categorized by income who responded with "Did not vote" in the 2000 presidential election. The table also gives the number of eligible voters in the city, categorized by income.

Income (percentile)	Percent Who "Did Not Vote"	Eligible Voters
0–16	52	4,000
17–33	31	11,000
34–67	30	17,500
68–95	14	12,500
96–100	12	5,000

If an eligible voter from this city who had voted in the election is selected at random, what is the probability that this person had an income in the 17–33 percentile?

Source: The National Election Studies

42. VOTER TURNOUT BY PROFESSION   The following table gives the percent of eligible voters grouped according to profession who responded with "Voted" in the 2000 presidential election. The table also gives the percent of people in a survey categorized by their profession.

Profession	Percent Who Voted	Percent in Each Profession
Professionals	84	12
White collar	73	24
Blue collar	66	32
Unskilled	57	10
Farmers	68	8
Housewives	66	14

If an eligible voter who participated in the survey and voted in the election is selected at random, what is the probability that this person is a housewife?

Source: The National Election Studies

7.6 Solutions to Self-Check Exercises

1. By Bayes' theorem, we have, using the probabilities given in the tree diagram,

$$P(B|D) = \frac{P(B) \cdot P(D|B)}{P(A) \cdot P(D|A) + P(B) \cdot P(D|B) + P(C) \cdot P(D|C)}$$

$$= \frac{\left(\frac{1}{2}\right)\left(\frac{1}{2}\right)}{\left(\frac{1}{4}\right)\left(\frac{1}{3}\right) + \left(\frac{1}{2}\right)\left(\frac{1}{2}\right) + \left(\frac{1}{4}\right)\left(\frac{1}{4}\right)} = \frac{12}{19}$$

2. Let R and D, respectively, denote the event that the Republican and the Democratic candidate won the presidential election. Then, $P(R) = \frac{3}{5}$ and $P(D) = \frac{2}{5}$. Also, let C denote the event that research for the new manned bomber would continue. These data may be exhibited as in the accompanying tree diagram:

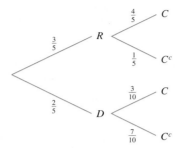

Using Bayes' theorem, we find that the probability that the Republican candidate had won the election is given by

$$P(R|C^c) = \frac{P(R) \cdot P(C^c|R)}{P(R) \cdot P(C^c|R) + P(D) \cdot P(C^c|D)}$$

$$= \frac{\left(\frac{3}{5}\right)\left(\frac{1}{5}\right)}{\left(\frac{3}{5}\right)\left(\frac{1}{5}\right) + \left(\frac{2}{5}\right)\left(\frac{7}{10}\right)} = \frac{3}{10}$$

CHAPTER 7 **Summary of Principal Formulas and Terms**

 FORMULAS

1. Probability of an event in a uniform sample space	$P(E) = \dfrac{n(E)}{n(S)}$
2. Probability of the union of two mutually exclusive events	$P(E \cup F) = P(E) + P(F)$
3. Addition rule	$P(E \cup F) = P(E) + P(F) - P(E \cap F)$
4. Rule of complements	$P(E^c) = 1 - P(E)$
5. Conditional probability	$P(B|A) = \dfrac{P(A \cap B)}{P(A)}$
6. Product rule	$P(A \cap B) = P(A) \cdot P(B|A)$
7. Test for independence	$P(A \cap B) = P(A) \cdot P(B)$

TERMS

experiment (380)
outcome (380)
sample point (380)
sample space (380)
event (380)
finite sample space (380)
union of two events (381)
intersection of two events (381)

complement of an event (381)
mutually exclusive events (382)
relative frequency (389)
empirical probability (389)
probability of an event (389)
elementary (simple) event (389)
probability distribution (389)
probability function (389)

uniform sample space (390)
addition principle (391)
conditional probability (415)
finite stochastic process (419)
independent events (422)
Bayes' theorem (430)

CHAPTER 7 **Concept Review Questions**

Fill in the blanks.

1. An activity with observable results is called a/an _____; an outcome of an experiment is called a _____ point, and the set consisting of all possible sample points of an experiment is called a sample _____; a subset of a sample space of an experiment is called a/an _____.

2. The events E and F are mutually exclusive if $E \cap F =$ _____.

3. A sample space in which the outcomes are equally likely is called a/an _____ sample space; if such a space contains n simple events, then the probability of each simple event is _____.

4. The probability of the event B given that the event A has already occurred is called the _____ probability of B given A.

5. If the outcome of one event does not depend on the other, then the two events are said to be _____.

6. The probability of an event after the outcomes of an experiment have been observed is called a/an _____ _____.

CHAPTER 7 **Review Exercises**

1. Let E and F be two mutually exclusive events and suppose $P(E) = .4$ and $P(F) = .2$. Compute
   a. $P(E \cap F)$ **b.** $P(E \cup F)$
   c. $P(E^c)$ **d.** $P(E^c \cap F^c)$
   e. $P(E^c \cup F^c)$

2. Let E and F be two events of an experiment with sample space S. Suppose $P(E) = .3$, $P(F) = .2$, and $P(E \cap F) = .15$. Compute
   a. $P(E \cup F)$
   b. $P(E^c \cap F^c)$
   c. $P(E^c \cap F)$

3. A die is loaded, and it was determined that the probability distribution associated with the experiment of casting the die and observing which number falls uppermost is given by

Simple Event	Probability
{1}	.20
{2}	.12
{3}	.16
{4}	.18
{5}	.15
{6}	.19

   a. What is the probability of the number being even?
   b. What is the probability of the number being either a 1 or a 6?
   c. What is the probability of the number being less than 4?

4. An urn contains six red, five black, and four green balls. If two balls are selected at random without replacement from the urn, what is the probability that a red ball and a black ball will be selected?

5. QUALITY CONTROL The quality-control department of Starr Communications, the manufacturer of video-game cartridges, has determined from records that 1.5% of the cartridges sold have video defects, 0.8% have audio defects, and 0.4% have both audio and video defects. What is the probability that a cartridge purchased by a customer
   a. Will have a video or audio defect?
   b. Will not have a video or audio defect?

6. Let E and F be two events and suppose $P(E) = .35$, $P(F) = .55$, and $P(E \cup F) = .70$. Find $P(E|F)$.

The accompanying tree diagram represents an experiment consisting of two trials. In Exercises 7–11, use the diagram to find the given probability.

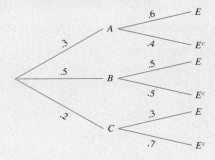

7. $P(A \cap E)$ 8. $P(B \cap E)$

9. $P(C \cap E)$ 10. $P(A|E)$

11. $P(E)$

12. An experiment consists of tossing a fair coin three times and observing the outcomes. Let A be the event that at least one

head is thrown and B the event that at most two tails are thrown.
  a. Find $P(A)$.   **b.** Find $P(B)$.
  c. Are A and B independent events?

13. QUALITY CONTROL   In a group of 20 ballpoint pens on a shelf in the stationery department of Metro Department Store, 2 are known to be defective. If a customer selects 3 of these pens, what is the probability that
  a. At least 1 is defective?
  b. No more than 1 is defective?

14. Five people are selected at random. What is the probability that none of the people in this group were born on the same day of the week?

15. A pair of fair dice is cast. What is the probability that the sum of the numbers falling uppermost is 8 if it is known that the two numbers are different?

16. A fair die is cast three times. What is the probability that it shows an even number in the first toss, an odd number in the second toss, and a 1 on the third toss? Assume that the outcomes of the tosses are independent.

17. A fair die is cast, a fair coin is tossed, and a card is drawn from a standard deck of 52 playing cards. Assuming these events are independent, what is the probability that the number falling uppermost on the die is a 6, the coin shows a tail, and the card drawn is a face card?

Three cards are drawn at random without replacement from a standard deck of 52 playing cards. In Exercises 18–22, find the probability of each of the given events.

18. All three cards are aces.

19. All three cards are face cards.

20. The second and third cards are red.

21. The second card is black, given that the first card was red.

22. The second card is a club, given that the first card was black.

23. AIRFONE USAGE   The number of planes in the fleets of five leading airlines that contain Airfones is shown in the accompanying table:

Airline	Planes with Airfones	Size of Fleet
A	50	295
B	40	325
C	31	167
D	29	50
E	25	248

  a. If a plane is selected at random from airline A, what is the probability that it contains an Airfone?
  b. If a plane is selected at random from the entire fleet of the five airlines, what is the probability that it contains an Airfone?

24. FLEX-TIME   Of 320 male and 280 female employees at the home office of Gibraltar Insurance Company, 160 of the men and 190 of the women are on flex-time (flexible working hours). Given that an employee selected at random from this group is on flex-time, what is the probability that the employee is a man?

25. QUALITY CONTROL   In a manufacturing plant, three machines, A, B, and C, produce 40%, 35%, and 25%, respectively, of the total production. The company's quality-control department has determined that 1% of the items produced by machine A, 1.5% of the items produced by machine B, and 2% of the items produced by machine C are defective. If an item is selected at random and found to be defective, what is the probability that it was produced by machine B?

26. COLLEGE ADMISSIONS   Applicants who wish to be admitted to a certain professional school in a large university are required to take a screening test that was devised by an educational testing service. From past results, the testing service has estimated that 70% of all applicants are eligible for admission and that 92% of those who are eligible for admission pass the exam, whereas 12% of those who are ineligible for admission pass the exam. Using these results, what is the probability that an applicant for admission
  a. Passed the exam?
  b. Passed the exam but was actually ineligible?

27. COMMUTING TIMES   Bill commutes to work in the business district of Boston. He takes the train $\frac{3}{5}$ of the time and drives $\frac{2}{5}$ of the time (when he visits clients). If he takes the train, then he gets home by 6:30 p.m. 85% of the time; if he drives, then he gets home by 6:30 p.m. 60% of the time. If Bill gets home by 6:30 p.m., what is the probability that he drove to work?

28. CUSTOMER SURVEYS   The sales department of Thompson Drug Company released the accompanying data concerning the sales of a certain pain reliever manufactured by the company:

Pain Reliever	Percent of Drug Sold	Percent of Group Sold in Extra-Strength Dosage
Group I (capsule form)	.57	.38
Group II (tablet form)	.43	.31

If a customer purchased the extra-strength dosage of this drug, what is the probability that it was in capsule form?

CHAPTER 7   Before Moving On . . .

1. Let $S = \{s_1, s_2, s_3, s_4, s_5, s_6\}$ be the sample space associated with an experiment having the following probability distribution:

Outcome	s_1	s_2	s_3	s_4	s_5	s_6
Probability	$\frac{1}{12}$	$\frac{2}{12}$	$\frac{3}{12}$	$\frac{2}{12}$	$\frac{3}{12}$	$\frac{1}{12}$

Find the probability of the event $A = \{s_1, s_3, s_6\}$.

2. A card is drawn from a well-shuffled 52-card deck. What is the probability that the card drawn is a deuce or a face card?

3. Let E and F be events of an experiment with sample space S. Suppose $P(E) = .5$, $P(F) = .6$, and $P(E \cap F) = .2$. Compute:
a. $P(E \cup F)$ **b.** $P(E \cap F^c)$

4. Suppose A and B are independent events with $P(A) = .3$ and $P(B) = .6$. Find $P(A \cup B)$.

5. The accompanying tree diagram represents a two-stage experiment. Use the diagram to find $P(A|D)$.

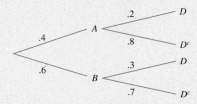

8 Probability Distributions and Statistics

© Andy Sacks/Stone/Getty Images

Where do we go from here? Some students in the top 10% of this senior class will further their education at one of the campuses of the State University system. In Example 3, page 500, we will determine the minimum grade point average a senior needs to be eligible for admission to a state university.

STATISTICS IS THAT branch of mathematics concerned with the collection, analysis, and interpretation of data. In Sections 8.1–8.3, we take a look at descriptive statistics; here, our interest lies in the description and presentation of data in the form of tables and graphs. In the rest of the chapter, we briefly examine inductive statistics, and we see how mathematical tools such as those developed in Chapter 7 may be used in conjunction with these data to help us draw certain conclusions and make forecasts.

8.1   Distensions of Random Variables

▬ Random Variables

In many situations, it is desirable to assign numerical values to the outcomes of an experiment. For example, if an experiment consists of casting a die and observing the face that lands uppermost, then it is natural to assign the numbers 1, 2, 3, 4, 5, and 6, respectively, to the outcomes *one, two, three, four, five,* and *six* of the experiment. If we let X denote the outcome of the experiment, then X assumes one of the numbers. Because the values assumed by X depend on the outcomes of a chance experiment, the outcome X is referred to as a random variable.

> **Random Variable**
>
> A random variable is a rule that assigns a number to each outcome of a chance experiment.

More precisely, a random variable is a function with domain given by the set of outcomes of a chance experiment and range contained in the set of real numbers.

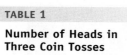

EXAMPLE 1   A coin is tossed three times. Let the random variable X denote the number of heads that occur in the three tosses.
   a. List the outcomes of the experiment; that is, find the domain of the function X.
   b. Find the value assigned to each outcome of the experiment by the random variable X.
   c. Find the event comprising the outcomes to which a value of 2 has been assigned by X. This event is written $(X = 2)$ and is the event comprising the outcomes in which two heads occur.

Solution
   a. From the results of Example 3, Section 7.1 (page 382), we see that the set of outcomes of the experiment is given by the sample space

$$S = \{\text{HHH, HHT, HTH, THH, HTT, THT, TTH, TTT}\}$$

   b. The outcomes of the experiment are displayed in the first column of Table 1. The corresponding value assigned to each such outcome by the random variable X (the number of heads) appears in the second column.
   c. With the aid of Table 1, we see that the event $(X = 2)$ is given by the set

$$\{\text{HHT, HTH, THH}\}$$

EXAMPLE 2   A coin is tossed repeatedly until a head occurs. Let the random variable Y denote the number of coin tosses in the experiment. What are the values of Y?

Solution   The outcomes of the experiment make up the infinite set

$$S = \{\text{H, TH, TTH, TTTH, TTTTH, \ldots}\}$$

These outcomes of the experiment are displayed in the first column of Table 2. The corresponding values assumed by the random variable Y (the number of tosses) appear in the second column.

TABLE 1

Number of Heads in Three Coin Tosses

Outcome	Value of X
HHH	3
HHT	2
HTH	2
THH	2
HTT	1
THT	1
TTH	1
TTT	0

TABLE 2

Number of Coin Tosses before Heads Appear

Outcome	Value of Y
H	1
TH	2
TTH	3
TTTH	4
TTTTH	5
⋮	⋮

APPLIED EXAMPLE 3   Product Reliability A disposable flashlight is turned on until its battery runs out. Let the random variable Z denote the length (in hours) of the life of the battery. What values may Z assume?

Solution   The values assumed by Z may be any nonnegative real numbers; that is, the possible values of Z comprise the interval $0 \le Z < \infty$.

One advantage of working with random variables rather than working directly with the outcomes of an experiment is that random variables are functions that may be added, subtracted, and multiplied. Because of this, results developed in the field of algebra and other areas of mathematics may be used freely to help us solve problems in probability and statistics.

A random variable is classified into three categories depending on the set of values it assumes. A random variable is called **finite discrete** if it assumes only finitely many values. For example, the random variable X of Example 1 is finite discrete since it may assume values only from the finite set $\{0, 1, 2, 3\}$ of numbers. Next, a random variable is said to be **infinite discrete** if it takes on infinitely many values, which may be arranged in a sequence. For example, the random variable Y of Example 2 is infinite discrete since it assumes values from the set $\{1, 2, 3, 4, 5, \ldots\}$, which has been arranged in the form of an infinite sequence. Finally, a random variable is called **continuous** if the values it may assume comprise an interval of real numbers. For example, the random variable Z of Example 3 is continuous since the values it may assume comprise the interval of nonnegative real numbers. For the remainder of this section, *all random variables will be assumed to be finite discrete*.

Probability Distributions of Random Variables

In Section 7.2, we learned how to construct the probability distribution for an experiment. There, the probability distribution took the form of a table that gave the probabilities associated with the outcomes of an experiment. Since the random variable associated with an experiment is related to the outcomes of the experiment, it is clear that we should be able to construct a probability distribution associated with the *random variable* rather than one associated with the outcomes of the experiment. Such a distribution is called the **probability distribution of a random variable** and may be given in the form of a formula or displayed in a table that gives the distinct (numerical) values of the random variable X and the probabilities associated with these values. Thus, if $x_1, x_2, \ldots, x_n$ are the values assumed by the random variable X with associated probabilities $P(X = x_1), P(X = x_2), \ldots, P(X = x_n)$, respectively, then the required probability distribution of the random variable X, where $p_i = P(X = x_1)$, $i = 1, 2, \ldots, n$, may be expressed in the form of the table shown in Table 3.

In the next several examples we illustrate the construction of probability distributions.

EXAMPLE 4   Find the probability distribution of the random variable associated with the experiment of Example 1.

Solution   From the results of Example 1, we see that the values assumed by the random variable X are 0, 1, 2, and 3, corresponding to the events of 0, 1, 2, and

TABLE 3

Probability Distribution for the Random Variable X

x	$P(X = x)$
x_1	p_1
x_2	p_2
x_3	p_3
.	.
.	.
.	.
x_n	p_n

TABLE 4

Probability Distribution

x	$P(X = x)$
0	$\frac{1}{8}$
1	$\frac{3}{8}$
2	$\frac{3}{8}$
3	$\frac{1}{8}$

TABLE 5

Probability Distribution for the Random Variable That Gives the Sum of the Faces of Two Dice

x	$P(X = x)$
2	$\frac{1}{36}$
3	$\frac{2}{36}$
4	$\frac{3}{36}$
5	$\frac{4}{36}$
6	$\frac{5}{36}$
7	$\frac{6}{36}$
8	$\frac{5}{36}$
9	$\frac{4}{36}$
10	$\frac{3}{36}$
11	$\frac{2}{36}$
12	$\frac{1}{36}$

TABLE 6

Probability Distribution

x	$P(X = x)$
0	.03
1	.15
2	.27
3	.20
4	.13
5	.10
6	.07
7	.03
8	.02

3 heads occurring, respectively. Referring to Table 1 once again, we see that the outcome associated with the event ($X = 0$) is given by the set {TTT}. Consequently, the probability associated with the random variable X when it assumes the value 0 is given by

$$P(X = 0) = \frac{1}{8} \qquad \text{Note that } n(S) = 8.$$

Next, observe that the event ($X = 1$) is given by the set {HTT, THT, TTH}, so

$$P(X = 1) = \frac{3}{8}$$

In a similar manner, we may compute $P(X = 2)$ and $P(X = 3)$, which gives the probability distribution shown in Table 4. ∎

EXAMPLE 5   Let X denote the random variable that gives the sum of the faces that fall uppermost when two fair dice are cast. Find the probability distribution of X.

Solution   The values assumed by the random variable X are 2, 3, 4, . . . , 12, corresponding to the events $E_2, E_3, E_4, \ldots, E_{12}$ (see Example 4, Section 7.1). Next, the probabilities associated with the random variable X when X assumes the values 2, 3, 4, . . . , 12 are precisely the probabilities $P(E_2), P(E_3), \ldots, P(E_{12})$, respectively, and may be computed in much the same way as the solution to Example 3, Section 7.2. Thus,

$$P(X = 2) = P(E_2) = \frac{1}{36}$$

$$P(X = 3) = P(E_3) = \frac{2}{36}$$

and so on. The required probability distribution of X is given in Table 5. ∎

EXAMPLE 6   Waiting Lines   The following data give the number of cars observed waiting in line at the beginning of 2-minute intervals between 3 and 5 p.m. on a certain Friday at the drive-in teller of Westwood Savings Bank and the corresponding frequency of occurrence. Find the probability distribution of the random variable X, where X denotes the number of cars observed waiting in line.

Cars	0	1	2	3	4	5	6	7	8
Frequency of Occurrence	2	9	16	12	8	6	4	2	1

Solution   Dividing each number in the last row of the given table by 60 (the sum of these numbers) gives the respective probabilities (here, we use the relative frequency interpretation of probability) associated with the random variable X when X assumes the values 0, 1, 2, . . . , 8. For example,

$$P(X = 0) = \frac{2}{60} \approx .03$$

$$P(X = 1) = \frac{9}{60} = .15$$

and so on. The resulting probability distribution is shown in Table 6. ∎

Histograms

A probability distribution of a random variable may be exhibited graphically by means of a **histogram.** To construct a histogram of a particular probability distribution, first locate the values of the random variable on a number line. Then, above each such number, erect a rectangle with width 1 and height equal to the probability associated with that value of the random variable. For example, the histogram of the probability distribution appearing in Table 4 is shown in Figure 1. The histograms of the probability distributions of Examples 5 and 6 are constructed in a similar manner and are displayed in Figures 2 and 3, respectively.

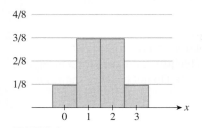

FIGURE 1
Histogram showing the probability distribution for the number of heads occurring in three coin tosses

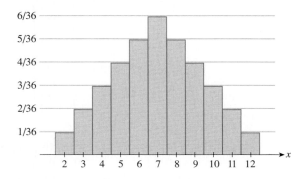

FIGURE 2
Histogram showing the probability distribution for the sum of the faces of two dice

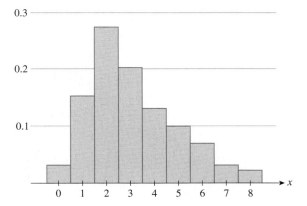

FIGURE 3
Histogram showing the probability distribution for the number of cars waiting in line

Observe that in each histogram, the area of a rectangle associated with a value of a random variable X gives precisely the probability associated with the value of X. This follows because each such rectangle, by construction, has width 1 and height corresponding to the probability associated with the value of the random variable. Another consequence arising from the method of construction of a histogram is that *the probability associated with more than one value of the random variable X is given by the sum of the areas of the rectangles associated with those values of X.* For example, in the coin-tossing experiment of Example 1, the event of obtaining at least two heads, which corresponds to the event $(X = 2)$ or $(X = 3)$, is given by

$$P(X = 2) + P(X = 3)$$

and may be obtained from the histogram depicted in Figure 1 by adding the areas associated with the values 2 and 3, respectively, of the random variable X. We obtain

$$P(X = 2) + P(X = 3) = (1)\left(\frac{3}{8}\right) + (1)\left(\frac{1}{8}\right) = \frac{1}{2}$$

This result provides us with a method of computing the probabilities of events directly from the knowledge of a histogram of the probability distribution of the random variable associated with the experiment.

EXAMPLE 7 Suppose the probability distribution of a random variable X is represented by the histogram shown in Figure 4. Identify that part of the histogram whose area gives the probability $P(10 \le X \le 20)$. Do not evaluate the result.

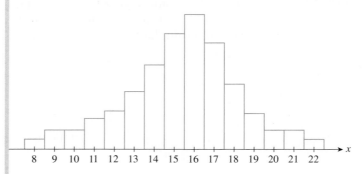

Solution The event $(10 \le X \le 20)$ is the event comprising outcomes related to the values 10, 11, 12, . . . , 20 of the random variable X. The probability of this event $P(10 \le X \le 20)$ is therefore given by the shaded area of the histogram in Figure 5.

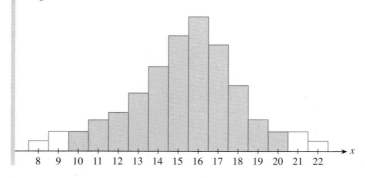

FIGURE 5
$P(10 \le X \le 20)$

8.1 Self-Check Exercises

1. Three balls are selected at random without replacement from an urn containing four black balls and five white balls. Let the random variable X denote the number of black balls drawn.
   a. List the outcomes of the experiment.
   b. Find the value assigned to each outcome of the experiment by the random variable X.
   c. Find the event comprising the outcomes to which a value of 2 has been assigned by X.

2. The following data, extracted from the records of Dover Public Library, give the number of books borrowed by the library's members over a 1-mo period:

Books	0	1	2	3	4	5	6	7	8
Frequency of Occurrence	780	300	412	205	98	54	57	30	6

   a. Find the probability distribution of the random variable X, where X denotes the number of books checked out over a 1-mo period by a randomly chosen member.
   b. Draw the histogram representing this probability distribution.

Solutions to Self-Check Exercises 8.1 can be found on page 451.

8.1 Concept Questions

1. What is a *random variable*? Give an example.

2. Give an example of (a) a *finite discrete random variable*, (b) an *infinite discrete random variable*, and (c) a *continuous random variable*.

3. Suppose you are given the probability distribution for a random variable X. Explain how you would construct a histogram for this probability distribution. What does the area of each rectangle in the histogram represent?

8.1 Exercises

1. Three balls are selected at random without replacement from an urn containing four green balls and six red balls. Let the random variable X denote the number of green balls drawn.
  a. List the outcomes of the experiment.
  b. Find the value assigned to each outcome of the experiment by the random variable X.
  c. Find the event comprising the outcomes to which a value of 3 has been assigned by X.

2. A coin is tossed four times. Let the random variable X denote the number of tails that occur.
  a. List the outcomes of the experiment.
  b. Find the value assigned to each outcome of the experiment by the random variable X.
  c. Find the event comprising the outcomes to which a value of 2 has been assigned by X.

3. A die is cast repeatedly until a 6 falls uppermost. Let the random variable X denote the number of times the die is cast. What are the values that X may assume?

4. Cards are selected one at a time without replacement from a well-shuffled deck of 52 cards until an ace is drawn. Let X denote the random variable that gives the number of cards drawn. What values may X assume?

5. Let X denote the random variable that gives the sum of the faces that fall uppermost when two fair dice are cast. Find $P(X = 7)$.

6. Two cards are drawn from a well-shuffled deck of 52 playing cards. Let X denote the number of aces drawn. Find $P(X = 2)$.

In Exercises 7–12, give the range of values that the random variable X may assume and classify the random variable as finite discrete, infinite discrete, or continuous.

7. X = The number of times a die is thrown until a 2 appears

8. X = The number of defective watches in a sample of eight watches

9. X = The distance a commuter travels to work

10. X = The number of hours a child watches television on a given day

11. X = The number of times an accountant takes the CPA examination before passing

12. X = The number of boys in a four-child family

13. The probability distribution of the random variable X is shown in the accompanying table:

x	−10	−5	0	5	10	15	20
P(X = x)	.20	.15	.05	.1	.25	.1	.15

Find
  a. $P(X = -10)$ **b.** $P(X \geq 5)$
  c. $P(-5 \leq X \leq 5)$ **d.** $P(X \leq 20)$

14. The probability distribution of the random variable X is shown in the accompanying table:

x	−5	−3	−2	0	2	3
P(X = x)	.17	.13	.33	.16	.11	.10

Find
  a. $P(X \leq 0)$ **b.** $P(X \leq -3)$
  c. $P(-2 \leq X \leq 2)$

15. Suppose a probability distribution of a random variable X is represented by the accompanying histogram. Shade that part of the histogram whose area gives the probability $P(17 \leq X \leq 20)$.

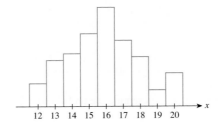

16. EXAMS   An examination consisting of ten true-or-false questions was taken by a class of 100 students. The probability distribution of the random variable X, where X denotes the number of questions answered correctly by a randomly chosen student, is represented by the accompanying histogram. The rectangle with base centered on the number 8 is missing. What should be the height of this rectangle?

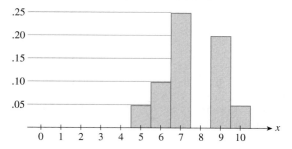

17. Two dice are cast. Let the random variable X denote the number that falls uppermost on the first die and let Y denote the number that falls uppermost on the second die.
a. Find the probability distributions of X and Y.
b. Find the probability distribution of $X + Y$.

18. DISTRIBUTION OF FAMILIES BY SIZE   A survey was conducted by the Public Housing Authority in a certain community among 1000 families to determine the distribution of families by size. The results follow:

Family Size	2	3	4	5	6	7	8
Frequency of Occurrence	350	200	245	125	66	10	4

a. Find the probability distribution of the random variable X, where X denotes the number of persons in a randomly chosen family.
b. Draw the histogram corresponding to the probability distribution found in part (a).

19. WAITING LINES   The accompanying data were obtained in a study conducted by the manager of SavMore Supermarket. In this study, the number of customers waiting in line at the express checkout at the beginning of each 3-min interval between 9 a.m. and 12 noon on Saturday was observed.

Customers	0	1	2	3	4
Frequency of Occurrence	1	4	2	7	14

Customers	5	6	7	8	9	10
Frequency of Occurrence	8	10	6	3	4	1

a. Find the probability distribution of the random variable X, where X denotes the number of customers observed waiting in line.
b. Draw the histogram representing the probability distribution.

20. MONEY MARKET RATES   The rates paid by 30 financial institutions on a certain day for money market deposit accounts are shown in the accompanying table:

Rate, %	6	6.25	6.55	6.56
Institutions	1	7	7	1

Rate, %	6.58	6.60	6.65	6.85
Institutions	1	8	3	2

Let the random variable X denote the interest paid by a randomly chosen financial institution on its money market deposit accounts and find the probability distribution associated with these data.

21. TELEVISION PILOTS   After the private screening of a new television pilot, audience members were asked to rate the new show on a scale of 1 to 10 (10 being the highest rating). From a group of 140 people, the accompanying responses were obtained:

Rating	1	2	3	4	5	6	7	8	9	10
Frequency of Occurrence	1	4	3	11	23	21	28	29	16	4

Let the random variable X denote the rating given to the show by a randomly chosen audience member. Find the probability distribution associated with these data.

22. U.S. POPULATION BY AGE   The following table gives the 2002 age distribution of the U.S. population:

Age (in years)	Under 5	5–19	20–24	25–44	45–64	65 and over
Number (in thousands)	19,527	59,716	18,611	83,009	66,088	33,590

Let the random variable X denote a randomly chosen age group within the population. Find the probability distribution associated with these data.

Source: U.S. Census Bureau

In Exercises 23 and 24, determine whether the statement is true or false. If it is true, explain why it is true. If it is false, give an example to show why it is false.

23. Suppose X is a finite discrete random variable assuming the values $x_1, x_2, \ldots, x_n$ and associated probabilities $p_1, p_2, \ldots, p_n$, then $p_1 + p_2 + \cdots + p_n = 1$.

24. The area of a histogram associated with a probability distribution is a number between 0 and 1.

8.1 Solutions to Self-Check Exercises

1. a. Using the accompanying tree diagram, we see that the outcomes of the experiment are

$$S = \{BBB, BBW, BWB, BWW,$$
$$WBB, WBW, WWB, WWW\}$$

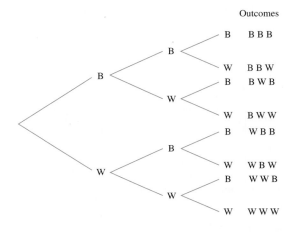

Outcomes

B B B B
W B B W
B B W B
W B W W
B W B B
W W B W
B W W B
W W W W

b. Using the results of part (a), we obtain the values assigned to the outcomes of the experiment as follows:

Outcome	BBB	BBW	BWB	BWW
Value	3	2	2	1

Outcome	WBB	WBW	WWB	WWW
Value	2	1	1	0

c. The required event is {BBW, BWB, WBB}.

2. a. We divide each number in the bottom row of the given table by 1942 (the sum of these numbers) to obtain the probabilities associated with the random variable X when X takes on the values 0, 1, 2, 3, 4, 5, 6, 7, and 8. For example,

$$P(X = 0) = \frac{780}{1942} \approx .402$$

$$P(X = 1) = \frac{300}{1942} \approx .154$$

The required probability distribution and histogram follow:

x	0	1	2	3	4
$P(X = x)$	.402	.154	.212	.106	.050

x	5	6	7	8
$P(X = x)$	.028	.029	.015	.003

b.

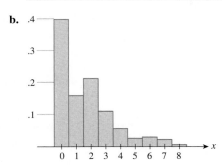

USING TECHNOLOGY

■ Graphing a Histogram

Graphing Utility

A graphing utility can be used to plot the histogram for a given set of data, as illustrated by the following example.

 APPLIED EXAMPLE 1 A survey of 90,000 households conducted in 1995 revealed the following percent of women who wear the given shoe size.

Shoe Size	<5	5–5$\frac{1}{2}$	6–6$\frac{1}{2}$	7–7$\frac{1}{2}$	8–8$\frac{1}{2}$	9–9$\frac{1}{2}$	10–10$\frac{1}{2}$	>10$\frac{1}{2}$
Women, %	1	5	15	27	29	14	7	2

Source: Footwear Market Insights survey

(continued)

a. Plot a histogram for the given data.

b. What percent of women in the survey wear size 7–$7\frac{1}{2}$ or 8–$8\frac{1}{2}$ shoes?

Solution

a. Let X denote the random variable taking on the values 1 through 8, where 1 corresponds to a shoe size less than 5, 2 corresponds to a shoe size of 5–$5\frac{1}{2}$, and so on. Entering the values of X as $x_1 = 1$, $x_2 = 2, \ldots, x_8 = 8$ and the corresponding values of Y as $y_1 = 1$, $y_2 = 5, \ldots, y_8 = 2$, and then using the **DRAW** function from the Statistics menu, we draw the histogram shown in Figure T1.

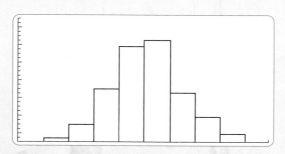

FIGURE T1
The histogram for the given data, using the viewing window $[0, 9] \times [0, 35]$

b. The probability that a woman participating in the survey wears size 7–$7\frac{1}{2}$ or 8–$8\frac{1}{2}$ shoes is given by

$$P(X = 4) + P(X = 5) = .27 + .29 = .56$$

which tells us that 56% of the women wear either size 7–$7\frac{1}{2}$ or 8–$8\frac{1}{2}$ shoes. ∎

Excel

Excel can be used to plot the histogram for a given set of data, as illustrated by the following example.

 APPLIED EXAMPLE 2   A survey of 90,000 households conducted in 1995 revealed the following percent of women who wear the given shoe size.

Shoe Size	<5	5–$5\frac{1}{2}$	6–$6\frac{1}{2}$	7–$7\frac{1}{2}$	8–$8\frac{1}{2}$	9–$9\frac{1}{2}$	10–$10\frac{1}{2}$	$>10\frac{1}{2}$
Women, %	1	5	15	27	29	14	7	2

Source: Footwear Market Insights survey

a. Plot a histogram for the given data.

b. What percent of women in the survey wear size 7–$7\frac{1}{2}$ or 8–$8\frac{1}{2}$ shoes?

Solution

a. Let X denote the random variable taking on the values 1 through 8, where 1 corresponds to a shoe size less than 5, 2 corresponds to a shoe size of 5–$5\frac{1}{2}$, and so on. Next, enter the given data in columns A and B onto a spreadsheet, as shown in Figure T2. Highlight the data in column B and select $\boxed{\Sigma}$ from the toolbar. The

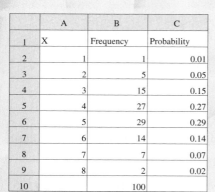

	A	B	C
	X	Frequency	Probability
1			
2	1	1	0.01
3	2	5	0.05
4	3	15	0.15
5	4	27	0.27
6	5	29	0.29
7	6	14	0.14
8	7	7	0.07
9	8	2	0.02
10		100	

FIGURE T2
Completed spreadsheet for Example 2

Note: Boldfaced words/characters enclosed in a box (for example, $\boxed{\text{Enter}}$) indicate an action (click, select, or press) is required. Words/characters printed blue (for example, Chart sub-type:) indicate words/characters that appear on the screen. Words/characters printed in a typewriter font (for example, = (−2/3)*A2+2) indicate words/characters that need to be typed and entered.

sum of the numbers in this column (100) will appear in cell B10. In cell C2, type
=B2/100 and then press [Enter]. To extend the formula to cell C9, move the
pointer to the small black box at the lower right corner of cell C2. Drag the black
+ that appears (at the lower right corner of cell C2) through cell C9 and then
release it. The probability distribution shown in cells C2 to C9 will then appear
on your spreadsheet. Then highlight the data in the Probability column and select
[Chart Wizard] from the toolbar. Select [Column] under Chart type: and click
[Next] twice. Under the Titles tab, enter Histogram, X, and Probability
in the appropriate boxes. Under the Legend tab, click the [Show legend] box to
delete the check mark. Then click [Finish]. The histogram shown in Figure T3
will appear.

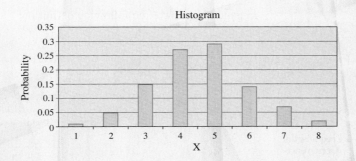

FIGURE T3
The histogram for the random
variable X

b. The probability that a woman participating in the survey wears size 7–7$\frac{1}{2}$ or
8–8$\frac{1}{2}$ shoes is given by

$$P(X = 4) + P(X = 5) = .27 + .29 = .56$$

which tells us that 56% of the women wear either size 7–7$\frac{1}{2}$ or 8–8$\frac{1}{2}$ shoes. ∎

TECHNOLOGY EXERCISES

1. Graph the histogram associated with the data given in Table 1, page 444. Compare your graph with that given in Figure 1, page 447.

2. Graph the histogram associated with the data given in Exercise 18, page 450.

3. Graph the histogram associated with the data given in Exercise 19, page 450.

4. Graph the histogram associated with the data given in Exercise 21, page 450.

8.2   Expected Value

Mean

The average value of a set of numbers is a familiar notion to most people. For example, to compute the average of the four numbers

$$12, \quad 16, \quad 23, \quad 37$$

we simply add these numbers and divide the resulting sum by 4, giving the required average as

$$\frac{12 + 16 + 23 + 37}{4} = \frac{88}{4} = 22$$

In general, we have the following definition:

> **Average, or Mean**
> The **average**, or **mean**, of the n numbers
> $$x_1, x_2, \ldots, x_n$$
> is $\bar{x}$ (read "x bar"), where
> $$\bar{x} = \frac{x_1 + x_2 + \cdots + x_n}{n}$$

TABLE 7

Cars	Frequency of Occurrence
0	2
1	9
2	16
3	12
4	8
5	6
6	4
7	2
8	1

APPLIED EXAMPLE 1   Waiting Times   Refer to Example 6, Section 8.1. Find the average number of cars waiting in line at the bank's drive-in teller at the beginning of each 2-minute interval during the period in question.

Solution   The number of cars, together with its corresponding frequency of occurrence, are reproduced in Table 7. Observe that the number 0 (of cars) occurs twice, the number 1 occurs 9 times, and so on. There are altogether

$$2 + 9 + 16 + 12 + 8 + 6 + 4 + 2 + 1 = 60$$

numbers to be averaged. Therefore, the required average is given by

$$\frac{(0 \cdot 2) + (1 \cdot 9) + (2 \cdot 16) + (3 \cdot 12) + (4 \cdot 8) + (5 \cdot 6) + (6 \cdot 4) + (7 \cdot 2) + (8 \cdot 1)}{60} \approx 3.1 \qquad \textbf{(1)}$$

or approximately 3.1 cars.

Expected Value

Let's reconsider the expression in Equation (1) that gives the average of the frequency distribution shown in Table 7. Dividing each term by the denominator, the expression may be rewritten in the form

$$0 \cdot \left(\frac{2}{60}\right) + 1 \cdot \left(\frac{9}{60}\right) + 2 \cdot \left(\frac{16}{60}\right) + 3 \cdot \left(\frac{12}{60}\right) + 4 \cdot \left(\frac{8}{60}\right) + 5 \cdot \left(\frac{6}{60}\right)$$

$$+ 6 \cdot \left(\frac{4}{60}\right) + 7 \cdot \left(\frac{2}{60}\right) + 8 \cdot \left(\frac{1}{60}\right)$$

Observe that each term in the sum is a product of two factors; the first factor is the value assumed by the random variable X, where X denotes the number of cars waiting in line, and the second factor is just the probability associated with that value of the random variable. This observation suggests the following general method for calculating the expected value (that is, the average, or mean) of a random variable X that assumes a finite number of values from the knowledge of its probability distribution.

> **Expected Value of a Random Variable X**
> Let X denote a random variable that assumes the values $x_1, x_2, \ldots , x_n$ with associated probabilities $p_1, p_2, \ldots , p_n$, respectively. Then the **expected value** of X, $E(X)$, is given by
>
> $$E(X) = x_1 p_1 + x_2 p_2 + \cdots + x_n p_n \qquad (2)$$

Note The numbers $x_1, x_2, \ldots , x_n$ may be positive, zero, or negative. For example, such a number will be positive if it represents a profit and negative if it represents a loss. ■

APPLIED EXAMPLE 2 **Waiting Times** Re-solve Example 1 by using the probability distribution associated with the experiment, reproduced in Table 8.

Solution Let X denote the number of cars waiting in line. Then, the average number of cars waiting in line is given by the expected value of X—that is, by

$$E(X) = (0)(.03) + (1)(.15) + (2)(.27) + (3)(.20) + (4)(.13)$$
$$+ (5)(.10) + (6)(.07) + (7)(.03) + (8)(.02)$$
$$= 3.1 \text{ cars}$$

which agrees with the earlier result. ■

TABLE 8

Probability Distribution

x	$P(X = x)$
0	.03
1	.15
2	.27
3	.20
4	.13
5	.10
6	.07
7	.03
8	.02

The expected value of a random variable X is a measure of the central tendency of the probability distribution associated with X. In repeated trials of an experiment with random variable X, the average of the observed values of X gets closer and closer to the expected value of X as the number of trials gets larger and larger. Geometrically, the expected value of a random variable X has the following simple interpretation: If a laminate is made of the histogram of a probability distribution associated with a random variable X, then the expected value of X corresponds to the point on the base of the laminate at which the latter will balance perfectly when the point is directly over a fulcrum (Figure 6).

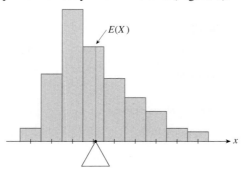

FIGURE 6
Expected value of a random variable X

TABLE 9

Probability Distribution

x	$P(X = x)$
2	$\frac{1}{36}$
3	$\frac{2}{36}$
4	$\frac{3}{36}$
5	$\frac{4}{36}$
6	$\frac{5}{36}$
7	$\frac{6}{36}$
8	$\frac{5}{36}$
9	$\frac{4}{36}$
10	$\frac{3}{36}$
11	$\frac{2}{36}$
12	$\frac{1}{36}$

FIGURE 7
Histogram showing the probability distribution for the sum of the faces of two dice

EXAMPLE 3   Let X denote the random variable that gives the sum of the faces that fall uppermost when two fair dice are cast. Find the expected value, $E(X)$, of X.

Solution   The probability distribution of X, reproduced in Table 9, was found in Example 5, Section 8.1. Using this result, we find

$$E(X) = 2\left(\frac{1}{36}\right) + 3\left(\frac{2}{36}\right) + 4\left(\frac{3}{36}\right) + 5\left(\frac{4}{36}\right) + 6\left(\frac{5}{36}\right) + 7\left(\frac{6}{36}\right)$$
$$+ 8\left(\frac{5}{36}\right) + 9\left(\frac{4}{36}\right) + 10\left(\frac{3}{36}\right) + 11\left(\frac{2}{36}\right) + 12\left(\frac{1}{36}\right)$$
$$= 7$$

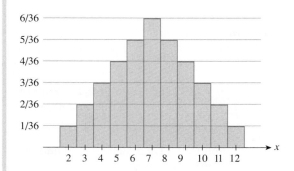

Note that, because of the symmetry of the histogram of the probability distribution with respect to the vertical line $x = 7$, the result could have been obtained by merely inspecting Figure 7. ∎

The next example shows how we can use the concept of expected value to help us make the best investment decision.

APPLIED EXAMPLE 4   Expected Profit   A group of private investors intends to purchase one of two motels currently being offered for sale in a certain city. The terms of sale of the two motels are similar, although the Regina Inn has 52 rooms and is in a slightly better location than the Merlin Motor Lodge, which has 60 rooms. Records obtained for each motel reveal that the occupancy rates, with corresponding probabilities, during the May–September tourist season are as shown in the following tables:

Regina Inn					
Occupancy Rate	.80	.85	.90	.95	1.00
Probability	.19	.22	.31	.23	.05

Merlin Motor Lodge						
Occupancy Rate	.75	.80	.85	.90	.95	1.00
Probability	.35	.21	.18	.15	.09	.02

The average profit per day for each occupied room at the Regina Inn is $10, whereas the average profit per day for each occupied room at the Merlin Motor Lodge is $9.

a. Find the average number of rooms occupied per day at each motel.

b. If the investors' objective is to purchase the motel that generates the higher daily profit, which motel should they purchase? (Compare the expected daily profit of the two motels.)

Solution

a. Let X denote the occupancy rate at the Regina Inn. Then the average daily occupancy rate at the Regina Inn is given by the expected value of X—that is, by

$$E(X) = (.80)(.19) + (.85)(.22) + (.90)(.31)$$
$$+ (.95)(.23) + (1.00)(.05)$$
$$= .8865$$

The average number of rooms occupied per day at the Regina is

$$(.8865)(52) \approx 46.1$$

or approximately 46.1 rooms. Similarly, letting Y denote the occupancy rate at the Merlin Motor Lodge, we have

$$E(Y) = (.75)(.35) + (.80)(.21) + (.85)(.18) + (.90)(.15)$$
$$+ (.95)(.09) + (1.00)(.02)$$
$$= .8240$$

The average number of rooms occupied per day at the Merlin is

$$(.8240)(60) \approx 49.4$$

or approximately 49.4 rooms.

b. The expected daily profit at the Regina is given by

$$(46.1)(10) = 461$$

or $461. The expected daily profit at the Merlin is given by

$$(49.4)(9) \approx 445$$

or $445. From these results we conclude that the investors should purchase the Regina Inn, which is expected to yield a higher daily profit. ▨

APPLIED EXAMPLE 5   Raffles   The Island Club is holding a fund-raising raffle. Ten thousand tickets have been sold for $2 each. There will be a first prize of $3000, 3 second prizes of $1000 each, 5 third prizes of $500 each, and 20 consolation prizes of $100 each. Letting X denote the net winnings (that is, winnings less the cost of the ticket) associated with the tickets, find $E(X)$. Interpret your results.

Solution   The values assumed by X are $(0 - 2)$, $(100 - 2)$, $(500 - 2)$, $(1000 - 2)$, and $(3000 - 2)$—that is, -2, 98, 498, 998, and 2998—which correspond, respectively, to the value of a losing ticket, a consolation prize, a third prize, and so on.

TABLE 10	
Probability Distribution for a Raffle	
x	$P(X = x)$
-2	.9971
98	.0020
498	.0005
998	.0003
2998	.0001

The probability distribution of X may be calculated in the usual manner and appears in Table 10. Using the table, we find

$$E(X) = (-2)(.9971) + 98(.0020) + 498(.0005)$$
$$+ 998(.0003) + 2998(.0001)$$
$$= -0.95$$

This expected value gives the long-run average loss (negative gain) of a holder of one ticket; that is, if one participated in such a raffle by purchasing one ticket each time, in the long run, one may expect to lose, on the average, 95 cents per raffle. ∎

APPLIED EXAMPLE 6 Roulette In the game of roulette as played in Las Vegas casinos, the wheel is divided into 38 compartments numbered 1 through 36, 0, and 00. One-half of the numbers 1 through 36 are red, the other half black, and 0 and 00 are green (Figure 8). Of the many types of bets that may be placed, one type involves betting on the outcome of the color of the winning number. For example, one may place a certain sum of money on *red*. If the winning number is red, one wins an amount equal to the bet placed and loses the bet otherwise. Find the expected value of the winnings on a $1 bet placed on *red*.

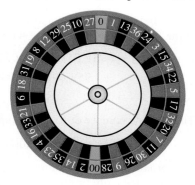

FIGURE 8
Roulette wheel

Solution Let X be a random variable whose values are 1 and -1, which correspond to a win and a loss. The probabilities associated with the values 1 and -1 are $\frac{18}{38}$ and $\frac{20}{38}$, respectively. Therefore, the expected value is given by

$$E(X) = 1\left(\frac{18}{38}\right) + (-1)\left(\frac{20}{38}\right) = -\frac{2}{38}$$
$$\approx -0.053$$

Thus, if one places a $1 bet on *red* over and over again, one may expect to lose, on the average, approximately 5 cents per bet in the long run. ∎

Examples 5 and 6 illustrate games that are not "fair." Of course, most participants in such games are aware of this fact and participate in them for other reasons. In a fair game, neither party has an advantage, a condition that translates into the condition that $E(X) = 0$, where X takes on the values of a player's winnings.

APPLIED EXAMPLE 7 Fair Games Mike and Bill play a card game with a standard deck of 52 cards. Mike selects a card from a well-shuffled deck and receives A dollars from Bill if the card selected is a diamond; otherwise, Mike pays Bill a dollar. Determine the value of A if the game is to be fair.

Solution Let X denote a random variable whose values are associated with Mike's winnings. Then X takes on the value A with probability $P(X = A) = \frac{1}{4}$ (since there are 13 diamonds in the deck) if Mike wins and takes on the value -1 with probability $P(X = -1) = \frac{3}{4}$ if Mike loses. Since the game is to be a fair one, the expected value $E(X)$ of Mike's winnings must be equal to zero; that is,

$$E(X) = A\left(\frac{1}{4}\right) + (-1)\left(\frac{3}{4}\right) = 0$$

Solving this equation for A gives $A = 3$. Thus, the card game will be fair if Bill makes a $3 payoff for a winning bet of $1 placed by Mike. ∎

Odds

In everyday parlance, the probability of the occurrence of an event is often stated in terms of the *odds in favor of* (or *odds against*) the occurrence of the event. For example, one often hears statements such as "the odds that the Dodgers will win the World Series this season are 7 to 5" and "the odds that it will not rain tomorrow are 3 to 2." We will return to these examples later. But first, let us look at a definition that ties together these two concepts.

Odds in Favor Of and Odds Against

If $P(E)$ is the probability of an event E occurring, then

1. The odds in favor of E occurring are

$$\frac{P(E)}{1 - P(E)} = \frac{P(E)}{P(E^c)} \qquad [P(E) \neq 1] \qquad \textbf{(3a)}$$

2. The odds against E occurring are

$$\frac{1 - P(E)}{P(E)} = \frac{P(E^c)}{P(E)} \qquad [P(E) \neq 0] \qquad \textbf{(3b)}$$

Notes

1. The odds in favor of the occurrence of an event are given by the ratio of the probability of the event occurring to the probability of the event not occurring. The odds against the occurrence of an event are given by the reciprocal of the odds in favor of the occurrence of the event.

2. Whenever possible, odds are expressed as ratios of whole numbers. If the odds in favor of E are a/b, we say the odds in favor of E are a to b. If the odds against E occurring are b/a, we say the odds against E are b to a. ∎

EXAMPLE 8 Find the odds in favor of winning a bet on *red* in American roulette. What are the odds against winning a bet on *red*?

Solution The probability of winning a bet here—the probability that the ball lands in a red compartment—is given by $P = \frac{18}{38}$. Therefore, using Formula (3a), we see that the odds in favor of winning a bet on *red* are

$$\frac{P(E)}{1 - P(E)} = \frac{\frac{18}{38}}{1 - \frac{18}{38}}$$ 　　　　E, event of winning a bet on *red*

$$= \frac{\frac{18}{38}}{\frac{38 - 18}{38}}$$

$$= \frac{18}{38} \cdot \frac{38}{20}$$

$$= \frac{18}{20} = \frac{9}{10}$$

or 9 to 10. Next, using (3b), we see that the odds against winning a bet on *red* are $\frac{10}{9}$, or 10 to 9.   ∎

Now, suppose the odds in favor of the occurrence of an event are a to b. Then, (3a) gives

$$\frac{a}{b} = \frac{P(E)}{1 - P(E)}$$

$$a[1 - P(E)] = bP(E)$$ 　　　　Cross-multiply

$$a - aP(E) = bP(E)$$

$$a = (a + b)P(E)$$

$$P(E) = \frac{a}{a + b}$$

which leads us to the following result.

Probability of an Event (Given the Odds)

If the odds in favor of an event E occurring are a to b, then the probability of E occurring is

$$P(E) = \frac{a}{a + b} \tag{4}$$

Formula (4) is often used to determine subjective probabilities, as the next example shows.

EXAMPLE 9   Consider each of the following statements.
a. "The odds that the Dodgers will win the World Series this season are 7 to 5."
b. "The odds that it will not rain tomorrow are 3 to 2."

Express each of these odds as a probability of the event occurring.

Solution
a. Using Formula (4) with $a = 7$ and $b = 5$ gives the required probability as
$$\frac{7}{7 + 5} = \frac{7}{12} \approx .5833$$
b. Here, the event is that it will not rain tomorrow. Using (4) with $a = 3$ and $b = 2$, we conclude that the probability that it will not rain tomorrow is
$$\frac{3}{3 + 2} = \frac{3}{5} = .6$$ ∎

TITLE   Senior Project Director
INSTITUTION   GfK ARBOR

GfK ARBOR, LLC is a full-service, custom marketing research and consulting firm. We develop and apply advanced research methodologies and analyses to a wide array of marketing and marketing research problems. We have provided services to many of the largest corporations in the United States as well as many other countries around the world across a wide array of industries. Statistics play a big part in helping our clients find solutions to their marketing problems.

A manufacturer of a brand of juice wanted to know how their brand was performing overall and how various factors in the marketplace were affecting consumers' perceptions and usage of their brand as well as the juice category overall. This juice brand was beginning a new advertising campaign, introducing new packaging and flavors, and had reduced the juice content of their product.

A year-long tracking study was conducted over the Internet with a sample of juice users. Consumers were asked their opinions, perceptions, and consumption of various brands of juice.

Using statistics, we were able to determine with certainty whether the juice brand's overall performance had improved or declined over time. Opinion ratings and brand awareness as well as consumption levels from before the new advertising campaign, packaging, flavors, and product formulation occurred were compared to levels after these events. Oftentimes we see movement or change over time, but these changes or differences must be statistically significant for us to say there has in fact been a change.

The information obtained from this research gave the juice manufacturer direction for future advertising and the ability to tailor their advertising to include elements that appeal to consumers and motivate them to purchase this particular brand of juice. The manufacturer was able to make informed decisions about whether or not to continue with the new packaging and flavors and if modifications were needed. We were able to determine that the reduced juice content did not have detrimental effects on consumer opinions or consumption of the brand.

EXPLORE & DISCUSS

In the movie *Casino*, the executive of the Tangiers Casino, Sam Rothstein (Robert DeNiro), fired the manager of the slot machines in the casino after three gamblers hit three "million dollar" jackpots in a span of 20 minutes. Rothstein claimed that it was a scam and that somebody had gotten into those machines to set the wheels. He was especially annoyed at the slot machine manager's assertion that there was no way to determine this. According to Rothstein the odds of hitting a jackpot in a four-wheel machine is 1 in $1\frac{1}{2}$ million, and the probability of hitting three jackpots in a row is "in the billions." "It cannot happen! It will not happen!" To see why Mr. Rothstein was so indignant, find the odds of hitting the jackpots in three of the machines in quick succession and comment on the likelihood of this happening.

Median and Mode

In addition to the mean, there are two other measures of central tendency of a set of numerical data: the median and the mode of a set of numbers.

Median
The median of a set of numbers arranged in increasing or decreasing order is the middle number if there is an odd number of entries, and the mean of the two middle numbers if there is an even number of entries.

APPLIED EXAMPLE 10   Commuting Times

a. The times, in minutes, Susan took to go to work on nine consecutive working days were

$$46 \quad 42 \quad 49 \quad 40 \quad 52 \quad 48 \quad 45 \quad 43 \quad 50$$

What is the median of her morning commute times?

b. The times, in minutes, Susan took to return home from work on eight consecutive working days were

$$37 \quad 36 \quad 39 \quad 37 \quad 34 \quad 38 \quad 41 \quad 40$$

What is the median of her evening commute times?

Solution

a. Arranging the numbers in increasing order, we have

$$40 \quad 42 \quad 43 \quad 45 \quad 46 \quad 48 \quad 49 \quad 50 \quad 52$$

Here we have an odd number of entries with the middle number equal to 46, and this gives the required median.

b. Arranging the numbers in increasing order, we have

$$34 \quad 36 \quad 37 \quad 37 \quad 38 \quad 39 \quad 40 \quad 41$$

Here the number of entries is even, and the required median is

$$\frac{37 + 38}{2} = 37.5$$

Mode

The **mode** of a set of numbers is the number in the set that occurs most frequently.

Note

A set may have no mode, a unique mode, or more than one mode.

EXAMPLE 11   Find the mode, if it has one, of the set of numbers.

a. 1, 2, 3, 4, 6   **b.** 2, 3, 3, 4, 6, 8   **c.** 2, 3, 3, 3, 4, 4, 4, 8

Solution

a. The set has no mode because there isn't a number that occurs more frequently than the others.

b. The mode is 3 because it occurs more frequently than the others.

c. The modes are 3 and 4 because each number occurs three times.

Finally, observe that of the three measures of central tendency of a set of numerical data, only the mean is suitable in work that requires mathematical computations.

8.2 Self-Check Exercises

1. Find the expected value of a random variable X having the following probability distribution:

x	-4	-3	-1	0	1	2
$P(X = x)$	.10	.20	.25	.10	.25	.10

2. The developer of Shoreline Condominiums has provided the following estimate of the probability that 20, 25, 30, 35, 40, 45, or 50 of the townhouses will be sold within the first month they are offered for sale:

Units	20	25	30	35	40	45	50
Probability	.05	.10	.30	.25	.15	.10	.05

How many townhouses can the developer expect to sell within the first month they are put on the market?

Solutions to Self-Check Exercises 8.2 can be found on page 467.

8.2 Concept Questions

1. What is the *expected value* of a random variable? Give an example.

2. What is a *fair game*? Is the game of roulette as played in American casinos a fair game? Why?

3. a. If the probability of an event E occurring is $P(E)$, what are the odds in favor of E occurring?
   b. If the odds in favor of an event occurring are a to b, what is the probability of E occurring?

8.2 Exercises

1. During the first year at a university that uses a 4-point grading system, a freshman took ten 3-credit courses and received two As, three Bs, four Cs, and one D.
   a. Compute this student's grade point average.
   b. Let the random variable X denote the number of points corresponding to a given letter grade. Find the probability distribution of the random variable X and compute $E(X)$, the expected value of X.

2. Records kept by the chief dietitian at the university cafeteria over a 30-wk period show the following weekly consumption of milk (in gallons):

Milk	200	205	210	215	220
Weeks	3	4	6	5	4

Milk	225	230	235	240
Weeks	3	2	2	1

   a. Find the average number of gallons of milk consumed per week in the cafeteria.
   b. Let the random variable X denote the number of gallons of milk consumed in a week at the cafeteria. Find the probability distribution of the random variable X and compute $E(X)$, the expected value of X.

3. Find the expected value of a random variable X having the following probability distribution:

x	-5	-1	0	1	5	8
$P(X = x)$	.12	.16	.28	.22	.12	.1

4. Find the expected value of a random variable X having the following probability distribution:

x	0	1	2	3	4	5
$P(X = x)$	$\frac{1}{8}$	$\frac{1}{4}$	$\frac{3}{16}$	$\frac{1}{4}$	$\frac{1}{16}$	$\frac{1}{8}$

5. The daily earnings X of an employee who works on a commission basis are given by the following probability distribution. Find the employee's expected earnings.

x (in \$)	0	25	50	75
$P(X = x)$	.07	.12	.17	.14

x (in \$)	100	125	150
$P(X = x)$	.28	.18	.04

6. In a four-child family, what is the expected number of boys? (Assume that the probability of a boy being born is the same as the probability of a girl being born.)

7. Based on past experience, the manager of the VideoRama Store has compiled the following table, which gives the probabilities that a customer who enters the VideoRama Store will buy 0, 1, 2, 3, or 4 videocassettes. How many videocassettes can a customer entering this store be expected to buy?

Video-cassettes	0	1	2	3	4
Probability	.42	.36	.14	.05	.03

8. If a sample of three batteries is selected from a lot of ten, of which two are defective, what is the expected number of defective batteries?

9. AUTO ACCIDENTS The number of accidents that occur at a certain intersection known as "Five Corners" on a Friday afternoon between the hours of 3 p.m. and 6 p.m., along with the corresponding probabilities, are shown in the following table. Find the expected number of accidents during the period in question.

Accidents	0	1	2	3	4
Probability	.935	.03	.02	.01	.005

10. EXPECTED DEMAND The owner of a newsstand in a college community estimates the weekly demand for a certain magazine as follows:

Quantity Demanded	10	11	12	13	14	15
Probability	.05	.15	.25	.30	.20	.05

Find the number of issues of the magazine that the newsstand owner can expect to sell per week.

11. EXPECTED PRODUCT RELIABILITY A bank has two automatic tellers at its main office and two at each of its three branches. The number of machines that break down on a given day, along with the corresponding probabilities, are shown in the following table:

Machines That Break Down	0	1	2	3	4
Probability	.43	.19	.12	.09	.04

Machines That Break Down	5	6	7	8
Probability	.03	.03	.02	.05

Find the expected number of machines that will break down on a given day.

12. EXPECTED SALES The management of the Cambridge Company has projected the sales of its products (in millions of dollars) for the upcoming year, with the associated probabilities shown in the following table:

Sales	20	22	24	26	28	30
Probability	.05	.1	.35	.3	.15	.05

What does the management expect the sales to be next year?

13. INTEREST RATE PREDICTION A panel of 50 economists was asked to predict the average prime interest rate for the upcoming year. The results of the survey follow:

Interest Rate, %	4.9	5.0	5.1	5.2	5.3	5.4
Economists	3	8	12	14	8	5

Based on this survey, what does the panel expect the average prime interest rate to be next year?

14. UNEMPLOYMENT RATES A panel of 64 economists was asked to predict the average unemployment rate for the upcoming year. The results of the survey follow:

Unemployment Rate, %	4.5	4.6	4.7	4.8	4.9	5.0	5.1
Economists	2	4	8	20	14	12	4

Based on this survey, what does the panel expect the average unemployment rate to be next year?

15. LOTTERIES In a lottery, 5000 tickets are sold for $1 each. One first prize of $2000, 1 second prize of $500, 3 third prizes of $100, and 10 consolation prizes of $25 are to be awarded. What are the expected net earnings of a person who buys one ticket?

16. LIFE INSURANCE PREMIUMS A man wishes to purchase a 5-yr term-life insurance policy that will pay the beneficiary $20,000 in the event that the man's death occurs during the next 5 yr. Using life insurance tables, he determines that the probability that he will live another 5 yr is .96. What is the minimum amount that he can expect to pay for his premium?
Hint: The minimum premium occurs when the insurance company's expected profit is zero.

17. LIFE INSURANCE PREMIUMS A woman purchased a $10,000, 1-yr term-life insurance policy for $130. Assuming that the probability that she will live another year is .992, find the company's expected gain.

18. LIFE INSURANCE POLICIES As a fringe benefit, Dennis Taylor receives a $25,000 life insurance policy from his employer. The probability that Dennis will live another year is .9935. If he purchases the same coverage for himself, what is the minimum amount that he can expect to pay for the policy?

19. EXPECTED PROFIT A buyer for Discount Fashions, an outlet for women's apparel, is considering buying a batch of clothing for $64,000. She estimates that the company will be able to sell it for $80,000, $75,000, or $70,000 with probabilities of .30, .60, and .10, respectively. Based on these estimates, what will be the company's expected gross profit?

20. INVESTMENT ANALYSIS The proprietor of Midland Construction Company has to decide between two projects. He estimates that the first project will yield a profit of $180,000 with a probability of .7 or a profit of $150,000 with a probability of .3; the second project will yield a profit of $220,000 with a probability of .6 or a profit of $80,000 with a probability of .4. Which project should the proprietor choose if he wants to maximize his expected profit?

21. **CABLE TELEVISION** The management of MultiVision, a cable TV company, intends to submit a bid for the cable television rights in one of two cities, A or B. If the company obtains the rights to city A, the probability of which is .2, the estimated profit over the next 10 yr is $10 million; if the company obtains the rights to city B, the probability of which is .3, the estimated profit over the next 10 yr is $7 million. The cost of submitting a bid for rights in city A is $250,000 and that of city B is $200,000. By comparing the expected profits for each venture, determine whether the company should bid for the rights in city A or city B.

22. **EXPECTED AUTO SALES** Roger Hunt intends to purchase one of two car dealerships currently for sale in a certain city. Records obtained from each of the two dealers reveal that their weekly volume of sales, with corresponding probabilities, are as follows:

Dahl Motors

Cars Sold/Week	5	6	7	8
Probability	.05	.09	.14	.24

Cars Sold/Week	9	10	11	12
Probability	.18	.14	.11	.05

Farthington Auto Sales

Cars Sold/Week	5	6	7	8	9	10
Probability	.08	.21	.31	.24	.10	.06

The average profit/car at Dahl Motors is $362, and the average profit/car at Farthington Auto Sales is $436.
a. Find the average number of cars sold each week at each dealership.
b. If Roger's objective is to purchase the dealership that generates the higher weekly profit, which dealership should he purchase? (Compare the expected weekly profit for each dealership.)

23. **EXPECTED HOME SALES** Sally Leonard, a real estate broker, is relocating in a large metropolitan area where she has received job offers from realty company A and realty company B. The number of houses she expects to sell in a year at each firm and the associated probabilities are shown in the following tables:

Company A

Houses Sold	12	13	14	15	16
Probability	.02	.03	.05	.07	.07

Houses Sold	17	18	19	20
Probability	.16	.17	.13	.11

Houses Sold	21	22	23	24
Probability	.09	.06	.03	.01

Company B

Houses Sold	6	7	8	9	10
Probability	.01	.04	.07	.06	.11

Houses Sold	11	12	13	14
Probability	.12	.19	.17	.13

Houses Sold	15	16	17	18
Probability	.04	.03	.02	.01

The average price of a house in the locale of company A is $154,000, whereas the average price of a house in the locale of company B is $237,000. If Sally will receive a 3% commission on sales at both companies, which job offer should she accept to maximize her expected yearly commission?

24. **INVESTMENT ANALYSIS** Bob, the proprietor of Midway Lumber, bases his projections for the annual revenues of the company on the performance of the housing market. He rates the performance of the market as very strong, strong, normal, weak, and very weak. For the next year, Bob estimates that the probabilities for these outcomes are .18, .27, .42, .10, and .03, respectively. He also thinks that the revenues corresponding to these outcomes are $20, $18.8, $16.2, $14, and $12 million, respectively. What is Bob's expected revenue for next year?

25. **REVENUE PROJECTION** Maria sees the growth of her business for the upcoming year as being tied to the gross domestic product (GDP). She believes that her business will grow (or contract) at the rate of 5%, 4.5%, 3%, 0%, or −0.5% per year if the GDP grows (or contracts) at the rate of between 2 and 2.5%, between 1.5 and 2%, between 1 and 1.5%, between 0 and 1%, and between −1 and 0%, respectively. Maria has decided to assign a probability of .12, .24, .40, .20, and .04, respectively, to each outcome. At what rate does Maria expect her business to grow next year?

26. **WEATHER PREDICTIONS** Suppose the probability that it will rain tomorrow is .3.
a. What are the odds that it will rain tomorrow?
b. What are the odds that it will not rain tomorrow?

27. **ROULETTE** In American roulette, as described in Example 6, a player may bet on a split (two adjacent numbers). In this case, if the player bets $1 and either number comes up, the player wins $17 and gets his $1 back. If neither comes up, he loses $1 bet. Find the expected value of the winnings on a $1 bet placed on a split.

28. **ROULETTE** If a player placed a $1 bet on *red* and a $1 bet on *black* in a single play in American roulette, what would be the expected value of his winnings?

29. **ROULETTE** In European roulette, the wheel is divided into 37 compartments numbered 1 through 36 and 0. (In American roulette there are 38 compartments numbered 1 through 36, 0, and 00.) Find the expected value of the winnings on a $1 bet placed on *red* in European roulette.

30. The probability of an event E occurring is .8. What are the odds in favor of E occurring? What are the odds against E occurring?

31. The probability of an event E not occurring is .6. What are the odds in favor of E occurring? What are the odds against E occurring?

32. The odds in favor of an event E occurring are 9 to 7. What is the probability of E occurring?

33. The odds against an event E occurring are 2 to 3. What is the probability of E not occurring?

34. ODDS   Carmen, a computer sales representative, feels that the odds are 8 to 5 that she will clinch the sale of a mini-computer to a certain company. What is the (subjective) probability that Carmen will make the sale?

35. SPORTS   Steffi feels that the odds in favor of her winning her tennis match tomorrow are 7 to 5. What is the (subjective) probability that she will win her match tomorrow?

36. SPORTS   If a sports forecaster states that the odds of a certain boxer winning a match are 4 to 3, what is the (subjective) probability that the boxer will win the match?

37. ODDS   Bob, the proprietor of Midland Lumber, feels that the odds in favor of a business deal going through are 9 to 5. What is the (subjective) probability that this deal will not materialize?

38. ROULETTE
   a. Show that for any number c,
$$E(cX) = cE(X)$$
   b. Use this result to find the expected loss if a gambler bets $300 on red in a single play in American roulette.
   Hint: Use the results of Example 6.

39. EXAM SCORES   In an examination given to a class of 20 students, the following test scores were obtained:

40   45   50   50   55   60   60   75   75   80
80   85   85   85   85   90   90   95   95   100

   a. Find the mean, or average, score, the mode, and the median score.
   b. Which of these three measures of central tendency do you think is the least representative of the set of scores?

40. WAGE RATES   The frequency distribution of the hourly wage rates (in dollars) among blue-collar workers in a certain factory is given in the following table. Find the mean (or average) wage rate, the mode, and the median wage rate of these workers.

Wage Rate	10.70	10.80	10.90	11.00	11.10	11.20
Frequency	60	90	75	120	60	45

41. WAITING TIMES   Refer to Example 6, Section 8.1. Find the median of the number of cars waiting in line at the bank's drive-in teller at the beginning of each 2-minute interval during the period in question. Compare you answer to the mean obtained in Example 1, Section 8.2.

42. SAN FRANCISCO WEATHER   The normal daily minimum temperature in degrees Fahrenheit for the months of January through December in San Francisco follows:

46.2   48.4   48.6   49.2   50.7   52.5
53.1   54.2   55.8   54.8   51.5   47.2

Find the average and the median daily temperature in San Francisco for these months.

Source: San Francisco Convention and Visitors Bureau

43. WEIGHT OF POTATO CHIPS   The weights, in ounces, of ten packages of potato chips are

16.1   16   15.8   16   15.9   16.1   15.9   16   16   16.2

Find the average, the median, and the mode of these weights.

44. BOSTON WEATHER   The relative humidity, in percent, in the morning for the months of January through December in Boston follows:

68   67   69   69   71   73
74   76   79   77   74   70

Find the average, the median, and the mode of these humidity readings.

Source: National Weather Service Forecast Office

In Exercises 45 and 46, determine whether the statement is true or false. If it is true, explain why it is true. If it is false, give an example to show why it is false.

45. A game between two persons is fair if the expected value to both persons is zero.

46. If the odds in favor of an event E occurring are a to b, then the probability of E^c occurring is $b/(a + b)$.

8.2 Solutions to Self-Check Exercises

1. $E(X) = (-4)(.10) + (-3)(.20) + (-1)(.25)$

$\quad\quad\quad + (0)(.10) + (1)(.25) + (2)(.10)$

$\quad\quad = -0.8$

2. Let X denote the number of townhouses that will be sold within 1 mo of being put on the market. Then, the number of townhouses the developer expects to sell within 1 mo is given by the expected value of X—that is, by

$$E(X) = 20(.05) + 25(.10) + 30(.30) + 35(.25)$$
$$+ 40(.15) + 45(.10) + 50(.05)$$
$$= 34.25$$

or 34 townhouses.

8.3 Variance and Standard Deviation

▬ Variance

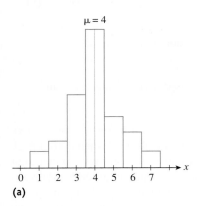

$\mu = 4$

(a)

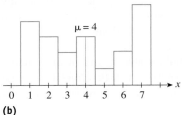

$\mu = 4$

(b)

FIGURE 9
The histograms of two probability distributions

The mean, or expected value, of a random variable enables us to express an important property of the probability distribution associated with the random variable in terms of a single number. But the knowledge of the location, or central tendency, of a probability distribution alone is usually not enough to give a reasonably accurate picture of the probability distribution. Consider, for example, the two probability distributions whose histograms appear in Figure 9. Both distributions have the same expected value, or mean, $\mu = 4$ (the Greek letter μ is read "mu"). Note that the probability distribution with the histogram shown in Figure 9a is closely concentrated about its mean μ, whereas the one with the histogram shown in Figure 9b is widely dispersed or spread about its mean.

As another example, suppose David Horowitz, host of the popular television show *The Consumer Advocate*, decides to demonstrate the accuracy of the weights of two popular brands of potato chips. Ten packages of potato chips of each brand are selected at random and weighed carefully. The results are as follows:

Weight in Ounces									
Brand A 16.1	16	15.8	16	15.9	16.1	15.9	16	16	16.2
Brand B 16.3	15.7	15.8	16.2	15.9	16.1	15.7	16.2	16	16.1

In Example 3, we will verify that the mean weights for each of the two brands is 16 ounces. However, a cursory examination of the data now shows that the weights of the brand B packages exhibit much greater dispersion about the mean than those of brand A.

One measure of the degree of dispersion, or spread, of a probability distribution about its mean is given by the variance of the random variable associated with the probability distribution. A probability distribution with a small spread about its mean will have a small variance, whereas one with a larger spread will have a larger variance. Thus, the variance of the random variable associated with the probability distribution whose histogram appears in Figure 9a is smaller than the variance of the random variable associated with the probability distribution whose histogram is shown in Figure 9b (see Example 1). Also, as we will see in Example 3, the variance of the random variable associated with the weights of the brand A potato chips is smaller than that of the random variable associated with the weights of the brand B potato chips. (This observation was made earlier.)

We now define the variance of a random variable.

Variance of a Random Variable *X*

Suppose a random variable has the probability distribution

x	x_1	x_2	x_3	$\cdots$	x_n
$P(X = x)$	p_1	p_2	p_3	$\cdots$	p_n

and expected value

$$E(X) = \mu$$

Then the **variance** of the random variable X is

$$\text{Var}(X) = p_1(x_1 - \mu)^2 + p_2(x_2 - \mu)^2 + \cdots + p_n(x_n - \mu)^2 \qquad (5)$$

Let's look a little closer at Equation (5). First, note that the numbers

$$x_1 - \mu, \quad x_2 - \mu, \ldots, \quad x_n - \mu \qquad (6)$$

measure the **deviations** of $x_1, x_2, \ldots, x_n$ from μ, respectively. Thus, the numbers

$$(x_1 - \mu)^2, \quad (x_2 - \mu)^2, \ldots, \quad (x_n - \mu)^2 \qquad (7)$$

measure the squares of the deviations of $x_1, x_2, \ldots, x_n$ from μ, respectively. Next, by multiplying each of the numbers in (7) by the probability associated with each value of the random variable X, the numbers are weighted accordingly so that their sum is a measure of the variance of X about its mean. An attempt to define the variance of a random variable about its mean in a similar manner using the deviations in (6) rather than their squares would not be fruitful since some of the deviations may be positive whereas others may be negative and (because of cancellations) the sum will not give a satisfactory measure of the variance of the random variable.

EXAMPLE 1   Find the variance of the random variable X and of the random variable Y whose probability distributions are shown in the following table. These are the probability distributions associated with the histograms shown in Figure 9a–b.

x	$P(X = x)$	y	$P(Y = y)$
1	.05	1	.2
2	.075	2	.15
3	.2	3	.1
4	.375	4	.15
5	.15	5	.05
6	.1	6	.1
7	.05	7	.25

Solution The mean of the random variable X is given by

$$\mu_X = (1)(.05) + (2)(.075) + (3)(.2) + (4)(.375) + (5)(.15)$$
$$+ (6)(.1) + (7)(.05)$$
$$= 4$$

Therefore, using Equation (5) and the data from the probability distribution of X, we find that the variance of X is given by

$$\text{Var}(X) = (.05)(1 - 4)^2 + (.075)(2 - 4)^2 + (.2)(3 - 4)^2$$
$$+ (.375)(4 - 4)^2 + (.15)(5 - 4)^2$$
$$+ (.1)(6 - 4)^2 + (.05)(7 - 4)^2$$
$$= 1.95$$

Next, we find that the mean of the random variable Y is given by

$$\mu_Y = (1)(.2) + (2)(.15) + (3)(.1) + (4)(.15) + (5)(.05)$$
$$+ (6)(.1) + (7)(.25)$$
$$= 4$$

and so the variance of Y is given by

$$\text{Var}(Y) = (.2)(1 - 4)^2 + (.15)(2 - 4)^2 + (.1)(3 - 4)^2$$
$$+ (.15)(4 - 4)^2 + (.05)(5 - 4)^2$$
$$+ (.1)(6 - 4)^2 + (.25)(7 - 4)^2$$
$$= 5.2$$

Note that $\text{Var}(X)$ is smaller than $\text{Var}(Y)$, which confirms the earlier observations about the spread, or dispersion, of the probability distribution of X and Y, respectively.

Standard Deviation

Because Equation (5), which gives the variance of the random variable X, involves the squares of the deviations, the unit of measurement of $\text{Var}(X)$ is the square of the unit of measurement of the values of X. For example, if the values assumed by the random variable X are measured in units of a gram, then $\text{Var}(X)$ will be measured in units involving the *square* of a gram. To remedy this situation, one normally works with the square root of $\text{Var}(X)$ rather than $\text{Var}(X)$ itself. The former is called the standard deviation of X.

Standard Deviation of a Random Variable X

The **standard deviation** of a random variable X, σ (pronounced "sigma"), is defined by

$$\sigma = \sqrt{\text{Var}(X)}$$
$$= \sqrt{p_1(x_1 - \mu)^2 + p_2(x_2 - \mu)^2 + \cdots + p_n(x_n - \mu)^2} \qquad \textbf{(8)}$$

where $x_1, x_2, \ldots, x_n$ denote the values assumed by the random variable X and $p_1 = P(X = x_1), p_2 = P(X = x_2), \ldots, p_n = P(X = x_n)$.

EXAMPLE 2   Find the standard deviations of the random variables X and Y of Example 1.

Solution   From the results of Example 1, we have $\text{Var}(X) = 1.95$ and $\text{Var}(Y) = 5.2$. Taking their respective square roots, we have

$$\sigma_X = \sqrt{1.95}$$
$$\approx 1.40$$
$$\sigma_Y = \sqrt{5.2}$$
$$\approx 2.28$$

APPLIED EXAMPLE 3   **Packaging**   Let X and Y denote the random variables whose values are the weights of the brand A and brand B potato chips, respectively (see page 467). Compute the means and standard deviations of X and Y and interpret your results.

Solution   The probability distributions of X and Y may be computed from the given data as follows:

Brand A			Brand B		
x	Relative Frequency of Occurrence	$P(X = x)$	y	Relative Frequency of Occurrence	$P(Y = y)$
15.8	1	.1	15.7	2	.2
15.9	2	.2	15.8	1	.1
16.0	4	.4	15.9	1	.1
16.1	2	.2	16.0	1	.1
16.2	1	.1	16.1	2	.2
			16.2	2	.2
			16.3	1	.1

The means of X and Y are given by

$$\mu_X = (.1)(15.8) + (.2)(15.9) + (.4)(16.0) + (.2)(16.1)$$
$$+ (.1)(16.2)$$
$$= 16$$
$$\mu_Y = (.2)(15.7) + (.1)(15.8) + (.1)(15.9) + (.1)(16.0)$$
$$+ (.2)(16.1) + (.2)(16.2) + (.1)(16.3)$$
$$= 16$$

Therefore,

$$\text{Var}(X) = (.1)(15.8 - 16)^2 + (.2)(15.9 - 16)^2 + (.4)(16 - 16)^2$$
$$+ (.2)(16.1 - 16)^2 + (.1)(16.2 - 16)^2$$
$$= 0.012$$
$$\text{Var}(Y) = (.2)(15.7 - 16)^2 + (.1)(15.8 - 16)^2 + (.1)(15.9 - 16)^2$$
$$+ (.1)(16 - 16)^2 + (.2)(16.1 - 16)^2 + (.2)(16.2 - 16)^2$$
$$+ (.1)(16.3 - 16)^2$$
$$= 0.042$$

so that the required standard deviations are

$$\sigma_X = \sqrt{\text{Var}(X)}$$
$$= \sqrt{0.012}$$
$$\approx 0.11$$
$$\sigma_Y = \sqrt{\text{Var}(Y)}$$
$$= \sqrt{0.042}$$
$$\approx 0.20$$

The mean of X and that of Y are both equal to 16. Therefore, the average weight of a package of potato chips of either brand is 16 ounces. However, the standard deviation of Y is greater than that of X. This tells us that the weights of the packages of brand B potato chips are more widely dispersed about the common mean of 16 than are those of brand A. ◼

EXPLORE & DISCUSS

Suppose the mean weight of m packages of brand A potato chips is μ_1 and the standard deviation from the mean of their weight distribution is σ_1. Also suppose the mean weight of n packages of brand B potato chips is μ_2 and the standard deviation from the mean of their weight distribution is σ_2.

1. Show that the mean of the weights of packages of brand A and brand B combined is

$$\mu = \frac{m\mu_1 + n\mu_2}{m + n}$$

2. If $\mu_1 = \mu_2$, show that the standard deviation from the mean of the combined-weight distribution is

$$\sigma = \left[\frac{m\sigma_1^2 + n\sigma_2^2}{m + n}\right]^{1/2}$$

3. Refer to Example 3, page 470. Using the results of parts 1 and 2, find the mean and the standard deviation of the combined-weight distribution.

◼ Chebychev's Inequality

The standard deviation of a random variable X may be used in statistical estimations. For example, the following result, derived by the Russian mathematician P. L. Chebychev (1821–1894), gives the proportion of the values of X lying within k standard deviations of the expected value of X.

Chebychev's Inequality

Let X be a random variable with expected value μ and standard deviation σ. Then, the probability that a randomly chosen outcome of the experiment lies between $\mu - k\sigma$ and $\mu + k\sigma$ is at least $1 - (1/k^2)$; that is,

$$P(\mu - k\sigma \leq X \leq \mu + k\sigma) \geq 1 - \frac{1}{k^2} \tag{9}$$

FIGURE 10
At least 75% of the outcomes fall within this interval

FIGURE 11
At least 89% of the outcomes fall within this interval

To shed some light on this result, let's take $k = 2$ in Inequality (9) and compute

$$P(\mu - 2\sigma \leq X \leq \mu + 2\sigma) \geq 1 - \frac{1}{2^2} = 1 - \frac{1}{4} = .75$$

This tells us that at least 75% of the outcomes of the experiment lie within 2 standard deviations of the mean (Figure 10). Taking $k = 3$ in Formula (9), we have

$$P(\mu - 3\sigma \leq X \leq \mu + 3\sigma) \geq 1 - \frac{1}{3^2} = 1 - \frac{1}{9} = \frac{8}{9} \approx .89$$

This tells us that at least 89% of the outcomes of the experiment lie within 3 standard deviations of the mean (Figure 11).

EXAMPLE 4 A probability distribution has a mean of 10 and a standard deviation of 1.5. Use Chebychev's inequality to estimate the probability that an outcome of the experiment lies between 7 and 13.

Solution Here, $\mu = 10$ and $\sigma = 1.5$. Next, to determine the value of k, note that $\mu - k\sigma = 7$ and $\mu + k\sigma = 13$. Substituting the appropriate values for μ and σ, we find $k = 2$. Using Chebychev's Inequality (9), we see that the probability that an outcome of the experiment lies between 7 and 13 is given by

$$P(7 \leq X \leq 13) \geq 1 - \left(\frac{1}{2^2} \right)$$

$$= \frac{3}{4}$$

—that is, at least 75%.   ◼

Note The results of Example 4 tell us that at least 75% of the outcomes of the experiment lie between $10 - 2\sigma$ and $10 + 2\sigma$—that is, between 7 and 13.   ◼

APPLIED EXAMPLE 5   Industrial Accidents Great Northwest Lumber Company employs 400 workers in its mills. It has been estimated that X, the random variable measuring the number of mill workers who have industrial accidents during a 1-year period, is distributed with a mean of 40 and a standard deviation of 6. Using Chebychev's Inequality (9), estimate the probability that the number of workers who will have an industrial accident over a 1-year period is between 30 and 50, inclusive.

Solution Here, $\mu = 40$ and $\sigma = 6$. We wish to estimate $P(30 \leq X \leq 50)$. To use Chebychev's Inequality (9), we first determine the value of k from the equation

$$\mu - k\sigma = 30 \quad \text{or} \quad \mu + k\sigma = 50$$

Since $\mu = 40$ and $\sigma = 6$ in this case, we see that k satisfies

$$40 - 6k = 30 \quad \text{and} \quad 40 + 6k = 50$$

from which we deduce that $k = \frac{5}{3}$. Thus, the probability that the number of mill workers who will have an industrial accident during a 1-year period is between 30 and 50 is given by

$$P(30 \leq X \leq 50) \geq 1 - \frac{1}{(\frac{5}{3})^2}$$

$$= \frac{16}{25}$$

—that is, at least 64%.

8.3 Self-Check Exercises

1. Compute the mean, variance, and standard deviation of the random variable X with probability distribution as follows:

x	-4	-3	-1	0	2	5
$P(X = x)$	.1	.1	.2	.3	.1	.2

2. James recorded the following travel times (the length of time in minutes it took him to drive to work) on 10 consecutive days:

   55   50   52   48   50   52   46   48   50   51

Calculate the mean and standard deviation of the random variable X associated with these data.

Solutions to Self-Check Exercises 8.3 can be found on page 477.

8.3 Concept Questions

1. a. What is the *variance* of a random variable X?
   b. What is the *standard deviation* of a random variable X?

2. What does Chebychev's inequality measure?

8.3 Exercises

In Exercises 1–6, the probability distribution of a random variable X is given. Compute the mean, variance, and standard deviation of X.

1.

x	1	2	3	4
$P(X = x)$	.4	.3	.2	.1

2.

x	-4	-2	0	2	4
$P(X = x)$	.1	.2	.3	.1	.3

3.

x	-2	-1	0	1	2
$P(X = x)$	1/16	4/16	6/16	4/16	1/16

4.

x	10	11	12	13	14	15
$P(X = x)$	1/8	2/8	1/8	2/8	1/8	1/8

5.

x	430	480	520	565	580
$P(X = x)$	.1	.2	.4	.2	.1

6.

x	-198	-195	-193	-188	-185
$P(X = x)$	.15	.30	.10	.25	.20

7. The following histograms represent the probability distributions of the random variables X and Y. Determine by inspection which probability distribution has the larger variance.

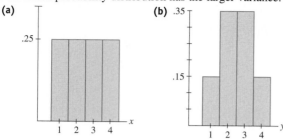

8. The following histograms represent the probability distributions of the random variables X and Y.

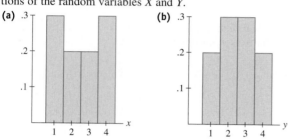

Determine by inspection which probability distribution has the larger variance.

In Exercises 9 and 10, find the variance of the probability distribution for the histogram.

9.

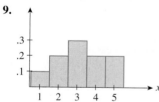

10.

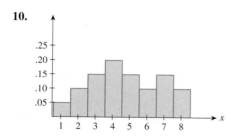

11. An experiment consists of casting an eight-sided die (numbered 1 through 8) and observing the number that appears uppermost. Find the mean and variance of this experiment.

12. DRIVING AGE REQUIREMENTS   The minimum age requirement for a regular driver's license differs from state to state. The frequency distribution for this age requirement in the 50 states is given in the following table:

Minimum Age	15	16	17	18	19	21
Frequency of Occurrence	1	15	4	28	1	1

a. Describe a random variable X that is associated with these data.
b. Find the probability distribution for the random variable X.
c. Compute the mean, variance, and standard deviation of X.

13. BIRTHRATES   The birthrates in the United States for the years 1991–2000 are given in the following table. (The birthrate is the number of live births/1000 population.)

Year	1991	1992	1993	1994
Birthrate	16.3	15.9	15.5	15.2

Year	1995	1996	1997
Birthrate	14.8	14.7	14.5

Year	1998	1999	2000
Birthrate	14.6	14.5	14.7

a. Describe a random variable X that is associated with these data.
b. Find the probability distribution for the random variable X.
c. Compute the mean, variance, and standard deviation of X.
Source: National Center for Health Statistics

14. INVESTMENT ANALYSIS   Paul Hunt is considering two business ventures. The anticipated returns (in thousands of dollars) of each venture are described by the following probability distributions:

Venture A

Earnings	Probability
−20	.3
40	.4
50	.3

Venture B

Earnings	Probability
−15	.2
30	.5
40	.3

a. Compute the mean and variance for each venture.
b. Which investment would provide Paul with the higher expected return (the greater mean)?
c. In which investment would the element of risk be less (that is, which probability distribution has the smaller variance)?

15. **INVESTMENT ANALYSIS** Rosa Walters is considering investing $10,000 in two mutual funds. The anticipated returns from price appreciation and dividends (in hundreds of dollars) are described by the following probability distributions:

Mutual Fund A

Returns	Probability
−4	.2
8	.5
10	.3

Mutual Fund B

Returns	Probability
−2	.2
6	.4
8	.4

a. Compute the mean and variance associated with the returns for each mutual fund.
b. Which investment would provide Rosa with the higher expected return (the greater mean)?
c. In which investment would the element of risk be less (that is, which probability distribution has the smaller variance)?

16. The distribution of the number of chocolate chips (x) in a cookie is shown in the following table. Find the mean and the variance of the number of chocolate chips in a cookie.

x	0	1	2
$P(X = x)$	.01	.03	.05

x	3	4	5
$P(X = x)$	.11	.13	.24

x	6	7	8
$P(X = x)$	.22	.16	.05

17. Formula (5) can also be expressed in the form

$$\text{Var}(X) = (p_1 x_1^2 + p_2 x_2^2 + \cdots + p_n x_n^2) - \mu^2$$

Find the variance of the distribution of Exercise 1, using this formula.

18. Find the variance of the distribution of Exercise 16, using the formula

$$\text{Var}(X) = (p_1 x_1^2 + p_2 x_2^2 + \cdots + p_n x_n^2) - \mu^2$$

19. **HOUSING PRICES** A survey was conducted by the market research department of the National Real Estate Company among 500 prospective buyers in a large metropolitan area to determine the maximum price a prospective buyer would be willing to pay for a house. From the data collected, the distribution that follows was obtained. Compute the mean, variance, and standard deviation of the maximum price (in thousands of dollars) that these buyers were willing to pay for a house.

Maximum Price Considered, x	$P(X = x)$
180	$\frac{10}{500}$
190	$\frac{20}{500}$
200	$\frac{75}{500}$
210	$\frac{85}{500}$
220	$\frac{70}{500}$
250	$\frac{90}{500}$
280	$\frac{90}{500}$
300	$\frac{55}{500}$
350	$\frac{5}{500}$

20. **VOLKSWAGEN'S REVENUE** The revenue of Volkswagen (in billions of Euros) for the five quarters beginning with the first quarter of 2003 are summarized in the following table:

	2003				**2004**
Quarter	Q1	Q2	Q3	Q4	Q1
Revenue	20.6	22.1	21.3	21.8	21.9

Find the average quarterly revenue of Volkswagen for the five quarters in question. What is the standard deviation?
Source: Company Reports

21. **EXAM SCORES** The following table gives the scores of 30 students in a mathematics examination:

Scores	90–99	80–89	70–79	60–69	50–59
Students	4	8	12	4	2

Find the mean and the standard deviation of the distribution of the given data.
Hint: Assume that all scores lying within a group interval take the midvalue of that group.

22. **MARITAL STATUS OF MEN** The number of married men (in thousands) between the ages of 20 and 44 in the United States in 1998 is given in the following table:

Age	20–24	25–29	30–34	35–39	40–44
Men	1332	4219	6345	7598	7633

Find the mean and the standard deviation of the given data.
Hint: See the hint in Exercise 21.
Source: U.S. Census Bureau

23. **HOURS WORKED IN SOME COUNTRIES** The average hours worked per year per worker in the United States and five European countries in 2002 is given in the following table:

Country	U.S.	Spain	Great Britain	France	West Germany	Norway
Average Hours Worked, per Worker	1815	1807	1707	1545	1428	1342

Find the average of the average hours worked, per worker, in 2002 for workers in the six countries. What is the standard deviation?

Source: Office of Economic Cooperation and Development

24. **AMERICANS WITHOUT HEALTH INSURANCE** The number of Americans without health insurance, in millions, from 1995 through 2002 is summarized in the following table:

Year	1995	1996	1997	1998	1999	2000	2001	2002
Americans	40.7	41.8	43.5	44.5	40.2	39.9	41.2	43.6

Find the average number of Americans without health insurance in the period from 1995 through 2002. What is the standard deviation?

Source: U.S. Census Bureau

25. **ACCESS TO CAPITAL** One of the key determinants to economic growth is access to capital. Using 54 variables to create an index of 1–7, 7 being best possible access to capital, Milken Institute ranked the following as the top ten nations (although, technically Hong Kong is not a nation) by the ability of their entrepreneurs to get access to capital:

Country	Hong Kong	Netherlands	U.K.	Singapore	Switzerland
Index	5.7	5.59	5.57	5.56	5.55

Country	U.S.	Australia	Finland	Germany	Denmark
Index	5.55	5.31	5.24	5.23	5.22

Find the mean of the indices at the top ten nations. What is the standard deviation?

Source: Milken Institute

26. **ACCESS TO CAPITAL** Refer to Exercise 25. Milken Institute also ranked the following as the ten worst-performing nations by the ability of their entrepreneurs to get access to capital:

Country	Peru	Mexico	Bulgaria	Brazil	Indonesia
Index	3.76	3.7	3.66	3.5	3.46

Country	Colombia	Turkey	Argentina	Venezuela	Russia
Index	3.46	3.43	3.2	2.88	2.19

Find the mean of the indices at the ten worst-performing nations. What is the standard deviation?

Source: Milken Institute

27. **SALES OF VEHICLES** The seasonally adjusted annualized sales rate for U.S. cars and light trucks, in millions of units, for May 2003 through April 2004 are given in the following tables:

2003

M	J	J	A	S	O	N	D
16.5	16.5	17.0	18.5	17.0	16.0	17.0	18.0

2004

J	F	M	A
16.3	16.5	16.8	16.5

What is the average monthly seasonally adjusted annualized sales rate for U.S. motor vehicles for the period in question? What is the standard deviation, and what does it say about the monthly sales?

Source: Autodata

28. **ELECTION TURNOUT** The percent of the voting age population who cast ballots in presidential election years from 1932 through 2000 are given in the following table:

Election Year	1932	1936	1940	1944	1948	1952	1956	1960	1964
Turnout, %	53	57	59	56	51	62	59	59	62

Election Year	1968	1972	1976	1980	1984	1988	1992	1996	2000
Turnout %	61	55	54	53	53	50	55	49	51

Find the mean and the standard deviation of the given data.

Source: Federal Election Commission

29. A probability distribution has a mean of 42 and a standard deviation of 2. Use Chebychev's inequality to estimate the probability that an outcome of the experiment lies between
   a. 38 and 46.   b. 32 and 52.

30. A probability distribution has a mean of 20 and a standard deviation of 3. Use Chebychev's inequality to estimate the probability that an outcome of the experiment lies between
   a. 15 and 25.   b. 10 and 30.

31. A probability distribution has a mean of 50 and a standard deviation of 1.4. Use Chebychev's inequality to find the value of c that guarantees that the probability is at least 96% that an outcome of the experiment lies between $50 - c$ and $50 + c$.

32. Suppose X is a random variable with mean μ and standard deviation σ. If a large number of trials is observed, at least what percentage of these values is expected to lie between $\mu - 2\sigma$ and $\mu + 2\sigma$?

33. **PRODUCT RELIABILITY** The expected lifetime of the deluxe model hair dryer produced by Roland Electric has a mean life of 24 mo and a standard deviation of 3 mo. Find the probability that one of these hair dryers will last between 20 and 28 mo.

34. **PRODUCT RELIABILITY** A Christmas tree light has an expected life of 200 hr and a standard deviation of 2 hr.
   a. Estimate the probability that one of these Christmas tree lights will last between 190 and 210 hr.
   b. Suppose 150,000 of these Christmas tree lights are used by a large city as part of its Christmas decorations. Estimate the number of lights that will require replacement between 180 and 220 hr of use.

35. **STARTING SALARIES** The mean annual starting salary of a new graduate in a certain profession is $42,000 with a standard deviation of $500. What is the probability that the starting

salary of a new graduate in this profession will be between $40,000 and $44,000?

36. QUALITY CONTROL   Sugar packaged by a certain machine has a mean weight of 5 lb and a standard deviation of 0.02 lb. For what values of c can the manufacturer of the machinery claim that the sugar packaged by this machine has a weight between $5 - c$ and $5 + c$ lb with probability at least 96%?

In Exercises 37 and 38, determine whether the statement is true or false. If it is true, explain why it is true. If it is false, give an example to show why it is false.

37. Both the variance and the standard deviation of a random variable measure the spread of a probability distribution.

38. Chebychev's inequality is useless when $k \le 1$.

8.3 Solutions to Self-Check Exercises

1. The mean of the random variable X is

$$\mu = (-4)(.1) + (-3)(.1) + (-1)(.2)$$
$$+ (0)(.3) + (2)(.1) + (5)(.2)$$
$$= 0.3$$

The variance of X is

$$\mathrm{Var}(X) = (.1)(-4 - 0.3)^2 + (.1)(-3 - 0.3)^2$$
$$+ (.2)(-1 - 0.3)^2 + (.3)(0 - 0.3)^2$$
$$+ (.1)(2 - 0.3)^2 + (.2)(5 - 0.3)^2$$
$$= 8.01$$

The standard deviation of X is

$$\sigma = \sqrt{\mathrm{Var}(X)} = \sqrt{8.01} \approx 2.83$$

2. We first compute the probability distribution of X from the given data as follows:

x	Relative Frequency of Occurrence	$P(X = x)$
46	1	.1
48	2	.2
50	3	.3
51	1	.1
52	2	.2
55	1	.1

The mean of X is

$$\mu = (.1)(46) + (.2)(48) + (.3)(50)$$
$$+ (.1)(51) + (.2)(52) + (.1)(55)$$
$$= 50.2$$

The variance of X is

$$\mathrm{Var}(X) = (.1)(46 - 50.2)^2 + (.2)(48 - 50.2)^2$$
$$+ (.3)(50 - 50.2)^2 + (.1)(51 - 50.2)^2$$
$$+ (.2)(52 - 50.2)^2 + (.1)(55 - 50.2)^2$$
$$= 5.76$$

from which we deduce the standard deviation

$$\sigma = \sqrt{5.76}$$
$$= 2.4$$

USING TECHNOLOGY

■ Finding the Mean and Standard Deviation

The calculation of the mean and standard deviation of a random variable is facilitated by the use of a graphing utility.

EXAMPLE 1 A survey conducted in 1995 of the Fortune 1000 companies revealed the following age distribution of the company directors:

Age	20–25	25–30	30–35	35–40	40–45	45–50	50–55
Directors	1	6	28	104	277	607	1142

Age	55–60	60–65	65–70	70–75	75–80	80–85	85–90
Directors	1413	1424	494	159	62	31	5

Source: Directorship

a. Plot a histogram for the given data.
b. Find the mean age and the standard deviation of the company directors.

Solution

a. Let X denote the random variable taking on the values 1 through 14, where 1 corresponds to the age bracket 20–25, 2 corresponds to the age bracket 25–30, and so on. Entering the values of X as $x_1 = 1$, $x_2 = 2, \ldots, x_{14} = 14$ and the corresponding values of Y as $y_1 = 1$, $y_2 = 6, \ldots, y_{14} = 5$, and then using the **DRAW** function from the Statistics menu of a graphing utility, we obtain the histogram shown in Figure T1.

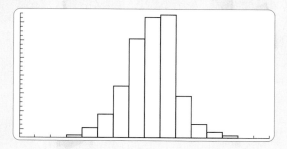

FIGURE T1
The histogram for the given data, using the viewing window [0, 16] × [0, 1500]

b. Using the appropriate function from the Statistics menu, we find that $\bar{x} = 7.9193$ and $\sigma x = 1.6378$; that is, the mean of X is $\mu \approx 7.9$, and the standard deviation is $\sigma \approx 1.6$. Thus, the average age of the directors is in the 55- to 60-year old bracket. ■

TECHNOLOGY EXERCISES

1. a. Graph the histogram associated with the random variable X in Example 1, page 468.
   b. Find the mean and the standard deviation for these data.

2. a. Graph the histogram associated with the random variable Y in Example 1, page 468.
   b. Find the mean and the standard deviation for these data.

3. a. Graph the histogram associated with the data given in Exercise 12, page 474.
   b. Find the mean and the standard deviation for these data.

4. a. Graph the histogram associated with the data given in Exercise 16, page 475.
   b. Find the mean and the standard deviation for these data.

5. A sugar refiner uses a machine to pack sugar in 5-lb cartons. To check the machine's accuracy, cartons are selected at random and weighed. The results follow:

4.98	5.02	4.96	4.97	5.03
4.96	4.98	5.01	5.02	5.06
4.97	5.04	5.04	5.01	4.99
4.98	5.04	5.01	5.03	5.05
4.96	4.97	5.02	5.04	4.97
5.03	5.01	5.00	5.01	4.98

   a. Describe a random variable X that is associated with these data.
   b. Find the probability distribution for the random variable X.
   c. Compute the mean and standard deviation of X.

6. The scores of 25 students in a mathematics examination follow:

90	85	74	92	68	94	66
87	85	70	72	68	73	72
69	66	58	70	74	88	90
98	71	75	68

   a. Describe a random variable X that is associated with these data.
   b. Find the probability distribution for the random variable X.
   c. Compute the mean and standard deviation of X.

7. HEIGHTS OF WOMEN The following data, obtained from the records of the Westwood Health Club, give the heights (to the nearest inch) of 200 female members of the club:

Height	62	$62\frac{1}{2}$	63	$63\frac{1}{2}$	64	$64\frac{1}{2}$	65	$65\frac{1}{2}$	66
Frequency	2	3	4	8	11	20	32	30	18

Height	$66\frac{1}{2}$	67	$67\frac{1}{2}$	68	$68\frac{1}{2}$	69	$69\frac{1}{2}$	70	$70\frac{1}{2}$	71
Frequency	18	16	8	10	5	5	4	3	2	1

   a. Plot a histogram for the given data.
   b. Find the mean and the standard deviation (from the mean).

8. AGE DISTRIBUTION IN A TOWN The following table gives the distribution of the ages (in years) of the residents (in hundreds) of the town of Monroe under the age of 40:

Age	0–3	4–7	8–11	12–15	16–19
Residents	30	42	50	60	50

Age	20–23	24–27	28–31	32–35	36–39
Residents	41	50	45	42	34

Let X denote the random variable taking on the values 1 through 10, where 1 corresponds to the range 0–3, . . . , and 10 corresponds to the range 36–39.
   a. Plot a histogram for the given data.
   b. Find the mean and the standard deviation.

8.4 The Binomial Distribution

▬ Bernoulli Trials

An important class of experiments have (or may be viewed as having) two outcomes. For example, in a coin-tossing experiment, the two outcomes are *heads* and *tails*. In the card game played by Mike and Bill (Example 7, Section 8.2), one may view the selection of a diamond as a *win* (for Mike) and the selection of a card of another suit as a *loss* for Mike. For a third example, consider the experiment in which a person is inoculated with a flu vaccine. Here, the vaccine may be classified as being "effective" or "ineffective" with respect to that particular person.

In general, experiments with two outcomes are called **Bernoulli trials,** or **binomial trials.** It is standard practice to label one of the outcomes of a binomial trial a *success* and the other a *failure*. For example, in a coin-tossing experiment, the outcome a *head* may be called a success, in which case the outcome a *tail* is called a failure. Note that by using the terms *success* and *failure* in this way, we depart from their usual connotations.

A sequence of Bernoulli (binomial) trials is called a binomial experiment. More precisely, we have the following definition:

Binomial Experiment

A **binomial experiment** has the following properties:

1. The number of trials in the experiment is fixed.
2. There are two outcomes of the experiment: "success" and "failure."
3. The probability of success in each trial is the same.
4. The trials are independent of each other.

In a binomial experiment, it is customary to denote the probability of a success by the letter p and the probability of a failure by the letter q. Because the event of a success and the event of a failure are complementary events, we have the relationship

$$p + q = 1$$

or, equivalently,

$$q = 1 - p$$

The properties of a binomial experiment are illustrated in the following example.

EXAMPLE 1 A fair die is cast four times. Compute the probability of obtaining exactly one 6 in the four throws.

Solution There are four trials in this experiment. Each trial consists of casting the die once and observing the face that lands uppermost. We may view each trial as an experiment with two outcomes: a success (S) if the face that lands uppermost is a 6 and a failure (F) if it is any of the other five numbers. Letting p and q

denote the probability of success and failure, respectively, of a single trial of the experiment, we find that

$$p = \frac{1}{6} \quad \text{and} \quad q = 1 - \frac{1}{6} = \frac{5}{6}$$

Furthermore, we may assume that the trials of this experiment are independent. Thus, we have a binomial experiment.

With the aid of the multiplication principle, we see that the experiment has 2^4, or 16, outcomes. We can obtain these outcomes by constructing the tree diagram associated with the experiment (see Table 11, where the outcomes are listed according to the number of successes). From the table, we see that the event of obtaining exactly one success in four trials is given by

$$E = \{\text{SFFF, FSFF, FFSF, FFFS}\}$$

with probability given by

$$P(E) = P(\text{SFFF}) + P(\text{FSFF}) + P(\text{FFSF}) + P(\text{FFFS}) \tag{10}$$

TABLE 11

0 Success	1 Success	2 Successes	3 Successes	4 Successes
FFFF	SFFF	SSFF	SSSF	SSSS
	FSFF	SFSF	SSFS	
	FFSF	SFFS	SFSS	
	FFFS	FSSF	FSSS	
		FSFS		
		FFSS		

Since the trials (throws) are independent, the terms on the right-hand side of Equation (10) may be computed as follows:

$$P(\text{SFFF}) = P(\text{S})P(\text{F})P(\text{F})P(\text{F}) = p \cdot q \cdot q \cdot q = pq^3$$
$$P(\text{FSFF}) = P(\text{F})P(\text{S})P(\text{F})P(\text{F}) = q \cdot p \cdot q \cdot q = pq^3$$
$$P(\text{FFSF}) = P(\text{F})P(\text{F})P(\text{S})P(\text{F}) = q \cdot q \cdot p \cdot q = pq^3$$
$$P(\text{FFFS}) = P(\text{F})P(\text{F})P(\text{F})P(\text{S}) = q \cdot q \cdot q \cdot p = pq^3$$

Therefore, upon substituting these values in (10), we obtain

$$P(E) = pq^3 + pq^3 + pq^3 + pq^3 = 4pq^3$$
$$= 4\left(\frac{1}{6}\right)\left(\frac{5}{6}\right)^3 \approx .386$$

▬ Probabilities in Bernoulli Trials

Let's reexamine the computations performed in the last example. There it was found that the probability of obtaining exactly one success in a binomial experiment with four independent trials with probability of success in a single trial p is given by

$$P(E) = 4pq^3 \qquad (\text{where } q = 1 - p) \tag{11}$$

Observe that the coefficient 4 of pq^3 appearing in Equation (11) is precisely the number of outcomes of the experiment with exactly one success and three failures, the outcomes being

$$SFFF \quad FSFF \quad FFSF \quad FFFS$$

Another way of obtaining this coefficient is to think of the outcomes as arrangements of the letters S and F. Then, the number of ways of selecting one position for S from four possibilities is given by

$$C(4, 1) = \frac{4!}{1!(4 - 1)!}$$
$$= 4$$

Next, observe that, because the trials are independent, each of the four outcomes of the experiment has the same probability, given by

$$pq^3$$

where the exponents 1 and 3 of p and q, respectively, correspond to exactly one success and three failures in the trials that make up each outcome.

As a result of the foregoing discussion, we may write (11) as

$$P(E) = C(4, 1)pq^3 \tag{12}$$

We are also in a position to generalize this result. Suppose that in a binomial experiment the probability of success in any trial is p. What is the probability of obtaining exactly x successes in n independent trials? We start by counting the number of outcomes of the experiment, each of which has exactly x successes. Now, one such outcome involves x successive successes followed by $(n - x)$ failures—that is,

$$\underbrace{SS \cdots S}_{x} \underbrace{FF \cdots F}_{n - x} \tag{13}$$

The other outcomes, each of which has exactly x successes, are obtained by rearranging the Ss (x of them) and Fs ($n - x$ of them). But there are $C(n, x)$ ways of arranging these letters. Next, arguing as in Example 1, we see that each such outcome has probability given by

$$p^x q^{n-x}$$

For example, for the outcome (13), we find

$$P(\underbrace{SS \cdots S}_{x} \underbrace{FF \cdots F}_{(n-x)}) = \underbrace{P(S)P(S) \cdots P(S)}_{x} \underbrace{P(F)\,P(F) \cdots P(F)}_{(n-x)}$$

$$= \underbrace{pp \cdots p}_{x} \underbrace{qq \cdots q}_{n-x}$$

$$= p^x q^{n-x}$$

Let's state this important result formally:

Computation of Probabilities in Bernoulli Trials

In a binomial experiment in which the probability of success in any trial is p, the probability of exactly x successes in n independent trials is given by

$$C(n, x)p^x q^{n-x}$$

If we let X be the random variable that gives the number of successes in a binomial experiment, then the probability of exactly x successes in n independent trials may be written

$$P(X = x) = C(n, x)p^x q^{n-x} \qquad (x = 0, 1, 2, \ldots, n) \tag{14}$$

The random variable X is called a **binomial random variable,** and the probability distribution of X is called a **binomial distribution.**

EXAMPLE 2   A fair die is cast five times. If a 1 or a 6 lands uppermost in a trial, then the throw is considered a success. Otherwise, the throw is considered a failure.
a. Find the probability of obtaining exactly 0, 1, 2, 3, 4, and 5 successes, respectively, in this experiment.
b. Using the results obtained in the solution to part (a), construct the binomial distribution for this experiment and draw the histogram associated with it.

Solution
a. This is a binomial experiment with X, the binomial random variable, taking on each of the values 0, 1, 2, 3, 4, and 5 corresponding to exactly 0, 1, 2, 3, 4, and 5 successes, respectively, in five trials. Since the die is fair, the probability of a 1 or a 6 landing uppermost in any trial is given by $p = 2/6 = 1/3$, from which it also follows that $q = 1 - p = 2/3$. Finally, $n = 5$ since there are five trials (throws of the die) in this experiment. Using Equation (14), we find that the required probabilities are

$$P(X = 0) = C(5, 0)\left(\frac{1}{3}\right)^0\left(\frac{2}{3}\right)^5 = \frac{5!}{0!5!} \cdot 1 \cdot \frac{32}{243} \approx .132$$

$$P(X = 1) = C(5, 1)\left(\frac{1}{3}\right)^1\left(\frac{2}{3}\right)^4 = \frac{5!}{1!4!} \cdot \frac{16}{243} \approx .329$$

$$P(X = 2) = C(5, 2)\left(\frac{1}{3}\right)^2\left(\frac{2}{3}\right)^3 = \frac{5!}{2!3!} \cdot \frac{8}{243} \approx .329$$

$$P(X = 3) = C(5, 3)\left(\frac{1}{3}\right)^3\left(\frac{2}{3}\right)^2 = \frac{5!}{3!2!} \cdot \frac{4}{243} \approx .165$$

$$P(X = 4) = C(5, 4)\left(\frac{1}{3}\right)^4\left(\frac{2}{3}\right)^1 = \frac{5!}{4!1!} \cdot \frac{2}{243} \approx .041$$

$$P(X = 5) = C(5, 5)\left(\frac{1}{3}\right)^5\left(\frac{2}{3}\right)^0 = \frac{5!}{5!0!} \cdot \frac{1}{243} \approx .004$$

b. Using these results, we find the required binomial distribution associated with this experiment given in Table 12. Next, we use this table to construct the histogram associated with the probability distribution (Figure 12). ■

EXAMPLE 3   A fair die is cast five times. If a 1 or a 6 lands uppermost in a trial, then the throw is considered a success. Use the results from Example 2 to answer the following questions:
a. What is the probability of obtaining 0 or 1 success in the experiment?
b. What is the probability of obtaining at least 1 success in the experiment?

TABLE 12

Probability Distribution

x	$P(X = x)$
0	.132
1	.329
2	.329
3	.165
4	.041
5	.004

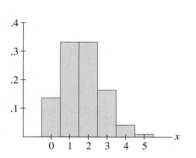

FIGURE 12
The probability of the number of successes in five throws

Solution   Interpreting the probability associated with the random variable X, when X assumes the value $X = a$, as the area of the rectangle centered about $X = a$ (Figure 12), or otherwise, we find that

a. The probability of obtaining 0 or 1 success in the experiment is given by

$$P(X = 0) + P(X = 1) = .132 + .329 = .461$$

b. The probability of obtaining at least 1 success in the experiment is given by

$$P(X = 1) + P(X = 2) + P(X = 3) + P(X = 4) + P(X = 5)$$
$$= .329 + .329 + .165 + .041 + .004$$
$$= .868$$

EXPLORE & DISCUSS

Consider the equation

$$P(X = x) = C(n, x)p^x q^{n-x}$$

for the binomial distribution.

1. Construct the histogram with $n = 5$ and $p = .2$; the histogram with $n = 5$ and $p = .5$; and the histogram with $n = 5$ and $p = .8$.

2. Comment on the shape of the histograms and give an interpretation.

The following formulas (which we state without proof) will be useful in solving problems involving binomial experiments.

Mean, Variance, and Standard Deviation of a Random Variable X

If X is a binomial random variable associated with a binomial experiment consisting of n trials with probability of success p and probability of failure q, then the **mean** (expected value), **variance,** and **standard deviation** of X are

$$\mu = E(X) = np \qquad\qquad \textbf{(15a)}$$
$$\text{Var}(X) = npq \qquad\qquad \textbf{(15b)}$$
$$\sigma_X = \sqrt{npq} \qquad\qquad \textbf{(15c)}$$

EXAMPLE 4   For the experiment in Examples 2 and 3, compute the mean, the variance, and the standard deviation of X, (a) using Formulas (15a), (15b), and (15c) and (b) using the definition of each term (Sections 8.2 and 8.3).

Solution

a. We use (15a), (15b), and (15c), with $p = \frac{1}{3}$, $q = \frac{2}{3}$, and $n = 5$, obtaining

$$\mu = E(X) = (5)\left(\frac{1}{3}\right) = \frac{5}{3} \approx 1.67$$

$$\text{Var}(X) = (5)\left(\frac{1}{3}\right)\left(\frac{2}{3}\right) = \frac{10}{9} \approx 1.11$$

$$\sigma_X = \sqrt{\text{Var}(X)} = \sqrt{1.11} \approx 1.05$$

We leave it to you to interpret the results.

b. Using the definition of expected value (Section 8.2) and the values of the probability distribution shown in Table 12, we find that

$$\mu = E(X) = (0)(.132) + (1)(.329) + (2)(.329)$$
$$+ (3)(.165) + (4)(.041) + (5)(.004)$$
$$\approx 1.67$$

which agrees with the result obtained in part (a). Next, using the definition of variance and $\mu = 1.67$, we find that

$$\text{Var}(X) = (.132)(-1.67)^2 + (.329)(-0.67)^2 + (.329)(0.33)^2$$
$$+ (.165)(1.33)^2 + (.041)(2.33)^2 + (.004)(3.33)^2$$
$$\approx 1.11$$
$$\sigma_X \approx \sqrt{\text{Var}(X)}$$
$$\approx \sqrt{1.11} \approx 1.05$$

which again agrees with the earlier results.   ■

We close this section by looking at several examples involving binomial experiments. In working through these examples, you may use a calculator, or you may consult Table 1, Appendix C.

APPLIED EXAMPLE 5   Quality Control   A division of Solaron manufactures photovoltaic cells to use in the company's solar energy converters. It is estimated that 5% of the cells manufactured are defective. If a random sample of 20 is selected from a large lot of cells manufactured by the company, what is the probability that it will contain at most 2 defective cells?

Solution   We may view this as a binomial experiment. To see this, first note that a fixed number of trials ($n = 20$) correspond to the selection of exactly 20 photovoltaic cells. Second, observe that there are exactly two outcomes in the experiment, defective ("success") and nondefective ("failure"). Third, the probability of success in each trial is .05 ($p = .05$) and the probability of failure in each trial is .95 ($q = .95$). This assumption is justified by virtue of the fact that the lot from which the cells are selected is "large," so the removal of a few cells will not appreciably affect the percentage of defective cells in the lot in each successive trial. Finally, the trials are independent of each other, once again because of the lot size.

Letting X denote the number of defective cells, we find that the probability of finding at most 2 defective cells in the sample of 20 is given by

$$P(X = 0) + P(X = 1) + P(X = 2)$$
$$= C(20, 0)(.05)^0(.95)^{20} + C(20, 1)(.05)^1(.95)^{19}$$
$$+ C(20, 2)(.05)^2(.95)^{18}$$
$$\approx .3585 + .3774 + .1887$$
$$= .9246$$

Thus, for lots of photovoltaic cells manufactured by Solaron, approximately 92% of the samples will have at most 2 defective cells; equivalently, approximately 8% of the samples will contain more than 2 defective cells.   ■

APPLIED EXAMPLE 6   Success of Heart Transplants The probability that a heart transplant performed at the Medical Center is successful (that is, the patient survives 1 year or more after undergoing such an operation) is .7. Of six patients who have recently undergone such an operation, what is the probability that 1 year from now

a. None of the heart recipients will be alive?
b. Exactly three will be alive?
c. At least three will be alive?
d. All will be alive?

Solution   Here, $n = 6$, $p = .7$, and $q = .3$. Let X denote the number of successful operations. Then,

a. The probability that no heart recipients will be alive after 1 year is given by

$$P(X = 0) = C(6, 0)(.7)^0(.3)^6$$

$$= \frac{6!}{0!6!} \cdot 1 \cdot (.3)^6$$

$$\approx .0007$$

b. The probability that exactly three will be alive after 1 year is given by

$$P(X = 3) = C(6, 3)(.7)^3(.3)^3$$

$$= \frac{6!}{3!3!} (.7)^3(.3)^3$$

$$\approx .19$$

c. The probability that at least three will be alive after 1 year is given by

$$P(X = 3) + P(X = 4) + P(X = 5) + P(X = 6)$$
$$= C(6, 3)(.7)^3(.3)^3 + C(6, 4)(.7)^4(.3)^2$$
$$+ C(6, 5)(.7)^5(.3)^1 + C(6, 6)(.7)^6(.3)^0$$
$$= \frac{6!}{3!3!} (.7)^3(.3)^3 + \frac{6!}{4!2!} (.7)^4(.3)^2 + \frac{6!}{5!1!} (.7)^5(.3)^1$$
$$+ \frac{6!}{6!0!} (.7)^6 \cdot 1$$
$$\approx .93$$

d. The probability that all will be alive after 1 year is given by

$$P(X = 6) = C(6, 6)(.7)^6(.3)^0 = \frac{6!}{6!0!} (.7)^6$$

$$\approx .12$$

APPLIED EXAMPLE 7   Quality Control PAR Bearings manufactures ball bearings packaged in lots of 100 each. The company's quality-control department has determined that 2% of the ball bearings manufactured do not meet the specifications imposed by a buyer. Find the average number of ball bearings per package that fail to meet with the specification imposed by the buyer.

Solution The experiment under consideration is binomial. The average number of ball bearings per package that fail to meet with the specifications is therefore given by the expected value of the associated binomial random variable. Using (15a), we find that

$$\mu = E(X) = np = (100)(.02) = 2$$

substandard ball bearings in a package of 100. ∎

8.4 Self-Check Exercises

1. A binomial experiment consists of four independent trials. The probability of success in each trial is .2.
   a. Find the probability of obtaining exactly 0, 1, 2, 3, and 4 successes, respectively, in this experiment.
   b. Construct the binomial distribution and draw the histogram associated with this experiment.
   c. Compute the mean and the standard deviation of the random variable associated with this experiment.

2. A survey shows that 60% of the households in a large metropolitan area have microwave ovens. If ten households are selected at random, what is the probability that five or fewer of these households have microwave ovens?

Solutions to Self-Check Exercises 8.4 can be found on page 490.

8.4 Concept Questions

1. Suppose that you are given a Bernoulli experiment.
   a. How many outcomes are there in the experiment?
   b. Can the number of trials in the experiment vary, or is it fixed?
   c. Are the trials in the experiment dependent?
   d. If the probability of success in any trial is p, what is the probability of exactly x successes in n independent trials?

2. Give the formula for the mean, variance, and standard deviation of X, where X is a binomial random variable associated with a binomial experiment consisting of n trials with probability of success p and probability of failure q.

8.4 Exercises

In Exercises 1–6, determine whether the experiment is a binomial experiment. Justify your answer.

1. Casting a fair die 3 times and observing the number of times a 6 is thrown

2. Casting a fair die and observing the number of times the die is thrown until a 6 appears uppermost

3. Casting a fair die 3 times and observing the number that appears uppermost

4. A card is selected from a deck of 52 cards, and its color is observed. A second card is then drawn (without replacement), and its color is observed.

5. Recording the number of accidents that occur at a given intersection on 4 clear days and 1 rainy day

6. Recording the number of hits a baseball player, whose batting average is .325, gets after being up to bat 5 times

In Exercises 7–10, find $C(n, x)p^x q^{n-x}$ for the values of n, x, and p.

7. $n = 4, x = 2, p = \dfrac{1}{3}$ 8. $n = 6, x = 4, p = \dfrac{1}{4}$

9. $n = 5, x = 3, p = .2$ 10. $n = 6, x = 5, p = .4$

In Exercises 11–16, use the formula $C(n, x)p^x q^{n-x}$ to determine the probability of the event.

11. The probability of exactly no successes in five trials of a binomial experiment in which $p = \frac{1}{3}$

12. The probability of exactly three successes in six trials of a binomial experiment in which $p = \frac{1}{2}$

13. The probability of at least three successes in six trials of a binomial experiment in which $p = \frac{1}{2}$

14. The probability of no successful outcomes in six trials of a binomial experiment in which $p = \frac{1}{3}$

15. The probability of no failures in five trials of a binomial experiment in which $p = \frac{1}{3}$

16. The probability of at least one failure in five trials of a binomial experiment in which $p = \frac{1}{3}$

17. A fair die is cast four times. Calculate the probability of obtaining exactly two 6's.

18. Let X be the number of successes in five independent trials of a binomial experiment in which the probability of success is $p = \frac{2}{5}$. Find:
    a. $P(X = 4)$ **b.** $P(2 \leq X \leq 4)$

19. A binomial experiment consists of five independent trials. The probability of success in each trial is .4.
    a. Find the probability of obtaining exactly 0, 1, 2, 3, 4, and 5 successes, respectively, in this experiment.
    b. Construct the binomial distribution and draw the histogram associated with this experiment.
    c. Compute the mean and the standard deviation of the random variable associated with this experiment.

20. Let the random variable X denote the number of girls in a five-child family. If the probability of a female birth is .5,
    a. Find the probability of 0, 1, 2, 3, 4, and 5 girls in a five-child family.
    b. Construct the binomial distribution and draw the histogram associated with this experiment.
    c. Compute the mean and the standard deviation of the random variable X.

21. The probability that a fuse produced by a certain manufacturing process will be defective is $\frac{1}{50}$. Is it correct to infer from this statement that there is at most 1 defective fuse in each lot of 50 produced by this process? Justify your answer.

22. **Sports** If the probability that a certain tennis player will serve an ace is $\frac{1}{4}$, what is the probability that he will serve exactly two aces out of five serves?

23. **Sports** If the probability that a certain tennis player will serve an ace is .15, what is the probability that she will serve at least two aces out of five serves?

24. **Sales Predictions** From experience, the manager of Kramer's Book Mart knows that 40% of the people who are browsing in the store will make a purchase. What is the probability that among ten people who are browsing in the store, at least three will make a purchase?

25. **Customer Services** Mayco, a mail-order department store, has six telephone lines available for customers who wish to place their orders. If the probability that during business hours any one of the six telephone lines is engaged is $\frac{1}{4}$, find the probability that when a customer calls to place an order all six lines will be in use.

26. **Restaurant Violations of the Health Code** Suppose 30% of the restaurants in a certain part of a town are in violation of the health code. If a health inspector randomly selects five of the restaurants for inspection, what is the probability that
    a. None of the restaurants are in violation of the health code?
    b. Just one of the restaurants is in violation of the health code?
    c. At least two of the restaurants are in violation of the health code?

27. **Advertisements** An advertisement for brand A chicken noodle soup claims that 60% of all consumers prefer brand A over brand B, the chief competitor's product. To test this claim, David Horowitz, host of *The Consumer Advocate*, selected ten people at random from the audience. After tasting both soups, each person was asked to state his or her preference. Assuming the company's claim is correct, find the probability that
    a. The company's claim was supported by the experiment; that is, six or more people stated a preference for brand A.
    b. The company's claim was not supported by the experiment; that is, fewer than six people stated a preference for brand A.

28. **Voters** In a certain congressional district, it is known that 40% of the registered voters classify themselves as conservatives. If ten registered voters are selected at random from this district, what is the probability that four of them will be conservatives?

29. **Violations of the Building Code** Suppose that one third of the new buildings in a town are in violation of the building code. If a building inspector inspects five of the buildings, find the probability that
    a. The first three buildings will pass the inspection and the remaining two will fail the inspection.
    b. Just three of the buildings will pass inspection.

30. **Exams** A biology quiz consists of eight multiple-choice questions. Five must be answered correctly to receive a passing grade. If each question has five possible answers, of

which only one is correct, what is the probability that a student who guesses at random on each question will pass the examination?

31. **BLOOD TYPES**   It is estimated that one third of the general population has blood type A^+. If a sample of nine people is selected at random, what is the probability that
    a. Exactly three of them have blood type A^+?
    b. At most three of them have blood type A^+?

32. **EXAMS**   A psychology quiz consists of ten true-or-false questions. If a student knows the correct answer to six of the questions but determines the answers to the remaining questions by flipping a coin, what is the probability that she will obtain a score of at least 90%?

33. **QUALITY CONTROL**   The probability that a DVD player produced by VCA Television is defective is estimated to be .02. If a sample of ten players is selected at random, what is the probability that the sample contains
    a. No defectives?   **b.**   At most two defectives?

34. **QUALITY CONTROL**   As part of its quality-control program, the video cartridges produced by Starr Communications are subjected to a final inspection before shipment. A sample of six cartridges is selected at random from each lot of cartridges produced, and the lot is rejected if the sample contains one or more defective cartridges. If 1.5% of the cartridges produced by Starr is defective, find the probability that a shipment will be accepted.

35. **ROBOT RELIABILITY**   An automobile-manufacturing company uses ten industrial robots as welders on its assembly line. On a given working day, the probability that a robot will be inoperative is .05. What is the probability that on a given working day
    a. Exactly two robots are inoperative?
    b. More than two robots are inoperative?

36. **ENGINE FAILURES**   The probability that an airplane engine will fail in a transcontinental flight is .001. Assuming that engine failures are independent of each other, what is the probability that, on a certain transcontinental flight, a four-engine plane will experience
    a. Exactly one engine failure?
    b. Exactly two engine failures?
    c. More than two engine failures? (*Note:* In this event, the airplane will crash!)

37. **QUALITY CONTROL**   The manager of Toy World has decided to accept a shipment of electronic games if none of a random sample of 20 is found to be defective.

a. What is the probability that he will accept the shipment if 10% of the electronic games is defective?
b. What is the probability that he will accept the shipment if 5% of the electronic games is defective?

38. **QUALITY CONTROL**   Refer to Exercise 37. If the manager's criterion for accepting shipment is that there be no more than 1 defective electronic game in a random sample of 20, what is the probability that he will accept the shipment if 10% of the electronic games is defective?

39. **QUALITY CONTROL**   Refer to Exercise 37. If the manager of the store changes his sample size to 10 and decides to accept shipment if none of the games is defective, what is the probability that he will accept the shipment if 10% of the games is defective?

40. How many times must a person toss a coin if the chances of obtaining at least one head are 99% or better?

41. **DRUG TESTING**   A new drug has been found to be effective in treating 75% of the people afflicted by a certain disease. If the drug is administered to 500 people who have this disease, what are the mean and the standard deviation of the number of people for whom the drug can be expected to be effective?

42. **COLLEGE GRADUATES**   At a certain university the probability that an entering freshman will graduate within 4 yr is .6. From an incoming class of 2000 freshmen, find
    a. The expected number of students who will graduate within 4 yr.
    b. The standard deviation of the number of students who will graduate within 4 yr.

In Exercises 43–46, determine whether the statement is true or false. If it is true, explain why it is true. If it is false, give an example to show why it is false.

43. In a binomial experiment, the number of outcomes of the experiment may be any finite number.

44. In a binomial experiment with $n = 3$, $P(X = 1 \text{ or } 2) = 3pq$.

45. If the probability that a batter gets a hit is $\frac{1}{4}$, then the batter is sure to get a hit if she bats four times.

46. The histogram associated with a binomial distribution is symmetric with respect to $x = \frac{n}{2}$ if $p = \frac{1}{2}$.

8.4   Solutions to Self-Check Exercises

1. a. We use Formula (14) with $n = 4$, $p = .2$, and $q = 1 - .2 = .8$, obtaining

$$P(X = 0) = C(4, 0)(.2)^0(.8)^4$$

$$= \frac{4!}{0!4!} \cdot 1 \cdot (.8)^4 \approx .410$$

$$P(X = 1) = C(4, 1)(.2)^1(.8)^3$$

$$= \frac{4!}{1!3!}(.2)(.8)^3 \approx .410$$

$$P(X = 2) = C(4, 2)(.2)^2(.8)^2$$

$$= \frac{4!}{2!2!}(.2)^2(.8)^2 \approx .154$$

$$P(X = 3) = C(4, 3)(.2)^3(.8)^1$$

$$= \frac{4!}{3!1!}(.2)^3(.8) \approx .026$$

$$P(X = 4) = C(4, 4)(.2)^4(.8)^0$$

$$= \frac{4!}{4!0!}(.2)^4 \cdot 1 \approx .002$$

b. The required binomial distribution and histogram are as follows:

x	$P(X = x)$
0	.410
1	.410
2	.154
3	.026
4	.002

c. The mean is

$$\mu = E(X) = np = (4)(.2)$$

$$= 0.8$$

and the standard deviation is

$$\sigma = \sqrt{npq} = \sqrt{(4)(.2)(.8)}$$

$$= 0.8$$

2. This is a binomial experiment with $n = 10$, $p = .6$, and $q = .4$. Let X denote the number of households that have microwave ovens. Then, the probability that five or fewer households have microwave ovens is given by

$$P(X = 0) + P(X = 1) + P(X = 2) + P(X = 3)$$
$$+ P(X = 4) + P(X = 5)$$
$$= C(10, 0)(.6)^0(.4)^{10} + C(10, 1)(.6)^1(.4)^9$$
$$+ C(10, 2)(.6)^2(.4)^8 + C(10, 3)(.6)^3(.4)^7$$
$$+ C(10, 4)(.6)^4(.4)^6 + C(10, 5)(.6)^5(.4)^5$$
$$\approx 0 + .002 + .011 + .042 + .111 + .201$$
$$\approx .37$$

8.5   The Normal Distribution

■ Probability Density Functions

The probability distributions discussed in the preceding sections were all associated with finite random variables—that is, random variables that take on finitely many values. Such probability distributions are referred to as *finite probability distributions*. In this section, we consider probability distributions associated with a continuous random variable—that is, a random variable that may take on any value lying in an interval of real numbers. Such probability distributions are called **continuous probability distributions.**

Unlike a finite probability distribution, which may be exhibited in the form of a table, a continuous probability distribution is defined by a function f whose domain coincides with the interval of values taken on by the random variable associated with the experiment. Such a function f is called the **probability density function** associated with the probability distribution and has the following properties:

1. $f(x)$ is nonnegative for all values of x.
2. The area of the region between the graph of f and the x-axis is equal to 1 (Figure 13).

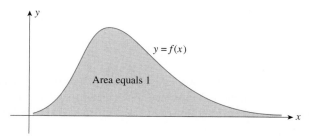

FIGURE 13
A probability density function

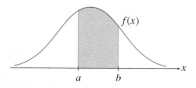

FIGURE 14
$P(a < X < b)$ is given by the area of the shaded region.

Now suppose we are given a continuous probability distribution defined by a probability density function f. Then, the probability that the random variable X assumes a value in an interval $a < x < b$ is given by the area of the region between the graph of f and the x-axis from $x = a$ to $x = b$ (Figure 14). We denote the value of this probability by $P(a < X < b)$.* Observe that Property 2 of the probability density function states that the probability that a continuous random variable takes on a value lying in its range is 1, a certainty, which is expected. Note the analogy between the areas under the probability density curves and the histograms associated with finite probability distributions (see Section 8.1).

■ Normal Distributions

The mean μ and the standard deviation σ of a continuous probability distribution have roughly the same meanings as the mean and standard deviation of a finite probability distribution. Thus, the mean of a continuous probability distribution is a measure of the central tendency of the probability distribution, and the standard deviation of the probability distribution measures its spread about its mean. Both of these numbers will play an important role in the following discussion.

For the remainder of this section, we will discuss a special class of continuous probability distributions known as **normal distributions.** The normal distribution is without doubt the most important of all the probability distributions. Many phenomena, such as the heights of people in a given population, the weights of newborn infants, the IQs of college students, the actual weights of 16-ounce packages of cereals, and so on, have probability distributions that are normal. The normal distribution also provides us with an accurate approximation to the distributions of many random variables associated with random-sampling problems. In fact, in the next section we will see how a normal distribution may be used to approximate a binomial distribution under certain conditions.

The graph of a normal distribution, which is bell shaped, is called a **normal curve** (Figure 15).

---

*Because the area under one point of the graph of f is equal to zero, we see that $P(a < X < b) = P(a < X \le b) = P(a \le X < b) = P(a \le X \le b)$.

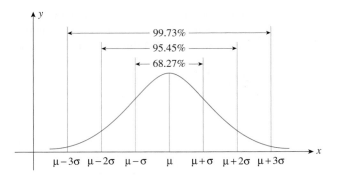

FIGURE 15
A normal curve

The normal curve (and therefore the corresponding normal distribution) is completely determined by its mean μ and standard deviation σ. In fact, the normal curve has the following characteristics, described in terms of these two parameters:*

1. The curve has a peak at $x = \mu$.
2. The curve is symmetric with respect to the vertical line $x = \mu$.
3. The curve always lies above the x-axis but approaches the x-axis as x extends indefinitely in either direction.
4. The area under the curve is 1.
5. For any normal curve, 68.27% of the area under the curve lies within 1 standard deviation of the mean (that is, between $\mu - \sigma$ and $\mu + \sigma$), 95.45% of the area lies within 2 standard deviations of the mean, and 99.73% of the area lies within 3 standard deviations of the mean.

Figure 16 shows two normal curves with different means μ_1 and μ_2 but the same deviation. Next, Figure 17 shows two normal curves with the same mean but different standard deviations σ_1 and σ_2. (Which number is smaller?)

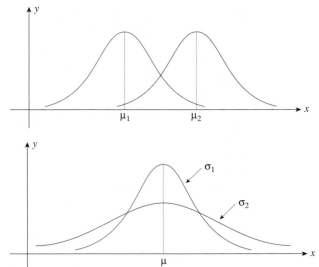

FIGURE 16
Two normal curves that have the same standard deviation but different means

FIGURE 17
Two normal curves that have the same mean but different standard deviations

---

*The probability density function associated with this normal curve is given by

$$y = \frac{1}{\sigma\sqrt{2\pi}}\, e^{-(1/2)[(x-\mu)/\sigma]^2}$$

but the direct use of this formula will not be required in our discussion of the normal distribution.

In general, the mean μ of a normal distribution determines where the center of the curve is located, whereas the standard deviation σ of a normal distribution determines the sharpness (or flatness) of the curve.

As this discussion reveals, there are infinitely many normal curves corresponding to different choices of the parameters μ and σ, which characterize such curves. Fortunately, any normal curve may be transformed into any other normal curve (as we will see later), so in the study of normal curves it suffices to single out one such particular curve for special attention. The normal curve with mean $\mu = 0$ and standard deviation $\sigma = 1$ is called the **standard normal curve.** The corresponding distribution is called the **standard normal distribution.** The random variable itself is called the **standard normal variable** and is commonly denoted by Z.

EXPLORING WITH TECHNOLOGY

Consider the probability density function

$$f(x) = \frac{1}{\sqrt{2\pi}} e^{-x^2/2}$$

which is the formula given in the footnote on page 492 with $\mu = 0$ and $\sigma = 1$.

1. Use a graphing utility to plot the graph of f, using the viewing window $[-4, 4] \times [0, 0.5]$.

2. Use the numerical integration function of a graphing utility to find the area of the region under the graph of f on the intervals $[-1, 1]$, $[-2, 2]$, and $[-3, 3]$ and thus verify Property 5 of normal distributions for the special case where $\mu = 0$ and $\sigma = 1$.

■ Computations of Probabilities Associated with Normal Distributions

Areas under the standard normal curve have been extensively computed and tabulated. Table 2, Appendix C, gives the areas of the regions under the standard normal curve to the left of the number z; these areas correspond, of course, to probabilities of the form $P(Z < z)$ or $P(Z \le z)$. The next several examples illustrate the use of this table in computations involving the probabilities associated with the standard normal variable.

EXAMPLE 1 Let Z be the standard normal variable. By first making a sketch of the appropriate region under the standard normal curve, find the values of
a. $P(Z < 1.24)$ **b.** $P(Z > 0.5)$ **c.** $P(0.24 < Z < 1.48)$
d. $P(-1.65 < Z < 2.02)$

Solution
a. The region under the standard normal curve associated with the probability $P(Z < 1.24)$ is shown in Figure 18. To find the area of the required region using Table 2, Appendix C, we first locate the number 1.2 in the column and the number 0.04 in the row, both headed by z, and read off the number 0.8925 appearing in the body of the table. Thus,

$$P(Z < 1.24) = .8925$$

FIGURE 18
$P(Z < 1.24)$

b. The region under the standard normal curve associated with the probability $P(Z > 0.5)$ is shown in Figure 19a. Observe, however, that the required area is, by virtue of the symmetry of the standard normal curve, equal to the shaded area shown in Figure 19b. Thus,

$$P(Z > 0.5) = P(Z < -0.5)$$
$$= .3085$$

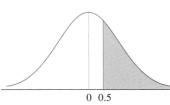 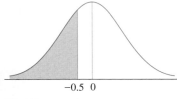

(a) $P(Z > 0.5)$ (b) $P(Z < -0.5)$

FIGURE 19

c. The probability $P(0.24 < Z < 1.48)$ is equal to the shaded area shown in Figure 20. But this area is obtained by subtracting the area under the curve to the left of $z = 0.24$ from the area under the curve to the left of $z = 1.48$; that is

$$P(0.24 < Z < 1.48) = P(Z < 1.48) - P(Z < 0.24)$$
$$= .9306 - .5948$$
$$= .3358$$

d. The probability $P(-1.65 < Z < 2.02)$ is given by the shaded area shown in Figure 21. We have

$$P(-1.65 < Z < 2.02) = P(Z < 2.02) - P(Z < -1.65)$$
$$= .9783 - .0495$$
$$= .9288$$

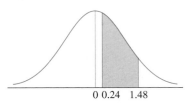

FIGURE 20
$P(0.24 < Z < 1.48)$

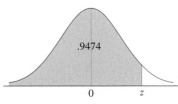

FIGURE 21
$P(-1.65 < Z < 2.02)$

EXAMPLE 2 Let Z be the standard normal variable. Find the value of z if z satisfies
a. $P(Z < z) = .9474$ **b.** $P(Z > z) = .9115$
c. $P(-z < Z < z) = .7888$

Solution
a. Refer to Figure 22. We want the value of Z such that the area of the region under the standard normal curve and to the left of $Z = z$ is .9474. Locating the number .9474 in Table 2, Appendix C, and reading back, we find that $z = 1.62$.
b. Since $P(Z > z)$, or equivalently, the area of the region to the right of z is greater than 0.5, z must be negative (Figure 23). Therefore $-z$ is positive. Furthermore, the area of the region to the right of z is the same as the area of the region to the left of $-z$. Therefore,

$$P(Z > z) = P(Z < -z)$$
$$= .9115$$

Looking up the table, we find $-z = 1.35$, so $z = -1.35$.

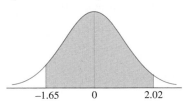

FIGURE 22
$P(Z < z) = .9474$

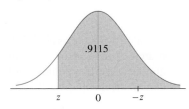

FIGURE 23
$P(Z > z) = .9115$

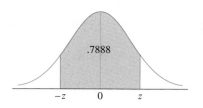

FIGURE 24
$P(-z < Z < z) = .7888$

c. The region associated with $P(-z < Z < z)$ is shown in Figure 24. Observe that by symmetry the area of this region is just double that of the area of the region between $Z = 0$ and $Z = z$; that is,

$$P(-z < Z < z) = 2P(0 < Z < z)$$

$$P(0 < Z < z) = P(Z < z) - \frac{1}{2}$$

(Figure 25). Therefore,

$$\frac{1}{2} P(-z < Z < z) = P(Z < z) - \frac{1}{2}$$

or, solving for $P(Z < z)$,

$$P(Z < z) = \frac{1}{2} + \frac{1}{2} P(-z < Z < z)$$

$$= \frac{1}{2} (1 + .7888)$$

$$= .8944$$

Consulting the table, we find $z = 1.25$.

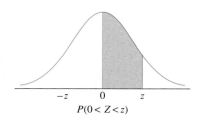

$P(0 < Z < z)$ $=$ $P(Z < z)$ $-$ $\frac{1}{2}$

FIGURE 25

We now turn our attention to the computation of probabilities associated with normal distributions whose means and standard deviations are not necessarily equal to 0 and 1, respectively. As mentioned earlier, any normal curve may be transformed into the standard normal curve. In particular, it may be shown that if X is a normal random variable with mean μ and standard deviation σ, then it can be transformed into the standard normal random variable Z by means of the substitution

$$Z = \frac{X - \mu}{\sigma}$$

The area of the region under the normal curve (with random variable X) between $x = a$ and $x = b$ is *equal* to the area of the region under the standard normal curve between $z = (a - \mu)/\sigma$ and $z = (b - \mu)/\sigma$. In terms of probabilities associated with these distributions, we have

$$P(a < X < b) = P\left(\frac{a - \mu}{\sigma} < Z < \frac{b - \mu}{\sigma}\right) \tag{16}$$

(Figure 26). Similarly, we have

$$P(X < b) = P\left(Z < \frac{b - \mu}{\sigma}\right) \tag{17}$$

$$P(X > a) = P\left(Z > \frac{a - \mu}{\sigma}\right) \tag{18}$$

Thus, with the help of Equations (16)–(18), computations of probabilities associated with any normal distribution may be reduced to the computations of areas of regions under the standard normal curve.

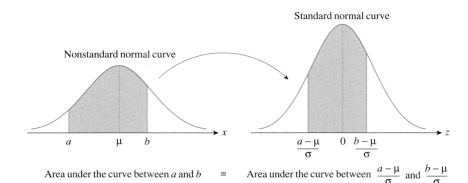

FIGURE 26

Area under the curve between a and b = Area under the curve between $\dfrac{a-\mu}{\sigma}$ and $\dfrac{b-\mu}{\sigma}$

EXAMPLE 3 Suppose X is a normal random variable with $\mu = 100$ and $\sigma = 20$. Find the values of

a. $P(X < 120)$ **b.** $P(X > 70)$ **c.** $P(75 < X < 110)$

Solution

a. Using Formula (17) with $\mu = 100$, $\sigma = 20$, and $b = 120$, we have

$$P(X < 120) = P\left(Z < \frac{120 - 100}{20}\right)$$
$$= P(Z < 1) = .8413 \qquad \text{Use the table of values of Z.}$$

b. Using Formula (18) with $\mu = 100$, $\sigma = 20$, and $a = 70$, we have

$$P(X > 70)$$
$$= P\left(Z > \frac{70 - 100}{20}\right)$$
$$= P(Z > -1.5) = P(Z < 1.5) = .9332 \qquad \text{Use the table of values of Z.}$$

c. Using Formula (16) with $\mu = 100$, $\sigma = 20$, $a = 75$, and $b = 110$, we have

$$P(75 < X < 110)$$
$$= P\left(\frac{75 - 100}{20} < Z < \frac{110 - 100}{20}\right)$$
$$= P(-1.25 < Z < 0.5)$$
$$= P(Z < 0.5) - P(Z < -1.25) \qquad \text{See Figure 27.}$$
$$= .6915 - .1056 = .5859 \qquad \text{Use the table of values of Z.}$$

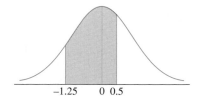

FIGURE 27

8.5 Self-Check Exercises

1. Let Z be a standard normal variable.
  a. Find the value of $P(-1.2 < Z < 2.1)$ by first making a sketch of the appropriate region under the standard normal curve.
  b. Find the value of z if z satisfies $P(-z < Z < z) = .8764$.

2. Let X be a normal random variable with $\mu = 80$ and $\sigma = 10$. Find the values of
  a. $P(X < 100)$ **b.** $P(X > 60)$ **c.** $P(70 < X < 90)$

Solutions to Self-Check Exercises 8.5 can be found on page 498.

8.5 Concept Questions

1. Consider the following normal curve with mean μ and standard deviation σ:

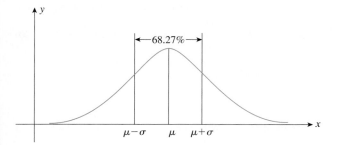

  a. What is the x-coordinate of the peak of the curve?
  b. What can you say about the symmetry of the curve?
  c. Does the curve always lie above the x-axis? What happens to the curve as x extends indefinitely to the left or right?
  d. What is the value of the area under the curve?
  e. Between what values does 68.27% of the area under the curve lie?

2. a. What is the difference between a *normal curve* and a *standard normal curve*?
  b. If X is a normal random variable with mean μ and standard deviation σ, write $P(a < X < b)$ in terms of the probabilities associated with the standard normal random variable Z.

8.5 Exercises

In Exercises 1–6, find the value of the probability of the standard normal variable Z corresponding to the shaded area under the standard normal curve.

1. $P(Z < 1.45)$

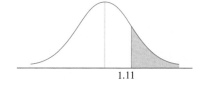

1.45

2. $P(Z > 1.11)$

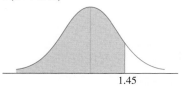

1.11

3. $P(Z < -1.75)$

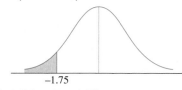

−1.75

4. $P(0.3 < Z < 1.83)$

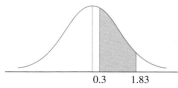

0.3   1.83

5. $P(-1.32 < Z < 1.74)$

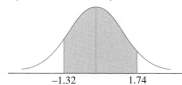

−1.32   1.74

6. $P(-2.35 < Z < -0.51)$

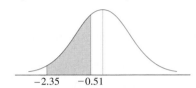

-2.35   -0.51

In Exercises 7–14, (a) make a sketch of the area under the standard normal curve corresponding to the probability and (b) find the value of the probability of the standard normal variable Z corresponding to this area.

7. $P(Z < 1.37)$

8. $P(Z > 2.24)$

9. $P(Z < -0.65)$

10. $P(0.45 < Z < 1.75)$

11. $P(Z > -1.25)$

12. $P(-1.48 < Z < 1.54)$

13. $P(0.68 < Z < 2.02)$

14. $P(-1.41 < Z < -0.24)$

15. Let Z be the standard normal variable. Find the values of z if z satisfies
  a. $P(Z < z) = .8907$ **b.** $P(Z < z) = .2090$

16. Let Z be the standard normal variable. Find the values of z if z satisfies
  a. $P(Z > z) = .9678$ **b.** $P(-z < Z < z) = .8354$

17. Let Z be the standard normal variable. Find the values of z if z satisfies
  a. $P(Z > -z) = .9713$ **b.** $P(Z < -z) = .9713$

18. Suppose X is a normal random variable with $\mu = 380$ and $\sigma = 20$. Find the value of
  a. $P(X < 405)$ **b.** $P(400 < X < 430)$ **c.** $P(X > 400)$

19. Suppose X is a normal random variable with $\mu = 50$ and $\sigma = 5$. Find the value of
  a. $P(X < 60)$ **b.** $P(X > 43)$ **c.** $P(46 < X < 58)$

20. Suppose X is a normal random variable with $\mu = 500$ and $\sigma = 75$. Find the value of
  a. $P(X < 750)$ **b.** $P(X > 350)$ **c.** $P(400 < X < 600)$

8.5   Solutions to Self-Check Exercises

1. a. The probability $P(-1.2 < Z < 2.1)$ is given by the shaded area in the accompanying figure:

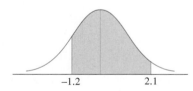

-1.2   2.1

We have

$$P(-1.2 < Z < 2.1) = P(Z < 2.1) - P(Z < -1.2)$$
$$= .9821 - .1151$$
$$= .867$$

b. The region associated with $P(-z < Z < z)$ is shown in the accompanying figure:

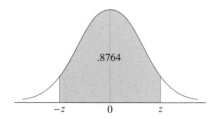

.8764

-z   0   z

Observe that we have the following relationship:

$$P(Z < z) = \frac{1}{2}[1 + P(-z < Z < z)]$$

(see Example 2c). With $P(-z < Z < z) = .8764$, we find that

$$P(Z < z) = \frac{1}{2}(1 + .8764)$$
$$= .9382$$

Consulting the table, we find $z = 1.54$.

2. Using the transformation (16) and the table of values of Z, we have

a. $P(X < 100) = P\left(Z < \dfrac{100 - 80}{10}\right)$
  $= P(Z < 2)$
  $= .9772$

b. $P(X > 60) = P\left(Z > \dfrac{60 - 80}{10}\right)$
  $= P(Z > -2)$
  $= P(Z < 2)$
  $= .9772$

c. $P(70 < X < 90) = P\left(\dfrac{70 - 80}{10} < Z < \dfrac{90 - 80}{10}\right)$
  $= P(-1 < Z < 1)$
  $= P(Z < 1) - P(Z < -1)$
  $= .8413 - .1587$
  $= .6826$

8.6 Applications of the Normal Distribution

■ Applications Involving Normal Random Variables

In this section we look at some applications involving the normal distribution.

APPLIED EXAMPLE 1 Birth Weights of Infants The medical records of infants delivered at the Kaiser Memorial Hospital show that the infants' birth weights in pounds are normally distributed with a mean of 7.4 and a standard deviation of 1.2. Find the probability that an infant selected at random from among those delivered at the hospital weighed more than 9.2 pounds at birth.

Solution Let X be the normal random variable denoting the birth weights of infants delivered at the hospital. Then, the probability that an infant selected at random has a birth weight of more than 9.2 pounds is given by $P(X > 9.2)$. To compute $P(X > 9.2)$, we use Formula (18), Section 8.5, with $\mu = 7.4$, $\sigma = 1.2$, and $a = 9.2$. We find that

$$P(X > 9.2) = P\left(Z > \frac{9.2 - 7.4}{1.2}\right) \qquad P(X > a) = P\left(Z > \frac{a - \mu}{\sigma}\right)$$
$$= P(Z > 1.5)$$
$$= P(Z < -1.5)$$
$$= .0668$$

Thus, the probability that an infant delivered at the hospital weighs more than 9.2 pounds is .0668. ■

APPLIED EXAMPLE 2 Packaging Idaho Natural Produce Corporation ships potatoes to its distributors in bags whose weights are normally distributed with a mean weight of 50 pounds and standard deviation of 0.5 pound. If a bag of potatoes is selected at random from a shipment, what is the probability that it weighs
a. More than 51 pounds?
b. Less than 49 pounds?
c. Between 49 and 51 pounds?

Solution Let X denote the weight of potatoes packed by the company. Then, the mean and standard deviation of X are $\mu = 50$ and $\sigma = 0.5$, respectively.
a. The probability that a bag selected at random weighs more than 51 pounds is given by

$$P(X > 51) = P\left(Z > \frac{51 - 50}{0.5}\right) \qquad P(X > a) = P\left(Z > \frac{a - \mu}{\sigma}\right)$$
$$= P(Z > 2)$$
$$= P(Z < -2)$$
$$= .0228$$

b. The probability that a bag selected at random weighs less than 49 pounds is given by

$$P(X < 49) = P\left(Z < \frac{49 - 50}{0.5}\right) \qquad P(X < b) = P\left(Z < \frac{b - \mu}{\sigma}\right)$$

$$= P(Z < -2)$$

$$= .0228$$

c. The probability that a bag selected at random weighs between 49 and 51 pounds is given by

$$P(49 < X < 51) \qquad\qquad P(a < X < b) =$$

$$= P\left(\frac{49 - 50}{0.5} < Z < \frac{51 - 50}{0.5}\right) \qquad P\left(\frac{a - \mu}{\sigma} < Z < \frac{b - \mu}{\sigma}\right)$$

$$= P(-2 < Z < 2)$$

$$= P(Z < 2) - P(Z < -2)$$

$$= .9772 - .0228$$

$$= .9544$$ ∎

VIDEO   **APPLIED EXAMPLE 3   College Admissions**   The grade point average (GPA) of the senior class of Jefferson High School is normally distributed with a mean of 2.7 and a standard deviation of 0.4 point. If a senior in the top 10% of his or her class is eligible for admission to any of the nine campuses of the State University system, what is the minimum GPA that a senior should have to ensure eligibility for admission to the State University system?

Solution   Let X denote the GPA of a randomly selected senior at Jefferson High School and let x denote the minimum GPA to ensure his or her eligibility for admission to the university. Since only the top 10% is eligible for admission, x must satisfy the equation

$$P(X \geq x) = .1$$

Using Formula (18), Section 8.5, with $\mu = 2.7$ and $\sigma = 0.4$, we find that

$$P(X \geq x) = P\left(Z \geq \frac{x - 2.7}{0.4}\right) = .1 \qquad P(X > a) = P\left(Z > \frac{a - \mu}{\sigma}\right)$$

But, this is equivalent to the equation

$$P\left(Z \leq \frac{x - 2.7}{0.4}\right) = .9 \qquad \text{Why?}$$

Consulting Table 2, Appendix C, we find that

$$\frac{x - 2.7}{0.4} = 1.28$$

Upon solving for x, we obtain

$$x = (1.28)(0.4) + 2.7$$

$$\approx 3.2$$

Thus, to ensure eligibility for admission to one of the nine campuses of the State University system, a senior at Jefferson High School should have a minimum of 3.2 GPA. ∎

Approximating Binomial Distributions

As mentioned in the last section, one important application of the normal distribution is that it provides us with an accurate approximation of other continuous probability distributions. We now show how a binomial distribution may be approximated by a suitable normal distribution. This technique leads to a convenient and simple solution to certain problems involving binomial probabilities.

Recall that a binomial distribution is a probability distribution of the form

$$P(X = x) = C(n, x)p^x q^{n-x} \qquad x = 0, 1, 2, \ldots, n \qquad \textbf{(19)}$$

(See Section 8.4.) For small values of n, the arithmetic computations of the binomial probabilities may be done with relative ease. However, if n is large, then the work involved becomes prodigious, even when tables of $P(X = x)$ are available. For example, if $n = 50$, $p = .3$, and $q = .7$, then the probability of ten or more successes is given by

$$P(X \geq 10) = P(X = 10) + P(X = 11) + \cdots + P(X = 50)$$

$$= \frac{50!}{10!40!}(.3)^{10}(.7)^{40} + \frac{50!}{11!39!}(.3)^{11}(.7)^{39} + \cdots + \frac{50!}{50!0!}(.3)^{50}(.7)^0$$

To see how the normal distribution helps us in such situations, let's consider a coin-tossing experiment. Suppose a fair coin is tossed 20 times and we wish to compute the probability of obtaining ten or more heads. The solution to this problem may be obtained of course by computing

$$P(X \geq 10) = P(X = 10) + P(X = 11) + \cdots + P(X = 20)$$

The inconvenience of this approach for solving the problem at hand has already been pointed out. As an alternative solution, let's begin by interpreting the solution in terms of finding the area of suitable rectangles of the histogram for the distribution associated with the problem. Using Formula (19) we compute the probability of obtaining exactly x heads in 20 coin tosses. The results lead to the binomial distribution displayed in Table 13.

Using the data from the table, we next construct the histogram for the distribution (Figure 28). The probability of obtaining ten or more heads in 20 coin tosses

TABLE 13

Probability Distribution

x	$P(X = x)$
0	.0000
1	.0000
2	.0002
3	.0011
4	.0046
5	.0148
6	.0370
7	.0739
8	.1201
9	.1602
10	.1762
11	.1602
12	.1201
.	.
.	.
.	.
20	.0000

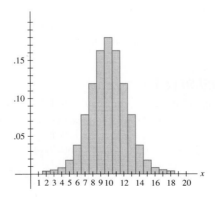

FIGURE 28
Histogram showing the probability of obtaining x heads in 20 coin tosses

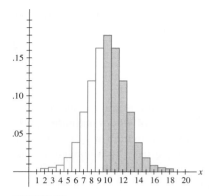

FIGURE 29
The shaded area gives the probability of obtaining ten or more heads in 20 coin tosses.

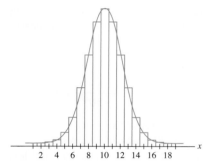

FIGURE 30
Normal curve superimposed on the histogram for a binomial distribution

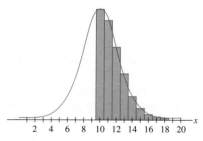

FIGURE 31
$P(X \geq 10)$ is approximated by the area under the normal curve.

is equal to the sum of the areas of the shaded rectangles of the histogram of the binomial distribution shown in Figure 29.

Next, observe that the shape of the histogram suggests that the binomial distribution under consideration may be approximated by a suitable normal distribution. Since the mean and standard deviation of the binomial distribution are given by

$$\mu = np$$
$$= (20)(.5) = 10$$
$$\sigma = \sqrt{npq}$$
$$= \sqrt{(20)(.5)(.5)}$$
$$= 2.24$$

respectively (see Section 8.4), the natural choice of a normal curve for this purpose is one with a mean of 10 and standard deviation of 2.24. Figure 30 shows such a normal curve superimposed on the histogram of the binomial distribution.

The good fit suggests that the sum of the areas of the rectangles representing $P(X \geq 10)$, the probability of obtaining ten or more heads in 20 coin tosses, may be approximated by the area of an appropriate region under the normal curve. To determine this region, let's note that the base of the portion of the histogram representing the required probability extends from $x = 9.5$ on, since the base of the leftmost rectangle is centered at $x = 10$ and the base of each rectangle has length 1 (Figure 31). Therefore, the required region under the normal curve should also have $x \geq 9.5$. Letting Y denote the continuous normal variable, we have

$$P(X \geq 10) \approx P(Y \geq 9.5)$$
$$= P(Y > 9.5)$$
$$= P\left(Z > \frac{9.5 - 10}{2.24}\right) \qquad P(X > a) = P\left(Z > \frac{a - \mu}{\sigma}\right)$$
$$= P(Z > -0.22)$$
$$= P(Z < 0.22)$$
$$= .5871 \qquad\qquad \text{Use the table of values of } Z.$$

The exact value of $P(X \geq 10)$ may be found by computing

$$P(X = 10) + P(X = 11) + \cdots + P(X = 20)$$

in the usual fashion and is equal to .5881. Thus, the normal distribution with suitably chosen mean and standard deviation does provide us with a good approximation of the binomial distribution.

In the general case, the following result, which is a special case of the *central limit theorem,* guarantees the accuracy of the approximation of a binomial distribution by a normal distribution under certain conditions.

THEOREM 1

Suppose we are given a binomial distribution associated with a binomial experiment involving n trials, each with a probability of success p and probability of failure q. Then, if n is large and p is not close to 0 or 1, the binomial distribution may be approximated by a normal distribution with

$$\mu = np \quad \text{and} \quad \sigma = \sqrt{npq}$$

Note   It can be shown that if both np and nq are greater than 5, then the error resulting from this approximation is negligible.   ■

■ Applications Involving Binomial Random Variables

Next, we look at some applications involving binomial random variables.

APPLIED EXAMPLE 4   Quality Control   An automobile manufacturer receives the microprocessors used to regulate fuel consumption in its automobiles in shipments of 1000 each from a certain supplier. It has been estimated that, on the average, 1% of the microprocessors manufactured by the supplier are defective. Determine the probability that more than 20 of the microprocessors in a single shipment are defective.

Solution   Let X denote the number of defective microprocessors in a single shipment. Then, X has a binomial distribution with $n = 1000$, $p = .01$, and $q = .99$, so

$$\mu = (1000)(.01) = 10$$
$$\sigma = \sqrt{(1000)(.01)(.99)}$$
$$\approx 3.15$$

Approximating the binomial distribution by a normal distribution with a mean of 10 and a standard deviation of 3.15, we find that the probability that more than 20 microprocessors in a shipment are defective is given by

$$P(X > 20) \approx P(Y > 20.5) \qquad \text{Where } Y \text{ denotes the normal random variable}$$
$$= P\left(Z > \frac{20.5 - 10}{3.15}\right) \qquad P(X > a) = P\left(Z > \frac{a - \mu}{\sigma}\right)$$
$$= P(Z > 3.33)$$
$$= P(Z < -3.33)$$
$$= .0004$$

In other words, approximately 0.04% of the shipments containing 1000 microprocessors each will contain more than 20 defective units.   ■

APPLIED EXAMPLE 5   Heart Transplant Survival Rate   The probability that a heart transplant performed at the Medical Center is successful (that is, the patient survives 1 year or more after undergoing the surgery) is .7. Of 100 patients who have undergone such an operation, what is the probability that
a. Fewer than 75 will survive 1 year or more after the operation?
b. Between 80 and 90, inclusive, will survive 1 year or more after the operation?

Solution   Let X denote the number of patients who survive 1 year or more after undergoing a heart transplant at the Medical Center. Then, X is a binomial random variable. Also, $n = 100$, $p = .7$, and $q = .3$, so

$$\mu = (100)(.7) = 70$$
$$\sigma = \sqrt{(100)(.7)(.3)}$$
$$\approx 4.58$$

Approximating the binomial distribution by a normal distribution with a mean of 70 and a standard deviation of 4.58, we find, upon letting Y denote the associated normal random variable:

a. The probability that fewer than 75 patients will survive 1 year or more is given by

$$P(X < 75) \approx P(Y < 74.5) \qquad \text{Why?}$$

$$= P\left(Z < \frac{74.5 - 70}{4.58}\right) \qquad P(X < b) = P\left(Z < \frac{b - \mu}{\sigma}\right)$$

$$= P(Z < 0.98)$$

$$= .8365$$

b. The probability that between 80 and 90, inclusive, of the patients will survive 1 year or more is given by

$$P(80 \leq X \leq 90)$$

$$\approx P(79.5 < Y < 90.5)$$

$$= P\left(\frac{79.5 - 70}{4.58} < Z < \frac{90.5 - 70}{4.58}\right) \qquad \begin{aligned} &P(a < X < b) \\ &= P\left(\frac{a - \mu}{\sigma} < Z < \frac{b - \mu}{\sigma}\right) \end{aligned}$$

$$= P(2.07 < Z < 4.48)$$

$$= P(Z < 4.48) - P(Z < 2.07)$$

$$= 1 - .9808 \qquad \text{Note: } P(Z < 4.48) \approx 1$$

$$= .0192$$

8.6   Self-Check Exercises

1. The serum cholesterol levels in milligrams/decaliter (mg/dL) in a current Mediterranean population are found to be normally distributed with a mean of 160 and a standard deviation of 50. Scientists at the National Heart, Lung, and Blood Institute consider this pattern ideal for a minimal risk of heart attacks. Find the percent of the population having blood cholesterol levels between 160 and 180 mg/dL.

2. It has been estimated that 4% of the luggage manufactured by The Luggage Company fails to meet the standards established by the company and is sold as "seconds" to discount and outlet stores. If 500 bags are produced, what is the probability that more than 30 will be classified as "seconds"?

Solutions to Self-Check Exercises 8.6 can be found on page 506.

8.6   Concept Questions

1. What does the central limit theorem allow us to do?

2. Suppose a binomial distribution is associated with a binomial experiment involving n trials, each with a probability of success p and probability of failure q, and n and p satisfy the other conditions given in the central limit theorem. What are the formulas for μ and σ that can be used to approximate this binomial distribution by a normal distribution?

8.6 Exercises

1. MEDICAL RECORDS   The medical records of infants delivered at Kaiser Memorial Hospital show that the infants' lengths at birth (in inches) are normally distributed with a mean of 20 and a standard deviation of 2.6. Find the probability that an infant selected at random from among those delivered at the hospital measures
   a. More than 22 in.   **b.** Less than 18 in.
   c. Between 19 and 21 in.

2. FACTORY WORKERS' WAGES   According to the data released by the Chamber of Commerce of a certain city, the weekly wages of factory workers are normally distributed with a mean of $600 and a standard deviation of $50. What is the probability that a worker selected at random from the city makes a weekly wage
   a. Of less than $600?   **b.** Of more than $760?
   c. Between $550 and $650?

3. PRODUCT RELIABILITY   TKK Products manufactures electric light bulbs in the 50-, 60-, 75-, and 100-watt range. Laboratory tests show that the lives of these light bulbs are normally distributed with a mean of 750 hr and a standard deviation of 75 hr. What is the probability that a TKK light bulb selected at random will burn
   a. For more than 900 hr?
   b. For less than 600 hr?
   c. Between 750 and 900 hr?
   d. Between 600 and 800 hr?

4. EDUCATION   On the average, a student takes 100 words/minute midway through an advanced court reporting course at the American Institute of Court Reporting. Assuming that the dictation speeds of the students are normally distributed and that the standard deviation is 20 words/minute, what is the probability that a student randomly selected from the course can take dictation at a speed
   a. Of more than 120 words/minute?
   b. Between 80 and 120 words/minute?
   c. Of less than 80 words/minute?

5. IQs   The IQs of students at Wilson Elementary School were measured recently and found to be normally distributed with a mean of 100 and a standard deviation of 15. What is the probability that a student selected at random will have an IQ
   a. Of 140 or higher?   **b.** Of 120 or higher?
   c. Between 100 and 120?   **d.** Of 90 or less?

6. PRODUCT RELIABILITY   The tread lives of the Super Titan radial tires under normal driving conditions are normally distributed with a mean of 40,000 mi and a standard deviation of 2000 mi. What is the probability that a tire selected at random will have a tread life of more than 35,000 mi? If four new tires are installed in a car and they experience even wear, determine the probability that all four tires still have useful tread lives after 35,000 mi of driving.

7. FEMALE FACTORY WORKERS' WAGES   According to data released by the Chamber of Commerce of a certain city, the weekly wages (in dollars) of female factory workers are normally distributed with a mean of 575 and a standard deviation of 50. Find the probability that a female factory worker selected at random from the city makes a weekly wage of $550 to $650.

8. CIVIL SERVICE EXAMS   To be eligible for further consideration, applicants for certain Civil Service positions must first pass a written qualifying examination on which a score of 70 or more must be obtained. In a recent examination it was found that the scores were normally distributed with a mean of 60 points and a standard deviation of 10 points. Determine the percentage of applicants who passed the written qualifying examination.

9. WARRANTIES   The general manager of the Service Department of MCA Television has estimated that the time that elapses between the dates of purchase and the dates on which the 19-in. sets manufactured by the company first require service is normally distributed with a mean of 22 mo and a standard deviation of 4 mo. If the company gives a 1-yr warranty on parts and labor for these sets, determine the percentage of sets manufactured and sold by the company that may require service before the warranty period runs out.

10. GRADE DISTRIBUTIONS   The scores on an Economics examination are normally distributed with a mean of 72 and a standard deviation of 16. If the instructor assigns a grade of A to 10% of the class, what is the lowest score a student may have and still obtain an A?

11. GRADE DISTRIBUTIONS   The scores on a Sociology examination are normally distributed with a mean of 70 and a standard deviation of 10. If the instructor assigns A's to 15%, B's to 25%, C's to 40%, D's to 15%, and F's to 5% of the class, find the cutoff points for these grades.

12. HIGHWAY SPEEDS   The speeds (in mph) of motor vehicles on a certain stretch of Route 3A as clocked at a certain place along the highway are normally distributed with a mean of 64.2 mph and a standard deviation of 8.44 mph. What is the probability that a motor vehicle selected at random is traveling at
   a. More than 65 mph?   **b.** Less than 60 mph?
   c. Between 65 and 70 mph?

In Exercises 13–24, use the appropriate normal distributions to approximate the resulting binomial distributions.

13. A coin is weighted so that the probability of obtaining a head in a single toss is .4. If the coin is tossed 25 times, what is the probability of obtaining
   a. Fewer than 10 heads?
   b. Between 10 and 12 heads, inclusive?
   c. More than 15 heads?

14. A fair coin is tossed 20 times. What is the probability of obtaining
   a. Fewer than 8 heads?   **b.** More than 6 heads?
   c. Between 6 and 10 heads inclusive?

15. SPORTS   A marksman's chance of hitting a target with each of his shots is 60%. If he fires 30 shots, what is the probability of his hitting the target
   a. At least 20 times?   **b.** Fewer than 10 times?
   c. Between 15 and 20 times, inclusive?

16. SPORTS   A basketball player has a 75% chance of making a free throw. What is the probability of her making 100 or more free throws in 120 trials?

17. QUALITY CONTROL   The manager of C & R Clothiers, a major manufacturer of men's shirts, has determined that 3% of C & R's shirts do not meet with company standards and are sold as "seconds" to discount and outlet stores. What is the probability that in a day's production of 200 dozen shirts, less than 10 dozen will be classified as "seconds"?

18. TELEMARKETING   Jorge sells magazine subscriptions over the phone. He estimates that the probability of his making a sale with each attempt is .12. What is the probability of Jorge making more than 10 sales if he makes 80 calls?

19. INDUSTRIAL ACCIDENTS   Colorado Mining and Mineral has 800 employees engaged in its mining operations. It has been estimated that the probability of a worker meeting with an accident during a 1-yr period is .1. What is the probability that more than 70 workers will meet with an accident during the 1-yr period?

20. QUALITY CONTROL   PAR Bearings is the principal supplier of ball bearings for the Sperry Gyroscope Company. It has been determined that 6% of the ball bearings shipped are rejected because they fail to meet tolerance requirements. What is the probability that a shipment of 200 ball bearings contains more than 10 rejects?

21. DRUG TESTING   An experiment was conducted to test the effectiveness of a new drug in treating a certain disease. The drug was administered to 50 mice that had been previously exposed to the disease. It was found that 35 mice subsequently recovered from the disease. It was determined that the natural recovery rate from the disease is 0.5.
   a. Determine the probability that 35 or more of the mice not treated with the drug would recover from the disease.
   b. Using the results obtained in part (a), comment on the effectiveness of the drug in the treatment of the disease.

22. LOAN DELINQUENCIES   The manager of Madison Finance Company has estimated that, because of a recession year, 5% of its 400 loan accounts will be delinquent. If the manager's estimate is correct, what is the probability that 25 or more of the accounts will be delinquent?

23. CRUISE SHIP BOOKINGS   Because of late cancellations, Neptune Lines, an operator of cruise ships, has a policy of accepting more reservations than there are accommodations available. From experience, 8% of the bookings for the 90-day around-the-world cruise on the S.S. *Drion,* which has accommodations for 2000 passengers, are subsequently canceled. If the management of Neptune Lines has decided, for public relations reasons, that a person who has made a reservation should have a probability of .99 of obtaining accommodation on the ship, determine the largest number of reservations that should be taken for a cruise on the S.S. *Drion.*

24. THEATER BOOKINGS   Preview Showcase, a research firm, screens pilots of new TV shows before a randomly selected audience and then solicits their opinions of the shows. Based on past experience, 20% of those who get complimentary tickets are "no-shows." The theater has a seating capacity of 500. Management has decided, for public relations reasons, that a person who has been solicited for a screening should have a probability of .99 of being seated. How many tickets should the company send out to prospective viewers for each screening?

8.6   Solutions to Self-Check Exercises

1. Let X be the normal random variable denoting the serum cholesterol levels in mg/dL in the current Mediterranean population under consideration. Then, the percentage of the population having blood cholesterol levels between 160 and 180 mg/dL is given by $P(160 < X < 180)$. To compute $P(160 < X < 180)$, we use Formula (16), Section 8.5, with $\mu = 160$, $\sigma = 50$, $a = 160$, and $b = 180$. We find

$$P(160 < X < 180) = P\left(\frac{160 - 160}{50} < Z < \frac{180 - 160}{50}\right)$$
$$= P(0 < Z < 0.4)$$
$$= P(Z < 0.4) - P(Z < 0)$$
$$= .6554 - .5000$$
$$= .1554$$

Thus, approximately 15.5% of the population has blood cholesterol levels between 160 and 180 mg/dL.

2. Let X denote the number of substandard bags in the production. Then, X has a binomial distribution with $n = 500$, $p = .04$, $q = .96$, so

$$\mu = (500)(.04) = 20$$
$$\sigma = \sqrt{(500)(.04)(.96)} = 4.38$$

Approximating the binomial distribution by a normal distribution with a mean of 20 and standard deviation of 4.38, we find that the probability that more than 30 bags in the production of 500 will be substandard is given by

$$P(X > 30) \approx P(Y > 30.5) \qquad \text{Where } Y \text{ denotes the normal random variable}$$

$$= P\left(Z > \frac{30.5 - 20}{4.38}\right)$$
$$= P(Z > 2.40)$$
$$= P(Z < -2.40)$$
$$= .0082$$

or approximately 0.8%.

CHAPTER 8 Summary of Principal Formulas and Terms

 ### FORMULAS

1. Mean of n numbers	$\bar{x} = \dfrac{x_1 + x_2 + \cdots + x_n}{n}$
2. Expected value	$E(X) = x_1 p_1 + x_2 p_2 + \cdots + x_n p_n$
3. Odds in favor of E occurring	$\dfrac{P(E)}{P(E^c)}$
4. Odds against E occurring	$\dfrac{P(E^c)}{P(E)}$
5. Probability of an event occurring given the odds	$\dfrac{a}{a + b}$
6. Variance of a random variable	$\begin{aligned}\text{Var}(X) = {}& p_1(x_1 - \mu)^2 \\ &+ p_2(x_2 - \mu)^2 + \cdots \\ &+ p_n(x_n - \mu)^2\end{aligned}$
7. Standard deviation of a random variable	$\sigma = \sqrt{\text{Var}(X)}$
8. Chebychev's inequality	$P(\mu - k\sigma \le X \le \mu + k\sigma)$ $\ge 1 - \dfrac{1}{k^2}$
9. Probability of x successes in n Bernoulli trials	$C(n, x)p^x q^{n-x}$
10. Binomial random variable: Mean Variance Standard deviation	$\mu = E(X) = np$ $\text{Var}(X) = npq$ $\sigma_x = \sqrt{npq}$

TERMS

random variable (444)

finite discrete random variable (445)

infinite discrete random variable (445)

continuous random variable (445)

probability distribution of a random variable (445)

histogram (447)

average (mean) (454)

expected value (455)

median (461)

mode (462)

variance (468)

standard deviation (469)

Bernoulli (binomial) trial (480)

binomial experiment (480)

binomial random variable (483)

binomial distribution (483)

probability density function (491)

normal distribution (491)

CHAPTER 8   Concept Review Questions

Fill in the blanks.

1. A rule that assigns a number to each outcome of a chance experiment is called a/an _____ variable.

2. If a random variable assumes only finitely many values, then it is called _____ discrete; if it takes on infinitely many values that can be arranged in a sequence, then it is called _____ discrete; if it takes on all real numbers in an interval, then it is said to be _____.

3. The expected value of a random variable X is given by the _____ of the products of the values assumed by the random variable and its associated probabilities. For example, if X assumes the values -2, 3, and 4 with associated probabilities $\frac{1}{2}$, $\frac{1}{4}$, and $\frac{1}{4}$, then its expected value is _____.

4. **a.** If the probability of an event E occurring is $P(E)$, then the odds in favor of E occurring are _____.
   b. If the odds in favor of an event E occurring are a to b, then the probability of E occurring is _____.

5. If a random variable X takes on the values $x_1, x_2, \ldots, x_n$ with probabilities $p_1, p_2, \ldots, p_n$ and has a mean of μ, then the variance of X is _____. The standard deviation of X is _____.

6. In a binomial experiment, the number of trials is _____, there are exactly _____ outcomes of the experiment, the probability of "success" in each trial is the _____, and the trials are _____ of each other.

7. A probability distribution that is associated with a continuous random variable is called a/an _____ probability distribution. Such a probability distribution is defined by a/an _____ _____ _____ whose domain is the _____ of values taken on by the random variable associated with the experiment.

8. A binomial distribution may be approximated by a/an _____ distribution with $\mu = np$ and $\sigma = \sqrt{npq}$ if n is _____ and p is not close to _____ or _____.

CHAPTER 8   Review Exercises

1. Three balls are selected at random without replacement from an urn containing three white balls and four blue balls. Let the random variable X denote the number of blue balls drawn.
   a. List the outcomes of this experiment.
   b. Find the value assigned to each outcome of this experiment by the random variable X.
   c. Find the probability distribution of the random variable associated with this experiment.
   d. Draw the histogram representing this distribution.

2. A man purchased a $25,000, 1-yr term-life insurance policy for $375. Assuming that the probability that he will live for another year is .989, find the company's expected gain.

3. The probability distribution of a random variable X is shown in the accompanying table:

x	$P(X = x)$
0	.1
1	.1
2	.2
3	.3
4	.2
5	.1

   a. Compute $P(1 \leq X \leq 4)$.
   b. Compute the mean and standard deviation of X.

4. A binomial experiment consists of four trials in which the probability of success in any one trial is $\frac{2}{5}$.
   a. Construct the probability distribution for the experiment.
   b. Compute the mean and standard deviation of the probability distribution.

In Exercises 5–8, let Z be the standard normal variable. Make a rough sketch of the appropriate region under the standard normal curve and find the probability.

5. $P(Z < 0.5)$

6. $P(Z < -0.75)$

7. $P(-0.75 < Z < 0.5)$

8. $P(-0.42 < Z < 0.66)$

In Exercises 9–12, let Z be the standard normal variable. Find z if z satisfies the value.

9. $P(Z < z) = .9922$

10. $P(Z < z) = .1469$

11. $P(Z > z) = .9788$

12. $P(-z < Z < z) = .8444$

In Exercises 13–16, let X be a normal random variable with $\mu = 10$ and $\sigma = 2$. Find the value of the probability.

13. $P(X < 11)$

14. $P(X > 8)$

15. $P(7 < X < 9)$

16. $P(6.5 < X < 11.5)$

17. **SPORTS** If the probability that a bowler will bowl a strike is .7, what is the probability that he will get exactly two strikes in four attempts? At least two strikes in four attempts?

18. **HEIGHTS OF WOMEN** The heights of 4000 women who participated in a recent survey were found to be normally distributed with a mean of 64.5 in. and a standard deviation of 2.5 in. What percentage of these women have heights of 67 in. or greater?

19. **HEIGHTS OF WOMEN** Refer to Exercise 18. Use Chebychev's inequality to estimate the probability that the height of a woman who participated in the survey will fall within 2 standard deviations of the mean—that is, that her height will be between 59.5 and 69.5 in.

20. **MARITAL STATUS OF WOMEN** The number of single women (in thousands) between the ages of 20 and 44 in the United States in 1998 is given in the following table:

Age	20–24	25–29	30–34	35–39	40–44
Women	6178	3689	2219	1626	1095

Find the mean and the standard deviation of the given data. Hint: Assume that all scores lying within a group interval take the midvalue of that group.

Source: U.S. Census Bureau

21. **QUALITY CONTROL** The proprietor of a hardware store will accept a shipment of ceramic wall tiles if no more than 2 of a random sample of 20 are found to be defective. What is the probability that he will accept shipment if exactly 10% of the tiles in a certain shipment is defective?

22. **DRUG EFFECTIVENESS** An experimental drug has been found to be effective in treating 15% of the people afflicted by a certain disease. If the drug is administered to 800 people who have this disease, what are the mean and standard deviation of the number of people for whom the drug can be expected to be effective?

23. **QUALITY CONTROL** Dayton Iron Works manufactures steel rods to a specification of 1-in. diameter. These rods are accepted by the buyer if they fall within the tolerance limits of 0.995 and 1.005. Assuming that the diameter of the rods is normally distributed about a mean of 1 in. and a standard deviation of 0.002 in., estimate the percentage of rods that will be rejected by the buyer.

24. **COIN TOSSES** A coin is biased so that the probability of it landing heads is .6. If the coin is tossed 100 times, what is the probability that heads will appear more than 50 times in the 100 tosses?

25. **QUALITY CONTROL** A division of Solaron Corporation manufactures photovoltaic cells for use in the company's solar energy converters. It is estimated that 5% of the cells manufactured is defective. In a batch of 200 cells manufactured by the company, what is the probability that it will contain at most 20 defective units?

CHAPTER 8 **Before Moving On . . .**

1. The values taken on by a random variable X and the frequency of their occurrence are shown in the following table. Find the probability distribution of X.

x	−3	−2	0	1	2	3
Frequency of Occurrence	4	8	20	24	16	8

2. The probability distribution of the random variable X is shown in the following table. Find (a) $P(X \le 0)$ and (b) $P(-4 \le X \le 1)$.

x	−4	−3	−1	0	1	3
$P(X = x)$	.06	.14	.32	.28	.12	.08

3. Find the mean, variance, and standard deviation of a random variable X having the following probability distribution:

x	−3	−1	0	1	3	5
$P(X = x)$	.08	.24	.32	.16	.12	.08

4. A binomial experiment consists of four independent trials. The probability of success in each trial is 0.3.
   a. Find the probability of obtaining 0, 1, 2, 3, and 4 successes, respectively.
   b. Compute the mean and standard deviation of the random variable associated with this experiment.

5. Let X be a normal random variable with $\mu = 60$ and $\sigma = 5$. Find the values of (a) $P(X < 70)$, (b) $P(X > 50)$, and (c) $P(50 < X < 70)$.

6. A fair coin is tossed 30 times. Using the appropriate normal distribution to approximate a binomial distribution, find the probability of obtaining (a) fewer than 10 heads, (b) between 12 and 16 heads, inclusive, and (c) more than 20 heads.

Markov Chains and the Theory of Games

Will the flowers be red? A certain species of plant produces red, pink, or white flowers, depending on its genetic makeup. In Example 4, page 538, we will show that if the offspring of two plants are crossed successively with plants of a certain genetic makeup only, then in the long run all the flowers produced by the plants will be red.

© Joseph Sohm; Chromosohm/Corbis

IN THIS CHAPTER, we look at two important applications of mathematics that are based primarily on matrix theory and the theory of probability. Both of these applications, *Markov chains* and the *theory of games*, are relatively recent developments in the field of mathematics and have wide applications in many practical areas.

9.1 Markov Chains

■ Transitional Probabilities

A finite stochastic process, you may recall, is an experiment consisting of a finite number of stages in which the outcomes and associated probabilities at each stage depend on the outcomes and associated probabilities of the *preceding stages*. In this chapter, we are concerned with a special class of stochastic processes—namely, those in which the probabilities associated with the outcomes at any stage of the experiment depend only on the outcomes of the *preceding stage*. Such a process is called a **Markov process,** or a **Markov chain,** named after the Russian mathematician A. A. Markov (1856–1922).

The outcome at any stage of the experiment in a Markov process is called the **state** of the experiment. In particular, the outcome at the current stage of the experiment is called the **current state** of the process. Here is a typical problem involving a Markov chain:

Starting from one state of a process (the current state), determine the probability that the process will be at a particular state at some future time.

APPLIED EXAMPLE 1　Common Stocks　An analyst at Weaver and Kline, a stock brokerage firm, observes that the closing price of the preferred stock of an airline company over a short span of time depends only on its previous closing price. At the end of each trading day, he makes a note of the stock's performance for that day, recording the closing price as "higher," "unchanged," or "lower" according to whether the stock closes higher, unchanged, or lower than the previous day's closing price. This sequence of observations may be viewed as a Markov chain. ■

The transition from one state to another in a Markov chain may be studied with the aid of tree diagrams, as in the next example.

APPLIED EXAMPLE 2　Common Stocks　Refer to Example 1. If on a certain day the stock's closing price is higher than that of the previous day, then the probability that it closes higher, unchanged, or lower on the next trading day is .2, .3, and .5, respectively. Next, if the stock's closing price is unchanged from the previous day, then the probability that it closes higher, unchanged, or lower on the next trading day is .5, .2, and .3, respectively. Finally, if the stock's closing price is lower than that of the previous day, then the probability that it closes higher, unchanged, or lower on the next trading day is .4, .4, and .2, respectively. With the aid of tree diagrams, describe the transition between states and the probabilities associated with these transitions.

Solution　The Markov chain being described has three states: higher, unchanged, and lower. If the current state is higher, then the transition to the other states from this state may be displayed by constructing a tree diagram in which the associated probabilities are shown on the appropriate limbs (Figure 1). Tree diagrams describing the transition from each of the other two possible current states, unchanged and lower, to the other states are constructed in a similar manner.

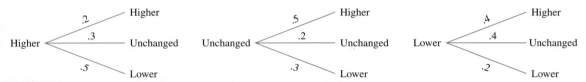

FIGURE 1
Tree diagrams showing transition probabilities between states

The probabilities encountered in this example are called **transition probabilities** because they are associated with the transition from one state to the next in the Markov process. These transition probabilities may be conveniently represented in the form of a matrix. Suppose for simplicity that we have a Markov chain with three possible outcomes at each stage of the experiment. Let's refer to these outcomes as state 1, state 2, and state 3. Then the transition probabilities associated with the transition from state 1 to each of the states 1, 2, and 3 in the next phase of the experiment are precisely the respective conditional probabilities that the outcome is state 1, state 2, and state 3 *given* that the outcome state 1 has occurred. In short, the desired transition probabilities are $P(\text{state } 1|\text{state } 1)$, $P(\text{state } 2|\text{state } 1)$, and $P(\text{state } 3|\text{state } 1)$, respectively. Thus, we write

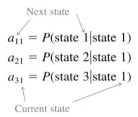

Note that the first subscript in this notation refers to the state in the next stage of the experiment, and the second subscript refers to the current state. Using a tree diagram, we have the following representation:

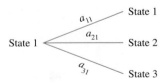

Similarly, the transition probabilities associated with the transition from state 2 and state 3 to each of the states 1, 2, and 3 are

$$a_{12} = P(\text{state } 1|\text{state } 2) \quad \text{and} \quad a_{13} = P(\text{state } 1|\text{state } 3)$$
$$a_{22} = P(\text{state } 2|\text{state } 2) \qquad\qquad a_{23} = P(\text{state } 2|\text{state } 3)$$
$$a_{32} = P(\text{state } 3|\text{state } 2) \qquad\qquad a_{33} = P(\text{state } 3|\text{state } 3)$$

These observations lead to the following matrix representation of the transition probabilities:

Current state

| | State 1 | State 2 | State 3 |

Next state

$$\begin{array}{c} \text{State 1} \\ \text{State 2} \\ \text{State 3} \end{array} \begin{bmatrix} a_{11} & a_{12} & a_{13} \\ a_{21} & a_{22} & a_{23} \\ a_{31} & a_{32} & a_{33} \end{bmatrix}$$

EXAMPLE 3   Use a matrix to represent the transition probabilities obtained in Example 2.

Solution   There are three states at each stage of the Markov chain under consideration. Letting state 1, state 2, and state 3 denote the states "higher," "unchanged," and "lower," respectively, we find that

$$a_{11} = .2 \quad a_{21} = .3 \quad a_{31} = .5$$

and so on, so the required matrix representation is given by

$$T = \begin{bmatrix} .2 & .5 & .4 \\ .3 & .2 & .4 \\ .5 & .3 & .2 \end{bmatrix}$$

The matrix obtained in Example 3 is a transition matrix. In the general case, we have the following definition:

Transition Matrix

A **transition matrix** associated with a Markov chain with n states is an $n \times n$ matrix T with entries a_{ij} $(1 \le i \le n; 1 \le j \le n)$

Current state

| | State 1 | State 2 | $\cdots$ | State j | $\cdots$ | State n |

$$T = \begin{array}{c} \text{Next} \\ \text{state} \end{array} \begin{array}{c} \text{State 1} \\ \text{State 2} \\ \vdots \\ \text{State } i \\ \vdots \\ \text{State } n \end{array} \begin{bmatrix} a_{11} & a_{12} & \cdots & a_{1j} & \cdots & a_{1n} \\ a_{21} & a_{22} & \cdots & a_{2j} & \cdots & a_{2n} \\ \vdots & \vdots & & \vdots & & \vdots \\ a_{i1} & a_{i2} & \cdots & a_{ij} & \cdots & a_{in} \\ \vdots & \vdots & & \vdots & & \vdots \\ a_{n1} & a_{n2} & \cdots & a_{nj} & \cdots & a_{nn} \end{bmatrix}$$

having the following properties:

1. $a_{ij} \ge 0$ for all i and j.
2. The sum of the entries in each column of T is 1.

Since $a_{ij} = P(\text{state } i | \text{state } j)$ is the probability of the occurrence of an event, it must be nonnegative, and this is precisely what Property 1 implies. Property 2 follows from the fact that the transition from any one of the current states must terminate in one of the n states in the next stage of the experiment. Any square matrix satisfying properties 1 and 2 is referred to as a **stochastic matrix.**

EXPLORE & DISCUSS

Let

$$A = \begin{bmatrix} p & q \\ 1 - p & 1 - q \end{bmatrix} \quad \text{and} \quad B = \begin{bmatrix} r & s \\ 1 - r & 1 - s \end{bmatrix}$$

be two 2×2 stochastic matrices, where $0 \le p \le 1$, $0 \le q \le 1$, $0 \le r \le 1$, and $0 \le s \le 1$.

1. Show that AB is a 2×2 stochastic matrix.
2. Use the result of part (a) to explain why $A^2, A^3, \ldots, A^n$, where n is a positive integer, are also 2×2 stochastic matrices.

One advantage in representing the transition probabilities in the form of a matrix is that we may use the results from matrix theory to help us solve problems involving Markov processes, as we will see in the next several sections.

Next, for simplicity, let's consider the following Markov process where each stage of the experiment has precisely two possible states.

APPLIED EXAMPLE 4 Urban–Suburban Population Flow Because of the continued successful implementation of an urban renewal program, it is expected that each year 3% of the population currently residing in the city will move to the suburbs and 6% of the population currently residing in the suburbs will move into the city. At present, 65% of the total population of the metropolitan area lives in the city itself, while the remaining 35% lives in the suburbs. Assuming that the total population of the metropolitan area remains constant, what will be the distribution of the population 1 year from now?

Solution This problem may be solved with the aid of a tree diagram and the techniques of Chapter 7. The required tree diagram describing this process is shown in Figure 2. Using the method of Section 7.5, we find the probability that a person selected at random will be a city dweller 1 year from now is given by

$$(.65)(.97) + (.35)(.06) = .6515$$

In a similar manner, we find that the probability that a person selected at random will reside in the suburbs 1 year from now is given by

$$(.65)(.03) + (.35)(.94) = .3485$$

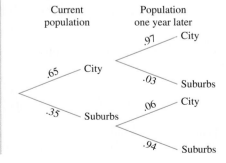

FIGURE 2
Tree diagram showing a Markov process with two states: living in the city and living in the suburbs

Thus, the population of the area 1 year from now may be expected to be distributed as follows: 65.15% living in the city and 34.85% residing in the suburbs.

Let's reexamine the solution to this problem. As noted earlier, the process under consideration may be viewed as a Markov chain with two possible states at each stage of the experiment: "living in the city" (state 1) and "living in the suburbs" (state 2). The transition matrix associated with this Markov chain is

$$T = \begin{matrix} \text{State 1} \\ \text{State 2} \end{matrix} \begin{bmatrix} .97 & .06 \\ .03 & .94 \end{bmatrix} \qquad \text{Transition matrix}$$

with column headings State 1, State 2.

Next, observe that the initial (current) probability distribution of the population may be summarized in the form of a column vector of dimension 2 (that is, a 2×1 matrix). Thus,

$$X_0 = \begin{matrix} \text{State 1} \\ \text{State 2} \end{matrix} \begin{bmatrix} .65 \\ .35 \end{bmatrix} \qquad \text{Initial-state matrix}$$

Using the results of Example 4, we may write the population distribution 1 year later as

$$X_1 = \begin{matrix} \text{State 1} \\ \text{State 2} \end{matrix} \begin{bmatrix} .6515 \\ .3485 \end{bmatrix} \qquad \text{Distribution after 1 year}$$

You may now verify that

$$TX_0 = \begin{bmatrix} .97 & .06 \\ .03 & .94 \end{bmatrix} \begin{bmatrix} .65 \\ .35 \end{bmatrix} = \begin{bmatrix} .6515 \\ .3485 \end{bmatrix} = X_1$$

so this problem may be solved using matrix multiplication.

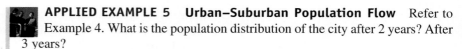 **APPLIED EXAMPLE 5   Urban–Suburban Population Flow**   Refer to Example 4. What is the population distribution of the city after 2 years? After 3 years?

Solution   Let X_2 be the column vector representing the probability population distribution of the metropolitan area after 2 years. We may view X_1, the vector representing the probability population distribution of the metropolitan area after 1 year, as representing the "initial" probability distribution in this part of our calculation. Thus,

$$X_2 = TX_1 = \begin{bmatrix} .97 & .06 \\ .03 & .94 \end{bmatrix} \begin{bmatrix} .6515 \\ .3485 \end{bmatrix} = \begin{bmatrix} .6529 \\ .3471 \end{bmatrix}$$

The vector representing the probability distribution of the metropolitan area after 3 years is given by

$$X_3 = TX_2 = \begin{bmatrix} .97 & .06 \\ .03 & .94 \end{bmatrix} \begin{bmatrix} .6529 \\ .3471 \end{bmatrix} = \begin{bmatrix} .6541 \\ .3459 \end{bmatrix}$$

That is, after 3 years, the population will be distributed as follows: 65.41% will live in the city and 34.59% will live in the suburbs.

Distribution Vectors

Observe that, in the foregoing computations, we have $X_1 = TX_0$, $X_2 = TX_1 = T^2X_0$, and $X_3 = TX_2 = T^3X_0$. These results are easily generalized. To see this, suppose we have a Markov process in which there are n possible states at each stage of the experiment. Suppose further that the probability of the system being in state 1, state 2, ..., state n, initially, is given by $p_1, p_2, \ldots, p_n$, respectively. This distribution may be represented as an n-dimensional vector

$$X_0 = \begin{bmatrix} p_1 \\ p_2 \\ \vdots \\ p_n \end{bmatrix}$$

called a **distribution vector.** If T represents the $n \times n$ transition matrix associated with the Markov process, then the probability distribution of the system after m observations is given by

$$X_m = T^m X_0 \tag{1}$$

VIDEO

APPLIED EXAMPLE 6   Taxicab Movement   To keep track of the location of its cabs, Zephyr Cab has divided a town into three zones: zone I, zone II, andzone III. Zephyr's management has determined from company records that of the passengers picked up in zone I, 60% are discharged in the same zone, 30% are discharged in zone II, and 10% are discharged in zone III. Of those picked up in zone II, 40% are discharged in zone I, 30% are discharged in zone II, and 30% are discharged in zone III. Of those picked up in zone III, 30% are discharged in zone I, 30% are discharged in zone II, and 40% are discharged in zone III. At the beginning of the day, 80% of the cabs are in zone I, 15% are in zone II, 5% are in zone III, and a taxi without a passenger will cruise within the zone it is currently in until a pickup is made.

a. Find the transition matrix for the Markov chain that describes the successive locations of a cab.
b. What is the distribution of the cabs after all of them have made one pickup and discharge?
c. What is the distribution of the cabs after all of them have made two pickups and discharges?

Solution   Let zone I, zone II, and zone III correspond to state 1, state 2, and state 3 of the Markov chain.

a. The required transition matrix is given by

$$T = \begin{bmatrix} .6 & .4 & .3 \\ .3 & .3 & .3 \\ .1 & .3 & .4 \end{bmatrix}$$

b. The initial distribution vector associated with the problem is

$$X_0 = \begin{bmatrix} .8 \\ .15 \\ .05 \end{bmatrix}$$

If X_1 denotes the distribution vector after one observation—that is, after all the cabs have made one pickup and discharge—then

$$X_1 = TX_0$$

$$= \begin{bmatrix} .6 & .4 & .3 \\ .3 & .3 & .3 \\ .1 & .3 & .4 \end{bmatrix} \begin{bmatrix} .8 \\ .15 \\ .05 \end{bmatrix} = \begin{bmatrix} .555 \\ .3 \\ .145 \end{bmatrix}$$

That is, 55.5% of the cabs are in zone I, 30% are in zone II, and 14.5% are in zone III.

c. Let X_2 denote the distribution vector after all the cabs have made two pickups and discharges. Then

$$X_2 = TX_1$$

$$= \begin{bmatrix} .6 & .4 & .3 \\ .3 & .3 & .3 \\ .1 & .3 & .4 \end{bmatrix} \begin{bmatrix} .555 \\ .3 \\ .145 \end{bmatrix} = \begin{bmatrix} .4965 \\ .3 \\ .2035 \end{bmatrix}$$

That is, 49.65% of the cabs are in zone I, 30% are in zone II, and 20.35% are in zone III. You should verify that the same result may be obtained by computing $T^2 X_0$. ◼

Note In this simplified model, we do not take into consideration variable demand and variable delivery time. ◼

9.1 Self-Check Exercises

1. Three supermarkets serve a certain section of a city. During the upcoming year, supermarket A is expected to retain 80% of its customers, lose 5% of its customers to supermarket B, and lose 15% to supermarket C. Supermarket B is expected to retain 90% of its customers and lose 5% of its customers to each of supermarkets A and C. Supermarket C is expected to retain 75% of its customers, lose 10% to supermarket A, and lose 15% to supermarket B. Construct the transition matrix for the Markov chain that describes the change in the market share of the three supermarkets.

2. Refer to Self-Check Exercise 1. Currently the market shares of supermarket A, supermarket B, and supermarket C are 0.4, 0.3, and 0.3, respectively.
   a. Find the initial distribution vector for this Markov chain.
   b. What share of the market will be held by each supermarket after 1 yr? Assuming that the trend continues, what will be the market share after 2 yr?

Solutions to Self-Check Exercises 9.1 can be found on page 522.

9.1 Concept Questions

1. What is a *finite stochastic process*? What can you say about the finite stochastic processes in a Markov chain?

2. Define the following terms for a Markov chain:
   a. *State*
   b. *Current state*
   c. *Transition probabilities*

3. Consider a transition matrix T for a Markov chain with entries a_{ij}, where $1 \leq i \leq n; \ 1 \leq j \leq n$.
   a. If there are n states associated with the Markov chain, what is the size of the matrix T?
   b. Describe the probability that each entry represents. Can an entry be negative?
   c. What is the sum of the entries in each column of T?

9.1 Exercises

In Exercises 1–10, determine which of the matrices are stochastic.

1. $\begin{bmatrix} .4 & .7 \\ .6 & .3 \end{bmatrix}$

2. $\begin{bmatrix} .8 & .2 \\ .3 & .7 \end{bmatrix}$

3. $\begin{bmatrix} \frac{1}{4} & \frac{1}{8} \\ \frac{3}{4} & \frac{7}{8} \end{bmatrix}$

4. $\begin{bmatrix} \frac{1}{3} & 0 & \frac{1}{2} \\ \frac{1}{2} & 1 & 0 \\ \frac{1}{4} & 0 & \frac{1}{2} \end{bmatrix}$

5. $\begin{bmatrix} .3 & .2 & .4 \\ .4 & .7 & .3 \\ .3 & .1 & .2 \end{bmatrix}$

6. $\begin{bmatrix} \frac{1}{3} & \frac{1}{4} & \frac{1}{2} \\ \frac{1}{3} & 0 & -\frac{1}{2} \\ \frac{1}{4} & \frac{3}{4} & \frac{1}{2} \end{bmatrix}$

7. $\begin{bmatrix} .1 & .4 & .3 \\ .7 & .2 & .1 \\ .2 & .4 & .6 \end{bmatrix}$

8. $\begin{bmatrix} 1 & 0 & 0 \\ 0 & 0 & 1 \\ 0 & 1 & 0 \end{bmatrix}$

9. $\begin{bmatrix} .2 & .3 \\ .3 & .1 \\ .5 & .6 \end{bmatrix}$

10. $\begin{bmatrix} .5 & .2 & .3 \\ .2 & .3 & .2 \\ .3 & .4 & .1 \\ 0 & .1 & .4 \end{bmatrix}$

11. The transition matrix for a Markov process is given by

$$T = \begin{matrix} & \text{State} \\ & \begin{matrix} 1 & \ 2 \end{matrix} \\ \begin{matrix} \text{State } 1 \\ \text{State } 2 \end{matrix} & \begin{bmatrix} .3 & .6 \\ .7 & .4 \end{bmatrix} \end{matrix}$$

a. What does the entry $a_{11} = .3$ represent?
b. Given that the outcome state 1 has occurred, what is the probability that the next outcome of the experiment will be state 2?

c. If the initial-state distribution vector is given by

$$X_0 = \begin{matrix} \text{State } 1 \\ \text{State } 2 \end{matrix} \begin{bmatrix} .4 \\ .6 \end{bmatrix}$$

find TX_0, the probability distribution of the system after one observation.

12. The transition matrix for a Markov process is given by

$$T = \begin{matrix} & \text{State} \\ & \begin{matrix} 1 & \ \ 2 \end{matrix} \\ \begin{matrix} \text{State } 1 \\ \text{State } 2 \end{matrix} & \begin{bmatrix} \frac{1}{6} & \frac{2}{3} \\ \frac{5}{6} & \frac{1}{3} \end{bmatrix} \end{matrix}$$

a. What does the entry $a_{22} = \frac{1}{3}$ represent?
b. Given that the outcome state 1 has occurred, what is the probability that the next outcome of the experiment will be state 2?
c. If the initial-state distribution vector is given by

$$X_0 = \begin{matrix} \text{State } 1 \\ \text{State } 2 \end{matrix} \begin{bmatrix} \frac{1}{4} \\ \frac{3}{4} \end{bmatrix}$$

find TX_0, the probability distribution of the system after one observation.

13. The transition matrix for a Markov process is given by

$$T = \begin{matrix} & \text{State} \\ & \begin{matrix} 1 & \ \ 2 \end{matrix} \\ \begin{matrix} \text{State } 1 \\ \text{State } 2 \end{matrix} & \begin{bmatrix} .6 & .2 \\ .4 & .8 \end{bmatrix} \end{matrix}$$

and the initial-state distribution vector is given by

$$X_0 = \begin{matrix} \text{State 1} \\ \text{State 2} \end{matrix} \begin{bmatrix} .5 \\ .5 \end{bmatrix}$$

Find TX_0 and interpret your result with the aid of a tree diagram.

14. The transition matrix for a Markov process is given by

$$X_0 = \begin{matrix} & \text{State} \\ & 1 \quad\ 2 \end{matrix}$$
$$\begin{matrix} \text{State 1} \\ \text{State 2} \end{matrix} \begin{bmatrix} \frac{1}{2} & \frac{3}{4} \\ \frac{1}{2} & \frac{1}{4} \end{bmatrix}$$

and the initial-state distribution vector is given by

$$X_0 = \begin{matrix} \text{State 1} \\ \text{State 2} \end{matrix} \begin{bmatrix} \frac{1}{3} \\ \frac{2}{3} \end{bmatrix}$$

Find TX_0 and interpret your result with the aid of a tree diagram.

In Exercises 15–18, find X_2 (the probability distribution of the system after two observations) for the distribution vector X_0 and the transition matrix T.

15. $X_0 = \begin{bmatrix} .6 \\ .4 \end{bmatrix}$, $T = \begin{bmatrix} .4 & .8 \\ .6 & .2 \end{bmatrix}$

16. $X_0 = \begin{bmatrix} \frac{1}{2} \\ \frac{1}{2} \\ 0 \end{bmatrix}$, $T = \begin{bmatrix} \frac{1}{2} & \frac{1}{3} & \frac{1}{2} \\ 0 & \frac{1}{3} & \frac{1}{4} \\ \frac{1}{2} & \frac{1}{3} & \frac{1}{4} \end{bmatrix}$

17. $X_0 = \begin{bmatrix} \frac{1}{4} \\ \frac{1}{2} \\ \frac{1}{4} \end{bmatrix}$, $T = \begin{bmatrix} \frac{1}{4} & \frac{1}{4} & \frac{1}{2} \\ \frac{1}{4} & \frac{1}{2} & \frac{1}{2} \\ \frac{1}{2} & \frac{1}{4} & 0 \end{bmatrix}$

18. $X_0 = \begin{bmatrix} .25 \\ .40 \\ .35 \end{bmatrix}$, $T = \begin{bmatrix} .1 & .1 & .3 \\ .8 & .7 & .2 \\ .1 & .2 & .5 \end{bmatrix}$

19. PSYCHOLOGY EXPERIMENTS   A psychologist conducts an experiment in which a mouse is placed in a T-maze, where it has a choice at the T-junction of turning left and receiving a reward (cheese) or turning right and receiving a mild electric shock (see accompanying figure). At the end of each trial, a record is kept of the mouse's response. It is observed that the mouse is as likely to turn left (state 1) as right (state 2) during the first trial. In subsequent trials, however, the observation is made that if the mouse had turned left in the previous trial, then on the next trial the probability that it will turn left is .8, whereas the probability that it will turn right is .2. If the mouse had turned right in the previous trial, then the probability that it will turn right on the next trial is .1, whereas the probability that it will turn left is .9.

a. Using a tree diagram, describe the transitions between states and the probabilities associated with these transitions.

b. Represent the transition probabilities obtained in part (a) in terms of a matrix.

c. What is the initial-state probability vector?

d. Use the results of parts (b) and (c) to find the probability that a mouse will turn left on the second trial.

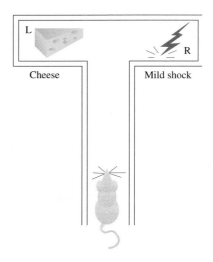

20. SMALL-TOWN REVIVAL   At the beginning of 1990, the population of a certain state was 55.4% rural and 44.6% urban. Based on past trends, it is expected that 10% of the population currently residing in the rural areas will move into the urban areas, while 17% of the population currently residing in the urban areas will move into the rural areas in the next decade. What was the population distribution in that state at the beginning of 2000?

21. POLITICAL POLLS   Morris Polling conducted a poll 6 mo before an election in a state in which a Democrat and a Republican were running for governor and found that 60% of the voters intended to vote for the Republican and 40% intended to vote for the Democrat. In a poll conducted 3 mo later, it was found that 70% of those who had earlier stated a preference for the Republican candidate still maintained that preference, whereas 30% of these voters now preferred the Democratic candidate. Of those who had earlier stated a preference for the Democrat, 80% still maintained that preference, whereas 20% now preferred the Republican candidate.

a. If the election were held at this time, who would win?

b. Assuming that this trend continues, which candidate is expected to win the election?

22. COMMUTER TRENDS   Within a large metropolitan area, 20% of the commuters currently use the public transportation system, whereas the remaining 80% commute via automobile. The city has recently revitalized and expanded its public transportation system. It is expected that 6 mo from now

30% of those who are now commuting to work via automobile will switch to public transportation, and 70% will continue to commute via automobile. At the same time, it is expected that 20% of those now using public transportation will commute via automobile and 80% will continue to use public transportation.

a. Construct the transition matrix for the Markov chain that describes the change in the mode of transportation used by these commuters.

b. Find the initial distribution vector for this Markov chain.

c. What percentage of the commuters are expected to use public transportation 6 mo from now?

23. Refer to Example 6. If the initial distribution vector for the location of the taxis is

$$X_0 = \begin{array}{c} \text{Zone I} \\ \text{Zone II} \\ \text{Zone III} \end{array} \begin{bmatrix} .6 \\ .2 \\ .2 \end{bmatrix}$$

what will be the distribution after all of them have made one pickup and discharge?

24. **URBAN–SUBURBAN POPULATION FLOW** Refer to Example 4. If the initial probability distribution is

$$X_0 = \begin{array}{c} \text{City} \\ \text{Suburb} \end{array} \begin{bmatrix} .80 \\ .20 \end{bmatrix}$$

what will be the population distribution of the city after 1 yr? After 2 yr?

25. **MARKET SHARE** At a certain university, three bookstores—the University Bookstore, the Campus Bookstore, and the Book Mart—currently serve the university community. From a survey conducted at the beginning of the fall quarter, it was found that the University Bookstore and the Campus Bookstore each had 40% of the market, whereas the Book Mart had 20% of the market. Each quarter the University Bookstore retains 80% of its customers but loses 10% to the Campus Bookstore and 10% to the Book Mart. The Campus Bookstore retains 75% of its customers but loses 10% to the University Bookstore and 15% to the Book Mart. The Book Mart retains 90% of its customers but loses 5% to the University Bookstore and 5% to the Campus Bookstore. If these trends continue, what percent of the market will each store have at the beginning of the second quarter? The third quarter?

26. **MARKET SHARE OF AUTO MANUFACTURERS** In a study of the domestic market share of the three major automobile manufacturers A, B, and C in a certain country, it was found that their current market shares were 60%, 30%, and 10%, respectively. Furthermore, it was found that of the customers who bought a car manufactured by A, 75% would again buy a car manufactured by A, 15% would buy a car manufactured by B, and 10% would buy a car manufactured by C. Of the customers who bought a car manufactured by B, 90% would again buy a car manufactured by B, whereas 5% each

would buy cars manufactured by A and C, respectively. Finally, of the customers who bought a car manufactured by C, 85% would again buy a car manufactured by C, 5% would buy a car manufactured by A, and 10% would buy a car manufactured by B. Assuming that these sentiments reflect the buying habits of customers in the future, determine the market share that will be held by each manufacturer after the next two model years.

27. **COLLEGE MAJORS** Records compiled by the Admissions Office at a state university indicating the percentage of students that change their major each year are shown in the following transition matrix. Of the freshmen now at the university, 30% have chosen their major field in Business, 30% in the Humanities, 20% in Education, and 20% in the Natural Sciences and other fields. Assuming that this trend continues, find the percent of these students that will be majoring in each of the given areas in their senior year.
Hint: Find $T^3 X_0$.

	Bus.	Hum.	Educ.	Nat. Sc. and others
Business	.80	.10	.20	.10
Humanities	.10	.70	.10	.05
Education	.05	.10	.60	.05
Nat. Sci. and others	.05	.10	.10	.80

28. **HOMEOWNERS' CHOICE OF ENERGY** A study conducted by the Urban Energy Commission in a large metropolitan area indicates the probabilities that homeowners within the area will use certain heating fuels or solar energy during the next 10 yr as the major source of heat for their homes. The transition matrix representing the transition probabilities from one state to another is

	Elec.	Gas	Oil	Solar
Electricity	.70	0	0	0
Natural gas	.15	.90	.20	.05
Fuel oil	.05	.02	.75	0
Solar energy	.10	.08	.05	.95

Among homeowners within the area, 20% currently use electricity, 35% use natural gas, 40% use oil, and 5% use solar energy as the major source of heat for their homes. What is the expected distribution of the homeowners that will be using each type of heating fuel or solar energy within the next decade?

In Exercises 29 and 30, determine whether the statement is true or false. If it is true, explain why it is true. If it is false, give an example to show why it is false.

29. A Markov chain is a process in which the outcomes at any stage of the experiment depend on the outcomes of the preceding stages.

30. The sum of the entries in each column of a transition matrix must not exceed 1.

9.1   Solutions to Self-Check Exercises

1. The required transition matrix is

$$T = \begin{bmatrix} .80 & .05 & .10 \\ .05 & .90 & .15 \\ .15 & .05 & .75 \end{bmatrix}$$

2. a. The initial distribution vector is

$$X_0 = \begin{bmatrix} .4 \\ .3 \\ .3 \end{bmatrix}$$

b. The vector representing the market share of each supermarket after 1 yr is

$$X_1 = TX_0$$
$$= \begin{bmatrix} .80 & .05 & .10 \\ .05 & .90 & .15 \\ .15 & .05 & .75 \end{bmatrix} \begin{bmatrix} .4 \\ .3 \\ .3 \end{bmatrix} = \begin{bmatrix} .365 \\ .335 \\ .3 \end{bmatrix}$$

That is, after 1 yr supermarket A will command a 36.5% market share, supermarket B will have a 33.5% share, and supermarket C will have a 30% market share.

The vector representing the market share of the supermarkets after 2 yr is

$$X_2 = TX_1$$
$$= \begin{bmatrix} .80 & .05 & .10 \\ .05 & .90 & .15 \\ .15 & .05 & .75 \end{bmatrix} \begin{bmatrix} .365 \\ .335 \\ .3 \end{bmatrix} = \begin{bmatrix} .3388 \\ .3648 \\ .2965 \end{bmatrix}$$

That is, 2 yr later the market shares of supermarkets A, B, and C will be 33.88%, 36.48%, and 29.65%, respectively.

USING TECHNOLOGY

■ Finding Distribution Vectors

Since the computation of the probability distribution of a system after a certain number of observations involves matrix multiplication, a graphing utility may be used to facilitate the work.

EXAMPLE 1   Consider the problem posed in Example 6, page 517, where

$$T = \begin{bmatrix} .6 & .4 & .3 \\ .3 & .3 & .3 \\ .1 & .3 & .4 \end{bmatrix} \quad \text{and} \quad X_0 = \begin{bmatrix} .8 \\ .15 \\ .05 \end{bmatrix}$$

Verify that

$$X_2 = \begin{bmatrix} .4965 \\ .3 \\ .2035 \end{bmatrix}$$

as obtained in that example.

Solution   First, we enter the matrix X_0 as the matrix A and the matrix T as the matrix B. Then, performing the indicated multiplication, we find that

$$B^2 * A = \begin{bmatrix} .4965 \\ .3 \\ .2035 \end{bmatrix}$$

That is,

$$X_2 = T^2 X_0 = \begin{bmatrix} .4965 \\ .3 \\ .2035 \end{bmatrix}$$

as was to be shown.

TECHNOLOGY EXERCISES

In Exercises 1–2, find X_5 (the probability distribution of the system after five observations) for the distribution vector X_0 and the transition matrix T.

1. $X_0 = \begin{bmatrix} .2 \\ .3 \\ .2 \\ .1 \\ .2 \end{bmatrix}$, $T = \begin{bmatrix} .2 & .2 & .3 & .2 & .1 \\ .1 & .2 & .1 & .2 & .1 \\ .3 & .4 & .1 & .3 & .3 \\ .2 & .1 & .2 & .2 & .2 \\ .2 & .1 & .3 & .1 & .3 \end{bmatrix}$

2. $X_0 = \begin{bmatrix} .1 \\ .2 \\ .2 \\ .3 \\ .2 \end{bmatrix}$, $T = \begin{bmatrix} .3 & .2 & .1 & .3 & .1 \\ .2 & .1 & .2 & .1 & .2 \\ .1 & .2 & .3 & .2 & .2 \\ .1 & .3 & .2 & .3 & .2 \\ .3 & .2 & .2 & .1 & .3 \end{bmatrix}$

3. Refer to Exercise 26 on page 521. Using the same data, determine the market share that will be held by each manufacturer five model years after the study began.

4. Refer to Exercise 25 on page 521. Using the same data, determine the expected percent of the market that each store will have at the beginning of the fourth quarter.

9.2 Regular Markov Chains

■ Steady-State Distribution Vectors

In the last section, we derived a formula for computing the likelihood that a physical system will be in any one of the possible states associated with each stage of a Markov process describing the system. In this section we use this formula to help us investigate the long-term trends of certain Markov processes.

APPLIED EXAMPLE 1   Educational Status of Women   A survey conducted by the National Commission on the Educational Status of Women reveals that 70% of the daughters of women who have completed 2 or more years of college have also completed 2 or more years of college, whereas 20% of the daughters of women who have had less than 2 years of college have completed 2 or more years of college. If this trend continues, determine, in the long run, the percentage of women in the population who will have completed at least 2 years of college given that currently only 20% of the women have completed at least 2 years of college.

Solution   This problem may be viewed as a Markov process with two possible states: "completed 2 or more years of college" (state 1) and "completed less than 2 years of college" (state 2). The transition matrix associated with this Markov chain is given by

$$T = \begin{bmatrix} .7 & .2 \\ .3 & .8 \end{bmatrix}$$

The initial distribution vector is given by

$$X_0 = \begin{bmatrix} .2 \\ .8 \end{bmatrix}$$

To study the long-term trend pertaining to this particular aspect of the educational status of women, let's compute X_1, X_2, . . . , the distribution vectors associated with the Markov process under consideration. These vectors give the percentage of women with 2 or more years of college and that of women with less than 2 years of college after one generation, after two generations, and so on. With the aid of Formula (1), Section 9.1, we find (to four decimal places)

After one generation $X_1 = TX_0 = \begin{bmatrix} .7 & .2 \\ .3 & .8 \end{bmatrix} \begin{bmatrix} .2 \\ .8 \end{bmatrix} = \begin{bmatrix} .3 \\ .7 \end{bmatrix}$

After two generations $X_2 = TX_1 = \begin{bmatrix} .7 & .2 \\ .3 & .8 \end{bmatrix} \begin{bmatrix} .3 \\ .7 \end{bmatrix} = \begin{bmatrix} .35 \\ .65 \end{bmatrix}$

After three generations $X_3 = TX_2 = \begin{bmatrix} .7 & .2 \\ .3 & .8 \end{bmatrix} \begin{bmatrix} .35 \\ .65 \end{bmatrix} = \begin{bmatrix} .375 \\ .625 \end{bmatrix}$

Proceeding further, we obtain the following sequence of vectors:

$$X_4 = \begin{bmatrix} .3875 \\ .6125 \end{bmatrix}$$

$$X_5 = \begin{bmatrix} .3938 \\ .6062 \end{bmatrix}$$

$$X_6 = \begin{bmatrix} .3969 \\ .6031 \end{bmatrix}$$

$$X_7 = \begin{bmatrix} .3984 \\ .6016 \end{bmatrix}$$

$$X_8 = \begin{bmatrix} .3992 \\ .6008 \end{bmatrix}$$

$$X_9 = \begin{bmatrix} .3996 \\ .6004 \end{bmatrix}$$

After ten generations $X_{10} = \begin{bmatrix} .3998 \\ .6002 \end{bmatrix}$

From the results of these computations, we see that as m increases, the probability distribution vector X_m approaches the probability distribution vector

$$\begin{bmatrix} .4 \\ .6 \end{bmatrix} \quad \text{or} \quad \begin{bmatrix} \frac{2}{5} \\ \frac{3}{5} \end{bmatrix}$$

Such a vector is called the limiting, or steady-state, distribution vector for the system. We interpret these results in the following way: Initially, 20% of the women in the population have completed 2 or more years of college, whereas 80% have completed less than 2 years of college. After one generation, the former has increased to 30% of the population, and the latter has dropped to 70% of the population. The trend continues, and eventually, 40% of all women in future generations will have completed 2 or more years of college, whereas 60% will have completed less than 2 years of college. ∎

To explain the foregoing result, let's analyze Formula (1), Section 9.1, more closely. Now, the initial distribution vector X_0 is a constant; that is, it remains fixed throughout our computation of $X_1, X_2, \ldots$. It appears reasonable, therefore, to conjecture that this phenomenon is a result of the behavior of the powers, T^m, of the transition matrix T. Pursuing this line of investigation, we compute

$$T^2 = \begin{bmatrix} .7 & .2 \\ .3 & .8 \end{bmatrix} \begin{bmatrix} .7 & .2 \\ .3 & .8 \end{bmatrix} = \begin{bmatrix} .55 & .3 \\ .45 & .7 \end{bmatrix}$$

$$T^3 = \begin{bmatrix} .7 & .2 \\ .3 & .8 \end{bmatrix} \begin{bmatrix} .55 & .3 \\ .45 & .7 \end{bmatrix} = \begin{bmatrix} .475 & .35 \\ .525 & .65 \end{bmatrix}$$

Proceeding further, we obtain the following sequence of matrices:

$$T^4 = \begin{bmatrix} .4375 & .375 \\ .5625 & .625 \end{bmatrix} \qquad T^5 = \begin{bmatrix} .4188 & .3875 \\ .5813 & .6125 \end{bmatrix}$$

$$T^6 = \begin{bmatrix} .4094 & .3938 \\ .5906 & .6062 \end{bmatrix} \qquad T^7 = \begin{bmatrix} .4047 & .3969 \\ .5953 & .6031 \end{bmatrix}$$

$$T^8 = \begin{bmatrix} .4023 & .3984 \\ .5977 & .6016 \end{bmatrix} \qquad T^9 = \begin{bmatrix} .4012 & .3992 \\ .5988 & .6008 \end{bmatrix}$$

$$T^{10} = \begin{bmatrix} .4006 & .3996 \\ .5994 & .6004 \end{bmatrix} \qquad T^{11} = \begin{bmatrix} .4003 & .3998 \\ .5997 & .6002 \end{bmatrix}$$

These results show that the powers T^m of the transition matrix T tend toward a fixed matrix as m gets larger and larger. In this case, the "limiting matrix" is the matrix

$$L = \begin{bmatrix} .40 & .40 \\ .60 & .60 \end{bmatrix} \quad \text{or} \quad \begin{bmatrix} \frac{2}{5} & \frac{2}{5} \\ \frac{3}{5} & \frac{3}{5} \end{bmatrix}$$

Such a matrix is called the steady-state matrix for the system. Thus, as suspected, the long-term behavior of a Markov process such as the one in this example depends on the behavior of the limiting matrix of the powers of the transition matrix—the steady-state matrix for the system. In view of this, the long-term (steady-state) distribution vector for this problem may be found by computing the product

$$LX_0 = \begin{bmatrix} .40 & .40 \\ .60 & .60 \end{bmatrix} \begin{bmatrix} .2 \\ .8 \end{bmatrix} = \begin{bmatrix} .40 \\ .60 \end{bmatrix}$$

which agrees with the result obtained earlier.

Next, since the transition matrix T in this situation seems to have a stabilizing effect over the long term, we are led to wonder whether the steady state would be reached regardless of the initial state of the system. To answer this question, suppose the initial distribution vector is

$$X_0 = \begin{bmatrix} p \\ 1 - p \end{bmatrix}$$

Then, as before, the steady-state distribution vector is given by

$$LX_0 = \begin{bmatrix} .40 & .40 \\ .60 & .60 \end{bmatrix} \begin{bmatrix} p \\ 1 - p \end{bmatrix} = \begin{bmatrix} .40 \\ .60 \end{bmatrix}$$

Thus, the steady state is reached regardless of the initial state of the system!

Regular Markov Chains

The transition matrix T of Example 1 has several important properties, which we emphasized in the foregoing discussion. First, the sequence $T, T^2, T^3, \ldots$ approaches a steady-state matrix in which the rows of the limiting matrix are all equal and all entries are positive. A matrix T having this property is given the special name regular Markov chain.

> **Regular Markov Chain**
> A stochastic matrix T is a **regular Markov chain** if the sequence
>
> $$T, T^2, T^3, \ldots$$
>
> approaches a steady-state matrix in which the rows of the limiting matrix are all equal and all the entries are positive.

It can be shown that *a stochastic matrix T is regular if and only if some power of T has entries that are all positive.* Second, as in the case of Example 1, a Markov chain with a regular transition matrix has a steady-state distribution vector whose elements coincide with those of a row (since they are all the same) of the steady-state matrix; thus, this steady-state distribution vector is always reached regardless of the initial distribution vector.

We will return to computations involving regular Markov chains, but for the moment let's see how one may determine whether a given matrix is indeed regular.

EXAMPLE 2   Determine which of the following matrices are regular.

a. $\begin{bmatrix} .7 & .2 \\ .3 & .8 \end{bmatrix}$ **b.** $\begin{bmatrix} .4 & 1 \\ .6 & 0 \end{bmatrix}$ **c.** $\begin{bmatrix} 0 & 1 \\ 1 & 0 \end{bmatrix}$

Solution

a. Since all the entries of the matrix are positive, the given matrix is regular. Note that this is the transition matrix of Example 1.

b. In this case, one of the entries of the given matrix is equal to zero. Let's compute

$$\begin{bmatrix} .4 & 1 \\ .6 & 0 \end{bmatrix}^2 = \begin{bmatrix} .4 & 1 \\ .6 & 0 \end{bmatrix}\begin{bmatrix} .4 & 1 \\ .6 & 0 \end{bmatrix} = \begin{bmatrix} .76 & .4 \\ .24 & .6 \end{bmatrix}$$

↑
All entries are positive.

Since the second power of the matrix has entries that are all positive, we conclude that the given matrix is in fact regular.

c. Denote the given matrix by A. Then,

$$A = \begin{bmatrix} 0 & 1 \\ 1 & 0 \end{bmatrix}$$

$$A^2 = \begin{bmatrix} 0 & 1 \\ 1 & 0 \end{bmatrix}\begin{bmatrix} 0 & 1 \\ 1 & 0 \end{bmatrix} = \begin{bmatrix} 1 & 0 \\ 0 & 1 \end{bmatrix}$$

$$A^3 = \begin{bmatrix} 0 & 1 \\ 1 & 0 \end{bmatrix}\begin{bmatrix} 1 & 0 \\ 0 & 1 \end{bmatrix} = \begin{bmatrix} 0 & 1 \\ 1 & 0 \end{bmatrix}$$

Not all entries are positive.

Observe that $A^3 = A$. It therefore follows that $A^4 = A^2$, $A^5 = A$, and so on. In other words, any power of A must coincide with either A or A^2. Since not all entries of A and A^2 are positive, the same is true of any power of A. We conclude accordingly that the given matrix is not regular. ∎

We now return to the study of regular Markov chains. In Example 1 we found the steady-state distribution vector associated with a regular Markov chain by studying the limiting behavior of a sequence of distribution vectors. Alternatively, as pointed out in the subsequent discussion, the steady-state distribution vector may also be obtained by first determining the steady-state matrix associated with the regular Markov chain.

Fortunately, there is a relatively simple procedure for finding the steady-state distribution vector associated with a regular Markov process. It does not involve the rather tedious computations required to obtain the sequences in Example 1. The procedure follows.

> **EXPLORE & DISCUSS**
>
> Find the set of all 2×2 stochastic matrices with elements that are either 0 or 1.

Finding the Steady-State Distribution Vector

Let T be a regular stochastic matrix. Then the steady-state distribution vector X may be found by solving the vector equation

$$TX = X$$

together with the condition that the sum of the elements of the vector X be equal to 1.

A justification of the foregoing procedure is given in Exercise 29.

VIDEO

EXAMPLE 3   Find the steady-state distribution vector for the regular Markov chain whose transition matrix is

$$T = \begin{bmatrix} .7 & .2 \\ .3 & .8 \end{bmatrix}$$ See Example 1.

Solution   Let

$$X = \begin{bmatrix} x \\ y \end{bmatrix}$$

be the steady-state distribution vector associated with the Markov process, where the numbers x and y are to be determined. The condition $TX = X$ translates into the matrix equation

$$\begin{bmatrix} .7 & .2 \\ .3 & .8 \end{bmatrix} \begin{bmatrix} x \\ y \end{bmatrix} = \begin{bmatrix} x \\ y \end{bmatrix}$$

or, equivalently, the system of linear equations

$$0.7x + 0.2y = x$$
$$0.3x + 0.8y = y$$

But each of the equations that make up this system of equations is equivalent to the single equation

$$0.3x - 0.2y = 0 \qquad \begin{matrix} 0.7x - x + 0.2y = 0 \\ 0.3x + 0.8y - y = 0 \end{matrix}$$

Next, the condition that the sum of the elements of X add up to 1 gives

$$x + y = 1$$

Thus, the fulfillment of the two conditions simultaneously implies that x and y are the solutions of the system

$$0.3x - 0.2y = 0$$
$$x + \quad y = 1$$

Solving the first equation for x, we obtain

$$x = \frac{2}{3} y$$

which, upon substitution into the second, yields

$$\frac{2}{3} y + y = 1$$

$$y = \frac{3}{5}$$

Thus, $x = \frac{2}{5}$, and the required steady-state distribution vector is given by

$$X = \begin{bmatrix} \frac{2}{5} \\ \frac{3}{5} \end{bmatrix}$$

which agrees with the result obtained earlier.   ∎

APPLIED EXAMPLE 4 Taxicab Movement In Example 6, Section 9.1, we showed that the transition matrix that described the movement of taxis from zone to zone was given by the regular stochastic matrix

$$T = \begin{bmatrix} .6 & .4 & .3 \\ .3 & .3 & .3 \\ .1 & .3 & .4 \end{bmatrix}$$

Use this information to determine the long-term distribution of the taxis in the three zones.

Solution Let

$$X = \begin{bmatrix} x \\ y \\ z \end{bmatrix}$$

be the steady-state distribution vector associated with the Markov process under consideration, where x, y, and z are to be determined. The condition $TX = X$ translates into the matrix equation

$$\begin{bmatrix} .6 & .4 & .3 \\ .3 & .3 & .3 \\ .1 & .3 & .4 \end{bmatrix} \begin{bmatrix} x \\ y \\ z \end{bmatrix} = \begin{bmatrix} x \\ y \\ z \end{bmatrix}$$

or, equivalently, the system of linear equations

$$0.6x + 0.4y + 0.3z = x$$
$$0.3x + 0.3y + 0.3z = y$$
$$0.1x + 0.3y + 0.4z = z$$

The system simplifies into

$$4x - 4y - 3z = 0$$
$$3x - 7y + 3z = 0$$
$$x + 3y - 6z = 0$$

Since $x + y + z = 1$ as well, we are required to solve the system

$$x + y + z = 1$$
$$4x - 4y - 3z = 0$$
$$3x - 7y + 3z = 0$$
$$x + 3y - 6z = 0$$

Using the Gauss–Jordan elimination procedure of Chapter 2, we find that

$$x = \frac{33}{70} \qquad y = \frac{3}{10} \qquad z = \frac{8}{35}$$

or $x \approx 0.47$, $y = 0.30$, and $z \approx 0.23$. Thus, in the long run, 47% of the taxis will be in zone I, 30% in zone II, and 23% in zone III.

9.2   Self-Check Exercises

1. Find the steady-state distribution vector for the regular Markov chain whose transition matrix is

$$T = \begin{bmatrix} .5 & .8 \\ .5 & .2 \end{bmatrix}$$

2. Three supermarkets serve a certain section of a city. During the year, supermarket A is expected to retain 80% of its customers, lose 5% of its customers to supermarket B, and lose 15% to supermarket C. Supermarket B is expected to retain 90% of its customers and lose 5% to each of supermarket A and supermarket C. Supermarket C is expected to retain 75% of its customers, lose 10% to supermarket A, and lose 15% to supermarket B. If these trends continue, what will be the market share of each supermarket in the long run?

Solutions to Self-Check Exercises 9.2 can be found on page 532.

9.2   Concept Questions

1. Explain (a) a *steady-state distribution vector*, (b) a *steady-state matrix*, and (c) a *regular Markov chain*.

2. How do you find the steady-state distribution vector given a regular stochastic matrix T associated with the Markov process?

9.2   Exercises

In Exercises 1–8, determine which of the matrices are regular.

1. $\begin{bmatrix} \frac{2}{5} & \frac{3}{4} \\ \frac{3}{5} & \frac{1}{4} \end{bmatrix}$ 2. $\begin{bmatrix} 0 & .3 \\ 1 & .7 \end{bmatrix}$

3. $\begin{bmatrix} 1 & .8 \\ 0 & .2 \end{bmatrix}$ 4. $\begin{bmatrix} \frac{1}{3} & 0 \\ \frac{2}{3} & 1 \end{bmatrix}$

5. $\begin{bmatrix} \frac{1}{2} & \frac{3}{4} & 0 \\ \frac{1}{2} & 0 & \frac{1}{2} \\ 0 & \frac{1}{4} & \frac{1}{2} \end{bmatrix}$ 6. $\begin{bmatrix} 1 & .3 & .1 \\ 0 & .4 & .8 \\ 0 & .3 & .1 \end{bmatrix}$

7. $\begin{bmatrix} .7 & .2 & .3 \\ .3 & .8 & .3 \\ 0 & 0 & .4 \end{bmatrix}$ 8. $\begin{bmatrix} 0 & 0 & \frac{1}{4} \\ 1 & 0 & 0 \\ 0 & 1 & \frac{3}{4} \end{bmatrix}$

In Exercises 9–16, find the steady-state vector for the transition matrix.

9. $\begin{bmatrix} \frac{1}{3} & \frac{1}{4} \\ \frac{2}{3} & \frac{3}{4} \end{bmatrix}$ 10. $\begin{bmatrix} \frac{4}{5} & \frac{3}{5} \\ \frac{1}{5} & \frac{2}{5} \end{bmatrix}$

11. $\begin{bmatrix} .5 & .2 \\ .5 & .8 \end{bmatrix}$ 12. $\begin{bmatrix} .9 & 1 \\ .1 & 0 \end{bmatrix}$

13. $\begin{bmatrix} 0 & \frac{1}{8} & 1 \\ 1 & \frac{5}{8} & 0 \\ 0 & \frac{1}{4} & 0 \end{bmatrix}$ 14. $\begin{bmatrix} .6 & .3 & 0 \\ .4 & .4 & .6 \\ 0 & .3 & .4 \end{bmatrix}$

15. $\begin{bmatrix} .2 & 0 & .3 \\ 0 & .6 & .4 \\ .8 & .4 & .3 \end{bmatrix}$ 16. $\begin{bmatrix} .1 & .2 & .3 \\ .1 & .2 & .3 \\ .8 & .6 & .4 \end{bmatrix}$

17. **PSYCHOLOGY EXPERIMENTS**   A psychologist conducts an experiment in which a mouse is placed in a T-maze, where it has a choice at the T-junction of turning left and receiving a reward (cheese) or turning right and receiving a mild shock. At the end of each trial a record is kept of the mouse's response. It is observed that the mouse is as likely to turn left (state 1) as right (state 2) during the first trial. In subsequent trials, however, the observation is made that if the mouse had turned left in the previous trial, then the probability that it will turn left in the next trial is .8, whereas the probability that it will turn right is .2. If the mouse had turned right in the previous trial, then the probability that it will turn right in the next trial is .1, whereas the probability that it will turn left is .9. In the long run, what percentage of the time will the mouse turn left at the T-junction?

18. **COMMUTER TRENDS**   Within a large metropolitan area, 20% of the commuters currently use the public transportation system, whereas the remaining 80% commute via automobile. The city has recently revitalized and expanded its public transportation system. It is expected that 6 mo from now 30% of those who are now commuting to work via automobile will switch to public transportation, and 70% will continue to commute via automobile. At the same time, it is

expected that 20% of those now using public transportation will commute via automobile, and 80% will continue to use public transportation. In the long run, what percent of the commuters will be using public transportation?

19. **ONE- AND TWO-INCOME FAMILIES** From data compiled over a 10-yr period by Manpower, Inc., in a statewide study of married couples in which at least one spouse was working, the following transition matrix was constructed. It gives the transitional probabilities for one and two wage earners among married couples.

		Current State	
		1 Wage Earner	2 Wage Earners
Next	1 Wage Earner	.72	.12
State	2 Wage Earners	.28	.88

At the present time, 48% of the married couples (in which at least one spouse is working) have one wage earner, and 52% have two wage earners. Assuming that this trend continues, what will be the distribution of one- and two-wage earner families among married couples in this area 10 years from now? Over the long run?

20. **PROFESSIONAL WOMEN** From data compiled over a 5-yr period by *Women's Daily* in a study of the number of women in the professions, the following transition matrix was constructed. It gives the transitional probabilities for the number of men and women in the professions.

		Current State	
		Men	Women
Next State	Men	.95	.04
	Women	.05	.96

As of the beginning of 1986, 52.9% of professional jobs were held by men. If this trend continues, what percent of professional jobs will be held by women in the long run?

21. **BUYING TRENDS OF HOME BUYERS** From data collected by the Association of Realtors of a certain city, the following transition matrix was obtained. The matrix describes the buying pattern of home buyers who buy single-family homes (S) or condominiums (C).

		Current State	
		S	C
Next State	S	.85	.35
	C	.15	.65

Currently, 80% of the homeowners live in single-family homes, whereas 20% live in condominiums. If this trend continues, what will be the percent of homeowners in this city who will own single-family homes and condominiums 2 years from now? In the long run?

22. **HOMEOWNERS' CHOICE OF ENERGY** A study conducted by the Urban Energy Commission in a large metropolitan area indicates the probabilities that homeowners within the area will use certain heating fuels or solar energy during the next 10 yr as the major source of heat for their homes. The following transition matrix represents the transition probabilities from one state to another:

	Elec.	Gas	Oil	Solar
Electricity	.70	0	.10	0
Natural gas	.15	.90	.10	.05
Fuel oil	.05	.02	.75	.05
Solar energy	.10	.08	.05	.90

Among the homeowners within the area, 20% currently use electricity, 35% use natural gas, 40% use oil, and 5% use solar energy as their major source of heat for their homes. In the long run, what percent of homeowners within the area will be using solar energy as their major source of heating fuel?

23. **NETWORK NEWS VIEWERSHIP** A television poll was conducted among regular viewers of the national news in a certain region where the three national networks share the same time slot for the evening news. Results of the poll indicate that 30% of the viewers watch the ABC evening news, 40% watch the CBS evening news, and 30% watch the NBC evening news. Furthermore, it was found that of those viewers who watched the ABC evening news during 1 wk, 80% would again watch the ABC evening news during the next week, 10% would watch the CBS news, and 10% would watch the NBC news. Of those viewers who watched the CBS evening news during 1 wk, 85% would again watch the CBS evening news during the next week, 10% would watch the ABC news, and 5% would watch the NBC news. Of those viewers who watched the NBC evening news during 1 wk, 85% would again watch the NBC news during the next week, 10% would watch ABC, and 5% would watch CBS.

a. What share of the audience consisting of regular viewers of the national news will each network command after 2 wk?

b. In the long run, what share of the audience will each network command?

24. **NETWORK NEWS VIEWERSHIP** Refer to Exercise 23. If the initial distribution vector is

$$X_0 = \begin{matrix} \text{ABC} \\ \text{CBS} \\ \text{NBC} \end{matrix} \begin{bmatrix} .40 \\ .40 \\ .20 \end{bmatrix}$$

what share of the audience will each network command in the long run?

25. **GENETICS** In a certain species of roses, a plant with genotype (genetic makeup) *AA* has red flowers, a plant with genotype *Aa* has pink flowers, and a plant with genotype *aa* has white flowers, where *A* is the dominant gene and *a* is the recessive gene for color. If a plant with one genotype is crossed with another plant, then the color of the offspring's flowers is determined by the genotype of the parent plants. If a plant of

each genotype is crossed with a pink-flowered plant, then the transition matrix used to determine the color of the offspring's flowers is given by

$$
\begin{array}{c}
 \\
 \\
\text{Offspring}\quad
\begin{array}{c}
\text{Red } (AA) \\
\text{Pink } (Aa) \text{ or } (aA) \\
\text{White } (aa)
\end{array}
\end{array}
\overset{\begin{array}{c}\text{Parent}\\ \text{Red}\quad\text{Pink}\quad\text{White}\end{array}}{
\begin{bmatrix}
\frac{1}{2} & \frac{1}{4} & 0 \\
\frac{1}{2} & \frac{1}{2} & \frac{1}{2} \\
0 & \frac{1}{4} & \frac{1}{2}
\end{bmatrix}}
$$

If the offspring of each generation are crossed only with pink-flowered plants, in the long run what percentage of the plants will have red flowers? Pink flowers? White flowers?

26. **MARKET SHARE OF AUTO MANUFACTURERS** In a study of the domestic market share of the three major automobile manufacturers A, B, and C in a certain country, it was found that of the customers who bought a car manufactured by A, 75% would again buy a car manufactured by A, 15% would buy a car manufactured by B, and 10% would buy a car manufactured by C. Of the customers who bought a car manufactured by B, 90% would again buy a car manufactured by B, whereas 5% each would buy cars manufactured by A and C, respectively. Finally, of the customers who bought a car manufactured by C, 85% would again buy a car manufactured by C, 5% would buy a car manufactured by A, and 10% would buy a car manufactured by B. Assuming that these sentiments reflect the buying habits of customers in the future model years, determine the market share that will be held by each manufacturer in the long run.

In Exercises 27 and 28, determine whether the statement is true or false. If it is true, explain why it is true. If it is false, give an example to show why it is false.

27. A stochastic matrix T is a regular Markov chain if the powers of T approach a fixed matrix whose rows are all equal.

28. To find the steady-state distribution vector X, we solve the system

$$
\begin{cases}
TX = X \\
x_1 + x_2 + \cdots + x_n = 1
\end{cases}
$$

where T is the regular stochastic matrix associated with the Markov process and

$$
X = \begin{bmatrix} x_1 \\ x_2 \\ \vdots \\ x_n \end{bmatrix}
$$

29. Let T be a regular stochastic matrix. Show that the steady-state distribution vector X may be found by solving the vector equation $TX = X$ together with the condition that the sum of the elements of X be equal to 1.
    Hint: Take the initial distribution to be X, the steady-state distribution vector. Then, when n is large, $X \approx T^n X$. (Why?) Multiply both sides of the last equation by T (on the left) and consider the resulting equation when n is large.

9.2   Solutions to Self-Check Exercises

1. Let

$$
X = \begin{bmatrix} x \\ y \end{bmatrix}
$$

be the steady-state distribution vector associated with the Markov process, where the numbers x and y are to be determined. The condition $TX = X$ translates into the matrix equation

$$
\begin{bmatrix} .5 & .8 \\ .5 & .2 \end{bmatrix}\begin{bmatrix} x \\ y \end{bmatrix} = \begin{bmatrix} x \\ y \end{bmatrix}
$$

which is equivalent to the system of linear equations

$$
0.5x + 0.8y = x
$$
$$
0.5x + 0.2y = y
$$

Each equation in the system is equivalent to the equation

$$
0.5x - 0.8y = 0
$$

Next, the condition that the sum of the elements of X adds up to 1 gives

$$
x + y = 1
$$

Thus, the fulfillment of the two conditions simultaneously implies that x and y are the solutions of the system

$$
0.5x - 0.8y = 0
$$
$$
x + \quad y = 1
$$

Solving the first equation for x, we obtain

$$
x = \frac{8}{5}y
$$

which, upon substitution into the second, yields

$$
\frac{8}{5}y + y = 1
$$
$$
y = \frac{5}{13}
$$

Therefore, $x = \dfrac{8}{13}$, and the required steady-state distribution vector is

$$\begin{bmatrix} \frac{8}{13} \\ \frac{5}{13} \end{bmatrix}$$

2. The transition matrix for the Markov process under consideration is

$$T = \begin{bmatrix} .80 & .05 & .10 \\ .05 & .90 & .15 \\ .15 & .05 & .75 \end{bmatrix}$$

Now, let

$$X = \begin{bmatrix} x \\ y \\ z \end{bmatrix}$$

be the steady-state distribution vector associated with the Markov process under consideration, where x, y, and z are to be determined. The condition $TX = X$ is

$$\begin{bmatrix} .80 & .05 & .10 \\ .05 & .90 & .15 \\ .15 & .05 & .75 \end{bmatrix} \begin{bmatrix} x \\ y \\ z \end{bmatrix} = \begin{bmatrix} x \\ y \\ z \end{bmatrix}$$

or, equivalently, the system of linear equations

$$\begin{aligned} 0.80x + 0.05y + 0.10z &= x \\ 0.05x + 0.90y + 0.15z &= y \\ 0.15x + 0.05y + 0.75z &= z \end{aligned}$$

This system simplifies into

$$\begin{aligned} 4x - y - 2z &= 0 \\ x - 2y + 3z &= 0 \\ 3x + y - 5z &= 0 \end{aligned}$$

Since $x + y + z = 1$ as well, we are required to solve the system

$$\begin{aligned} 4x - y - 2z &= 0 \\ x - 2y + 3z &= 0 \\ 3x + y - 5z &= 0 \\ x + y + z &= 1 \end{aligned}$$

Using the Gauss–Jordan elimination procedure, we find

$$x = \frac{1}{4} \qquad y = \frac{1}{2} \qquad z = \frac{1}{4}$$

Therefore, in the long run, supermarkets A and C will each have one-quarter of the customers, and supermarket B will have half the customers.

USING TECHNOLOGY

■ Finding the Long-Term Distribution Vector

The problem of finding the long-term distribution vector for a regular Markov chain ultimately rests on the problem of solving a system of linear equations. As such, the **rref** or equivalent function of a graphing utility proves indispensable, as the following example shows.

EXAMPLE 1 Find the steady-state distribution vector for the regular Markov chain whose transition matrix is

$$T = \begin{bmatrix} .4 & .2 & .1 \\ .3 & .4 & .5 \\ .3 & .4 & .4 \end{bmatrix}$$

Solution Let

$$\begin{bmatrix} x \\ y \\ z \end{bmatrix}$$

(*continued*)

be the steady-state distribution vector, where x, y, and z are to be determined. The condition $TX = X$ translates into the matrix equation

$$\begin{bmatrix} .4 & .2 & .1 \\ .3 & .4 & .5 \\ .3 & .4 & .4 \end{bmatrix} \begin{bmatrix} x \\ y \\ z \end{bmatrix} = \begin{bmatrix} x \\ y \\ z \end{bmatrix}$$

or, equivalently, the system of linear equations

$$\begin{aligned} 0.4x + 0.2y + 0.1z &= x \\ 0.3x + 0.4y + 0.5z &= y \\ 0.3x + 0.4y + 0.4z &= z \end{aligned}$$

Since $x + y + z = 1$, we are required to solve the system

$$\begin{aligned} -0.6x + 0.2y + 0.1z &= 0 \\ 0.3x - 0.6y + 0.5z &= 0 \\ 0.3x + 0.4y - 0.6z &= 0 \\ x + \quad y + \quad z &= 1 \end{aligned}$$

Entering this system into the graphing calculator as the augmented matrix

$$A = \begin{bmatrix} -.6 & .2 & .1 & | & 0 \\ .3 & -.6 & .5 & | & 0 \\ .3 & .4 & -.6 & | & 0 \\ 1 & 1 & 1 & | & 1 \end{bmatrix}$$

and then using the **rref** function, we obtain the equivalent system (to two decimal places)

$$\begin{bmatrix} 1 & 0 & 0 & | & .20 \\ 0 & 1 & 0 & | & .42 \\ 0 & 0 & 1 & | & .38 \\ 0 & 0 & 0 & | & 0 \end{bmatrix}$$

Therefore, $x \approx 0.20$, $y \approx 0.42$, and $z \approx 0.38$, and so the required steady-state distribution vector is

$$\begin{bmatrix} .20 \\ .42 \\ .38 \end{bmatrix}$$

TECHNOLOGY EXERCISES

In Exercises 1 and 2, find the steady-state vector for the matrix T.

1.
$$T = \begin{bmatrix} .2 & .2 & .3 & .2 & .1 \\ .1 & .2 & .1 & .2 & .1 \\ .3 & .4 & .1 & .3 & .3 \\ .2 & .1 & .2 & .2 & .2 \\ .2 & .1 & .3 & .1 & .3 \end{bmatrix}$$

2.
$$T = \begin{bmatrix} .3 & .2 & .1 & .3 & .1 \\ .2 & .1 & .2 & .1 & .2 \\ .1 & .2 & .3 & .2 & .2 \\ .1 & .3 & .2 & .3 & .2 \\ .3 & .2 & .2 & .1 & .3 \end{bmatrix}$$

3. Verify that the steady-state vector for Example 4, page 529, is

$$X = \begin{bmatrix} .47 \\ .30 \\ .23 \end{bmatrix}$$

9.3 Absorbing Markov Chains

Absorbing Markov Chains

In this section, we investigate the long-term trends of a certain class of Markov chains that involve transition matrices that are not regular. In particular, we study Markov chains in which the transition matrices, known as absorbing stochastic matrices, have the special properties to be described presently.

Consider the stochastic matrix

$$\begin{bmatrix} 1 & 0 & .2 & 0 \\ 0 & 1 & .3 & 1 \\ 0 & 0 & .5 & 0 \\ 0 & 0 & 0 & 0 \end{bmatrix}$$

associated with a Markov process. Interpreting it in the usual fashion, we see that after one observation, the probability is 1 (a certainty) that an object previously in state 1 will remain in state 1. Similarly, we see that an object previously in state 2 must remain in state 2. Next, we find that an object previously in state 3 has a probability of .2 of going to state 1, a probability of .3 of going to state 2, a probability of .5 of remaining in state 3, and no chance of going to state 4. Finally, an object previously in state 4 must, after one observation, end up in state 2.

This stochastic matrix exhibits certain special characteristics. First, as observed earlier, an object in state 1 or state 2 must stay in state 1 or state 2, respectively. Such states are called absorbing states. In general, an **absorbing state** is one from which it is impossible for an object to leave. To identify the absorbing states of a stochastic matrix, we simply examine each column of the matrix. If column i has a 1 in the a_{ii} position (that is, on the main diagonal of the matrix) and zeros elsewhere in that column, then and only then is state i an absorbing state.

Second, observe that states 3 and 4, although not absorbing states, have the property that an object in each of these states has a possibility of going to an absorbing state. For example, an object currently in state 3 has a probability of .2 of ending up in state 1, an absorbing state, and an object in state 4 must end up in state 2, also an absorbing state, after one transition.

Absorbing Stochastic Matrix

An **absorbing stochastic matrix** has the following properties:

1. There is at least one absorbing state.
2. It is possible to go from any nonabsorbing state to an absorbing state in one or more stages.

A Markov chain is said to be an **absorbing Markov chain** if the transition matrix associated with the process is an absorbing stochastic matrix.

VIDEO

EXAMPLE 1   Determine whether the following matrices are absorbing stochastic matrices.

a. $\begin{bmatrix} .7 & 0 & .1 & 0 \\ 0 & 1 & .5 & 0 \\ .3 & 0 & .2 & 0 \\ 0 & 0 & .2 & 1 \end{bmatrix}$ **b.** $\begin{bmatrix} 1 & 0 & 0 & 0 \\ 0 & 1 & 0 & 0 \\ 0 & 0 & .5 & .4 \\ 0 & 0 & .5 & .6 \end{bmatrix}$

Solution

a. States 2 and 4 are both absorbing states. Furthermore, even though state 1 is not an absorbing state, there is a possibility (with probability .3) that an object may go from this state to state 3. State 3 itself is nonabsorbing, but an object in that state has a probability of .5 of going to the absorbing state 2 and a probability of .2 of going to the absorbing state 4. Thus, the given matrix is an absorbing stochastic matrix.

b. States 1 and 2 are absorbing states. However, it is impossible for an object to go from the nonabsorbing states 3 and 4 to either or both of the absorbing states. Thus, the given matrix is not an absorbing stochastic matrix.   ∎

Given an absorbing stochastic matrix, it is always possible, by suitably reordering the states if necessary, to rewrite it so that the absorbing states appear first. Then, the resulting matrix can be partitioned into four submatrices,

$$\begin{array}{cc} \text{Absorbing} & \text{Nonabsorbing} \end{array}$$
$$\left[\begin{array}{c:c} I & S \\ \hdashline O & R \end{array}\right]$$

where I is an identity matrix whose order is determined by the number of absorbing states and O is a zero matrix. The submatrices R and S correspond to the nonabsorbing states. As an example, the absorbing stochastic matrix of Example 1(a)

$$\begin{array}{c} \\ 1 \\ 2 \\ 3 \\ 4 \end{array} \begin{array}{cccc} 1 & 2 & 3 & 4 \\ \end{array} \begin{bmatrix} .7 & 0 & .1 & 0 \\ 0 & 1 & .5 & 0 \\ .3 & 0 & .2 & 0 \\ 0 & 0 & .2 & 1 \end{bmatrix}$$

may be written as

$$\begin{array}{c} \\ 4 \\ 2 \\ 1 \\ 3 \end{array} \begin{array}{cccc} 4 & 2 & 1 & 3 \\ \end{array} \left[\begin{array}{cc:cc} 1 & 0 & 0 & .2 \\ 0 & 1 & 0 & .5 \\ \hdashline 0 & 0 & .7 & .1 \\ 0 & 0 & .3 & .2 \end{array}\right]$$

upon reordering the states as indicated.

APPLIED EXAMPLE 2   Gambler's Ruin   John has decided to risk $2 in the following game of chance. He places a $1 bet on each repeated play of the game in which the probability of his winning $1 is .4, and he continues to play until he has accumulated a total of $3 or he has lost all of his money. Write the transition matrix for the related absorbing Markov chain.

Solution   There are four states in this Markov chain, which correspond to John accumulating a total of $0, $1, $2, and $3. Since the first and last states listed are absorbing states, we will list these states first, resulting in the transition matrix

$$
\begin{array}{c}
\overbrace{}^{\text{Absorbing}}\ \overbrace{}^{\text{Nonabsorbing}} \\
\ \ \$0\ \ \$3\ \ \$1\ \ \$2 \\
\begin{array}{c}
\$0 \\ \$3 \\ \$1 \\ \$2
\end{array}
\left[
\begin{array}{cccc}
1 & 0 & .6 & 0 \\
0 & 1 & 0 & .4 \\
0 & 0 & 0 & .6 \\
0 & 0 & .4 & 0
\end{array}
\right]
\end{array}
$$

which is constructed as follows: Since the state "$0" is an absorbing state, we see that $a_{11} = 1$, $a_{21} = a_{31} = a_{41} = 0$. Similarly, the state "$3" is an absorbing state, so $a_{22} = 1$, and $a_{12} = a_{32} = a_{42} = 0$. To construct the column corresponding to the nonabsorbing state "$1," we note that there is a probability of .6 (John loses) in going from an accumulated amount of $1 to $0, so $a_{13} = .6$; $a_{23} = a_{33} = 0$ because it is not feasible to go from an accumulated amount of $1 to either an accumulated amount of $3 or $1 in one transition (play). Finally, there is a probability of .4 (John wins) in going from an accumulated amount of $1 to an accumulated amount of $2, so $a_{43} = .4$. The last column of the transition matrix is constructed by reasoning in a similar manner. ■

The following question arises in connection with the last example: If John continues to play the game as originally planned, what is the probability that he will depart from the game victorious—that is, leave with an accumulated amount of $3?

To answer this question, we have to look at the long-term trend of the relevant Markov chain. Taking a cue from our work in the last section, we may compute the powers of the transition matrix associated with the Markov chain. Just as in the case of regular stochastic matrices, it turns out that the powers of an absorbing stochastic matrix approach a steady-state matrix. However, instead of demonstrating this, we use the following result, which we state without proof, for computing the steady-state matrix:

Finding the Steady-State Matrix for an Absorbing Stochastic Matrix

Suppose an absorbing stochastic matrix A has been partitioned into submatrices

$$
A = \left[
\begin{array}{c|c}
I & S \\
\hline
O & R
\end{array}
\right]
$$

Then the *steady-state matrix* of A is given by

$$
\left[
\begin{array}{c|c}
I & S(I - R)^{-1} \\
\hline
O & O
\end{array}
\right]
$$

where the order of the identity matrix appearing in the expression $(I - R)^{-1}$ is chosen to have the same order as R.

APPLIED EXAMPLE 3 Gambler's Ruin (continued) Refer to Example 2. If John continues to play the game until he has accumulated a sum of $3 or until he has lost all of his money, what is the probability that he will accumulate $3?

Solution   The transition matrix associated with the Markov process is (see Example 2)

$$A = \begin{bmatrix} 1 & 0 & \vdots & .6 & 0 \\ 0 & 1 & \vdots & 0 & .4 \\ \hdashline 0 & 0 & \vdots & 0 & .6 \\ 0 & 0 & \vdots & .4 & 0 \end{bmatrix}$$

We need to find the steady-state matrix of A. In this case,

$$R = \begin{bmatrix} 0 & .6 \\ .4 & 0 \end{bmatrix} \quad \text{and} \quad S = \begin{bmatrix} .6 & 0 \\ 0 & .4 \end{bmatrix}$$

so

$$I - R = \begin{bmatrix} 1 & 0 \\ 0 & 1 \end{bmatrix} - \begin{bmatrix} 0 & .6 \\ .4 & 0 \end{bmatrix} = \begin{bmatrix} 1 & -.6 \\ -.4 & 1 \end{bmatrix}$$

Using the formula in Section 2.6 for finding the inverse of a 2×2 matrix, we find that

$$(I - R)^{-1} = \begin{bmatrix} 1.32 & .79 \\ .53 & 1.32 \end{bmatrix}$$

and so

$$S(I - R)^{-1} = \begin{bmatrix} .6 & 0 \\ 0 & .4 \end{bmatrix} \begin{bmatrix} 1.32 & .79 \\ .53 & 1.32 \end{bmatrix} = \begin{bmatrix} .79 & .47 \\ .21 & .53 \end{bmatrix}$$

Therefore, the required steady-state matrix of A is given by

$$\begin{bmatrix} I & \vdots & S(I-R)^{-1} \\ \hdashline O & \vdots & O \end{bmatrix} = \begin{array}{c} \\ \$0 \\ \$3 \\ \$1 \\ \$2 \end{array} \begin{bmatrix} \overset{\$0}{1} & \overset{\$3}{0} & \vdots & \overset{\$1}{.79} & \overset{\$2}{.47} \\ 0 & 1 & \vdots & .21 & .53 \\ \hdashline 0 & 0 & \vdots & 0 & 0 \\ 0 & 0 & \vdots & 0 & 0 \end{bmatrix}$$

From this result we see that starting with \$2, the probability is .53 that John will leave the game with an accumulated amount of \$3—that is, that he wins \$1. ∎

Our last example shows an application of Markov chains in the field of genetics.

APPLIED EXAMPLE 4   Genetics   In a certain species of flowers, a plant of genotype (genetic makeup) AA has red flowers, a plant of genotype Aa has pink flowers, and a plant of genotype aa has white flowers, where A is the dominant gene and a is the recessive gene for color. If a plant of one genotype is crossed with another plant, then the color of the offspring's flowers is determined by the genotype of the parent plants. If the offspring are crossed successively with plants of genotype AA only, show that in the long run all the flowers produced by the plants will be red.

Solution   First, let's construct the transition matrix associated with the resulting Markov chain. In crossing a plant of genotype AA with another of the same genotype AA, the offspring will inherit one dominant gene from each parent and thus

VIDEO

EXPLORE & DISCUSS

Consider the stochastic matrix

$$A = \begin{bmatrix} 1 & 0 & a \\ 0 & 1 & b \\ 0 & 0 & 1-a-b \end{bmatrix}$$

where a and b satisfy $0 < a < 1$, $0 < b < 1$, and $0 < a + b < 1$.

1. Find the steady-state matrix.

2. What is the probability that state 3 will be absorbed in state 2?

will have genotype *AA*. Therefore, the probabilities of the offspring being genotype *AA*, *Aa*, and *aa* are 1, 0, and 0, respectively.

Next, in crossing a plant of genotype *AA* with one of genotype *Aa*, the probability of the offspring having genotype *AA* (inheriting an *A* gene from the first parent and an *A* from the second) is $\frac{1}{2}$; the probability of the offspring having genotype *Aa* (inheriting an *A* gene from the first parent and an *a* gene from the second parent) is $\frac{1}{2}$; finally, the probability of the offspring being of genotype *aa* is 0 since this is clearly impossible.

A similar argument shows that in crossing a plant of genotype *AA* with one of genotype *aa*, the probabilities of the offspring having genotype *AA*, *Aa*, and *aa* are 0, 1, and 0, respectively.

The required transition matrix is thus given by

$$
\begin{array}{c}
\text{Absorbing state} \\
\downarrow \\
\begin{array}{ccc} AA & Aa & aa \end{array} \\
T = \begin{array}{c} AA \\ Aa \\ aa \end{array}
\begin{bmatrix} 1 & \frac{1}{2} & 0 \\ 0 & \frac{1}{2} & 1 \\ 0 & 0 & 0 \end{bmatrix}
\end{array}
$$

Observe that the state *AA* is an absorbing state. Furthermore, it is possible to go from each of the other two nonabsorbing states to the absorbing state *AA*. Thus, the Markov chain is an absorbing Markov chain. To determine the long-term effects of this experiment, let's compute the steady-state matrix of *T*. Partitioning *T* in the usual manner, we find

$$
T = \begin{bmatrix} 1 & \frac{1}{2} & 0 \\ \hline 0 & \frac{1}{2} & 1 \\ 0 & 0 & 0 \end{bmatrix}
$$

so that

$$
R = \begin{bmatrix} \frac{1}{2} & 1 \\ 0 & 0 \end{bmatrix} \quad \text{and} \quad S = \begin{bmatrix} \frac{1}{2} & 0 \end{bmatrix}
$$

Next, we compute

$$
I - R = \begin{bmatrix} 1 & 0 \\ 0 & 1 \end{bmatrix} - \begin{bmatrix} \frac{1}{2} & 1 \\ 0 & 0 \end{bmatrix} = \begin{bmatrix} \frac{1}{2} & -1 \\ 0 & 1 \end{bmatrix}
$$

and, using the formula for finding the inverse of a 2 × 2 matrix in Section 2.6,

$$
(I - R)^{-1} = \begin{bmatrix} 2 & 2 \\ 0 & 1 \end{bmatrix}
$$

Thus,

$$
S(I - R)^{-1} = \begin{bmatrix} \frac{1}{2} & 0 \end{bmatrix} \begin{bmatrix} 2 & 2 \\ 0 & 1 \end{bmatrix} = \begin{bmatrix} 1 & 1 \end{bmatrix}
$$

Therefore, the steady-state matrix of *T* is given by

$$
\begin{bmatrix} I & S(I-R)^{-1} \\ \hline O & O \end{bmatrix} =
\begin{array}{c}
\begin{array}{ccc} AA & Aa & aa \end{array} \\
\begin{array}{c} AA \\ Aa \\ aa \end{array}
\begin{bmatrix} 1 & 1 & 1 \\ \hline 0 & 0 & 0 \\ 0 & 0 & 0 \end{bmatrix}
\end{array}
$$

Interpreting the steady-state matrix of T, we see that the long-term result of crossing the offspring with plants of genotype AA only leads to the absorbing state AA. In other words, such a procedure will result in the production of plants that will bear only red flowers, as we set out to demonstrate.   ∎

9.3   Self-Check Exercises

1. Let

$$T = \begin{bmatrix} .2 & 0 & 0 \\ .3 & 1 & .6 \\ .5 & 0 & .4 \end{bmatrix}$$

   a. Show that T is an absorbing stochastic matrix.
   b. Rewrite T so that the absorbing states appear first, partition the resulting matrix, and identify the submatrices R and S.
   c. Compute the steady-state matrix of T.

2. There is a trend toward increased use of computer-aided transcription (CAT) and electronic recording (ER) as alternatives to manual transcription (MT) of court proceedings by court stenographers in a certain state. Suppose the following sto-

chastic matrix is the transition matrix associated with the Markov process:

$$T = \begin{array}{c} \\ \text{CAT} \\ \text{ER} \\ \text{MT} \end{array} \begin{array}{c} \text{CAT} \quad \text{ER} \quad \text{MT} \\ \begin{bmatrix} 1 & .3 & .2 \\ 0 & .6 & .3 \\ 0 & .1 & .5 \end{bmatrix} \end{array}$$

Determine the probability that a court now using electronic recording or manual transcribing of its proceedings will eventually change to CAT.

Solutions to Self-Check Exercises 9.3 can be found on page 542.

9.3   Concept Questions

1. What is an *absorbing stochastic matrix*?

2. Suppose the absorbing stochastic matrix A has been partitioned into submatrices

$$\left[\begin{array}{c|c} I & S \\ \hline O & R \end{array} \right]$$

Write the expression representing the steady-state matrix of A. (Assume that I and R have the same order.)

9.3   Exercises

In Exercises 1–8, determine whether the matrix is an absorbing stochastic matrix.

1. $\begin{bmatrix} \frac{2}{5} & 0 \\ \frac{3}{5} & 1 \end{bmatrix}$ 2. $\begin{bmatrix} 1 & 0 \\ 0 & 1 \end{bmatrix}$

3. $\begin{bmatrix} 1 & .5 & 0 \\ 0 & 0 & 1 \\ 0 & .5 & 0 \end{bmatrix}$ 4. $\begin{bmatrix} 1 & 0 & 0 \\ 0 & .7 & .2 \\ 0 & .3 & .8 \end{bmatrix}$

5. $\begin{bmatrix} \frac{1}{8} & 0 & 0 \\ \frac{1}{4} & 1 & 0 \\ \frac{5}{8} & 0 & 1 \end{bmatrix}$ 6. $\begin{bmatrix} 1 & 0 & 0 & 0 \\ 0 & \frac{5}{8} & 0 & \frac{1}{6} \\ 0 & \frac{1}{8} & 1 & 0 \\ 0 & \frac{1}{4} & 0 & \frac{5}{6} \end{bmatrix}$

7. $\begin{bmatrix} 1 & 0 & .3 & 0 \\ 0 & 1 & .2 & 0 \\ 0 & 0 & .1 & .5 \\ 0 & 0 & .4 & .5 \end{bmatrix}$ 8. $\begin{bmatrix} 1 & 0 & 0 & 0 \\ 0 & 1 & 0 & 0 \\ 0 & 0 & .2 & .6 \\ 0 & 0 & .8 & .4 \end{bmatrix}$

In Exercises 9–14, rewrite each absorbing stochastic matrix so that the absorbing states appear first, partition the resulting matrix, and identify the submatrices R and S.

9. $\begin{bmatrix} .6 & 0 \\ .4 & 1 \end{bmatrix}$ 10. $\begin{bmatrix} \frac{1}{4} & 0 & 0 \\ \frac{1}{4} & 1 & 0 \\ \frac{1}{2} & 0 & 1 \end{bmatrix}$

11. $\begin{bmatrix} 0 & .2 & 0 \\ .5 & .4 & 0 \\ .5 & .4 & 1 \end{bmatrix}$

12. $\begin{bmatrix} .5 & 0 & .3 \\ 0 & 1 & .1 \\ .5 & 0 & .6 \end{bmatrix}$

13. $\begin{bmatrix} .4 & .2 & 0 & 0 \\ .2 & .3 & 0 & 0 \\ 0 & .3 & 1 & 0 \\ .4 & .2 & 0 & 1 \end{bmatrix}$

14. $\begin{bmatrix} .1 & 0 & 0 & 0 \\ .2 & 1 & 0 & .2 \\ .3 & 0 & 1 & 0 \\ .4 & 0 & 0 & .8 \end{bmatrix}$

In Exercises 15–24, compute the steady-state matrix of each stochastic matrix.

15. $\begin{bmatrix} .55 & 0 \\ .45 & 1 \end{bmatrix}$

16. $\begin{bmatrix} \frac{3}{5} & 0 \\ \frac{2}{5} & 1 \end{bmatrix}$

17. $\begin{bmatrix} 1 & .2 & .3 \\ 0 & .4 & .2 \\ 0 & .4 & .5 \end{bmatrix}$

18. $\begin{bmatrix} \frac{1}{5} & 0 & 0 \\ 0 & 1 & \frac{3}{8} \\ \frac{4}{5} & 0 & \frac{5}{8} \end{bmatrix}$

19. $\begin{bmatrix} \frac{1}{2} & 0 & \frac{1}{3} & 0 \\ \frac{1}{2} & 1 & 0 & 0 \\ 0 & 0 & \frac{2}{3} & 0 \\ 0 & 0 & 0 & 1 \end{bmatrix}$

20. $\begin{bmatrix} 1 & \frac{1}{8} & \frac{1}{3} & 0 \\ 0 & \frac{1}{8} & 0 & 0 \\ 0 & \frac{1}{4} & \frac{2}{3} & 0 \\ 0 & \frac{1}{2} & 0 & 1 \end{bmatrix}$

21. $\begin{bmatrix} 1 & 0 & \frac{1}{4} & \frac{1}{3} \\ 0 & 1 & \frac{1}{4} & \frac{1}{3} \\ 0 & 0 & \frac{1}{2} & 0 \\ 0 & 0 & 0 & \frac{1}{3} \end{bmatrix}$

22. $\begin{bmatrix} 1 & 0 & .2 & .1 \\ 0 & 1 & .4 & .2 \\ 0 & 0 & 0 & .4 \\ 0 & 0 & .4 & .3 \end{bmatrix}$

23. $\begin{bmatrix} 1 & 0 & 0 & .2 & .1 \\ 0 & 1 & 0 & .1 & .2 \\ 0 & 0 & 1 & .3 & .1 \\ 0 & 0 & 0 & .2 & .2 \\ 0 & 0 & 0 & .2 & .4 \end{bmatrix}$

24. $\begin{bmatrix} 1 & 0 & \frac{1}{4} & \frac{1}{3} & 0 \\ 0 & 1 & 0 & \frac{1}{3} & \frac{1}{2} \\ 0 & 0 & \frac{1}{4} & \frac{1}{3} & 0 \\ 0 & 0 & \frac{1}{2} & 0 & \frac{1}{2} \\ 0 & 0 & 0 & 0 & 0 \end{bmatrix}$

25. GASOLINE CONSUMPTION   As more and more old cars are taken off the road and replaced by late models that use unleaded fuel, the consumption of leaded gasoline will continue to drop. Suppose the transition matrix

$$A = \begin{array}{c} \\ \text{L} \\ \text{UL} \end{array} \overset{\begin{array}{cc} \text{L} & \text{UL} \end{array}}{\begin{bmatrix} .80 & 0 \\ .20 & 1 \end{bmatrix}}$$

describes this Markov process, where L denotes leaded gasoline and UL denotes unleaded gasoline.

a. Show that A is an absorbing stochastic matrix and rewrite it so that the absorbing state appears first. Partition the resulting matrix and identify the submatrices R and S.

b. Compute the steady-state matrix of A and interpret your results.

26. Diane has decided to play the following game of chance. She places a $1 bet on each repeated play of the game in which

the probability of her winning $1 is .5. She has further decided to continue playing the game until she has either accumulated a total of $3 or has lost all her money. What is the probability that Diane will eventually leave the game a winner if she started with a capital of $1? Of $2?

27. Refer to Exercise 26. Suppose Diane has decided to stop playing only after she has accumulated a sum of $4 or has lost all her money. All other conditions being the same, what is the probability that Diane will leave the game a winner if she started with a capital of $1? Of $2? Of $3?

28. USE OF AUTOMATED OFFICE EQUIPMENT   Because of the proliferation of more affordable automated office equipment, more and more companies are turning to them as replacements for obsolete equipment. The following transition matrix describes the Markov process. Here, E stands for electric typewriters, W stands for electric typewriters with some form of word processing capabilities, and C stands for computers with word processing software.

$$A = \begin{array}{c} \\ \text{E} \\ \text{W} \\ \text{C} \end{array} \overset{\begin{array}{ccc} \text{E} & \text{W} & \text{C} \end{array}}{\begin{bmatrix} .10 & 0 & 0 \\ .70 & .60 & 0 \\ .20 & .40 & 1 \end{bmatrix}}$$

a. Show that A is an absorbing stochastic matrix and rewrite it so that the absorbing state appears first. Partition the resulting matrix and identify the submatrices R and S.

b. Compute the steady-state matrix of A and interpret your results.

29. EDUCATION RECORDS   The registrar of Computronics Institute has compiled the following statistics on the progress of the school's students in their 2-yr computer programming course leading to an associate degree: Of beginning students in a particular year, 75% successfully complete their first year of study and move on to the second year, whereas 25% drop out of the program; of second-year students in a particular year, 90% go on to graduate at the end of the year, whereas 10% drop out of the program.

a. Construct the transition matrix associated with this Markov process.

b. Compute the steady-state matrix.

c. Determine the probability that a beginning student enrolled in the program will complete the course successfully.

30. EDUCATION RECORDS   The registrar of a law school has compiled the following statistics on the progress of the school's students working toward the LLB degree: Of the first-year students in a particular year, 85% successfully complete their course of studies and move on to the second year, whereas 15% drop out of the program; of the second-year students in a particular year, 92% go on to the third year, whereas 8% drop out of the program; of the third-year students in a particular year, 98% go on to graduate at the end of the year, whereas 2% drop out of the program.

a. Construct the transition matrix associated with the Markov process.
b. Find the steady-state matrix.
c. Determine the probability that a beginning law student enrolled in the program will go on to graduate.

In Exercises 31 and 32, determine whether the statement is true or false. If it is true, explain why it is true. If it is false, give an example to show why it is false.

31. An absorbing stochastic matrix need not contain an absorbing state.

32. In partitioning an absorbing matrix into subdivisions,

$$A = \left[\begin{array}{c|c} I & S \\ \hline O & R \end{array} \right]$$

the identity matrix I is chosen to have the same order as R.

33. GENETICS   Refer to Example 4. If the offspring are crossed successively with plants of genotype aa only, show that in the long run all the flowers produced by the plants will be white.

9.3   Solutions to Self-Check Exercises

1. a. State 2 is an absorbing state. States 1 and 3 are not absorbing, but each has a possibility (with probability .3 and .6) that an object may go from these states to state 2. Therefore, the matrix T is an absorbing stochastic matrix.

 b. Denoting the states as indicated, we rewrite

$$\begin{array}{c} \\ 1 \\ 2 \\ 3 \end{array} \begin{array}{ccc} 1 & 2 & 3 \\ \left[\begin{array}{ccc} .2 & 0 & 0 \\ .3 & 1 & .6 \\ .5 & 0 & .4 \end{array} \right] \end{array}$$

in the form

$$\begin{array}{c} \\ 2 \\ 3 \\ 1 \end{array} \begin{array}{ccc} 2 & 3 & 1 \\ \left[\begin{array}{cc|c} 1 & .6 & .3 \\ 0 & .4 & .5 \\ \hline 0 & 0 & .2 \end{array} \right] \end{array}$$

We see that

$$S = [.6 \quad .3] \quad \text{and} \quad R = \left[\begin{array}{cc} .4 & .5 \\ 0 & .2 \end{array} \right]$$

 c. We complete

$$I - R = \left[\begin{array}{cc} 1 & 0 \\ 0 & 1 \end{array} \right] - \left[\begin{array}{cc} .4 & .5 \\ 0 & .2 \end{array} \right] = \left[\begin{array}{cc} .6 & -.5 \\ 0 & .8 \end{array} \right]$$

and, using the formula for finding the inverse of a 2×2 matrix in Section 2.6,

$$(I - R)^{-1} = \left[\begin{array}{cc} 1.67 & 1.04 \\ 0 & 1.25 \end{array} \right]$$

and so

$$S(I - R)^{-1} = [.6 \quad .3] \left[\begin{array}{cc} 1.67 & 1.04 \\ 0 & 1.25 \end{array} \right] = [1 \quad 1]$$

Therefore, the steady-state matrix of T is

$$\left[\begin{array}{c|cc} 1 & 1 & 1 \\ \hline 0 & 0 & 0 \\ 0 & 0 & 0 \end{array} \right]$$

2. We want to compute the steady-state matrix of T. Note that T is in the form

$$\left[\begin{array}{c|c} I & S \\ \hline O & R \end{array} \right]$$

where

$$S = [.3 \quad .2] \quad \text{and} \quad R = \left[\begin{array}{cc} .6 & .3 \\ .1 & .5 \end{array} \right]$$

We compute

$$I - R = \left[\begin{array}{cc} 1 & 0 \\ 0 & 1 \end{array} \right] - \left[\begin{array}{cc} .6 & .3 \\ .1 & .5 \end{array} \right] = \left[\begin{array}{cc} .4 & -.3 \\ -.1 & .5 \end{array} \right]$$

and, using the inverse formula in Section 2.6,

$$(I - R)^{-1} = \left[\begin{array}{cc} 2.94 & 1.76 \\ 0.59 & 2.36 \end{array} \right]$$

so

$$S(I - R)^{-1} = [.3 \quad .2] \left[\begin{array}{cc} 2.94 & 1.76 \\ 0.59 & 2.36 \end{array} \right] = [1 \quad 1]$$

Therefore, the steady-state matrix of T is

$$\begin{array}{c} \\ \text{CAT} \\ \text{ER} \\ \text{MT} \end{array} \begin{array}{ccc} \text{CAT} & \text{ER} & \text{MT} \\ \left[\begin{array}{c|cc} 1 & 1 & 1 \\ \hline 0 & 0 & 0 \\ 0 & 0 & 0 \end{array} \right] \end{array}$$

Interpreting the steady-state matrix of T, we see that in the long run all courts in this state will use computer-aided transcription.

9.4 Game Theory and Strictly Determined Games

The theory of games is a relatively new branch of mathematics and owes much of its development to John von Neumann (1903–1957), one of the mathematical giants of the twentieth century. John Harsanyi, John Nash, and Reinhard Selten won the Nobel Prize in Economics in 1994 for their work in this field. Basically, the theory of games combines matrix methods with the theory of probability to determine the optimal strategies to be employed by two or more opponents involved in a competitive situation, with each opponent seeking to maximize his or her "gains," or, equivalently, to minimize his or her "losses." As such, the players may be poker players, managers of rival corporations seeking to extend their share of the market, campaign managers, or generals of opposing armies, to name a few.

For simplicity, we limit our discussion to games with two players. Such games are, naturally enough, called two-person games.

Two-Person Games

EXAMPLE 1 Coin-Matching Game Richie and Chuck play a coin-matching game in which each player selects a side of a penny without prior knowledge of the other's choice. Then, upon a predetermined signal, both players disclose their choices simultaneously. Chuck agrees to pay Richie a sum of $3 if both choose heads; if Richie chooses heads and Chuck chooses tails, then Richie pays Chuck $6; if Richie chooses tails and Chuck chooses heads, then Chuck pays Richie $2; finally, if both Richie and Chuck choose tails, then Chuck pays Richie $1. In this game, the objective of each player is to discover a strategy that will ensure that his winnings are maximized (equivalently, that his losses are minimized).

The coin-matching game is an example of a **zero-sum game**—that is, a game in which the payoff to one party results in an equal loss to the other. For such games, the sum of the payments made by both players at the end of each play adds up to zero.

To facilitate the analysis of the problem, we represent the given data in the form of a matrix:

$$
\begin{array}{cc}
 & \begin{array}{cc} C\text{'s moves} \\ \text{Heads} \quad \text{Tails} \end{array} \\
R\text{'s moves} \begin{array}{c} \text{Heads} \\ \text{Tails} \end{array} & \begin{bmatrix} 3 & -6 \\ 2 & 1 \end{bmatrix}
\end{array}
$$

Each row of the matrix corresponds to one of the two possible moves by Richie (referred to as the row player, R), whereas each column corresponds to one of the two possible moves by Chuck (referred to as the column player, C). Each entry in the matrix represents the payoff from C to R. For example, the entry $a_{11} = 3$ represents a $3 payoff from Chuck to Richie (C to R) when Richie chooses to play row 1 (heads) and Chuck chooses to play column 1 (heads). On the other hand, the entry $a_{12} = -6$ represents (because it's negative) a $6 payoff to C (from R) when R chooses to play row 1 (heads) and C chooses to play column 2 (tails). (Interpret the meaning of $a_{21} = 2$ and $a_{22} = 1$ for yourself.)

More generally, suppose we are given a two-person game with two players R and C. Furthermore, suppose that R has m possible moves $R_1, R_2, \ldots, R_m$ and that

C has n possible moves $C_1, C_2, \ldots, C_n$. Then, we can represent the game in terms of an $m \times n$ matrix in which each row of the matrix represents one of the m possible moves of R and each column of the matrix represents one of the n possible moves of C:

$$
\begin{array}{c}
\\
\\
R\text{'s moves}
\end{array}
\begin{array}{c}
 \\
R_1 \\
R_2 \\
\vdots \\
R_i \\
\vdots \\
R_m
\end{array}
\begin{bmatrix}
a_{11} & a_{12} & \cdots & a_{1j} & \cdots & a_{1n} \\
a_{21} & a_{22} & \cdots & a_{2j} & \cdots & a_{2n} \\
\vdots & \vdots & & \vdots & & \vdots \\
a_{i1} & a_{i2} & \cdots & a_{ij} & \cdots & a_{in} \\
\vdots & \vdots & & \vdots & & \vdots \\
a_{m1} & a_{m2} & \cdots & a_{mj} & \cdots & a_{mn}
\end{bmatrix}
$$

C's moves
$C_1 \quad C_2 \quad \cdots \quad C_j \quad \cdots \quad C_n$

The entry a_{ij} in the ith row and jth column of the (payoff) matrix represents the payoff to R when R chooses move R_i and C chooses move C_j. In this context, note that a payoff to R means, in actuality, a payoff to C in the event that the value of a_{ij} is negative.

EXAMPLE 2   The payoff matrix associated with a game is given by

$$
R\text{'s moves}
\begin{array}{c}
R_1 \\
R_2
\end{array}
\begin{bmatrix}
1 & -2 & 3 \\
4 & -5 & -1
\end{bmatrix}
$$

C's moves
$C_1 \quad C_2 \quad C_3$

Give an interpretation of this payoff matrix.

Solution   In this two-person game, player R has two possible moves, whereas player C has three possible moves. The payoffs are determined as follows: If R chooses R_1, then

R wins 1 unit if C chooses C_1.

R loses 2 units if C chooses C_2.

R wins 3 units if C chooses C_3.

If R chooses R_2, then

R wins 4 units if C chooses C_1.

R loses 5 units if C chooses C_2.

R loses 1 unit if C chooses C_3.

Optimal Strategies

Let's return to the payoff matrix of Example 1 and see how it may be used to help us determine the "best" strategy for each of the two players R and C. For convenience, this matrix is reproduced here:

$$
R\text{'s moves}
\begin{array}{c}
R_1 \\
R_2
\end{array}
\begin{bmatrix}
3 & -6 \\
2 & 1
\end{bmatrix}
$$

C's moves
$C_1 \quad C_2$

Let's first consider the game from R's point of view. Since the entries in the payoff matrix represent payoffs to him, his initial reaction might be to seek out the largest entry in the matrix and consider the row containing such an entry as a possible move. Thus, he is led to the consideration of R_1 as a possible move.

Let's examine this choice a little more closely. To be sure, R would realize the largest possible payoff to himself (\$3) if C chose C_1; but if C chose C_2, then R would lose \$6! Since R does not know beforehand what C's move will be, a more prudent approach on his part would be to assume that no matter what row he chooses, C will counter with a move (column) that will result in the smallest payoff to him. To maximize the payoff to himself under these circumstances, R would then select from among the moves (rows) the one in which the smallest payoff is as large as possible. This strategy for R, called, for obvious reasons, the **maximin strategy,** may be summarized as follows:

Maximin Strategy

1. For each row of the payoff matrix, find the smallest entry in that row.
2. Choose the row for which the entry found in step 1 is as large as possible. This row constitutes R's "best" move.

For the problem under consideration, we can organize our work as follows:

$$\begin{bmatrix} 3 & -6 \\ 2 & 1 \end{bmatrix} \begin{matrix} \text{Row} \\ \text{minima} \\ -6 \\ \textcircled{1} \end{matrix} \quad \leftarrow \text{Larger of the row minima}$$

From these results, it is seen that R's "best" move is row 2. By choosing this move R stands to win at least \$1.

Next, let's consider the game from C's point of view. His objective is to minimize the payoff to R. This is accomplished by choosing the column whose largest payoff is as small as possible. This strategy for C, called the **minimax strategy,** may be summarized as follows:

Minimax Strategy

1. For each column of the payoff matrix, find the largest entry in that column.
2. Choose the column for which the entry found in step 1 is as small as possible. This column constitutes C's "best" move.

We can organize the work involved in determining C's "best" move as follows:

$$\begin{bmatrix} 3 & -6 \\ 2 & 1 \end{bmatrix}$$

$$\text{Column maxima} \quad 3 \quad \textcircled{1}$$

$$\uparrow$$

Smaller of the column maxima

From these results, we see that C's "best" move is column 2. By choosing this move, C stands to lose at most \$1.

EXAMPLE 3   For the game with the following payoff matrix, determine the maximin and minimax strategies for each player.

$$\begin{bmatrix} -3 & -2 & 4 \\ -2 & 0 & 3 \\ 6 & -1 & 1 \end{bmatrix}$$

Solution   We determine the minimum of each row and the maximum of each column of the payoff matrix and then display these numbers by circling the largest of the row minima and the smallest of the column maxima:

From these results, we conclude that the maximin strategy (for the row player) is to play row 3, whereas the minimax strategy (for the column player) is to play column 2.   ∎

EXAMPLE 4   Determine the maximin and minimax strategies for each player in a game whose payoff matrix is given by

$$\begin{bmatrix} 3 & 4 & -4 \\ 2 & -1 & -3 \end{bmatrix}$$

Solution   Proceeding as in the last example, we obtain the following:

from which we conclude that the maximin strategy for the row player is to play row 2, whereas the minimax strategy for the column player is to play column 3.   ∎

In arriving at the maximin and minimax strategies for the respective players, we have assumed that both players always act rationally, with the knowledge that their opponents will always act rationally. This means that each player adopts a strategy of always making the same move and assumes that his opponent is always going to counter that move with a move that will maximize the payoff to the opponent. Thus, each player adopts the pure strategy of always making the move that will minimize the payoff his opponent can receive and thereby maximize the payoff to himself.

This raises the following question: Suppose a game is played repeatedly and one of the players realizes that the opponent is employing his maximin (or minimax) strategy. Can this knowledge be used to the player's advantage? To obtain a partial answer to this question, let's consider the game posed in Example 3. There, the minimax strategy for the column player is to play column 2. Suppose, in repeated plays of the game, a player consistently plays that column and this strategy becomes known to the row player. The row player may then change the strategy from playing row 3 (the maximin strategy) to playing row 2, thereby reducing losses from 1 unit to zero. Thus, at least for this game, the knowledge that a player is employing the maximin or minimax strategy can be used to the opponent's advantage.

There is, however, a class of games in which the knowledge that a player is using the maximin (or minimax) strategy proves of no help to the opponent. Consider, for example, the game of Example 4. There, the row player's maximin strategy is to play row 2, and the column player's minimax strategy is to play column 3. Suppose, in repeated plays of the game, R (the row player) has discovered that C (the column player) consistently chooses to play column 3 (the minimax strategy). Can this knowledge be used to R's advantage? Now, other than playing row 2 (the maximin strategy), R may choose to play row 1. But if R makes this choice, then he would lose 4 units instead of 3 units! Clearly, in this case, the knowledge that C is using the minimax strategy cannot be used to advantage. R's optimal (best) strategy is the maximin strategy.

Optimal Strategy

The **optimal strategy** in a game is the strategy that is most profitable to a particular player.

Next, suppose that in repeated plays of the game, C has discovered that R consistently plays row 2 (his optimal strategy). Can this knowledge be used to C's advantage? Another glance at the payoff matrix reveals that by playing column 1, C stands to lose 2 units, and by playing column 2, he stands to win 1 unit, as compared with winning 3 units by playing column 3, as called for by the minimax strategy. Thus, as in the case of R, C does not benefit from knowing his opponent's move. Furthermore, his optimal strategy coincides with the minimax strategy.

This game, in which the row player cannot benefit from knowing that his opponent is using the minimax strategy and the column player cannot benefit from knowing that his opponent is using the maximin strategy, is said to be strictly determined.

Strictly Determined Game

A **strictly determined game** is characterized by the following properties:

1. There is an entry in the payoff matrix that is *simultaneously* the smallest entry in its row and the largest entry in its column. This entry is called the **saddle point** for the game.
2. The optimal strategy for the row player is precisely the maximin strategy and is the row containing the saddle point. The optimal strategy for the column player is the minimax strategy and is the column containing the saddle point.

The saddle point of a strictly determined game is also referred to as the **value of the game.** If the value of a strictly determined game is positive, then the game favors the row player. If the value is negative, it favors the column player. If the value of the game is zero, the game is called a **fair game.**

Returning to the coin-matching game discussed earlier, we conclude that Richie's optimal strategy consists of playing row 2 repeatedly, whereas Chuck's optimal strategy consists of playing column 2 repeatedly. Furthermore, the value of the game is 1, implying that the game favors the row player, Richie.

EXAMPLE 5   A two-person, zero-sum game is defined by the payoff matrix

$$A = \begin{bmatrix} 1 & 2 & -3 \\ -1 & 2 & -2 \\ 2 & 3 & -4 \end{bmatrix}$$

a. Show that the game is strictly determined and find the saddle point(s) for the game.
b. What is the optimal strategy for each player?
c. What is the value of the game? Does the game favor one player over the other?

Solution

a. First, we determine the minimum of each row and the maximum of each column of the payoff matrix A and display these minima and maxima as follows:

$$\begin{array}{ccc} & & \text{Row} \\ & & \text{minima} \end{array}$$

$$\begin{bmatrix} 1 & 2 & -3 \\ -1 & 2 & \boxed{-2} \\ 2 & 3 & -4 \end{bmatrix} \begin{array}{l} -3 \\ -2 \quad \leftarrow \text{Largest of the row minima} \\ -4 \end{array}$$

$$\begin{array}{llll} \text{Column maxima} & 2 & 3 & -2 \\ & & & \uparrow \\ & & & \text{Smallest of the column maxima} \end{array}$$

From these results, we see that the circled entry, -2, is simultaneously the smallest entry in its row and the largest entry in its column. Therefore, the game is strictly determined, with the entry $a_{23} = -2$ as its saddle point.

b. From these results, we see that the optimal strategy for the row player is to make the move represented by the second row of the matrix, and the optimal strategy for the column player is to make the move represented by the third column.

c. The value of the game is -2, which implies that if both players adopt their best strategy, the column player will win 2 units in a play. Consequently, the game favors the column player.   ∎

A game may have more than one saddle point, as the next example shows.

EXAMPLE 6   A two-person, zero-sum game is defined by the payoff matrix

$$A = \begin{bmatrix} 4 & 5 & 4 \\ -2 & -5 & -3 \\ 4 & 6 & 8 \end{bmatrix}$$

a. Show that the game is strictly determined and find the saddle points for the game.

b. Discuss the optimal strategies for the players.

c. Does the game favor one player over the other?

Solution

a. Proceeding as in the previous example, we obtain the following information:

$$
\begin{array}{ccc}
 & & \text{Row} \\
 & & \text{minima}
\end{array}
$$

$$
\begin{bmatrix}
\boxed{④} & 5 & 4 \\
-2 & -5 & -3 \\
\boxed{④} & 6 & 8
\end{bmatrix}
\begin{array}{l}
4 \longleftarrow \\
-5 \quad \text{Largest of the row minima} \\
4 \longleftarrow
\end{array}
$$

Column maxima   4   6   8

↑

Smallest of the column maxima

We see that each of the circled entries, 4, is simultaneously the smallest entry in the row and the largest entry in the column containing it. Therefore, the game is strictly determined, and in this case it has two saddle points: the entry $a_{11} = 4$ and the entry $a_{31} = 4$. In general, it can be shown that every saddle point of a payoff matrix must have the same value.

b. Since the game has two saddle points, both lying in the first column and in the first and third rows of the payoff matrix, we see that the row player's optimal strategy consists of playing either row 1 or row 3 consistently, whereas the column player's optimal strategy consists of playing column 1 repeatedly.

c. The value of the game is 4, which implies that it favors the row player. ∎

APPLIED EXAMPLE 7   Bidding for Rights   Two television subscription companies, UBS and Telerama, are planning to extend their operations to a certain city. Each has the option of making its services available to prospective subscribers with a special introductory subscription rate. It is estimated that if both UBS and Telerama offer the special subscription rate, each will get 50% of the market, whereas if UBS offers the special subscription rate and Telerama does not, UBS will get 70% of the market. If Telerama offers the special subscription rate and UBS does not, it is estimated that UBS will get 40% of the market. If both companies elect not to offer the special subscription rate, it is estimated that UBS will get 60% of the market.

a. Construct the payoff matrix for the game.

b. Show that the game is strictly determined.

c. Determine the optimal strategy for each company and find the value of the game.

Solution

a. The required payoff matrix is given by

		Telerama	
		Intro. rate	Usual rate
UBS	Intro. rate	.50	.70
	Usual rate	.40	.60

b. The entry $a_{11} = .50$ is the smaller entry in its row and the larger entry in its column. Therefore, the entry $a_{11} = .50$ is a saddle point of the game and the game is strictly determined.

c. The entry $a_{11} = .50$ is the only saddle point of the game, so UBS's optimal strategy is to choose row 1, and Telerama's optimal strategy is to choose column 1. In other words, both companies should offer their potential customers their respective introductory subscription rates. ∎

EXPLORE & DISCUSS

A two-person, zero-sum game is defined by the payoff matrix

$$A = \begin{bmatrix} a & a \\ c & d \end{bmatrix}$$

1. Show that the game is strictly determined.

2. What can you say about the game with the following payoff matrix?

$$B = \begin{bmatrix} a & b \\ c & c \end{bmatrix}$$

9.4 Self-Check Exercises

1. A two-person, zero-sum game is defined by the payoff matrix

$$A = \begin{bmatrix} -2 & 1 & 3 \\ 3 & 2 & 2 \\ 2 & -1 & 4 \end{bmatrix}$$

a. Show that the game is strictly determined and find the saddle point(s) for the game.
b. What is the optimal strategy for each player?
c. What is the value of the game? Does the game favor one player over the other?

2. The management of Delta Corporation, a construction and development company, is deciding whether to go ahead with the construction of a large condominium complex. A financial analysis of the project indicates that if Delta goes ahead with the development and the home mortgage rate drops 1 point or more by next year, when the complex is expected to be completed, it will stand to make a profit of $750,000. If Delta goes ahead with the development and the mortgage rate stays within 1 point of the current rate by next year, it will stand to make a profit of $600,000. If Delta goes ahead with the development and the mortgage rate increases 1 point or more by next year, it will stand to make a profit of $350,000. If Delta does not go ahead with the development and the mortgage rate drops 1 point or more by next year, it will stand to make a profit of $400,000. If Delta does not go ahead with the development and the mortgage rate stays within 1 point of the current rate by next year, it will stand to make a profit of $350,000. Finally, if Delta does not go ahead with the development and the mortgage rate increases 1 point or more by next year, it stands to make $250,000.

a. Represent this information in the form of a payoff matrix.
b. Assuming that the home mortgage rate trend is volatile over the next year, determine whether or not Delta should go ahead with the project.

Solutions to Self-Check Exercises 9.4 can be found on page 553.

9.4 Concept Questions

1. **a.** What is the maximin strategy for the row player in a two-person game that is represented by a payoff matrix?
   b. What is the minimax strategy for the column player in a two-person game that is represented by a payoff matrix?

2. **a.** How do you find the saddle point in the payoff matrix for a strictly determined game?
   b. What is the optimal strategy for the row player in a strictly determined game? The column player?

9.4 Exercises

In Exercises 1–8, determine the maximin and minimax strategies for each two-person, zero-sum matrix game.

1. $\begin{bmatrix} 2 & 3 \\ 4 & 1 \end{bmatrix}$

2. $\begin{bmatrix} -1 & 3 \\ 2 & 5 \end{bmatrix}$

3. $\begin{bmatrix} 1 & 3 & 2 \\ 0 & -1 & 4 \end{bmatrix}$

4. $\begin{bmatrix} 1 & 4 & -2 \\ 4 & 6 & -3 \end{bmatrix}$

5. $\begin{bmatrix} 3 & 2 & 1 \\ 1 & -2 & 3 \\ 6 & 4 & 1 \end{bmatrix}$

6. $\begin{bmatrix} 1 & 4 \\ 2 & -2 \\ 3 & 0 \end{bmatrix}$

7. $\begin{bmatrix} 4 & 2 & 1 \\ 1 & 0 & -1 \\ 2 & 1 & 3 \end{bmatrix}$

8. $\begin{bmatrix} -1 & 1 & 2 \\ 3 & 1 & 1 \\ -1 & 1 & 2 \\ 3 & 2 & -1 \end{bmatrix}$

In Exercises 9–18, determine whether the two-person, zero-sum matrix game is strictly determined. If a game is strictly determined,
a. Find the saddle point(s) of the game.
b. Find the optimal strategy for each player.
c. Find the value of the game.
d. Determine whether the game favors one player over the other.

9. $\begin{bmatrix} 2 & 3 \\ 1 & -4 \end{bmatrix}$

10. $\begin{bmatrix} 1 & 0 \\ 0 & -1 \end{bmatrix}$

11. $\begin{bmatrix} 1 & 3 & 2 \\ -1 & 4 & -6 \end{bmatrix}$

12. $\begin{bmatrix} 3 & 2 \\ -1 & -2 \\ 4 & 1 \end{bmatrix}$

13. $\begin{bmatrix} 1 & 3 & 4 & 2 \\ 0 & 2 & 6 & -4 \\ -1 & -3 & -2 & 1 \end{bmatrix}$

14. $\begin{bmatrix} 2 & 4 & 2 \\ 0 & 3 & 0 \\ -1 & -2 & 1 \end{bmatrix}$

15. $\begin{bmatrix} 1 & 2 \\ 0 & 3 \\ -1 & 2 \\ 2 & -2 \end{bmatrix}$

16. $\begin{bmatrix} -1 & 2 & 4 \\ 2 & 3 & 5 \\ 0 & 1 & -3 \\ -2 & 4 & -2 \end{bmatrix}$

17. $\begin{bmatrix} 1 & -1 & 3 & 2 \\ 1 & 0 & 2 & 2 \\ -2 & 2 & 3 & -1 \end{bmatrix}$

18. $\begin{bmatrix} 3 & -1 & 0 & -4 \\ 2 & 1 & 0 & 2 \\ -3 & 1 & -2 & 1 \\ -1 & -1 & -2 & 1 \end{bmatrix}$

19. Robin and Cathy play a game of matching fingers. On a predetermined signal, both players simultaneously extend 1, 2, or 3 fingers from a closed fist. If the sum of the number of fingers extended is even, then Robin receives an amount in dollars equal to that sum from Cathy. If the sum of the number of fingers extended is odd, then Cathy receives an amount in dollars equal to that sum from Robin.
   a. Construct the payoff matrix for the game.
   b. Find the maximin and the minimax strategies for Robin and Cathy, respectively.
   c. Is the game strictly determined?
   d. If the answer to part (c) is yes, what is the value of the game?

20. **MANAGEMENT DECISIONS** Brady's, a conventional department store, and ValueMart, a discount department store, are each considering opening new stores at one of two possible sites: the Civic Center and North Shore Plaza. The strategies available to the management of each store are given in the following payoff matrix, where each entry represents the amounts (in hundreds of thousands of dollars) either gained or lost by one business from or to the other as a result of the sites selected.

$$\begin{array}{cc} & \begin{matrix} \text{ValueMart} \\ \text{Center} \quad \text{Plaza} \end{matrix} \\ \text{Brady's} \begin{matrix} \text{Civic Center} \\ \text{North Shore Plaza} \end{matrix} & \begin{bmatrix} 2 & -2 \\ 3 & -4 \end{bmatrix} \end{array}$$

a. Show that the game is strictly determined.

b. What is the value of the game?

c. Determine the best strategy for the management of each store (that is, determine the ideal locations for each store).

21. FINANCIAL ANALYSIS   The management of Acrosonic is faced with the problem of deciding whether to expand the production of its line of electrostatic loudspeaker systems. It has been estimated that an expansion will result in an annual profit of $200,000 for Acrosonic if the general economic climate is good. On the other hand, an expansion during a period of economic recession will cut its annual profit to $120,000. As an alternative, Acrosonic may hold the production of its electrostatic loudspeaker systems at the current level and expand its line of conventional loudspeaker systems. In this event, the company is expected to make a profit of $50,000 in an expanding economy (because many potential customers will be expected to buy electrostatic loudspeaker systems from other competitors) and a profit of $150,000 in a recessionary economy.

a. Construct the payoff matrix for this game.

Hint: The row player is the management of the company and the column player is the economy.

b. Should management recommend expanding the company's line of electrostatic loudspeaker systems?

22. FINANCIAL ANALYSIS   The proprietor of Belvedere's is faced with the problem of deciding whether to expand his restaurant facilities now or to wait until some future date to do so. If he expands the facilities now and the economy experiences a period of growth during the coming year, he will make a net profit of $442,000; if he expands now and a period of zero growth follows, then he will make a net profit of $40,000; and if he expands now and an economic recession follows, he will suffer a net loss of $108,000. If he does not expand the restaurant now and the economy experiences a period of growth during the coming year, he will make a net profit of $280,000; if he does not expand now and a period of zero growth follows, he will make a net profit of $190,000. Finally, if he does not expand now and an economic recession follows, he will make a net profit of $100,000.

a. Represent this information in the form of a payoff matrix.

b. Assuming that the state of the economy during the coming year is uncertain, determine whether the owner of the restaurant should expand his facilities at this time.

23. MARKET SHARE   Roland's Barber Shop and Charley's Barber Shop are both located in the business district of a certain town. Roland estimates that if he raises the price of a haircut by $1, he will increase his market share by 3% if Charley raises his price by the same amount; he will decrease his market share by 1% if Charley holds his price at the same level; and he will decrease his market share by 3% if Charley lowers his price by $1. If Roland keeps his price the same, he will increase his market share by 2% if Charley raises his price by $1; he will keep the same market share if Charley holds the price at the same level; and he will decrease his market share by 2% if Charley lowers his price by $1. Finally, if Roland lowers the price he charges by $1, his market share will increase by 5% if Charley raises his prices by the same amount; he will increase his market share by 2% if Charley holds his price at the same level; and he will increase his market share by 1% if Charley lowers his price by $1.

a. Construct the payoff matrix for this game.

b. Show that the game is strictly determined.

c. If neither party is willing to lower the price he charges for a haircut, show that both should keep their present price structures.

In Exercises 24–26, determine whether the statement is true or false. If it is true, explain why it is true. If it is false, give an example to show why it is false.

24. In a zero-sum game, the payments made by the players at the end of each play add up to zero.

25. In a strictly determined game, the value of the game is given by the saddle point of the game.

26. If the value of a strictly determined game is not negative, it favors the row player.

9.4   Solutions to Self-Check Exercises

1. a. Displaying the minimum of each row and the maximum of each column of the payoff matrix A, we obtain

$$\begin{bmatrix} -2 & 1 & 3 \\ 3 & ②\ & 2 \\ 2 & -1 & 4 \end{bmatrix} \begin{matrix} -2 \\ 2 \\ -1 \end{matrix}$$

Row minima

$\leftarrow$ Largest of the row minima

Column maxima   3   2   4

↑
Smallest of the column maxima

From these results, we see that the circled entry, 2, is simultaneously the smallest entry in its row and the largest entry in its column. Therefore, the game is strictly determined, with the entry $a_{22} = 2$ as its saddle point.

b. From these results, we see that the optimal strategy for the row player is to make the move represented by the second row of the matrix, and the optimal strategy for the column player is to make the move represented by the second column.

c. The value of the game is 2, which implies that if both players adopt their best strategy, the row player will win 2 units in a play. Consequently, the game favors the row player.

2. a. We may review this situation as a game in which the row player is Delta and the column player is the home mortgage rate. The required payoff matrix is

Mortgage rate

		Decrease	Steady	Increase
Delta	Project	750	600	350
Corp.	No project	400	350	250

(All figures are in thousands of dollars.)

b. From part (a), the payoff matrix under consideration is

$$\begin{bmatrix} 750 & 600 & 350 \\ 400 & 350 & 250 \end{bmatrix}$$

Proceeding in the usual manner, we find

$$\begin{bmatrix} 750 & 600 & ③50 \\ 400 & 350 & 250 \end{bmatrix} \begin{matrix} 350 \\ 250 \end{matrix}$$

Row minima

$\leftarrow$ Larger of the row minima

Column maxima   750   600   350

↑
Smallest of the column maxima

From these results, we see that the entry $a_{13} = 350$ is a saddle point and the game is strictly determined. We can also conclude that the company should go ahead with the project.

9.5   Games with Mixed Strategies

In Section 9.4, we discussed strictly determined games and found that the optimal strategy for the row player is to select the row containing a saddle point for the game, and the optimal strategy for the column player is to select the column containing a saddle point. Furthermore, in repeated plays of the game, each player's optimal strategy consists of making the same move over and over again, since the discovery of the opponent's optimal strategy cannot be used to advantage. Such strategies are called **pure strategies.** In this section, we look at games that are not strictly determined and the strategies associated with such games.

Mixed Strategies

As a simple example of a game that is not strictly determined, let's consider the following slightly modified version of the coin-matching game played by Richie and Chuck (see Example 1, Section 9.4). Suppose Richie wins $3 if both parties choose heads and $1 if both choose tails and loses $2 if one chooses heads and the other tails. Then, the payoff matrix for this game is given by

$$
\begin{array}{cc}
 & \begin{array}{cc} C\text{'s moves} \\ \begin{array}{cc} C_1 & C_2 \\ \text{(heads)} & \text{(tails)} \end{array} \end{array} \\
R\text{'s moves} \quad \begin{array}{c} R_1 \text{ (heads)} \\ R_2 \text{ (tails)} \end{array} & \begin{bmatrix} 3 & -2 \\ -2 & 1 \end{bmatrix}
\end{array}
$$

A quick examination of this matrix reveals that it contains no entry that is simultaneously the smallest entry in its row and the largest entry in its column; that is, the game has no saddle point and is therefore not strictly determined. What strategy might Richie adopt for the game? Offhand, it would seem that he should consistently select row 1 since he stands to win $3 by playing this row and only $1 by playing row 2, at a risk, in either case, of losing $2. However, if Chuck discovers that Richie is playing row 1 consistently, he would counter this strategy by playing column 2, causing Richie to lose $2 on each play! In view of this, Richie is led to consider a strategy whereby he chooses row 1 some of the time and row 2 at other times. A similar analysis of the game from Chuck's point of view suggests that he might consider choosing column 1 some of the time and column 2 at other times. Such strategies are called **mixed strategies.**

From a practical point of view, there are many ways in which a player may choose moves in a game with mixed strategies. For example, in the game just mentioned, if Richie decides to play heads half the time and tails the other half of the time, he could toss an unbiased coin before each move and let the outcome of the toss determine which move he should make. Another more general but less practical way of deciding on the choice of a move is: Having determined beforehand the proportion of the time row 1 is to be chosen (and therefore the proportion of the time row 2 is to be chosen), Richie might construct a spinner (Figure 3) in which the areas of the two sectors reflect these proportions and let the move be decided by the outcome of a spin. These two methods for determining a player's move in a game with mixed strategies guarantee that the strategy will not fall into a pattern that will be discovered by the opponent.

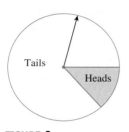

FIGURE 3

From a mathematical point of view, we may describe the mixed strategy of a row player in terms of a row vector whose dimension coincides with the number of possible moves the player has. For example, if Richie had decided on a strategy in which he chose to play row 1 half the time and row 2 the other half of the time, then this strategy is represented by the row vector

$$[.5 \quad .5]$$

Similarly, the mixed strategy for a column vector may be represented by a column vector of appropriate dimension. For example, returning to our illustration, suppose Chuck has decided that 20% of the time he will choose column 1 and 80% of the time he will choose column 2. This strategy is represented by the column vector

$$\begin{bmatrix} .2 \\ .8 \end{bmatrix}$$

Expected Value of a Game

For the purpose of comparing the merits of a player's different mixed strategies in a game, it is convenient to introduce a number called the expected value of a game. The **expected value of a game** measures the average payoff to the row player when both players adopt a particular set of mixed strategies. We now explain this notion using a 2×2 matrix game whose payoff matrix has the general form

$$A = \begin{bmatrix} a_{11} & a_{12} \\ a_{21} & a_{22} \end{bmatrix}$$

Suppose in repeated plays of the game the row player R adopts the mixed strategy

$$P = [p_1 \quad p_2]$$

That is, the player selects row 1 with probability p_1 and row 2 with probability p_2, and the column player C adopts the mixed strategy

$$Q = \begin{bmatrix} q_1 \\ q_2 \end{bmatrix}$$

That is, the column player selects column 1 with probability q_1 and column 2 with probability q_2. Now, in each play of the game, there are four possible outcomes that may be represented by the ordered pairs

(row 1, column 1)

(row 1, column 2)

(row 2, column 1)

(row 2, column 2)

where the first number of each ordered pair represents R's selection and the second number of each ordered pair represents C's selection. Since the choice of moves is made by one player without knowing the other's choice, each pair of events (for example, the events "row 1" and "column 1") constitutes a pair of independent events. Therefore, the probability of R choosing row 1 and C choosing column 1, $P(\text{row } 1, \text{column } 1)$, is given by

$$P(\text{row } 1, \text{column } 1) = P(\text{row } 1) \cdot P(\text{column } 1)$$
$$= p_1 q_1$$

In a similar manner, we compute the probability of each of the other three outcomes. These calculations, together with the payoffs associated with each of the four possible outcomes, may be summarized as follows:

Outcome	Probability	Payoff
(row 1, column 1)	$p_1 q_1$	a_{11}
(row 1, column 2)	$p_1 q_2$	a_{12}
(row 2, column 1)	$p_2 q_1$	a_{21}
(row 2, column 2)	$p_2 q_2$	a_{22}

Then, the *expected payoff E* of the game is the sum of the products of the payoffs and the corresponding probabilities (see Section 8.2). Thus,

$$E = p_1 q_1 a_{11} + p_1 q_2 a_{12} + p_2 q_1 a_{21} + p_2 q_2 a_{22}$$

In terms of the matrices P, A, and Q, we have the following relatively simple expression for E—namely,

$$E = PAQ$$

which you may verify (Exercise 22). This result may be generalized as follows:

Expected Value of a Game

Let

$$P = [p_1 p_2 \cdots p_m] \quad \text{and} \quad Q = \begin{bmatrix} q_1 \\ q_2 \\ \vdots \\ q_n \end{bmatrix}$$

be the vectors representing the mixed strategies for the row player R and the column player C, respectively, in a game with an $m \times n$ payoff matrix

$$A = \begin{bmatrix} a_{11} & a_{12} & \cdots & a_{1n} \\ a_{21} & a_{22} & \cdots & a_{2n} \\ \vdots & \vdots & & \vdots \\ a_{m1} & a_{m2} & \cdots & a_{mn} \end{bmatrix}$$

Then the expected value of a game is given by

$$E = PAQ = [p_1 p_2 \cdots p_m] \begin{bmatrix} a_{11} & a_{12} & \cdots & a_{1n} \\ a_{21} & a_{22} & \cdots & a_{2n} \\ \vdots & \vdots & & \vdots \\ a_{m1} & a_{m2} & \cdots & a_{mn} \end{bmatrix} \begin{bmatrix} q_1 \\ q_2 \\ \vdots \\ q_n \end{bmatrix}$$

We now look at several examples involving the computation of the expected value of a game.

EXAMPLE 1 Consider a coin-matching game played by Richie and Chuck with a payoff matrix given by

$$A = \begin{bmatrix} 3 & -2 \\ -2 & 1 \end{bmatrix}$$

Compute the expected payoff of the game if Richie adopts the mixed strategy P and Chuck adopts the mixed strategy Q, where

a. $P = [.5 \quad .5]$ and $Q = \begin{bmatrix} .5 \\ .5 \end{bmatrix}$

b. $P = [.8 \quad .2]$ and $Q = \begin{bmatrix} .1 \\ .9 \end{bmatrix}$

Solution

a. We compute

$$E = PAQ = \begin{bmatrix} .5 & .5 \end{bmatrix} \begin{bmatrix} 3 & -2 \\ -2 & 1 \end{bmatrix} \begin{bmatrix} .5 \\ .5 \end{bmatrix}$$

$$= \begin{bmatrix} .5 & -.5 \end{bmatrix} \begin{bmatrix} .5 \\ .5 \end{bmatrix}$$

$$= 0$$

Thus, in repeated plays of the game, it may be expected in the long run that the game will end in a draw.

b. We compute

$$E = PAQ = \begin{bmatrix} .8 & .2 \end{bmatrix} \begin{bmatrix} 3 & -2 \\ -2 & 1 \end{bmatrix} \begin{bmatrix} .1 \\ .9 \end{bmatrix}$$

$$= \begin{bmatrix} 2 & -1.4 \end{bmatrix} \begin{bmatrix} .1 \\ .9 \end{bmatrix}$$

$$= -1.06$$

That is, in the long run Richie may be expected to lose $1.06 on the average in each play. ■

EXAMPLE 2 The payoff matrix for a certain game is given by

$$A = \begin{bmatrix} 1 & -2 \\ -1 & 2 \\ 3 & -3 \end{bmatrix}$$

a. Find the expected payoff to the row player if the row player R uses her maximin pure strategy and the column player C uses her minimax pure strategy.
b. Find the expected payoff to the row player if R uses her maximin strategy 50% of the time and chooses each of the other two rows 25% of the time, while C chooses each column 50% of the time.

Solution

a. The maximin and minimax strategies for the row and column players, respectively, may be found using the method of the last section. Thus,

Row
minima

$$\begin{bmatrix} 1 & -2 \\ -1 & 2 \\ 3 & -3 \end{bmatrix} \begin{matrix} -2 \\ \text{\textcircled{-1}} \quad \leftarrow \text{Largest of the row minima} \\ -3 \end{matrix}$$

Column maxima   3   ②
                      ↑
          Smaller of the column maxima

From these results, we see that R's optimal pure strategy is to choose row 2, whereas C's optimal pure strategy is to choose column 2. Furthermore, if both players use their respective optimal strategy, then the expected payoff to R is 2 units.

b. In this case, R's mixed strategy may be represented by the row vector

$$P = [.25 \quad .50 \quad .25]$$

and C's mixed strategy may be represented by the column vector

$$Q = \begin{bmatrix} .5 \\ .5 \end{bmatrix}$$

The expected payoff to the row player will then be given by

$$E = PAQ = [.25 \quad .50 \quad .25] \begin{bmatrix} 1 & -2 \\ -1 & 2 \\ 3 & -3 \end{bmatrix} \begin{bmatrix} .5 \\ .5 \end{bmatrix}$$

$$= [.25 \quad .50 \quad .25] \begin{bmatrix} -.5 \\ .5 \\ 0 \end{bmatrix}$$

$$= .125$$

In the last section, we studied optimal strategies associated with strictly determined games and found them to be precisely the maximin and minimax pure strategies adopted by the row and column players. We now look at optimal mixed strategies associated with matrix games that are not strictly determined. In particular, we state, without proof, the optimal mixed strategies to be adopted by the players in a 2×2 matrix game.

As we saw earlier, a player in a nonstrictly determined game should adopt a mixed strategy since a pure strategy will soon be detected by the opponent, who may then use this knowledge to his advantage in devising a counterstrategy. Since there are infinitely many mixed strategies for each player in such a game, the question arises as to how an optimal mixed strategy may be discovered for each player. Recall that an optimal mixed strategy for a player is one in which the row player seeks to maximize his expected payoff and the column player simultaneously seeks to minimize it.

More precisely, the optimal mixed strategy for the row player is arrived at using the following argument: The row player anticipates that any mixed strategy he adopts will be met by a counterstrategy by the column player that will minimize the row player's payoff. Consequently, the row player adopts the mixed strategy for which the expected payoff to the row player (when the column player uses his best counterstrategy) is maximized.

Similarly, the optimal mixed strategy for the column player is arrived at using the following argument: The column player anticipates that the row player will choose a counterstrategy that will maximize the row player's payoff regardless of whatever mixed strategy he (the column player) chooses. Consequently, the column player adopts the mixed strategy for which the expected payoff to the row player (who will use his best counterstrategy) is minimized.

Without going into unnecessary details, let's note that the problem of finding the optimal mixed strategies for the players in a nonstrictly determined game is equivalent to one of solving the related linear programming problem. However, for a 2 × 2 nonstrictly determined game, the optimal mixed strategies for the players may be found by employing the formulas contained in the following result, which we state without proof.

Optimal Strategies for Nonstrictly Determined Games
Let

$$\begin{bmatrix} a & b \\ c & d \end{bmatrix}$$

be the payoff matrix for a nonstrictly determined game. Then, the **optimal mixed strategy for the row player** is given by

$$P = [p_1 \quad p_2] \tag{2a}$$

where

$$p_1 = \frac{d - c}{a + d - b - c} \quad \text{and} \quad p_2 = 1 - p_1$$

and the **optimal mixed strategy for the column player** is given by

$$Q = \begin{bmatrix} q_1 \\ q_2 \end{bmatrix} \tag{2b}$$

where

$$q_1 = \frac{d - b}{a + d - b - c} \quad \text{and} \quad q_2 = 1 - q_1$$

Furthermore, the **value of the game** is given by the expected value of the game $E = PAQ$, where P and Q are the optimal mixed strategies for the row and column players, respectively. Thus,

$$E = PAQ$$
$$= \frac{ad - bc}{a + d - b - c} \tag{2c}$$

The next example illustrates the use of these formulas in finding the optimal mixed strategies and in finding the value of a 2 × 2 (nonstrictly determined) game.

EXAMPLE 3 Coin-Matching Game (continued) Consider the coin-matching game played by Richie and Chuck with the payoff matrix

$$A = \begin{bmatrix} 3 & -2 \\ -2 & 1 \end{bmatrix} \qquad \text{See Example 1.}$$

a. Find the optimal mixed strategies for both Richie and Chuck.
b. Find the value of the game. Does it favor one player over the other?

Solution

a. The game under consideration has no saddle point and is accordingly non-strictly determined. Using Formula (2a) with $a = 3$, $b = -2$, $c = -2$, and $d = 1$, we find that

$$p_1 = \frac{d - c}{a + d - b - c} = \frac{1 - (-2)}{3 + 1 - (-2) - (-2)} = \frac{3}{8}$$

$$p_2 = 1 - p_1$$

$$= 1 - \frac{3}{8}$$

$$= \frac{5}{8}$$

so Richie's optimal mixed strategy is given by

$$P = [p_1 \quad p_2]$$

$$= \left[\tfrac{3}{8} \quad \tfrac{5}{8}\right]$$

Using (2b), we find that

$$q_1 = \frac{d - b}{a + d - b - c} = \frac{1 - (-2)}{3 + 1 - (-2) - (-2)} = \frac{3}{8}$$

$$q_2 = 1 - q_1$$

$$= 1 - \frac{3}{8}$$

$$= \frac{5}{8}$$

giving Chuck's optimal mixed strategy as

$$Q = \begin{bmatrix} \tfrac{3}{8} \\ \tfrac{5}{8} \end{bmatrix}$$

b. The value of the game may be found by computing the matrix product PAQ, where P and Q are the vectors found in part (a). Equivalently, using (2c) we find that

$$E = \frac{ad - bc}{a + d - b - c}$$

$$= \frac{(3)(1) - (-2)(-2)}{3 + 1 - (-2) - (-2)}$$

$$= -\frac{1}{8}$$

Since the value of the game is negative, we conclude that the coin-matching game with the particular given payoff matrix favors Chuck (the column player) over Richie. Over the long run, in repeated plays of the game, where each player uses his optimal strategy, Chuck is expected to win $\frac{1}{8}$, or 12.5¢, on the average per play. ∎

APPLIED EXAMPLE 4   Investment Strategies As part of their investment strategy, the Carringtons have earmarked $40,000 for short-term investments in the stock market and the money market. The performance of the investments depends on the prime rate (that is, the interest rate that banks charge their best customers). An increase in the prime rate generally favors their investment in the money market, whereas a decrease in the prime rate generally favors their investment in the stock market. Suppose the following payoff matrix gives the expected percentage increase or decrease in the value of each investment for each state of the prime rate:

$$
\begin{array}{cc}
 & \begin{array}{cc} \text{Prime rate} & \text{Prime rate} \\ \text{up} & \text{down} \end{array} \\
\begin{array}{l} \text{Money market investment} \\ \text{Stock market investment} \end{array} & \begin{bmatrix} 15 & 10 \\ -5 & 25 \end{bmatrix}
\end{array}
$$

a. Determine the optimal investment strategy for the Carringtons' short-term investment of $40,000.
b. What short-term profit can the Carringtons expect to make on their investments?

Solution

a. We treat the problem as a matrix game in which the Carringtons are the row player. Letting $p = [p_1 \quad p_2]$ denote their optimal strategy, we find that

$$
p_1 = \frac{d - c}{a + d - b - c} = \frac{25 - (-5)}{15 + 25 - 10 - (-5)} = \frac{30}{35} = \frac{6}{7}
$$

$$
p_2 = 1 - p_1 = 1 - \frac{6}{7} = \frac{1}{7}
$$

Thus, the Carringtons should put $(6/7)(40,000)$, or approximately $34,300, into the money market and $(1/7)(\$40,000)$, or approximately $5700, into the stock market.

b. The expected value of the game is given by

$$
E = \frac{ad - bc}{a + d - b - c}
$$

$$
= \frac{(15)(25) - (10)(-5)}{15 + 25 - 10 - (-5)} = \frac{425}{35}
$$

$$
\approx 12.14
$$

Thus, the Carringtons can expect to make a short-term profit of 12.14% on their total investment of 40,000—that is, a profit of $(0.1214)(40,000)$, or $4856. ∎

EXPLORE & DISCUSS

A two-person, zero-sum game is defined by the payoff matrix

$$
A = \begin{bmatrix} x & 1 - x \\ 1 - x & x \end{bmatrix}
$$

1. For what value(s) of x is the game strictly determined? For what value(s) of x is the game not strictly determined?

2. What is the value of the game?

9.5 Self-Check Exercises

1. The payoff matrix for a game is given by

$$A = \begin{bmatrix} 2 & 3 & -1 \\ -3 & 2 & -2 \\ 3 & -2 & 2 \end{bmatrix}$$

   a. Find the expected payoff to the row player if the row player R uses the maximin pure strategy and the column player C uses the minimax pure strategy.
   b. Find the expected payoff to the row player if R uses the maximin strategy 40% of the time and chooses each of the other two rows 30% of the time, while C uses the minimax strategy 50% of the time and chooses each of the other two columns 25% of the time.
   c. Which pair of strategies favors the row player?

2. A farmer has allocated 2000 acres of her farm for planting two crops. Crop A is more susceptible to frost than crop B. If there is no frost in the growing season, then she can expect to make $40/acre from crop A and $25/acre from crop B. If there is mild frost, the expected profits are $20/acre from crop A and $30/acre from crop B. How many acres of each crop should the farmer cultivate in order to maximize her profits? What profit could she expect to make using this optimal strategy?

Solutions to Self-Check Exercises 9.5 can be found on page 565.

9.5 Concept Questions

1. What does the *expected value of a game* measure?

2. Suppose

$$\begin{bmatrix} a & b \\ c & d \end{bmatrix}$$

   is the payoff matrix for a nonstrictly determined game.
   a. What is the optimal mixed strategy for the column player?
   b. What is the optimal mixed strategy for the row player?
   c. What is the value of the game?

9.5 Exercises

In Exercises 1–6, find the expected payoff E of each game whose payoff matrix and strategies P and Q (for the row and column players, respectively) are given.

1. $\begin{bmatrix} 3 & 1 \\ -4 & 2 \end{bmatrix}, P = \begin{bmatrix} \frac{1}{2} & \frac{1}{2} \end{bmatrix}, Q = \begin{bmatrix} \frac{3}{5} \\ \frac{2}{5} \end{bmatrix}$

2. $\begin{bmatrix} -1 & 4 \\ 3 & -2 \end{bmatrix}, P = [.8 \quad .2], Q = \begin{bmatrix} .6 \\ .4 \end{bmatrix}$

3. $\begin{bmatrix} -4 & 3 \\ 2 & 1 \end{bmatrix}, P = \begin{bmatrix} \frac{1}{3} & \frac{2}{3} \end{bmatrix}, Q = \begin{bmatrix} \frac{3}{4} \\ \frac{1}{4} \end{bmatrix}$

4. $\begin{bmatrix} 1 & 2 \\ -3 & 1 \end{bmatrix}, P = \begin{bmatrix} \frac{3}{5} & \frac{2}{5} \end{bmatrix}, Q = \begin{bmatrix} \frac{1}{3} \\ \frac{2}{3} \end{bmatrix}$

5. $\begin{bmatrix} 2 & 0 & -2 \\ 1 & -1 & 3 \\ 2 & 1 & -4 \end{bmatrix}, P = [.2 \quad .6 \quad .2], Q = \begin{bmatrix} .2 \\ .6 \\ .2 \end{bmatrix}$

6. $\begin{bmatrix} 1 & -4 & 2 \\ 2 & 1 & -1 \\ 2 & -2 & 0 \end{bmatrix}, P = [.2 \quad .3 \quad .5], Q = \begin{bmatrix} .6 \\ .2 \\ .2 \end{bmatrix}$

7. The payoff matrix for a game is given by

$$\begin{bmatrix} 1 & -2 \\ -2 & 3 \end{bmatrix}$$

   Compute the expected payoffs of the game for the pairs of strategies in parts (a–d). Which of these strategies is most advantageous to R?

   a. $P = [1 \quad 0], Q = \begin{bmatrix} 1 \\ 0 \end{bmatrix}$

   b. $P = [0 \quad 1], Q = \begin{bmatrix} 1 \\ 0 \end{bmatrix}$

c. $P = \begin{bmatrix} \frac{1}{2} & \frac{1}{2} \end{bmatrix}$, $Q = \begin{bmatrix} \frac{1}{2} \\ \frac{1}{2} \end{bmatrix}$

d. $P = \begin{bmatrix} .5 & .5 \end{bmatrix}$, $Q = \begin{bmatrix} .8 \\ .2 \end{bmatrix}$

8. The payoff matrix for a game is

$$\begin{bmatrix} 3 & 1 & 1 \\ 0 & 2 & 0 \\ -1 & 0 & 2 \end{bmatrix}$$

Compute the expected payoffs of the game for the pairs of strategies in parts (a–d). Which of these strategies is most advantageous to R?

a. $P = \begin{bmatrix} \frac{1}{3} & \frac{1}{3} & \frac{1}{3} \end{bmatrix}$, $Q = \begin{bmatrix} \frac{1}{3} \\ \frac{1}{3} \\ \frac{1}{3} \end{bmatrix}$

b. $P = \begin{bmatrix} \frac{1}{4} & \frac{1}{2} & \frac{1}{4} \end{bmatrix}$, $Q = \begin{bmatrix} \frac{1}{8} \\ \frac{3}{8} \\ \frac{1}{2} \end{bmatrix}$

c. $P = \begin{bmatrix} .4 & .3 & .3 \end{bmatrix}$, $Q = \begin{bmatrix} .6 \\ .2 \\ .2 \end{bmatrix}$

d. $P = \begin{bmatrix} .1 & .5 & .4 \end{bmatrix}$, $Q = \begin{bmatrix} .3 \\ .3 \\ .4 \end{bmatrix}$

9. The payoff matrix for a game is

$$\begin{bmatrix} -3 & 3 & 2 \\ -3 & 1 & 1 \\ 1 & -2 & 1 \end{bmatrix}$$

a. Find the expected payoff to the row player if the row player R uses the maximin pure strategy and the column player C uses the minimax pure strategy.

b. Find the expected payoff to the row player if R uses the maximin strategy 50% of the time and chooses each of the other two rows 25% of the time, while C uses the minimax strategy 60% of the time and chooses each of the other columns 20% of the time.

c. Which of these strategies favors the row player?

10. The payoff matrix for a game is

$$\begin{bmatrix} 4 & -3 & 3 \\ -4 & 2 & 1 \\ 3 & -5 & 2 \end{bmatrix}$$

a. Find the expected payoff to the row player if the row player R uses the maximin pure strategy and the column player C uses the minimax pure strategy.

b. Find the expected payoff to the row player if R uses the maximin strategy 40% of the time and chooses each of the other two rows 30% of the time, while C uses the minimax strategy 50% of the time and chooses each of the other columns 25% of the time.

c. Which of these strategies favors the row player?

In Exercises 11–16, find the optimal strategies, P and Q, for the row and column players, respectively. Also compute the expected payoff E of each matrix game and determine which player it favors, if any, if the row and column players use their optimal strategies.

11. $\begin{bmatrix} 4 & 1 \\ 2 & 3 \end{bmatrix}$

12. $\begin{bmatrix} 2 & 5 \\ 3 & -6 \end{bmatrix}$

13. $\begin{bmatrix} -1 & 2 \\ 1 & -3 \end{bmatrix}$

14. $\begin{bmatrix} -1 & 3 \\ 2 & 0 \end{bmatrix}$

15. $\begin{bmatrix} -2 & -6 \\ -8 & -4 \end{bmatrix}$

16. $\begin{bmatrix} 2 & 5 \\ -2 & 4 \end{bmatrix}$

17. Consider the coin-matching game played by Richie and Chuck (see Examples 1 and 3) with the payoff matrix

$$A = \begin{bmatrix} 4 & -2 \\ -2 & 1 \end{bmatrix}$$

a. Find the optimal strategies for Richie and Chuck.

b. Find the value of the game. Does it favor one player over the other?

18. INVESTMENT STRATEGIES As part of their investment strategy, the Carringtons have decided to put $100,000 into stock market investments and also into purchasing precious metals. The performance of the investments depends on the state of the economy in the next year. In an expanding economy, it is expected that their stock market investment will outperform their investment in precious metals, whereas an economic recession will have precisely the opposite effect. Suppose the following payoff matrix gives the expected percent increase or decrease in the value of each investment for each state of the economy:

	Expanding economy	Economic recession
Stock market investment	20	−5
Commodity investment	10	15

a. Determine the optimal investment strategy for the Carringtons' investment of $100,000.

b. What profit can the Carringtons expect to make on their investments over the year if they use their optimal investment strategy?

19. INVESTMENT STRATEGIES The Maxwells have decided to invest $40,000 in the common stocks of two companies listed on the New York Stock Exchange. One of the companies derives its revenue mainly from its worldwide operation of a chain of hotels, whereas the other company is a major brewery in the country. It is expected that if the economy is in a state of growth, then the hotel stock should outperform the brewery stock; however, the brewery stock is expected to hold its own better than the hotel stock in a recessionary period. Suppose the following payoff matrix gives the expected percent increase or decrease in the value of each investment for each state of the economy:

	Expanding economy	Economic recession
Investment in hotel stock	$\begin{bmatrix} 25 \\ 10 \end{bmatrix}$	$\begin{bmatrix} -5 \\ 15 \end{bmatrix}$
Investment in brewery stock		

a. Determine the optimal investment strategy for the Maxwells' investment of $40,000.
b. What profit can the Maxwells expect to make on their investments if they use their optimal investment strategy?

20. CAMPAIGN STRATEGIES Bella Robinson and Steve Carson are running for a seat in the U.S. Senate. If both candidates campaign only in the major cities of the state, then Robinson is expected to get 60% of the votes; if both candidates campaign only in the rural areas, then Robinson is expected to get 55% of the votes; if Robinson campaigns exclusively in the city and Carson campaigns exclusively in the rural areas, then Robinson is expected to get 40% of the votes; finally, if Robinson campaigns exclusively in the rural areas and Carson campaigns exclusively in the city, then Robinson is expected to get 45% of the votes.
a. Construct the payoff matrix for the game and show that it is not strictly determined.
b. Find the optimal strategy for both Robinson and Carson.

21. ADVERTISEMENTS Two dentists, Lydia Russell and Jerry Carlton, are planning to establish practices in a newly developed community. Both have allocated approximately the same total budget for advertising in the local newspaper and for the distribution of fliers announcing their practices. Because of the location of their offices, Russell is expected to get 48% of the business if both dentists advertise only in the local newspaper; if both dentists advertise through fliers, then Russell is expected to get 45% of the business; if Russell advertises exclusively in the local newspaper and

Carlton advertises exclusively through fliers, then Russell is expected to get 65% of the business. Finally, if Russell advertises through fliers exclusively and Carlton advertises exclusively in the local newspaper, then Russell is expected to get 50% of the business.
a. Construct the payoff matrix for the game and show that it is not strictly determined.
b. Find the optimal strategy for both Russell and Carlton.

22. Let

$$\begin{bmatrix} a_{11} & a_{12} \\ a_{21} & a_{22} \end{bmatrix}$$

be the payoff matrix with a 2×2 matrix game. Assume that either the row player uses the optimal mixed strategy $P = [p_1 \quad p_2]$, where

$$p_1 = \frac{d - c}{a + d - b - c} \quad \text{and} \quad p_2 = 1 - p_1$$

or the column player uses the optimal mixed strategy

where

$$Q = \begin{bmatrix} q_1 \\ q_2 \end{bmatrix}$$

$$q_1 = \frac{d - b}{a + d - b - c} \quad \text{and} \quad q_2 = 1 - q_1$$

Show by direct computation that the expected value of the game is given by $E = PAQ$.

23. Let

$$\begin{bmatrix} a & b \\ c & d \end{bmatrix}$$

be the payoff matrix associated with a nonstrictly determined 2×2 matrix game. Prove that the expected payoff of the game is given by

$$E = \frac{ad - bc}{a + d - b - c}$$

Hint: Compute $E = PAQ$, where P and Q are the optimal strategies for the row and column players, respectively.

9.5 Solutions to Self-Check Exercises

1. a. From the following calculations,

$$
\begin{bmatrix} 2 & 3 & -1 \\ -3 & 2 & -2 \\ 3 & -2 & 2 \end{bmatrix}
\begin{matrix} \text{Row} \\ \text{minima} \\ \boxed{-1} \\ -3 \\ -2 \end{matrix}
\leftarrow \begin{matrix} \text{Largest of the} \\ \text{row minima} \end{matrix}
$$

Column maxima   3   3   ②

↑
Smallest of the column maxima

we see that R's optimal pure strategy is to choose row 1, whereas C's optimal pure strategy is to choose column 3. Furthermore, if both players use their respective optimal strategies, then the expected payoff to R is -1 unit.

b. R's mixed strategy may be represented by the row vector

$$ P = [.4 \quad .3 \quad .3] $$

and C's mixed strategy may be represented by the column vector

$$ Q = \begin{bmatrix} .25 \\ .25 \\ .50 \end{bmatrix} $$

The expected payoff to the row player will then be given by

$$
E = PAQ = [.4 \quad .3 \quad .3]
\begin{bmatrix} 2 & 3 & -1 \\ -3 & 2 & -2 \\ 3 & -2 & 2 \end{bmatrix}
\begin{bmatrix} .25 \\ .25 \\ .50 \end{bmatrix}
$$

$$
= [.4 \quad .3 \quad .3]
\begin{bmatrix} .75 \\ -1.25 \\ 1.25 \end{bmatrix}
$$

$$ = .3 $$

c. From the results of parts (a) and (b), we see that the mixed strategies of part (b) will be better for R.

2. We may view this problem as a matrix game with the farmer as the row player and the weather as the column player. The payoff matrix for the game is

$$
\begin{matrix} & \text{No} & \text{Mild} \\ & \text{frost} & \text{frost} \end{matrix}
$$
$$
\begin{matrix} \text{Crop A} \\ \text{Crop B} \end{matrix}
\begin{bmatrix} 40 & 20 \\ 25 & 30 \end{bmatrix}
$$

The game under consideration has no saddle point and is accordingly nonstrictly determined. Letting $p = [p_1 \quad p_2]$ denote the farmer's optimal strategy and using the formula for determining the optimal mixed strategies for a 2×2 game with $a = 40$, $b = 20$, $c = 25$, and $d = 30$, we find

$$ p_1 = \frac{d - c}{a + d - b - c} = \frac{30 - 25}{40 + 30 - 20 - 25} = \frac{5}{25} = \frac{1}{5} $$

$$ p_2 = 1 - p_1 = 1 - \frac{1}{5} = \frac{4}{5} $$

Therefore, the farmer should cultivate $\left(\frac{1}{5}\right)(2000)$, or 400, acres of crop A and 1600 acres of crop B. By using her optimal strategy, the farmer can expect to realize a profit of

$$ E = \frac{ad - bc}{a + d - b - c} $$

$$ = \frac{(40)(30) - (20)(25)}{40 + 30 - 20 - 25} $$

$$ = 28 $$

or \$28/acre—that is, a total profit of $(28)(2000)$, or \$56,000.

CHAPTER 9 **Summary of Principal Formulas and Terms**

FORMULAS

1. Steady-state matrix for an absorbing stochastic matrix	If $A = \begin{bmatrix} I & S \\ \hline O & R \end{bmatrix}$ then the steady-state matrix of A is $\begin{bmatrix} I & S(I - R)^{-1} \\ \hline O & O \end{bmatrix}$
2. Expected value of a game	$E = PAQ =$ $[p_1 p_2 \cdots p_m] \begin{bmatrix} a_{11} & a_{12} & \cdots & a_{1n} \\ a_{21} & a_{22} & \cdots & a_{2n} \\ \vdots & \vdots & & \vdots \\ a_{m1} & a_{m2} & \cdots & a_{mn} \end{bmatrix} \begin{bmatrix} q_1 \\ q_2 \\ \vdots \\ q_n \end{bmatrix}$
3. Optimal strategy for a nonstrictly determined game	$P = [p_1 \quad p_2]$, where $p_1 = \dfrac{d - c}{a + d - b - c}$ and $p_2 = 1 - p_1$ and $Q = \begin{bmatrix} q_1 \\ q_2 \end{bmatrix}$ where $q_1 = \dfrac{d - b}{a + d - b - c}$ and $q_2 = 1 - q_1$ The expected value of the game is $E = PAQ$ $= \dfrac{ad - bc}{a + d - b - c}$

TERMS

CHAPTER 9 Concept Review Questions

Fill in the blanks.

1. A Markov chain is a stochastic process in which the _____ associated with the outcomes at any stage of the experiment depend only on the outcomes of the _____ stage.

2. The outcome at any stage of the experiment in a Markov process is called the _____ of the experiment; the outcome at the current stage of the experiment is called the current _____.

3. The probabilities in a Markov chain are called _____ probabilities because they are associated with the transition from one state to the next in the Markov process.

4. A transition matrix associated with a Markov chain with n states is a/an _____ matrix T with entries satisfying the following conditions: (a) all entries are _____ and (b) the sum of the entries in each column of T is _____.

5. If the probability distribution vector X_N associated with a Markov process approaches a fixed vector as N gets larger and larger, then the latter is called the steady-state _____ vector for the system. To find this vector, we are led to finding the limit of T^m, which (if it exists) is called the _____ matrix.

6. A stochastic matrix T is a/an _____ Markov chain if T^n approaches a steady-state matrix in which the _____ of the limiting matrix are all _____ and all the entries are _____. To find the steady-state distribution vector X, we solve the vector equation _____ together with the condition that the sum of the _____ of the vector X is equal to _____.

7. In an absorbing stochastic matrix, (a) there is at least one _____ state, a state in which it is impossible for an object to _____, and (b) it is possible to go from any nonabsorbing state to an absorbing state in one or more _____.

8. **a.** A game in which the payoff to one party results in an equal loss to the other is called a/an _____ game.
   b. The strategy employed by the row player in which he or she selects from among the rows the one in which the smallest payoff is as large as possible is called the _____ strategy. The strategy in which C chooses from among the columns, the one in which the largest payoff is as small as possible, is called the _____ strategy.

9. The strategy that is most profitable to a particular player is called the _____ strategy.

10. In a strictly determined game, an entry in the payoff matrix that is simultaneously the smallest entry in the row and the largest entry in the column is called a/an _____ _____; the optimal strategy for the row player in a strictly determined game is the _____ strategy and is the _____ containing the _____ point; the optimal strategy for the column player is the _____ strategy and is the _____ containing the _____ point.

CHAPTER 9 Review Exercises

In Exercises 1–4, determine which of the following are regular stochastic matrices.

1. $\begin{bmatrix} 1 & -2 \\ 0 & -8 \end{bmatrix}$

2. $\begin{bmatrix} .3 & 1 \\ .7 & 0 \end{bmatrix}$

3. $\begin{bmatrix} \frac{1}{2} & 0 & \frac{1}{3} \\ 0 & 0 & \frac{1}{3} \\ \frac{1}{2} & 1 & \frac{1}{3} \end{bmatrix}$

4. $\begin{bmatrix} .3 & 0 & .5 \\ .2 & 1 & 0 \\ .1 & 0 & .5 \end{bmatrix}$

In Exercises 5 and 6, find X_2, (the probability distribution of the system after two observations) for the distribution vector X_0 and the transition matrix T.

5. $X_0 = \begin{bmatrix} \frac{1}{2} \\ \frac{1}{2} \\ 0 \end{bmatrix}$, $T = \begin{bmatrix} 0 & \frac{1}{4} & \frac{3}{5} \\ \frac{2}{5} & \frac{1}{2} & \frac{1}{5} \\ \frac{3}{5} & \frac{1}{4} & \frac{1}{5} \end{bmatrix}$

6. $X_0 = \begin{bmatrix} .35 \\ .25 \\ .40 \end{bmatrix}$, $T = \begin{bmatrix} .2 & .1 & .3 \\ .5 & .4 & .4 \\ .3 & .5 & .3 \end{bmatrix}$

In Exercises 7–10, determine whether the matrix is an absorbing stochastic matrix.

7. $\begin{bmatrix} 1 & .6 & .1 \\ 0 & .2 & .6 \\ 0 & .2 & .3 \end{bmatrix}$

8. $\begin{bmatrix} .3 & .2 & .1 \\ .7 & .5 & .3 \\ 0 & .3 & .6 \end{bmatrix}$

9. $\begin{bmatrix} .32 & .22 & .44 \\ .68 & .78 & .56 \\ 0 & 0 & 0 \end{bmatrix}$

10. $\begin{bmatrix} .31 & .35 & 0 \\ .32 & .40 & 0 \\ .37 & .25 & 1 \end{bmatrix}$

In Exercises 11–14, find the steady-state matrix for the transition matrix.

11. $\begin{bmatrix} .6 & .3 \\ .4 & .7 \end{bmatrix}$ 12. $\begin{bmatrix} .5 & .4 \\ .5 & .6 \end{bmatrix}$

13. $\begin{bmatrix} .6 & .4 & .3 \\ .2 & .2 & .2 \\ .2 & .4 & .5 \end{bmatrix}$ 14. $\begin{bmatrix} .1 & .2 & .6 \\ .3 & .4 & .2 \\ .6 & .4 & .2 \end{bmatrix}$

15. **URBANIZATION OF FARMLAND**   A study conducted by the State Department of Agriculture in a Sunbelt state reveals an increasing trend toward urbanization of the farmland within the state. Ten years ago, 50% of the land within the state was used for agricultural purposes (A), 15% had been urbanized (U), and the remaining 35% was neither agricultural nor urban (N). Since that time, 10% of the agricultural land has been converted to urban land, 5% has been used for other purposes, and the remaining 85% is still agricultural. Of the urban land, 95% has remained urban, whereas 5% of it has been used for nonagricultural purposes. Of the land that was neither agricultural nor urban, 10% has been converted to agricultural land, 5% has been urbanized, and the remaining 85% remains unchanged.

   a. Construct the transition matrix for the Markov chain that describes the shift in land use within the state.

   b. Find the probability vector describing the distribution of land within the state 10 yr ago.

   c. Assuming that this trend continues, find the probability vector describing the distribution of land within the state 10 yr from now.

16. **AUTOMOBILE SURVEY**   *Auto Trend* magazine conducted a survey among automobile owners in a certain area of the country to determine what type of car they now own and what type of car they expect to own 4 yr from now. For purposes of classification, automobiles mentioned in the survey were placed into three categories: large, intermediate, and small. Results of the survey follow:

			Present car	
		Large	Intermediate	Small
Future car	Large	.3	.1	.1
	Intermediate	.3	.5	.2
	Small	.4	.4	.7

Assuming that these results indicate the long-term buying trend of car owners within the area, what will be the distribution of cars (relative to size) in this area over the long run?

In Exercises 17–20, determine whether each game within the given payoff matrix is strictly determined. If so, give the optimal pure strategies for the row player and the column player and also give the value of the game.

17. $\begin{bmatrix} 1 & 2 \\ 3 & 5 \\ 4 & 6 \end{bmatrix}$ 18. $\begin{bmatrix} 1 & 0 & 3 \\ 2 & -1 & -2 \end{bmatrix}$

19. $\begin{bmatrix} 1 & 3 & 6 \\ -2 & 4 & 3 \\ -5 & -4 & -2 \end{bmatrix}$ 20. $\begin{bmatrix} 4 & 3 & 2 \\ -6 & 3 & -1 \\ 2 & 3 & 4 \end{bmatrix}$

In Exercises 21–24, find the expected payoff E of each game whose payoff matrix and strategies P and Q (for the row and column players, respectively) are given.

21. $\begin{bmatrix} 4 & 8 \\ 6 & -12 \end{bmatrix}, P = \begin{bmatrix} \frac{1}{2} & \frac{1}{2} \end{bmatrix}, Q = \begin{bmatrix} \frac{1}{4} \\ \frac{3}{4} \end{bmatrix}$

22. $\begin{bmatrix} 3 & 0 & -3 \\ 2 & 1 & 2 \end{bmatrix}, P = \begin{bmatrix} \frac{1}{3} & \frac{2}{3} \end{bmatrix}, Q = \begin{bmatrix} \frac{1}{3} \\ \frac{1}{3} \\ \frac{1}{3} \end{bmatrix}$

23. $\begin{bmatrix} 3 & -1 & 2 \\ 1 & 2 & 4 \\ -2 & 3 & 6 \end{bmatrix}, P = \begin{bmatrix} .2 & .4 & .4 \end{bmatrix}, Q = \begin{bmatrix} .2 \\ .6 \\ .2 \end{bmatrix}$

24. $\begin{bmatrix} 2 & -2 & 3 \\ 1 & 2 & -1 \\ -1 & 2 & 3 \end{bmatrix}, P = \begin{bmatrix} .2 & .4 & .4 \end{bmatrix}, Q = \begin{bmatrix} .3 \\ .3 \\ .4 \end{bmatrix}$

In Exercises 25–28, find the optimal strategies, P and Q, for the row player and the column player, respectively. Also compute the expected payoff E of each matrix game if the row and column players adopt their optimal strategies and determine which player it favors, if any.

25. $\begin{bmatrix} 1 & -2 \\ 0 & 3 \end{bmatrix}$ 26. $\begin{bmatrix} 4 & -7 \\ -5 & 6 \end{bmatrix}$

27. $\begin{bmatrix} 3 & -6 \\ 1 & 2 \end{bmatrix}$ 28. $\begin{bmatrix} 12 & 10 \\ 6 & 14 \end{bmatrix}$

29. **PRICING PRODUCTS**   Two competing music stores, Disco-Mart and Stereo World, each have the option of selling a certain popular compact disc (CD) label at a price of either $7 or $8/CD. If both sell the label at the same price, they are each expected to get 50% of the business. If DiscoMart sells the label at $7/CD and Stereo World sells the label at $8/CD, DiscoMart is expected to get 70% of the business; if DiscoMart sells the label at $8/CD and Stereo World sells the label at $7/CD, DiscoMart is expected to get 40% of the business.

a. Represent this information in the form of a payoff matrix.

b. Determine the optimal price that each company should sell the compact disc label for to ensure that it captures the largest possible expected market share.

30. MAXIMIZING PRODUCTION The management of a division of National Motor Corporation that produces compact and subcompact cars has estimated that the quantity demanded of their compact models is 1500 units/week if the price of oil increases at a higher than normal rate, whereas the quantity demanded of their subcompact models is 2500 units/week under similar conditions. However, the quantity demanded of their compact models and subcompact models is 3000 units and 2000 units/week, respectively, if the price of oil increases at a normal rate. Determine the percents of compact and subcompact cars the division should plan to manufacture to maximize the expected number of cars demanded each week.

CHAPTER 9 Before Moving On . . .

1. The transition matrix for a Markov process is

$$T = \begin{array}{c} \text{State} \\ \begin{array}{c} \text{State 1} \\ \text{State 2} \end{array} \begin{bmatrix} .3 & .4 \\ .7 & .6 \end{bmatrix} \end{array}$$

and the initial-state distribution vector is

$$X_0 = \begin{array}{c} \text{State 1} \\ \text{State 2} \end{array} \begin{bmatrix} .6 \\ .4 \end{bmatrix}$$

Find X_2.

2. Find the steady-state vector for the transition matrix

$$T = \begin{bmatrix} \frac{1}{3} & \frac{1}{4} \\ \frac{2}{3} & \frac{3}{4} \end{bmatrix}$$

3. Compute the steady-state matrix of the absorbing stochastic matrix

$$\begin{bmatrix} \frac{1}{3} & 0 & 0 \\ 0 & 1 & \frac{1}{4} \\ \frac{2}{3} & 0 & \frac{3}{4} \end{bmatrix}$$

4. A two-person, zero-sum game is defined by the matrix

$$A = \begin{bmatrix} 2 & 3 & -1 \\ -1 & 2 & -3 \\ 3 & 4 & -2 \end{bmatrix}$$

a. Show that the game is strictly determined and find the saddle point for the game.

b. What is the optimal strategy for each player?

c. What is the value of the game? Does the game favor one player over the other?

5. The payoff matrix for a certain game is

$$A = \begin{bmatrix} 2 & -1 \\ 3 & 2 \\ -3 & 4 \end{bmatrix}$$

a. Find the expected payoff to the row player if the row player R uses her maximin pure strategy and the column player C uses his minimax pure strategy.

b. Find the expected payoff to the row player if R uses her maximin strategy 40% of the time and chooses each of the other two rows 30% of the time, while C chooses the optimal strategy 60% of the time.

6. The payoff matrix for a certain game is

$$\begin{bmatrix} 3 & 1 \\ -2 & 2 \end{bmatrix}$$

a. Find the optimal strategies, P and Q, for the row and column players, respectively.

b. Find the expected payoff E of the game and determine which player it favors, if any, if the row and column players use their optimal strategies.

A | Introduction to Logic

ONE OF THE earliest records of man's efforts to understand the reasoning process can be found in Aristotle's work *Organon* (third century B.C.), in which he presented his ideas on logical arguments. Up until the present day, Aristotelian logic has been the basis for traditional logic, the systematic study of valid inferences.

The field of mathematics known as symbolic logic, in which symbols are used to replace ordinary language, had its beginnings in the 18th and 19th centuries with the works of the German mathematician Gottfried Wilhelm Leibniz (1646–1716) and the English mathematician George Boole (1815–1864). Boole introduced algebraic-type operations to the field of logic, thereby providing us with a systematic method of combining statements. Boolean algebra is the basis of modern-day computer technology, as well as being central to the study of pure mathematics.

In this appendix, we introduce the fundamental concepts of symbolic logic. Beginning with the definitions of statements and their truth values, we proceed to a discussion of the combination of statements and valid arguments. In Section A.6, we present an application of symbolic logic that is widely used in computer technology—switching networks.

A.1   Propositions and Connectives

We use deductive reasoning in many of the things we do. Whether in a formal debate or in an expository article, we use language to express our thoughts. In English, sentences are used to express assertions, questions, commands, and wishes. In our study of logic, we will be concerned with only one type of sentence, the declarative sentence.

> **Proposition**
>
> A **proposition**, or *statement*, is a declarative sentence that can be classified as either true or false, but not both.

Commands, requests, questions, or exclamations are examples of sentences that are not propositions.

EXAMPLE 1   Which of the following are propositions?

a. Toronto is the capital of Ontario.
b. Close the door!
c. There are 100 trillion connections among the neurons in the human brain.
d. Who is the chief justice of the Supreme Court?
e. The new television sit-com is a successful show.
f. The 2000 Summer Olympic Games were held in Montreal.
g. $x + 3 = 8$
h. How wonderful!
i. Seven is either an odd number or it is even.

Solution   Statements (a), (c), (e), (f), and (i) are propositions. Statement (c) would be difficult to verify, but the validity of the statement can at least theoretically be determined. Statement (e) is a proposition if we assume that the meaning of the term *successful* has been defined. For example, one might use the Nielsen ratings to determine the success of a new show. Statement (f) is a proposition that is false.

Statements (b), (d), (g), and (h) are *not* propositions. Statement (b) is a command, (d) is a question, and (h) is an exclamation. Statement (g) is an open sentence that cannot be classified as true or false. For example, if $x = 5$, then $5 + 3 = 8$ and the sentence is true. On the other hand, if $x = 4$, then $4 + 3 \neq 8$ and the sentence is false. ∎

Having considered what is meant by a proposition, we now discuss the ways in which propositions may be combined. For example, the two propositions

a.   Toronto is the capital of Ontario

and

b.   Toronto is the largest city in Canada

may be joined to form the proposition

c.   Toronto is the largest city in Canada and is the capital of Ontario.

Propositions (a) and (b) are called **prime,** or *simple*, **propositions** because they are simple statements expressing a single complete thought. In the ensuing discussion, we use the lowercase letters p, q, r, and so on to denote prime propositions.

Propositions that are combinations of two or more propositions, such as proposition (c), are called **compound propositions.** The words used to combine propositions are called **logical connectives.** The connectives that we will consider are given in Table 1 along with their symbols. We discuss the first three in this section and the remaining two in Section A.3.

TABLE 1

Name	Logical Connective	Symbol
Conjunction	and	$\wedge$
Negation	not	$\sim$
Disjunction	or	$\vee$
Conditional	if . . . then	$\rightarrow$
Biconditional	if and only if	$\leftrightarrow$

Conjunction

A **conjunction** is a statement of the form "*p* and *q*" and is represented symbolically by

$$p \wedge q$$

The conjunction $p \wedge q$ is true if *both* p and q are true; it is false otherwise.

We have already encountered a conjunction in an earlier example. The two propositions

p: Toronto is the capital of Ontario

q: Toronto is the largest city in Canada

were combined to form the conjunction

$p \wedge q$: Toronto is the capital of Ontario and is the largest city in Canada.

Disjunction

A **disjunction** is a proposition of the form "*p* or *q*" and is represented symbolically by

$$p \vee q$$

The disjunction $p \vee q$ is false if *both* statements p and q are false; it is true in all other cases.

The use of the word *or* in this definition is meant to convey the meaning "one or the other, or both." Since this disjunction is true when *both* p and q are true, as well as when either p or q is true, it is sometimes referred to as an **inclusive disjunction.**

EXAMPLE 2   Consider the propositions

p: Dorm residents can purchase meal plans.

q: Dorm residents can purchase à la carte cards.

The disjunction is

$p \vee q$: Dorm residents can purchase meal plans or à la carte cards.   ■

The disjunction in Example 2 is true when dorm residents can purchase either meal plans or à la carte cards or both meal plans and à la carte cards.

Exclusive Disjunction
An **exclusive disjunction** is a proposition of the form "p or q" and is denoted by

$$p \veebar q$$

The disjunction $p \veebar q$ is true if *either* p or q is true.

In contrast to an inclusive disjunction, an exclusive disjunction is *false* when both p and q are true. In Example 2 the exclusive disjunction would not be true if dorm residents could buy both meal plans and à la carte cards. The difference between exclusive and inclusive disjunctions is further illustrated in the next example.

EXAMPLE 3   Consider the propositions

p: The base price of each condominium unit includes a private deck.

q: The base price of each condominium unit includes a private patio.

Find the exclusive disjunction $p \veebar q$.

Solution   The exclusive disjunction is

$p \veebar q$: The base price of each condominium unit includes either a private deck or a patio.

Observe that this statement is not true when both p and q are true. In other words, the base price of each condominium unit does not include both a private deck and a patio.   ■

In our everyday use of language, the meaning of the word *or* is not always clear. In legal documents the two cases are distinguished by the words *and/or* (inclusive) and *either/or* (exclusive). In mathematics we use the word *or* in the inclusive sense, unless otherwise specified.

Negation
A **negation** is a proposition of the form "not p" and is represented symbolically by

$$\sim p$$

The proposition $\sim p$ is true if p is false and vice versa.

EXAMPLE 4 Form the negation of the proposition

p: Prices rose on the New York Stock Exchange today.

Solution The negation is

$\sim p$: Prices did not rise on the New York Stock Exchange today. ∎

EXAMPLE 5 Consider the two propositions

p: The birthrate declined in the United States last year.
q: The population of the United States increased last year.

Write the following statements in symbolic form:

a. Last year the birthrate declined and the population increased in the United States.
b. Either the birthrate declined or the population increased in the United States last year.
c. It is not true that the birthrate declined and the population increased in the United States last year.

Solution The symbolic form of each statement is given by

a. $p \wedge q$ **b.** $p \veebar q$ **c.** $\sim(p \wedge q)$ ∎

A.1 Exercises

In Exercises 1–14, determine whether the statement is a proposition.

1. The defendant was convicted of grand larceny.

2. Who won the 2004 presidential election?

3. The first month of the year is February.

4. The number 2 is odd.

5. $x - 1 \geq 0$

6. Keep off the grass!

7. Coughing may be caused by a lack of water in the air.

8. The first McDonald's fast-food restaurant was opened in California.

9. Exercise protects men and women from sudden heart attacks.

10. If voters do not pass the proposition, then taxes will be increased.

11. Don't swim immediately after eating!

12. What ever happened to Baby Jane?

13. $\dfrac{x^2 + 1}{x^2 + 1} = 1$

14. Several major corporations will enter or expand operations into the home-security products market this year.

In Exercises 15–20, identify the logical connective that is used in the statement.

15. Housing starts in the United States did not increase last month.

16. Mel Bieber's will is valid if and only if he was of sound mind and memory when he made the will.

17. Americans are saving less and spending more this year.

18. If you traveled on your job and weren't reimbursed, then you can deduct these expenses from your taxable income.

19. Both loss of appetite and irritability are symptoms of mental stress.

20. Prices for many imported goods have either stayed flat or dropped slightly this year.

In Exercises 21–26, state the negation of the proposition.

21. New orders for manufactured goods fell last month.

22. For many men, housecleaning is not a familiar task.

23. Drinking during pregnancy affects the size and weight of babies.

24. Not all patients suffering from influenza lose weight.

25. The commuter airline industry is now undergoing a shakeup.

26. The Dow Jones Industrial Average registered its fourth consecutive decline today.

27. Let p and q denote the propositions

p: Domestic car sales increased over the past year.

q: Foreign car sales decreased over the past year.

Express the following compound propositions in words:
a. $p \vee q$ **b.** $p \wedge q$ **c.** $p \veebar q$
d. $\sim p$ **e.** $\sim p \vee q$ **f.** $\sim p \vee \sim q$

28. Let p and q denote the propositions

p: Every employee is required to be fingerprinted.

q: Every employee is required to take an oath of allegiance.

Express the following compound propositions in words:
a. $p \vee q$ **b.** $p \wedge q$ **c.** $\sim p \vee \sim q$
d. $\sim p \wedge \sim q$ **e.** $p \vee \sim q$

29. Let p and q denote the propositions

p: The doctor recommended surgery to treat his hyperthyroidism.

q: The doctor recommended radioactive iodine to treat his hyperthyroidism.

a. State the exclusive disjunction for these propositions in words.
b. State the inclusive disjunction for these propositions in words.

30. Let p and q denote the propositions

p: The investment newsletter recommended buying bond mutual funds.

q: The investment newsletter recommended buying stock mutual funds.

a. State the exclusive disjunction for these propositions in words.
b. State the inclusive disjunction for these propositions in words.

31. Let p and q denote the propositions

p: The SAT verbal scores improved in this school district last year.

q: The SAT math scores improved in this school district last year.

Express each of the following statements symbolically.
a. The SAT verbal scores and the SAT math scores improved in this school district last year.
b. Either the SAT verbal scores or the SAT math scores improved in this school district last year.
c. Neither the SAT verbal scores nor the SAT math scores improved in this school district last year.
d. It is not true that the SAT math scores did not improve in this school district last year.

32. Let p and q denote the propositions

p: Laura purchased a VHS videocassette recorder.

q: Laura did not purchase a DVD player.

Express each of the following statements symbolically.
a. Laura purchased either a VHS or a DVD player.
b. Laura purchased a VHS and a DVD player.
c. It is not true that Laura purchased a DVD player.
d. Laura purchased neither a VHS nor a DVD player.

33. Let p, q, and r denote the propositions

p: The popularity of prime-time soaps increased this year.

q: The popularity of prime-time situation comedies increased this year.

r: The popularity of prime-time detective shows decreased this year.

Express each of the following propositions in words.
a. $\sim p \wedge \sim q$ **b.** $\sim p \vee r$
c. $\sim(\sim r) \vee \sim q$ **d.** $\sim p \veebar \sim q$

A.2   Truth Tables

A primary objective of studying logic is to determine whether a given proposition is true or false. In other words, we wish to determine the **truth value** of a given statement.

Consider, for example, the conjunction $p \wedge q$ formed from the prime propositions p and q. There are four possible true values for the given conjunction—namely,

1. p is true and q is true.
2. p is true and q is false.
3. p is false and q is true.
4. p is false and q is false.

By definition the conjunction is true when both p and q are true, and it is false otherwise. We can summarize this information in the form of a truth table, as shown in Table 2.

In a similar manner we can construct the truth tables for inclusive disjunction, exclusive disjunction, and negation (Tables 3a–c).

TABLE 2

p	q	$p \wedge q$
T	T	T
T	F	F
F	T	F
F	F	F

Conjunction

TABLE 3

p	q	$p \vee q$		p	q	$p \veebar q$		p	$\sim p$
T	T	T		T	T	F		T	F
T	F	T		T	F	T		F	T
F	T	T		F	T	T			
F	F	F		F	F	F			

(a) Inclusive disjunction   **(b)** Exclusive disjunction   **(c)** Negation

In general, when we are given a compound proposition, we are concerned with the problem of finding every possible combination of truth values associated with the compound proposition. To systematize this procedure, we construct a truth table exhibiting all possible truth values for the given proposition. The next several examples illustrate this method.

EXAMPLE 1   Construct the truth table for the proposition $\sim p \vee q$.

Solution   The truth table is constructed in the following manner.

1. The two prime propositions p and q are placed at the head of the first two columns (Table 4).
2. The two propositions containing the connectives, $\sim p$ and $\sim p \vee q$, are placed at the head of the next two columns.
3. The possible truth values for p are entered in the column headed by p, and then the possible truth values for q are entered in the column headed by q. Notice that the *possible T values are always exhausted first*.
4. The possible truth values for the negation $\sim p$ are entered in the column headed by $\sim p$, and then the possible truth values for the disjunction $\sim p \vee q$ are entered in the column headed by $\sim p \vee q$.   ■

TABLE 4

p	q	$\sim p$	$\sim p \vee q$
T	T	F	T
T	F	F	F
F	T	T	T
F	F	T	T

EXAMPLE 2   Construct the truth table for the proposition $\sim p \vee (p \wedge q)$.

Solution   Proceeding as in the previous example, we construct the truth table as shown in Table 5. Since the given proposition is the disjunction of the propositions $\sim p$ and $(p \wedge q)$, columns are introduced for $\sim p$, $p \wedge q$, and $\sim p \vee (p \wedge q)$.

TABLE 5

p	q	$\sim p$	$p \wedge q$	$\sim p \vee (p \wedge q)$
T	T	F	T	T
T	F	F	F	F
F	T	T	F	T
F	F	T	F	T

The following is an example involving three prime propositions.

EXAMPLE 3   Construct the truth table for the proposition $(p \vee q) \vee (r \wedge \sim p)$.

Solution   Following the method of the previous example, we construct the truth table as shown in Table 6.

TABLE 6

p	q	r	$p \vee q$	$\sim p$	$r \wedge \sim p$	$(p \vee q) \vee (r \wedge \sim p)$
T	T	T	T	F	F	T
T	T	F	T	F	F	T
T	F	T	T	F	F	T
T	F	F	T	F	F	T
F	T	T	T	T	T	T
F	T	F	T	T	F	T
F	F	T	F	T	T	T
F	F	F	F	T	F	F

Observe that the truth table in Example 3 contains eight rows, whereas the truth tables in Examples 1 and 2 contain four rows. In general, *if a compound proposition P(p, q, . . .) contains the n prime propositions p, q, . . . , then the corresponding truth table contains 2^n rows.* For example, since the compound proposition in Example 3 contains three prime propositions, its corresponding truth table contains 2^3, or 8, rows.

A.2   Exercises

In Exercises 1–18, construct a truth table for each compound proposition.

1. $p \vee \sim q$

2. $\sim p \wedge \sim q$

3. $\sim(\sim p)$

4. $\sim(p \wedge q)$

5. $p \vee \sim p$

6. $\sim(\sim p \vee \sim q)$

7. $\sim p \wedge (p \vee q)$

8. $(p \vee \sim q) \wedge q$

9. $(p \vee q) \wedge (p \wedge \sim q)$

10. $(p \vee q) \wedge \sim p$

11. $(p \vee q) \wedge \sim(p \vee q)$

12. $(p \vee q) \vee (\sim p \wedge q)$

13. $(p \vee q) \wedge (p \vee r)$

14. $p \wedge (q \vee r)$

15. $(p \wedge q) \vee \sim r$

16. $(\sim p \vee q) \wedge \sim r$

17. $(p \wedge \sim q) \vee (p \wedge r)$

18. $\sim(p \wedge q) \vee (q \wedge r)$

19. If a compound proposition consists of the prime propositions p, q, r, and s, how many rows does its corresponding truth table contain?

A.3   The Conditional and the Biconditional Connectives

In this section, we introduce two other connectives: the conditional and the biconditional. We also discuss three variations of conditional statements: the inverse, the contrapositive, and the converse.

We often use expressions of the form

If it rains, *then* the baseball game will be postponed

to specify the conditions under which a statement will be true. The "if . . . then" statement is the building block on which deductive reasoning is based, and it is important to understand its use in forming logical proofs.

> **Conditional Statement**
>
> A **conditional statement** is a proposition of the form "if p, then q" and is represented symbolically by
>
> $$p \rightarrow q$$
>
> The connective "if . . . then" is called the **conditional connective**; the proposition p, the **hypothesis**; and the proposition q, the **conclusion**. A conditional statement is false if the hypothesis is true and the conclusion is false; it is true in all other cases.

TABLE 7

p	q	$p \rightarrow q$
T	T	T
T	F	F
F	T	T
F	F	T

The truth table determined by the conditional $p \rightarrow q$ is shown in Table 7.

One question that is often asked is: Why is a conditional statement true when its hypothesis is false? We can answer this question by considering the following conditional statement made by a mother to her son:

If you do your homework, then you may watch TV.

Think of the statement consisting of the two prime propositions

p: You do your homework

q: You may watch TV

as a *promise* made by the mother to her son. Four cases arise:

1. The son does his homework, and his mother lets him watch TV.
2. The son does his homework, and his mother does not let him watch TV.
3. The son does not do his homework, and his mother lets him watch TV.
4. The son does not do his homework, and his mother does not let him watch TV.

In case 1, p and q are both true, the promise has been kept and, consequently, $p \rightarrow q$ is true. In case 2, p is true and q is false, and the mother has broken her promise. Therefore, $p \rightarrow q$ is false. In cases 3 and 4, p is not true, and the promise is *not* broken. Thus, $p \rightarrow q$ is regarded as a true statement. In other words, the conditional statement is regarded as false only if the "promise" is broken.

There are several equivalent expressions for the conditional connective "if . . . then." Among them are

1. p implies q
2. p only if q

3. q if p

4. q whenever p

5. Suppose p, then q

Care should be taken not to confuse the conditional "q if p" with the conditional $q \rightarrow p$, because the two statements have quite different meanings. For example, the conditional $p \rightarrow q$ formed from the two propositions

p: There is a fire

q: You call the fire department

is

> If there is a fire, then you call the fire department.

The conditional $q \rightarrow p$ is

> If you call the fire department, then there is a fire.

Obviously, the two statements have quite different meanings.

We refer to statements that are variations of the conditional $p \rightarrow q$ as **logical variants**. We define the three logical variants of the conditional $p \rightarrow q$ as follows:

1. The **converse** is a compound statement of the form "if q, then p" and is represented symbolically by

$$q \rightarrow p$$

2. The **contrapositive** is a compound statement of the form "if not q, then not p" and is represented symbolically by

$$\sim q \rightarrow \sim p$$

3. The **inverse** is a compound statement of the form "if not p, then not q" and is represented symbolically by

$$\sim p \rightarrow \sim q$$

EXAMPLE 1   Given the two propositions

p: You vote in the presidential election

q: You are a registered voter

a. State the conditional $p \rightarrow q$.
b. State the converse, the contrapositive, and the inverse of $p \rightarrow q$.

Solution

a. The conditional is "If you vote in the presidential election, then you are a registered voter."
b. The converse is "If you are a registered voter, then you vote in the presidential election." The contrapositive is "If you are not a registered voter, then you do not vote in the presidential election." The inverse is "If you do not vote in the presidential election, then you are not a registered voter." ◼

The truth table for the conditional $p \rightarrow q$ and its three logical variants is shown in Table 8. Notice that the conditional $p \rightarrow q$ and its contrapositive $\sim q \rightarrow \sim p$ have identical truth tables. In other words, the conditional statement and its contrapositive have the same meaning.

TABLE 8

p	q	Conditional $p \to q$	Converse $q \to p$	$\sim p$	$\sim q$	Contrapositive $\sim q \to \sim p$	Inverse $\sim p \to \sim q$
T	T	T	T	F	F	T	T
T	F	F	T	F	T	F	T
F	T	T	F	T	F	T	F
F	F	T	T	T	T	T	T

Logical Equivalence

Two propositions p and q are logically equivalent, denoted by

$$p \Leftrightarrow q$$

if they have identical truth tables.

Referring once again to the truth table in Table 8, we see that $p \to q$ is logically equivalent to its contrapositive. Similarly, the converse of a conditional statement is logically equivalent to the inverse.

It is sometimes easier to prove the contrapositive of a conditional statement than it is to prove the conditional itself. We may use this to our advantage in establishing a proof, as shown in the next example.

EXAMPLE 2   Prove that if n^2 is an odd number, then n is an odd number.

Solution   Let

p: n^2 is odd.

q: n is an odd number.

Since $p \to q$ is logically equivalent to $\sim q \to \sim p$, it suffices to prove $\sim q \to \sim p$. Thus, we wish to prove that if n is not an odd number, then n^2 is an even number. If n is even, then $n = 2k$, where k is an integer. Therefore,

$$n^2 = (2k)(2k) = 2(2k^2)$$

Since $2(2k^2)$ is a multiple of 2, it is an even number and consequently not odd. Thus, we have shown that the contrapositive $\sim q \to \sim p$ is true, and it follows that the conditional $p \to q$ is also true.   ■

We now turn our attention to the last of the five basic connectives.

Biconditional Propositions

Statements of the form "p if and only if q" are called biconditional propositions and are represented symbolically by

$$p \leftrightarrow q$$

The connective "if and only if" is called the *biconditional connective*. The biconditional $p \leftrightarrow q$ is true whenever p and q are *both true* or *both false*.

TABLE 9

p	q	$p \leftrightarrow q$
T	T	T
T	F	F
F	T	F
F	F	T

The truth table for $p \leftrightarrow q$ is shown in Table 9.

EXAMPLE 3 Let

p: Mark is going to the senior prom.

q: Linda is going to the senior prom.

State the biconditional $p \leftrightarrow q$.

Solution The required statement is

Mark is going to the senior prom,

if and only if,

Linda is going to the senior prom.

As suggested by its name, the biconditional statement is actually composed of two conditional statements. For instance, an *equivalent expression* for the biconditional statement in Example 3 is given by the two conditional statements

Mark is going to the senior prom
if Linda is going to the senior prom.

Linda is going to the senior prom
if Mark is going to the senior prom.

Thus, "*p* if *q* and *q* if *p*" is equivalent to "*p* if and only if *q*."

The equivalent expressions that are most commonly used in forming conditional and biconditional statements are summarized in Table 10. The words *necessary* and *sufficient* occur frequently in mathematical proofs. When we say that p is **sufficient** for q we mean that *when p is true, q is also true*—that is, $p \rightarrow q$. In like manner, when we say that p is **necessary** for q, we mean that *if p is not true, then q is not true*—that is, $\sim p \rightarrow \sim q$. But this last expression is logically equivalent to $q \rightarrow p$. Thus, when we say that p is necessary and sufficient for q, it follows that $p \rightarrow q$ and $q \rightarrow p$. Example 4 illustrates the use of these expressions.

TABLE 10

		Equivalent Form
Conditional statement	If p, then q	p is sufficient for q p only if q q is necessary for p q, if p
Biconditional statement	p if and only if q	If p then q; if q then p p is necessary and sufficient for q

EXAMPLE 4 Let

p: The stock market will go up.

q: Interest rates decrease.

Represent the following statements symbolically:

a. If interest rates decrease, then the stock market will go up.
b. If interest rates do not decrease, then the stock market will not go up.

c. The stock market will go up if and only if interest rates decrease.

d. Decreasing interest rates is a sufficient condition for the stock market to go up.

e. Decreasing interest rates is a necessary and sufficient condition for a rising stock market.

Solution

a. $q \to p$ b. $\sim q \to \sim p$ or $p \to q$ c. $p \leftrightarrow q$

d. $q \to p$ e. $q \leftrightarrow p$ ■

Just as there is an order of precedence when we use the arithmetical operations $\times$, $\div$, $+$, and $-$, we must also observe an order of precedence when we use the logical connectives. The following list dictates the order of precedence for logical connectives.

Order of Precedence of Logical Connectives

$$\sim, \quad \wedge, \quad \vee, \quad \to, \leftrightarrow$$

Thus, the connective $\sim$ should be applied first, followed by $\wedge$, and so on. For example, using the order of precedence, we see that $\sim p \vee q \to p \wedge r \vee s$ is $[(\sim p) \vee q] \to [(p \wedge r) \vee s]$.

Also, as in the case of arithmetical operations, parentheses take the highest priority in the order of precedence.

A.3   Exercises

In Exercises 1–4, write the converse, the contrapositive, and the inverse of the conditional statement.

1. $p \to \sim q$

2. $\sim p \to \sim q$

3. $q \to p$

4. $\sim p \to q$

In Exercises 5 and 6, refer to the following propositions p and q:

p: It is snowing

q: The temperature is below freezing.

5. Express the conditional and the biconditional of p and q in words.

6. Express the converse, contrapositive, and inverse of the conditional $p \to q$ in words.

In Exercises 7 and 8, refer to the following propositions p and q:

p: The company's union and management reach a settlement.

q: The workers will not strike.

7. Express the conditional and biconditional of p and q in words.

8. Express the converse, contrapositive, and inverse of the conditional $p \to q$ in words.

In Exercises 9–12, determine whether the statement is true or false.

9. A conditional proposition and its converse are logically equivalent.

10. The converse and the inverse of a conditional proposition are logically equivalent.

11. A conditional proposition and its inverse are logically equivalent.

12. The converse and the contrapositive of a conditional proposition are logically equivalent.

13. Consider the conditional statement "If the owner lowers the selling price of the house, then I will buy it." Under what conditions is the conditional statement false?

14. Consider the biconditional statement "I will buy the house if and only if the owner lowers the selling price." Under what conditions is the biconditional statement false?

In Exercises 15–28, construct a truth table for the compound proposition.

15. $\sim(p \to q)$

16. $\sim(q \to \sim p)$

17. $\sim(p \to q) \land p$

18. $(p \to q) \lor (q \to p)$

19. $(p \to \sim q) \veebar \sim p$

20. $(p \to q) \land (\sim p \lor q)$

21. $(p \to q) \leftrightarrow (\sim q \to \sim p)$

22. $(p \to q) \leftrightarrow (\sim p \lor q)$

23. $(p \land q) \to (p \lor q)$

24. $\sim q \to (\sim p \land \sim q)$

25. $(p \lor q) \to \sim r$

26. $(p \leftrightarrow q) \lor r$

27. $p \to (q \lor r)$

28. $[(p \to q) \lor (q \to r)] \to (p \to r)$

In Exercises 29–36, determine whether the compound proposition is logically equivalent.

29. $\sim p \lor q; p \to q$

30. $\sim(p \lor q); \sim p \land \sim q$

31. $q \to p; \sim p \to \sim q$

32. $\sim p \to q; \sim p \lor q$

33. $p \land q; p \to \sim q$

34. $\sim(p \land \sim q); \sim p \lor q$

35. $(p \to q) \to r; (p \lor q) \lor r$

36. $p \lor (q \land r); (p \lor q) \land (p \lor r)$

37. Let p and q denote the following propositions:

p: Taxes are increased.

q: The federal deficit increases.

Represent the following statements symbolically.
a. If taxes are increased, then the federal deficit will not increase.
b. If taxes are not increased, then the federal deficit will increase.
c. The federal deficit will not increase if and only if taxes are increased.
d. Increased taxation is a sufficient condition for halting the growth of the federal deficit.
e. Increased taxation is a necessary and sufficient condition for halting the growth of the federal deficit.

38. Let p and q denote the following propositions:

p: The unemployment rate decreases.

q: Consumer confidence will improve.

Represent the following statements symbolically.
a. If the unemployment rate does not decrease, consumer confidence will not improve.
b. Consumer confidence will improve if and only if the unemployment rate does not decrease.
c. A decreasing unemployment rate is a sufficient condition for consumer confidence to improve.
d. A decreasing unemployment rate is a necessary and sufficient condition for consumer confidence to improve.

A.4   Laws of Logic

Just as the laws of algebra enable us to perform operations with real numbers, the **laws of logic** provide us with a systematic method of connecting statements.

> **Laws of Logic**
> Let p, q, and r be any three propositions. Then
>
> | | |
> |---|---|
> | **1.** $p \land p \Leftrightarrow p$ | *Idempotent law for conjunction* |
> | **2.** $p \lor p \Leftrightarrow p$ | *Idempotent law for disjunction* |
> | **3.** $(p \land q) \land r \Leftrightarrow p \land (q \land r)$ | *Associative law for conjunction* |
> | **4.** $(p \lor q) \lor r \Leftrightarrow p \lor (q \lor r)$ | *Associative law for disjunction* |
> | **5.** $p \land q \Leftrightarrow q \land p$ | *Commutative law for conjunction* |
> | **6.** $p \lor q \Leftrightarrow q \lor p$ | *Commutative law for disjunction* |
> | **7.** $p \land (q \lor r) \Leftrightarrow (p \land q) \lor (p \land r)$ | *Distributive law for conjunction* |
> | **8.** $p \lor (q \land r) \Leftrightarrow (p \lor q) \land (p \lor r)$ | *Distributive law for disjunction* |
> | **9.** $\sim(p \lor q) \Leftrightarrow \sim p \land \sim q$ | *De Morgan's law* |
> | **10.** $\sim(p \land q) \Leftrightarrow \sim p \lor \sim q$ | *De Morgan's law* |

To verify any of these laws, we need only construct a truth table to show that the given statements are logically equivalent. We illustrate this procedure in the next example.

EXAMPLE 1 Prove the distributive law for conjunction.

Solution We wish to prove that $p \wedge (q \vee r)$ is logically equivalent to $(p \wedge q) \vee (p \wedge r)$. This is easily done by constructing the associated truth table, as shown in Table 11. Since the entries in the last two columns are the same, we conclude that $p \wedge (q \vee r) \Leftrightarrow (p \wedge q) \vee (p \wedge r)$.

TABLE 11

p	q	r	$q \vee r$	$p \wedge q$	$p \wedge r$	$p \wedge (q \vee r)$	$(p \wedge q) \vee (p \wedge r)$
T	T	T	T	T	T	T	T
T	T	F	T	T	F	T	T
T	F	T	T	F	T	T	T
T	F	F	F	F	F	F	F
F	T	T	T	F	F	F	F
F	T	F	T	F	F	F	F
F	F	T	T	F	F	F	F
F	F	F	F	F	F	F	F

The proofs of the other laws are left to you as an exercise.

De Morgan's laws, Laws 9 and 10, are useful in forming the negation of a statement. For example, if we wish to state the negation of the proposition

Steve plans to major in business administration or economics

we can represent the statement symbolically by $p \vee q$, where the prime propositions are

p: Steve plans to major in business administration.

q: Steve plans to major in economics.

Then, the negation of $p \vee q$ is $\sim(p \vee q)$. Using De Morgan's laws, we have

$$\sim(p \vee q) \Leftrightarrow \sim p \wedge \sim q$$

Thus, the required statement is

Steve does not plan to major in business administration, *and* he does not plan to major in economics.

Up to this point we have considered propositions that have both true and false entries in their truth tables. Some statements have the property that they are always true; other propositions have the property that they are always false.

Tautologies and Contradictions

A tautology is a statement that is always true.

A contradiction is a statement that is always false.

EXAMPLE 2   Show that

a. $p \lor \sim p$ is a tautology.   **b.** $p \land \sim p$ is a contradiction.

Solution

a. The truth table associated with $p \lor \sim p$ is shown in Table 12a. Since all the entries in the last column are Ts, the proposition is a tautology.

b. The truth table for $p \land \sim p$ is shown in Table 12b. Since all the entries in the last column are Fs, the proposition is a contradiction.

TABLE 12

p	$\sim p$	$p \lor \sim p$		p	$\sim p$	$p \lor \sim p$
T	F	T		T	F	F
F	T	T		F	T	F

(a)   (b)

In Section A.3, we defined logically equivalent statements as statements that have identical truth tables. The definition of equivalence may also be restated in terms of a tautology.

> **Logical Equivalence**
> Two propositions p and q are **logically equivalent** if the biconditional $p \leftrightarrow q$ is a tautology.

We can see why these two definitions are really the same by looking more closely at the truth table for $p \leftrightarrow q$ (Table 13a).

TABLE 13

	p	q	$p \leftrightarrow q$		p	q	$p \leftrightarrow q$
Case 1	T	T	T		T	T	T
Case 2	T	F	F		F	F	T
Case 3	F	T	F				
Case 4	F	F	T				

(a)   (b)

If the biconditional is always true, then the second and third cases shown in the truth table are excluded and we are left with the truth table in Table 13b. Notice that the entries in each row of the p and q columns are identical. In other words, p and q have identical truth values and hence are logically equivalent.

In addition to the ten laws stated earlier, we have the following four laws involving tautologies and contradictions.

> **Laws of Logic**
> Let t be a tautology and c a contradiction. Then
>
> **11.** $p \lor \sim p \Leftrightarrow t$ **12.** $p \land \sim p \Leftrightarrow c$
> **13.** $p \lor t \Leftrightarrow t$ **14.** $p \land t \Leftrightarrow p$

It now remains only to show how these laws are used to simplify proofs.

EXAMPLE 3   Using the laws of logic, show that

$$p \vee (\sim p \wedge q) \Leftrightarrow (p \vee q)$$

Solution

$$
\begin{aligned}
p \vee (\sim p \wedge q) &\Leftrightarrow (p \vee \sim p) \wedge (p \vee q) && \text{By Law 8} \\
&\Leftrightarrow t \wedge (p \vee q) && \text{By Law 11} \\
&\Leftrightarrow p \vee q && \text{By Law 14}
\end{aligned}
$$

EXAMPLE 4   Using the laws of logic, show that

$$\sim(p \vee q) \vee (\sim p \wedge q) \Leftrightarrow \sim p$$

Solution

$$
\begin{aligned}
\sim(p \vee q) &\vee (\sim p \wedge q) \\
&\Leftrightarrow (\sim p \wedge \sim q) \vee (\sim p \wedge q) && \text{By Law 9} \\
&\Leftrightarrow \sim p \wedge (\sim q \vee q) && \text{By Law 7} \\
&\Leftrightarrow \sim p \wedge t && \text{By Law 11} \\
&\Leftrightarrow \sim p && \text{By Law 14}
\end{aligned}
$$

A.4   Exercises

1. Prove the idempotent law for conjunction,
   $p \wedge p \Leftrightarrow p$.

2. Prove the idempotent law for disjunction,
   $p \vee p \Leftrightarrow p$.

3. Prove the associative law for conjunction,
   $(p \wedge q) \wedge r \Leftrightarrow p \wedge (q \wedge r)$.

4. Prove the associative law for disjunction,
   $(p \vee q) \vee r \Leftrightarrow p \vee (q \vee r)$.

5. Prove the commutative law for conjunction,
   $p \wedge q \Leftrightarrow q \wedge p$.

6. Prove the commutative law for disjunction,
   $p \vee q \Leftrightarrow q \vee p$.

7. Prove the distributive law for disjunction,
   $p \vee (q \wedge r) \Leftrightarrow (p \vee q) \wedge (p \vee r)$.

8. Prove De Morgan's laws
   a. $\sim(p \vee q) \Leftrightarrow \sim p \wedge \sim q$ **b.** $\sim(p \wedge q) \Leftrightarrow \sim p \vee \sim q$

In Exercises 9–18, determine whether the statement is a tautology, a contradiction, or neither.

9. $(p \rightarrow q) \leftrightarrow (\sim p \vee q)$ 10. $(p \veebar q) \wedge (p \leftrightarrow q)$

11. $p \rightarrow (p \vee q)$ 12. $(p \rightarrow q) \vee (q \rightarrow p)$

13. $(p \rightarrow q) \leftrightarrow (\sim q \rightarrow \sim p)$

14. $p \wedge (p \rightarrow q) \rightarrow q$

15. $(p \rightarrow q) \wedge (\sim q) \rightarrow (\sim p)$

16. $[(p \rightarrow q) \wedge (q \rightarrow r)] \rightarrow (p \rightarrow r)$

17. $[(p \rightarrow q) \vee (q \rightarrow r)] \rightarrow (p \rightarrow r)$

18. $[p \wedge (q \vee r)] \leftrightarrow [(p \wedge q) \vee (p \wedge r)]$

19. Let p and q denote the statements

    p: The candidate opposes changes in the Social Security system.

    q: The candidate supports the ERA.

    Use De Morgan's laws to state the negation of $p \wedge q$ and the negation of $p \vee q$.

20. Let p and q denote the statements

    p: The recycling bill was passed by the voters.

    q: The tax on oil and hazardous materials was not approved by the voters.

    Use De Morgan's laws to state the negation of $p \wedge q$ and the negation of $p \vee q$.

In Exercises 21–26, use the laws of logic to prove the propositions.

21. $[p \wedge (q \vee \sim q) \vee (p \wedge q)] \Leftrightarrow p \vee (p \wedge q)$

22. $p \vee (\sim p \wedge \sim q) \Leftrightarrow p \vee \sim q$

23. $(p \wedge \sim q) \vee (p \wedge \sim r) \Leftrightarrow p \wedge (\sim q \vee \sim r)$

24. $(p \vee q) \vee \sim q \Leftrightarrow t$

25. $p \wedge [\sim(q \wedge r)] \Leftrightarrow (p \wedge \sim q) \vee (p \wedge \sim r)$

26. $p \vee (q \vee r) \Leftrightarrow r \vee (q \vee p)$

A.5 Arguments

In this section, we discuss arguments and the methods used to determine the validity of arguments.

> **Argument**
> An **argument** or **proof** consists of a set of propositions $p_1, p_2, \ldots, p_n$, called the *premises*, and a proposition q, called the *conclusion*. An argument is *valid* if and only if the conclusion is true whenever the premises are all true. An argument that is *not* valid is called a *fallacy*, or an *invalid* argument.

The next example illustrates the form in which an argument is presented. Notice that the premises are written separately above the horizontal line and the conclusion is written below the line.

EXAMPLE 1 An argument is presented in the following way:

Premises: If Pam studies diligently, she passes her exams.

Pam studies diligently.

Conclusion: Pam passes her exams. ■

To determine the validity of an argument, we first write the argument in symbolic form and then construct the associated truth table containing the prime propositions, the premises, and the conclusion. We then *check the rows in which the premises are all true. If the conclusion in each of these rows is also true, then the argument is valid. Otherwise, it is a fallacy.*

EXAMPLE 2 Determine the validity of the argument in Example 1.

Solution The symbolic form of the argument is

$$p \to q$$
$$\underline{p}$$
$$\therefore q$$

Observe that the conclusion is preceded by the symbol $\therefore$, which is used to represent the word *therefore*. The truth table associated with this argument is shown in Table 14. Observe that only row 1 contains true values for both premises. Since the conclusion is also true in this row, we conclude that the argument is valid.

TABLE 14

Propositions		Premises		Conclusion
p	q	$p \to q$	p	q
T	T	T	T	T
T	F	F	T	F
F	T	T	F	T
F	F	T	F	F

EXAMPLE 3   Determine the validity of the argument

> If Michael is overtired, then he is grumpy.
>
> Michael is grumpy.
> _____
>
> Therefore, Michael is overtired.

Solution   The argument is written symbolically as follows:

$$p \to q$$
$$\underline{q \qquad}$$
$$\therefore p$$

The associated truth table is shown in Table 15. Observe that the entries for the premises in the third row are both true, but the corresponding entry for the conclusion is false. We conclude that the argument is a fallacy.

TABLE 15

p	q	$p \to q$	q	p
T	T	T	T	T
T	F	F	F	T
F	T	T	T	F
F	F	T	F	F

When we consider the validity of an argument, we are concerned only with the *form* of the argument and not the truth or falsity of the premises. In other words, the conclusion of an argument may follow validly from the premises, but the premises themselves may be false. The next example demonstrates this point.

EXAMPLE 4   Determine the validity of the argument

> Reggie is a wealthy man.
>
> Wealthy men are happy.
> _____
>
> Therefore, Reggie is happy.

Solution   The symbolic form of the argument is

$$p \to q$$
$$\underline{q \to r}$$
$$\therefore p \to r$$

The associated truth table is shown in Table 16. Since the conclusion is true in each of the rows that contain true values for both premises, we conclude that the argument is valid. Note that the validity of the argument is not affected by the truth or falsity of the premise "Wealthy men are happy."

TABLE 16

p	q	r	p → q	q → r	p → r	
T	T	T	T	T	T	(✓)
T	T	F	T	F	F	
T	F	T	F	T	T	
T	F	F	F	T	F	
F	T	T	T	T	T	(✓)
F	T	F	T	F	T	
F	F	T	T	T	T	(✓)
F	F	F	T	T	T	(✓)

A question that may already have arisen in your mind is: How does one determine the validity of an argument if the associated truth table does not contain true values for all the premises? The next example provides the answer to this question.

EXAMPLE 5   Determine the validity of the argument

> The door is locked.
>
> The door is unlocked.
> _____
>
> Therefore, the door is locked.

Solution   The symbolic form of the argument is

$$p$$
$$\sim p$$
$$\therefore p$$

TABLE 17

p	~p	p
T	F	T
F	T	F

The associated truth table is shown in Table 17. Observe that no rows contain true values for both premises. Nevertheless, the argument is considered to be valid since the condition that the conclusion is true whenever the premises are all true is not violated. Again, we remind you that the validity of an argument is not determined by the truth or falsity of its premises or conclusion, but rather it is determined only by the form of the argument.

The next proposition provides us with an alternative method for determining the validity of an argument.

Propositions
Suppose an argument consists of the premises $p_1, p_2, \ldots, p_n$ and conclusion q. Then, the argument is valid if and only if the proposition $[p_1 \wedge p_2 \wedge \ldots \wedge p_n] \rightarrow q$ is a tautology.

To prove this proposition, we must show that (a) if an argument is valid, then the given proposition is a tautology, and (b) if the given proposition is a tautology, then the argument is valid.

Proof of (a) Since the argument is valid, it follows that q is true whenever all of the premises $p_1, p_2, \ldots, p_n$ are true. But the conjunction $[p_1 \wedge p_2 \wedge \ldots \wedge p_n]$ is true only when all the premises are true (by the definition of conjunction). Hence, q is true whenever the conjunction is true, and we conclude that the conditional $[p_1 \wedge p_2 \wedge \ldots \wedge p_n] \to q$ is true. (Recall that a conditional statement is always true when its hypotheses are false. Hence, to show that a conditional is a tautology, we need only prove that its conclusion is always true when its hypotheses are true.)

Proof of (b) Since the given proposition is a tautology, q is always true. Therefore, q is true when the conjunction

$$(p_1 \wedge p_2 \wedge \ldots \wedge p_n)$$

is true. But the conjunction is true only when all of the premises $p_1, p_2, \ldots, p_n$ are true. Hence, q is true whenever the premises are all true, and the argument is valid.

Having proved this proposition, we can now use it to determine the validity of an argument. If we are given an argument, we construct a truth table for $(p_1 \wedge p_2 \wedge \ldots \wedge p_n) \to q$. If the truth table contains all the true values in its last column, then the argument is valid; otherwise it is invalid. This method of proof is illustrated in the next example.

EXAMPLE 6 Determine the validity of the argument

You return the book on time, or you will have to pay a fine.

You do not return the book on time.

Therefore, you will have to pay a fine.

Solution The symbolic form of the argument is

$$p \vee q$$
$$\underline{\sim p}$$
$$\therefore q$$

Following the method described, we construct the truth table for

$$[(p \vee q) \wedge \sim p] \to q$$

as shown in Table 18. Since the entries in the last column are all T's, the proposition is a tautology, and we conclude that the argument is valid.

TABLE 18

p	q	$p \vee q$	$\sim p$	$(p \vee q) \wedge \sim p$	$[(p \vee q) \wedge \sim p] \to q$
T	T	T	F	F	T
T	F	T	F	F	T
F	T	T	T	T	T
F	F	F	T	F	T

By familiarizing ourselves with a few of the most commonly used argument forms, we can simplify the problem of determining the validity of an argument. Some of the argument forms most commonly used are the following:

1. Modus ponens (a manner of affirming), or rule of detachment

$$p \to q$$
$$\underline{p \qquad\quad}$$
$$\therefore q$$

2. Modus tollens (a manner of denying)

$$p \to q$$
$$\underline{\sim q \qquad\quad}$$
$$\therefore \sim p$$

3. Law of syllogisms

$$p \to q$$
$$\underline{q \to r \quad}$$
$$\therefore p \to r$$

The truth tables verifying modus ponens and the law of syllogisms have already been constructed (see Tables 14 and 16, respectively). The verification of modus tollens is left to you as an exercise. The next two examples illlustrate the use of these argument forms.

EXAMPLE 7 Determine the validity of the argument

If the battery is dead, the car will not start.
The car starts.
_____
Therefore, the battery is not dead.

Solution As before, we express the argument in symbolic form. Thus,

$$p \to q$$
$$\underline{\sim q \qquad\quad}$$
$$\therefore \sim p$$

Identifying the form of this argument as modus tollens, we conclude that the given argument is valid. ▪

EXAMPLE 8 Determine the validity of the argument

If mortgage rates are lowered, housing sales increase.
If housing sales increase, then the prices of houses increase.
_____
Therefore, if mortgage rates are lowered, the prices of houses increase.

Solution The symbolic form of the argument is

$$p \to q$$
$$\underline{q \to r \quad}$$
$$\therefore p \to r$$

Using the law of syllogisms, we conclude that the argument is valid. ▪

A.5 Exercises

In Exercises 1–16, determine whether the argument is valid.

1. $p \rightarrow q$
$\dfrac{q \rightarrow r}{\therefore p \rightarrow r}$

2. $p \vee q$
$\dfrac{\sim p}{\therefore q}$

3. $p \wedge q$
$\dfrac{\sim p}{\therefore q}$

4. $p \rightarrow q$
$\dfrac{\sim q}{\therefore \sim p}$

5. $p \rightarrow q$
$\dfrac{\sim p}{\therefore \sim q}$

6. $p \rightarrow q$
$\dfrac{q \wedge r}{\therefore p \vee r}$

7. $p \leftrightarrow q$
$\dfrac{q}{\therefore p}$

8. $p \wedge q$
$\dfrac{\sim p \rightarrow \sim q}{\therefore p \wedge \sim q}$

9. $p \rightarrow q$
$\dfrac{q \rightarrow p}{\therefore p \leftrightarrow q}$

10. $p \rightarrow \sim q$
$\dfrac{p \veebar q}{\therefore \sim q}$

11. $p \leftrightarrow q$
$\dfrac{q \leftrightarrow r}{\therefore p \leftrightarrow r}$

12. $p \rightarrow q$
$\dfrac{q \leftrightarrow r}{\therefore p \wedge r}$

13. $p \veebar r$
$\dfrac{q \wedge r}{\therefore p \rightarrow r}$

14. $p \leftrightarrow q$
$q \vee r$
$\dfrac{\sim r}{\therefore p \rightarrow \sim r}$

15. $p \leftrightarrow q$
$q \vee r$
$\dfrac{\sim p}{\therefore \sim p \rightarrow \sim r}$

16. $p \rightarrow q$
$r \rightarrow q$
$\dfrac{p \wedge q}{\therefore p \vee r}$

In Exercises 17–22, represent the argument symbolically and determine whether it is a valid argument.

17. If Carla studies, then she passes her exams.

Carla did not study.

Therefore, Carla did not pass her exams.

18. If Tony is wealthy, he is either intelligent or a good businessman.

Tony is intelligent and he is not a good businessman.

Therefore, Tony is not wealthy.

19. Steve will attend the matinee and/or the evening show.

If Steve doesn't go to the matinee show, then he will not go to the evening show.

Therefore, Steve will attend the matinee show.

20. If Mary wins the race, then Stacy loses the race.

Neither Mary nor Linda won the race.

Therefore, Stacy won the race.

21. If mortgage rates go up, then housing prices will go up.

If housing prices go up, more people will rent houses.

More people are renting houses.

Therefore, mortgage rates went up.

22. If taxes are cut, then retail sales increase.

If retail sales increase, then the unemployment rate will decrease.

If the unemployment rate decreases, then the incumbent will win the election.

Therefore, if taxes are not cut, the incumbent will not win the election.

23. Given the statement "All good cooks prepare gourmet food," which of the following conclusions follow logically?
   a. If George prepares gourmet food, then he is a good cook.
   b. If Brenda does not prepare gourmet food, then she is not a good cook.
   c. Everyone who prepares gourmet food is a good cook.
   d. Some people who prepare gourmet food are not good cooks.

24. What conclusion can be drawn from the following statements?

His date is pretty, or she is tall and skinny.

If his date is tall, then she is a brunette.

His date is not a brunette.

25. Show that modus tollens is a valid form of argument.

A.6   Applications of Logic to Switching Networks

In this section, we see how the principles of logic can be used in the design and analysis of switching networks. A **switching network** is an arrangement of wires and switches connecting two terminals. These networks are used extensively in digital computers.

A switch may be open or closed. If a switch is closed, current will flow through the wire. If it is open, no current will flow through the wire. Because a switch has exactly two states, it can be represented by a proposition p that is true if the switch is closed and false if the switch is open.

Now let's consider a circuit with two switches p and q. If the circuit is connected as shown in Figure 1, the switches p and q are said to be **in series**.

FIGURE 1
Two switches connected in series

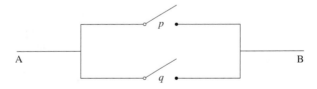

For such a network, current will flow from A to B if and only if both p and q are closed, but no current will flow if one or more of the switches is open. Thinking of p and q as propositions, we have the truth table shown in Table 19. (Recall that T corresponds to the situation in which the switch is closed.) From the truth table we see that two switches p and q connected in series are analogous to the conjunction $p \wedge q$ of the two propositions p and q.

If a circuit is connected as shown in Figure 2, the switches p and q are said to be connected **in parallel.**

TABLE 19		
p	**q**	
T	T	T
T	F	F
F	T	F
F	F	F

FIGURE 2
Two switches connected in parallel

For such a network, current will flow from A to B if and only if one or more of the switches p or q is closed. Once again, thinking of p and q as propositions, we have the truth table shown in Table 20. From the truth table, we conclude that two switches p and q connected in parallel are analogous to the inclusive disjunction $p \vee q$ of the two propositions p and q.

TABLE 20		
p	**q**	
T	T	T
T	F	T
F	T	T
F	F	F

EXAMPLE 1   Find a logic statement that represents the network shown in Figure 3. By constructing the truth table for this logic statement, determine the conditions under which current will flow from A to B in the network.

FIGURE 3
We want to determine when the current will flow from A to B.

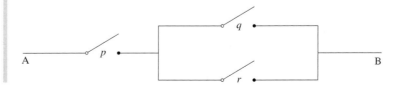

Solution The required logic statement is $p \wedge (q \vee r)$. Next, we construct the truth table for the logic statement $p \wedge (q \vee r)$ (Table 21).

TABLE 21

p	q	r	$q \vee r$	$p \wedge (q \vee r)$
T	T	T	T	T
T	T	F	T	T
T	F	T	T	T
T	F	F	F	F
F	T	T	T	F
F	T	F	T	F
F	F	T	T	F
F	F	F	F	F

From the truth table, we conclude that current will flow from A to B if and only if one of the following conditions is satisfied:

1. p, q, and r are all closed.
2. p and q are closed, but r is open.
3. p and r are closed, but q is open.

In other words, current will flow from A to B if and only if p is closed and either q or r or both q and r are closed.

Before looking at some additional examples, let's remark that $\sim p$ represents a switch that is open when p is closed and vice versa. Furthermore, a circuit that is always closed is represented by a tautology, $p \vee \sim p$, whereas a circuit that is always open is represented by a contradiction, $p \wedge \sim p$. (Why?)

EXAMPLE 2 Given the logic statement $(p \vee q) \wedge (r \vee s \vee \sim p)$, draw the corresponding network.

Solution Recalling that the disjunction $p \vee q$ of the propositions p and q represents two switches p and q connected in parallel and the conjunction $p \wedge q$ represents two switches connected in series, we obtain the following network (Figure 4).

FIGURE 4
The network that corresponds to the logic statement
$(p \vee q) \wedge (r \vee s \vee \sim p)$

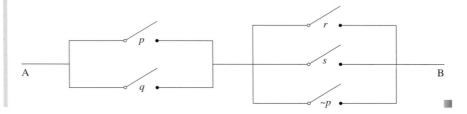

When used in conjunction with the laws of logic, the theory of networks developed so far is a useful tool in network analysis. In particular, network analysis enables us to find equivalent, and often simpler, networks, as the following example shows.

EXAMPLE 3   Find a logic statement representing the network shown in Figure 5. Also, find a simpler but equivalent network.

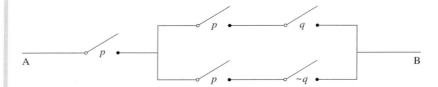

Solution   The logic statement corresponding to the given network is $p \wedge [(p \wedge q) \vee (p \wedge \sim q)]$. Next, using the rules of logic to simplify this statement, we obtain

$$p \wedge [(p \wedge q) \vee (p \wedge \sim q)] \Leftrightarrow p \wedge [p \wedge (q \vee \sim q)] \quad \text{Distributive law}$$
$$\Leftrightarrow p \wedge p \qquad\qquad \text{Tautology}$$
$$\Leftrightarrow p$$

Thus, the network is equivalent to the one shown in Figure 6.   ■

FIGURE 6
The network is equivalent to the one
shown in Figure 5.

A.6   Exercises

In Exercises 1–5, find a logic statement corresponding to the network. Determine the conditions under which current will flow from A to B.

1.

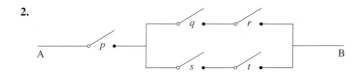

2.

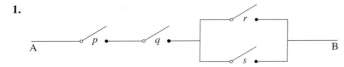

3.

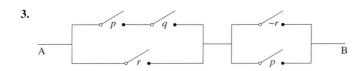

4.

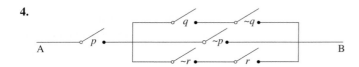

5.

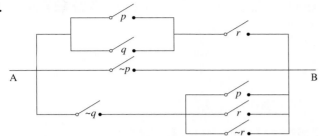

In Exercises 6–11, draw the network corresponding to the logic statement.

6. $p \vee (q \wedge r)$ **7.** $(p \wedge q) \wedge r$

8. $[p \vee (q \wedge r)] \wedge \sim q$ **9.** $(p \vee q) \vee [r \wedge (\sim r \vee \sim p)]$

10. $(\sim p \vee q) \wedge (p \vee \sim q)$

11. $(p \wedge q) \vee [(r \vee \sim q) \wedge (s \vee \sim p)]$

In Exercises 12–15, find a logic statement corresponding to the network. Then find a simpler but equivalent network.

12.

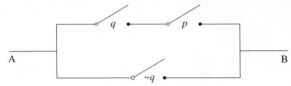

13.

14.

15.

16. A hallway light is to be operated by two switches, one located at the bottom of the staircase and the other located at the top of the staircase. Design a suitable network.
Hint: Let p and q be the switches. Construct a truth table for the associated propositions.

B   The System of Real Numbers

In this appendix, we briefly review the system of real numbers. This system consists of a set of objects called real numbers together with two operations, addition and multiplication, that enable us to combine two or more real numbers to obtain other real numbers. These operations are subjected to certain rules that we will state after first recalling the set of real numbers.

The set of real numbers may be constructed from the set of natural (also called counting) numbers

$$N = \{1, 2, 3, \ldots\}$$

by adjoining other objects (numbers) to it. Thus, the set

$$W = \{0, 1, 2, 3, \ldots\}$$

obtained by adjoining the single number 0 to N is called the set of whole numbers. By adjoining *negatives* of the numbers 1, 2, 3, ... to the set W of whole numbers, we obtain the set of integers

$$I = \{\ldots, -3, -2, -1, 0, 1, 2, 3, \ldots\}$$

Next, consider the set

$$Q = \left\{\frac{a}{b} \,\middle|\, a \text{ and } b \text{ are integers with } b \neq 0\right\}$$

Now, the set I of integers is contained in the set Q of rational numbers. To see this, observe that each integer may be written in the form a/b with $b = 1$, thus qualifying as a member of the set Q. The converse, however, is false, for the rational numbers (fractions) such as

$$\frac{1}{2}, \quad \frac{23}{25}, \quad \text{and so on}$$

are clearly not integers.

The sets N, W, I, and Q constructed thus far have

$$N \subset W \subset I \subset Q$$

That is, N is a proper subset of W, W is a proper subset of I, and so on.

Finally, consider the set Ir of all numbers that cannot be expressed in the form a/b, where a, b are integers $(b \neq 0)$. The members of this set, called the set of irrational numbers, include $\sqrt{2}$, 3, π, and so on. The set

$$R = Q \cup Ir$$

which is the set of all rational and irrational numbers, is called the set of real numbers (Figure 1).

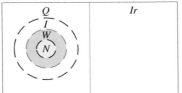

Q = Rationals	
I = Integers	
W = Whole numbers	
N = Natural numbers	
Ir = Irrationals	

FIGURE 1
The set of all real numbers consists of the set of rational numbers plus the set of irrational numbers.

Note the following important representation of real numbers: Every real number has a decimal representation; a rational number has a representation in terms of a repeated decimal. For example,

$$\frac{1}{7} = 0.142857142857142857\ldots$$

Note that the block of integers 142857 repeats.

On the other hand, the irrational number $\sqrt{2}$ has a representation in terms of a non-repeating decimal. Thus,

$$\sqrt{2} = 1.41421\ldots$$

As mentioned earlier, any two real numbers may be combined to obtain another real number. The operation of *addition*, written $+$, enables us to combine any two numbers a and b to obtain their sum, denoted by $a + b$. Another operation, called *multiplication*, and written $\cdot$, enables us to combine any two real numbers a and b to form their product, the number $a \cdot b$, or, written more simply, ab. These two operations are subjected to the following rules of operation: Given any three real numbers a, b, and c, we have

I. Under addition
   1. $a + b = b + a$ Commutative law of addition
   2. $a + (b + c) = (a + b) + c$ Associative law of addition
   3. $a + 0 = a$ Identity law of addition
   4. $a + (-a) = 0$ Inverse law of addition

II. Under multiplication
   1. $ab = ba$ Commutative law of multiplication
   2. $a(bc) = (ab)c$ Associative law of multiplication
   3. $a \cdot 1 = a$ Identity law of multiplication
   4. $a(1/a) = 1 \quad (a \neq 0)$ Inverse law of multiplication

III. Under addition and multiplication
   1. $a(b + c) = ab + ac$ Distributive law for multiplication with respect to addition

C   Tables

TABLE 1

Binomial Probabilities

n	x	p 0.05	0.1	0.2	0.3	0.4	0.5	0.6	0.7	0.8	0.9	0.95
2	0	0.902	0.810	0.640	0.490	0.360	0.250	0.160	0.090	0.040	0.010	0.002
	1	0.095	0.180	0.320	0.420	0.480	0.500	0.480	0.420	0.320	0.180	0.095
	2	0.002	0.010	0.040	0.090	0.160	0.250	0.360	0.490	0.640	0.810	0.902
3	0	0.857	0.729	0.512	0.343	0.216	0.125	0.064	0.027	0.008	0.001	
	1	0.135	0.243	0.384	0.441	0.432	0.375	0.288	0.189	0.096	0.027	0.007
	2	0.007	0.027	0.096	0.189	0.288	0.375	0.432	0.441	0.384	0.243	0.135
	3		0.001	0.008	0.027	0.064	0.125	0.216	0.343	0.512	0.729	0.857
4	0	0.815	0.656	0.410	0.240	0.130	0.062	0.026	0.008	0.002		
	1	0.171	0.292	0.410	0.412	0.346	0.250	0.154	0.076	0.026	0.004	
	2	0.014	0.049	0.154	0.265	0.346	0.375	0.346	0.265	0.154	0.049	0.014
	3		0.004	0.026	0.076	0.154	0.250	0.346	0.412	0.410	0.292	0.171
	4			0.002	0.008	0.026	0.062	0.130	0.240	0.410	0.656	0.815
5	0	0.774	0.590	0.328	0.168	0.078	0.031	0.010	0.002			
	1	0.204	0.328	0.410	0.360	0.259	0.156	0.077	0.028	0.006		
	2	0.021	0.073	0.205	0.309	0.346	0.312	0.230	0.132	0.051	0.008	0.001
	3	0.001	0.008	0.051	0.132	0.230	0.312	0.346	0.309	0.205	0.073	0.021
	4			0.006	0.028	0.077	0.156	0.259	0.360	0.410	0.328	0.204
	5				0.002	0.010	0.031	0.078	0.168	0.328	0.590	0.774
6	0	0.735	0.531	0.262	0.118	0.047	0.016	0.004	0.001			
	1	0.232	0.354	0.393	0.303	0.187	0.094	0.037	0.010	0.002		
	2	0.031	0.098	0.246	0.324	0.311	0.234	0.138	0.060	0.015	0.001	
	3	0.002	0.015	0.082	0.185	0.276	0.312	0.276	0.185	0.082	0.015	0.002
	4		0.001	0.015	0.060	0.138	0.234	0.311	0.324	0.246	0.098	0.031
	5			0.002	0.010	0.037	0.094	0.187	0.303	0.393	0.354	0.232
	6				0.001	0.004	0.016	0.047	0.118	0.262	0.531	0.735
7	0	0.698	0.478	0.210	0.082	0.028	0.008	0.002				
	1	0.257	0.372	0.367	0.247	0.131	0.055	0.017	0.004			
	2	0.041	0.124	0.275	0.318	0.261	0.164	0.077	0.025	0.004		
	3	0.004	0.023	0.115	0.227	0.290	0.273	0.194	0.097	0.029	0.003	
	4		0.003	0.029	0.097	0.194	0.273	0.290	0.227	0.115	0.023	0.004
	5			0.004	0.025	0.077	0.164	0.261	0.318	0.275	0.124	0.041
	6				0.004	0.017	0.055	0.131	0.247	0.367	0.372	0.257
	7					0.002	0.008	0.028	0.082	0.210	0.478	0.698

TABLE 1 (continued)

n	x	p 0.05	0.1	0.2	0.3	0.4	0.5	0.6	0.7	0.8	0.9	0.95
8	0	0.663	0.430	0.168	0.058	0.017	0.004	0.001				
	1	0.279	0.383	0.336	0.198	0.090	0.031	0.008	0.001			
	2	0.051	0.149	0.294	0.296	0.209	0.109	0.041	0.010	0.001		
	3	0.005	0.033	0.147	0.254	0.279	0.219	0.124	0.047	0.009		
	4		0.005	0.046	0.136	0.232	0.273	0.232	0.136	0.046	0.005	
	5			0.009	0.047	0.124	0.219	0.279	0.254	0.147	0.033	0.005
	6			0.001	0.010	0.041	0.109	0.209	0.296	0.294	0.149	0.051
	7				0.001	0.008	0.031	0.090	0.198	0.336	0.383	0.279
	8					0.001	0.004	0.017	0.058	0.168	0.430	0.663
9	0	0.630	0.387	0.134	0.040	0.010	0.002					
	1	0.299	0.387	0.302	0.156	0.060	0.018	0.004				
	2	0.063	0.172	0.302	0.267	0.161	0.070	0.021	0.004			
	3	0.008	0.045	0.176	0.267	0.251	0.164	0.074	0.021	0.003		
	4	0.001	0.007	0.066	0.172	0.251	0.246	0.167	0.074	0.017	0.001	
	5		0.001	0.017	0.074	0.167	0.246	0.251	0.172	0.066	0.007	0.001
	6			0.003	0.021	0.074	0.164	0.251	0.267	0.176	0.045	0.008
	7				0.004	0.021	0.070	0.161	0.267	0.302	0.172	0.063
	8					0.004	0.018	0.060	0.156	0.302	0.387	0.299
	9						0.002	0.010	0.040	0.134	0.387	0.630
10	0	0.599	0.349	0.107	0.028	0.006	0.001					
	1	0.315	0.387	0.268	0.121	0.040	0.010	0.002				
	2	0.075	0.194	0.302	0.233	0.121	0.044	0.011	0.001			
	3	0.010	0.057	0.201	0.267	0.215	0.117	0.042	0.009	0.001		
	4	0.001	0.011	0.088	0.200	0.251	0.205	0.111	0.037	0.006		
	5		0.001	0.026	0.103	0.201	0.246	0.201	0.103	0.026	0.001	
	6			0.006	0.037	0.111	0.205	0.251	0.200	0.088	0.011	0.001
	7			0.001	0.009	0.042	0.117	0.215	0.267	0.201	0.057	0.010
	8				0.001	0.011	0.044	0.121	0.233	0.302	0.194	0.075
	9					0.002	0.010	0.040	0.121	0.268	0.387	0.315
	10						0.001	0.006	0.028	0.107	0.349	0.599
11	0	0.569	0.314	0.086	0.020	0.004						
	1	0.329	0.384	0.236	0.093	0.027	0.005	0.001				
	2	0.087	0.213	0.295	0.200	0.089	0.027	0.005	0.001			
	3	0.014	0.071	0.221	0.257	0.177	0.081	0.023	0.004			
	4	0.001	0.016	0.111	0.220	0.236	0.161	0.070	0.017	0.002		
	5		0.002	0.039	0.132	0.221	0.226	0.147	0.057	0.010		
	6			0.010	0.057	0.147	0.226	0.221	0.132	0.039	0.002	
	7			0.002	0.017	0.070	0.161	0.236	0.220	0.111	0.016	0.001
	8				0.004	0.023	0.081	0.177	0.257	0.221	0.071	0.014
	9				0.001	0.005	0.027	0.089	0.200	0.295	0.213	0.087
	10					0.001	0.005	0.027	0.093	0.236	0.384	0.329
	11						0.004	0.020	0.086	0.314	0.569	

TABLE 1 (*continued*)

n	x	0.05	0.1	0.2	0.3	0.4	0.5	0.6	0.7	0.8	0.9	0.95
12	0	0.540	0.282	0.069	0.014	0.002						
	1	0.341	0.377	0.206	0.071	0.017	0.003					
	2	0.099	0.230	0.283	0.168	0.064	0.016	0.002				
	3	0.017	0.085	0.236	0.240	0.142	0.054	0.012	0.001			
	4	0.002	0.021	0.133	0.231	0.213	0.121	0.042	0.008	0.001		
	5		0.004	0.053	0.158	0.227	0.193	0.101	0.029	0.003		
	6			0.016	0.079	0.177	0.226	0.177	0.079	0.016		
	7			0.003	0.029	0.101	0.193	0.227	0.158	0.053	0.004	
	8			0.001	0.008	0.042	0.121	0.213	0.231	0.133	0.021	0.002
	9				0.001	0.012	0.054	0.142	0.240	0.236	0.085	0.017
	10					0.002	0.016	0.064	0.168	0.283	0.230	0.099
	11						0.003	0.017	0.071	0.206	0.377	0.341
	12							0.002	0.014	0.069	0.282	0.540
13	0	0.513	0.254	0.055	0.010	0.001						
	1	0.351	0.367	0.179	0.054	0.011	0.002					
	2	0.111	0.245	0.268	0.139	0.045	0.010	0.001				
	3	0.021	0.100	0.246	0.218	0.111	0.035	0.006	0.001			
	4	0.003	0.028	0.154	0.234	0.184	0.087	0.024	0.003			
	5		0.006	0.069	0.180	0.221	0.157	0.066	0.014	0.001		
	6		0.001	0.023	0.103	0.197	0.209	0.131	0.044	0.006		
	7			0.006	0.044	0.131	0.209	0.197	0.103	0.023	0.001	
	8			0.001	0.014	0.066	0.157	0.221	0.180	0.069	0.006	
	9				0.003	0.024	0.087	0.184	0.234	0.154	0.028	0.003
	10				0.001	0.006	0.035	0.111	0.218	0.246	0.100	0.021
	11					0.001	0.010	0.045	0.139	0.268	0.245	0.111
	12						0.002	0.011	0.054	0.179	0.367	0.351
	13							0.001	0.010	0.055	0.254	0.513
14	0	0.488	0.229	0.044	0.007	0.001						
	1	0.359	0.356	0.154	0.041	0.007	0.001					
	2	0.123	0.257	0.250	0.113	0.032	0.006	0.001				
	3	0.026	0.114	0.250	0.194	0.085	0.022	0.003				
	4	0.004	0.035	0.172	0.229	0.155	0.061	0.014	0.001			
	5		0.008	0.086	0.196	0.207	0.122	0.041	0.007			
	6		0.001	0.032	0.126	0.207	0.183	0.092	0.023	0.002		
	7			0.009	0.062	0.157	0.209	0.157	0.062	0.009		
	8			0.002	0.023	0.092	0.183	0.207	0.126	0.032	0.001	
	9				0.007	0.041	0.122	0.207	0.196	0.086	0.008	
	10				0.001	0.014	0.061	0.155	0.229	0.172	0.035	0.004
	11					0.003	0.022	0.085	0.194	0.250	0.114	0.026
	12					0.001	0.006	0.032	0.113	0.250	0.257	0.123
	13						0.001	0.007	0.041	0.154	0.356	0.359
	14							0.001	0.007	0.044	0.229	0.488

TABLE 1 (*continued*)

						p						
n	x	0.05	0.1	0.2	0.3	0.4	0.5	0.6	0.7	0.8	0.9	0.95
15	0	0.463	0.206	0.035	0.005							
	1	0.366	0.343	0.132	0.031	0.005						
	2	0.135	0.267	0.231	0.092	0.022	0.003					
	3	0.031	0.129	0.250	0.170	0.063	0.014	0.002				
	4	0.005	0.043	0.188	0.219	0.127	0.042	0.007	0.001			
	5	0.001	0.010	0.103	0.206	0.186	0.092	0.024	0.003			
	6		0.002	0.043	0.147	0.207	0.153	0.061	0.012	0.001		
	7			0.014	0.081	0.177	0.196	0.118	0.035	0.003		
	8			0.003	0.035	0.118	0.196	0.177	0.081	0.014		
	9			0.001	0.012	0.061	0.153	0.207	0.147	0.043	0.002	
	10				0.003	0.024	0.092	0.186	0.206	0.103	0.010	0.001
	11				0.001	0.007	0.042	0.127	0.219	0.188	0.043	0.005
	12					0.002	0.014	0.063	0.170	0.250	0.129	0.031
	13						0.003	0.022	0.092	0.231	0.267	0.135
	14							0.005	0.031	0.132	0.343	0.366
	15								0.005	0.035	0.206	0.463

TABLE 2

The Standard Normal Distribution

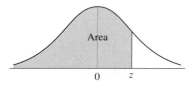

$$F_z(z) = P[Z \leq z]$$

z	0.00	0.01	0.02	0.03	0.04	0.05	0.06	0.07	0.08	0.09
−3.4	0.0003	0.0003	0.0003	0.0003	0.0003	0.0003	0.0003	0.0003	0.0003	0.0002
−3.3	0.0005	0.0005	0.0005	0.0004	0.0004	0.0004	0.0004	0.0004	0.0004	0.0003
−3.2	0.0007	0.0007	0.0006	0.0006	0.0006	0.0006	0.0006	0.0005	0.0005	0.0005
−3.1	0.0010	0.0009	0.0009	0.0009	0.0008	0.0008	0.0008	0.0008	0.0007	0.0007
−3.0	0.0013	0.0013	0.0013	0.0012	0.0012	0.0011	0.0011	0.0011	0.0010	0.0010
−2.9	0.0019	0.0018	0.0017	0.0017	0.0016	0.0016	0.0015	0.0015	0.0014	0.0014
−2.8	0.0026	0.0025	0.0024	0.0023	0.0023	0.0022	0.0021	0.0021	0.0020	0.0019
−2.7	0.0035	0.0034	0.0033	0.0032	0.0031	0.0030	0.0029	0.0028	0.0027	0.0026
−2.6	0.0047	0.0045	0.0044	0.0043	0.0041	0.0040	0.0039	0.0038	0.0037	0.0036
−2.5	0.0062	0.0060	0.0059	0.0057	0.0055	0.0054	0.0052	0.0051	0.0049	0.0048
−2.4	0.0082	0.0080	0.0078	0.0075	0.0073	0.0071	0.0069	0.0068	0.0066	0.0064
−2.3	0.0107	0.0104	0.0102	0.0099	0.0096	0.0094	0.0091	0.0089	0.0087	0.0084
−2.2	0.0139	0.0136	0.0132	0.0129	0.0125	0.0122	0.0119	0.0116	0.0113	0.0110
−2.1	0.0179	0.0174	0.0170	0.0166	0.0162	0.0158	0.0154	0.0150	0.0146	0.0143
−2.0	0.0228	0.0222	0.0217	0.0212	0.0207	0.0202	0.0197	0.0192	0.0188	0.0183
−1.9	0.0287	0.0281	0.0274	0.0268	0.0262	0.0256	0.0250	0.0244	0.0239	0.0233
−1.8	0.0359	0.0352	0.0344	0.0336	0.0329	0.0322	0.0314	0.0307	0.0301	0.0294
−1.7	0.0446	0.0436	0.0427	0.0418	0.0409	0.0401	0.0392	0.0384	0.0375	0.0367
−1.6	0.0548	0.0537	0.0526	0.0516	0.0505	0.0495	0.0485	0.0475	0.0465	0.0455
−1.5	0.0668	0.0655	0.0643	0.0630	0.0618	0.0606	0.0594	0.0582	0.0571	0.0559
−1.4	0.0808	0.0793	0.0778	0.0764	0.0749	0.0735	0.0722	0.0708	0.0694	0.0681
−1.3	0.0968	0.0951	0.0934	0.0918	0.0901	0.0885	0.0869	0.0853	0.0838	0.0823
−1.2	0.1151	0.1131	0.1112	0.1093	0.1075	0.1056	0.1038	0.1020	0.1003	0.0985
−1.1	0.1357	0.1335	0.1314	0.1292	0.1271	0.1251	0.1230	0.1210	0.1190	0.1170
−1.0	0.1587	0.1562	0.1539	0.1515	0.1492	0.1469	0.1446	0.1423	0.1401	0.1379
−0.9	0.1841	0.1814	0.1788	0.1762	0.1736	0.1711	0.1685	0.1660	0.1635	0.1611
−0.8	0.2119	0.2090	0.2061	0.2033	0.2005	0.1977	0.1949	0.1922	0.1894	0.1867
−0.7	0.2420	0.2389	0.2358	0.2327	0.2296	0.2266	0.2236	0.2206	0.2177	0.2148
−0.6	0.2743	0.2709	0.2676	0.2643	0.2611	0.2578	0.2546	0.2514	0.2483	0.2451
−0.5	0.3085	0.3050	0.3015	0.2981	0.2946	0.2912	0.2877	0.2843	0.2810	0.2776
−0.4	0.3446	0.3409	0.3372	0.3336	0.3300	0.3264	0.3228	0.3192	0.3156	0.3121
−0.3	0.3821	0.3783	0.3745	0.3707	0.3669	0.3632	0.3594	0.3557	0.3520	0.3483
−0.2	0.4207	0.4168	0.4129	0.4090	0.4052	0.4013	0.3974	0.3936	0.3897	0.3859
−0.1	0.4602	0.4562	0.4522	0.4483	0.4443	0.4404	0.4364	0.4325	0.4286	0.4247
−0.0	0.5000	0.4960	0.4920	0.4880	0.4840	0.4801	0.4761	0.4721	0.4681	0.4641

TABLE 2 (*continued*)

$$F_z(z) = P[Z \leq z]$$

z	0.00	0.01	0.02	0.03	0.04	0.05	0.06	0.07	0.08	0.09
0.0	0.5000	0.5040	0.5080	0.5120	0.5160	0.5199	0.5239	0.5279	0.5319	0.5359
0.1	0.5398	0.5438	0.5478	0.5517	0.5557	0.5596	0.5636	0.5675	0.5714	0.5753
0.2	0.5793	0.5832	0.5871	0.5910	0.5948	0.5987	0.6026	0.6064	0.6103	0.6141
0.3	0.6179	0.6217	0.6255	0.6293	0.6331	0.6368	0.6406	0.6443	0.6480	0.6517
0.4	0.6554	0.6591	0.6628	0.6664	0.6700	0.6736	0.6772	0.6808	0.6844	0.6879
0.5	0.6915	0.6950	0.6985	0.7019	0.7054	0.7088	0.7123	0.7157	0.7190	0.7224
0.6	0.7257	0.7291	0.7324	0.7357	0.7389	0.7422	0.7454	0.7486	0.7517	0.7549
0.7	0.7580	0.7611	0.7642	0.7673	0.7704	0.7734	0.7764	0.7794	0.7823	0.7852
0.8	0.7881	0.7910	0.7939	0.7967	0.7995	0.8023	0.8051	0.8078	0.8106	0.8133
0.9	0.8159	0.8186	0.8212	0.8238	0.8264	0.8289	0.8315	0.8340	0.8365	0.8389
1.0	0.8413	0.8438	0.8461	0.8485	0.8508	0.8531	0.8554	0.8577	0.8599	0.8621
1.1	0.8643	0.8665	0.8686	0.8708	0.8729	0.8749	0.8770	0.8790	0.8810	0.8830
1.2	0.8849	0.8869	0.8888	0.8907	0.8925	0.8944	0.8962	0.8980	0.8997	0.9015
1.3	0.9032	0.9049	0.9066	0.9082	0.9099	0.9115	0.9131	0.9147	0.9162	0.9177
1.4	0.9192	0.9207	0.9222	0.9236	0.9251	0.9265	0.9278	0.9292	0.9306	0.9319
1.5	0.9332	0.9345	0.9357	0.9370	0.9382	0.9394	0.9406	0.9418	0.9429	0.9441
1.6	0.9452	0.9463	0.9474	0.9484	0.9495	0.9505	0.9515	0.9525	0.9535	0.9545
1.7	0.9554	0.9564	0.9573	0.9582	0.9591	0.9599	0.9608	0.9616	0.9625	0.9633
1.8	0.9641	0.9649	0.9656	0.9664	0.9671	0.9678	0.9686	0.9693	0.9699	0.9706
1.9	0.9713	0.9719	0.9726	0.9732	0.9738	0.9744	0.9750	0.9756	0.9761	0.9767
2.0	0.9772	0.9778	0.9783	0.9788	0.9793	0.9798	0.9803	0.9808	0.9812	0.9817
2.1	0.9821	0.9826	0.9830	0.9834	0.9838	0.9842	0.9846	0.9850	0.9854	0.9857
2.2	0.9861	0.9864	0.9868	0.9871	0.9875	0.9878	0.9881	0.9884	0.9887	0.9890
2.3	0.9893	0.9896	0.9898	0.9901	0.9904	0.9906	0.9909	0.9911	0.9913	0.9916
2.4	0.9918	0.9920	0.9922	0.9925	0.9927	0.9929	0.9931	0.9932	0.9934	0.9936
2.5	0.9938	0.9940	0.9951	0.9943	0.9945	0.9946	0.9948	0.9949	0.9951	0.9952
2.6	0.9953	0.9955	0.9956	0.9957	0.9959	0.9960	0.9961	0.9962	0.9963	0.9964
2.7	0.9965	0.9966	0.9967	0.9968	0.9969	0.9970	0.9971	0.9972	0.9973	0.9974
2.8	0.9974	0.9975	0.9976	0.9977	0.9977	0.9978	0.9979	0.9979	0.9980	0.9981
2.9	0.9981	0.9982	0.9982	0.9983	0.9984	0.9984	0.9985	0.9985	0.9986	0.9986
3.0	0.9987	0.9987	0.9987	0.9988	0.9988	0.9989	0.9989	0.9989	0.9990	0.9990
3.1	0.9990	0.9991	0.9991	0.9991	0.9992	0.9992	0.9992	0.9992	0.9993	0.9993
3.2	0.9993	0.9993	0.9994	0.9994	0.9994	0.9994	0.9994	0.9995	0.9995	0.9995
3.3	0.9995	0.9995	0.9995	0.9996	0.9996	0.9996	0.9996	0.9996	0.9996	0.9997
3.4	0.9997	0.9997	0.9997	0.9997	0.9997	0.9997	0.9997	0.9997	0.9997	0.9998

Answers to Odd-Numbered Exercises

CHAPTER 1

Exercises 1.1, page 7

1. $(3, 3)$; Quadrant I   3. $(2, -2)$; Quadrant IV

5. $(-4, -6)$; Quadrant III   7. A 9. E, F, and G

11. F 13–19. See the following figure.

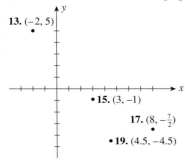

13. $(-2, 5)$

15. $(3, -1)$

17. $(8, -\frac{7}{2})$

19. $(4.5, -4.5)$

21. 5 23. $\sqrt{61}$ 25. $(-8, -6)$ and $(8, -6)$

29. $(x - 2)^2 + (y + 3)^2 = 25$ 31. $x^2 + y^2 = 25$

33. $(x - 2)^2 + (y + 3)^2 = 34$ 35. 3400 mi

37. Route 1   39. Model C

41. **a.** $d = \sqrt{1300}\, t$ **b.** 72.11 mi   43. False

Exercises 1.2, page 19

1. $\frac{1}{2}$ 3. Not defined   5. 5 7. $\frac{5}{6}$

9. $\dfrac{d - b}{c - a}\ (a \neq c)$ 11. **a.** 4 **b.** -8 13. Parallel

15. Perpendicular   17. $a = -5$ 19. $y = -3$

21. e   23. a   25. f   27. $y = 2x - 10$

29. $y = 2$ 31. $y = 3x - 2$ 33. $y = x + 1$

35. $y = 3x + 4$ 37. $y = 5$

39. $y = \frac{1}{2}x$; $m = \frac{1}{2}$; $b = 0$

41. $y = \frac{2}{3}x - 3$; $m = \frac{2}{3}$; $b = -3$

43. $y = -\frac{1}{2}x + \frac{7}{2}$; $m = -\frac{1}{2}$; $b = \frac{7}{2}$

45. $y = \frac{1}{2}x + 3$ 47. $y = -6$ 49. $y = b$

51. $y = \frac{2}{3}x - \frac{2}{3}$ 53. $k = 8$

55.

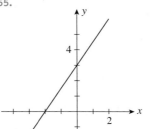

57.

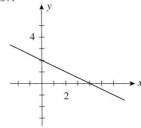

59.

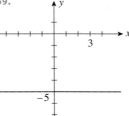

63. $y = -2x - 4$ 65. $y = \frac{1}{8}x - \frac{1}{2}$ 67. Yes

69. **a.**

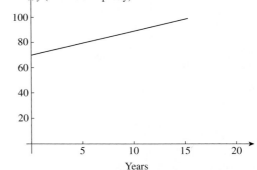

b. 1.9467; 70.082

c. The capacity utilization has been increasing by 1.9467% each year since 1990 when it stood at 70.082%.

d. Shortly after 2005

71. **a.** $y = 0.55x$ **b.** 2000 73. 84.8%

75. a and **b.**

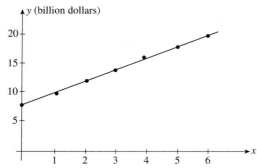

c. $y = 1.82x + 7.9$ **d.** $17 billion; same

77. a and **b.**

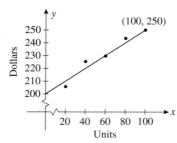

c. $y = \frac{1}{2}x + 200$ **d.** $227

79. a and **b.**

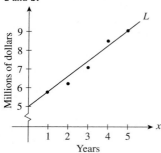

c. $y = 0.8x + 5$ **d.** $12.2 million

81. a. A family of parallel lines having slope m
   b. A family of straight lines that pass through the point $(0, b)$

83. False   **85.** True   **87.** True

Using Technology Exercises 1.2, page 27

Graphing Utility

1.

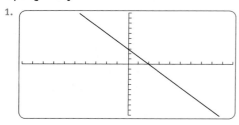

3.

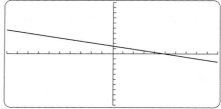

5.

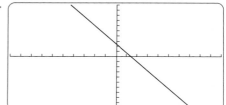

7. a.

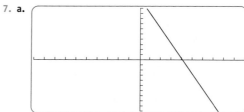

b.

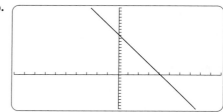

9. a.

b.

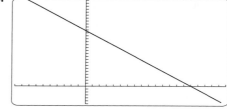

11.

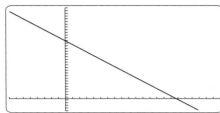

13.

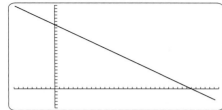

15.

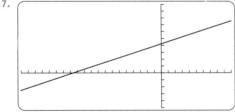

17.

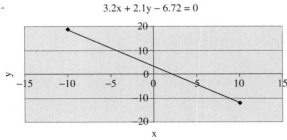

Excel

1.

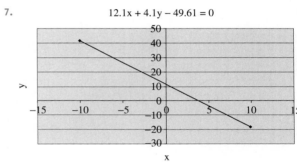

$3.2x + 2.1y - 6.72 = 0$

3.

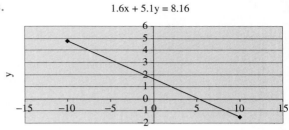

$1.6x + 5.1y = 8.16$

5.

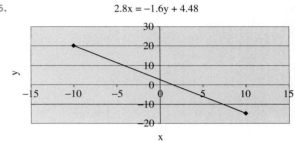

$2.8x = -1.6y + 4.48$

7.

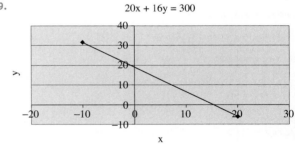

$12.1x + 4.1y - 49.61 = 0$

9.

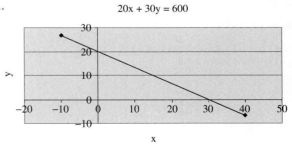

$20x + 16y = 300$

11.

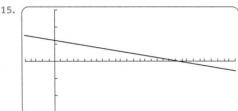

$20x + 30y = 600$

13.

$$22.4x + 16.1y - 352 = 0$$

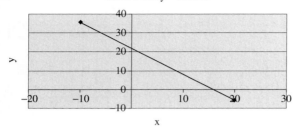

15.

$$1.2x + 20y = 24$$

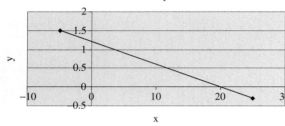

17.

$$-4x + 12y = 50$$

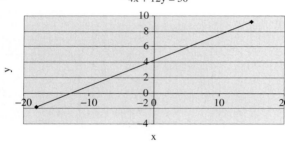

Exercises 1.3, page 35

1. Yes; $y = -\frac{2}{3}x + 2$ 3. Yes; $y = \frac{1}{2}x + 2$

5. Yes; $y = \frac{1}{2}x + \frac{9}{4}$ 7. No   9. No

11. **a.** $C(x) = 8x + 40,000$ **b.** $R(x) = 12x$
    c. $P(x) = 4x - 40,000$
    d. Loss of $8000; profit of $8000

13. $m = -1; b = 2$ 15. $900,000; $800,000

17. $6 billion; $43.5 billion; $81 billion

19. **a.** $y = 1.053x$ **b.** $652.86

21. $C(x) = 0.6x + 12,100; R(x) = 1.15x$;
    $P(x) = 0.55x - 12,100$

23. **a.** $12,000/yr   **b.** $V = 60,000 - 12,000t$
    c.

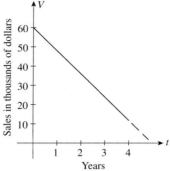

    d. $24,000

25. $900,000; $800,000

27. **a.** $m = a/1.7; b = 0$ **b.** 117.65 mg

29. $f(t) = 7.5t + 20$ $(0 \le t \le 6)$; 65 million

31. **a.** $F = \frac{9}{5}C + 32$ **b.** 68°F   **c.** 21.1°C   33. L_2

35. **a.**   **b.** 3000

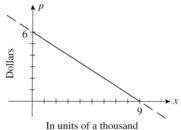

In units of a thousand

37. **a.**   **b.** 10,000

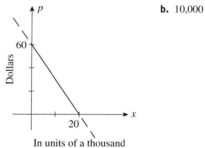

In units of a thousand

39. $p = -\frac{3}{40}x + 130$; $130; 1733

41. 2500 units

43. **a.**   **b.** 2667 units

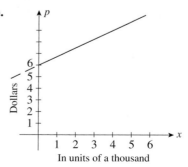

In units of a thousand

45. a.

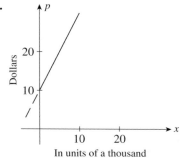

In units of a thousand

b. 2000 units

47. $p = \frac{1}{2}x + 40$ (x is measured in units of a thousand)

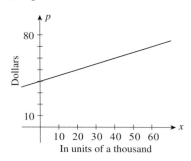

In units of a thousand

60,000 units

49. False

Using Technology Exercises 1.3, page 42

1. 2.2875   **3.** 2.880952381   **5.** 7.2851648352

7. 2.4680851064

Exercises 1.4, page 49

1. (2, 10)   **3.** $\left(4, \frac{2}{3}\right)$ **5.** $(-4, -6)$

7. 1000 units; $15,000   **9.** 600 units; $240

11. a.

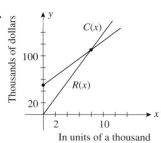

In units of a thousand

b. 8000 units; $112,000

c.

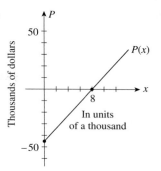

In units
of a thousand

d. (8000, 0)

13. 9259 units; $83,331

15. a. $C_1(x) = 18{,}000 + 15x$
$C_2(x) = 15{,}000 + 20x$

b.

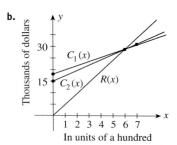

In units of a hundred

c. Machine II; machine II; machine I
d. ($1500); $1500; $4750

17. Middle of 2003

19. a.

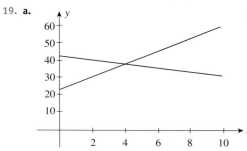

b. Feb. 2005

21. 8000 units; $9   **23.** 2000 units; $18

25. a. $p = -0.08x + 725$ **b.** $p = 0.09x + 300$
c. 2500 DVD players; $525

27. 300 fax machines; $600

29. a. $\dfrac{b - d}{c - a}; \dfrac{bc - ad}{c - a}$

b. If c is increased, x gets smaller and p gets larger.
c. If b is decreased, x decreases and p decreases.

31. True

33. **a.** $m_1 = m_2$ and $b_2 \neq b_1$
    b. $m_1 \neq m_2$
    c. $m_1 = m_2$ and $b_1 = b_2$

Using Technology Exercises 1.4, page 53

1. (0.6, 6.2)   3. (3.8261, 0.1304)

5. (386.9091, 145.3939)

7. **a.**

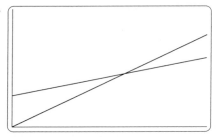

b. (3548.39, 27,996.77)

c.

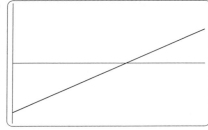

x-intercept: 3548

9. **a.** $C_1(x) = 34 + 0.18x$; $C_2(x) = 28 + 0.22x$
   b.

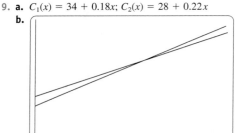

   c. (150, 61)

   d. If the distance driven is less than or equal to 150 mi, rent from Acme Truck Leasing; if the distance driven is more than 150 mi, rent from Ace Truck Leasing.

11. **a.** $p = -\frac{1}{10}x + 284$; $p = \frac{1}{60}x + 60$
    b.

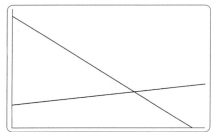

    (1920, 92)

    c. 1920/wk; $92/radio

Exercises 1.5, page 59

1. **a.** $y = 2.3x + 1.5$

   b.

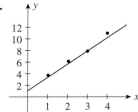

3. **a.** $y = -0.77x + 5.74$

   b.

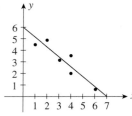

5. **a.** $y = 1.2x + 2$

   b.

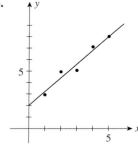

7. **a.** $y = 0.34x - 0.9$

   b.

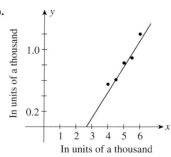

   c. 1276 applications

9. **a.** $y = -2.8x + 440$

b. 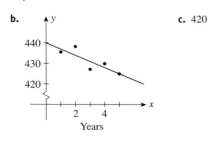   **c.** 420

11. **a.** $y = 2.8x + 17.6$ **b.** \$40,000,000

13. **a.** $y = 22.7t + 124.2$ **b.** \$260.4 billion

15. **a.** $y = 12.2x + 20.9$ **b.** 106.3 million

17. **a.** $y = 0.059x + 19.5$ **b.** 21.9 yr   **c.** 21.3 yr

19. **a.** $y = 1.757x + 7.914$ **b.** \$16.69 billion

21. **a.** $y = 10x + 90.34$ **b.** 150,340,000

23. **a.** $y = 46.61x + 495.04$ **b.** \$46.61/buyer

25. **a.** $y = 3.24x + 70.16$ **b.** \$96,080

27. False   29. True

Using Technology Exercises 1.5, page 65

1. $y = 3.8639 + 2.3596x$ 3. $y = 3.5525 - 1.1948x$

5. **a.** $y = 0.55x + 1.17$ **b.** \$5.57 billion

7. **a.** $y = 13.321x + 72.571$ **b.** 192 million tons

9. **a.** $y = 14.43x + 212.1$ **b.** 247 trillion cu ft

Chapter 1 Concept Review Questions, page 67

1. Ordered; abscissa (x-coordinate); ordinate (y-coordinate)

2. **a.** x-; y-;   **b.** third

3. $\sqrt{(c - a)^2 + (d - b)^2}$ 4. $(x - a^2) + (y - b^2) = r^2$

5. **a.** $\dfrac{y_2 - y_1}{x_2 - x_1}$ **b.** Undefined   **c.** 0   **d.** Positive

6. $m_1 = m_2$; $m_1 = -\dfrac{1}{m_2}$

7. **a.** $y - y_1 = m(x - x_1)$; point-slope
   b. $y = mx + b$; slope-intercept

8. **a.** $Ax + By + C = 0$ (A, B, not both zero)   **b.** $-\dfrac{a}{b}$

9. $mx + b$

10. **a.** Price; demanded; demand   **b.** Price; supplied; supply

11. Break-even   12. Demand; supply

Chapter 1 Review Exercises, page 68

1. 5   2. 5   3. 5   4. 2

5. $x = -2$ 6. $y = 4$

7. $x + 10y - 38 = 0$ 8. $y = -\frac{4}{5}x + \frac{12}{5}$

9. $5x - 2y + 18 = 0$ 10. $y = \frac{3}{4}x + \frac{11}{2}$

11. $y = -\frac{1}{2}x - 3$ 12. $\frac{3}{5}$; $-\frac{6}{5}$ 13. $3x + 4y - 18 = 0$

14. $y = -\frac{3}{5}x + \frac{12}{5}$ 15. $3x + 2y + 14 = 0$

16.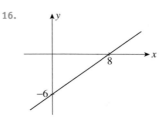

17.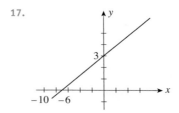

18. 60,000   19. **a.** $f(x) = x + 2.4$ **b.** \$5.4 million

21. **b.** ≈ 117 mg   22. **a.** \$200,000/yr   **b.** \$4,000,000

23. **a.** \$22,500/yr   **b.** $V = -22,500t + 300,000$

24. **a.** $6x + 30,000$ **b.** $10x$ **c.** $4x - 30,000$
    d. (\$6,000); \$2000; \$18,000

25. $p = -0.05x + 200$
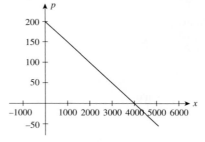

26. $p = \frac{1}{36}x + \frac{400}{9}$ 27. $(2, -3)$ 28. $(6, \frac{21}{2})$

29. $(2500, 50,000)$ 30. 6000; \$22

31. **a.** $y = 0.25x$ **b.** 1600   32. 600; \$80

Chapter 1 Before Moving On, page 69

1.

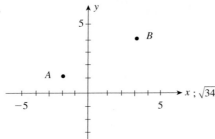

2. $y = 3x - 8$ 3. Yes

4. **a.** $18   **b.** $22,000   **c.** $15

5. $\left(1, \frac{4}{3}\right)$ 6. After 5 yr

CHAPTER 2

Exercises 2.1, page 77

1. Unique solution; $(2, 1)$ 3. No solution

5. Unique solution; $(3, 2)$

7. Infinitely many solutions; $\left(t, \frac{2}{5}t - 2\right)$; t, a parameter

9. Unique solution; $(1, -2)$

11. No solution   13. $k = -2$

15. $\begin{aligned} x + y &= 500 \\ 42x + 30y &= 18{,}600 \end{aligned}$ 17. $\begin{aligned} x + y &= 100 \\ 2.5x + 3y &= 280 \end{aligned}$

19. $\begin{aligned} x + y &= 1000 \\ .25x + .75y &= 650 \end{aligned}$

21. $\begin{aligned} .06x + .08y + .12z &= 21{,}600 \\ z &= 2x \\ .12z &= .08y \end{aligned}$

23. $\begin{aligned} 18x + 20y + 24z &= 26{,}400 \\ 4x + 4y + 3z &= 4{,}900 \\ 5x + 4y + 6z &= 6{,}200 \end{aligned}$

25. $\begin{aligned} 12{,}000x + 18{,}000y + 24{,}000z &= 1{,}500{,}000 \\ x &= 2y \\ x + y + z &= 100 \end{aligned}$

27. $\begin{aligned} 10x + 6y + 8z &= 100 \\ 10x + 12y + 6z &= 100 \\ 5x + 4y + 12z &= 100 \end{aligned}$

29. True

Exercises 2.2, page 91

1. $\begin{bmatrix} 2 & -3 & | & 7 \\ 3 & 1 & | & 4 \end{bmatrix}$

3. $\begin{bmatrix} 0 & -1 & 2 & | & 6 \\ 2 & 2 & -8 & | & 7 \\ 0 & 3 & 4 & | & 0 \end{bmatrix}$

5. $\begin{aligned} 3x + 2y &= -4 \\ x - y &= 5 \end{aligned}$ 7. $\begin{aligned} x + 3y + 2z &= 4 \\ 2x &= 5 \\ 3x - 3y + 2z &= 6 \end{aligned}$

9. Yes   11. No   13. Yes   15. No   17. No

19. $\begin{bmatrix} 1 & 2 & | & 4 \\ 0 & -5 & | & -10 \end{bmatrix}$ 21. $\begin{bmatrix} 1 & -2 & | & -3 \\ 0 & 16 & | & 20 \end{bmatrix}$

23. $\begin{bmatrix} 1 & 2 & 3 & | & 6 \\ 0 & -1 & -5 & | & -7 \\ 0 & -7 & -7 & | & -14 \end{bmatrix}$

25. $\begin{bmatrix} -6 & -11 & 0 & | & -5 \\ 2 & 4 & 1 & | & 3 \\ 1 & -2 & 0 & | & -10 \end{bmatrix}$

27. $\begin{bmatrix} 3 & 9 & | & 6 \\ 2 & 1 & | & 4 \end{bmatrix} \xrightarrow{\frac{1}{3}R_1} \begin{bmatrix} 1 & 3 & | & 2 \\ 2 & 1 & | & 4 \end{bmatrix}$

$\xrightarrow{R_2 - 2R_1} \begin{bmatrix} 1 & 3 & | & 2 \\ 0 & -5 & | & 0 \end{bmatrix} \xrightarrow{-\frac{1}{5}R_2}$

$\begin{bmatrix} 1 & 3 & | & 2 \\ 0 & 1 & | & 0 \end{bmatrix} \xrightarrow{R_1 - 3R_2} \begin{bmatrix} 1 & 0 & | & 2 \\ 0 & 1 & | & 0 \end{bmatrix}$

29. $\begin{bmatrix} 1 & 3 & 1 & | & 3 \\ 3 & 8 & 3 & | & 7 \\ 2 & -3 & 1 & | & -10 \end{bmatrix} \xrightarrow[R_3 - 2R_1]{R_2 - 3R_1}$

$\begin{bmatrix} 1 & 3 & 1 & | & 3 \\ 0 & -1 & 0 & | & -2 \\ 0 & -9 & -1 & | & -16 \end{bmatrix} \xrightarrow{-R_2}$

$\begin{bmatrix} 1 & 3 & 1 & | & 3 \\ 0 & 1 & 0 & | & 2 \\ 0 & -9 & -1 & | & -16 \end{bmatrix} \xrightarrow[R_3 + 9R_2]{R_1 - 3R_2}$

$\begin{bmatrix} 1 & 0 & 1 & | & -3 \\ 0 & 1 & 0 & | & 2 \\ 0 & 0 & -1 & | & 2 \end{bmatrix} \xrightarrow[-R_3]{R_1 + R_3}$

$\begin{bmatrix} 1 & 0 & 0 & | & -1 \\ 0 & 1 & 0 & | & 2 \\ 0 & 0 & 1 & | & -2 \end{bmatrix}$

31. $(2, 0)$ 33. $(-1, 2, -2)$

35. $(4, -2)$ 37. $(-1, 2)$

39. $\left(\frac{7}{9}, -\frac{1}{9}, -\frac{2}{3}\right)$ 41. $(19, -7, -15)$

43. $(3, 0, 2)$ 45. $(1, -2, 1)$

47. $(-20, -28, 13)$ 49. $(4, -1, 3)$

51. 300 acres of corn, 200 acres of wheat

53. In 100 lb of blended coffee, use 40 lb of the $2.50/lb coffee and 60 lb of the $3/lb coffee.

55. 200 children and 800 adults

57. $40,000 in a savings account, $120,000 in mutual funds, $80,000 in bonds

59. 400 bags of grade-A fertilizer; 600 bags of grade-B fertilizer; 300 bags of grade-C fertilizer

61. 60 compact, 30 intermediate, and 10 full-size cars

63. 4 oz of food I, 2 oz of food II, 6 oz of food III

65. 240 front orchestra seats, 560 rear orchestra seats, 200 front balcony seats

67. 7 days in London, 4 days in Paris, and 3 days in Rome

69. False

Using Technology Exercises 2.2, page 96

1. $x_1 = 3; x_2 = 1; x_3 = -1; x_4 = 2$

3. $x_1 = 5; x_2 = 4; x_3 = -3; x_4 = -4$

5. $x_1 = 1; x_2 = -1; x_3 = 2; x_4 = 0; x_5 = 3$

Exercises 2.3, page 104

1. **a.** One solution   **b.** $(3, -1, 2)$

3. **a.** One solution   **b.** $(2, 4)$

5. **a.** Infinitely many solutions
   b. $(4 - t, -2, t)$; t, a parameter

7. **a.** No solution

9. **a.** Infinitely many solutions
   b. $(2, -1, 2 - t, t)$; t, a parameter

11. **a.** Infinitely many solutions
    b. $(2 - 3s, 1 + s, s, t)$; s, t, parameters

13. $(2, 1)$ 15. No solution   17. $(1, -1)$

19. $(2 + 2t, t)$; t, a parameter

21. $(\frac{4}{3} - \frac{2}{3}t, t)$; t, a parameter

23. $(-2 + \frac{1}{2}s - \frac{1}{2}t, s, t)$; s, t, parameters

25. $(-1, \frac{17}{7}, \frac{23}{7})$

27. $(1 - \frac{1}{4}s + \frac{1}{4}t, s, t)$; s, t, parameters

29. No solution   31. $(2, -1, 4)$

33. $x = 20 + z, y = 40 - 2z$; 25 compact cars, 30 mid-sized cars, and 5 full-sized cars; 30 compact cars, 20 mid-sized cars, and 10 full-sized cars

37. **a.**
$$\begin{aligned}
x_1 - x_2 & & & = 200 \\
x_1 & & - x_5 & = 100 \\
-x_2 + x_3 & & + x_6 & = 600 \\
-x_3 + x_4 & & & = 200 \\
& x_4 - x_5 & + x_6 & = 700
\end{aligned}$$
b. $x_1 = s + 100; x_2 = s - 100; x_3 = s - t + 500;$
$x_4 = s - t + 700; x_5 = s; x_6 = t$
$(250, 50, 600, 800, 150, 50)$
$(300, 100, 600, 800, 200, 100)$
c. $(300, 100, 700, 900, 200, 0)$

39. $k = 6; (-2, 2)$ 41. False

Using Technology Exercises 2.3, page 107

1. $(1 + t, 2 + t, t)$; t, a parameter

3. $(-\frac{17}{7} + \frac{6}{7}t, 3 - t, -\frac{18}{7} + \frac{1}{7}t, t)$; t, a parameter

5. No solution

Exercises 2.4, page 115

1. $4 \times 4; 4 \times 3; 1 \times 5; 4 \times 1$ 3. $2; 3; 8$

5. $D; D^T = [1 \quad 3 \quad -2 \quad 0]$ 7. $3 \times 2; 3 \times 2; 3 \times 3; 3 \times 3$

9. $\begin{bmatrix} 1 & 6 \\ 6 & -1 \\ 2 & 2 \end{bmatrix}$ 11. $\begin{bmatrix} 1 & 1 & -4 \\ -1 & -8 & 1 \\ 6 & 3 & 1 \end{bmatrix}$

13. $\begin{bmatrix} 3 & 5 & 9 \\ 4 & 10 & 13 \end{bmatrix}$ 15. $\begin{bmatrix} 3 & -4 & -16 \\ 17 & -4 & 16 \end{bmatrix}$

17. $\begin{bmatrix} -1.9 & 3.0 & -0.6 \\ 6.0 & 9.6 & 1.2 \end{bmatrix}$

19. $\begin{bmatrix} \frac{7}{2} & 3 & -1 & \frac{10}{3} \\ -\frac{19}{6} & \frac{2}{3} & -\frac{17}{2} & \frac{23}{3} \\ \frac{29}{3} & \frac{17}{6} & -1 & -2 \end{bmatrix}$

21. $x = \frac{5}{2}, y = 7, z = 2,$ and $u = 3$

23. $x = 2, y = 2, z = -\frac{7}{3},$ and $u = 15$

31. $\begin{bmatrix} 3 \\ 2 \\ -1 \\ 5 \end{bmatrix}$ 33. $\begin{bmatrix} 1 & 3 & 0 \\ -1 & 4 & 1 \\ 2 & 2 & 0 \end{bmatrix}$

35. $\begin{bmatrix} 220 & 215 & 210 & 205 \\ 220 & 210 & 200 & 195 \\ 215 & 205 & 195 & 190 \end{bmatrix}$

37. **a.** $D = \begin{bmatrix} 2960 & 1510 & 1150 \\ 1100 & 550 & 490 \\ 1230 & 590 & 470 \end{bmatrix}$

b. $E = \begin{bmatrix} 3256 & 1661 & 1265 \\ 1210 & 605 & 539 \\ 1353 & 649 & 517 \end{bmatrix}$

39. Mass.
$$\begin{array}{c} \\ \text{Mass.} \\ \text{U.S.} \end{array}\begin{array}{ccc} 2000 & 2001 & 2002 \\ \left[\begin{array}{ccc} 6.88 & 7.05 & 7.18 \\ 4.13 & 4.09 & 4.06 \end{array}\right] \end{array}$$

41.
$$\begin{array}{c} \\ \text{Women} \\ \text{Men} \end{array}\begin{array}{ccc} \text{White} & \text{Black} & \text{Hispanic} \\ \left[\begin{array}{ccc} 81 & 76.1 & 82.2 \\ 76 & 69.9 & 75.9 \end{array}\right] \end{array}$$

$$\begin{array}{c} \\ \text{White} \\ \text{Black} \\ \text{Hispanic} \end{array}\begin{array}{cc} \text{Women} & \text{Men} \\ \left[\begin{array}{cc} 81 & 76 \\ 76.1 & 69.9 \\ 82.2 & 75.9 \end{array}\right] \end{array}$$

43. True   **45.** False

Using Technology Exercises 2.4, page 121

1. $\begin{bmatrix} 15 & 38.75 & -67.5 & 33.75 \\ 51.25 & 40 & 52.5 & -38.75 \\ 21.25 & 35 & -65 & 105 \end{bmatrix}$

3. $\begin{bmatrix} -5 & 6.3 & -6.8 & 3.9 \\ 1 & 0.5 & 5.4 & -4.8 \\ 0.5 & 4.2 & -3.5 & 5.6 \end{bmatrix}$

5. $\begin{bmatrix} 16.44 & -3.65 & -3.66 & 0.63 \\ 12.77 & 10.64 & 2.58 & 0.05 \\ 5.09 & 0.28 & -10.84 & 17.64 \end{bmatrix}$

7. $\begin{bmatrix} 22.2 & -0.3 & -12 & 4.5 \\ 21.6 & 17.7 & 9 & -4.2 \\ 8.7 & 4.2 & -20.7 & 33.6 \end{bmatrix}$

Exercises 2.5, page 129

1. 2×5; not defined   **3.** 1×1; 7×7

5. $n = s$; $m = t$ **7.** $\begin{bmatrix} -1 \\ 3 \end{bmatrix}$ **9.** $\begin{bmatrix} 9 \\ -10 \end{bmatrix}$

11. $\begin{bmatrix} 4 & -2 \\ 9 & 13 \end{bmatrix}$ **13.** $\begin{bmatrix} 2 & 9 \\ 5 & 16 \end{bmatrix}$ **15.** $\begin{bmatrix} 0.57 & 1.93 \\ 0.64 & 1.76 \end{bmatrix}$

17. $\begin{bmatrix} 6 & -3 & 0 \\ -2 & 1 & -8 \\ 4 & -4 & 9 \end{bmatrix}$ **19.** $\begin{bmatrix} 5 & 1 & -3 \\ 1 & 7 & -3 \end{bmatrix}$

21. $\begin{bmatrix} -4 & -20 & 4 \\ 4 & 12 & 0 \\ 12 & 32 & 20 \end{bmatrix}$ **23.** $\begin{bmatrix} 4 & -3 & 2 \\ 7 & 1 & -5 \end{bmatrix}$

27. $AB = \begin{bmatrix} 10 & 7 \\ 22 & 15 \end{bmatrix}$; $BA = \begin{bmatrix} 5 & 8 \\ 13 & 20 \end{bmatrix}$

31. $A = \begin{bmatrix} -2 & -1 \\ 5 & 2 \end{bmatrix}$ **33. a.** $A^T = \begin{bmatrix} 2 & 5 \\ 4 & -6 \end{bmatrix}$

35. $AX = B$, where $A = \begin{bmatrix} 2 & -3 \\ 3 & -4 \end{bmatrix}$, $X = \begin{bmatrix} x \\ y \end{bmatrix}$,

and $B = \begin{bmatrix} 7 \\ 8 \end{bmatrix}$

37. $AX = B$, where $A = \begin{bmatrix} 2 & -3 & 4 \\ 0 & 2 & -3 \\ 1 & -1 & 2 \end{bmatrix}$, $X = \begin{bmatrix} x \\ y \\ z \end{bmatrix}$,

and $B = \begin{bmatrix} 6 \\ 7 \\ 4 \end{bmatrix}$

39. $AX = B$, where $A = \begin{bmatrix} -1 & 1 & 1 \\ 2 & -1 & -1 \\ -3 & 2 & 4 \end{bmatrix}$, $X = \begin{bmatrix} x_1 \\ x_2 \\ x_3 \end{bmatrix}$,

and $B = \begin{bmatrix} 0 \\ 2 \\ 4 \end{bmatrix}$

41. a. $AB = \begin{bmatrix} 51,400 \\ 54,200 \end{bmatrix}$

b. The first entry shows that William's total stockholdings are $51,400; the second shows that Michael's stockholdings are $54,200.

43. a. $B = \begin{bmatrix} 20,000 \\ 22,000 \\ 25,000 \\ 30,000 \end{bmatrix}$

b. $7,160,000 in New York, $2,860,000 in Connecticut, and $1,430,000 in Massachusetts; the total profit is $11,450,000.

45. $\begin{array}{c} \quad\quad D \quad\quad R \quad\quad\; I \\ BA = [41,000 \quad 35,000 \quad 14,000] \end{array}$

47. $AB = \begin{bmatrix} 1575 & 1590 & 1560 & 975 \\ 410 & 405 & 415 & 270 \\ 215 & 205 & 225 & 155 \end{bmatrix}$

49. [277.60]; it represents Cindy's long-distance bill for phone calls to London, Tokyo, and Hong Kong.

51. a. $\begin{bmatrix} 8800 \\ 3380 \\ 1020 \end{bmatrix}$ **b.** $\begin{bmatrix} 8800 \\ 3380 \\ 1020 \end{bmatrix}$ **c.** $\begin{bmatrix} 17,600 \\ 6,760 \\ 2,040 \end{bmatrix}$

53. False   **55.** True

Using Technology Exercises 2.5, page 136

1. $\begin{bmatrix} 18.66 & 15.2 & -12 \\ 24.48 & 41.88 & 89.82 \\ 15.39 & 7.16 & -1.25 \end{bmatrix}$

3. $\begin{bmatrix} 20.09 & 20.61 & -1.3 \\ 44.42 & 71.6 & 64.89 \\ 20.97 & 7.17 & -60.65 \end{bmatrix}$

5. $\begin{bmatrix} 32.89 & 13.63 & -57.17 \\ -12.85 & -8.37 & 256.92 \\ 13.48 & 14.29 & 181.64 \end{bmatrix}$

7. $\begin{bmatrix} 128.59 & 123.08 & -32.50 \\ 246.73 & 403.12 & 481.52 \\ 125.06 & 47.01 & -264.81 \end{bmatrix}$

9. $\begin{bmatrix} 87 & 68 & 110 & 82 \\ 119 & 176 & 221 & 143 \\ 51 & 128 & 142 & 94 \\ 28 & 174 & 174 & 112 \end{bmatrix}$

$\begin{bmatrix} 113 & 117 & 72 & 101 & 90 \\ 72 & 85 & 36 & 72 & 76 \\ 81 & 69 & 76 & 87 & 30 \\ 133 & 157 & 56 & 121 & 146 \\ 154 & 157 & 94 & 127 & 122 \end{bmatrix}$

11. $\begin{bmatrix} 170 & 18.1 & 133.1 & -106.3 & 341.3 \\ 349 & 226.5 & 324.1 & 164 & 506.4 \\ 245.2 & 157.7 & 231.5 & 125.5 & 312.9 \\ 310 & 245.2 & 291 & 274.3 & 354.2 \end{bmatrix}$

Exercises 2.6, page 145

5. $\begin{bmatrix} 3 & -5 \\ -1 & 2 \end{bmatrix}$ 7. Does not exist

9. $\begin{bmatrix} 2 & -11 & -3 \\ 1 & -6 & -2 \\ 0 & -1 & 0 \end{bmatrix}$ 11. Does not exist

13. $\begin{bmatrix} -\frac{13}{10} & \frac{7}{5} & \frac{1}{2} \\ \frac{2}{5} & -\frac{1}{5} & 0 \\ -\frac{7}{10} & \frac{3}{5} & \frac{1}{2} \end{bmatrix}$

15. $\begin{bmatrix} 3 & 4 & -6 & 1 \\ -2 & -3 & 5 & -1 \\ -4 & -4 & 7 & -1 \\ -4 & -5 & 8 & -1 \end{bmatrix}$

17. **a.** $A = \begin{bmatrix} 2 & 5 \\ 1 & 3 \end{bmatrix}; X = \begin{bmatrix} x \\ y \end{bmatrix}; B = \begin{bmatrix} 3 \\ 2 \end{bmatrix}$

b. $x = -1; y = 1$

19. **a.** $A = \begin{bmatrix} 2 & -3 & -4 \\ 0 & 0 & -1 \\ 1 & -2 & 1 \end{bmatrix}; X = \begin{bmatrix} x \\ y \\ z \end{bmatrix}; B = \begin{bmatrix} 4 \\ 3 \\ -8 \end{bmatrix}$

b. $x = -1; y = 2; z = -3$

21. **a.** $A = \begin{bmatrix} 1 & 4 & -1 \\ 2 & 3 & -2 \\ -1 & 2 & 3 \end{bmatrix}; X = \begin{bmatrix} x \\ y \\ z \end{bmatrix}; B = \begin{bmatrix} 3 \\ 1 \\ 7 \end{bmatrix}$

b. $x = 1; y = 1; z = 2$

23. **a.** $A = \begin{bmatrix} 1 & 1 & -1 & 1 \\ 2 & 1 & 1 & 0 \\ 2 & 1 & 0 & 1 \\ 2 & -1 & -1 & 3 \end{bmatrix}; X = \begin{bmatrix} x_1 \\ x_2 \\ x_3 \\ x_4 \end{bmatrix}$

$B = \begin{bmatrix} 6 \\ 4 \\ 7 \\ 9 \end{bmatrix}$

b. $x_1 = 1; x_2 = 2; x_3 = 0; x_4 = 3$

25. **b.(i)** $x = \frac{24}{5}; y = \frac{23}{5}$ **(ii)** $x = \frac{2}{5}; y = \frac{9}{5}$

27. **b.(i)** $x = -1; y = 3; z = 2$
(ii) $x = 1; y = 8; z = -12$

29. **b.(i)** $x = -\frac{2}{17}; y = -\frac{10}{17}; z = -\frac{60}{17}$
(ii) $x = 1; y = 0; z = -5$

31. **b.(i)** $x_1 = 1; x_2 = -4; x_3 = 5; x_4 = -1$
(ii) $x_1 = 12; x_2 = -24; x_3 = 21; x_4 = -7$

33. **a.** $A^{-1} = \begin{bmatrix} -\frac{5}{2} & -\frac{3}{2} \\ 2 & 1 \end{bmatrix}$

35. **a.** $ABC = \begin{bmatrix} 4 & 10 \\ 2 & 3 \end{bmatrix}; A^{-1} = \begin{bmatrix} 3 & -5 \\ 1 & -2 \end{bmatrix};$

$B^{-1} = \begin{bmatrix} 1 & -3 \\ -1 & 4 \end{bmatrix}; C^{-1} = \begin{bmatrix} \frac{1}{8} & -\frac{3}{8} \\ \frac{1}{4} & \frac{1}{4} \end{bmatrix}$

37. **a.** 3214; 3929   **b.** 4286; 3571   **c.** 3929; 5357

39. **a.** 400 acres of soybeans; 300 acres of corn; 300 acres of wheat
b. 500 acres of soybeans; 400 acres of corn; 300 acres of wheat

41. **a.** $80,000 in high-risk stocks; $20,000 in medium-risk stocks; $100,000 in low-risk stocks
b. $88,000 in high-risk stocks; $22,000 in medium-risk stocks; $110,000 in low-risk stocks
c. $56,000 in high-risk stocks; $64,000 in medium-risk stocks, $120,000 in low-risk stocks

43. All values of k except $k = \frac{3}{2}; \dfrac{1}{3-2k}\begin{bmatrix} 3 & -2 \\ -k & 1 \end{bmatrix}$

45. True   47. True

Using Technology Exercises 2.6, page 152

1. $\begin{bmatrix} 0.36 & 0.04 & -0.36 \\ 0.06 & 0.05 & 0.20 \\ -0.19 & 0.10 & 0.09 \end{bmatrix}$

3. $\begin{bmatrix} 0.01 & -0.09 & 0.31 & -0.11 \\ -0.25 & 0.58 & -0.15 & -0.02 \\ 0.86 & -0.42 & 0.07 & -0.37 \\ -0.27 & 0.01 & -0.05 & 0.31 \end{bmatrix}$

5. $\begin{bmatrix} 0.30 & 0.85 & -0.10 & -0.77 & -0.11 \\ -0.21 & 0.10 & 0.01 & -0.26 & 0.21 \\ 0.03 & -0.16 & 0.12 & -0.01 & 0.03 \\ -0.14 & -0.46 & 0.13 & 0.71 & -0.05 \\ 0.10 & -0.05 & -0.10 & -0.03 & 0.11 \end{bmatrix}$

7. $x = 1.2; y = 3.6; z = 2.7$

9. $x_1 = 2.50; x_2 = -0.88; x_3 = 0.70; x_4 = 0.51$

Exercises 2.7, page 158

1. **a.** \$10 million   **b.** \$160 million
   c. Agricultural; manufacturing and transportation

3. $x = 23.75$ and $y = 21.25$

5. $x = 42.85$ and $y = 57.14$

9. **a.** \$318.2 million worth of agricultural goods and \$336.4 million worth of manufactured products
   b. \$198.2 million worth of agricultural products and \$196.4 million worth of manufactured goods

11. **a.** \$443.75 million, \$381.25 million, and \$281.25 million worth of agricultural products, manufactured goods, and transportation, respectively
    b. \$243.75 million, \$281.25 million, and \$221.25 million worth of agricultural products, manufactured goods, and transportation, respectively

13. \$45 million and \$75 million

15. \$34.4 million, \$33 million, and \$21.6 million

Using Technology Exercises 2.7, page 162

1. The final outputs of the first, second, third, and fourth industries are 602.62, 502.30, 572.57, and 523.46 units, respectively.

3. The final outputs of the first, second, third, and fourth industries are 143.06, 132.98, 188.59, and 125.53 units, respectively.

Chapter 2 Concept Review Questions, page 163

1. **a.** One; many; no   **b.** One; many; no   2. Equations

3. $R_i \longleftrightarrow R_j; cR_i; R_i + aR_j$; solution

4. **a.** Unique   **b.** No; infinitely many; unique

5. Size; entries   6. Size; corresponding

7. $m \times n; n \times m; a_{ji}$ 8. $cA; c$

9. **a.** Columns; rows   **b.** $m \times p$

10. **a.** $A(BC); AB + AC$ **b.** $n \times r$

11. $A^{-1}A; A^{-1}A$; singular   12. $A^{-1}B$

Chapter 2 Review Exercises, page 164

1. $\begin{bmatrix} 2 & 2 \\ -1 & 4 \\ 3 & 3 \end{bmatrix}$ 2. $\begin{bmatrix} -2 & 0 \\ -2 & 6 \end{bmatrix}$ 3. $[-6 \ -2]$ 4. $\begin{bmatrix} 17 \\ 13 \end{bmatrix}$

5. $x = 2; y = 3; z = 1; w = 3$ 6. $x = 2; y = -2$

7. $a = 3; b = 4; c = -2; d = 2; e = -3$

8. $x = -1; y = -2; z = 1$

9. $\begin{bmatrix} 8 & 9 & 11 \\ -10 & -1 & 3 \\ 11 & 12 & 10 \end{bmatrix}$ 10. $\begin{bmatrix} -1 & 7 & -3 \\ -2 & 5 & 11 \\ 10 & -8 & 2 \end{bmatrix}$

11. $\begin{bmatrix} 6 & 18 & 6 \\ -12 & 6 & 18 \\ 24 & 0 & 12 \end{bmatrix}$ 12. $\begin{bmatrix} -10 & 10 & -18 \\ 4 & 14 & 26 \\ 16 & -32 & -4 \end{bmatrix}$

13. $\begin{bmatrix} -11 & -16 & -15 \\ -4 & -2 & -10 \\ -6 & 14 & 2 \end{bmatrix}$ 14. $\begin{bmatrix} 5 & 20 & 19 \\ -2 & 20 & 8 \\ 26 & 10 & 30 \end{bmatrix}$

15. $\begin{bmatrix} -3 & 17 & 8 \\ -2 & 56 & 27 \\ 74 & 78 & 116 \end{bmatrix}$ 16. $\begin{bmatrix} \frac{3}{2} & -2 & -5 \\ \frac{11}{2} & -1 & 11 \\ \frac{7}{2} & -3 & 0 \end{bmatrix}$

17. $x = 1; y = -1$ 18. $x = -1; y = 3$

19. $x = 1; y = 2; z = 3$

20. $(2, 2t - 5, t); t$, a parameter   21. No solution

22. $x = 1; y = -1; z = 2; w = 2$

23. $x = 1; y = 0; z = 1$ 24. $x = 2; y = -1; z = 3$

25. $\begin{bmatrix} \frac{2}{5} & -\frac{1}{5} \\ -\frac{1}{5} & \frac{3}{5} \end{bmatrix}$ 26. $\begin{bmatrix} \frac{3}{4} & -\frac{1}{2} \\ -\frac{1}{8} & \frac{1}{4} \end{bmatrix}$

27. $\begin{bmatrix} -1 & 2 \\ 1 & -\frac{3}{2} \end{bmatrix}$ 28. $\begin{bmatrix} \frac{1}{4} & \frac{1}{2} \\ \frac{1}{8} & -\frac{1}{4} \end{bmatrix}$

29. $\begin{bmatrix} \frac{5}{4} & \frac{1}{4} & -\frac{7}{4} \\ -\frac{1}{4} & -\frac{1}{4} & \frac{3}{4} \\ -\frac{3}{4} & \frac{1}{4} & \frac{5}{4} \end{bmatrix}$ 30. $\begin{bmatrix} -\frac{1}{4} & \frac{1}{2} & -\frac{1}{4} \\ \frac{7}{8} & -\frac{3}{4} & -\frac{5}{8} \\ -\frac{1}{8} & \frac{1}{4} & \frac{3}{8} \end{bmatrix}$

31. $\begin{bmatrix} -\frac{1}{5} & \frac{2}{5} & 0 \\ \frac{2}{3} & -\frac{1}{3} & \frac{1}{3} \\ -\frac{1}{30} & \frac{1}{15} & -\frac{1}{6} \end{bmatrix}$ 32. $\begin{bmatrix} 0 & -\frac{1}{5} & \frac{2}{5} \\ -2 & 1 & 1 \\ -1 & \frac{1}{5} & \frac{3}{5} \end{bmatrix}$

33. $\begin{bmatrix} \frac{3}{2} & 1 \\ -\frac{7}{2} & -1 \end{bmatrix}$ 34. $\begin{bmatrix} \frac{11}{24} & -\frac{7}{8} \\ -\frac{1}{12} & \frac{1}{4} \end{bmatrix}$

35. $\begin{bmatrix} \frac{2}{5} & -\frac{3}{5} \\ \frac{1}{5} & \frac{1}{5} \end{bmatrix}$ 36. $\begin{bmatrix} \frac{4}{7} & -\frac{3}{7} \\ -\frac{3}{7} & \frac{4}{7} \end{bmatrix}$

37. $A^{-1} = \begin{bmatrix} \frac{2}{7} & \frac{3}{7} \\ \frac{1}{7} & -\frac{2}{7} \end{bmatrix}; x = -1; y = -2$

38. $A^{-1} = \begin{bmatrix} \frac{2}{5} & \frac{3}{10} \\ -\frac{1}{5} & \frac{1}{10} \end{bmatrix}; x = 2; y = 1$

39. $A^{-1} = \begin{bmatrix} 1 & -\frac{2}{5} & \frac{4}{5} \\ -1 & 1 & -1 \\ -\frac{1}{2} & \frac{3}{5} & -\frac{7}{10} \end{bmatrix}; x = 1; y = 2; z = 4$

40. $A^{-1} = \begin{bmatrix} 0 & \frac{1}{7} & \frac{2}{7} \\ -1 & -\frac{4}{7} & \frac{6}{7} \\ -\frac{1}{2} & -\frac{1}{2} & \frac{1}{2} \end{bmatrix}$; $x = 3$; $y = -1$; $z = 2$

41. $4330, $4000, and $5300

42. $2,300,000; $2,450,000; an increase of $150,000

43. 30 of each type

44. Houston: 100,000 gallons; Tulsa: 600,000 gallons

Chapter 2 Before Moving On, page 166

1. $\left(\frac{2}{3}, -\frac{2}{3}, \frac{5}{3}\right)$

2. **a.** $(2, -3, 1)$ **b.** No solution   **c.** $(2, 1 - 3t, t)$
   d. $(0, 0, 0, 0)$ **e.** $(2 + t, 3 - 2t, t)$

3. **a.** $(-1, 2)$ **b.** $\left(\frac{4}{7}, -\frac{5}{7} + 2t, t\right)$

4. **a.** $\begin{bmatrix} 3 & 1 & 4 \\ 5 & -2 & 6 \end{bmatrix}$ **b.** $\begin{bmatrix} 14 & 3 & 7 \\ 14 & 5 & 1 \end{bmatrix}$ **c.** $\begin{bmatrix} 0 & 5 & 3 \\ 4 & -1 & -11 \end{bmatrix}$

5. $\begin{bmatrix} 3 & -2 & -5 \\ -3 & 2 & 6 \\ -1 & 1 & 2 \end{bmatrix}$ 6. $(1, -1, 2)$

CHAPTER 3

Exercises 3.1, page 173

1.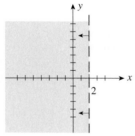

3.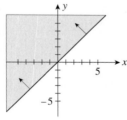

5.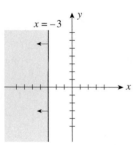

7.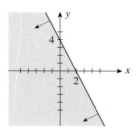

9.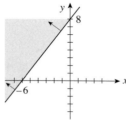

11. $x \geq 1$; $x \leq 5$; $y \geq 2$; $y \leq 4$

13. $2x - y \geq 2$; $5x + 7y \geq 35$; $x \leq 4$

15. $x - y \geq -10$; $7x + 4y \leq 140$; $x + 3y \geq 30$

17. $x + y \geq 7$; $x \geq 2$; $y \geq 3$; $y \leq 7$

19.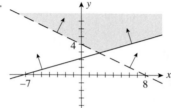
Unbounded

21.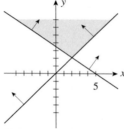
Unbounded

23.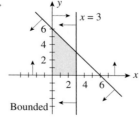
Bounded   No solution

25.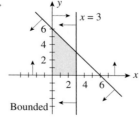
Bounded

27.

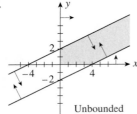

Unbounded

29.

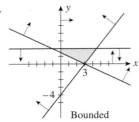

$3x - 7y = -24$

$x + 3y = 8$

Unbounded

31.

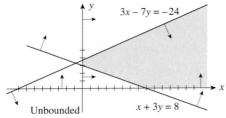

Bounded

33.

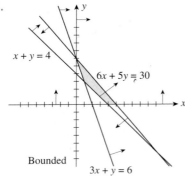

$x + y = 4$

$6x + 5y = 30$

$3x + y = 6$

Bounded

35.

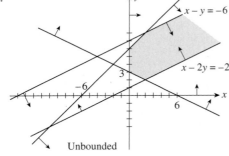

$x - y = -6$

$x - 2y = -2$

Unbounded

37. False   **39.** True

Exercises 3.2, page 181

1. Maximize $P = 3x + 4y$
  subject to $6x + 9y \le 300$
    $5x + 4y \le 180$
    $x \ge 0, y \ge 0$

3. Maximize $P = 2x + 1.5y$
  subject to $3x + 4y \le 1000$
    $6x + 3y \le 1200$
    $x \ge 0, y \ge 0$

5. Maximize $P = 0.1x + 0.12y$
  subject to $x + y \le 20$
    $x - 4y \ge 0$
    $x \ge 0, y \ge 0$

7. Maximize $P = 50x + 40y$
  subject to $\frac{1}{200} x + \frac{1}{200} y \le 1$
    $\frac{1}{100} x + \frac{1}{300} y \le 1$
    $x \ge 0, y \ge 0$

9. Minimize $C = 14{,}000x + 16{,}000y$
  subject to $50x + 75y \ge 650$
    $3000x + 1000y \ge 18{,}000$
    $x \ge 0, y \ge 0$

11. Minimize $C = 2x + 5y$
  subject to $30x + 25y \ge 400$
    $x + 0.5y \ge 10$
    $2x + 5y \ge 40$
    $x \ge 0, y \ge 0$

13. Minimize $C = 1000x + 800y$
  subject to $70{,}000x + 10{,}000y \ge 2{,}000{,}000$
    $40{,}000x + 20{,}000y \ge 1{,}400{,}000$
    $20{,}000x + 40{,}000y \ge 1{,}000{,}000$
    $x \ge 0, y \ge 0$

15. Maximize $P = 0.1x + 0.15y + 0.2z$
  subject to $x + y + z \le 2{,}000{,}000$
    $-2x - 2y + 8z \le 0$
    $-6x + 4y + 4z \le 0$
    $-10x + 6y + 6z \le 0$
    $x \ge 0, y \ge 0, z \ge 0$

17. Maximize $P = 18x + 12y + 15z$
  subject to $2x + y + 2z \le 900$
    $3x + y + 2z \le 1080$
    $2x + 2y + z \le 840$
    $x \ge 0, y \ge 0, z \ge 0$

19. Maximize $P = 26x + 28y + 24z$
  subject to $\frac{5}{4}x + \frac{3}{2}y + \frac{3}{2}z \le 310$
    $x + y + \frac{3}{4}z \le 205$
    $x + y + \frac{1}{2}z \le 190$
    $x \ge 0, y \ge 0, z \ge 0$

21. Minimize $C = 60x_1 + 60x_2 + 80x_3 + 80x_4 + 70x_5 + 50x_6$
subject to $x_1 + x_2 + x_3 \leq 300$
$x_4 + x_5 + x_6 \leq 250$
$x_1 + x_4 \geq 200$
$x_2 + x_5 \geq 150$
$x_3 + x_6 \geq 200$
$x_1 \geq 0, x_2 \geq 0, \ldots, x_6 \geq 0$

23. Maximize $P = x + 0.8y + 0.9z$
subject to $8x \qquad + 4z \leq 16{,}000$
$8x + 12y + 8z \leq 24{,}000$
$4y + 4z \leq \quad 5000$
$z \leq \quad\; 800$
$x \geq 0, y \geq 0, z \geq 0$

25. False

Exercises 3.3, page 192

1. Max: 35; min: 5   **3.** No max. value; min: 27

5. Max: 44; min: 15   **7.** $x = 0$; $y = 6$; $P = 18$

9. Any point (x, y) lying on the line segment joining $(\frac{5}{2}, 0)$ and $(1, 3)$; $P = 5$

11. $x = 0$; $y = 8$; $P = 64$

13. $x = 0$; $y = 4$; $P = 12$

15. $x = 2$; $y = 1$; $C = 10$

17. Any point (x, y) lying on the line segment joining $(20, 10)$ and $(40, 0)$; $C = 120$

19. $x = 14$; $y = 3$; $C = 58$

21. $x = 3$; $y = 3$; $C = 75$

23. $x = 15$; $y = 17.5$; $P = 115$

25. $x = 10$; $y = 38$; $P = 134$

27. Max: $x = 6$; $y = \frac{33}{2}$; $P = 258$
Min: $x = 15$; $y = 3$; $P = 186$

29. 20 product A, 20 product B; $140

31. 120 model A, 160 model B; $480

33. $16 million in homeowner loans, $4 million in auto loans; $2.08 million

35. 50 fully assembled units, 150 kits; $8500

37. Saddle Mine: 4 days; Horseshoe Mine: 6 days; $152,000

39. Infinitely many solutions; 10 oz of food A and 4 oz of food B or 20 oz of food A and 0 oz of food B, etc., with a minimum value of 40 mg of cholesterol

41. 30 in newspaper I, 10 in newspaper II; $38,000

43. 80 from I to A, 20 from I to B, 0 from II to A, 50 from II to B

45. a. $22,500 in growth stocks and $7500 in speculative stocks; maximum return; $5250

47. 750 urban, 750 suburban; $5475

49. False   **51. a.** True   **b.** True

55. a.

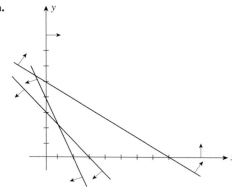

b. No solution

Exercises 3.4, page 209

1. c. $216   **3. a.** Between 750 and 1425
  b. It cannot be decreased by more than 60 units.

5. a. $x = 3$; $y = 2$; $P = 17$ **b.** $\frac{8}{3} \leq c \leq 8$
  c. $8 \leq b \leq 24$ **d.** $\frac{5}{4}$
  e. Both constraints are binding.

7. a. $x = 4$; $y = 0$; $C = 8$ **b.** $0 \leq c \leq \frac{5}{2}$
  c. $b \geq 3$ **d.** 2
  e. Constraint 1 is binding; constraint 2 is nonbinding.

9. a. $x = 3$; $y = 5$; $P = 27$ **b.** $2 \leq c \leq 5$
  c. $21 \leq b \leq 33$ **d.** $\frac{2}{3}$
  e. The first two constraints are binding; the third is nonbinding.

11. a. 20 units of each product   **b.** $\frac{8}{3} \leq c \leq 5$
  c. $216 \leq b \leq 405$ **d.** $\frac{8}{21}$

13. a. Operate Saddle Mine for 4 days, Horseshoe Mine for 6 days; minimum cost of $152,000
  b. $10{,}666\frac{2}{3} \leq c \leq 48{,}000$ **c.** $300 \leq b \leq 1350$
  d. $194.29

15. a. Produce 60 of each; maximum profit of $1320
  b. $8\frac{1}{3} \leq c \leq 15$ **c.** $1100 \leq b \leq 1633\frac{1}{3}$
  d. $0.55
  e. Constraints 1 and 2 are binding; constraint 3 is not.

17. a. 120 model A and 160 model B grates; maximum profit of $480
  b. $1.125 \leq c \leq 3$ **c.** $600 \leq b \leq 1100$
  d. $0.20
  e. Constraints 1 and 2 are binding; constraint 3 is nonbinding.

Chapter 3 Concept Review Questions, page 212

1. a. Half plane; line   **b.** $ax + by \leq c$; $ax + by = c$

2. a. Points; each   **b.** Bounded; enclosed

3. Objective function; maximized; minimized; linear; inequalities

4. **a.** Corner point   **b.** Line

5. Parameters; optimal

6. Resource; amount; value; improved; increased

Chapter 3 Review Exercises, page 213

1. Max: 18—any point (x, y) lying on the line segment joining $(0, 6)$ to $(3, 4)$; Min: 0

2. Max: 27; Min: 7   3. $x = 0$; $y = 4$; $P = 20$

4. $x = 0$; $y = 12$; $P = 36$

5. $x = 3$; $y = 4$; $C = 26$ 6. $x = 1.25$; $y = 1.5$; $C = 9.75$

7. $x = 3$; $y = 10$; $P = 29$ 8. $x = 8$; $y = 0$; $P = 48$

9. $x = 20$; $y = 0$; $C = 40$ 10. $x = 2$; $y = 6$; and $C = 14$

11. Max: $x = 22$; $y = 0$; $Q = 22$; Min: $x = 3$; $y = \frac{5}{2}$; $Q = \frac{11}{2}$

12. Max: $x = 12$; $y = 6$; $Q = 54$; Min: $x = 4$; $y = 0$; $Q = 8$

13. $40,000 in each company; $P = \$13,600$

14. 60 model A clock radios; 60 model B clock radios; $P = \$1320$

15. 93 model A, 180 model B; $P = \$456$

16. 600 to Warehouse I and 400 to Warehouse II

Chapter 3 Before Moving On, page 214

1. **a.**

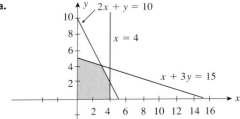

b.

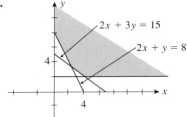

2. Min: $x = 3$, $y = 16$; $C = -7$ 3. Max: $x = 0$, $y = \frac{24}{7}$; $P = \frac{72}{7}$
   Max: $x = 28$, $y = 8$; $P = 76$

4. Min: $x = 0$, $y = 10$; $C = 10$

5. **a.** Max: $x = 4$, $y = 6$; $P = 26$ **b.** Between 1.5 and 4.5
   c. Between 8 and 24   **d.** $1.25
   e. Constraints 1 and 2 are binding.

CHAPTER 4

Exercises 4.1, page 233

1. In final form; $x = \frac{30}{7}$, $y = \frac{20}{7}$, $u = 0$, $v = 0$, and $P = \frac{220}{7}$

3. Not in final form; pivot element is $\frac{1}{2}$, lying in the first row, second column.

5. In final form; $x = \frac{1}{3}$, $y = 0$, $z = \frac{13}{3}$, $u = 0$, $v = 6$, $w = 0$, and $P = 17$

7. Not in final form; pivot element is 1, lying in the third row, second column.

9. In final form; $x = 30$, $y = 10$, $z = 0$, $u = 0$, $v = 0$, $P = 60$ and $x = 30$, $y = 0$, $z = 0$, $u = 10$, $v = 0$, $P = 60$

11. $x = 0$, $y = 4$, $u = 0$, $v = 1$, and $P = 16$

13. $x = 6$, $y = 3$, $u = 0$, $v = 0$, and $P = 96$

15. $x = 6$, $y = 6$, $u = 0$, $v = 0$, $w = 0$, and $P = 60$

17. $x = 0$, $y = 4$, $z = 4$, $u = 0$, $v = 0$, and $P = 36$

19. $x = 0$, $y = 3$, $z = 0$, $u = 90$, $v = 0$, $w = 75$, and $P = 12$

21. $x = 15$, $y = 3$, $z = 0$, $u = 2$, $v = 0$, $w = 0$, and $P = 78$

23. $x = \frac{5}{4}$, $y = \frac{15}{2}$, $z = 0$, $u = 0$, $v = \frac{15}{4}$, $w = 0$, and $P = 90$

25. $x = 2$, $y = 1$, $z = 1$, $u = 0$, $v = 0$, $w = 0$, and $P = 87$

29. No model A, 2500 model B; $100,000

31. 65 acres of crop A, 80 acres of crop B; $25,750

33. $62,500 in the money market fund, $125,000 in the international equity fund, $62,500 in the growth-and-income fund; $25,625

35. 22 min of morning advertising time, 3 min of evening advertising time; maximum exposure: 6.2 million viewers

37. 80 units of model A, 80 units of model B, and 60 units of model C; maximum profit: $5760; no

39. 9000 bottles of formula I, 7833 bottles of formula II, 6000 bottles of formula III; maximum profit: $4986.67; Yes, ingredients for 4167 bottles of formula II

41. Project A: $800,000, project B: $800,000, and project C: $400,000; $280,000

43. False   45. True

Using Technology Exercises 4.1, page 242

1. $x = 1.2$, $y = 0$, $z = 1.6$, $w = 0$, and $P = 8.8$

3. $x = 1.6$, $y = 0$, $z = 0$, $w = 3.6$, and $P = 12.4$

Exercises 4.2, page 252

1. $x = 4$, $y = 0$, and $C = -8$

3. $x = 4$, $y = 3$, and $C = -18$

5. $x = 0$, $y = 13$, $z = 18$, $w = 14$, and $C = -111$

7. $x = \frac{5}{4}$, $y = \frac{1}{4}$, $u = 2$, $v = 3$, and $C = P = 13$

9. $x = 5$, $y = 10$, $z = 0$, $u = 1$, $v = 2$, and $C = P = 80$

11. Maximize $P = 4u + 6v$
    subject to $u + 3v \le 2$
    $2u + 2v \le 5$; $x = 4$, $y = 0$, $C = 8$
    $u \ge 0, v \ge 0$

13. Maximize $P = 60u + 40v + 30w$
    subject to $6u + 2v + w \le 6$
    $u + v + w \le 4$; $x = 10$, $y = 20$, $C = 140$
    $u \ge 0, v \ge 0, w \ge 0$

15. Maximize $P = 10u + 20v$
    subject to $20u + v \le 200$
    $10u + v \le 150$; $x = 0$, $y = 0$, $z = 10$, $C = 1200$
    $u + 2v \le 120$
    $u \ge 0, v \ge 0$

17. Maximize $P = 10u + 24v + 16w$
    subject to: $u + 2v + w \le 6$
    $2u + v + w \le 8$; $x = 8$, $y = 0$, $z = 8$, $C = 80$
    $2u + v + w \le 4$
    $u \ge 0, v \ge 0, w \ge 0$

19. Maximize $P = 6u + 2v + 4w$
    subject to: $2u + 6v \le 30$
    $4u + 6w \le 12$; $x = \frac{1}{3}$, $y = \frac{4}{3}$, $z = 0$, $C = 26$
    $3u + v + 2w \le 20$
    $u \ge 0, v \ge 0, w \ge 0$

21. 2 type-A vessels; 3 type-B vessels; $250,000

23. 30 in newspaper I; 10 in newspaper II; $38,000

25. 8 oz of orange juice; 6 oz of pink grapefruit juice; 178 calories

27. True

Using Technology Exercises 4.2, page 259

1. $x = \frac{4}{3}$, $y = \frac{10}{3}$, $z = 0$, and $C = \frac{14}{3}$

3. $x = 0.9524$; $y = 4.2857$; $z = 0$; $C = 6.0952$

5. **a.** $x = 3$, $y = 2$, and $P = 17$ **b.** $\frac{8}{3} \le c_1 \le 8$; $\frac{3}{2} \le c_2 \le \frac{9}{2}$
   c. $8 \le b_1 \le 24$; $4 \le b_2 \le 12$ **d.** $\frac{5}{4}$; $\frac{1}{4}$
   e. Both constraints are binding.

7. **a.** $x = 4$, $y = 0$, and $C = 8$ **b.** $0 \le c_1 \le \frac{5}{2}$; $4 \le c_2 < \infty$
   c. $3 \le b_1 < \infty$; $-\infty < b_2 \le 4$ **d.** 2; 0
   e. Both constraints are binding.

Exercises 4.3, page 270

1. Maximize $P = -C = -2x + 3y$
   subject to $-3x - 5y \le -20$
   $3x + y \le 16$
   $-2x + y \le 1$
   $x \ge 0, y \ge 0$

3. Maximize $P = -C = -5x - 10y - z$
   subject to $-2x - y - z \le -4$
   $-x - 2y - 2z \le -2$
   $2x + 4y + 3z \le 12$
   $x \ge 0, y \ge 0, z \ge 0$

5. $x = 5$, $y = 2$, and $P = 9$

7. $x = 4$, $y = 0$, and $C = -8$

9. $x = 4$, $y = \frac{2}{3}$, and $P = \frac{20}{3}$

11. $x = 3$, $y = 2$, and $P = 7$

13. $x = 24$, $y = 0$, $z = 0$, and $P = 120$

15. $x = 0$, $y = 17$, $z = 1$, and $C = -33$

17. $x = \frac{46}{7}$, $y = 0$, $z = \frac{50}{7}$, and $P = \frac{142}{7}$

19. $x = 0$, $y = 0$, $z = 10$, and $P = 30$

21. 80 acres of crop A, 68 acres of crop B; $P = \$25,600$

23. $50 million worth of home loans, $10 million worth of commercial-development loans; maximum return: $4.6 million

25. 0 units of product A, 280 units of product B, 280 units of product C; $P = \$7560$

27. 10 oz of food A, 4 oz of food B, 40 mg of cholesterol; infinitely many solutions

Chapter 4 Concept Review Questions, page 274

1. Maximized; nonnegative; less than; equal to

2. Equations; slack variables; $-c_1x_1 - c_2x_2 - \cdots - c_nx_n + P = 0$; below; augmented

3. Minimized; nonnegative; greater than; equal to

4. Dual; objective; optimal value

Chapter 4 Review Exercises, page 274

1. $x = 3$, $y = 4$, $u = 0$, $v = 0$, and $P = 25$

2. $x = 3$, $y = 6$, $u = 4$, $v = 0$, $w = 0$, and $P = 36$

3. $x = \frac{56}{5}$, $y = \frac{2}{5}$, $z = 0$, $u = 0$, $v = 0$, and $P = 23\frac{3}{5}$

4. $x = 0$, $y = \frac{11}{3}$, $z = \frac{25}{6}$, $u = \frac{37}{6}$, $v = 0$, $w = 0$, and $P = \frac{119}{6}$

5. $x = \frac{3}{2}$, $y = 1$, $u = \frac{1}{4}$, $v = \frac{5}{4}$, and $C = \frac{13}{2}$

6. $x = \frac{32}{11}$, $y = \frac{36}{11}$, $u = \frac{2}{11}$, $v = 0$, and $C = \frac{104}{11}$

7. $x = \frac{3}{4}$, $y = 0$, $z = \frac{7}{4}$, $u = 6$, $v = 6$, and $C = 60$

8. $x = 0$, $y = 2$, $z = 0$, $u = 1$, $v = 0$, $w = 0$, and $C = 4$

9. $x = 45$, $y = 0$, $u = 0$, $v = 35$, and $P = 135$

10. $x = 4$, $y = 4$, $u = 2$, $v = 0$, $w = 0$, and $C = 20$

11. $x = 5$, $y = 2$, $u = 0$, $v = 0$, and $P = 16$

12. $x = 20$, $y = 25$, $z = 0$, $u = 0$, $v = 30$, $w = 10$, and $C = -160$

13. Saddle Mine: 4 days; Horseshoe Mine: 6 days; $152,000

14. $70,000 in blue-chip stocks; $0 in growth stocks; $30,000 in speculative stocks; maximum return: $13,000

15. 0 unit product A, 30 units product B, 0 unit product C; $P = $180

16. $50,000 in stocks, $100,000 in bonds, $50,000 in money market funds; $P = $21,500

Chapter 4 Before Moving On, page 275

1.

x	y	z	u	v	w	P	Constant
2	①	−1	1	0	0	0	3
1	−2	3	0	1	0	0	1
3	2	4	0	0	1	0	17
−1	−2	3	0	0	0	1	0

2. $x = 2$; $y = 0$; $z = 11$; $u = 2$; $v = 0$; $w = 0$; $P = 28$

3. Max: $x = 6$, $y = 2$; $P = 34$ 4. Min: $x = 3$, $y = 0$; $C = 3$

5. Max: $x = 10$, $y = 0$; $P = 20$

CHAPTER 5

Exercises 5.1, page 289

1. $80; $580   3. $836   5. $1000   7. 146 days

9. 10%/yr   11. $1718.19   13. $4974.47

15. $27,566.93   17. $261,751.04   19. $214,986.69

21. $10\frac{1}{4}$%/yr   23. 8.3%/yr   25. $29,277.61

27. $30,255.95   29. $6885.64   31. 6.08%/yr

33. 2.06 yr   35. 24%/yr   37. $123,600

39. 5%/yr   41. $705.28   43. $255,256

45. $2.58 million   47. $22,163.75   49. $26,267.49

51. a. $34,626.88   b. $33,886.16   c. $33,506.76

53. Acme Mutual Fund   55. $23,329.48   57. 4.2%

59. 5.75%   61. 5.12%/yr   63. $115.3 billion

65. Investment A   67. $33,885.14; $33,565.38

69. $80,000e^{(\sqrt{t/2} - 0.09t)}$; $151,718   73. True   75. True

Using Technology Exercises 5.1, page 296

1. $5872.78   3. $475.49   5. 8.95%/yr

7. 10.20%/yr   9. $29,743.30   11. $53,303.25

Exercises 5.2, page 304

1. $15,937.42   3. $54,759.35   5. $37,965.57

7. $137,209.97   9. $28,733.19   11. $15,558.61

13. $15,011.29   15. $109,658.91   17. $455.70

19. $44,526.45   21. Karen   23. $9850.12   25. $608.54

27. Between $167,712 and $203,390

29. Between $143,312 and $172,890   31. $17,887.62

33. False

Using Technology Exercises 5.2, page 308

1. $59,622.15   3. $8453.59   5. $35,607.23

7. $13,828.60

Exercises 5.3, page 315

1. $14,902.95   3. $444.24   5. $622.13

7. $731.79   9. $1491.19   11. $516.76

13. $172.95   15. $1957.36   17. $3450.87

19. $16,274.54   21. a. $212.27   b. $1316.36; $438.79

23. a. $387.21; $304.35   b. $1939.56; $2608.80

25. $1761.03; $41,833; $59,461; $124,853   27. $60,982.31

29. $3135.48   31. $242.23   33. $199.07

35. $2090.41; $4280.21   37. $33,835.20

39. $167,341.33   41. $152,018.20

43. a. $1681.71   b. $194,282.67   c. $1260.11   d. $421.60

45. $71,799

Using Technology Exercises 5.3, page 321

1. $3645.40   3. $18,443.75   5. $1863.99

7. $707.96   9. $18,288.92

Exercises 5.4, page 328

1. 30   3. −4.5   5. −3, 8, 19, 30, 41   7. $x + 6y$

9. 795   11. 792   13. 550

15. a. 275   b. −280   17. 37 wk   19. $7.90

21. b. $800   23. GP; 256; 508   25. Not a GP

27. GP; 1/3; $364\frac{1}{3}$ 29. 3; 0   31. 293,866

33. $38,209.67   35. Annual raise of 8%/yr

37. a. $20,113.57   b. $87,537.38   39. $25,165.82

41. $39,321.60; $110,678.40   43. True

Chapter 5 Concept Review Questions, page 331

1. **a.** Original; $P(1 + rt)$ **b.** Interest; $P(1 + i)^n$; $A(1 + i)^{-n}$

2. Simple; one; nominal; m; $\left(1 + \dfrac{r}{m}\right)^m - 1$

3. Annuity; ordinary annuity; simple annuity

4. **a.** $R\left[\dfrac{(1 + i)^n - 1}{i}\right]$ **b.** $R\left[\dfrac{1 - (1 + i)^{-n}}{i}\right]$

5. $\dfrac{Pi}{1 - (1 + i)^{-n}}$ 6. Future; $\dfrac{iS}{(1 + i)^n - 1}$

7. Constant d; $a(n - 1)d$; $\dfrac{n}{2}[2a + (n - 1)d]$

8. Constant r; ar^{n-1}; $\dfrac{a(1 - r^n)}{1 - r}$

Chapter 5 Review Exercises, page 332

1. **a.** $7320.50   **b.** $7387.28   **c.** $7422.53
   d. $7446.77

2. **a.** $19,859.95   **b.** $20,018.07   **c.** $20,100.14
   d. $20,156.03

3. **a.** 12%   **b.** 12.36%   **c.** 12.5509%   **d.** 12.6825%

4. **a.** 11.5%   **b.** 11.8306%   **c.** 12.0055%   **d.** 12.1259%

5. $30,000.29   6. $39,999.95   7. $5557.68

8. $23,221.71   9. $7861.70   10. $173,804.43

11. $694.49   12. $318.93   13. $332.73   14. $208.44

15. 7.179%   16. 9.563%   17. 12%/yr   18. $80,000

19. $2,592,702; $8,612,002   20. $5,491,922   21. $2982.73

22. $15,000   23. $5000   24. 7.6%   25. $218.64

26. $73,178.41   27. $13,026.89   28. $2000

29. **a.** $965.55   **b.** $227,598   **c.** $42,684

30. **a.** $1217.12   **b.** $99,081.60   **c.** $91,367

31. $19,573.56   32. $4727.67   33. $205.09; 20.27%/yr

34. $2203.83

Chapter 5 Before Moving On, page 334

1. $2540.47   2. 6.2%/yr   3. $569,565.47   4. $1213.28

5. $35.13   6. **a.** 210   **b.** 127.5

CHAPTER 6

Exercises 6.1, page 343

1. $\{x \mid x$ is a gold medalist in the 2002 Winter Olympic Games$\}$

3. $\{x \mid x$ is an integer greater than 2 and less than 8$\}$

5. $\{2, 3, 4, 5, 6\}$ 7. $\{-2\}$

9. **a.** True   **b.** False   11. **a.** False   **b.** False

13. True   15. **a.** True   **b.** False

17. **a.** and **b.**

19. **a.** $\varnothing$, $\{1\}$, $\{2\}$, $\{1, 2\}$
    b. $\varnothing$, $\{1\}$, $\{2\}$, $\{3\}$, $\{1, 2\}$, $\{1, 3\}$, $\{2, 3\}$, $\{1, 2, 3\}$
    c. $\varnothing$, $\{1\}$, $\{2\}$, $\{3\}$, $\{4\}$, $\{1, 2\}$, $\{1, 3\}$, $\{1, 4\}$, $\{2, 3\}$, $\{2, 4\}$, $\{3, 4\}$, $\{1, 2, 3\}$, $\{1, 2, 4\}$, $\{1, 3, 4\}$, $\{2, 3, 4\}$, $\{1, 2, 3, 4\}$

21. $\{1, 2, 3, 4, 6, 8, 10\}$

23. $\{$Jill, John, Jack, Susan, Sharon$\}$

25. **a.**

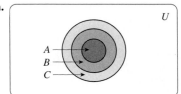

b.

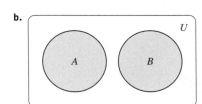

c.

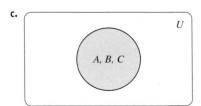

27. **a.**

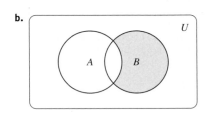

b.

29. a.

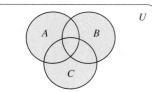

b.

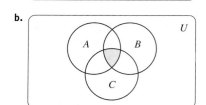

31. a.

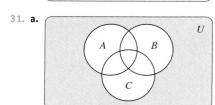

b.

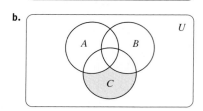

33. a. {2, 4, 6, 8, 10}   **b.** {1, 2, 4, 5, 6, 8, 9, 10}
   c. U

35. a. $C = \{1, 2, 4, 5, 8, 9\}$ **b.** $\varnothing$ **c.** U

37. a. Not disjoint   **b.** Disjoint

39. a. The set of all employees at Universal Life Insurance who do not
      drink tea
   b. The set of all employees at Universal Life Insurance who do not
      drink coffee

41. a. The set of all employees at Universal Life Insurance who drink
      tea but not coffee
   b. The set of all employees at Universal Life Insurance who drink
      coffee but not tea

43. a. The set of all employees in a hospital who are not doctors
   b. The set of all employees in a hospital who are not nurses

45. a. The set of all employees in a hospital who are female doctors
   b. The set of all employees in a hospital who are both doctors and
      administrators

47. a. $D \cap F$ **b.** $R \cap F^c \cap L^c$

49. a. B^c **b.** $A \cap B$ **c.** $A \cap B \cap C^c$

51. a. $A \cap B \cap C$; the set of tourists who have taken the underground,
      a cab, and a bus over a 1-wk period in London
   b. $A \cap C$; the set of tourists who have taken the underground and
      a bus over a 1-wk period in London
   c. B^c; the set of tourists who have not taken a cab over a 1-wk
      period in London

53. a.

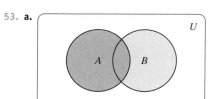

b.

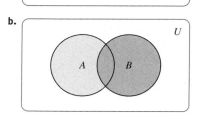

55.

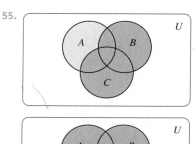

57.

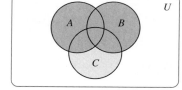

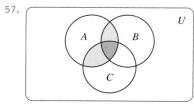

61. a. x, y, v, r, w, u **b.** v, r

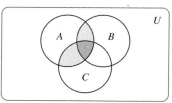

63. a. s, t, y **b.** t, z, w, x, s **65.** $A \subset C$

67. False   **69.** True   **71.** True

Exercises 6.2, page 349

3. a. 4   **b.** 5   **c.** 7   **d.** 2

7. 20

9. a. 140   **b.** 100   **c.** 60

11. 13   **13.** 0   **15.** 13   **17.** 61

19. a. 106   **b.** 64   **c.** 38   **d.** 14

21. a. 182   **b.** 118   **c.** 56   **d.** 18   **23.** 30

25. a. 64   **b.** 10   **27. a.** 36   **b.** 36   **29.** 5

31. a. 62   **b.** 33   **c.** 25   **d.** 38

33. a. 108   **b.** 15   **c.** 45   **d.** 12

35. a. 22   **b.** 80

37. True   **39.** True

Exercises 6.3, page 357

1. 12   **3.** 64   **5.** 24

7. 24   **9.** 60   **11.** 1 billion   **13.** 5^{50}

15. 30   **17.** 9990

19. a. 17,576,000   **b.** 17,576,000

21. 1024; 59,049   **23.** 2730

25. 217   **27.** True

Exercises 6.4, page 369

1. 360   **3.** 10   **5.** 120

7. 20   **9.** n **11.** 1

13. 35   **15.** 1   **17.** 84

19. $\dfrac{n(n-1)}{2}$ **21.** $\dfrac{n!}{2}$

23. Permutation   **25.** Combination

27. Permutation   **29.** Combination

31. $P(4, 4) = 24$ **33.** $P(4, 4) = 24$

35. $P(9, 9) = 362,880$ **37.** $C(12, 3) = 220$

39. 151,200   **41.** $C(12, 3) = 220$

43. $C(100, 3) = 161,700$ **45.** $P(6, 6) = 720$

47. $P(12, 6) = 665,280$

49. a. $P(10, 10) = 3,628,800$
 b. $P(3, 3)P(4, 4)P(3, 3)P(3, 3) = 5184$

51. a. $P(20, 20) = 20!$
 b. $P(5, 5)P[(4, 4)]^5 = 5!(4!)^5 = 955,514,880$

53. $P(2, 1)P(3, 1) = 6$

55. $C(3,3)[C(8, 6) + C(8, 7) + C(8, 8)] = 37$

57. a. $C(12, 3) = 220$ **b.** $C(11, 2)] = 55$
 c. $C(5, 1)C(7, 2) + C(5, 2)C(7, 1) + C(5, 3) = 185$

59. $P(7, 3) + C(7, 2)P(3, 2) = 336$

61. $[C(5, 1)C(3, 1)C(6, 2)][C(4, 1) + C(3, 1)] = 1575$

63. $C(10, 8) + C(10, 9) + C(10, 10) = 56$

65. $10C(4, 1) = 40$

67. $4C(13, 5) - 40 = 5108$

69. $13C(4, 3)12C(4, 2) = 3744$

71. $C(6, 2) = 15$

73. $C(12, 6) + C(12, 7) + C(12, 8) + C(12, 9) + C(12, 10) + C(12, 11) + C(12, 12) = 2510$

75. $4! = 24$ **79.** True   **81.** True

Using Technology Exercises 6.4, page 374

1. $1.307674368 \times 10^{12}$ **3.** $2.56094948229 \times 10^{16}$

5. 674,274,182,400   **7.** 133,784,560   **9.** 4,656,960

11. 658,337,004,000

Chapter 6 Concept Review Questions, page 375

1. Set; elements; set   **2.** Equal   **3.** Subset

4. a. No   **b.** All   **5.** Union; intersection

6. Complement   **7.** $A^C \cap B^C \cap C^C$

8. Permutation; combination

Chapter 6 Review Exercises, page 375

1. {3}   **2.** {A, E, H, L, S, T}

3. {4, 6, 8, 10}   **4.** {−4}   **5.** Yes   **6.** Yes   **7.** Yes

8. No

9.

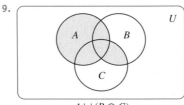

$A \cup (B \cap C)$

10.

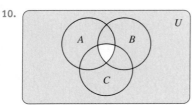

$(A \cap B \cap C)^C$

11.

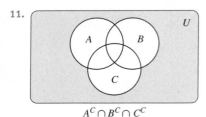

$A^C \cap B^C \cap C^C$

12.

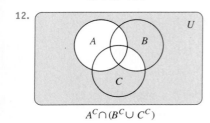

$A^C \cap (B^C \cup C^C)$

17. The set of all participants in a consumer-behavior survey who both avoided buying a product because it is not recyclable and boycotted a company's products because of its record on the environment.

18. The set of all participants in a consumer-behavior survey who avoided buying a product because it is not recyclable and/or voluntarily recycled their garbage.

19. The set of all participants in a consumer-behavior survey who both did not use cloth diapers rather than disposable diapers and voluntarily recycled their garbage.

20. The set of all participants in a consumer-behavior survey who did not boycott a company's products because of the company's record on the environment and/or who do not voluntarily recycle their garbage.

21. 150   **22.** 230   **23.** 270   **24.** 30   **25.** 70   **26.** 200

27. 190   **28.** 181,440   **29.** 120   **30.** 8400   **31.** None

32. a. 446   **b.** 377   **c.** 34   **33.** 720   **34.** 20

35. a. 50,400   **b.** 5040   **36. a.** 60   **b.** 125

37. 80   **38. a.** 1287   **b.** 288

39. 720   **40.** 1050   **41. a.** 5040   **b.** 3600

42. a. 487,635   **b.** 550   **c.** 341,055

43. a. $C(15, 4) = 1365$ **b.** $C(15, 4) - C(10, 4) = 1155$

Chapter 6 Before Moving On, page 377

1. a. $\{d, f, g\}$ **b.** $\{b, c, d, e, f, g\}$ **c.** $\{b, c, e\}$

2. 3   **3.** 360   **4.** 264   **5.** 200

CHAPTER 7

Exercises 7.1, page 385

1. $\{a, b, d, f\}$; $\{a\}$ **3.** $\{b, c, e\}$; $\{a\}$ **5.** No   **7.** S

9. $\varnothing$ **11.** Yes   **13.** Yes   **15.** $E \cup F$ **17.** G^c

19. $(E \cup F \cup G)^c$

21. a. $\{(2, 1), (3, 1), (4, 1), (5, 1), (6, 1), (3, 2), (4, 2), (5, 2), (6, 2),$ $(4, 3), (5, 3), (6, 3), (5, 4), (6, 4), (6, 5)\}$
    b. $\{(1, 2), (2, 4), (3, 6)\}$

23. $\varnothing, \{a\}, \{b\}, \{c\}, \{a, b\}, \{a, c\}, \{b, c\}, S$

25. a. $S = \{B, R\}$ **b.** $\varnothing, \{B\}, \{R\}, S$

27. a. $S = \{(H, 1), (H, 2), (H, 3), (H, 4), (H, 5), (H, 6), (T, 1),$ $(T, 2), (T, 3), (T, 4), (T, 5), (T, 6)\}$
    b. $\{(H, 2), (H, 4), (H, 6)\}$

29. a. No   **b.** No

31. $S = \{ddd, ddn, dnd, ndd, dnn, ndn, nnd, nnn\}$

33. a. $\{ABC, ABD, ABE, ACD, ACE, ADE, BCD, BCE, BDE, CDE\}$
    b. 6   **c.** 3   **d.** 6

35. a. E^c **b.** $E^c \cap F^c$ **c.** $E \cup F$
    d. $(E \cap F^c) \cup (E^c \cap F)$

37. a. $\{t \mid t > 0\}$ **b.** $\{t \mid 0 < t \le 2\}$ **c.** $\{t \mid t > 2\}$

39. a. $S = \{0, 1, 2, 3, \ldots, 10\}$ **b.** $E = \{0, 1, 2, 3\}$
    c. $F = \{5, 6, 7, 8, 9, 10\}$

41. a. $S = \{0, 1, 2, \ldots, 20\}$
    b. $E = \{0, 1, 2, \ldots, 9\}$ **c.** $F = \{20\}$

47. False

Exercises 7.2, page 393

1. $\{(H, H)\}, \{(H, T)\}, \{(T, H)\}, \{(T, T)\}$

3. $\{(D, m)\}, \{(D, f)\}, \{(R, m)\}, \{(R, f)\}, \{(I, m)\}, \{(I, f)\}$

5. $\{(1, i)\}, \{(1, d)\}, \{(1, s)\}, \{(2, i)\}, \{(2, d)\}, \{(2, s)\}, \ldots,$ $\{(5, i)\}, \{(5, d)\}, \{(5, s)\}$

7. $\{(A, Rh^+)\}, \{(A, Rh^-)\}, \{(B, Rh^+)\}, \{(B, Rh^-)\},$ $\{(AB, Rh^+)\}, \{(AB, Rh^-)\}, \{(O, Rh^+)\}, \{(O, Rh^-)\}$

9.

Grade	A	B	C	D	F
Probability	.10	.25	.45	.15	.05

11. a. $S = \{(0 < x \le 200), (200 < x \le 400),$
$(400 < x \le 600), (600 < x \le 800),$
$(800 < x \le 1000), (x > 1000)\}$

b.

Cars, x	Probability
$0 < x \le 200$	.075
$200 < x \le 400$	.1
$400 < x \le 600$	.175
$600 < x \le 800$	.35
$800 < x \le 1000$	.225
$x > 1000$	.075

13.

Opinion	Favor	Oppose	Don't know
Probability	.47	.46	.07

15.

Event	A	B	C	D	E
Probability of an Event	.026	.199	.570	.193	.012

17.

Figures Produced (in dozens)	30	31	32
Probability	.125	0	.1875

Figures Produced (in dozens)	33	34	35	36
Probability	.25	.1875	.125	.125

19. .469

21.

Income, $	0–24,999	25,000–49,999	50,000–74,999	75,000–99,999
Probability	.287	.293	.195	.102

Income, $	100,000–124,999	125,000–149,999	150,000–199,999	200,000 or more
Probability	.052	.025	.022	.024

23. a. .856   **b.** .144   **25.** .46

27. a. $\frac{1}{4}$ **b.** $\frac{1}{2}$ **c.** $\frac{1}{13}$ **29.** $\frac{3}{8}$

31. .95   **33. a.** .633   **b.** .276

35. There are two ways of obtaining a sum of 7.

37. No   **39.** No   **41. a.** $\frac{3}{8}$ **b.** $\frac{1}{2}$ **c.** $\frac{1}{4}$

43. .783   **45.** False

Exercises 7.3, page 403

1. $\frac{1}{2}$ **3.** $\frac{1}{36}$ **5.** $\frac{1}{9}$ **7.** $\frac{1}{52}$

9. $\frac{3}{13}$ **11.** $\frac{12}{13}$ **13.** .002; .998

15. $P(a) + P(b) + P(c) \ne 1$

17. Since the five events are not mutually exclusive, Property (3) cannot be used; that is, he could win more than one purse.

19. The two events are not mutually exclusive; hence, the probability of the given event is $\frac{1}{6} + \frac{1}{6} - \frac{1}{36} = \frac{11}{36}$.

21. $E^C \cap F^C = \{e\} \ne \varnothing$

23. $P(G \cup C)^C \ne 1 - P(G) - P(C)$; he has not considered the case in which a customer buys both glasses and contact lenses.

25. a. 0   **b.** .7   **c.** .8   **d.** .3

27. a. $\frac{1}{2}, \frac{3}{8}$ **b.** $\frac{1}{2}, \frac{5}{8}$ **c.** $\frac{1}{8}$ **d.** $\frac{3}{4}$ **29.** .33

31. a. .16   **b.** .38   **c.** .22   **33. a.** .38   **b.** .58

35. a. .2   **b.** .34   **37. a.** .68   **b.** .87

39. a. .90   **b.** .40   **c.** .40

41. a. .6   **b.** .332   **c.** .232   **d.** .6

45. True   **47.** False

Exercises 7.4, page 412

1. $\frac{1}{32}$ **3.** $\frac{31}{32}$

5. $P(A) = 13C(4, 2)/C(52, 2) \approx .0588$

7. $C(26, 2)/C(52, 2) = .245$

9. $[C(3, 2)C(5, 2)]/C(8, 4) = 3/7$

11. $[C(5, 3)C(3, 1)]/C(8, 4) = 3/7$ **13.** $[C(3, 2)C(1, 1)]/8 = 3/8$

15. 1/8   **17.** $C(10, 6)/2^{10} \approx .205$

19. a. $C(4, 2)/C(24, 2) \approx .022$
b. $1 - C(20, 2)/C(24, 2) \approx .312$

21. a. $C(6, 2)/C(80, 2) \approx .005$
b. $1 - C(74, 2)/C(80, 2) \approx .145$

23. a. .12; $C(98, 10)/C(100, 12) \approx .013$
b. .15; .015

25. $[C(12, 8)C(8, 2) + C(12, 9)C(8, 1) + C(12, 10)]/C(20, 10) \approx .085$

27. a. $\frac{3}{5}$ **b.** $C(3, 1)/C(5, 3) = .3$ **c.** $1 - C(3, 3)/C(5, 3) = .9$

29. $\frac{1}{729}$ **31.** .0001   **33.** .10   **35.** $40/C(52, 5) \approx .0000154$

37. $[4C(13, 5) - 40]/C(52, 5) \approx .00197$

39. $[13C(4, 3)12C(4, 2)]/C(52, 5) \approx .00144$

41. a. .618   **b.** .059   **43.** .03

Exercises 7.5, page 425

1. a. .4   **b.** .33   **3.** .3   **5.** Independent

7. Independent   **9. a.** .24   **b.** .76

11. a. .5   **b.** .4   **c.** .2   **d.** .35   **e.** No   **f.** No

13. a. .4   **b.** .3   **c.** .12   **d.** .30   **e.** Yes   **f.** Yes

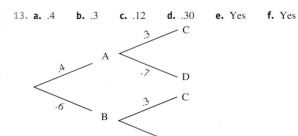

15. a. $\frac{1}{12}$ **b.** $\frac{1}{36}$ **c.** $\frac{1}{6}$ **d.** $\frac{1}{6}$ **e.** No

17. $\frac{4}{11}$ **19.** Independent   **21.** Not independent   **23.** .1875

25. a. $\frac{4}{9}$ **b.** $\frac{4}{9}$ **27. a.** $\frac{1}{21}$ **b.** $\frac{1}{3}$ **29.** .25   **31.** $\frac{1}{7}$

33. a.

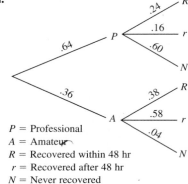

P = Professional
A = Amateur
R = Recovered within 48 hr
r = Recovered after 48 hr
N = Never recovered

b. .24   **c.** .40

35. a. .16   **b.** .424   **c.** .1696

37. a. .092   **b.** ≈ .008

39. a. .280; .390; .180; .643; .292   **b.** Not independent

41. Not independent   **43.** .0000068   **45. a.** $\frac{7}{10}$ **b.** $\frac{1}{5}$

47. 3   **51.** 1   **53.** True   **55.** True

Exercises 7.6, page 433

1.

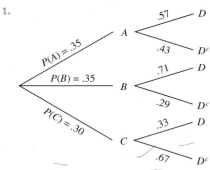

3. a. .45   **b.** .22   **5. a.** .48   **b.** .33
7. a. .08   **b.** .15   **c.** .35

9. a. $\frac{1}{12}$ **b.** $\frac{1}{4}$ **c.** $\frac{1}{18}$ **d.** $\frac{3}{14}$

11. $\frac{4}{17}$ **13.** .0784

15.

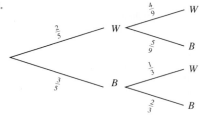

17. .53   **19.** .422   **21. a.** $\frac{3}{4}$ **b.** $\frac{2}{9}$ **23.** .856

25. a. .03   **b.** .29

27. .35   **29. a.** .30   **b.** .10

31. a. .513   **b.** .390   **33.** .65   **35.** .93

37. .3010   **39.** .3758   **41.** .2056

Chapter 7 Concept Review Questions, page 439

1. Experiment; sample; space; event   **2.** ∅   **3.** Uniform; $\frac{1}{n}$

4. Conditional   **5.** Independent   **6.** A posteriori probability

Chapter 7 Review Exercises, page 439

1. a. 0 **b.** .6 **c.** .6 **d.** .4 **e.** 1

2. a. .35   **b.** .65   **c.** .05

3. a. .49   **b.** .39   **c.** .48

4. $\frac{2}{7}$ **5. a.** .019   **b.** .981   **6.** .364

7. .18   **8.** .25   **9.** .06   **10.** .367   **11.** .49

12. a. $\frac{7}{8}$ **b.** $\frac{7}{8}$ **c.** No   **13. a.** .284   **b.** .984

14. .150   **15.** $\frac{2}{15}$ **16.** $\frac{1}{24}$ **17.** $\frac{1}{52}$ **18.** .00018

19. .00995   **20.** .2451   **21.** .510   **22.** .2451

23. a. .17   **b.** .16   **24.** .457   **25.** .368

26. a. .68   **b.** .053   **27.** .32   **28.** .619

Chapter 7 Before Moving On, page 441

1. $\frac{5}{12}$ **2.** $\frac{4}{13}$ **3. a.** .9   **b.** .3   **4.** .72   **5.** .3077

CHAPTER 8

Exercises 8.1, page 449

1. a. See part (b)

b.

Outcome	GGG	GGR	GRG	RGG
Value	3	2	2	2

Outcome	GRR	RGR	RRG	RRR
Value	1	1	1	0

c. {GGG}

3. Any positive integer   **5.** $\frac{1}{6}$

7. Any positive integer; infinite discrete

9. $0 \le x < \infty$; continuous

11. Any positive integer; infinite discrete

13. a. .20   **b.** .60   **c.** .30   **d.** 1

15.

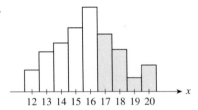

17. a.

x	1	2	3	4	5	6
$P(X = x)$	$\frac{1}{6}$	$\frac{1}{6}$	$\frac{1}{6}$	$\frac{1}{6}$	$\frac{1}{6}$	$\frac{1}{6}$

y	1	2	3	4	5	6
$P(Y = y)$	$\frac{1}{6}$	$\frac{1}{6}$	$\frac{1}{6}$	$\frac{1}{6}$	$\frac{1}{6}$	$\frac{1}{6}$

b.

x + y	2	3	4	5	6	7
$P(X + Y = x + y)$	$\frac{1}{36}$	$\frac{2}{36}$	$\frac{3}{36}$	$\frac{4}{36}$	$\frac{5}{36}$	$\frac{6}{36}$

x + y	8	9	10	11	12
$P(X + Y = x + y)$	$\frac{5}{36}$	$\frac{4}{36}$	$\frac{3}{36}$	$\frac{2}{36}$	$\frac{1}{36}$

19. a.

x	0	1	2	3	4
$P(X = x)$	.017	.067	.033	.117	.233

x	5	6	7	8	9	10
$P(X = x)$	.133	.167	.1	.05	.067	.017

b.

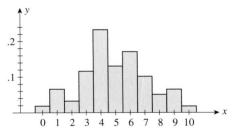

21.

x	1	2	3	4	5
$P(X = x)$	.007	.029	.021	.079	.164

x	6	7	8	9	10
$P(X = x)$	.15	.20	.207	.114	.029

23. True

Using Technology Exercises 8.1, page 453

Graphing Utility

1.

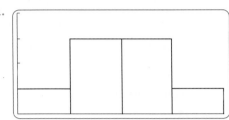

3.

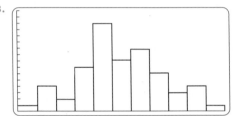

Excel

1.

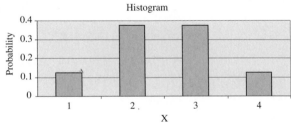

3.

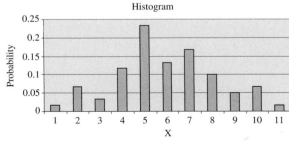

Exercises 8.2, page 463

1. a. 2.6

b.

x	0	1	2	3	4	
$P(X = x)$	0	.1	.4	.3	.2	; 2.6

3. 0.86   **5.** $78.50   **7.** 0.91

9. 0.12   **11.** 1.73   **13.** 5.16%

15. −39¢   **17.** $50   **19.** $12,000

21. City B   **23.** Company B   **25.** 2.86%

27. −5.3¢   **29.** −2.7¢   **31.** 2 to 3; 3 to 2

33. 0.4   **35.** .5833   **37.** ≈ .3571

39. a. Mean: 74; mode: 85; median: 80   **b.** Mode

41. 3; close   **43.** 16; 16; 16   **45.** True

Exercises 8.3, page 473

1. $\mu = 2$, Var$(X) = 1$, $\sigma = 1$

3. $\mu = 0$, Var$(X) = 1$, $\sigma = 1$

5. $\mu = 518$, Var$(X) = 1891$, $\sigma \approx 43.5$

7. Figure (a)   **9.** 1.56

11. $\mu = 4.5$, Var$(X) = 5.25$

13. a. Let X = the annual birthrate during the years 1991–2000

b.

x	14.5	14.6	14.7	14.8	15.2	15.5	15.9	16.3
$P(X = x)$	.2	.1	.2	.1	.1	.1	.1	.1

  c. $\mu \approx 15.07$, Var$(X) \approx 0.3621$, $\sigma \approx 0.6017$

15. a. Mutual fund A: $\mu = \$620$, Var$(X) = \$267,600$;
  Mutual fund B: $\mu = \$520$, Var$(X) = \$137,600$
  b. Mutual fund A   **c.** Mutual fund B

17. 1

19. $\mu = \$239,600$; Var$(X) = \$1,443,840,000$; $\sigma \approx \$37,998$

21. $\mu \approx 77.17$; $\sigma \approx 10.62$

23. $\mu = 1607.33$; $\sigma \approx 182.29$

25. $\mu = 5.452$; $\sigma \approx 0.1713$

27. 16.88 million/mo; 0.6841 million

29. a. At least .75   **b.** At least .96

31. $c = 7$ **33.** At least 7/16   **35.** .9375   **37.** True

Using Technology Exercises 8.3, page 479

1. a.

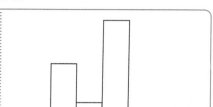

b. $\mu = 4$,
   $\sigma = 1.40$

3. a.

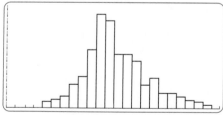

b. $\mu = 17.34$,
   $\sigma = 1.11$

5. a. Let X denote the random variable that gives the weight of a carton of sugar.

b.

x	4.96	4.97	4.98	4.99	5.00	5.01
$P(X = x)$	$\frac{3}{30}$	$\frac{4}{30}$	$\frac{4}{30}$	$\frac{1}{30}$	$\frac{1}{30}$	$\frac{5}{30}$

x	5.02	5.03	5.04	5.05	5.06
$P(X = x)$	$\frac{3}{30}$	$\frac{3}{30}$	$\frac{4}{30}$	$\frac{1}{30}$	$\frac{1}{30}$

c. $\mu \approx 5.00$; $\sigma \approx 0.03$

7. a.

b. 65.875; 1.73

Exercises 8.4, page 487

1. Yes

3. No. There are more than two outcomes to the experiment.

5. No. The probability of an accident on a clear day is not the same as the probability of an accident on a rainy day.

7. .296   **9.** .0512   **11.** .132

13. $\frac{21}{32}$ **15.** .0041   **17.** .116

19. a. $P(X = 0) \approx .08$; $P(X = 1) \approx .26$;
$P(X = 2) \approx .35$; $P(X) = 3) \approx .23$;
$P(X = 4) \approx .08$; $P(X = 5) \approx .01$

b.

x	0	1	2	3	4	5
$P(X = x)$	.08	.26	.35	.23	.08	.01

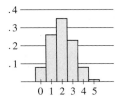

c. $\mu = 2$; $\sigma \approx 1.1$

21. No. The probability that at most 1 is defective is
$P(X = 0) + P(X = 1) = .74$.

23. .165   **25.** $\approx .0002$

27. a. $\approx .633$ **b.** $\approx .367$ **29. a.** $\approx .0329$ **b.** $\approx .3292$

31. a. $\approx .273$ **b.** $\approx .650$ **33. a.** $\approx .817$ **b.** $\approx .999$

35. a. $\approx .075$ **b.** $\approx .012$ **37. a.** $\approx .1216$ **b.** $\approx .3585$

39. $\approx .3487$ **41.** $\mu = 375$; $\sigma \approx 9.68$ **43.** False   **45.** False

Exercises 8.5, page 497

1. .9265   **3.** .0401   **5.** .8657

7. a.

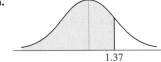

1.37

b. .9147

9. a.

−0.65

b. .2578

11. a.

−1.25

b. .8944

13. a.

0.68 2.02

b. .2266

15. a. 1.23   **b.** −0.81   **17. a.** 1.9   **b.** −1.9

19. a. .9772   **b.** .9192   **c.** .7333

Exercises 8.6, page 505

1. a. .2206   **b.** .2206   **c.** .2960

3. a. .0228   **b.** .0228   **c.** .4772   **d.** .7258

5. a. .0038   **b.** .0918   **c.** .4082   **d.** .2514

7. .6247   **9.** 0.62%

11. A: 80; B: 77; C: 73; D: 62; F: 54

13. a. .4207   **b.** .4254   **c.** .0125

15. a. .2877   **b.** .0008   **c.** .7287

17. .9265   **19.** .8686

21. a. .0037   **b.** The drug is very effective.

23. 2142

Chapter 8 Concept Review Questions, page 508

1. Random   **2.** Finite; infinite; continuous   **3.** Sum; .75

4. a. $\dfrac{P(E)}{P(E^C)}$ **b.** $\dfrac{a}{a + b}$

5. $p_1(x_1 - \mu)^2 + p_2(x_2 - \mu)^2 + \cdots + p_n(x_n - \mu)^2$; $\sqrt{\text{Var}(X)}$

6. Fixed; two; same; independent

7. Continuous; probability density function; set

8. Normal; large; 0; 1

Chapter 8 Review Exercises, page 508

1. a. {WWW, BWW, WBW, WWB, BBW, BWB, WBB, BBB}

b.

Outcome	WWW	BWW	WBW	WWB
Value of X	0	1	1	1

Outcome	BBW	BWB	WBB	BBB
Value of X	2	2	2	3

c.

x	0	1	2	3
$P(X = x)$	$\frac{1}{35}$	$\frac{12}{35}$	$\frac{18}{35}$	$\frac{4}{35}$

d.

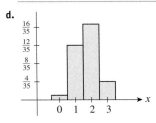

2. $100

3. a. .8   **b.** $\mu = 2.7$; $\sigma \approx 1.42$

4. a.

x	0	1	2	3	4
$P(X = x)$	.1296	.3456	.3456	.1536	.0256

b. $\mu = 1.6$; $V(x) = 0.96$; $\sigma \approx 0.9798$

5. .6915

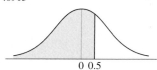

0 0.5

6. .2266

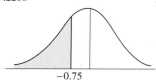

−0.75

7. .4649

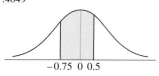

−0.75 0 0.5

8. .4082

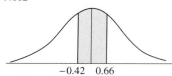

−0.42   0.66

9. 2.42   **10.** −1.05   **11.** −2.03   **12.** 1.42   **13.** .6915

14. .8413   **15.** .2417   **16.** .7333   **17.** .2646; .9163

18. 15.87%   **19.** At least .75   **20.** $\mu = 27.87$; $\sigma = 6.41$

21. .677   **22.** $\mu = 120$; $\sigma \approx 10.1$ **23.** 0.6%

24. .9738   **25.** .9997

Chapter 8 Before Moving On, page 509

1.

x	−3	−2	0	1	2	3
$P(X = x)$	.05	.1	.25	.3	.2	.1

2. a. .8   **b.** .92   **3.** 0.44; 4.0064; 2

4. a. .2401; .4116; .2646; .0756; .0081   **b.** .12; .917

5. a. .9772   **b.** .9772   **c.** .9544

6. a. .0228   **b.** .5086   **c.** .0228

CHAPTER 9

Exercises 9.1, page 519

1. Yes   **3.** Yes   **5.** No   **7.** Yes   **9.** No

11. a. Given that the outcome state 1 has occurred, the conditional probability that the outcome state 1 will occur is .3.

b. .7   **c.** $\begin{bmatrix} .48 \\ .52 \end{bmatrix}$

13. $TX_0 = \begin{bmatrix} .4 \\ .6 \end{bmatrix}$;

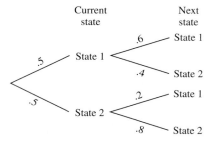

15. $X_2 = \begin{bmatrix} .576 \\ .424 \end{bmatrix}$ **17.** $X_2 = \begin{bmatrix} \frac{5}{16} \\ \frac{27}{64} \\ \frac{17}{64} \end{bmatrix}$

19. a.

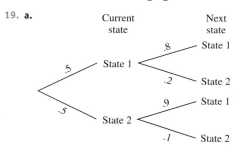

b. $T = \begin{matrix} & \begin{matrix} L & R \end{matrix} \\ \begin{matrix} L \\ R \end{matrix} & \begin{bmatrix} .8 & .9 \\ .2 & .1 \end{bmatrix} \end{matrix}$ **c.** $X_0 = \begin{matrix} L \\ R \end{matrix} \begin{bmatrix} .5 \\ .5 \end{bmatrix}$ **d.** .85

21. a. Vote is evenly split.   **b.** Democrat

23. 50% in zone I, 30% in zone II, 20% in zone III

25. University: 37%, Campus: 35%, Book Mart: 28%; University: 34.5%, Campus: 31.35%, Book Mart: 34.15%

27. Business: 36%, Humanities: 23.8%, Education: 15%, Natural Sciences and others: 25.1%

29. False

Using Technology Exercises 9.1, page 523

1. $X_5 = \begin{bmatrix} .204489 \\ .131869 \\ .261028 \\ .186814 \\ .2158 \end{bmatrix}$

3. Manufacturer A will have 23.95% of the market, manufacturer B will have 49.71% of the market share, and manufacturer C will have 26.34% of the market share.

Exercises 9.2, page 530

1. Regular   3. Not regular   5. Regular

7. Not regular   9. $\begin{bmatrix} \frac{3}{11} \\ \frac{8}{11} \end{bmatrix}$ 11. $\begin{bmatrix} \frac{2}{7} \\ \frac{5}{7} \end{bmatrix}$

13. $\begin{bmatrix} \frac{3}{13} \\ \frac{8}{13} \\ \frac{2}{13} \end{bmatrix}$ 15. $\begin{bmatrix} \frac{3}{19} \\ \frac{8}{19} \\ \frac{8}{19} \end{bmatrix}$ 17. 81.8%

19. 40.8% one wage earner and 59.2% two wage earners; 30% one wage earner and 70% two wage earners

21. 72.5% in single-family homes and 27.5% in condominiums; 70% in single-family homes and 30% in condominiums

23. **a.** 31.7% ABC, 37.35% CBS, 30.95% NBC
b. $33\frac{1}{3}$% ABC, $33\frac{1}{3}$% CBS, $33\frac{1}{3}$% NBC

25. 25% red, 50% pink, 25% white   27. False

Using Technology Exercises 9.2, page 534

1. $X_5 = \begin{bmatrix} .2045 \\ .1319 \\ .2610 \\ .1868 \\ .2158 \end{bmatrix}$

Exercises 9.3, page 540

1. Yes   3. Yes   5. Yes   7. Yes

9. $\begin{bmatrix} 1 & .4 \\ \hline 0 & .6 \end{bmatrix}$, $R = [.6]$, and $S = [.4]$

11. $\begin{bmatrix} 1 & .4 & .5 \\ \hline 0 & .4 & .5 \\ 0 & .2 & 0 \end{bmatrix}$, $R = \begin{bmatrix} .4 & .5 \\ .2 & 0 \end{bmatrix}$, and $S = [.4 \quad .5]$, or

$\begin{bmatrix} 1 & .5 & .4 \\ \hline 0 & 0 & .2 \\ 0 & .5 & .4 \end{bmatrix}$, $R = \begin{bmatrix} 0 & .2 \\ .5 & .4 \end{bmatrix}$, and $S = [.5 \quad .4]$

13. $\begin{bmatrix} 1 & 0 & .2 & .4 \\ 0 & 1 & .3 & 0 \\ \hline 0 & 0 & .3 & .2 \\ 0 & 0 & .2 & .4 \end{bmatrix}$,

$R = \begin{bmatrix} .3 & .2 \\ .2 & .4 \end{bmatrix}$ and $S = \begin{bmatrix} .2 & .4 \\ .3 & 0 \end{bmatrix}$, or

$\begin{bmatrix} 1 & 0 & .4 & .2 \\ 0 & 1 & 0 & .3 \\ \hline 0 & 0 & .4 & .2 \\ 0 & 0 & .2 & .3 \end{bmatrix}$,

$R = \begin{bmatrix} .4 & .2 \\ .2 & .3 \end{bmatrix}$, $S = \begin{bmatrix} .4 & .2 \\ 0 & .3 \end{bmatrix}$, and so forth

15. $\begin{bmatrix} 1 & 1 \\ \hline 0 & 0 \end{bmatrix}$ 17. $\begin{bmatrix} 1 & 1 & 1 \\ 0 & 0 & 0 \\ 0 & 0 & 0 \end{bmatrix}$

19. $\begin{bmatrix} 1 & 0 & 1 & 1 \\ 0 & 1 & 0 & 0 \\ 0 & 0 & 0 & 0 \\ 0 & 0 & 0 & 0 \end{bmatrix}$ 21. $\begin{bmatrix} 1 & 0 & \frac{1}{2} & \frac{1}{2} \\ 0 & 1 & \frac{1}{2} & \frac{1}{2} \\ 0 & 0 & 0 & 0 \\ 0 & 0 & 0 & 0 \end{bmatrix}$

23. $\begin{bmatrix} 1 & 0 & 0 & \frac{7}{22} & \frac{3}{11} \\ 0 & 1 & 0 & \frac{5}{22} & \frac{9}{22} \\ 0 & 0 & 1 & \frac{5}{11} & \frac{7}{22} \\ 0 & 0 & 0 & 0 & 0 \\ 0 & 0 & 0 & 0 & 0 \end{bmatrix}$

25. **a.**
$\begin{array}{c} \\ UL \\ L \end{array} \begin{array}{cc} UL & L \\ \end{array}$
$\begin{matrix} UL \\ L \end{matrix} \begin{bmatrix} 1 & .2 \\ \hline 0 & .8 \end{bmatrix}$, $R = [.8]$, and $S = [.2]$

b. $\begin{bmatrix} 1 & 1 \\ \hline 0 & 0 \end{bmatrix}$; eventually, only unleaded fuel will be used.

27. .25; .50; .75

29. **a.**
$\begin{array}{c} \\ D \\ G \\ 1 \\ 2 \end{array} \begin{array}{cccc} D & G & 1 & 2 \\ \end{array}$
$\begin{matrix} D \\ G \\ 1 \\ 2 \end{matrix} \begin{bmatrix} 1 & 0 & .25 & .1 \\ 0 & 1 & 0 & .9 \\ \hline 0 & 0 & 0 & 0 \\ 0 & 0 & .75 & 0 \end{bmatrix}$

b. $\begin{bmatrix} 1 & 0 & .325 & .1 \\ 0 & 1 & .675 & .9 \\ \hline 0 & 0 & 0 & 0 \\ 0 & 0 & 0 & 0 \end{bmatrix}$

c. .675

31. False

Exercises 9.4, page 551

1. R: row 1; C: column 2   **3.** R: row 1; C: column 1

5. R: row 1 or row 3; C: column 3

7. R: row 1 or row 3; C: column 2

9. Strictly determined;
  a. 2   **b.** R: row 1; C: column 1
  c. 2   **d.** Favors row player

11. Strictly determined;
  a. 1   **b.** R: row 1; C: column 1
  c. 1   **d.** Favors row player

13. Strictly determined;
  a. 1   **b.** R: row 1; C: column 1
  c. 1   **d.** Favors row player

15. Not strictly determined   **17.** Not strictly determined

19. a. $\begin{bmatrix} 2 & -3 & 4 \\ -3 & 4 & -5 \\ 4 & -5 & 6 \end{bmatrix}$

  b. Robin: row 1; Cathy: column 1 or column 2
  c. Not strictly determined
  d. Not strictly determined

21. a.

Economy

		Good	Recess.
Mgmt.	Expand	200,000	120,000
	Not exp.	50,000	150,000

  b. Yes

23. a.

Charley

		Raises	Holds	Lowers
Roland	Raises	3	−1	−3
	Holds	2	0	−2
	Lowers	5	2	1

25. True

Exercises 9.5, page 562

1. $\dfrac{3}{10}$ **3.** $\dfrac{5}{12}$ **5.** 0.16

7. a. 1   **b.** −2   **c.** 0
  d. $-\dfrac{3}{10}$; (a) is most advantageous

9. a. 1   **b.** $-\dfrac{7}{20}$ **c.** The first strategy

11. $P = \begin{bmatrix} \frac{1}{4} & \frac{3}{4} \end{bmatrix}$, $Q = \begin{bmatrix} \frac{1}{2} \\ \frac{1}{2} \end{bmatrix}$, and $E = 2.5$; favors row player

13. $P = \begin{bmatrix} \frac{4}{7} & \frac{3}{7} \end{bmatrix}$, $Q = \begin{bmatrix} \frac{5}{7} \\ \frac{2}{7} \end{bmatrix}$, and $E = -\frac{1}{7}$; favors column player

15. $P = \begin{bmatrix} \frac{1}{2} & \frac{1}{2} \end{bmatrix}$, $Q = \begin{bmatrix} \frac{1}{4} \\ \frac{3}{4} \end{bmatrix}$, and $E = -5$; favors column player

17. a. $P = \begin{bmatrix} \frac{1}{3} & \frac{2}{3} \end{bmatrix}$ and $Q = \begin{bmatrix} \frac{1}{3} \\ \frac{2}{3} \end{bmatrix}$

  b. $E = 0$; no

19. a. \$5714 in hotel stocks; \$34,286 in brewery stock
  b. \$4857

21. a.

C

		N	F	
R	N	.48	.65	C = Carlton; R = Russell
	F	.50	.45	N = local newspaper; F = flyer

  b. Russell's strategy: $P = \begin{bmatrix} .23 & .77 \end{bmatrix}$
  Carlton's strategy: $Q = \begin{bmatrix} .91 \\ .09 \end{bmatrix}$

Chapter 9 Concept Review Questions, page 567

1. Probabilities; preceding   **2.** State; state   **3.** Transition

4. $n \times n$; nonnegative; 1   **5.** Distribution; steady-state

6. Regular; rows; equal; positive; $TX = X$; elements; 1

7. Absorbing; leave; stages

8. a. Zero-sum   **b.** Maximin; minimax   **9.** Optimal

10. Saddle point; maximin; row; saddle; minimax; column; saddle

Chapter 9 Review Exercises, page 567

1. Not regular   **2.** Regular   **3.** Regular   **4.** Not regular

5. $\begin{bmatrix} .3675 \\ .36 \\ .2725 \end{bmatrix}$ **6.** $\begin{bmatrix} .1915 \\ .4215 \\ .387 \end{bmatrix}$

7. Yes   **8.** No   **9.** No   **10.** Yes

11. $\begin{bmatrix} \frac{3}{7} & \frac{3}{7} \\ \frac{4}{7} & \frac{4}{7} \end{bmatrix}$ **12.** $\begin{bmatrix} \frac{4}{9} & \frac{4}{9} \\ \frac{5}{9} & \frac{5}{9} \end{bmatrix}$

13. $\begin{bmatrix} .457 & .457 & .457 \\ .200 & .200 & .200 \\ .343 & .343 & .343 \end{bmatrix}$ **14.** $\begin{bmatrix} .323 & .323 & .323 \\ .290 & .290 & .290 \\ .387 & .387 & .387 \end{bmatrix}$

15. a.

	A	U	N	
A	.85	0	.10	A = Agriculture
U	.10	.95	.05	U = Urban
N	.05	.05	.85	N = Nonagricultural

  b. A $\begin{bmatrix} .50 \\ .15 \\ .35 \end{bmatrix}$ **c.** A $\begin{bmatrix} .424 \\ .262 \\ .314 \end{bmatrix}$
    U   U
    N   N

16. 12.5% large cars, 30.36% intermediate-sized cars, 57.14% small cars

17. Favors row player; strictly determined; R: row 3; C: column 1; value is 4.

18. Does not favor either player; strictly determined; R: row 1; C: column 2; value is 0.

19. Favors row player; strictly determined; R: row 1; C: column 1; value is 1.

20. Not strictly determined

21. $-\frac{1}{4}$ 22. $\frac{10}{9}$ 23. 2   24. 1.04

25. $P = \begin{bmatrix} \frac{1}{2} & \frac{1}{2} \end{bmatrix}$, $Q = \begin{bmatrix} \frac{5}{6} \\ \frac{1}{6} \end{bmatrix}$, and $E = \frac{1}{2}$; favors row player

26. $P = \begin{bmatrix} \frac{1}{2} & \frac{1}{2} \end{bmatrix}$, $Q = \begin{bmatrix} \frac{13}{22} \\ \frac{9}{22} \end{bmatrix}$, and $E = -\frac{1}{2}$, favors column player

27. $P = \begin{bmatrix} \frac{1}{10} & \frac{9}{10} \end{bmatrix}$, $Q = \begin{bmatrix} \frac{4}{5} \\ \frac{1}{5} \end{bmatrix}$, and $E = 1.2$; favors row player

28. $P = \begin{bmatrix} \frac{4}{5} & \frac{1}{5} \end{bmatrix}$, $Q = \begin{bmatrix} \frac{2}{5} \\ \frac{3}{5} \end{bmatrix}$, and $E = 10.8$; favors row player

29. **a.** $\begin{bmatrix} .5 & .7 \\ .4 & .5 \end{bmatrix}$ **b.** \$7

30. 25% compact models; 75% subcompact models

Chapter 9 Before Moving On, page 569

1. $\begin{bmatrix} .366 \\ .634 \end{bmatrix}$ 2. $\begin{bmatrix} \frac{3}{11} \\ \frac{8}{11} \end{bmatrix}$ 3. $\begin{bmatrix} 1 & 1 & 1 \\ 0 & 0 & 0 \\ 0 & 0 & 0 \end{bmatrix}$

4. **a.** -1 **b.** R: row 1; C; column 3   **c.** -1; column player

5. **a.** 3 units   **b.** 1.22 units

6. **a.** $P = \begin{bmatrix} \frac{2}{3} & \frac{1}{3} \end{bmatrix}$; $Q = \begin{bmatrix} \frac{1}{6} \\ \frac{5}{6} \end{bmatrix}$ **b.** $\frac{4}{3}$; row player

Appendix A

Exercises A.1, page 575

1. Yes   3. Yes   5. No

7. Yes   9. Yes   11. No

13. No   15. Negation   17. Conjunction

19. Conjunction

21. New orders for manufactured goods did not fall last month.

23. Drinking during pregnancy does not affect both the size and weight of babies.

25. The commuter airline industry is not now undergoing a shakeup.

27. **a.** Domestic car sales increased over the past year, and/or foreign car sales decreased over the past year.
 b. Domestic car sales increased over the past year, and foreign car sales decreased over the past year.
 c. Either domestic car sales increased over the past year or foreign car sales decreased over the past year.
 d. Domestic car sales did not increase over the past year.
 e. Domestic car sales did not increase over the past year, and/or foreign car sales decreased over the past year.

 f. Domestic car sales did not increase over the past year, and/or foreign car sales did not decrease over the past year.

29. **a.** Either the doctor recommended surgery to treat his hyperthyroidism or the doctor recommended radioactive iodine to treat his hyperthyroidism.
 b. The doctor recommended surgery to treat his hyperthyroidism, and/or the doctor recommended radioactive iodine to treat his hyperthyroidism.

31. **a.** $p \wedge q$ **b.** $p \veebar q$ **c.** $\sim(p \wedge q)$ **d.** $\sim(\sim q)$

33. **a.** Both the popularity of prime-time soaps and prime-time situation comedies did not increase this year.
 b. The popularity of prime-time soaps did not increase this year, and/or the popularity of prime-time detective shows decreased this year.
 c. The popularity of prime-time detective shows decreased this year, and/or the popularity of prime-time situation comedies did not increase this year.
 d. Either the popularity of prime-time soaps did not increase this year or the popularity of prime-time situation comedies did not increase this year.

Exercises A.2, page 578

1.

p	q	$\sim q$	$p \vee \sim q$
T	T	F	T
T	F	T	T
F	T	F	F
F	F	T	T

3.

p	$\sim p$	$\sim(\sim p)$
T	F	T
F	T	F

5.

p	$\sim p$	$p \vee \sim p$
T	F	T
F	T	T

7.

p	$\sim p$	q	$p \vee q$	$\sim p \wedge (p \vee q)$
T	F	T	T	F
T	F	F	T	F
F	T	T	T	T
F	T	F	F	F

9.

p	q	$\sim q$	$p \vee q$	$p \wedge \sim q$	$(p \vee q) \wedge (p \wedge \sim q)$
T	T	F	T	F	F
T	F	T	T	T	T
F	T	F	T	F	F
F	F	T	F	F	F

11.

p	q	$p \lor q$	$\sim(p \lor q)$	$(p \lor q) \land \sim(p \lor q)$
T	T	T	F	F
T	F	T	F	F
F	T	T	F	F
F	F	F	T	F

13.

p	q	r	$p \lor q$	$p \lor r$	$(p \lor q) \land (p \lor r)$
T	T	T	T	T	T
T	T	F	T	T	T
T	F	T	T	T	T
T	F	F	T	T	T
F	T	T	T	T	T
F	T	F	T	F	F
F	F	T	F	T	F
F	F	F	F	F	F

15.

p	q	r	$p \land q$	$\sim r$	$(p \land q) \lor \sim r$
T	T	T	T	F	T
T	T	F	T	T	T
T	F	T	F	F	F
T	F	F	F	T	T
F	T	T	F	F	F
F	T	F	F	T	T
F	F	T	F	F	F
F	F	F	F	T	T

17.

p	q	r	$\sim q$	$p \land \sim q$	$p \land r$	$(p \land \sim q) \lor (p \land r)$
T	T	T	F	F	T	T
T	T	F	F	F	F	F
T	F	T	T	T	T	T
T	F	F	T	T	F	T
F	T	T	F	F	F	F
F	T	F	F	F	F	F
F	F	T	T	F	F	F
F	F	F	T	F	F	F

19. 16 rows

Exercises A.3, page 583

1. $\sim q \to p$; $q \to \sim p$; $\sim p \to q$

3. $p \to q$; $\sim p \to \sim q$; $\sim q \to \sim p$

5. Conditional: If it is snowing, then the temperature is below freezing.
Biconditional: It is snowing if and only if the temperature is below freezing.

7. Conditional: If the company's union and management reach a settlement, then the workers will not strike.
Biconditional: The company's union and management will reach a settlement if and only if the workers do not strike.

9. False   **11.** False

13. It is false when I do not buy the house and the owner lowers the selling price.

15.

p	q	$p \to q$	$\sim(p \to q)$
T	T	T	F
T	F	F	T
F	T	T	F
F	F	T	F

17.

p	q	$p \to q$	$\sim(p \to q)$	$\sim(p \to q) \land p$
T	T	T	F	F
T	F	F	T	T
F	T	T	F	F
F	F	T	F	F

19.

p	q	$\sim p$	$\sim q$	$p \to \sim q$	$(p \to \sim q) \veebar \sim p$
T	T	F	F	F	F
T	F	F	T	T	T
F	T	T	F	T	F
F	F	T	T	T	F

21.

p	q	$\sim p$	$\sim q$	$p \to q$	$\sim q \to \sim p$	$(p \to q) \leftrightarrow (\sim q \to \sim p)$
T	T	F	F	T	T	T
T	F	F	T	F	F	T
F	T	T	F	T	T	F
F	F	T	T	T	T	T

23.

p	q	$p \land q$	$p \lor q$	$(p \land q) \to (p \lor q)$
T	T	T	T	T
T	F	F	T	T
F	T	F	T	T
F	F	F	F	T

25.

p	q	r	$p \lor q$	$\sim r$	$(p \lor q) \to \sim r$
T	T	T	T	F	F
T	T	F	T	T	T
T	F	T	T	F	F
T	F	F	T	T	T
F	T	T	T	F	F
F	T	F	T	T	T
F	F	T	F	F	T
F	F	F	F	T	T

27.

p	q	r	$q \lor r$	$p \to (q \lor r)$
T	T	T	T	T
T	T	F	T	T
T	F	T	T	T
T	F	F	F	F
F	T	T	T	T
F	T	F	T	T
F	F	T	T	T
F	F	F	F	T

29. Logically equivalent

31. Not logically equivalent

33. Not logically equivalent

35. Not logically equivalent

37. a. $p \rightarrow {\sim}q$ **b.** ${\sim}p \rightarrow q$ **c.** ${\sim}q \leftrightarrow p$

   d. $p \rightarrow {\sim}q$ **e.** $p \leftrightarrow {\sim}q$

Exercises A.4, page 587

1.

p	p	$p \wedge p$
T	T	T
F	F	F

3.

p	q	r	$p \wedge q$
T	T	T	T
T	T	F	T
T	F	T	F
T	F	F	F
F	T	T	F
F	T	F	F
F	F	T	F
F	F	F	F

$(p \wedge q) \wedge r$	$q \wedge r$	$p \wedge (q \wedge r)$
T	T	T
F	F	F
F	F	F
F	F	F
F	T	F
F	F	F
F	F	F
F	F	F

5.

p	q	$p \wedge q$	$q \wedge p$
T	T	T	T
T	F	F	F
F	T	F	F
F	F	F	F

7.

p	q	r	$q \wedge r$	$p \vee (q \wedge r)$
T	T	T	T	T
T	T	F	F	T
T	F	T	F	T
T	F	F	F	T
F	T	T	T	T
F	T	F	F	F
F	F	T	F	F
F	F	F	F	F

$p \vee q$	$p \vee r$	$(p \vee q) \wedge (p \vee r)$
T	T	T
T	T	T
T	T	T
T	T	T
T	T	T
T	F	F
F	T	F
F	F	F

9. Tautology   **11.** Tautology

13. Tautology

15. Tautology   **17.** Neither

19. ${\sim}(p \wedge q)$ The candidate does not oppose changes in the Social Security system, or the candidate does not support the ERA.

   ${\sim}(p \vee q)$: The candidate does not oppose changes in the Social Security system, and the candidate does not support the ERA.

21. $[p \wedge (q \vee {\sim}q) \vee (p \wedge q)]$
   $\Leftrightarrow p \wedge t \vee (p \wedge q)]$ By Law 11
   $\Leftrightarrow p \vee (p \wedge q)$ By Law 14

23. $(p \wedge {\sim}q) \vee (p \wedge {\sim}r)$
   $\Leftrightarrow p \wedge ({\sim}q \vee {\sim}r)$ By Law 7

25. $(p \wedge [{\sim}(q \wedge r)]$
   $\Leftrightarrow p \wedge ({\sim}q \vee {\sim}r)$ By Law 10
   $\Leftrightarrow (p \wedge {\sim}q) \vee (p \wedge {\sim}r)$ By Law 7

Exercises A.5, page 593

1. Valid   **3.** Valid   **5.** Invalid   **7.** Valid

9. Valid   **11.** Valid   **13.** Valid   **15.** Invalid

17. $p \rightarrow q$; invalid   **19.** $p \vee q$; valid
   ${\sim}p$ ${\sim}p \rightarrow {\sim}q$
   $\therefore {\sim}q$ $\therefore p$

21. $p \rightarrow q$; invalid   **23.** b
   $q \rightarrow r$
   r
   $\therefore p$

25.

p	q	$p \rightarrow q$	${\sim}q$	${\sim}p$
T	T	T	F	F
T	F	F	T	F
F	T	T	F	T
F	F	T	T	T

Exercises A.6, page 596

1. $p \wedge q \wedge (r \vee s)$

3. $[(p \wedge q) \vee r] \wedge ({\sim}r \vee p)$

5. $[(p \vee q) \wedge r] \vee ({\sim}p) \vee [{\sim}q \wedge (p \vee r \vee {\sim}r)]$

7.

9.

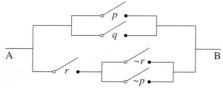

11.

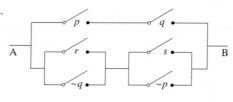

13. $p \wedge [\sim q \vee (\sim p \wedge q)]; p \wedge \sim q$

15. $p \wedge [\sim p \vee q \vee (q \wedge r)]; p \wedge q$

Index